Never at Rest

A Vulgar Mechanick can practice what he has been taught or seen done, but if he is in an error he knows not how to find it out and correct it, and if you put him out of his road, he is at a stand; Whereas he that is able to reason nimbly and judiciously about figure, force and motion, is never at rest till he gets over every rub.

Isaac Newton to Nathaniel Hawes
25 May 1694

NEVER AT REST

A Biography of Isaac Newton

RICHARD S. WESTFALL

Professor of History of Science
Indiana University

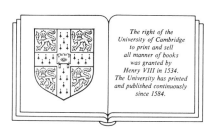

The right of the
University of Cambridge
to print and sell
all manner of books
was granted by
Henry VIII in 1534.
The University has printed
and published continuously
since 1584.

CAMBRIDGE UNIVERSITY PRESS

CAMBRIDGE

LONDON   NEW YORK   NEW ROCHELLE

MELBOURNE   SYDNEY

to

GLORIA

Published by the Press Syndicate of the University of Cambridge
The Pitt Building, Trumpington Street, Cambridge CB2 1RP
32 East 57th Street, New York, NY 10022, USA
296 Beaconsfield Parade, Middle Park, Melbourne 3206, Australia

First published 1980
Reprinted 1981, 1982
First paperback edition 1983
Reprinted 1984

Printed in the United States of America

Library of Congress Cataloging in Publication Data
Westfall, Richard S.
Never at rest.
Bibliography: p.
Includes index.
1. Newton, Isaac, Sir, 1642–1727.   2.   Physicists–
Great Britain–Biography.   I.   Title.
QC16.N7W35   509'.2'4   [B]   79-26294
ISBN 0 521 23143 4 hard covers
ISBN 0 521 27435 4 paperback

Contents

Preface to the paperback edition

Not a great deal of time has passed since I wrote a preface for the first printing of this book, and with that preface brought more than twenty years of work to a conclusion. After such a period I was, I confess, ready to move on to other matters, and with the exception of one Newtonian topic, I have done so. As a result I am not now prepared or inclined to revise the work in any fundamental way. Beyond correcting a few typographical and factual errors, I confine myself to indicating here some passages that reviews and recent publications have convinced me ought to be altered, and to mentioning the one topic on which my own research since completing the biography has led me to a deeper understanding.

In Chapter 2, pp. 57–58, I assert that Newton's secondary education did not include any significant mathematics. Recently, D. T. Whiteside has uncovered in Grantham a pocketbook dated 1654 that contains extensive "Notes for the Mathematicks." [See "Newton the Mathematician," in Z. Bechler, ed., *Contemporary Newtonian Research* (Dordrecht, 1982), pp. 110–11.] It appears to Whiteside to have been written in the hand of Henry Stokes, master of the grammar school of Grantham, which Newton began to attend during the year following the notes. Although most of the passage is devoted to elementary calculations, such as rules for determining the areas of fields, that a country gentleman might have found useful, not all of it remains at this level. Sixty-five pages on "The Measureing of Triangles [&] Circles" include instructions on how to calculate a table of sines (a task that Newton undertook, albeit briefly, at Cambridge), and a method for inscribing an equilateral septuagon in a circle (another topic that Newton encountered anew in his reading at Cambridge). Stokes even included the limits that Archimedes established for the value of pi, more than $3^{10}/_{71}$ and less than $3^{1}/_{7}$. As Whiteside does not fail to remark, this passage of the "Notes" inevitably reminds one of the geometric figures incised in the plaster at Woolsthorpe. The pocketbook strongly implies that my words on Newton's grammar school education need basic revision. It is highly probable that he was far from a novice in mathematics when he arrived at Cambridge, and the sudden immersion in mathematical study that began sometime during his undergraduate years as well as the burst of creativity that accompanied it, though not rendered one whit less stunning by this new information, do become considerably more comprehensible.

In his review of this book and in private communications with me, Whiteside has argued that in Chapter 4, the major discussion of Newton's mathematics (and I would add, in pp. 23–38 of Chapter 1, the discussion of seventeenth-century mathematics before Newton), I have neglected the distinction between indivisibles and infinitesimals, and that, as a result, the clarity of my exposition and some of the conclusions I draw suffer. I do not consider myself to be an historian of mathematics, and I am not now ready to plunge back into the demanding world of seventeenth-century mathematics in order to more fully clarify the matter in my own mind. Consequently, I am not able here to offer the revisions that are probably in order. Suffice it to say that in my experience, when Whiteside speaks on seventeenth-century mathematics, the wise attend. Hence I caution readers that in those passages, where something of importance to them involves indivisibles or infinitesimals, they would do well to consult the work of others, such as Whiteside's edition of the *Mathematical Papers,* his other writings, or the books I cite in the footnotes and bibliography.

Chapter 9 contains a discussion (pp. 351–6) of a theological work, *Theologiae gentilis origines philosophicae,* which Newton began to compose in the 1680s. The more I thought about it and about the fact that Newton continued to refer to it the rest of his life, the more important it seemed to me. About the time this biography first appeared, I became convinced that I had not sufficiently understood this manuscript; it constitutes the one Newtonian topic that I have explored further since publication of the book. The study confirmed my growing belief that it was the most important theological treatise Newton ever composed, even though he never put it in a finished form. One must not be misled by the earnest narratives, which ring so quaintly in our twentieth-century ears, of the lives of Noah and his offspring. These were the familiar scholarly themes of the day, and Newton could no more leap out of his own age than any of us can. He could, however, bend this accepted material to radical new purposes. It appears to me that the *Origines* can be adequately described only as the first of the deist tracts. I have discussed it at length and compared it to superficially similar works upon which he drew in an essay–"Isaac Newton's *Theologiae Gentilis Origines Philosophicae,*" in W. W. Wagar, ed., *The Secular Mind* (New York, 1982), pp. 15–34. To the best of my knowledge, this essay is the only discussion in print, beyond the brief passage in the present biography, of this important Newtonian manuscript, and I urge anyone interested in Newton's theological views to consult my essay.

I devote pp. 740–4 of Chapter 14 to the correction of the error in Proposition X, Book II, for the second edition of the *Principia,* remarking at one point that "One would like to know more about the circumstances." Volume 8 of Whiteside's edition of the *Mathematical Papers* has appeared since I wrote that line, and, as a result, we do know more about the circumstances. The treatment of that matter (pp. 312–424) is one of the *tours de force* of Whiteside's final volume. He assembles the papers on which Newton initially confirmed the validity of Bernoulli's objection, and he identifies, as Bernoulli was not able to do, the precise nature of the error. There follow the manuscripts of six successive attacks on the problem, together with seven other draft passages, leading up to Newton's successful location of the error and his correction of it. Indeed, as the manuscripts reveal, Newton even proceeded to an alternative demonstration of the correct result. Most interesting of all, Whiteside argues persuasively that only a computational error prevented Newton from realizing that he could have corrected the proposition within the framework of the original demonstration. Bernoulli seized on the radically amended proposition – for which he claimed credit, as the newly printed pages made its last-minute insertion clear – as evidence that Newton had not understood second derivatives when he composed the first edition of the *Principia.* That charge, endlessly repeated, figured prominently, much to Newton's embarrassment, in the priority dispute. This brief paragraph cannot begin to do justice to the richness of Whiteside's presentation, to which I refer readers wishing to know more about the incident.

In his review of the present work in the *American Historical Review,* 87 (1982), 1353, I. Bernard Cohen points out that the interpretations in the book are mine and are not necessarily shared by other Newtonian scholars. I never intended it otherwise. When I signed my name to the book, I understood that I was taking sole responsibility for its content, and I rather assumed that readers would understand the same. I have no privileged access to final truth, of course, and Professor Cohen may prove to be right about the interpretations he challenges in his review. Suffice it to say that I remain unrepentant as of this moment, and aside from the four passages above, the book continues to represent my view of Isaac Newton.

Preface

THE utility of biography, Dr. Johnson argued, rests on the fact that we can enter by sympathy into situations in which others have found themselves. Parallel circumstances to which we can conform our minds shape every life. Even the great are not removed from the factors common to all: "We are all prompted by the same motives, all deceived by the same fallacies, all animated by hope, obstructed by danger, entangled by desire, and seduced by pleasure." I must confess that twenty years devoted to the biography of Newton have not in my case confirmed Dr. Johnson's dictum. The more I have studied him, the more Newton has receded from me. It has been my privilege at various times to know a number of brilliant men, men whom I acknowledge without hesitation to be my intellectual superiors. I have never, however, met one against whom I was unwilling to measure myself, so that it seemed reasonable to say that I was half as able as the person in question, or a third or a fourth, but in every case a finite fraction. The end result of my study of Newton has served to convince me that with him there is no measure. He has become for me wholly other, one of the tiny handful of supreme geniuses who have shaped the categories of the human intellect, a man not finally reducible to the criteria by which we comprehend our fellow beings, those parallel circumstances of Dr. Johnson.

Why then, one might ask, am I attempting to write Newton's biography? My second prefatory confession is that increasingly I have asked the same question myself. Had I known, when in youthful self-confidence I committed myself to the task, that I would end up in similar self-doubt, surely I would never have set out. I did perceive that it would be a long and arduous task, though I was willing to undertake the labor. I thought at the time it would take ten years, not far short of eternity at that point in my life, though now, in its brevity, an indication of the chasm between expectation and reality. Perhaps even the prospect of twenty years would not have turned me back, but the other chasm, the unexpected gulf opening between me and my subject, would have been another matter. As I face the situation now, not in prospect but in retrospect, the lingering influence of the Puritan ethic makes the prospect of discarding the fruits of so much earnest toil abhorrent. For that matter, could any other potential biographer of Newton escape the same dilemma? Only another Newton could hope fully to enter into his being, and the economy of the human enterprise is such that a second Newton would not devote himself to the biography of the first. If history has a function – and my doubts have never extended to questioning that it has – perforce it must deal with the Newtons. Everyone who is informed agrees on the need for a new biography to replace Sir David Brewster's masterpiece, which is now one hundred and twenty-five years old. Others have aspired. It is not for me to decide whether I have succeeded where they failed. With all of the hesitations the paragraph above implies, I place the result of long years on the altar of history with the hope that it may add its bit to the understanding of the past to which the modern age has already contributed so much.

In writing Newton's biography, I have attempted, in accordance with my understanding of biography as a literary form, to avoid composing an essay on Newtonian science. At the same time I have sought to make Newton the scientist the central character of my drama. While he devoted himself extensively to other activities which a biography cannot ignore, from theology on one hand to administration of the Mint on the other, Newton holds our attention only because he was a scientist of transcendent importance. Hence I tend to think of my work as a scientific biography, that is, a biography in which Newton's scientific career furnishes the central theme. My goal has been to present his science, not as the finished product which has done so much to shape the whole of the modern intellect, but as the developing endeavor of a living man confronting it as problems still to be solved. Scientists and philosophers can probe the finished product. My interest in this biography centers exactly on what was not yet complete, the object of Newton's own activity, the substance of a life devoted to probing the unknown. I have tried to present his scientific endeavors in the context of his life, first in Woolsthorpe and Grantham, then in Cambridge, and finally in London. To an extent few others have equaled, however, Newton was a man of learning. Never fully at ease with others, he held his distance and lived largely in the setting of his own study. His books furnished the context of his life more than Cambridge or London did. A biographer ignores this truth about Newton at his own peril. I have done my best to keep it in mind and to present a picture of Newton in which the pursuit of truth, most importantly though not exclusively scientific truth, formed the essence of his life. To the extent that I have succeeded in this, the biography as a whole will also succeed.

The first volume of the Royal Society's edition of Newton's *Correspondence* appeared about the time I began serious work on the biography. Now all seven volumes are in print, and the footnotes in my book bear testimony to their indispensable aid. They are only part of the flood of Newtonian publications during the last two decades, on all of which my own work rests directly, to all of which I want to acknowledge a debt which is in fact beyond acknowledgment. The *Correspondence* has been necessary, as has the Whiteside edition of the *Mathematical Papers,* the final volume of which (to my loss I am sure) has yet to be published as I write this Preface. One of the features of my biography, which sets it apart from earlier ones, is the chronological account of Newton's mathematical activity. The account is my own, and I take full responsibility for it. It would have been impossible, however, without Whiteside's monument of scholarship.

Beyond these general publications are a number of others more restricted to single aspects of Newton's life: I. Bernard Cohen's *Papers & Letters* plus his edition of the *Principia* with variant readings undertaken jointly with the late Alexandre Koyré; A. R. and M. B. Hall's *Unpublished Scientific Papers;* John Herivel's *Background to Newton's 'Principia';* and (though it differs from the above in being formally a monograph) B. J. T. Dobbs's *Foundations of Newton's Alchemy.* The publication of these works has significantly expanded the opportunities of Newtonian scholarship. I am not the first to benefit from them, nor will I be the last. I regret that my work is done too soon to derive further benefit from the publication of optical papers that Alan Shapiro has undertaken.

During the time I have been at work on Newton, I have received assistance of many kinds from many sources. Grants from the National Science Foundation, the George A. and Eliza Gardner Howard Foundation, the American Council of Learned Societies, and the National Endowment for the Humanities; and sabbatical leaves from Indiana University have provided most of the time for study and writing, much of it in England, where the great bulk of Newton's papers exist. One of those years I had the privilege and advantage to be a Visiting Fellow of Clare Hall, Cambridge. The National Science Foundation and Indiana University have also helped to finance the acquisition of photocopies of Newton's papers. The staffs of many libraries have outdone themselves in kind assistance, most prominently (in proportion to my demands) the Cambridge University Library, the Trinity College Library, the Widener Library at Harvard, the Babson College Library, the Indiana University Library, and the Public Record Office. Most of the typing I owe to a succession of secretaries over the years in the Department of History and Philosophy of Science at Indiana University, but among them especially Karen Blaisdell. The help of Anita Guerrini in proofreading has been invaluable. I cannot sufficiently express my gratitude to those I have mentioned and to many others who have helped in less central ways. I can at least try to express it, and I do.

Nor can any author omit his family. By the time my children reached consciousness, I had embarked on the biography. Now I finish it as they complete their educations and set out on their own. The whole of their intimate experience of me has been flavored by the additional presence of Newton. I do not know if I would have been a more satisfactory father without the biography. Suffice it for them to realize that what they put up with over the years did in the end achieve some sort of conclusion, incarnation as a book. My wife of course endured more, and I thank her for enduring it with grace and understanding. In the end she discovered the only adequate defense – she is writing a book herself and may, if I improve, acknowledge my encouragement and support in her own preface.

R.S.W.

Acknowledgments

I WISH to acknowledge permission granted me by the Babson College Library to reproduce an alchemical diagram, a plan of the Jewish temple, and a scheme of the twelve gods of the ancient peoples, all found among the Grace K. Babson Collection; by the Trustees of the British Museum to reproduce a picture of the ivory bust by Le Marchand; by the University of California Press to reproduce six diagrams from their edition of the English translation of the *Principia;* by the Syndics of Cambridge University Library to reproduce eleven sketches and passages from the Portsmouth Papers; by Cambridge University Press to reproduce the picture of Croker's medal of Newton from John Craig, *The Mint* (Cambridge, 1953); by the Joseph Halle Schaffner Collection, University of Chicago Library, to reproduce a drawing of chemical furnaces; by Columbia University to reproduce an engraving of the Richter miniature and a lithograph of the Gandy portrait from the David Eugene Smith Collection, Rare Book and Manuscript Library; by the President and Fellows of Corpus Christi College, Oxford, to reproduce a drawing of the comet of 1680–1; by Lord Egremont and the Petworth Estate to reproduce the Kneller portrait of 1720; by the Bibliothèque Publique et Universitaire de Genève to reproduce the portrait of Nicolas Fatio de Duillier; by W. Heffer & Sons Ltd. of Cambridge, England, to reproduce the portrait of Newton in their possession; by the Jewish National and University Library to reproduce a scheme of the Revelation of St. John the Divine from the Yahuda Papers; by the Provost and Fellows of King's College, Cambridge, to reproduce a set of alchemical symbols from the Keynes Collection and an ivory plaque by Le Marchand; by the Trustees of the National Portrait Gallery to reproduce the Kneller portrait of 1702 and a portrait, artist unknown, of 1726; by Neale Watson Academic Publications, Inc., to reproduce four diagrams from Richard S. Westfall, *Force in Newton's Physics* (London, 1971); by the Warden and Fellows of New College, Oxford, to reproduce a drawing of Woolsthorpe, a picture of the house on St. Martin's Street, a drawing of the *experimentum crucis,* and a scheme of chronology from the New College MSS; by Lord Portsmouth and the Trustees of the Portsmouth Estates to reproduce two Kneller portraits as well as the Thornhill portrait of 1710; by the Royal Society to reproduce their drawing of the reflecting telescope, the sketch by Stukeley, an ivory plaque by Le Marchand, the Jervas portrait of 1703, the Vanderbank portrait of 1725, and the Vanderbank portrait of 1726; by Sotheby Parke Bernet & Co. to reproduce pictures of a bust and a plaque sculpted by Le Marchand; by the University of Texas to reproduce the Newton family tree; by the Master and Fellows of Trinity College, Cambridge, to reproduce the Thornhill portrait of 1710, the Murray portrait of 1718, the Vanderbank portrait of 1725, and the Seeman portrait of 1726; and by the Yale Medical Library to reproduce the drawing of Jupiter enthroned from their alchemical paper.

I wish further to acknowledge the permission and courtesy given me by the American Philosophical Society; Babson College (for the Grace K. Babson Collection); the Bodleian Library; the Syndics of the Cambridge University Library (for the Portsmouth Papers and other MSS); the University of Chicago Library (for the Joseph Halle Schaffner Collection); the William Andrews Clark Memorial Library (of UCLA, Los Angeles, California); the Rare Book and Manuscript Library of Columbia University (for the David Eugene Smith Collection); the Francis A. Countway Library of Medicine (Boston, Massachusetts); the Edinburgh University Library; the Emmanuel College, Cambridge, Library; the Syndics of the Fitzwilliam Museum, Cambridge; the Huntington Library (San Marino, California); the Jewish National and University Library (for the Yahuda MSS); the Provost and Fellows of King's College, Cambridge (for the Keynes MSS); the Pierpont Morgan Library; the Warden and Fellows of New College, Oxford; the Royal Society; the Smithsonian Institution Libraries (for the Dibner Collection); the Department of Special Collections of the Green Library, Stanford University (for the Newton Collection); the Controller of H.M. Stationery Office (for Crown-copyright records in the Public

Record Office); the University of Texas (for manuscripts in the Humanities Research Center, Austin); and the Master and Fellows of Trinity College, Cambridge, to cite manuscripts.

The University of California Press has allowed me to quote from the Cajori edition of Newton's *Principia:* Cambridge University Press to quote from I. Bernard Cohen and Alexandre Koyré, eds., *Isaac Newton's Philosophiae Naturalis Principia Mathematica;* from B. J. T. Dobbs, *The Foundations of Newton's Alchemy;* from A. R. and M. B. Hall, eds., *Unpublished Scientific Papers of Isaac Newton;* from H. W. Turnbull et al., eds., *The Correspondence of Isaac Newton;* and from D. T. Whiteside, ed., *The Mathematical Papers of Isaac Newton:* Dover Publications, Inc., to quote from their edition of Newton's *Opticks:* A. E. Gunther to quote from R. W. T. Gunther, *Early Science in Oxford:* Harvard University Press to quote from I. Bernard Cohen, ed., *Isaac Newton's Papers & Letters on Natural Philosophy: History of Science* to quote from Karen Figala, "Newton as Alchemist": Oxford University Press to quote from Mark Curtis, *Oxford and Cambridge in Transition;* from John Herivel, *The Background to Newton's 'Principia';* and from Frank Manuel, *The Religion of Isaac Newton: The Notes and Records of the Royal Society* to quote from J. E. McGuire and P. M. Rattansi, "Newton and the 'Pipes of Pan' " and from R. S. Westfall, "Short-writing and the State of Newton's Conscience, 1662": and Yale University Press to quote from Marjorie Hope Nicolson, *Conway Letters.* I gratefully acknowledge all of their kindnesses.

A note about dates

BECAUSE England had not yet adopted the Gregorian calendar (which it treated as a piece of popish superstition), it was ten days out of phase with the Continent before 1700, which England observed as a leap year, and eleven days out of phase after 28 February 1700. That is, 1 March in England was 11 March on the Continent before 1700 and 12 March beginning with 1700. I have not seen any advantage to this work in adopting the cumbersome notation 1/11 March, etc. Everywhere I have given dates as they were to the people involved, that is, English dates for Englishmen in England and Continental dates for men on the Continent, without any attempt to reduce the ones to the others. In the small number of cases where confusion might arise, I have included in parentheses O.S. (Old Style) for the Julian calendar and N.S. (New Style) for the Gregorian.

In England the new year began legally on 25 March. Some men adhered faithfully to legal practice; many wrote double years (e.g., 1671/2) during the period from 1 January to 25 March. Everywhere, except in quotations, I have given the year as though the new year began on 1 January.

Abbreviations used in footnotes

Add MS	Additional MS in the Cambridge University Library (for this book, that part of the Additional MSS constituting the Portsmouth Papers)
Babson MS	Newton manuscript in the library of Babson College, Babson Park, Mass.
Baily	Francis Baily, *An Account of the Revd John Flamsteed, the First Astronomer Royal* (London, 1835–7)
Burndy MS	Newton manuscript in the Dibner Collection, Smithsonian Institution Libraries
CM	Council Minutes of the Royal Society
Cohen	*Isaac Newton's Papers & Letters on Natural Philosophy*, ed. I. Bernard Cohen (Cambridge, Mass., 1958)
Comm epist	*Commercium epistolicum D. Johannis Collins, et aliorum de analysi promota* (London, 1713)
Corres	*The Correspondence of Isaac Newton*, ed. H. W. Turnbull, J. F. Scott, A. R. Hall, and Laura Tilling, 7 vols. (Cambridge, 1959–77)
CSPD	Calendar of State Papers Domestic
CTB	Calendar of Treasury Books
CTP	Calendar of Treasury Papers
Edleston	*Correspondence of Sir Isaac Newton and Professor Cotes*, ed. J. Edleston (London, 1850)
Halls	*Unpublished Scientific Papers of Isaac Newton*, ed. A. R. and Marie Boas Hall (Cambridge, 1962)
Herivel	J. W. Herivel, *The Background to Newton's 'Principia'* (Oxford, 1965)
Hiscock	W. G. Hiscock, ed., *David Gregory, Isaac Newton and Their Circle* (Oxford, 1937)
JB	Journal Book of the Royal Society
JBC	Journal Book (Copy) of the Royal Society
Keynes MS	Newton manuscript in the Keynes Collection in the library of King's College, Cambridge

Math	The Mathematical Papers of Isaac Newton, ed. D. T. Whiteside, 8 vols. (Cambridge, 1967–80)
Mint	Mint Papers in the Public Record Office
Opticks	Opticks, based on the 4th ed. (New York, 1952)
Prin	Mathematical Principles of Natural Philosophy, trans. Andrew Motte, rev. Florian Cajori (Berkeley, 1934)
Stukeley	William Stukeley, Memoirs of Sir Isaac Newton's Life, ed. A. Hastings White (London, 1936)
Var Prin	Isaac Newton's Philosophiae Naturalis Principia Mathematica, 3rd ed. with variant readings, 2 vols., eds. Alexandre Koyré and I. Bernard Cohen (Cambridge, 1972)
Villamil	Richard de Villamil, Newton: The Man (London, 1931)
Yahuda MS	Newton manuscript in Yahuda MS Var. 1 in the Jewish National and University Library, Jerusalem

1

The discovery of a new world

Convictus primus.	Secundus.	Quadrantarij.
M^r. Johes Smith.	Ric: Smith	Johes Bigge.
	Ed: Lowrey.	Isaac Newton.
Coll: Trin:	Johes Nowell	Josua Scargell.
	Tho: Ferrar.	Georg: Crosland.
	Barhamus Olyver.	Hen: Wright.
	Johes Doud.	Johes Tenant
	Johes Hawkins.	Eras: Sturton.[1]
	Johes Rowland.	
	Ed: Jolly.	

NEWTON'S name in the matriculation book of Cambridge University on 8 July 1661, together with those of sixteen other students recently admitted to Trinity College, bears witness to an event so obviously significant in his life (as it must have been for the other sixteen, and as similiar events have been for countless young men through eight centuries of Western history) that one flirts with banality even to mention it. He had left his home in the hamlet of Woolsthorpe in Lincolnshire some five weeks earlier, a raw provincial youth venturing more than ten miles from the place of his birth probably for the first time. He had been admitted to Trinity College on 5 June. As it turned out, the change in scene involved much more than the inevitable shattering of rural provincialism. In Cambridge, Newton discovered a new world. In one sense, of course, every youth who truly enters a university discovers a new world; such is the process of education, the opening of fresh horizons. Newton discovered a new world in a more concrete sense of the phrase, however. By 1661, the radical restructuring of natural philosophy that is called the scientific revolution was well advanced. Behind the familiar facade of nature, philosophers – we would call them scientists – had indeed discovered a new world, a quantitative world instead of the qualitative world of daily experience, mechanistic instead of organic, indefinite in extent instead of finite, an alien world frightening to many but in its challenge thrilling to some. In Cambridge, Newton discovered this discovery. It was by no means inevitable or even probable that he do so, for Cambridge University did not thrust the new world of scientific thought before its students. In all likelihood, the other sixteen young men who matriculated from Trinity that day in July never suspected

[1] Cambridge University Library, *Matriculations 1613–1702,* 8 July 1661.

1

its existence. The obscurity of their youth passed imperceptibly into the obscurity of their manhood, and no one today writes their biographies. Cambridge was a place where books were sold, however, and where libraries collected them. One who chose could encounter learning which the university itself did not foster; Newton chose, and with the choice determined his place in history. There were many facets to Newton's life that were not concerned with natural science, and a biography worthy of the name must present them. Nevertheless, the sole reason one undertakes to write a biography of Newton is the relation in which he stood to the new world of the scientific revolution.

As the heavens embrace our globe, so astronomy surrounded the scientific revolution. To date the beginning of an intellectual movement is always an arbitrary choice, but the succession of developments that followed the publication of Nicholas Copernicus's *De revolutionibus orbium coelestium* (*On the Revolutions of the Heavenly Spheres*) in 1543 has led historians almost unanimously to assign the birth of modern science to that year. Copernicus proposed a new solution to the central problem that had occupied astronomy for two thousand years, to account for the irregular motions of the planets against the unchanging background of the fixed stars.[2] Previous astronomy had started from the assumption, dictated by common sense and daily experience, that the earth is at rest. All the phenomena observed in the heavens, then, are real motions. Copernicus proposed instead that heavenly phenomena are in part mere appearances which arise from the motion of the earth.

To the earth Copernicus assigned two motions, a daily rotation on its axis and an annual revolution about the sun.[3] The daily rotation from west to east accounted of course for the apparent daily rotation of the heavens from east to west. More than four hundred years after Copernicus, we still find it convenient to speak of the rising and setting of the sun and moon; before Copernicus, they were held to do so literally. More important for astronomy, Copernicus boldly wrenched the earth from its moorings in the center of the universe, labeled it as one planet among others, and set it in motion with the others around the sun. He promoted the sun from the rank of planet and placed it at the center of the system. The annual orbit of the earth explained the most dramatic of the planetary phenomena, their periodic retrogressions in their normal progression from west to east among the fixed stars. To explain

[2] I use the phrase "new solution" in contrast to the prevailing astronomy. It is well known that Copernicus was not the first to propose a heliocentric system.

[3] Copernicus himself assigned a third motion to the earth, an annual conical motion of its axis. Kepler later pointed out that it was illusory.

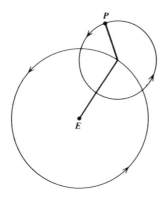

Figure 1.1. A major deferent on an epicycle whereby Ptolemaic astronomy accounted for the apparent retrogressions of the
planets.

these phenomena, Greek astronomers had developed the device of
the epicycle (Figure 1.1). The planets did not travel in circles about
the earth; rather they traveled in circles (epicycles) the centers of
which traveled in circles (deferents) about the earth. From our point
of view, this solution to the major problem of planetary orbits had
the effect of projecting the earth's annual orbit onto each planet's
motion – as epicycle in the cases of the superior planets (Mars,
Jupiter, and Saturn), as deferent in the cases of the inferior planets
(Mercury and Venus). The essence of Copernicus's revision of astronomy lay in his inauguration of our point of view and in his
demonstration that planetary retrogressions may be not real motions, but apparent motions deriving from the real motion of earth.

Unfortunately, the matter could not rest on this plane of simplicity. As ancient astronomers had established, one uniform circular
motion centered in the earth was unable to account for the positions
of the sun during a year, and two uniform circular motions, deferent and epicycle, were unable to account for the positions of the
planets. In all cases there were small deviations from the positions
predicted by the circles, to account for which astronomers had
resorted to other devices such as tiny epicycles and eccentric circles
(Figure 1.2). So also Copernicus found that a single circle could not
account for the motion of the earth or of any planet around the sun.
With the ancients he shared the conviction that the immutability
and perfection of the heavens require astronomy to confine itself to
combinations of the perfect figure, the circle. Hence he too had
recourse to tiny epicycles and eccentrics to account for the same
small deviations.

More than half a century later, Copernicus's greatest disciple,
Johannes Kepler, completed the structure of heliocentric astronomy

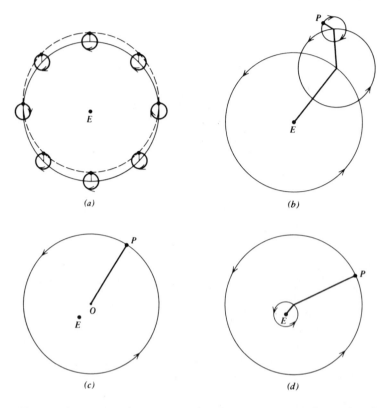

(a)   (b)

(c)   (d)

Figure 1.2. Various devices in Ptolemaic astronomy. *A* shows the effect of a minor epicycle with the same period as the deferent; *B,* an epicycle on a major epicycle; *C,* an eccentric; and
D, an eccentric on a deferent. The size of the minor epicycles and the amount of eccentricity in relation to the size
of the deferent circle are exaggerated considerably.

by abandoning the feature closest to Copernicus's heart, the perfect
circle. Employing the immense body of observations compiled by
the Danish astronomer, Tycho Brahe, Kepler concluded that Mars
travels about the sun in an ellipse, and he promptly generalized the
conclusion to all the planets. Published in his *Astronomia nova (A
New Astronomy, 1609),* the elliptical shape of the orbits is known as
Kepler's first law. In one stroke he swept away all the machinery of
epicycles and eccentrics. The minor deviations that had given rise to
them merged in the elegance of a single curved line. To the first law
he added a second, that the area of the ellipse swept out by the
radius vector of the moving planet is proportional to the time (Figure 1.3). With the second law he fulfilled a necessary demand of
every working astronomer in supplying a means by which to compute the location of a planet at any time. Ten years later, in the
Harmonices mundi (Harmonies of the World), Kepler added a third law

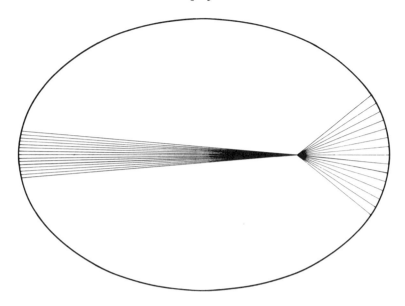

Figure 1.3. Kepler's second law, that the area swept out by the radius vector is proportional to time. The eccentricity of the el- lipse is greatly exaggerated.

which tied the planets together into an organized system by relating the mean radii and periods of their orbits: $T^2 \propto R^3$.

The same year that saw the publication of Kepler's *Astronomia nova* also witnessed the entry into astronomy of the instrument destined to become its principal tool. In 1609 Galileo Galilei first turned a telescope on the heavens, and the following year he began to publish his observations. Galileo was an ardent Copernican and attempted to turn his observations to the support of the Copernican system. He observed the rugged surface of the moon, which con- tradicted earlier notions of the crystalline perfection of the heavens. So did the spots on the sun, which formed and dissolved in rela- tively short periods of time, and which moved across its face, indi- cating that the sun turned on its axis. He discovered satellites of Jupiter, so that the earth ceased to be unique in its accompanying satellite. He observed the phases of Venus, which were incompat- ible with the geocentric system since they demonstrated that Venus circles the sun. All of this and especially the last lent support to heliocentric astronomy, though there was no way in which Galileo or anyone else could observe the motion of the earth or the central- ity of the sun through a telescope.[4]

---

[4] The literature on the development of heliocentric astronomy is immense. A brief account can be found in Thomas Kuhn, *The Copernican Revolution* (Cambridge, Mass., 1957), a longer account in Alexandre Koyré, *The Astronomical Revolution*, trans. R. E. W. Maddison (London, 1973).

When Newton arrived in Cambridge, the learned world had had a further half century to digest the work of Copernicus, Kepler, and Galileo. From the point of view of evidence, little had changed. Not until the nineteenth century did the Foucault pendulum furnish something like a direct demonstration that the earth turns on its axis, and the observation of stellar parallax directly confirm the annual motion. Nevertheless, by 1661 the debate on the heliocentric universe had been settled; those who mattered had surrendered to the irresistible elegance of Kepler's unencumbered ellipses, supported by the striking testimony of the telescope, whatever its ambiguities might be. For Newton, the heliocentric universe was never a matter in question.

But the heliocentric universe, especially in Kepler's formulation, offered as many challenges as conclusions. There had always been a tension within the geocentric system between physical and mathematical accounts. For cosmological purposes, Aristotle's homocentric spheres, which had solidified over the years into crystalline spheres, provided the physical picture of the universe. Working astronomers had meanwhile constructed planetary theory out of the various devices of deferents, epicycles, eccentrics, and equants. A final reconciliation between the two had been impossible, but astronomers had learned simply to tolerate the disparity. Heliocentric astronomy quickly abolished the crystalline spheres. In Kepler's view, Tycho's observations of comets had shattered the spheres; Kepler's own ellipses confirmed their demise. With the spheres went the very structure of the heavens, so that the new astronomy offered the improbable assertion that planets, without any means of support or guidance, whether visible or imagined, trace and retrace precisely defined ellipses in the immensity of space.

To this unavoidable problem Kepler offered a solution in terms of forces centered in the sun, one pushing the planets along in their paths, another controlling their distances from the sun. This celestial dynamics furnished the thread of Ariadne that Kepler's investigation successfully followed. Mathematically, it was consistent with his laws. It is one of the anomalies of seventeenth-century science, however, that a new science of mechanics rendered Kepler's dynamics obsolete within a generation, although we continue to accept as valid the laws that originally emerged from the dynamics. Descartes proposed a different physical theory of the cosmos. His vortex explained the gross phenomena of a heliocentric system; it carried all the planets in the same plane in the same direction around the sun. What the vortex could not produce were the exact mathematical relations of Kepler's laws – indeed not any one of the three.

Hence the old tension between the physical structure of the cosmos and the mathematical theory of astronomy reappeared in a new setting. One difference rendered the new version of the tension impossible to tolerate. In the ancient world, physics took precedence over astronomy. Astronomers might save the phenomena with ingenious constructions, but no one asserted their physical reality. Kepler did assert the reality of his ellipses as paths that planets follow through space; and in the end, the seventeenth century believed him. One way or another, the disparity between the mathematics of the heavens and their physics would have to be resolved; until it was, a gaping hole leered at natural philosophers from the very heart of their new science. Newton was one of those who quickly perceived this fact.

Astronomy, the science of the heavens, supplied the cosmic setting of the scientific revolution. The most important skirmishes of the revolution took place on earth. Early in the seventeenth century, Galileo laid the foundations of a new science of mechanics. It was impossible for there to be a radical restructuring of natural philosophy in which mechanics, the science of motion, would remain unmoved, for motion plays a central role in every conception of nature. Like astronomy, mechanics had a tradition stretching back to the ancient world; unlike astronomy, it had been a topic of extensive and productive discussion during the Middle Ages. In addition, it was inextricably entwined in the new astronomy. On the one hand, the received mechanics of Aristotle, even with its medieval modifications, refused to be reconciled to the assertion that the earth is in motion. On the other hand, in the climate of scientific thought as it existed in 1661, only the science of mechanics could resolve the cosmic problem posed by the new astronomy. If astronomy provided the setting of the scientific revolution, mechanics supplied its solid core.

More than any other man, Galileo created the new mechanics. A convinced Copernican, he devoted his energy not to the sort of technical details whereby Kepler brought astronomical theory into agreement with the observed positions of planets, but to the question of credence raised by the affront to common sense inherent in the proposition of a moving earth. The central argument in Galileo's *Dialogo sopra i due massimi sistemi del mondo (Dialogue on the Two Chief World Systems)* hinged on this point. As astronomy, the *Dialogue* was a fraud. It expounded a heliocentric system based on circular orbits which could not even approximate observed planetary positions. As mechanics, however, it showed that a moving earth can be compatible both with untutored daily experience and with carefully observed phenomena of motion on the earth. Central to Galileo's argument was the principle of inertia. Galileo did not in fact use the word "inertia"; for that matter, he did not enunciate the principle in the form we accept today. Nevertheless, he redefined the concept of motion in such a manner that we recognize in it the essential aspect of our principle of inertia.

To Aristotle, to move was to be moved. The motion of any body required a moving agent. Motion implied ontological change as well. The growth of an acorn, whereby it realizes its potential to be an oak, was motion. The education of a youth, whereby he realizes his potential to be a man of reason, was motion. Manifestly, both processes require a cause, that is, an agent or mover. Equally the motion of a heavy body falling, whereby it realizes its potential as a heavy body to be as close to the center of the universe as possible, appeared to require a moving agent. In the case of the heavy body, its fall was a natural motion caused by its nature as a heavy body once impediments to fall were removed. Bodies were also subject to violent motions in which external agents forced them to go where they had no inclination of their own; the moving agents were obvious enough in violent motions. Galileo set mechanics on a new course by redefining motion to eliminate most of its Aristotelian connotations. From the discussion of motion he cut away sprouting acorns and youths alearning. Local motion, which had been for Aristotle the simplest case and hence the example most suited to analysis, became for Galileo the sum total of the meaning of motion. And to motion in this sense, he insisted, a body is indifferent. A heavy body does not realize any potential when it falls; nothing it in changes. So also no violence is done to its nature when it is flung upward. Although he frequently used the phrase "natural motion," any real distinction between natural and violent motions disappeared from his mechanics. All motion is one and the same. It is not a process whereby potential is brought into realization. Motion is simply a state in which a body finds itself, a state to which it is indifferent.

Projectile motion had furnished the classic difficulty to Aristotelian mechanics, for a projectile continues to move after it has separated from its overt mover. Aristotle had solved the difficulty by arguing that the medium through which a projectile moves functions as mover and sustains the motion. Medieval philosophers had transferred the mover from the medium to the body itself. When it is placed in motion, they argued, a body acquires an impetus, an internal motive force, which sustains the motion. Galileo reduced projectile motion to its simplest case, a ball set rolling on a horizontal plane, and in his mind's eye he observed its motion. Since he required the frictionless plane of rational mechanics and a perfectly round ball, he had to observe it in his mind's eye. He concluded that under such ideal conditions a ball will continue to roll forever, as far as the plane continues. On real planes, of course, real balls come to rest, but the smoother the plane and the rounder the ball the longer it will roll. Motion is a state to which a body is indifferent. A ball on a horizontal plane can neither set itself in motion nor bring itself to rest. As Descartes, who shared this conception of motion, stated the case, philosophers had been asking the wrong question. They had asked what keeps a body in motion; one ought to ask instead what ever stops it.

With the new conception of motion, the principle of inertia, if I may speak loosely, the difficulties imagined to follow from the earth's motion dissolved away. Cannons fired east and west would carry equal ranges; cannons fired north and south would strike targets (or to speak more exactly perhaps of seventeenth-century gunnery, would miss them with normal facility). Objects dropped from towers would appear to observers to fall straight down since the observers are, of necessity, passengers on the same moving earth who participate in the common diurnal rotation. Galileo had more in mind than the question posed by astronomy, however. He proposed nothing less than a complete reconstruction of the science of motion – into a new science, as he proudly labeled it in his final book – and to reconstruct the science of motion completely was, in his view, to make it mathematical. Perhaps astronomy, which had always been a mathematical science of celestial motion, suggested the model. Galileo was accustomed to say that in the Copernican system the earth became a heavenly body. If the immutable heavens alone offer a subject proper to mathematics, the earth had been promoted into that class. One aspect of mechanics also offered a model of mathematical science. The balance, the lever, the inclined plane, the science of statics (as we would summarize them all), had invited mathematical treatment both in the ancient world and in medieval Europe. Archimedes especially, "divine Archimedes" as Galileo called him, had shown how the rigor of geometry could be applied even to a mundane science. Not the least of Galileo's self-esteem swelled from the realization that he had done what even divine Archimedes had not: To the mathematical science of bodies in equilibrium he had added a mathematical science of bodies in motion.

Galileo's new conception of motion supplied the cornerstone for the new structure. Defined mathematically, the motion of a perfect ball rolling on a frictionless horizontal plane is uniform motion. In equal periods of time it traverses equal distances. Bodies also move vertically, of course, and it had long been observed that when they fall, they move with increasing velocity. Medieval philosophers had even defined "uniformly difform motion," but they had analyzed it in the abstract without applying the definition to real motions. Changing the name to "uniformly accelerated motion," Galileo identified it with the motion of heavy bodies falling.[5] A body in

[5] As in the case of Copernicus, Galileo had a predecessor. Domingo de Soto, a Spanish scholastic of the mid-sixteenth century, had identified free fall as uniformly difform motion. He had not elaborated a full system of kinematics on the basis of his insight, however.

uniformly accelerated motion gains (or loses) equal increments of velocity in equal increments of time. Since he held, again in opposition to Aristotle, that all bodies are composed of the same matter, which is always heavy, he reasoned that bodies everywhere on the earth, whatever their size and whatever their substance, fall with a rate of acceleration common to them all. As in the case of uniform motion, the assertion supposed ideal conditions analogous to his frictionless planes. Since a medium such as air is always present, actual fall never realizes uniformly accelerated motion. As with horizontal motions again, the more one approximates ideal conditions, the more one approaches the defined motion.

The speed with which ordinary bodies fall, combined with the crudity of the instruments available to measure time, made it impossible to check this theory directly. Galileo recognized, however, that he could observe the identical phenomenon, slowed down to a measurable rate, on inclined planes. The established analysis of inclined planes even permitted him to calculate how much it is slowed down. Using a water clock to measure time, he corroborated the fact of uniformly accelerated motion in nature.

From the definition of uniformly accelerated motion Galileo proceeded to deduce the basic relations of kinematics that students still learn on their first introduction to mechanics, that a body falling from rest traverses distances proportional to the square of the time of the fall, that its velocity is proportional to the time of the fall and to the square root of the distance fallen. As a final tour de force, he demonstrated that a projectile, which (again under ideal conditions) moves with a motion compounded of uniform horizontal and uniformly accelerated vertical elements, must follow a parabolic trajectory.[6]

By 1661, the science of mechanics had assumed an anomalous stance. Though Galileo had not written in the Latin of the learned world but in Italian, a number of publicists had made his results available to the European scientific community. His relations of acceleration, velocity, distance and time both in uniform and in uniformly accelerated motion had become the common property of the science, accepted by all and questioned by none. Nevertheless,

---

[6] The literature on Galileo is at least as immense as that on the new astronomy. The most influential work has been Alexandre Koyré, *Etudes galiléennes* (Paris, 1939). Maurice Clavelin, *La Philosophie naturelle de Galilée* (Paris 1968), offers a more recent, excellent account. Stillman Drake has treated Galileo's mechanics in a series of articles too numerous to quote entirely here; among the most important are "The Concept of Inertia," *Saggi su Galileo Galilei* (Florence, 1967), pp. 3–14; "Galileo and the Law of Inertia," *American Journal of Physics, 32* (1964), 601–8; "Uniform Acceleration, Space, and Time," *British Journal for the History of Science, 5* (1970), 21–43. He has recently summarized his work on Galileo in *Galileo at Work: His Scientific Biography* (Chicago, 1978). *Galileo, Man of Science*, ed. Ernan McMullin (New York, 1967), contains articles which deal with every aspect and the major interpretations of Galileo's science.

the reconstruction of mechanics that Galileo had dared to promise had scarcely begun. The foundation on which his kinematics rested, the new conception of motion, presented difficulties of comprehension not easily surmounted. Pierre Gassendi had given a wonderfully clear statement of it; later in the same work, he had also stated the concept of impetus and apparently had failed to recognize that the two were incompatible. John Wallis would repeat that performance ten years hence. In fact, in 1661, only two significant figures had embraced the principle of inertia, René Descartes and Christiaan Huygens. Indeed, it was Descartes and not Galileo who stated the principle of inertia in the form we accept today, insisting on the rectilinear character of inertial motion. His writings were the primary vehicle by which the concept spread. In 1661, Huygens' work on mechanics still remained locked in his manuscripts. Other students of mechanics had derived Galileo's kinematic relations from concepts of impetus.[7]

In the meantime, two further issues of mechanics had been raised, both by Descartes. Partly for reasons connected with his philosophy as a whole, which allowed a body to influence only those other bodies which it touched, Descartes undertook to define the laws of impact. Almost no one accepted the laws as he gave them in his *Principia philosophiae (Principles of Philosophy, 1644)*, but by stating inadequate laws, he bequeathed a problem to the science of mechanics. In the same work, he attempted to define the mechanical elements of circular motion, which had only become a problem with the principle of rectilinear inertia. Once again, Descartes mangled the analysis and in so doing left a second problem. Both problems were attacked successfully by Christiaan Huygens in the late 1650s; but in 1661, he had not yet published either result. To one attracted to it, the science of mechanics at that time presented more challenges than conclusions.

Indeed no one, not even Huygens, had yet dreamed what implications lay hidden in those alluring kinematic equations of horizontal and vertical motion. In 1661, they remained as Galileo had presented them, brute facts of nature, enigmas to be probed, riddles to be explained. As yet, no causal dynamics made the uniformly accelerated motion of falling bodies appear as a natural result. When at length one was supplied, when the enigma was probed, not merely mechanics, but the entire body of natural philosophy shook with the consequences. One can scarcely speak of this as an unresolved problem in 1661, since even its possibility had yet to be recognized.

[7] See, for example, the discussion of Marcus Marci and of Evangelista Torricelli in Richard S. Westfall, *Force in Newton's Physics* (London, 1971), pp. 117–38. Later Claude-François Milliet de Chales, Giovanni Alfonso Borelli, and Edme Mariotte would carry through essentially identical analyses (*ibid.*, pp. 200–3, 213–30, 243–56).

Implicit it might be, but the challenge was there nevertheless, the greatest challenge of them all.

Like mechanics, the science of optics had its own tradition, one more medieval than ancient, and one which also received a powerful stimulus from astronomy in the seventeenth century. To observe the heavens was to employ the science of optics, whether one was conscious of it or not. The rectilinear propagation of light, which every observation assumed, was known to hold precisely only for celestial bodies directly overhead. Atmospheric refraction distorted the observed positions of every other body, though no one knew exactly how much. In observing eclipses, moreover, astronomers employed pinhole devices which introduced further problems in optics. Indeed, it was a problem introduced by the pinhole device which stimulated Kepler's initial interest in optics, and the subtitle of his great work of 1604, *Astronomiae pars optica (The Optical Part of Astronomy)*, stands as a permanent monument to the seminal role of astronomy in optics. Kepler's problem was the apparent shrinkage of the moon as it passed across the face of the sun. Having shown that the apparent shrinkage was a purely optical phenomenon generated by the pinhole device and quantitatively related to the size of the pinhole, Kepler extended his insight to the explanation of vision. The pupil of the eye became the pinhole opening of the astronomical device, but equipped with a focusing lens. The retina became the screen on which the image fell. The concept of the retinal image, enunciated by Kepler, offered a new answer to the central question of vision on which the entire history of optics had turned before the seventeenth century. When the telescope burst upon an unsuspecting world six years later, and the microscope shortly thereafter, optics assured itself an important role in seventeenth-century science.

Foremost among the achievements of optics after Kepler's seminal work was the discovery of the law of refraction, a subject explored by Kepler even before the telescope, but one made crucial by the instrument. Published by Descartes in 1637, the sine law of refraction complemented the law of reflection known since ancient times. Not long thereafter Francesco Maria Grimaldi discovered the phenomenon of diffraction, although his work was not published until 1665. Together, the two discoveries suggest the focus of seventeenth-century optical science, not the problem of sight which had occupied optics before, but the nature of light. By the seventeenth century, light was universally agreed to be an objective reality and not a power emitted by a seeing subject. It was a constituent aspect of nature which could be studied on the same terms with the rest of nature. The sine law of refraction added a new regularity to its behavior; the phenomenon of diffraction added mostly a puzzle.

Neither seemed to entail a definitive answer to the basic question now at the center of study: What is the nature of light? Descartes answered that it is a pressure transmitted instantaneously through a transparent medium. Gassendi, the reviver of atomism, replied alternatively that it is a stream of tiny particles moving with unimaginable speed. There were obvious problems with Gassendi's answer. Not only did it defy common sense, but it also appeared to contradict such phenomena as two men looking each other in the eye. On the other hand, very few liked Descartes's theory, which, in order to account for the geometrical regularities of light such as the very sine law he published, had resorted to the implausible argument that pressure, as a tendency to motion, obeys the same laws as motion itself. Nevertheless, Descartes's theory was capable of elaboration. His steady pressure might be converted to pulses of pressure, sometime called waves, and pulses or waves might overcome the difficulties presented by pressure alone. By 1661, at least two men were beginning to think in such terms: Robert Hooke and Christiaan Huygens. Neither had published a theory of light by that time, nor was it evident that such a theory would solve all the problems.

Moreover, the terms in which seventeenth-century science viewed nature as a whole had presented yet another problem to optics, the problem of color. Except in a small number of exceptional cases such as the rainbow (in which colors seemed to be associated more with the observer than with physical objects), colors had been placed among the properties of bodies. The new philosophy of nature asserted that all phenomena of colors are identical to the so-called exceptional cases. Size and shape alone are the real properties of bodies; colors exist only insofar as there are sentient subjects to observe them. From this point of view, colors were associated with light. Colors became a problem in optics. No one asserted that light itself is colored. Rather, as light itself, in falling on a retina, arouses sensations of sight, so also some aspect of light arouses sensations of color. Once again, Descartes was the man who posed the question; and once again, his proposed solution roused few enthusiasts. It remained to be seen what better answer could be offered. Like astronomy and mechanics, optics in 1661 was anything but a closed science.[8]

---

[8] Until recently, the history of optics has been studied much less extensively than the history of astronomy and mechanics. The established authority, Vasco Ronchi, *Storia della luce* (Bologna, 1939), only recently translated into English as *The Nature of Light: An Historical Survey,* trans. V. Barocas (London, 1970), has been superseded by several more recent works. For the history of optics through Kepler see David Lindberg, *Theories of Vision from Al-kindi to Kepler* (Chicago, 1976). The best works on optics in the seventeenth century are A. I. Sabra, *Theories of Light from Descartes to Newton* (London, 1967) and Alan E. Shapiro, "Kinematic Optics: A Study of the Wave Theory of Light in the Seventeenth Century," *Archive for History of Exact Sciences, 11* (1973), 134–266.

A common thread which insistently catches the twentieth-century eye runs through the astronomy, mechanics, and optics of the sixteenth and seventeenth centuries. Their achievements were expressed in mathematical terms, and what we seem to see is the establishment of the pattern that has increasingly dominated natural science ever since, the mathematization of nature. Perhaps that thread did not stand out so strikingly to an observer in 1661. Although seventeenth-century natural philosophers were as convinced as we are that a revolution was taking place in the study of nature, it was not the mathematical description of a handful of phenomena or the ultimate implications that might lie behind the possibility of such descriptions which they considered to be the revolution. They thought instead of the new philosophy of nature that had overthrown the Aristotelian categories that had dominated natural philosophy in the West for two thousand years. As far as men active in the study of nature were concerned, the word "overthrown" is not too strong. For them, Aristotelian philosophy was dead beyond resurrection. In its place stood a new philosophy for which the machine, not the organism, was the dominant analogy.

René Descartes, who contributed both to mechanics and to optics, contributed far more to the mechanical philosophy of nature. Although he was neither the first nor the sole philosopher in the seventeenth century to approach nature in mechanical terms, and although he and others could draw upon the inspiration of ancient atomism, Descartes was nevertheless the principal architect of the seventeenth century's mechanical philosophy. Hinted at in his *Discours de la méthode (Discourse on Method)* and its attendant essays (1637), given metaphysical underpinning in his *Meditationes de prima philosophia (Meditations on First Philosophy, 1641)*, Descartes's mechanical philosophy of nature was spelled out in full in his *Principia philosophiae (Principles of Philosophy, 1644)*.

Basic to it was the Cartesian dualism which attempted rigorously to separate mind or spirit from the operations of physical nature. The essence of mind (*res cogitans*) is the activity of thinking. In using the active participle to designate mind, Descartes deliberately specified that mind is the only locus of spontaneous activity in the universe. In making its activity thought alone, however, he equally indicated that mind or spirit does not play the role of an activating principle in physical nature. To be sure, the Creator summoned nature into being and continues to sustain it by His general concourse. He also decreed the laws by which nature operates, however, and He does not intervene to alter or obstruct their necessary operation. In addition, human souls can will motions for their bodies. Nevertheless, most of the functions of the human body, from digestion and growth on the one hand to reflex actions on the other, proceed independently of the human will. And the overwhelming majority of nature's phenomena, from the panorama of the heavenly vortices above us, through the realm of animal and vegetable life about us, to the motions of the particles of bodies below the threshold of perception, are never affected by an act of a human will. Physical nature, composed of extended matter (*res extensa*), is a machine which operates as machines must, according to the laws of mechanics. In the case of matter, Descartes consciously chose the passive participle, *extensa,* to express its nature. Wholly inert, shorn of any source of activity whereby it may initiate any change of motion, matter, together with the physical universe it composes, became the realm of pure physical necessity. Matter is divided into discrete particles which impinge on each other in their motions, producing the phenomena of nature. Though single particles may alter their motions in impact, coming to rest, speeding up, or changing direction, the total quantity of motion in the universe remains constant. This was the basic law of Descartes's mechanical universe; it insured that nonmechanical agents need not be introduced.

Although Descartes did not employ the analogy of the clock, later mechanical philosophers would make it the image of their conception of nature. "Clock" to them summoned up the image of the great cathedral clocks. Their descending weights powered the hands turning on their axles and a great deal more. Mechanical men rang bells. Saints appeared through the doors. Cocks crowed to tell the hour. One complex mechanism generated a multiplicity of operations that suggested the infinite phenomena of the infinitely complex machine we call nature, except that the cosmic clock made by the divine watchmaker required no winding up. The conservation of the total quantity of motion, and the principle of inertia on which it rested, insured its eternal operation.

If Descartes's mechanical philosophy of nature was the most prominent, it was not the only one. In England, Thomas Hobbes produced a similar, though less elaborately formulated philosophy. In France, the principal agent in the revival of atomism, Pierre Gassendi, offered another alternative. From a philosophical point of view concerned with the rigor of arguments, their differences were important. From the point of view of many practicing scientists, such as Robert Boyle, the extent of their agreement obscured serious attention to their differences. All agreed on some form of dualism which excluded from nature the possibility of what they called pejoratively "occult agents" and which presented natural phenomena as the necessary products of inexorable physical processes. All agreed that physical nature is composed of one common matter, qualitatively neutral and differentiated solely by the size, shape, and motion of the particles into which it is divided. All agreed that the program of natural philosophy lay in demonstrating that the phenomena of nature are produced by the mutual interplay of material particles which act on each other by direct contact alone. Thus the two competing conceptions of light in the later seventeenth century were the two possible conceptions allowed by the mechanical philosophy; light was either particles of matter in motion or pulses of motion transmitted through a material medium.

Descartes had defined the program of the new philosophy of nature in his sixth and final meditation. The first meditation, as he began the systematic search that led to a new foundation of certainty, had called the existence of the external world into doubt. We believe in it primarily on the evidence of our senses; but since our senses sometimes err, the existence of the external world cannot on their evidence claim the metaphysical certainty Descartes was seeking. By the time he reached the sixth meditation, Descartes was ready to replace the existence of the external world in the structure of certainty built on the rock of the *cogito*. The assertion of its existence now rested on a different foundation, not on the facile assumption of common sense buttressed by the unexamined perceptions of the senses, but on the evidence of necessary arguments from first principles. There is, he added – and the statement was the most important assertion in natural philosophy during the entire seventeenth century – no corresponding necessity that the external world be in any way similar to the one our senses depict. Our senses reveal a world of qualities. Reason tells us that quantity alone exists, particles of matter differentiated solely by their size, shape, and motion, which produce sensations of qualities when they impinge on our nerves. The program of the mechanical philosophy lay in demonstrating that assertion.

Aspects of the mechanical philosophy suggest an inherent harmony with the developments I have surveyed in astronomy, mechanics, and optics. The very word "mechanical" seems to incorporate the science of mechanics, and the program of the mechanical philosophy, to trace all phenomena to particles in motion, seems to demand the same. The assertions that quantity alone is real and that qualities are only sensations recall the mathematically formulated laws of physics. The harmony may be more apparent than real, however. The formalism of mathematical laws never satisfied the mechanical philosophy's demand that phenomena be explained in terms of particles in motion. Its explanations in turn refused to yield the mathematical laws. As we have seen, Descartes's vortices were incompatible with Kepler's laws, not just the ellipses, but all three laws. His explanation of gravity, like rival mechanical explanations, was incompatible with Galileo's kinematics of free fall. In order to derive the sine law of refraction, Descartes had to introduce arbitrary conditions which no one found believable; but when Pierre Fermat derived the law from the principle of least time, mechanical philosophers refused to have truck with what they considered occult notions. The words "quantitative" and "mechanical" should not mislead us into seeing the two dominant trends in seventeenth-century science as different facets of one program.

In 1661, to a young man attracted by the excitement of scientific discovery, the mechanical philosophy of nature possibly offered the more exciting prospect. Here no limiting horizons restricted one's attention to confined problems such as the shape of an orbit or the angle of refraction. If a new world had indeed been discovered, that world was found in the mechanical philosophy which took the world, the whole world, as its province. The aura of freshness still clung to it. The classic exposition was less than twenty years old; Gassendi's rival version, in its final statement, less than five. Its promise had ravished young Christiaan Huygens in the Netherlands, and in England, young Robert Boyle. Dissatisfied with the arid formulations of a tradition that seemed capable only of repeating insights now two thousand years old, they grasped eagerly at the promise of a completely new program which offered progress instead of repetition, true understanding of the depths of nature instead of superficial knowledge. It is a fact that no figure of importance in European science in the second half of the seventeenth century stood outside the precincts of the mechanical philosophy of nature.[9]

In 1661, nevertheless, to embrace the mechanical philosophy was also to face a series of problems which could not be avoided. For the philosophically serious, the rival mechanical systems demanded choices – the plenum or the void, the continuum or discrete particles. For the religiously serious, the exclusion of spirit from physical nature required that the full rigor of the system be compromised in the name of Christian sensibility. For the scientifically serious, the apparent incompatibility of mechanical explanations with rigorously formulated mathematical laws posed an equally serious problem. The mechanical philosophy excited the best minds of several generations, but its problems were acute. Their very acuteness suggests the possibility that disillusionment and revision might succeed the enthusiasm of youth.

---

[9] Of the vast literature on Descartes, the great majority is devoted to issues that fall outside his natural philosophy. The best expositions of the mechanical philosophy of nature are found elsewhere: in R. G. Collingwood, *The Idea of Nature* (Oxford, 1945); R. Harré, *Matter and Method* (London, 1964); Marie Boas [Hall], "The Establishment of the Mechanical Philosophy," *Osiris, 10* (1952), 412–541; E. J. Dijksterhuis, *The Mechanization of the World Picture*, trans. C. Dikshoorn (Oxford, 1961); R. Lenoble, *Mersenne ou la naissance du mécanisme* (Paris, 1943).

Standing largely outside the traditional boundaries of natural philosophy was another area of investigation, chemistry, which gradually asserted its role in an enterprise which was itself gradually changing from natural philosophy to natural science. Although the Aristotelian doctrine of four elements offered a possible foundation for chemical theory, an extensive Aristotelian chemistry had never been elaborated, and virtually no chemist in the seventeenth century was an Aristotelian. In the sixteenth century, Paracelsus had proposed a rival theory in which the four elements were supplemented by three principles—salt, sulfur, and mercury—which functioned as the primary agents in chemical explanation. Perhaps he merely adumbrated the theory and his followers elaborated it. By the dawn of the seventeenth century, in any case, Paracelsian chemistry dominated the field.

Chemistry was far more experimental than natural philosophy as a whole, with its background in the medieval universities and the ancient schools. The relative isolation of chemistry as a field of study stemmed partly from this fact. A student of mechanics worked at a desk with paper and pen. Until Tycho and the telescope brought about a new order, the same was largely true of an astronomer. The chemist, on the other hand, labored with crucibles and alembics at a hot furnace: "sooty empiric" the gentlemen scholars scornfully called him. One consequence of chemistry's experimental emphasis was an enormous expansion of its body of empirical knowledge. By the middle of the century, chemists controlled a corpus of experimental results whereby they were able to compound with assurance a range of substances wider by far than their predecessors had known. But chemistry lacked an adequate structure of theory which could give coherence and direction to the rapid growth of information. Perhaps the single-minded experimentation in chemistry stemmed partly from the failure of theory. Paracelsus's three principles were more a doctrine of substance than a chemical theory. Salt, sulphur, and mercury, as body, soul, and spirit, were the necessary constituents of any existent body. One might philosophize in this vein, but one could scarcely organize a body of experimental information effectively. The major chemical generalization of the century, the recognition that alkalis and acids neutralize each other, was announced by Johannes van Helmont, who was powerfully influenced by Paracelsian thought. The generalization can hardly be said to have grown naturally from the stock of Paracelsian theory. Quite the contrary, it could only be grafted onto that theory by brute force. Paracelsian chemists arbitrarily divided the principle salt to include, among others, acid salts and alkaline (or lixiviate) salts. Van Helmont's insight did not expand the explanatory power of Paracelsian theory; Paracelsian theory cast no illumination on van Helmont's insight.

In 1661, however, chemistry was undergoing what promised to be a momentous change. In that year Robert Boyle, who is sometimes called the father of chemistry, published his most famous, if not his most important book, *The Sceptical Chymist*. Boyle's work consisted of a sustained polemic against the Aristotelian concept of an element and the Paracelsian concept of a principle. He treated the two as brothers to be slain with the same sword. What he offered in their place was mechanical philosophy, that is, chemistry stated in terms of the mechanical philosophy of nature. Boyle's most famous sentence, his definition of an element, appeared in *The Sceptical Chymist*. It has been quoted and requoted by authors who have failed to notice that it merely repeated the traditional definition and that the entire work devoted itself to denying the possibility of elements in this sense, as Boyle explicitly stated in the final clause of the very sentence that contains the definition. Elements and principles do not exist. What does exist is the qualitatively neutral matter of the mechanical philosophy, divided into particles differentiated only by size, shape, and motion. From their various combinations arise all the appearances of the substances with which chemists deal.

To chemists Boyle offered full participation in the fraternity of natural philosophers. By mechanizing chemistry, he effectively obliterated the barriers that had separated their enterprise from the rest of natural philosophy. To the mechanical philosophy of nature he offered what chemistry alone could provide, an articulated science of matter, which surely had to be essential to a philosophy asserting that all phenomena result from particles of matter in motion. In 1661, Boyle's program, to which he and others would devote lifetimes of labor, was only a promise. He had shown that chemical phenomena could be expressed in mechanistic language. He had not yet shown that fuller understanding necessarily followed. If his program was still only a promise, nevertheless it was an exciting promise, especially to those who found the new mechanical approach to nature exciting.[10]

Even in the case of Boyle, chemistry at that time consisted of more than his program to mechanize it. Alchemy was still a living

[10] On Paracelsus and Paracelsian chemistry, see Walter Pagel, *Paracelsus. An Introduction to Philosophical Medicine in the Era of the Renaissance*, (Basel and New York, 1958), and *The Religious and Philosophical Aspects of van Helmont's Science and Medicine* (*Supplements to the Bulletin of the History of Medicine*, No. 2, Baltimore, 1944) and Allen Debus, *The English Paracelsians* (London, 1965). The best general expositions of seventeenth-century chemistry are Hélène Metzger, *Les doctrines chimiques en France du début du XVIIᵉ à la fin du XVIIIᵉ siècle* (Paris, 1923) and Marie Boas [Hall], *Robert Boyle and Seventeenth-Century Chemistry* (Cambridge, 1958).

enterprise and the sixteenth and seventeenth centuries witnessed its culmination in Western Europe. One of the greatest of English alchemists, who wrote under the pseudonym of Eirenaeus Philalethes (probably George Starkey), was still alive in 1661, and Elias Ashmole's collection of British alchemical writings, *Theatrum chemicum Britannicum*, had appeared only nine years earlier. Boyle himself had been introduced to chemistry by the alchemical circle in London surrounding Samuel Hartlib, and through a long career devoted to integrating chemistry into the mechanical philosophy he never ceased to search into the Art. Alchemy was alive in Cambridge as well, where Ezekiel Foxcroft, fellow of King's College, maintained connections with Hartlib's circle in London. Undoubtedly there were also other alchemists in Cambridge.

Alchemy was no one thing. By the seventeenth century it had been practiced in the West in a series of different cultures, from Hellenistic through Islamic and Latin-medieval to the European culture of the sixteenth and seventeenth centuries. Alchemy had had to adjust itself continually in order that different ages might comprehend it. Medieval alchemists expounded the Art in terms of Aristotle's four elements; alchemists of the late sixteenth century used Paracelsian salt, sulfur, and mercury. Hartlib's circle, influenced by the Platonist school of English philosophy, pursued a Neoplatonic alchemy. The varying philosophic costumes, however, always clothed an animistic philosophy of nature which appears contradictory in its very essence to the mechanical philosophy. Alchemists believed that life rather than mechanism stands at the very heart of nature. All things are generated by the conjunction of male and female; metals differ in no wise from the rest of nature. Like everything else, metals grow in the womb of the earth – rather, metal grows, for if we speak in strict terms, alchemy did not recognize more than one metal. That one, of course, was gold, the product that nature realizes when nothing interrupts her normal gestation. The other "metals," abortions of nature, are potential gold that has failed to reach maturity. Alchemists were trying to complete what nature had left incomplete. They were growing gold.

Composed of living things, nature to the alchemist abounded with centers of spontaneous activity, vital or active principles. These principles differed irreducibly from the inert matter of the mechanical philosophy which was able only to respond passively to external actions. Alchemists sought to extract the active principles of nature from the feculent dross which weighted them down. (The scatological imagery of alchemy was quite as explicit as the sexual.) They also sought to obtain pure matter or soil in which to plant the seeds. We must recall that every theory of generation until the late seventeenth century recognized a female semen. Conception and generation were held to occur when the male semen fermented with it. For the alchemist, the philosophic sulfur, inevitably the male principle in that age, the ultimate active agent, required the philosophic mercury, the female principle, the soil in which as seed it could be planted. Alchemists were forever separating and purifying. Purification – here was the basic need. Before the soul could shine the feces that fouled it had to be scrubbed away. Before it could fly, its burden of dross had to be removed. Before it could truly live, its encumbering flesh had to be purged. But once it was freed, it was "the subject of wonders," "the miracle of the world," "a most puissant & invincible king."[11]

Toward alchemy the seventeenth century was ambivalent.[12] To cite but one example, Ben Jonson's play *The Alchemist* testifies that some regarded alchemists as charlatans and rogues. By pretending to knowledge of how to make gold, they earned their way by gulling the gullible. Although there is no way to measure general opinion, it seems probable that a large number shared Jonson's attitude. Before we are tempted to see the beneficent influence of scientific skepticism already at work, we ought to recall that Geoffrey Chaucer expressed substantially the same opinion more than two centuries earlier. After all, gold has been a desirable commodity in all ages, and there are inherent dangers in repeatedly promising to make it unless one actually produces. The existence of skeptics is hardly surprising. The continued appeal of alchemy in all

---

[11] I have taken these particular expressions from Newton's alchemical papers: *Keynes MS* 40, f. 19ᵛ; *Keynes MS* 41, f. 15ᵛ; *Keynes MS* 40, f. 20. On alchemy in general, see Mircea Eliade, *The Forge and the Crucible*, trans. Stephen Corrin (New York, 1971); Arthur J. Hopkins, *Alchemy, Child of Greek Philosophy* (New York, 1934); John Read, *Prelude to Chemistry* (London, 1936); F. Sherwood Taylor, *The Alchemists, Founders of Modern Chemistry* (New York, 1949); Robert Multhauf, *The Origins of Chemistry* (London, 1966).

[12] A very small handful excepted, the twentieth century is not, and even to bring up alchemy is to court misunderstanding. Herbert Butterfield asserted, without much delicacy, that those who study alchemy in the twentieth century are "fabulous creatures themselves" who "seem sometimes to be under the wrath of God themselves . . . [and] seem to become tinctured with the kind of lunacy they set out to describe" (*The Origins of Modern Science*, rev. ed., New York, 1965, p. 141). Since I shall devote quite a few pages to Newton's alchemical interests, I feel the need to make a personal declaration. Even though I am sure that no fabulous creature recognizes the lunacy with which he is tinctured, I can only state my own perception of the situation. I am not myself an alchemist, nor do I believe in its premises. My modes of thought are so far removed from those of alchemy that I am constantly uneasy in writing on the subject, feeling that I have not fully penetrated an alien world of thought. Nevertheless, I have undertaken to write a biography of Newton, and my personal preferences cannot make more than a million words he wrote in the study of alchemy disappear. It is not inconceivable to most historians that twentieth-century criteria of rationality may not have prevailed in every age. Whether we like it or not, we have to conclude that anyone who devoted much of his time for nearly thirty years to alchemical study must have taken it very seriously – especially if he was Newton.

ages, but especially in the middle of the seventeenth century, when the movement of thought seemed set so decidedly against it, is what requires explanation.

I cannot pretend to give the explanation, either in general or in the particular case that concerns me here: I can only suggest several related speculations. Alchemy offered exactly what the mechanical philosophy of nature offered, though, to be sure, in a rather different package. That is, alchemy also promised to reveal the ultimate secrets of nature. Vulgar multipliers were one thing; there were Ben Jonsons aplenty to unmask them. Meanwhile serious men beyond suspicion of petty fraud found in alchemy something more precious than recipes to make gold, an all-embracing philosophy of irresistible attraction. In the seventeenth century alone, to name only a few, Michael Maier, Jean d'Espagnet, Johann Grasshoff, George Starkey, Alexandre Toussaint de Limojon de Saint Disdier, and Edmund Dickinson, who span the century from beginning to end, testify to that attraction. Are we to equate these men with Jonson's cheat? Is it impossible to imagine a grave young man, philosophically minded, desirous of understanding not limited problems but all that is – and perhaps convinced that he can – is it impossible to imagine the alchemical philosophers exciting him even as the mechanical philosophers did? In both philosophies, one had the pleasure all youths of spirit feel in rebelling against the established order. In 1661, the achievements of the mechanical philosophy could scarcely preclude the consideration of alternative philosophies.

If it resembled the mechanical philosophy in its scope, alchemy parted from it decidedly on a question of importance to most men of the seventeenth century. Without denying God as Creator, the mechanical philosophy denied the participation of spirit in the continuing operation of nature. Alchemy did not merely assert the participation of spirit; it asserted the primacy of spirit. All that happens in nature is the work of active principles, which passive matter serves as a mere vehicle. To the religiously sensitive, alchemy might have special attractions. Moreover, alchemy was pervaded with Christian symbolism. The delivery of active principle from its body of death, its renewal and exaltation, showed forth the rebirth of the soul in Christ in language that could not be mistaken. Even the resurrection of the body found its image in the purified *terra alba* prepared for the resurrected seed.

Nor was it evident to the seventeenth-century mind that the two philosophies of nature were mutually exclusive. The mechanical philosophy was a protean idiom. Those who spoke it were able to translate any theory into the language of particles in motion. On close examination, Descartes's three elements appear to be cousins german to the Paracelsian *tria prima*. His first element, the ceaselessly agitated subtle matter, might double very well for the philosophic sulfur, the ultimate active principle in the alchemical philosophy. Lines of demarcation that appear absolute to the twentieth-century mind were less clearly drawn in the seventeenth. It was perfectly possible to respond at once to the promise and excitement of both, perhaps to attempt to supply the defects of each from the achievements of the other.

Alchemy had the possible further attraction of the deliberately arcane. Where the mechanical philosophy sought to illuminate obscurity by the light of reason, alchemy employed obscurity to protect truth from contamination. Frequently alchemists themselves concealed their identity behind pseudonyms, such as Eirenaeus Philalethes. Alchemy did not speak in the ordinary language of ordinary mortals; it spoke in tongues, concealing its message in outlandish imagery comprehensible only to the initiated. Instead of a new program to correct the errors of centuries, it offered an old program, an ancient wisdom handed down generation by generation to a select few. An alchemist was not made; he was chosen. He was one of the elect, a member of the invisible fraternity of the sages of all generations. There is no need to contrast the twenteith century with the seventeenth in this case; the lure of the arcane is eternal. Nor is it impossible to imagine how a young man convinced of his genius might feel himself a member by right in the timeless circle of the elect.

The antipodes of alchemy with its eternal and exasperating secretiveness was mathematics, the very claim of which to be called knowledge rested on demonstrations open to all. Where the one made its way deviously with allusion and symbolism, the other proceeded in the cold light of rigorous logic. The diversity of the intellectual world of the seventeenth century has perhaps no better illustration than the coexistence of two such antithetical enquiries, both apparently in flourishing condition. Only to later ages would it be clear that seventeenth-century alchemy was the last blossom from a dying plant and seventeenth-century mathematics the first blooming of a hardy perennial. Whatever the state of alchemy, certainly it was manifest in 1661 that mathematics was a flourishing enterprise.

The very year 1661 marked a turning point in seventeenth-century mathematics. It witnessed the completion of the century's most influential mathematical publication, van Schooten's second Latin edition of Descartes's *Géométrie* with the addition of his own lengthy commentary and a number of ancillary treatises. No other work summed up the achievements of early seventeenth-century analysis so well. As a result, no work provided a more solid foundation for subsequent building.

The name "analysis," which seventeenth-century mathematicians gave to their work, both attaches it to and distinguishes it from ancient geometry. Part of the common lore of seventeenth-century mathematics concerned a method, now lost, by which the ancient geometers had made discoveries. The geometrical works that survived from the ancient world demonstrated the discoveries in the familiar language of synthetic geometry, starting from self-evident axioms and proceeding by ineluctable steps of geometrical logic. Seventeenth-century mathematicians set out to resurrect the method of discovery, or analysis, by which they were convinced the discoveries had originally been made. We still call what they created "analytic geometry." Part of the program was to show that the method provided a ready means of attack on knotty problems, such as drawing tangents to curves, and most importantly on the climactic problem of ancient geometry, the three- or four-line locus (or *locus solidus*). In actual fact, seventeenth-century analysis called upon a form of mathematics, algebra, which had scarcely been touched by the ancient world and was largely the creation of Western Europe during the previous century. The further it proceeded, the more analysis distinguished itself from classical geometry. As they began to sense the power of their new instrument, the analysts concerned themselves less and less with the question of demonstrative rigor. Rigor had provided the foundation of Greek geometry. Because of its rigor, geometry had offered the living example of the ideal of science in the original sense of the word, that which is truly known, in contrast to mere opinion, which cannot claim a similar degree of certainty. In the flush of excitement, analysts rushed forward into another newly discovered world and left to succeeding ages the delicate problem of demonstrating, in the full sense of the word, that the world they had discovered did indeed exist and was not mere rumor.

The problem of the three- and four-line locus had been posed some two thousand years before in the following terms: Having three or four lines given in position, find the locus of points from which the same number of lines can be drawn to the given lines, each making a given angle with each, and such that the product of the lengths of two of the lines shall bear a constant proportion to the square of the third (the three-line locus) or a constant proportion to the product of the other two (the four-line locus) (Figure 1.4). Ancient geometers had known that conic sections furnish the solution, but, at least according to Pappus's account, on which seventeenth-century analysts had to rely since the treatises them-

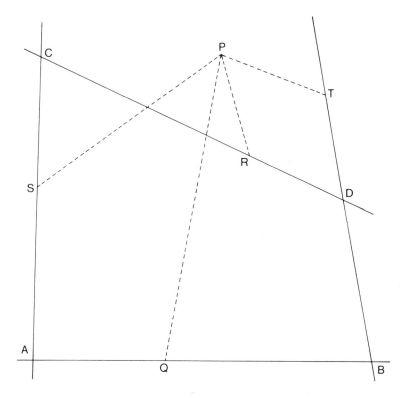

Figure 1.4. The four-line locus. The lines *AB, BD, CD,* and *AC* are given in position. Find the locus of points *P* from which lines *PQ, PR, PS,* and *PT* can be drawn to the four lines, each always making the same angle with the line it meets, such that *PQ* · *PR* is always in a given ratio to *PS* · *PT.* The locus is a conic that passes through the four intersections (*A, B, C, D,*) of the four given lines.

selves had been lost, they had been dissatisfied with the generality of their solution.

In his *Géométrie* (1637), Descartes applied algebraic techniques to the problem. Descartes's work started with the assertion that any problem in geometry can be reduced to terms in which knowledge of the lengths of certain lines suffices for its construction. The "lengths of certain lines" referred to what we now call the coordinates of a point. Descartes's coordinates look strange to anyone, familiar with "Cartesian coordinates," who expects to find curves graphed on *x* and *y* axes. Instead of starting with a system of coordinates and an equation, Descartes started with a diagram of the problem set. He chose two lines on the construction as his coordinates and expressed the other lines in terms of those two, the two variables, for which he employed letters from the end of the

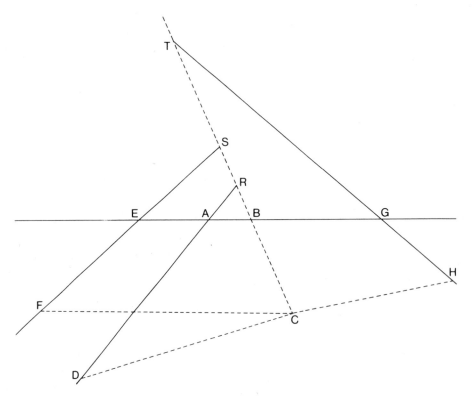

Figure 1.5. Descartes's diagram for the four-line locus. He set $AB =$ *x* and $BC = y$. In effect, *A* is the origin, and the coordinates of *C*, which is any point on the locus sought, are *x,y*. Manifestly the two coordinates are not perpendicular to each other as they are in what we today call Cartesian coordinates.

alphabet (Figure 1.5). Only a later generation would recognize fully the simplifying advantages of perpendicular coordinates. Meanwhile, Descartes was able to reduce the three- or four-line locus to a quadratic equation in two variables. He recognized that such an equation represents a conic, and he learned to distinguish the conics by the characteristics of their equations. The successful solution of Pappus's problem, as it was called, convinced Descartes of the universal utility of his method.

At much the same time, another French mathematician, Pierre Fermat, was attacking the same problem in much the same way. Fermat went beyond Descartes in his recognition that any equation in two unknowns defines a curve. He realized as well the desirability of perpendicular coordinates. Nevertheless, Fermat did not exert as much influence on the development of analytic geometry as Descartes. Most of his works remained unpublished, although they did circulate in manuscript. In contrast, Descartes had a disciple and publicist in Franz van Schooten, who translated his *Géométrie* into Latin, the international language of learning, and embellished it in successive editions with commentaries of his own and of a circle of students educated in the tradition of Cartesian geometry.

The full implications of the introduction of algebraic techniques into geometry appeared only with this second generation. The initial connection with the outlook of Greek geometry and absorption in its problems gradually receded, and the analytic consideration of algebraic equations in two variables which define curves gradually moved to the fore. Jan DeWitt's *Elementa curvarum (Elements of Curves)*, which was published in the second edition of Schooten's translation, has been called the first treatise in analytic geometry. The same claim has been made for John Wallis's *De sectionibus conicis (On Conic Sections)*, which appeared slightly earlier, in 1655. For my purposes the question is not worth resolving. As the title of Wallis's work and the content of DeWitt's testify, early analytic geometry was possible only because the Greek geometers had discovered and explored the conics. On the conics the first generation of analytic geometers concentrated almost exclusively, learning to express their characteristics in analytic form and extending knowledge of their properties. Fewer than five curves of the third degree had been studied by 1661, and in most cases their analyses had been defective. Nevertheless, in 1661, European mathematics possessed a new tool of immense power. The full extent of its power had yet to be explored.

Drawing tangents to curves presented one of the leading problems to which analysis addressed itself. Drawing a tangent is, of course, equivalent to finding the slope of a curve at any point, or what we now call differentiation. In his *Geometry*, Descartes had approached the problem indirectly by devising a method to find the normal to a curve at any point; since the tangent is perpendicular to the normal at the point of tangency, the position of the normal immediately defines the position of the tangent. To find the normal at any point C, Descartes constructed a circle, with its center P on the axis of the curve, that cuts the curve in two points, C and E. The simultaneous solution of the equations of the two curves gives the points of intersection. Now let the second intersection E coincide with C; that is, let the circle be tangent to the curve at C. Since two roots have coalesced, the equation for the intersection should have a double root at that point. Descartes proceeded then by purely algebraic means to establish the coordinates of the double root. Rather, since C is given, he established the coordinate on the

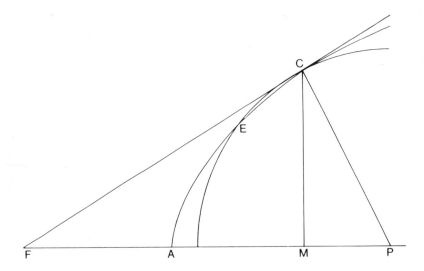

Figure 1.6. Descartes's method of finding tangents. By finding the double root when E and C merge into a single point, he found the length of the subnormal MP, in effect the location of P, the center on the axis AM of a circle that is tangent to the parabola AEC at point C. The tangent to the parabola at C is, of course, perpendicular to the normal PC.

axis of P, the center of the circle tangent to the curve at C (Figure 1.6).

In contrast to Descartes, Fermat attacked the problem of tangents directly by applying the techniques of analytic geometry to the ancient concept that a tangent is the longest line from a point to a convex curve. Given a curve and its equation – recall that with Fermat the curve was always a conic, a curve of second degree, as it nearly always was with Descartes – we wish to determine the tangent to any point B. Let E be the point, as yet undetermined, at which the tangent to B cuts the extended axis of the parabola DB. DC and CB are the x and y coordinates of the point B. (Actually, Fermat rejected Descartes's use of x and y as the symbols for variables, preferring Viète's earlier notation.) Let I be another point on the x axis a short distance removed from C. Because the curve is convex up, OI is longer than the ordinate to the curve at I. From the equation of the curve and the two similar triangles CBE, IOE, Fermat set up an inequality between $(OI^2)/(BC^2)$ and EI/EC. He then allowed I to coincide with C so that the ordinate to the curve at I coincided with OI. His inequality became an equation with a double root, from the solution of which EC, the length of the subtangent, emerged (Figure 1.7). Since he sought to determine the

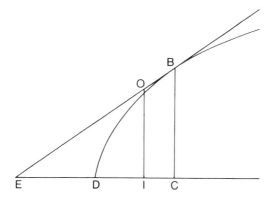

Figure 1.7. Fermat's method of finding tangents.

tangent directly, and since BC/EC defines the slope of the curve at B, Fermat's procedure looks more like differentiation than Descartes's determination of the tangent via the normal.

Schooten's second Latin edition of Descartes's *Geometry* contained two papers on maxima and minima, by Johann Hudde, setting forth a procedure for finding double roots, which figured both in Descartes's method and in Fermat's, which he had seen in manuscript. If the equation

$$x^n + a_1x^{n-1} + a_2x^{n-2} \cdots + a_n = 0$$

has a double root, Hudde demonstrated that the root also satisfies a second equation

$$ax^n + (a + b)a_1x^{n-1} + (a + 2b)a_2x^{n-2} \ldots + (a + nb)a_n = 0$$

which is obtained by multiplying the first equation by the terms of an arithmetic progression. Since the proposition is general, any arithmetic progression can be used, and in practice mathematicians usually chose a progression such that the most troublesome term in the original equation was multiplied by zero and eliminated. The implication of Hudde's rule appears most readily if the progression of exponents, n, $n - 1$, $n - 2$, . . . 1, 0, is used. Since the zero coefficient eliminates the constant term, an x can be canceled from each of the remaining terms, and the equation becomes

$$nx^{n-1} + (n - 1)a_1x^{n-2} + (n - 2)a_2x^{n-3} \ldots + a_{n-1} = 0$$

what we call the first derivative of the original equation, set equal to zero in order to find a maximum point, as in Fermat's method of tangents. (The equation in question is not that of the curve but that of EC, which becomes a maximum when EB is tangent to the curve.)

The problem of tangents, much in the air by the middle of the seventeenth century, also received a kinematic solution which reveals another facet of seventeenth-century mathematics. Both Torricelli and Roberval treated curves as the paths of points in motion. From this point of view, the instantaneous direction of the moving point at any position on the curve defines the tangent. Torricelli employed this method to determine tangents for parabolas of all degrees and for spirals. Roberval's method was similar enough to Torricelli's that a charge of plagiarism arose, one of the innumerable wrangles in seventeenth-century mathematics, when identical problems and identical considerations impelled mathematicians in identical directions. The concept of a curve as the line generated by the composition of two known motions did not spring newborn from the brains of seventeenth-century mathematicians. Nevertheless, the kinematic approach to mathematics most revealed its affinities with the developing science of mechanics. It also offered one possible road by which the new analysis could lead on into still unexplored territories.

During the early seventeenth century, mathematics concerned itself extensively with another question not initially an aspect of analysis, the problem of calculating areas under curves and volumes enclosed by curved surfaces. The use of infinitesimals found its way into mathematics primarily via this route. Kepler's *Steriometria doliorum (Measurement of the Volume of Casks, 1615)*, which summed infinitesimal layers to calculate the volumes, marked the beginning of the assault on such problems along this front. Twenty years later, Cavalieri's *Geometria indivisibilibus (Geometry by Means of Indivisibles, 1635)* attempted to place the use of indivisibles on a rigorous basis. By that time, the method was literally part of the atmosphere that mathematicians breathed, and even though Cavalieri first published general results for areas under the series of curves, $x = y$, $x = y^2$, $x = y^3$, . . . , $x = y^n$, Torricelli, Roberval, Pascal, Fermat, Gregory of St. Vincent, and Wallis either were already at work with similar methods and arriving at similar results, or soon would be.[13]

[13] Their ability in effect to integrate this series of equations does not contradict my earlier statement that analysis was confined almost entirely to second-degree equations before 1661. The curves that correspond to these equations of higher order were not studied. The results for the lower powers showed a consistent pattern and were simply generalized by induction. Even the curves that correspond to the equations of the lower powers, beyond the parabola $y = x^2$, were not studied; the "integrations" were carried through in isolation from any analysis of the curves. In the case of the simple parabola, mathematicians usually thought in terms of computing volumes rather than finding the area under a curve, but there was, of course, no equivalent analogy for the equations of higher order. Equations of more than second degree were also frequently used to expound Hudde's rule, but once again the curves that the equations define were not examined.

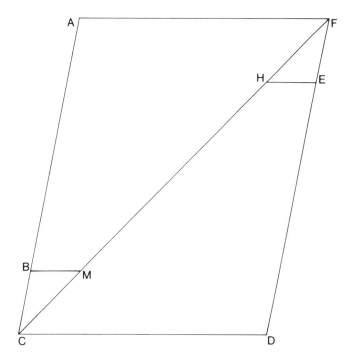

Figure 1.8. Cavalieri's calculation of the area of a triangle by means of infinitesimals. The diagram shows only one of the infinite corresponding pairs of lines in the two triangles.

Cavalieri utilized a triangle, the simplest of plane figures. On the base of the triangle, he constructed a parallelogram such that its diagonal, one side of the triangle, divided the parallelogram into two identical triangles. He then constructed lines parallel to the base and separated by equal distances such that the lines, infinite in number, filled the two triangles and, in effect, constituted their areas. Cavalieri was seeking a rigorous demonstration. He knew that he could neither add an infinite number of lines nor arrive at a two-dimensional area from such an addition. However, he could demonstrate the ratio between two infinite sums when every line (which he treated implicitly as an infinitesimal area) in one sum has the same ratio to a corresponding line in the other. In the case of the two triangles the ratio is equality. Every line *EH* in triangle *FDC* corresponds to an equal line *BM* in triangle *CAF*. Hence the two sums are equal; the area of triangle *FDC* equals the area of triangle *CAF*; and the area of each equals half the area of the parallelogram *AFDC* composed of the two (Figure 1.8). In modern notation, if we consider the diagonal as the line $x = y$ that divides a square, Cavalieri had demonstrated

$$\int_0^a x\, dy = \int_0^a y\, dy = \frac{a^2}{2}$$

Proceeding in an analogous way, he summed up ratios of squares on the infinitesimals and arrived at the conclusion that the volume of a pyramid (the sum of the squares of the infinitesimals in a triangle) is to the volume of the parallelepiped of equal height on the same base as 1 to 3. In our notation, the area under the curve $x = y^2$,

$$\int_0^a x\, dy = \int_0^a y^2\, dy = \frac{a^3}{3}$$

In a later work, he extended this result to all integral powers, $x = y^n$

$$\int_0^a x\, dy = \int_0^a y^n\, dy = \frac{a^{n+1}}{n+1}$$

By other routes other mathematicians arrived at the same result, which was an accepted common possession of the mathematical community by the mid-1650s. It was not, of course, expressed in our terminology, and the inverse relation of "integration" to "differentiation," which leaps from the page when the two are expressed in modern notation, had not been recognized.

Obviously, the equilateral hyperbola, $xy = 1$, or $y = 1/x$, could not fit into the general expression for areas, but in 1647 Gregory of St. Vincent successfully applied infinitesimals to the area under this curve as well (Figure 1.9). Although Gregory himself did not see the full implications of his work, others soon did. In our notation,

$$\int_a^b y\, dx = \int_a^b \frac{dx}{x} = \log b - \log a$$

In the seventeenth–century, when logarithms were still a new tool, this result, for the hyperbola $y = 1/(1 + x)$, provided a useful method, not to evaluate areas from logs, but to calculate logs from areas.

Although the method of infinitesimals led mathematics toward the concept of integration, it was inherently clumsy and crude. By 1661, its fertility was approaching exhaustion.

One of the distinctive new features in seventeenth-century mathematics, the use of infinite series, which began to appear prominently only in the middle of the century, developed primarily from attacks on problems of areas. Like most of seventeenth-century mathematics, this approach developed out of its counterpart in ancient geometry, in this case the method of exhaustion. When the new analysis was grafted onto the method of exhaustion, a wholly new branch of mathematics, infinite series, sprang forth.

Although precursors can be cited, such as Viète's expression for

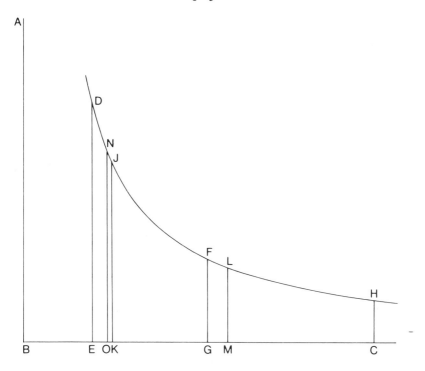

Figure 1.9. Gregory compared the areas *DEGF* and *FGCH* by means of infinitesimal segments equal in area.

the value of π, the prominent role of infinite series began with Gregory of St. Vincent's investigation of areas under the equilateral hyperbola. Once mathematicians recognized the logarithmic nature of the area function, they searched for means to calculate it by approximation, either by adding inscribed quadrangles of ever-decreasing size or by subtracting triangles of ever-decreasing size from the circumscribed quadrangle (Figure 1.10). In the first case, the area *ABCE* is approached by $BAFC + kFnd + mnpb + hklf + \ldots$ When $OA = AE = AB = 1$, and AB is divided into equal segments, the area $(= \log 2)$ equals $1/(1 \cdot 2) + 1/(3 \cdot 4) + 1/(5 \cdot 6) + 1/(7 \cdot 8) + \ldots$, a series that allows one to calculate the logarithm to whatever degree of accuracy he chooses by continuing the series. During the next quarter-century a whole series (not infinite to be sure but sufficiently long nevertheless) of series expansions of the logarithmic function, based on the area of the hyperbola, were developed as tools to simplify the calculation of logarithms.

Gregory's *Opus geometricum (Geometrical Work)* also considered infinite series by demonstrating that the sum of an infinite number of geometrically proportional indivisibles can approach a finite value as its limit. John Wallis's *Arithmetica infinitorum (Arithmetic of*

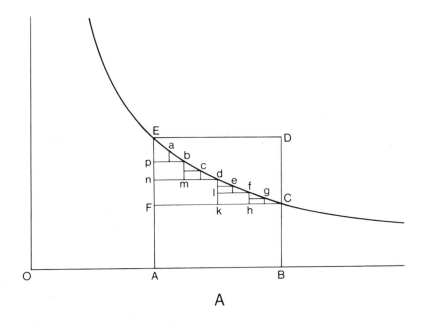

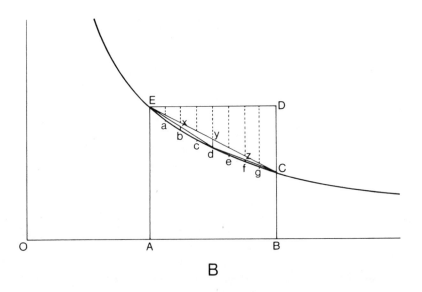

Figure 1.10. Two means of generating infinite series to compute the area of *ABCE* (= log 2, when *AB* = *AE* = *OA*). In diagram *A*, the area is approached by adding progressively smaller rectangles under the curve. In diagram *B*, the area is approached by subtracting progressively smaller triangles from the rectangle *ABDE*.

Infinites), published nine years after Gregory's work, revealed possibilities in the concept of infinite series that extended far beyond the logarithmic function. Casting caution to the winds and relying on his intuition of continuity in mathematical patterns as a substitute for demonstrative rigor, Wallis set out on a rather different path toward Cavalieri's goal, to determine the ratio of the sums of two infinite series of indivisibles as a means to calculate areas. Consider the parabola $y = x^2$. Let the base line of the area under the curve from $x = 0$ to $x = 1$ be divided into an infinite number of equal segments. Consider the length of the first ordinate to be 0^2, of the second 1^2, of the third 2^2, and so on to n^2. Now compare the sum of the ordinates to the sum of the ordinates (all equal to n^2) that make up the rectangle enclosing the segment of the curve being evaluated. Wallis proceeded by a rough induction. If the first two ordinates alone are considered,

$$\frac{0^2 + 1^2}{1^2 + 1^2} = \frac{1}{2} = \frac{1}{3} + \frac{1}{6} = \frac{1}{3} + \frac{1}{6 \cdot 1}$$

If three are considered,

$$\frac{0^2 + 1^2 + 2^2}{2^2 + 2^2 + 2^2} = \frac{5}{12} = \frac{1}{3} + \frac{1}{6 \cdot 2}$$

After a small number of similar calculations, all of which revealed the same pattern, Wallis was ready to generalize:

$$\frac{0^2 + 1^2 + 2^2 + 3^2 + \ldots n^2}{n^2 + n^2 + n^2 + n^2 + \ldots n^2} = \frac{1}{3} + \frac{1}{6n}$$

Since n is indefinitely large, the ratio of the area under the curve to the square enclosing it approaches the limiting value of 1/3 (Figure 1.11). In a similar manner, he determined that, for the curve $y = x^3$, the value of the ratio approaches 1/4 as a limit. Another pattern was now emerging, and Wallis did not hesitate to grasp it. For all positive integers k, the ratio of the area under the curve $y = x^k$ to the area of the enclosing square equals $1/(k + 1)$. This was hardly a startling result in 1656. Expressed in modern terminology, it was equivalent to the already established conclusion,

$$\int_0^1 x^k \, dx = \frac{1}{k + 1}$$

Undoubtedly the agreement with established results increased Wallis's confidence in his procedures. His further extension of them did break new ground. He generalized the results to fractional values of k, which he introduced into mathematical notation. He showed that the areas under binomials (such as $y = (a + x)^2$) can be evaluated by multiplying the binomial out and evaluating the terms one by one. Wallis's ultimate interest, however, lay in a much more

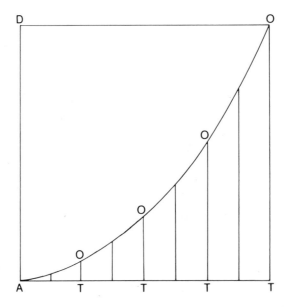

Figure 1.11. Wallis's demonstration of the ratio between the square
ATOD and the area *AOT* under the parabola.

difficult problem, the area under the circle. What he sought was the
ratio between the enclosing square and the quadrant of the circle y
$=(1 - x^2)^{1/2}$, which would be equivalent to the value of $4/\pi$. He
began by employing the method above to compose a table, for
integral values of p and n, of what we would now express as

$$\frac{1}{\int_0^1 (1 - x^{1/p})^n \, dx}$$

that is, of the ratio of the enclosing square to the area under the curve
between the y axis, and $x = 1$, where it cuts the x axis (Figure 1.12).

p				n			
	0	1	2	3	...		10
0	1	1	1	1	...		1
1	1	2	3	4	...		11
2	1	3	6	10	...		66
3	1	4	10	20	...		286
.	.	.	.	.	...		.
.	.	.	.	.	...		.
10	1	11	66	286	...		184756

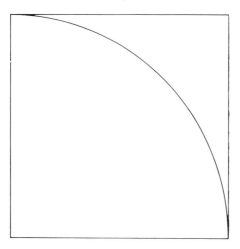

Figure 1.12. Wallis's table expressed the ratio of the area of the square to the area under the curve $y = (1 - x^{1/p})^n$ for values of p and n from 0 to 10. The diagram represents $p = \frac{1}{2}$, $n = \frac{1}{2}$, the equation for a circle of radius 1, from which Wallis established an infinite sequence for the value of π (or, more precisely, $4/\pi$) by interpolating into his table for integral values of p and n.

As Wallis recognized, the symmetry of the table repeats the pattern of the Pascalian triangle. He now sought to expand the table and, relying again on his faith in the continuity of patterns, to interpolate the values for 1/2, 3/2, 5/2 . . . Specifically, the function $p = 1/2$, $n = 1/2$ would be the value he sought,

$$\frac{1}{\int_0^1 (1 - x^2)^{1/2}\, dx}$$

He designated the value of this ratio by the symbol □. From the pattern in the rows and columns in which □ does not appear, Wallis was able to fill in the entire table. The row $p = 1/2$ contains the following values:

$$1, \qquad \square, \qquad \frac{3}{2}, \qquad \frac{4}{3}\square, \qquad \frac{15}{8}, \qquad \frac{24}{15}\square, \qquad \frac{105}{48} \quad \cdots$$

or

$$1, \qquad \square, \qquad \frac{3}{2}, \qquad \frac{4}{3}\square, \qquad \frac{3}{2}\cdot\frac{5}{4}, \qquad \frac{4}{3}\cdot\frac{6}{5}\square, \qquad \frac{3}{2}\cdot\frac{5}{4}\cdot\frac{7}{6} \quad \cdots$$

Since the alternate terms, in which □ does not appear, constantly increase, Wallis concluded that the terms with □ are also intermediate in value to the preceding and succeeding terms. By extending the row, he was able to establish ever narrower limits around the value of □. Hence □ (= $4/\pi$) approaches the limit, as the terms of the sequence are extended indefinitely,

$$\frac{3 \cdot 3 \cdot 5 \cdot 5 \cdot 7 \cdot 7 \cdot 9 \ldots}{2 \cdot 4 \cdot 4 \cdot 6 \cdot 6 \cdot 8 \cdot 8 \ldots}$$

Infinite series were still a recent innovation in 1661. The extent of the new possibilities they opened were as yet unprobed. As the example from Wallis illustrates, they had been used only to calculate certain values. From such cases, the concept of a limiting value which can be approached to any desired degree had emerged. No one had yet perceived how more flexible series composed of powers of a variable instead merely of numbers would extend the range of the new device.

Mathematics as a whole at that time consisted of innovations like infinite series not yet fully probed, and not yet bound together into a method or system that would reveal and exploit their mutual relations. Nevertheless, the very number of innovations was exciting. The world of mathematics was in ferment. On every side it was spilling out over its earlier boundaries.[14]

Let us try to place ourselves in the position of a young man in 1661, eager for knowledge, though of wholly untested capacity, as the new world of learning unrolled itself before his eyes. What an incredible challenge to the imagination – a world undreamed of in rural Lincolnshire, a world of many continents as extensive in their diversity as in their number. To the north lay the frigid lands of mathematics where one must breathe the bracing atmosphere of rigor. To the south lay the fetid tropical jungles of alchemy with their strange mythical fauna. Temperate lands for experimental investigation lay between. Manifestly, the very vastness of the new world placed it beyond the capacity of any one mind to grasp and to comprehend, finding an ordered cosmos where only chaos appeared. Perhaps. Perhaps not. Perhaps some rare individual, one of the intellectual supernovae who have burst intermittently but most infrequently into the visible heavens of a startled world, might

[14] As in the case of some of the earlier topics, there is a considerable body of literature on mathematics in the seventeenth century which I shall not attempt to summarize here. Indispensable though difficult is D. T. Whiteside, "Patterns of Mathematical Thought in the later Seventeenth Century," *Archive for History of Exact Sciences*, 1 (1961), 179–388. See also H. G. Zeuthen, *Geschichte der Mathematick im 16. und 17. Jahrhundert* (reprint ed., New York, 1966); J. E. Hofmann, *The History of Mathematics to 1800*, trans. F. Gaynor and H. O. Midoneck (Totowa, N.J., 1957); Carl Boyer, *History of Analytical Geometry* (New York, 1956), and *The Concepts of the Calculus* (New York, 1939); Michael S. Mahoney, *The Mathematical Career of Pierre de Fermat* (Princeton, 1973); and Margaret Baron, *The Origins of the Infinitesimal Calculus* (Oxford, 1969).

grapple effectively even with such a task. Other worlds new to Newton also opened themselves to his gaze in Cambridge, and his exploration of them played an important part in his life. Had he limited himself to them, however, his name would have passed long since into oblivion. As I said before, the only reason anyone writes a biography of Newton is because he chose to enter a world not only new to him as to all undergraduates, but new to man himself.

In Cambridge, Newton discovered that a new world had been discovered. He discovered as well something still more important. The early adventurers had only scouted its coasts. Vast continents remained to be explored.

2

A sober, silent, thinking lad

ISAAC Newton was born eighteen and a half years before he entered Cambridge, early on Christmas Day 1642, in the manor house of Woolsthorpe near the village of Colsterworth, seven miles south of Grantham in Lincolnshire. Since Galileo, on whose discoveries much of Newton's own career in science would squarely rest, had died that year, a significance attaches itself to 1642. I am far from the first to note it – and will be undoubtedly far from the last. Born in 1564, Galileo had lived nearly to eighty. Newton would live nearly to eighty-five. Between them they virtually spanned the entire scientific revolution, the central core of which their combined work constituted. In fact, only England's stiff-necked Protestantism permitted the chronological liaison. Because it considered that popery had fatally contaminated the Gregorian calendar, England was ten days out of phase with the Continent, where it was 4 January 1643 the day Newton was born. We can sacrifice the symbol without losing anything of substance. It matters only that he was born and at such a time that he could utilize Galileo's work.

Prior to Isaac, the Newton family was wholly without distinction and wholly without learning. Since it knew steady economic advance during the century prior to Isaac's birth, we may assume that it was not without diligence and not without the intelligence that can make diligence fruitful. In his old age, Newton may have entertained the idea that he was descended from a gentle Scottish family of East Lothian, one of whose members came to England with James I.[1] He settled in fact for an ancestry much more humble, however, and much closer to reality. According to his official pedigree, the family descended from a John Newton of Westby (a village about five miles southeast of Grantham), who lived in the mid-sixteenth century. Indeed there is a pedigree of another branch of the Newton family entered in the visitation of Lincolnshire of 1634 which traces the family to this same John Newton, who is said to be descended from the Newtons of Lancashire. It appears more probable that the family was indigenous to the area, taking its name from one of the numerous Newtons (from "new town") found nearby as they were found over all of England, which dated from

---

[1] The story is based on a conversation of James Gregory with Newton in 1725, repeated more than fifty years later via two intermediaries in a letter of Dr. Thomas Reid (see Appendix XXXII in David Brewster, *Memoirs of the Life, Writings, and Discoveries of Sir Isaac Newton,* 2nd ed., 2 vols. [Edinburgh, 1855], *2,* 537–45.)

the great medieval expansion of population in the eleventh and twelfth centuries.[2]

A Simon Newton, the first of the family to raise his head tentatively above rural anonymity, lived in Westby in 1524. Along with twenty-two other inhabitants of Westby, he had achieved the status of a taxpayer in the subsidy granted that year. Fourteen of the twenty-two, including Simon Newton, paid the minimum assessment of 4d. Eight others paid assessments ranging from 12d to 9s 6d, and one, Thomas Ellis, who was one of the richest men in Lincolnshire, paid over £16.[3] If the Newtons had risen above complete anonymity, clearly they did not rank very high in the social order, even in the village of Westby. Since the average village in that part of Lincolnshire consisted of about twenty-five or thirty households, his assessment may indicate that he and thirteen others occupied the lowest rung on the Westby ladder. They were climbing, however, and rather rapidly. When another subsidy was granted in 1544, only four men from Westby had the privilege to pay; two of them were Newtons. Simon Newton was gone, but John Newton, presumably the son of Simon, and another John Newton (the John Newton of Westby mentioned above), presumably his son, were now, after a man named Cony, the most flourishing inhabitants of Westby.[4] In his will of 1562, the younger John Newton still styled himself "husbandman"; twenty-one years later, his son, a third John, died a "yeoman," a step up the social ladder, and a brother William of the same generation also claimed that standing.[5]

Inevitably, Newton's pedigree has been worked out in considerable detail, first by Newton himself, later by the antiquarians whose attention the great attract. A list of his uncles, great-uncles, and the like, and the relationships in which they stood to him is of less interest than the implications wrapped up in the shift from husbandman to yeoman. In Lincolnshire, the sixteenth and seventeenth centuries witnessed a steady concentration of land and wealth with a consequent deepening of social and economic distinctions. The Newtons were among the minority who prospered. Westby is located on a limestone heath, the Kesteven plateau, a wedge of high ground thrust up toward Lincoln between the great fens to the east

---

[2] C.W. Foster, "Sir Isaac Newton's Family," *Reports and Papers of the Architectural Societies of the County of Lincoln, County of York, Archdeaconries of Northampton and Oakham, and County of Leicester, 39,* part I (1928), 4–5. Newton's own researches into the genealogy of his family, which also trace it back to the Newtons of Westby, are found in *Keynes MS* 112; *Babson MSS* 440, 441; and in an unnumbered manuscript in the Humanities Research Center, University of Texas.

[3] Foster, "Newton's Family," p. 5.

[4] *Ibid.,* p. 5.   [5] *Ibid.,* pp. 29, 36–7.

and the fenny bottom lands of the Trent valley to the west. The plateau had always presented itself as a likely highway to the north. The Romans had built Ermine Street along its back, and the Great North Road of medieval and early modern England followed the same route as far as Grantham, where it veered off to the west toward an easier passage over the Humber. Even today the main highway north near the eastern coast of England crosses the plateau along the same path. Woolsthorpe, where Newton was reared, lay less than a mile from a major thoroughfare of his day.

If the plateau was a natural highway, it was not a natural granary. The soil was thin and poor. Much of the arable land could sustain only a two-field rotation, which allowed it to stand fallow half the time. Enclosure here proceeded slowly, while large stretches of uncultivated waste were used in common as sheep walks. Wool from sheep was the foundation of the plateau's agricultural economy. In compensation for niggardly soil, the plateau bore a relatively sparse population. Those who would could prosper. The Newtons would.

The tale is told in the details of successive wills. From John Newton of Westby, who left a will when he died in 1562, each generation for a century left a considerably augmented estate. Rather, they left augmented estates. The Newtons were also a prolific clan (Figure 2.1). John Newton of Westby had eleven children, of whom ten survived. His son Richard, Isaac's great-grandfather, had seven children of whom five survived. Isaac's grandfather, Robert, had eleven, of whom six survived. There was no single inheritance which was augmented and passed on. The inheritance was continually being divided, but most of the segments took root and flourished. By the middle of the seventeenth century a considerable number of substantial yeomen named Newton were sprinkled over the area around Grantham, all of them descendants of John Newton of Westby, husbandman.[6] No doubt the fact that this John Newton married very well – Mary Nixe, the daughter of a prosperous yeoman – helped his position. He must also have known how to handle the dowry, however, for he was able to provide handsomely for three sons. The descendants of one of them, William, prospered even more than the rest; in 1661 one of his descendants, yet another John, pushed his way into the squirearchy as Sir John Newton, Bart.[7] In 1705, Isaac Newton anxiously pursued his son, also Sir John Newton, Bart., for corroboration of his pedigree. At about the time of his death, John Newton of Westby purchased an extensive farm of well over a hundred acres, including sixty acres of arable, in Woolsthorpe

---

[6] *Ibid., passim.*   [7] *Ibid.,* p. 10.

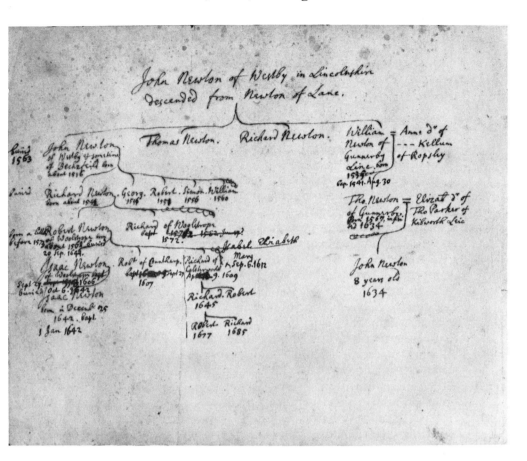

Figure 2.1. The Newton family tree as Newton himself drew it up for his coat of arms in 1705. (By permission of the University of Texas.)

for another son, Richard.[8] Woolsthorpe lay approximately three miles southwest of Westby, and Richard Newton was Isaac Newton's great-grandfather. To put the family's economic position in perspective, the average estate in the 1590s of peasants on the heath with property, that is, of the wealthier peasants, was about £49. The richest yeoman, as measured by his will, who died in Lincolnshire in the 1590s left personal property of nearly £400. Very few wills in that time left goods worth more than £100.[9] Richard Newton, whose father established him on a farm purchased for £40, left goods inventoried at £104; the inventory did

[8] *Ibid.*, p. 6.

[9] Joan Thirsk, *English Peasant Farming* (London, 1957), pp. 55, 103–4; M.W. Barley, *Lincolnshire and the Fens* (London, 1952), p. 103.

not include the land or the house.[10] It did include a flock of fifty sheep, well over the average number.[11] Sheep were the measure of wealth on the heath.

Not only did John Newton of Westby endow three sons magnificently by yeoman standards, but he also married a daughter to Henry Askew (or Ayscough) of Harlaxton.[12] The Ayscoughs were a prominent Lincolnshire family, though it is not clear what, if any, relation Henry Askew bore to the main stem of the family, whose seat lay well to the north. It was not the last alliance between the two families.

Robert Newton, Isaac's grandfather, was born about 1570. He inherited his father's property at Woolsthorpe to which he added the manor of Woolsthorpe by purchase in 1623. The manor was not in prosperous condition. It had changed hands by sale four times during the previous century.[13] Nevertheless its value was reckoned at £30 per year.[14] Added to the original estate, it gave the family a comfortable living indeed by yeoman standards of the day. Socially, it may have elevated Robert still further. He was now lord of a manor, legally entitled to exercise the powers of local authority, such as conducting court baron and court leet, which, as still operative elements of local administration, had jurisdiction over minor breaches of the peace and could levy fines but not imprison. The lord of a manor was no husbandman.[15] In December 1639, he settled the entire Woolsthorpe property on his eldest surviving son, Isaac, and Hannah Ayscough (or Askew), to whom he was betrothed. Isaac was hardly a young man. He had been born on 21 September 1606. Although Hannah Ayscough's age it not known, is seems likely that she was well beyond maidenhood herself; her parents had been married in 1609 and her brother William was probably the William Askue who matriculated in Cambridge from Trinity College early in 1630. Nevertheless, the couple did not marry at once, and there is every suggestion that they waited to obtain the inheritance first. After all, Robert Newton was nearly

---

[10] Foster, "Newton's Family," pp. 37–9.   [11] Thirsk, *Peasant Farming*, p. 71.

[12] Foster, "Newton's Family," p. 11.

[13] Turnor Papers, Lincolnshire Archives. I have not seen the Turnor Papers but I heard about them from Professor Gale Christianson of Indiana State University, who received his information from Mr. K. A. Baird of Colsterworth. As far as I know, Mr. Baird is the only person who has looked at the papers since the family deposited them in the archives nearly forty years ago. I draw my information from his unpublished report, which bears the mark of reliability.

[14] Foster, "Newton's Family," p. 13.

[15] In his notes intended for a life of Newton, John Conduitt, the husband of Newton's niece, Catherine Barton Conduitt, rather insisted on their position as lords of a manor. In the rolls, he said, the family styled themselves "Lords of the Mannour of Mortimer in the parishes or precincts of Wolstrope & Costerworth in the soak of Grantham in the County of Lincoln" (*Keynes MS* 130.3, sheet 2).

seventy. He obliged in the autumn of 1641; the following April they were united.[16]

The Ayscough match was another distinct step forward for the Newtons. Hannah was the daughter of James Ayscough, gentleman, of Market Overton, county Rutland. As marriage portion, she brought with her a property in Sewstern, Leicestershire, worth £50 per year.[17] It is difficult to imagine the match without Newton's recently purchased dignity of manorial lord. Hannah brought more than additional wealth. For the first time, the Newtons made contact with formal learning. Before 1642, no Newton in Isaac's branch of the family had been able to sign his own name. Their wills, drawn up by scriveners or curates, bore only their marks. Isaac Newton, the father of our subject, was unable to sign his name, and so was his brother who helped prepare the inventory of his possessions.[18] In contrast, one Ayscough at the very least was educated. William, Hannah's brother, M.A. Cambridge, 1637, pursued the calling for which learning was essential.[19] Ordained to the clergy of the Anglican Church, he was instituted to the rectory of Burton Coggles, two miles east of Colsterworth, in January of the year in which his sister married Isaac Newton.

The nearly exclusively masculine orientation of genealogical research in the past, buttressed by the analogous practice of patronymics, has focused attention entirely on the Newton family. Cannon Foster published an impeccable investigation of "Sir Isaac Newton's Family," in which Hannah Ayscough, daughter of James Ayscough and possibly related to the Askews of Harlaxton with whom the Newtons had mated once two generations earlier, sud-

---

[16] Foster, "Newton's Family," p. 15.

[17] A mistaken impression of distance and of farreaching connections may arise from the fact that three counties were involved. Since Woolsthorpe lies near the corner where the three meet, no more than five miles separated the three villages. Robert Newton had not reached any great distance to find a bride for his son. The match was distinctly a local affair.

[18] Foster, "Newton's Family," pp. 43–56. Although Richard, the brother, could not sign the inventory in 1642, he was able to sign his own will in 1659. Neither his son nor his wife were able to sign their wills at a later time, however.

[19] It is reasonable to question the extent of Hannah Ayscough Newton's literacy in the light both of her husband's illiteracy and of the general level of education among women at the time. She did sign her own will in 1672 (*ibid.*, p. 53). Moreover, one letter to her son Isaac has survived and is reproduced here in chapter 5 (*Corres 1, 2*). It is the letter of a woman for whom writing was clearly a heavy burden, two sentences long, or what would be two sentences if she had known how to punctuate and capitalize. It did not communicate anything beyond the fact that she had received a letter from him. There must have been other letters of this sort, it is true; indeed this one refers to an earlier one that Isaac had mentioned receiving. It seems very unlikely to me that there were many, and I find it quite impossible to imagine a profuse and revealing correspondence which would deepen our understanding of their relationship if it had survived.

denly appears as an adjunct to the Newtons. Perhaps we ought rather to view the match from the Ayscough perspective, but no dedicated genealogist has seen fit to investigate that family. We know only that James Ayscough, gentleman, of Market Overton had a daughter Hannah. The name raises the possibility of a blood relationship to the important Ayscoughs of northern Lincolnshire, but no available empirical evidence either supports or casts doubt on the speculation.[20] Whether they were connected or not, the genealogy of James Ayscough would undoubtedly be somewhat more impressive than that of Robert Newton, though his branch of the family would apparently show a recent decline in contrast to the Newtons' steady rise.[21] The intellectual history of England is not sprinkled with Ayscoughs, however, and anyone seeking to explain Isaac Newton's brilliance genetically is not likely to find it forecast among them any more than it is among the Newtons.

The overriding importance of the Ayscough contribution lies elsewhere. As it turned out, Isaac was reared entirely by the Ayscoughs. We can only speculate what would have happened had his father lived. The father was now the lord of a manor, as his own father had not been while he himself was being reared. Perhaps he would have seen the education of his son as a natural consequence of his position. However, his brother Richard, who was only a yeoman to be sure and not the lord of a manor, did not see fit to educate his son, who died illiterate.[22] Being reared as an Ayscough, Isaac met a different set of expectations. The presence of the Reverend William Ayscough only two miles to the east may have been the critical factor. At a later time, his intervention helped to direct Isaac toward the university. Whatever the individual roles, the Ayscoughs took it for granted that the boy would receive at least a basic education. We have some reason to doubt that the Newtons would have done so.

[20] William Stukeley, a younger contemporary of Newton's who collected a great deal of information about his life, took the connection for granted. He calls the Ayscoughs "a very antient and wealthy family in Lincolnshire, from a hamlet of that name, near Bedal, Yorkshire" (*Stukeley*, p. 34). In his memoir to Fontenelle, Conduitt probably drew upon Stukeley when he called them "an antient and hon^ble family in the County of Lincoln . . ." (*Keynes MS* 129A, p. 1).

[21] Conduitt said that the Ayscoughs were "formerly of great consideration in those parts," by which he meant the area about Colsterworth, since he connected them with Great Ponton, about four miles north. (*Keynes MS 130.2*, p. 9). The phrase suggests a gentle family fallen on hard times and forced into marriage with rising yeomen to recoup their fortunes.

[22] Foster, "Newton's Family," p. 55. Cf. n. 18 above. However, Isaac Newton *père* also called himself a yeoman in his will (*ibid.*, p. 45). Isaac Newton *fils*, the son of Hannah Ayscough, was the first in his line to call himself a gentleman (initially, as far as we know, in the visitation of Lincolnshire of 1666; *ibid.*, p. 3). Of course, he had commenced B.A. at Cambridge University, which automatically conferred gentle status.

Six months after his marriage, Isaac Newton died early in October, 1642. He left behind an estate and a pregnant widow but virtually no information about himself. We have only one brief description of him, from a century and a half after his death, by Thomas Maude, who claimed to have inquired diligently into Newton's ancestry among the descendants of his half brother and half sisters and around the parish of Colsterworth. According to Maude, Isaac Newton the father was "a wild, extravagant, and weak man . . . "[23] Such may have been the case; since Maude did not even get his name right, however, calling him John, we are scarcely compelled to accept the description. About his estate we are directly informed by his will. Since it defines the economic position of Isaac Newton (the son) at the time of his birth, it deserves some scrutiny. In addition to his extensive lands and the manor house, Isaac Newton, senior, left goods and chattels valued at £459 12s 4d. His flock of sheep, the ultimate measure of wealth in those parts, numbered 234, which compares with an average flock of about 35. He owned apparently 46 head of cattle (divided among three categories which are partly illegible on the document and hard to interpret in any case), also several times the average. In his barns were malt, oats, corn (probably barley, the staple crop of the heath), and hay valued at nearly £140. Since the inventory was made in October, these items undoubtedly represent the harvest of 1642. By putting oats (£1 15s) in a separate category, but coupling corn and hay (£130) in another, the men who drew up the inventory made it difficult to interpret. The oats and hay would have been fodder for the winter; surely the corn would not have been. The cattle (worth £101), and the sheep (worth £80) would have consumed the fodder during the approaching winter, so that it does not constitute a final product of the estate. Part of the final product was wool, and the inventory includes wool valued at £15. It is unlikely that the 1642 clip, from June, would still have been on hand; £15 is too small a sum in any case, since the annual clip averaged between a fourth and a third of the value of the flock. The estate also included, of course, extensive agricultural equipment and furnishings for the house. It included as well rights to graze sheep on the common.[24] The value of such rights is impossible to estimate, but when wool is king, grazing rights are gold. Like fodder, of course, they would only be means to the annual product, however. At this remove it is impossible to determine the total annual value of the estate. An estimate of at least £150 per year does not

---

[23] Thomas Maude, *Wensley-Dale; or, Rural Contemplations*, 3rd ed. (London, 1780), p. 29n.

[24] Newton still possessed these rights, plus some more attached to property acquired later, in 1712 when he wrote to Henry Ingle of Colsterworth about an agreement to enclose the commons (*Corres 5*, 346–7).

seem unreasonable.[25] We should add that the inventory may have
been lower than the long-term average value of the estate. The
1620s had been a hard decade, and probated inventories throughout
the 1630s were lower in consequence. They did not fully recover
their former level until about 1660. Newton's mother reserved the
income from the paternal estate to Isaac when she remarried; the
dowry lands in Sewstern appear to have been included. In addition,
her second husband settled a further piece of land on him.[26] Ultimately Newton inherited the entire paternal estate together with the
land from his stepfather and some additional properties purchased
by his mother. I have summarized the estate in financial terms
because that was its only meaning in Newton's life. At one time the
family intended that he manage it. This was not to be, however,
and the estate functioned in his life only as financial security. Whatever problems might await the child still unborn when the inventory was made, poverty was not likely to be among them.

The only child of Isaac Newton was born three months after his
father's death in the manor house at Woolsthorpe early Christmas
morning. The posthumous offspring, a son, was named after his
father, Isaac. Already fatherless, apparently premature, the baby
was so tiny that no one expected it to survive. Over eighty years

---

25 The will and inventory are in Foster, "Newton's Family," pp. 45–7. Stukeley stated that
    the manor of Woolsthorpe, Newton's paternal estate, was worth £30 per annum; in
    addition he had an estate at Sewstern from his mother that was worth £50 per annum
    (*Stukeley*, p. 36). Stukeley appears merely to have repeated the assessment of the manor,
    which was not the entire estate at Woolsthorpe. Thomas Maude, whose source was
    apparently the profligate Reverend Benjamin Smith, Newton's half-nephew, stated that
    Newton's whole estate at Woolsthorpe was worth about £105 per annum at his death.
    (Maude, *Wensley-Dale*, p. 29n). In 1694 Newton told David Gregory that the estate he
    inherited was worth £100 a year. Among the Turnor Papers there is an undated and
    unsigned list entitled "A valuation of the lands of the general persons heare underwritten
    as followeth," which appears to list the property owners of the parish. The valuation
    estimates the estate of Newton's mother at £70, a figure which would not include the
    dowry in Sewstern or the estate of the Reverend Barnabas Smith in North Witham.
    (Even so, she was the wealthiest person in the parish by far, with more than one-third of
    the total valuation.) According to the family tradition of the Ayscoughs, as written down
    by William Ayscough's granddaughter, the estate was worth £120 per annum (I.H.
    [James Hutton], "New Anecdotes of Sir Isaac Newton," *Annual Register, 19* [1776], 24).
    Only if we take seriously the limitation to Woolsthorpe, which would exclude both
    Hannah Newton's dowry in Sewstern (the produce of which was probably in the inven-
    tory), the property from Barnabas Smith, and later additions in Buckminster, do I find it
    possible to reconcile these figures with the facts of the inventory of his father's will. The
    estate could have declined in value, but there was no general decline in agricultural values
    in Lincolnshire during that period. In attempting to assess the estate, I have relied heavily
    on comparative information in Thirsk, *Peasant Farming, passim*. In 1733 the estate that
    Newton inherited, given as about 270 acres, sold for £1600 (Turnor Papers).
26 Foster, "Newton's Family," p. 17.

later, Newton told John Conduitt, the husband of his niece, the family legend about his birth.

> Sr I. N. told me that he had been told that when he was born he was so little they could put him into a quart pot & so weakly that he was forced to have a bolster all round his neck to keep it on his shoulders & so little likely to live that when two women were sent to Lady Pakenham at North Witham for something for him they sate down on a stile by the way & said there was no occasion for making haste for they were sure the child would be dead before they could get back.[27]

Apparently his life hung in the balance at least a week. He was not baptized until 1 January 1643.

We expect little information on the following years, and we are not disappointed. The great Civil War raged through England, but it is highly unlikely that its anxieties printed themselves on the infant's psyche. Before he was a year old the parliamentary forces had secured Lincolnshire; troops may have passed up and down the Great North Road within sight of the manor house. They would have been bodies of men without meaning to the tiny boy; the only man over whom the household could have worried was already dead.

An event of overwhelming importance shattered the security of his childhood immediately following his third birthday. Conduitt obtained an account of it from a Mrs. Hatton, née Ayscough:

> Mr Smith a neighbouring Clergyman, who had a very good Estate, had lived a Batchelor till he was pretty old, & one of his parishioners adviseing him to marry He said he did not know where to meet with a good wife: the man answered, the widow Newton is an extraordinary good woman: but saith Mr Smith, how do I know she will have me. & I don't care to ask & be denyed. But if you will go & ask her, I will pay you for your day's work. He went accordingly. Her answer was, She would be advised by her Bro: Ayscough. Upon which Mr Smith sent the same person to Mr Ayscough on ye same errand, who, upon consulting with his Sister, treated with Mr Smith: who gave her son Isaac a parcell of Land, being one of the terms insisted upon by the widow if she married him.[28]

Barnabas Smith was the rector of North Witham, the next village south along the Witham, a mile and a half away. Born in 1582, he had matriculated at Oxford in 1597, commencing B.A. (as graduation was called at the time) in 1601 and proceeding M.A. in 1604.

[27] Conduitt's memorandum of a conversation with Newton, 31 Aug. 1726 (*Keynes MS* 130.10). As he worked the memorandum into his projected life of Newton, he amended it in two successive drafts, each time for the worse (*Keynes MS* 130.3, sheets 5 and 6; *Keynes MS* 130.2 pp. 15–16). The progressive deterioration of the story makes a depressing warning of the dangers of attempted literary elegance.   [28] *Keynes MS* 125.

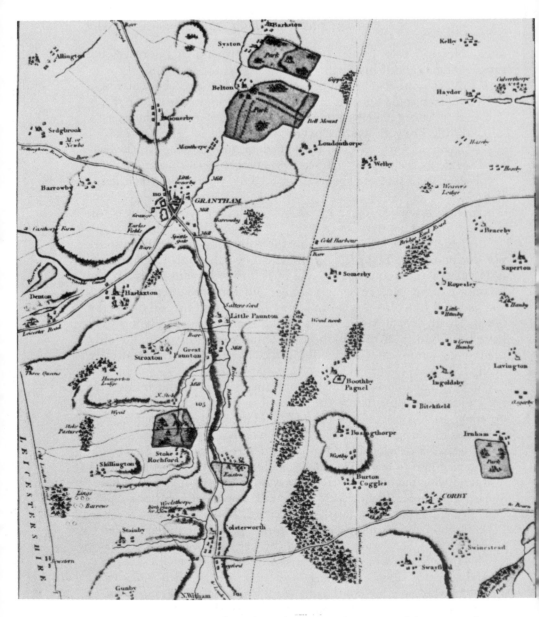

Figure 2.2. A map from the early nineteenth century of the corner of
Lincolnshire where Newton was born and reared. Only
two places connected with Newton and his family are not
included on the map. Buckminster, the location of two
fields that his mother willed to him, ought to appear near
the left-hand edge of the map, about a mile northwest of
Sewstern and directly west of Stainby. If the map extended
1¼ inches farther south, Market Overton, the original
home of his mother, would appear, 3¼ miles directly
south of Sewstern. (From Edmund Turnor, *Collections for
the History of the Town and Soke of Grantham,* London, 1806.)

"Pretty old," as Mrs. Hatton's account has it, rather understates the matter; he was sixty-three years old when he added "Smith" to Hannah Ayscough Newton's lengthening string of names. Nor had he lived a bachelor. He had buried a wife the previous June, and he had not allowed much grass to grow over her grave before he mended his single estate.[29]

We do not know a great deal about the rector of North Witham. To start with the best, he owned books. Newton's room at Woolsthorpe contained, on the shelves that Newton had built for them, two or three hundred books, mostly editions of the Fathers and theological treatises, which had belonged to his stepfather.[30] Purchasing books with intent to study is, of course, not the only way to obtain them. One might inherit a theological library, for example, if one's father was a clergyman as Barnabas Smith's father was. At any rate he had the books. He may even have read a bit in them. In a huge notebook, which he began in 1612, Smith entered a grandly conceived set of theological headings, and under the headings a few pertinent passages culled from his reading. If these notes represent the sum total of his lifetime assault on his library, it is not surprising that he left no reputation for learning. Such an expanse of blank paper was not to be discarded in the seventeenth century. Newton called it his "Waste Book," and what Barnabas Smith had once intended as a theological commonplace book witnessed the birth of the calculus and Newton's first steps in mechanics.[31] Possibly the library started Newton's theological voyage to lands his stepfather would not have recognized.

Smith must have been vigorous, not to say lusty; though already sixty-three when he married Hannah Newton, he fathered three children before he died at seventy-one. No surviving story suggests that he concerned himself much with the likelihood that the three children would soon be left without a father, even as another boy had been. Beyond the books and the vigor, little else about him sounds attractive. He occupied the rectory in North Witham because his father, the rector of South Witham, had bought it for him in 1610 by purchasing the next presentation from Sir Henry Pakenham, who controlled it. In the following year, a visitation by the Bishop of Lincoln reported that the Reverend Mr. Smith was of good behavior, was nonresident, and was not hospitable.[32] In effect, Barnabas Smith's father had purchased a comfortable annuity for his son. He received the income from North Witham for over forty years. For the first thirty, as far as we know, he conformed without protest to the ever more Arminian policies of the estab-

[29] Foster, "Newton's Family," p. 17. [30] *Stukeley,* p. 16.
[31] *Add MS* 4004. [32] Foster, "Newton's Family" p. 17.

lished church. With the Civil War came the Puritans and the Covenant. The Reverend Mr. Smith remained undisturbed in his living. The second Civil War brought the Independents and the Engagement. By now, large numbers of steadfast Anglican clergy had preferred ejection to conformity, and many were suffering real deprivation, but not the Reverend Mr. Smith. When he died in 1653, he had grasped his living firmly through all the upheavals – a pliable man, obviously, more concerned with the benefice than with principles. Although they had never met, John Milton knew him well.

> Anow of such as for their bellies sake,
> Creep and intrude, and climb into the fold?
> Of other care they little reck'ning make,
> Than how to scramble to the shearers feast,
> And shove away the worthy bidden guest;
> Blind mouthes![33]

Not that the living of North Witham was Barnabas Smith's primary means of support. He had an independent income of about £500 per annum – "wch in those days was a plentiful estate . . . ," said Conduitt, in his single essay at understatement.[34] For Newton, his stepfather's wealth meant in the end a significant increment to his own possessions. As Mrs. Hatton's account states, part of the marriage settlement was a parcel of land for him, increasing his paternal estate. Years later, Newton inherited from his mother additional lands that she had purchased for him, undoubtedly from the estate of her second husband.[35] The will of Newton's uncle, Richard Newton, suggests an economic status similar to that of Newton's father. The will of Hannah Ayscough Newton Smith bespeaks a wholly different level. The Ayscough marriage had been a step upward for the Newtons, more in status than in wealth. The Smith marriage brought a major increase in wealth. In return, it deprived Newton of a mother. His stepfather had no intention of taking the three-year-

---

[33] "Lycidas," in *The Works of John Milton,* ed. Frank A. Patterson (18 vols.; New York, 1931–8), *1,* Pt. 1, 76–83.

[34] *Keynes MS* 130.2, p. 11. According to the Ayscough family tradition, Smith was "a rich old batchelor, who had a good estate." In the context this meant a much better estate than the Newton estate of £120 per annum (I. H. [James Hutton], "New Anecdotes," p. 25).

[35] The two wills of Barnabas and Hannah Smith are in Foster, "Newton's Family," pp. 50–4. Beside some specific bequests (which include four silver vessels and a house in Lincoln), Barnabas Smith left all his lands to his son Benjamin, £500 to each daughter, and all the rest to his wife. All the rest must have been considerable, since Hannah Smith, when she died, was able to leave an additional £50 to Benjamin (in addition to a property in Sewstern and an annuity of £40 per year that she had already given him), an additional £80 to her married daughter Mary and her family, an additional £300 to her unmarried daughter Hannah, and to Isaac two fields in neighboring Buckminister and a house in Woolsthorpe that she had purchased, plus the uncatalogued goods and chattels, most of which went to Isaac.

old boy with the mother. Isaac was left in Woolsthorpe with his grandmother Ayscough. The Reverend Mr. Smith did have the house rebuilt for them; he could afford it.

The loss of his mother must have been a traumatic event in the life of the fatherless boy of three. There was a grandmother to replace her, to be sure; but significantly, Newton never recorded any affectionate recollection of her whatever. Even her death went unnoticed. As we shall see, Newton was a tortured man, an extremely neurotic personality who teetered always, at least through middle age, on the verge of breakdown. No one has to stretch his credulity excessively to believe that the second marriage and departure of his mother could have contributed enormously to the inner torment of the boy already perhaps bewildered by the realization that he, unlike others, had no father.[36] Moreover, there is reason to think that Isaac Newton and Barnabas Smith never learned to love each other. Nine years after his stepfather's death, when Newton was moved to draw up a list of his sins, he included, "Threatning my father and mother Smith to burne them and the house over them."[37] Probably every boy has angry confrontations with his parents, when puerile threats are screamed in frustration. Nevertheless, the scene must have etched itself deeply on Newton's consciousness if he recalled it nine years later. For Barnabas Smith's part, his actions speak clearly enough. For more than seven and a half years, until he died, while the child of three grew to be a boy of ten, he did not take him to live in the rectory in North Witham.

The manor house of Woolsthorpe stands on the west side of the small valley of the river Witham, a string down the Kesteven pla-

---

36 In his psychoanalysis of Newton (*A Portrait of Isaac Newton* [Cambridge, Mass., 1968]), Professor Manuel makes the remarriage of his mother the critical episode in Newton's entire life. Already lacking a father, whom he had never known, he was now bereft of the mother he had possessed exclusively. According to Manuel's interpretation, the sense of deprivation dominated his life. He had been robbed of his most precious possession. He spent the rest of his life finding surrogates on whom to vent the rage that he had not been able, in his infantile weakness, to pour out on its real object, Barnabas Smith, the man who had ravished his mother. Robert Hooke, John Flamsteed, Gottfried Wilhelm Leibniz, the unfortunate counterfeiters at the turn of the century – all of them suffered for the crime of Barnabas Smith. Manuel's analysis is subtle, complex, and ingenious. Is it also true? It appears to me that we lack entirely any means of knowing. It is plausible; it is equally plausible that it is miguided. I am unable to see how empirical evidence can be used to decide on it, one way or the other. I trust it is clear that I am not offering an alternative analysis. It would confront exactly the same problem of confirmation, and in any case, I have no qualifications whatever to offer one. Manuel's portrait, as distinct from his analysis, is a vivid and insightful account of established facts of Newton's life. No biographer of Newton can afford to ignore it. It does not attempt to give an account of his scientific career.

37 Richard S. Westfall, "Short-writing and the State of Newton's Conscience, 1662," *Notes and Records of the Royal Society, 18* (1963), 13.

The paternal House of Sir Isaac Newton, in which he was born, the 25th day of Decr 1642, at Woolsthorpe Lincolnshire.

Figure 2.3. An eighteenth-century drawing of Newton's home at Woolsthorpe. (Courtesy of the Warden and Fellows of New College, Oxford.)

teau beaded with villages, leading toward the town of Grantham. Built of the gray limestone which also builds the plateau, the house forms a squat letter **T**, with the kitchen in the stem, and the main hall and a parlor in the cross-stroke (Figure 2.3). The entrance, somewhat off-center between the hall and the parlor, faces the stairway which leads upstairs to two bedrooms. Here Newton was born, and here was the room he occupied while he grew to adolescence. Beyond the fact that he attended day schools in the neighboring villages of Skillington and Stoke, we know little about his youth. The area was liberally sprinkled with aunts, uncles, and cousins of varying degrees. Wills tell us of two uncles Newton, one living in Colsterworth and one in Counthorpe three miles away, both with children apparently not far removed from Isaac in age. Three married aunts, all with children, lived in neighboring Skillington. There were also Dentons, Vincents, and Welbys who were more distantly related on the Newton side of the family. At least some connection with them was maintained; in the affidavit that accompanied and justified his pedigree in 1705, he stated that his grandmother Ayscough "frequently conversed with [his] great Uncle, Richard Newton" at Woolsthorpe.[38] There were Ayscoughs as well. His grandmother had grown up in the area, and besides her daughter Hannah there was another married daughter Sarah, not to mention the Reverend William Ayscough two miles away. Nevertheless Isaac's boyhood appears to have been lonely. He formed no bond with any of his numerous relations that can be traced in his later life. The lonely boyhood was the first chapter in a long career of isolation.

In August 1653, the Reverend Barnabas Smith died, and Newton's mother returned to Woolsthorpe to live. Perhaps the period that followed was a joyful interlude for the boy of ten to whom a mother had been restored. Perhaps some bitterness tinged his joy as a half brother and two half sisters shared her attention, one an infant not yet a year old and another, just two, possibly dominating it. The fact is we do not know. We know only that the interlude was short. In less than two years, Isaac was sent off to grammar school in Grantham.

Despite its name, the Free Grammar School of King Edward VI of Grantham was an institution some three hundred years old when Newton enrolled in it. The upheavals of the Reformation had led to its reestablishment under a new charter as a Grammar School of King Edward VI in the mid-sixteenth century. Undoubtedly the reformation of learning associated with the Renaissance introduced more profound changes, but about them as they applied specifically to the grammar school in Grantham we know almost nothing. The school had an honorable reputation. More than a century earlier, William Cecil had studied there, more recently the Cambridge Platonist, Henry More, whom Newton would know at the university. The current master, Mr. Stokes, was considered to be a good schoolmaster, we are told by Newtonian lore, which begins now to accumulate. Presumably this means that the Reverend William Ayscough investigated him and approved.

By Newton's own testimony, he entered the school in Grantham

[38] Foster, "Newton's Family," p. 60.

when he was twelve.[39] The considerable number of anecdotes
about this period of his life concentrate on his academic progress
and on extracurricular recreations. By telling us nothing at all about
the nature of his studies, they leave us to assume that he studied
what every boy in grammar school at that time studied as a matter
of course. Latin and more Latin, with a bit of Greek toward the end
and no arithmetic or mathematics worth mentioning – such was the
standard curriculum of the English grammar school of the day, and
such we must assume Newton confronted in the school at Grantham which was respected, and which Mr. Stokes, reputed to be a
good schoolmaster, ran.[40] A product of Renaissance classicism, the

39 I am relying on Conduitt's memorandum of a conversation of 21 Aug. 1726 (*Keynes MS*
    130.10, p. 2). There has been considerable disagreement among biographers over the
    chronology of Newton's period in Grantham. Since this memorandum appears to be the
    most authoritative source, and since it is consistent with other established dates, I accept its
    testimony. It is necessary to note, however, and all the more so since I shall be citing this
    memorandum frequently, that one cannot accept it without reservations. In 1726, Newton
    was a very old man talking about events that had happened more than sixty years before. In
    another memorandum of a conversation a year and a half earlier, on 7 March 1725, Condu-
    itt remarked that Newton's head was clearer and his memory stronger than he had known
    him for some time. To say the least, this casts some cloud over the memorandum of 31
    August 1726. Nevertheless, we have no better evidence. According to it, Newton went to
    the grammar school in Grantham at the age of twelve – i.e., sometime in 1655. Possibly he
    arrived just in time to meet one of the illustrious alumni of the grammar school in which he
    was enrolling, Henry More, who had established himself as a prominent philosopher
    among the school of Platonists in Cambridge. More was in Grantham in the latter part of
    April that year. When he visited Grantham a year earlier, he stayed in the home of Mr.
    Clark where Newton lodged; he may have done so also in 1655 (see More to Anne
    Conway, 8 May 1654 and 18 April 1655 in Marjorie Hope Nicholson, ed., *Conway Letters.
    The Correspondence of Anne, Viscountess Conway, Henry More, and Their Friends* [New
    Haven, 1930], pp. 98, 107–8). Whether or not they met now, the two would have occasion
    to meet in the future. Newton was in Grantham initially four and a half years; thus he
    would have been there in 1658, the day of Cromwell's death and the great storm which
    figures in one anecdote. He returned to Woolsthorpe for three-quarters of a year, and then
    came back to Grantham for three-quarters of a year, after which he enrolled in Cambridge.
    Projecting backwards from his admission to Trinity, he came back to Grantham about the
    beginning of September 1660; he returned to Woolsthorpe earlier, about the beginning of
    December 1659; and he went originally to Grantham in the late spring of 1655. Perhaps all
    of these dates should be set back by some indeterminate but necessarily fairly short interval
    to allow for a pause in Woolsthorpe between grammar school and the university, although
    there is no mention of such. There is evidence that we must set the dates back a bit over a
    month at the least. On 28 Oct. 1659, Newton was fined by the manor court of Colster-
    worth and must therefore have been called home from school before the beginning of
    December. The exact dates are of no great significance; we would not know anything more
    even if they could be established.
40 Foster Watson, *The English Grammar Schools to 1660: Their Curriculum and Practice* (Cam-
    bridge, 1908), *passim. Villamil* contains two catalogues of Newton's library which indicate
    that he had a considerable collection of the classics, including most of the works that
    Watson indicates were basic to the curriculum of the grammar schools. Many of them
    were purchased after his grammar-school days, however, since they were post-1661
    editions. Moreover, some of the pre-1661 editions came into Newton's hands after he

grammar school of the seventeenth century had as its primary goal the development of proficiency in Latin, not merely the ability to read it, but also to write and to speak it. Latin was the sole path that led on to higher learning. Still the *lingua franca* of European scholarship, Latin was the means of communicating with the learned world beyond the borders of England. A generation earlier and some hundred miles to the south, John Wallis's education, as he remembered it late in his life, left him above all "well grounded in the Technical part of Grammar . . . " Wallis made it clear that he meant Latin grammar. Of arithmetic and mathematics Wallis got none at all; he learned of their very existence from his brother, who was preparing for a trade. Moreover, of the more than two hundred students with him in Emmanuel College, Cambridge, he did not know of two, and perhaps there were none at all, who had more mathematics than he.[41]

John Milton's tract, *Of Education,* written in the 1640s, told a similar story. Strangely, for one of the leading classical scholars of England, he summarized the grammar schools as "those Grammatick flats and shallows where [the students] stuck unreasonably to learn a few words with lamentable construction . . . "[42] To be sure, Milton did not want less Latin in the schools; he merely assumed that everyone could rush through it at his pace. The very silence on such a vital question by the collectors of anecdotes strongly suggests that Newton's education differed in no way from the ordinary one and some of the earliest surviving fragments of Newtoniana confirm this. In 1659 he purchased a small pocketbook (or notebook as we would say), dating his signature on the first page below a Latin couplet with "Martij 19, 1659." If one assumes this means 1659/60, it belongs to the period when he was at Woolsthorpe. He devoted most of the notebook to "Utilissimum prosodiae supplementum."[43] Further, in the Keynes Collection in King's College there is an edition of Pindar with Newton's signature and the date 1659, and the Babson Collection has his copy of Ovid's *Metamorphoses* dated that year.[44]

arrived in Cambridge. For example, his copy of Cicero's *Epistolae ad familiares,* a basic work in the grammar-school curriculum, contains the name of a man from Pembroke Hall, and his *Opera* of Macrobius has on its fly leaf the inscription "Johannes Laughton, 1667." Both of these works are in the partial reassembly of Newton's library in Trinity College, Cambridge. I have not bothered to look through every book in the collection for evidence of when he obtained it.

41 Wallis to Thomas Smith, 29 Jan. 1697; Thomas Hearne, *Works,* 4 vols. (London, 1810), *3,* cxliv, cxlvii–cxlviii.   42 *Of Education,* in *Works, 4,* 279.

43 Trinity College, Cambridge, *MS* R.4:48ᶜ.

44 A. N. L. Munby, "The Keynes Collection of the Works of Sir Isaac Newton at King's College, Cambridge," *Notes and Records of the Royal Society, 10,* (1952), 49; *A Descriptive Catalogue of the Grace K. Babson Collection of the Works of Sir Isaac Newton,* comp. Henry P. Macomber (New York, 1950), p. 188.

The reader in the twentieth century, surrounded by the achievements of modern mathematics and the material culture it has generated, can scarcely believe that the man who would discover the calculus four years after he left grammar school was probably not even introduced there to the already thriving mathematical culture out of which the calculus would come. Neither is there any suggestion that he studied natural philosophy. Nevertheless, the grammar school in Grantham served Newton well. Without exception, the mathematical works on which he fed a few years hence were written in Latin, as were most of his sources in natural philosophy. Later still, he could communicate with European science because he wrote Latin as readily as English. A little arithmetic, which he could have absorbed in a day's time anyway, would scarcely have compensated for a deficiency in Latin.

One other important feature of the grammar school in the seventeenth century was the Bible. Studied in the classical tongues, it both supported the basic curriculum and reinforced the Protestant faith of England. In Newton's case, biblical study may have joined with the Reverend Smith's library to launch his voyage over strange theological waters.

In Grantham, Newton lodged with the apothecary Mr. Clark, whose house stood on the High Street next to the George Inn. Also living in the house were three stepchildren of Mr. Clark, named Storer from his wife's first husband, a girl whose first name has been lost, and two boys, Edward and Arthur.[45] It seems clear that Newton did not get along with the boys. Among the incidents that he remembered uncomfortably in 1662 were "Stealing cherry cobs from Eduard Storer" and "Denying that I did so." He also recalled "Peevishness at Master Clarks for a piece of bread and butter."[46] As far as we know, Newton had grown up in relative isolation with his grandmother. He was different from other boys, and it is not surprising if he was unable to get along with them easily. As they came to recognize his intellectual superiority, the boys in the school apparently hated him. Years later there was only one, Chrichloe, whom he remembered with pleasure. Stukeley gathered that the

---

[45] In his list of sins of 1662, Newton mentioned Edward and Arthur Storer. Years later, Arthur Storer communicated observations of a comet to him from Maryland, and Edward Storer was his tenant at Woolsthorpe. The two brothers were nephews of Humphrey Babington, fellow of Trinity College and later rector of Boothby Pagnell, between Colsterworth and Grantham and somewhat to the east. Stukeley's anecdotes about Newton in Grantham include a "Miss Storey," the stepdaughter of Mr. Clark and the niece of Humphrey Babington. I assume that Stukeley's Miss Storey was really Miss Storer, and that Edward and Arthur were her brothers and hence also residents of the house.

[46] Westfall, "Short-writing," pp. 13–14.

others found him too cunning, able to get the better of them with his greater quickness. Perhaps it was one such incident, hardly calculated to endear him to boys already hostile, that Newton recorded in 1662. "Putting a pin in John Keys hat on Thy day to pick him."[47]

The stories that Stukeley collected in Grantham in the 1720s stressed the fact that Newton preferred the company of girls. For Miss Storer, who was several years his junior, and her friends he made doll furniture, delighting in his skill with tools. Indeed, as the two grew older, something of a romance apparently developed between Newton and Miss Storer. It was the first and last romantic connection with a woman in his life. The romance of an adolescent boy who prefers the company of girls is not likely to endure. This one did not. Though Newton remembered Mrs. Vincent (her married name) as one of his two friends in Grantham, it was only Mrs. Vincent who told of the romance. For the most part, he kept his own company. He was always "a sober, silent, thinking lad," Mrs. Vincent recalled, [and] "never was known scarce to play with the boys abroad."[48]

Early in Newton's stay in Grantham, a crisis occurred which burned deeply into his memory. He had not even had time to assert his intellectual prowess. Whether because he was ill-prepared by the village schools, or because he was alone again and frightened, he had been placed in the lowest form, and even there he stood next to the bottom. On the way to school one morning, the boy next above him kicked him in the belly, hard. It must have been Arthur Storer.[49] Boys will be boys, but even among boys a vicious kick in the stomach requires some provocation. Already there may have been one too many peevish scenes over bread and butter and cherry cobs and all the rest one can imagine. Though he played with the girls, Newton knew what he had to do:

> as soon as the school was over he challenged the boy to fight, & they went out together into the Church yard, the schoolmaster's son came to them whilst they were fighting & clapped one on the back & winked at the other to encourage them both. Tho Sr Isaac was not so lusty as his antagonist he had so much more spirit & resolution that he beat him till he declared he would fight no more, upon wch the schoolmaster's son bad him use him like a Coward, & rub his nose

---

[47] *Ibid.*, p. 13. John Keys attended the school in Grantham. If Newton put the incident on the list in 1662, it cannot have passed for a joke though it may have been an attempt at one.

[48] *Stukeley*, pp. 45–6. Cf. pp. 23, 46.

[49] One of the sins that Newton listed in 1662 was "Beating Arthur Storer" (Westfall, "Short-writing," p. 14). It seems probable that the fight he remembered seventy years later was the same one he remembered seven years later – the only fight in the list.

against the wall & accordingly Sr Isaac pulled him along by the ears & thrust his face against the side of the Church.[50]

Not content with beating him physically, he insisted on worsting him academically as well; once on his way, he rose to be first in the school. As he rose, he left his trail behind him, his name carved on every bench he occupied. The benches do not survive, but a stone windowsill still bears one of his signatures.[51]

By the time Stukeley was collecting anecdotes, Newton's genius was taken for granted. What everyone in Grantham remembered about him were "his strange inventions and extraordinary inclination for mechanical works." He filled his room in the garret of Clark's house with tools, spending all the money his mother gave him on them. While the other boys played their games, he made things from wood, not just doll furniture for the girls, but also and especially models. A windmill was built north of Grantham while he was there. Although water wheels were common in the area, windmills were not, and the inhabitants of Grantham used to walk out to watch its construction for diversion. Only the schoolboy Newton inspected it so closely that he could build a model of it, as good a piece of workmanship as the original and one which worked when he set it on the roof. He went the original one better. He equipped his model with a treadmill run by a mouse which was urged on either by tugs on a string tied to its tail or by corn placed above it to the front. Newton called the mouse his miller. He made a little vehicle for himself, a four-wheeled cart run by a crank which he turned as he sat in it. He made a lantern of "crimpled paper" to light his way to school on dark winter mornings, which he could simply fold up and put in his pocket for the day. The lantern had other possibilities; attached to the tail of a kite at night, it "wonderfully affrighted all the neighboring inhabitants for some time, and caus'd not a little discourse on market days, among the country people, when over their mugs of ale." By good fortune, Grantham was not burned to the ground.[52]

[50] *Keynes MS* 130.2, pp. 17–18. I have quoted Conduitt's original version (corresponding to *Keynes MS* 130.3, sheet 6); he later crossed out much of the account of the fight, including the role of the schoolmaster's son. Manuel makes a good deal of this incident. Arthur Storer is the first surrogate Barnabas Smith, and his final humiliation is the image of the fate of future antagonists. As it appears to me, he rather slurs over the role of the schoolmaster's son. If I am correct in identifying this fight with the beating Newton listed in 1662, his remorse does not seem to me to be compatible with Manuel's interpretation.

[51] It is reproduced in J. A. Holden, "Newton and his Homeland – the Haunts of his Youth," in *Isaac Newton, 1642–1727, a Memorial Volume*, ed. W. J. Greenstreet (London, 1927), opp. p. 142.

[52] *Stukeley*, pp. 38–42; *Keynes MS* 130.2, pp. 22–3, 28–9. Conduitt's account derives from Stukeley's.

Newton spent so much time at building that he frequently neglected his school work and fell behind, whereupon he turned to his books and quickly leaped ahead once more. Stokes remonstrated gently, but nothing could make him give up his mechanical contrivances. He could not leave them alone even on the Sabbath, although it filled him with remorse.[53] We know now that Newton found many of these contrivances in a book called *The Mysteries of Nature and Art* by John Bate. In another notebook from Grantham, with the information that he purchased it for 2 1/2d in 1659, Newton took down extensive notes from Bate, on drawing, catching birds, making various-colored inks, and the like. Although they do not appear in his notes, most of his devices remembered in Grantham, including a windmill, were described in the book.[54] Perhaps Newton's adolescent genius shrinks a little in the light of Bate's book. His genius is scarcely in doubt, however, and the fact is that he found a book which fed his natural interests.

There is a touch of whimsy in some of these stories, wholly unexpected because wholly absent from the rest of his life. At this distance, there appears as well a pathetic attempt to ingratiate himself with his schoolfellows by such means. He made lanterns for

---

[53] "Making a mousetrap on Thy day," "Contriving of the chimes on Thy day," "Twisting a cord on Sunday morning" (Westfall, "Short-writing," pp. 13–14).

[54] The notebook is now in the Pierpont Morgan Library in New York. On the flyleaf is the inscription "Isacus Newton hunc librum possidet. teste Eduardo Secker. pret. 2ᵈ ob. 1659." Professor Andrade identified the notes with Bate's work (E. N. da C. Andrade, "Newton's Early Notebook," *Nature, 135* [1935], 360). Its connection with Bate's book is discussed more fully in G. L. Huxley, "Newton's Boyhood Interests," *Harvard Library Bulletin, 13* (1959), 348–54. As with the other notebook, now in Trinity College, not everything in this one dates from the Grantham-Woolsthorpe period. After entering notes from Bate at one end, he entered extensive alphabetical lists of words under various headings such as "Artes, Trades, & Sciences," "Birdes," "Cloathes," and so on from the other. These lists, based on Francis Gregory's *Nomenclatura brevis anglo-latino,* were the example of a prominent characteristic of Newton, his desire, perhaps even compulsion, to organize and categorize information. There were to be many analogous exercises in the future, ranging from his "Quaestiones quaedam Philosophicae" based on his reading of the mechanical philosophers beginning in 1664, to his *Index chemicus,* begun in the early 1680s, which organized the results of his extensive reading in alchemical literature. Like the notes from Bate, these lists appear to date from the Grantham-Woolsthorpe period. Manuel makes extensive use of Newton's additions to the lists he found in Gregory. On the unused pages in the center of the notebook, Newton later entered fairly extensive astronomical tables, on such things as the rising and setting of the sun, its altitude during the year, and eclipses of the moon, and an ecclesiastical calendar which begins with 1662. There is a description of how to make a sundial. Three pages describe technical details of the Copernican system, and several pages are devoted to elementary plane and spherical trigonometry. He also outlined a phonetic system. All of these notes in the center of the notebook are in a later hand, some as late as 1664, I think. Dates of the Cambridge terms, which are found with the astronomical tables, and the calendar beginning with 1662 tend also to place them in his undergraduate period.

them also, and who can doubt that they participated in the artificial meteor? When they flew kites, Newton investigated their properties to determine their ideal proportions and the best points to attach the strings. Apparently his efforts were in vain; he only convinced them of his greater ingenuity and completed their alienation. As Conduitt says, even when he played with the boys, he was always exercising his mind. Ordinary boys must have found him disconcerting. He told the Earl of Pembroke that the first experiment he ever made was on the day of Cromwell's death, when a great storm swept over England. By jumping first with the wind and then against it, and comparing his leaps with those of a calm day, he measured "the vis of the storm." When the boys were puzzled by his saying that the storm was a foot stronger than any he had known before, he showed them the marks that measured his leaps.[55] According to one version of the story, he craftily used the wind to win a jumping contest–again the superior cunning which made him suspect.

Similar stories of mechanical models are told of Robert Hooke's boyhood. In both cases, manual skill served them well in constructing equipment for experiments. Far more important, however, is the testimony of such stories to the pervasive image of the machine in the seventeenth-century mind. Already that image had reshaped the conception of nature. The pursuits of his boyhood prepared Newton to embrace the mechanical philosophy as soon as he met it.

There were also other recreations in Grantham. Among them were sundials. Apparently dials had attracted his attention even earlier; there is one mounted in the Colsterworth church supposedly cut by Newton when he was nine.[56] Sundials involved much more than skill with tools. They presented an intellectual challenge. He filled poor Clark's house with dials–his own room, other rooms, the entry, wherever the sun came. He drove pegs into the walls to mark the hours, half-hours, and even quarter-hours, and tied strings with running balls to them to measure the shadows on successive days. By keeping a sort of almanac, he learned to distinguish the periods of the sun so that he could tell the equinoxes and solstices and even the days of the month. In the end the family and the neighbors came to consult "Isaac's dials."[57] Thus did the majesty of the heavens and the uniformity of nature spread themselves unforgettably before him. According to Conduitt, he was still watching the sun at the end of his life. He observed the shadows in

[55] *Keynes MS* 130.2, pp. 21–2. In the little book in which he collected anecdotes, Conduitt referred this one to Pembroke (*Keynes MS* 130.6, book 2).

[56] Foster, "Newton's Family," p. 21.

[57] *Stukeley*, p. 43; *Keynes MS* 130.2, p. 24. Again Conduitt's account derived from Stukeley's.

every room he frequented, and if asked, would look at the shadows instead of the clock to give the time.[58]

Living in an apothecary shop, he also interested himself in the composition of medicines. It was his first introduction to chemistry, which would occupy more of his time than the heavens.[59]

He became proficient in drawing as well, and once more Clark's house bore the brunt of his enthusiasm. A later occupant of the garret room testified that the walls were covered with charcoal drawings of birds, beasts, men, ships, and plants. He also drew portraits of Charles I, John Donne, and the schoolmaster Stokes. A few circles and triangles also appeared on the walls – more of a forecast of the Newton we know than all of the portraits and birds and ships together. And on nearly every board, testifying to his identity like the desks in the school, stood the name "Isaac Newton," carved and therefore indelible.[60]

What with carvings and drawings and sundials, pokings about in the shop, and peevish scenes over bread, the apothecary Clark may have looked forward to the departure of his precocious guest. That came late in 1659. Newton was turning seventeen. It was time that he face the realities of life and learn to manage his estate. With that end in view, his mother called him home to Woolsthorpe. From the beginning the attempt was a disaster. As the hero-worshipping Conduitt has it, his mind could not brook such "low employments." His mother appointed a trusty servant to teach him about the farm. Set to watch the sheep, he would build model waterwheels in a brook, both overshot and undershot, with proper dams and sluices. The sheep meanwhile would stray into the neighbors' corn, and his mother would have to pay damages. The records on the manor court of Colsterworth show that on 28 October 1659 Newton was fined 3s 4d "for suffering his sheep to break ye stubbs on 23 ouf loes [loose? i.e., unenclosed] Furlongs," as well as 1s each on two other counts, "for suffering his swine to trespass in ye corn fields," and "for suffering his fence belonging to his yards to be out of repair."[61] On market days, when he and the servant went to town to sell the produce of the farm and to purchase supplies, Newton would bribe the servant to drop him off beyond the first corner; he would spend the day building gadgets or with a book until the servant picked him up on the way home. If perhaps he

[58] *Keynes MS* 130.2, p. 24.   [59] *Keynes MS* 130.2, p. 21.

[60] *Stukeley*, p. 43; *Keynes MS* 130.2, p. 20. Conduitt's account derives from Stukeley's. Newton also made markings on the walls and windows of the Manor House at Woolsthorpe, many of them intersecting circles (H. W. Robinson, "Note on Some Recently Discovered Geometrical Drawings in the Stonework of Woolsthorpe Manor House," *Notes and Records of the Royal Society, 5* [1947], 35–6).

[61] Turnor Papers, Lincolnshire Archives.

went to town, he would run directly to his old room at Clark's
where a stock of books awaited, and again the servant had to conduct the business. Going home to Woolsthorpe from Grantham,
one had to mount Spittlegate hill immediately south of town. It
was customary to dismount and lead one's horse up the steep hill.
On one occasion, Newton became so lost in thought that he forget
to remount at the top and led the horse all the way home; on
another occasion (or perhaps in another version of the same story),
the horse slipped his bridle and went home while Isaac walked on,
bridle in hand, unaware that the horse was gone. Apparently the
servant stomached all of this. When Newton even forgot his meals,
however, he despaired of ever teaching him.[62]

Meanwhile, two other men were viewing Mrs. Smith's efforts
from a different perspective. Her brother, the Reverend William
Ayscough, had taken the young man's measure, and he urged his
sister to send him back to school to prepare for the university. The
schoolmaster, Mr. Stokes, was if anything more insistent. He remonstrated with Newton's mother on what a loss it was to bury
such talent in rural pursuits, all the more so since the attempt was
bound to fail. He even offered to remit the forty-shilling fee paid by
boys not residents of Grantham, and he took Newton to board in
his own home. Apparently Clark had had enough. In the autumn of
1660, as Charles II was learning to accustom himself to the perquisites of the throne, a more momentous event took place to the
north. Isaac Newton returned to grammar school in Grantham,
with the university in prospect beyond.[63]

The evidence available indicates that the nine months at home
were a nightmare.[64] The list of sins in 1662 suggests constant tension: "Refusing to go to the close at my mothers command."

---

[62] *Stukeley,* pp. 48–50; *Keynes MS* 130.2, pp. 29–31. Conduitt's account in this case appears
to have some additional material beyond Stukeley's.

[63] *Stukeley,* p. 51; *Keynes MS* 130.2, p. 32. In Stukeley's account, Stokes was the key figure;
Conduitt introduced William Ayscough as well.

[64] Manuel does not deal with this episode explicitly. For myself, I find it hard to reconcile
with Manuel's insistence on Newton's fixation on his mother and her central importance
in his creative life. Manuel suggests that the return to Woolsthorpe in 1665, at the time of
the plague, provided the psychic stimulus for the great discoveries of the *annus mirabilis.*
He even speaks of Newton's being tied to his mother's umbilical cord. Newton could
have spent his life with his mother in Woolsthorpe; in fact, such would have been his
easiest course, and he had to provoke a domestic crisis in order to escape. Later, he could
undoubtedly have returned from Cambridge to a vicarage or rectory in the vicinity, since
a career in the church was the goal and actual practice of the overwhelming majority of
university undergraduates in the seventeenth century. Instead, the excitement of learning
seems to have swept every other factor in his life before it. Let me make it clear that the
excitement of learning (under different rubrics) plays a major role in Manuel's analysis
also. Nevertheless, I find it hard, as I say, to reconcile the episode of 1660 with his
analysis.

"Striking many." "Peevishness with my mother." "With my sister." "Punching my sister." "Falling out with the servants." "Calling Derothy Rose a jade."[65] He must have been insufferable. In Grantham, he had begun to sample how delicious learning could be. His inescapably intellectual nature had set him apart from the other boys, but he had no more been able to deny his nature to win their favor than a lion can give up his mane. Just as he had begun to commit himself to learning, however, he had been called back to the farm to spend his life herding sheep and shoveling dung. Everything within him rebelled against his fate, and fortune was on his side. By the intervention of Stokes and William Ayscough he was to feast on learning after all. His excitement still permeates Conduitt's account, blurred neither by sixty-five years nor by Conduitt's attempt at grandiloquence.

> His genius now began to mount upwards apace & to shine out with more strength, & as he told me himself, he excelled particularly in making verses . . . In everything he undertook he discovered an application equal to the pregnancy of his parts & exceeded the most sanguine expectations his master had conceived of him.[66]

When Newton was ready finally to leave, Stokes set his favorite disciple before the school, and with tears in his eyes made a speech in his praise, urging the others to follow his example. According to Stukeley, from whom Conduitt got the story, there were tears in the eyes of the other boys as well.[67] We can imagine!

The school boys at Grantham were not the only group to whom Newton was a stranger and an enigma. To the servants at Woolsthorpe he was simply beyond comprehension. Surly on the one hand, inattentive on the other, not able even to remember his meals, he appeared both foolish and lazy in their eyes. They "rejoic'd at parting with him, declaring, he was fit for nothing but the 'Versity."[68]

---

[65] Westfall, "Short-writing," pp. 13–14.
[66] *Keynes MS* 130.2, p. 32–3.   [67] *Stukeley*, p. 51.   [68] *Stukeley*, p. 51.

3

The solitary scholar

NEWTON set out for Cambridge early in June. There was no greater watershed in his life. Although he would return to Woolsthorpe infrequently during the next eighteen years, with two extended visits during the plague, spiritually he now left it, and what a later commentator has called the idiocy of rural life, once and for all. Three short years would put him beyond any possibility of return, though three more years, perhaps somewhat longer, had to pass before a permanent stay in Cambridge was assured. His accounts show that he stopped at Sewstern, presumably to check on his property there; and after spending a second night at Stilton as he skirted the Great Fens, he arrived at Cambridge on the fourth of June and presented himself at Trinity College the following day.[1] If the procedures set forth in the statutes were followed, the senior dean and the head lecturer of the college examined him to determine if he was fit to hear lectures. He was admitted – although there is no record whatever of anything but the verdict, one feels constrained to add "forthwith." He purchased a lock for his desk, a quart bottle and ink to fill it, a notebook, a pound of candles, and a chamber pot, and was ready for whatever Cambridge might offer.[2]

Fifty years later, the German traveler Zacharius von Uffenbach found the city of Cambridge "no better than a village . . . one of the sorriest places in the world."[3] Uffenbach was not viewing Cambridge against the background of Colsterworth and Grantham, however. To the young man venturing forth from rural Lincolnshire, the town of five or six thousand inhabitants must have seemed a metropolis, especially when Sturbridge Fair, the greatest fair in England, located where coastal vessels could penetrate no farther inland through the waterways of the Fens, descended on the neighboring common in the late summer. In addition to the town, there was of course the university with its imposing colleges drawn up in ranks along the Cam. Even the worldly Uffenbach exempted them from his judgment of Cambridge. To young Isaac Newton, they must have appeared grand beyond compare. They appear so still to older and more experienced visitors in the twentieth century.

---

[1] Trinity College, Cambridge, MS R.4.48ᶜ.   [2] *Ibid.*

[3] J. E. B. Mayor, *Cambridge under Queen Anne* (Cambridge, 1870), pp. 123, 198. Uffenbach's account of Cambridge is one of three pieces that compose Mayor's volume. In his notes to Uffenbach, Mayor quotes very similar judgments of the town by John Evelyn in 1654 and by Edward Ward about 1700 (p.410).

Admission to a college was not tantamount to admission to the university. Many delayed matriculation in the university; a considerable number who had no interest in a degree, to which alone matriculation was relevant, managed to avoid it altogether. Newton did intend to take a degree. On 8 July, together with a number of students recently admitted to Trinity and to other colleges, he duly swore that he would preserve the privileges of the university as much as in him lay,[4] that he would save harmless its state, honor, and dignity as long as he lived, and that he would defend the same by his vote and counsel; and to testify to the same he paid his fee and saw his name entered in the university's matriculation book. He was now a full-fledged member of the university.

Most of those who took the matriculation oath added their own silent gloss to the duties they swore to so solemnly. They would preserve and defend the privileges and honor of the university to the extent that they might thereby exploit the venerable institution to their own profit. In 1661, Cambridge was more than four hundred years old. Organized initially by a migration from Oxford on the occasion of a crisis between the university and the town, Cambridge had lived in the shadow of the senior university until the great expansion of higher education under Elizabeth and James. Not only had Cambridge multiplied four- or fivefold in size at that time, until it surpassed Oxford and reached a maximum of more than three thousand souls in the early 1620s, but Cambridge outstripped Oxford intellectually as well. The heart of English Puritanism, it was the point of ferment in English intellectual life in the early seventeenth century. Expansion was not without its perils, to be sure. The primary pressure behind it was the perception that university education and a university degree were means toward preferment by the state, especially in its department of ecclesiastical affairs, the Anglican church. As both the monarchy and influential peers quickly realized, the two universities collected together one of the largest reservoirs of patronage in the realm. No church universal now stood between the university and secular powers as a buffer to protect it from their pressures. As Crown and peers on the one hand sought to control the university and the patronage it embodied, so on the other lesser figures flocked to receive their largess. The established curriculum, dating from the medieval university, was increasingly irrelevant to the new function of the university, but piecemeal changes had not added up to a coherent alternative. By the second half of the century, Cambridge University, like Oxford, drifted toward the status of a degree mill exploited without conscience by those fortunate enough

---

[4] *Keynes MS* 116. This interesting phrase later made its way into Definition III of the *Principia*, where Newton defined what he meant by the *vis insita* or *vis inertiae* of matter.

to gain access to it. Such an atmosphere might impinge in various ways on a youth who came in quest, not of a place, but of knowledge.

In 1660, there were additional perils for an institution recognized by all as the home of English Puritanism. Even Cambridge had not passed untouched through the upheavals of the great Civil War and the Interregnum. In comparison with Oxford, however, the blows it sustained had been mild. With the Restoration, all the conditions reversed themselves. Puritanism had lost more than political power. It had also lost confidence in its mission, and Cambridge scrambled to efface the memory of the sins of yore. Town and university together were ostentatious in proclaiming Charles II, to the extent that the *Parliamentary Intelligencer* singled their celebration out as "very remarkable both for the manner and continuance." On 10 May, the heads of the colleges were summoned with their scholars to the public schools, where university lectures and exercises were held – all in academic gowns, the vice-chancellor and doctors in scarlet, the regents, non-regents, and bachelors with their hoods turned, the scholars in caps. A procession with music advanced to the cross on Market Hill where the proclamation was read, and on to the Rose Tavern where it was read again. A band played a great while from the roof of King's College Chapel, and bonfires lit the night. On 11 May, the mayor accompanied by the recorder and aldermen, all in their scarlet gowns, and by the freemen of the town mounted on horseback proclaimed Charles in the name of the town no less than seven times, and bonfires burned again. On 12 May, soldiers fired a volley from the roof of King's College Chapel. An effigy of Cromwell hung from a gibbet in the market. The round undergraduate cap favored by the Puritans disappeared in favor of the square cap, which they had considered Romish; James Duport, a royalist, congratulated the Puritans on squaring the circle.[5] More difficult matters remained to be settled in private. The Commonwealth had sold the royal fee-farm rents, part of the Crown's estate, to various corporations, and Cambridge had been among the buyers. Although other corporations surrendered their purchases to the Crown, the university dallied with the notion of requesting that the king graciously permit it to keep the rents. A more prudent counsel urged that the university court royal favor by returning the fee-farm rents at once without waiting to be commanded. The

---

[5] The account of the proclamation of Charles II is taken from James Bass Mullinger, *The University of Cambridge*, 3 vols. (Cambridge, 1873–1911), *3*, 554–5, and J. E. Foster, ed., *The Diary of Samuel Newton, Alderman of Cambridge (1662–1717)* (Cambridge, 1890), p. 1. The anecdote about Duport is found in Mullinger, *The University of Cambridge*, *3*, 555; since Duport had retained his fellowship and his perquisites throughout the Civil War and the Interregnum, his sarcasm strikes a singularly sour note.

more prudent counsel, which prevailed, belonged to the master of Corpus Christi College, Dr. Love, who found himself shortly thereafter elevated to the deanery of Ely.[6] His advancement was a practical homily, the lesson of which was lost on few. At least in the immediate aftermath of the Restoration, Cambridge prospered without apparent injury. Nearly all the heads of colleges were replaced; a mastership was too juicy a morsel of preferment to be left untouched. Fellows ejected during the Civil War and the Commonwealth were restored, but the years had taken their toll so that the changes were few. With the Act of Uniformity in 1662, an additional small number departed. Meanwhile Cambridge expanded rapidly during the 1660s, admitting an average of more than three hundred new students each year and approaching again its maximum size of the 1620s. The changeover emphasized anew the position of the university in the network of royal patronage and preferment. When Samuel Pepys, who had matriculated in Cambridge in 1650, visited it in 1660, he found the "old preciseness" had almost ceased to exist.[7]

During the century since 1560, while Cambridge University, from one point of view, flourished as it had never flourished before, from another point of view it almost ceased to exist. To impose discipline on the universities, and to control the religious groups on either extreme, Puritans on one side, Catholics on the other, governmental policy under the Tudors had forced all the students into colleges. Long before 1600, the medieval hostels had disappeared. Increasingly the colleges usurped the educational duties of the university, so that by 1660 the university had little function beyond the conferment of degrees. Among the colleges at Cambridge, none was more important than the College of the Holy and Undivided Trinity, founded by Henry VIII in 1546, which Newton entered in 1661. Along with King's College Chapel, its great court, virtually the entire college when Newton arrived, furnished one of the two imposing spectacles of the university, "the fairest sight in Cambridge," in the words of one observer.[8] Together with its neighbor of equal size, St. John's College, it dominated the university; the two constituted a good third of the entire institution. Intellectually, it dominated the university by itself. Three of the five Regius Professorships were attached to it. During

---

[6] Mullinger, *The University of Cambridge, 3,* 561–2.

[7] *The Diary of Samuel Pepys,* ed. Robert Latham and William Matthews, 8 vols. continuing (Berkeley and Los Angeles, 1970–), *1, 67.* Note that Pepys's observation preceded the Restoration, although it was in the air, and his account of his visit to Cambridge makes it clear that everyone there was already adjusting to it.

[8] Cited in W. W. Rouse Ball, *Notes on the History of Trinity College Cambridge* (London, 1899), p. 75.

the reigns of Elizabeth and James, Trinity had furnished a greater number of bishops to the Anglican church than any other college in Oxford or Cambridge. In 1600, both archbishops and seven bishops were Trinity men. Six of the translators of the Authorized Version came from the college. Like the university as a whole, Trinity expanded anew and seemed to flourish in the decade following the Restoration – a community of more than four hundred men, fellows, scholars, students, clerks, choristers, servants, and twenty almsmen supported by the college under the terms of its charter.

There is nothing surprising in the fact that Newton chose to enter Trinity, "the famousest College in the University," in the opinion of John Strype, the future ecclesiastical historian, who was an undergraduate in Jesus College at the time.[9] As it happens, personal factors probably influenced Newton's choice more than the reputation of the college. The Reverend William Ayscough, his uncle, was a Trinity man, and according to the account that Conduitt later got from Mrs. Hatton, née Ayscough, the Reverend Mr. Ayscough persuaded Newton's mother to send him to Trinity.[10] Stukeley heard in Grantham that Humphrey Babington, the brother of Mrs. Clark and fellow of Trinity, was responsible. The doctor, Stukeley wrote, "is said to have had a particular kindness for him, which probably was owing to his own ingenuity."[11] There is some evidence to suggest a connection between Newton and Babington. "Mr Babingtons Woman," one of the bedmakers and chambermaids allowed to work in the college, appeared twice in the accounts Newton kept as a student, and he later indicated that he spent some of his time when he was home during the plague at neighboring Boothby Pagnell, where Babington held the rectory.[12] As a fellow with considerable seniority such that in 1667 he became one of the eight senior fellows who, along with the master, controlled the college (and reaped its ripest rewards), and furthermore as a man who had demonstrated his access to royal favor with two letters mandate immediately after the Restoration, Babington would be a powerful ally for a young man otherwise without connections. Both the nature of the college and the nature of Newton's studies made a powerful ally desirable at the least, and perhaps indispensable. For whatever reason, on 5 June 1661, the famousest college in the university, quite unaware, admitted its famousest student.

[9]   In a letter to his mother in 1662; printed in Charles Henry Cooper, *Annals of Cambridge,* 5 vols. (Cambridge, 1842–1908), *3,* 504.   [10] *Keynes MS* 125.

[11] Stukeley to Conduitt, 1 July 1727; *Keynes MS* 136, p. 7.

[12] Trinity College, MS R.4.48^c; Fitzwilliam notebook (Fitzwilliam Museum, Cambridge). In the university setting, "Mr." was the contraction for "Magister," Master of Arts. Babington was created Doctor of Divinity in 1669; hence the title "Doctor" by which Stukeley referred to him.

Newton entered Trinity as a subsizar, a poor student who earned his keep by performing menial tasks for the fellows, fellow commoners (very wealthy students who paid for privileges such as eating at high table), and pensioners (the merely affluent). Sizar and subsizar were terms peculiar to Cambridge; the corresponding Oxonian word, servitor, expressed their position unambiguously. So did the statutes of Trinity College, which called them "scholares pauperes, qui nominentur Sizatores" and introduced the definition of their status by reference to the requirement laid on Christians to support paupers.[13] The statutes allowed for thirteen sizars supported by the college, three to serve the master and ten for the ten fellows most senior; they also defined subsizars as students admitted in the same manner and subject to the same rules as sizars, but paying to hear lectures (at a rate lower than pensioners) and paying for their own food. That is, subsizars apparently were to be servants like sizars but not supported by the college – servants of fellows, of fellow commoners, and of pensioners, according to whatever arrangements they might make.[14] Essentially identical in status, sizar and subsizar stood at the bottom of the Cambridge social structure, which repeated the distinctions of English society.

Like those of servants everywhere, the duties of the sizars were menial. They functioned as valets to fellows and to other students, rousing them for morning chapel, cleaning their boots, dressing their hair, and carrying their orders from the buttery. In the hall, they waited on tables. Half a century earlier, an Oxford student, in justifying his expenditure on a poor student to his skeptical father, chose to pose a rhetorical question: "should I have carried wood, and dust, and emptied chamber pots . . . ?"[15] A Trinity Conclusion Book illustrated their duties indirectly, as it attempted to grapple with a somewhat more pressing problem in 1661. "Ordered that ye Woemen of ye Colledge dismiss their young maid servants, with y^e

[13] The Statutes of Trinity College from the year 1560 are found in Appendix B of the *Fourth Report from the Select Committee on Education* (London, 1818). I cite from p. 375.

[14] I am unable to determine that there was ever any consistently maintained distinction between sizars and subsizars in Trinity. Although the statutes made provision for subsizars, the first student admitted explicitly as a subsizar was Joseph Halsey in 1645 (W. W. Rouse Ball and J. A. Venn, ed., *Admissions to Trinity College,* 5 vols. [London, 1911–16]). From that time, subsizars outnumbered sizars in the admissions book. Nevertheless, it is quite impossible to reconcile the number admitted explicitly as sizars with the thirteen supported by the college. Perhaps the distinction, once it began to be made, was between servants of fellows and servants of wealthy students, but the Trinity records of admissions, unlike those of St. John's College (*Admissions to the College of St. John the Evangelist in the University of Cambridge* [Cambridge, 1882]), did not list sizars as the servants of specific students.

[15] Brian Twyne to his father, 1597; cited in Mark H. Curtis, *Oxford and Cambridge in Transition, 1558–1642* (Oxford, 1959), p. 56.

soonest, sometimes this quarter. That they carry about no burning turfe, that they go not for commons to yᵉ kitchen or for bear & bread to yᵉ Buttrie for Fellowes, Scholars, or others, but that these be done by Sizars." The sizars who waited on table dined after the hall was cleared on whatever the fellows left. Though the statutes specified that subsizars were to pay for their commons, they also specified that they were not to enter the hall with the other students. An order of 1699 implied that they were even expected to subsist on less.[16] They sat in their separate place in the college chapel. In some colleges, special gowns visibly set them apart; Queen's and Corpus Christi had separate sections for them in their registers so that even their names might not mingle with the others.[17] The language in which they were discussed almost summons up the image of chattel slavery. "Let the Master have three," said the Trinity statutes of the thirteen sizars. In 1707, the fellows of Emmanuel College disputed the services of the sizars "belonging to the eight senior fellows . . ."[18] Small wonder that the historian of Balliol College states that a servitor (the Oxford equivalent of the sizar) was "a social pariah with whom men of ordinary good sense and good feeling hardly cared to be seen walking and conversing in public."[19]

If all this was true, why was Newton a sizar? Only one possible answer presents itself. His mother, who had begrudged him further education in the first place, and (by one account) had sent him back to grammar school only when the forty-shilling fee was remitted, now begrudged him an allowance at the university that she could have afforded easily. Though her income probably exceeded £700 per annum, Newton's accounts seem to indicate that he received at

---

[16] 14 Oct. 1661 (Master's Old Conclusion Book, 1607–1673, p. 272). *Fourth Report*, p. 375. Point 2 of the order of 1699 specified that the weekly quantum of every scholar and pensioner in the Lower Butteries was not to exceed 4s, and that of every sizar and subsizar not to exceed 2s 6d (Conclusion Book, 1646–1811, p. 202).

[17] David Arthur Cressy, "Education and Literacy in London and East Anglia, 1580–1700" (dissertation, Cambridge University, 1972), pp. 242–3. In his account of his life as a student in Cambridge in 1667–8, Roger North, who was a fellow commoner, indicated that he envied the common scholars for their games and their freedom to ramble, but he was "tied up by quality from mixing with them . . ." (Roger North, *The Lives of the Right Hon. Francis North, Baron Guilford; the Hon. Sir Dudley North; and the Hon. and Rev. Dr. John North. Together with the Autobiography of the Author*, ed. Augustus Jessopp, 3 vols. [London, 1890], *3*, 14).

[18] *Fourth Report*, p. 375. The example from Emmanuel is cited in E. S. Shuckburgh, *Emmanuel College* (London, 1904), pp. 121–2. Thomas Baker, an early eighteenth-century antiquarian of Cambridge, indicated that the admission records of St. John's contained subsizars as early as 1572 along with the fellow or master "to whom the sizar belongs . . ." (*History of the College of St. John the Evangelist, Cambridge*, ed. J. E. B. Mayor, 2 vols. [Cambridge, 1869], *1*, 551).

[19] H. W. Carless Davis, *A History of Balliol College*, rev. R. H. C. Davis and Richard Hunt (Oxford, 1963), p. 135.

most £10 per annum.[20] There is a further possibility not inconsistent with the above. Newton may have gone to Trinity specifically as Humphrey Babington's sizar, perhaps to attend to the interests of Babington, who at that time was resident in Trinity only about four or five weeks a year. The payments mentioned above to "Mr Babingtons Woman" would fit into such a hypothesis. In the eighteenth century, the Ayscough family tradition recorded the story

[20] See especially the accounts in the Fitzwilliam notebook (Fitzwilliam Museum) for the years 1665, 1666, and 1667. They appear to sum his receipts from his mother (about £10 per annum) and his total expenditures over about three years. To be sure, these were highly unusual years when he was home much of the time. However, the years include two payments of £5 each to his tutor. Since they are the only such payments recorded (though one must add that we do not have all of Newton's accounts as a student), they may have been for the B.A. (paid in May 1665) and the M.A. (paid apparently in March 1666). Such a rate seems reasonable. His tutor took fifty-three pupils during the four years 1661–5, when Newton was an undergraduate. At £5 per sizar, and therefore at least at £10 per pensioner, Pulleyn would have made out quite handsomely during that period from undergraduates alone. Newton's total expenditures between May 1665, when he received £10, and March 1667, when he received £10 again, were £5 /1s/2d plus £10 to his tutor. The period coincides pretty closely with the plague years; at most, Newton was in Cambridge for six months between May 1665 and March 1667. His earlier accounts in the Trinity notebook (MS R.4.48^c) are consistent with this level of expenditure (cf. especially the summations). In this notebook Newton recorded his expenses for the trip to Cambridge and for settling in at the college. After a number of other items, including a payment to Agatha, a chambermaid who appears to have been paid at most on a quarterly basis, he summed his expenditures to £3 /5s/6d: "Habui 4. 0 0 Habeo 0. 14. 6." Unfortunately there is no date, but the clear indication is that he set out from home with £4 and that he stretched it to last for a considerable period. Inside the front cover of the notebook are two lists of expenditures that appear to be dated: "12.7." and "8.11." The second list, covering the four months July–November if that is what the otherwise unintelligible numbers mean, totals 9s/10d.

In considering Newton's status as a sizar, one must keep his mother's financial status in perspective. In his *Natural and Political Observations and Conclusions upon the State and Condition of England* written in 1696, Gregory King put the average annual income of baronets, the second highest category in his breakdown of the English social hierarchy, at £880, and he put the average annual income of knights, the third category, at £650. He also estimated that there were 160 households of temporal lords, the highest category, 800 households of baronets, and 600 of knights. If King's estimates were near the mark, and if the reports of the Reverend Smith's estate were accurate, Hannah Smith's household thirty-five years earlier must have been among the fifteen hundred wealthiest in all of England (Gregory King, *Two Tracts,* ed. George E. Barnett [Baltimore, 1936], p. 31). Even if we assume that both were badly in error, she was still a wealthy woman by the standards of the age. A century after Newton's student days, James Boswell gave an account of Peregrine Langton, who lived in a village in Lincolnshire on an annuity of £200 per annum (plus, perhaps, some small additional rents). Langton was able to maintain a household consisting of his sister and his niece as well as himself with four servants. He had·a postchaise and three horses. He entertained frequently and well. To be sure, the thrust of Boswell's account was meant to emphasize Langton's excellent management. Nevertheless, Newton's mother apparently had at least three times his income, and prices had certainly not declined during the intervening century. Despite her wealth she forced her son to be a sizar.

that "the pecuniary aid of some neighboring gentlemen" enabled Newton to study at Trinity.[21] As the rector of Boothby Pagnell, Babington might fit that description. At a later time it does appear that Babington's support (that is, his influence not his money) may have been crucial to Newton.

We cannot avoid a further question. What impact, if any, did his status as sizar have on Newton? He was, after all, heir to the lordship of a manor. If the manor itself was not grand, his family's economic status, thanks to the fortune of Barnabas Smith, ranked above that of most gentry. Newton was used to being served, not to serving. His own record, drawn up in 1662, indicates that he had used the servants at Woolsthorpe harshly, and they, for their part, had rejoiced to see him leave. It is hard to imagine that he did not find menial status galling.[22] His status probably reinforced his natural propensity to isolation. Already in Grantham Newton had found it impossible to get along with his fellow students. If he thought he was escaping them to study with a superior breed in Cambridge, he was mistaken. The same boys were there; the names were all that differed. Only now he was their servant, carrying their bread and beer from the buttery and emptying their chamber pots. The one surviving anecdote concerned with his relations with other students suggests that the isolation and alienation of Grantham had traveled with Newton to Cambridge, intensified perhaps by his menial status. Well over half a century later, Nicholas Wickins, the son of Newton's chamber-fellow, John Wickins, repeated what his father had told him about their meeting.

> My Father's Intimacy with Him came by meer accident My Father's first Chamber-fellow being very disagreeable to him, he retired one day into ye Walks, where he found Mr Newton solitary and dejected; Upon entering into discourse they found their cause of Retirem^t ye same, & thereupon agreed to shake off their present disorderly Companions & Chum together, wch they did as soon as conveniently they could, & so continued as long as my Father staid at College.[23]

Since Wickins entered Trinity in January 1663, the encounter above occurred at least eighteen months after Newton's admission. I am inclined to think that the walks of Trinity had frequently known a

---

[21] I. H. [James Hutton], "New Anecdotes," p. 24. Hutton, who did not appear to know that Newton had been a sizar, was sufficiently offended by the story that he eliminated it from his mother's account and referred to it only in a footnote.

[22] Professor Manuel does not consider this issue in his idyllic account of Newton's relations with his mother. Is it possible that he did not hold her responsible, by her stinginess, for his position?

[23] Wickins to Robert Smith, 16 Jan. 1728; *Keynes MS* 137. Newton's accounts show a payment of 1s "To a porter when I removed to another chamber" (Trinity College, MS R.4.48^c).

solitary figure during those eighteen months, as they would for thirty-five years more. With the exception of Wickins, Newton formed no single friendship that played a perceptible role in his life from among his fellow students, though he would live on in Trinity with some of them until 1696, and even his relation with Wickins was ambiguous.[24] Correspondingly, when Newton became England's most famous philosopher, none of his fellow students left any recorded mention that they had once known him. The sober, silent, thinking lad of Grantham had become the solitary and dejected scholar of Cambridge.

Significantly, I am inclined to think, Wickins was a pensioner. Trinity was less rigid in segregating sizars than some of the colleges. It did not prescribe separate gowns for them, and the possibility of a sizar "chumming" (that is, sharing a chamber) with a pensioner existed. At first blush, it might appear that Newton was more apt to find congenial companions among the other sizars. By and large, they were the serious students. Whereas only 30 percent of the gentlemen who entered Cambridge continued to the degree, roughly four out of five sizars commenced B.A.[25] On the whole, however, they were a plodding group, narrowly vocational in outlook, lower-class youths grimly intent on ecclesiastical preferment as the means to advancement.[26] Since he had entered Trinity at

[24] To one of them, Francis Aston, he did write a letter in 1669 (*Corres 1*, 9–11).

[25] I have conflated two sets of statistics from Cressy in this statement. One shows that sizars were the most reliable in matriculating. Between 1635 and 1700, 82 percent of sizars matriculated in the university, whereas 71 percent of pensioners did so and only 49 percent of fellow commoners. Two-thirds of matriculated students proceeded to the B.A. as opposed to about 50 percent of all students admitted. The second table, based on a systematic sample of one in five undergraduates who entered Caius in the seventeenth century, correlates the social origin of students with the percentage who graduated. Whereas only 30 percent of the sons of gentlemen graduated, 68 percent of the sons of clergy and professional men did, 79 percent of the sons of tradesmen, and 82 percent of the sons of yeomen and husbandmen. Most sizars come from the last two categories (Cressy, "Education and Literacy," pp. 223, 238).

[26] Cf. brief summaries of the careers of a random sampling of sizars in Trinity at Newton's time. Without exception, they pursued careers in the church. It is also worth noting that only one out of ten got a scholarship from the college and that none at all got a fellowship.

James Paston, admitted subsizar 30 Jan. 1660. B.A. 1665, M.A. 1668. Ordained deacon 1665. Held a rectory in Suffolk from 1667 to 1722, to which he added a second in 1681. His son, James Paston, attended Caius as a pensioner.

Richard Howard, admitted subsizar 5 March 1660. B.A. 1664. Ordained deacon 1665. Held a vicarage in Kent from 1672 to 1682.

Thomas Perkins, admitted sizar 7 May 1660. B.A. 1664. Ordained deacon and priest 1664. Held various church livings in Hertfordshire, Surrey, and Essex before his death in 1686.

Nicholas Pollard, admitted sizar 12 June 1660. B.A. 1664. M.A. 1668. Rector of Barton St. Mary, Norfolk, in 1667 and subsequently held two other livings in succession.

Robert Grace, admitted sizar 13 June 1660. B.A. 1664. M.A. 1668. Ordained deacon in 1664. Vicar of Shenstone, Staffordshire, in 1665.

eighteen, Newton was at least one year older than the average and perhaps two, another factor which separated him from them. Genius of Newton's order does not readily find companionship in any society in any age. He was perhaps even less apt to find it among the sizars of Restoration Cambridge. As in Grantham, he was unable to conceal his brilliance. "When he was young & first at university," his niece Catherine Conduitt told her husband, "he played at drafts & if any gave him first move sure to beat them."[27]

The society of pensioners would at least mitigate the stigma of menial status. Wickins is not the only evidence that he tried to assimilate himself with the pensioners. He made loans to a number of them, Henry Jermin, Barnham Oliver, and Francis Wilford, though to be sure most of his extensive business in usury was conducted among his fellow sizars.[28] His list of offenses in 1662 included "Using Wilfords towel to spare my own" and "Helping Pettit to make his water watch at 12 of the clock on Saturday night."[29] Like Wilford, Pettit was a pensioner; though the two instances differ in tone, they both testify to Newton's being in the company of pensioners. His accounts contain the heading "Supersedens" with several entries of 6d following it. The only meaning I can attach to it is a payment to sit above his station in the hall. His

Job Grace, admitted sizar 13 June 1660. Scholarship in 1664. B.A. 1664. M.A. 1668. Held two church livings and was treasurer of the Cathedral of Lichfield.

Valentine Booth, admitted sizar 1 Feb. 1661. B.A. 1665. Probably the vicar of Stanford, Leicestershire, 1668.

Robert Bond, admitted sizar 21 May 1661. B.A. 1665. M.A. 1668. Ordained priest 1667. Rector at Layer Breton, Essex, from 1677 to 1688, when he died.

Humphrey Pagett, admitted subsizar 19 March 1662. B.A. 1666. M.A. 1669. Rector of Peckleton, Leicestershire, 1671. Buried there 1708.

John Baldocke, admitted subsizar 11 April 1662. B.A. 1666. M.A. 1669. Vicar of Littlebury, Essex, 1669–72, followed by a succession of other church livings, until his death in 1709. [27] *Keynes MS* 130.6, Book 2.

[28] Newton kept close track of his loans. An **X** through the line recording one signified that it had been paid. The Trinity notebook (MS R.4.48ᶜ), from his early years at Cambridge, has a loan of 10s to Guy without an **X**. When Newton toted up his financial status in March 1667, at least three years later, he carried over an unpaid loan of 10s to Guy (now Mʳ Guy). There were no subsequent loans to Guy.

The records give no indication that Newton took interest. The sheer extent of his lending activity suggests that he did, however, as does also his self-condemnation in 1662 for setting his heart too much on money (Richard S. Westfall, "Short-writing and the State of Newton's Conscience, 1662," *Notes and Records of the Royal Society, 18* [1963], 13). Perhaps his activity in lending served further to isolate him. His supply of money told the other sizars in unmistakable terms that he was not one of them. In any case, the student usurer has never been a popular figure, no matter how essential he may be. James Duport, the prominent Trinity tutor, would not have approved of Newton's practice under any circumstances. One of his rules of conduct was directly relevant: "62 Doe you neither lend, nor borrow any things of any Scholler or other" ("Rules to be observed by . . . schollers in the University"; Trinity College, MS O. 10A. 33, p. 8).

[29] Westfall, "Short-writing," pp. 14–15.

accounts regularly showed payments to bedmakers and chamber-maids, which suggests that he tried to imitate the style of the pensioners. So also his expenditure for clothes was out of keeping with the stories of hand-me-downs and rags in which sizars supposedly dressed. As far as our evidence reveals, Newton found no more companionship among the pensioners than among the sizars – with the exception of Wickins, of course. The others may well have regarded him as a climber, a strange figure for other reasons as well, tolerated only to the extent that he always had money to loan. Hannah Smith's niggardly allowance, not enough to make him a pensioner but apparently more than the sizars had, gave the initial impulse to Newton's isolation in Cambridge. It was probably inevitable in any case.

In the summer of 1662, Newton underwent some sort of religious crisis. At least he felt impelled to examine the state of his conscience at Whitsunday, to draw up a list of his sins before that date, and to start a list of those committed thereafter. His earnestness did not survive long enough to extend the second list very far. Lest it fall under the wrong eyes, he recorded his sins in cipher, using Shelton's system of shortwriting just as Samuel Pepys was using it at the same time for a livelier and more revealing record.[30] Many of the incidents that Newton remembered with shame belonged to Grantham and to Woolsthorpe, but some of them belonged to Cambridge: "Having uncleane thoughts words and actions and dreamese." He had not kept the Lord's day as he ought: "Making pies on Sunday night"; "Squirting water on Thy day"; "Swimming in a kimnel [a tub] on Thy day";[31] "Idle discourse on

[30] Newton seems to have met Shelton's system while he was in Grantham. The notebook in the Morgan Library has "A remedy for a Ague" in short-writing, apparently an exercise in using the system. It follows the notes on Bate's book, and its title and first few words, written in English, are in his grammar-school hand. The first page of the notebook in Trinity College (MS R.4.48ᶜ), which also dates to his grammar-school days, has a sentence about the blood relationship of his grandmother Ayscough and Mr. Beaumont's father – again apparently practice since it contains nothing to be concealed, and again suggestive of Lincolnshire days. His interest in short-writing probably led him on to his interest in a phonetic alphabet (Cf. the Morgan notebook and Ralph W. V. Elliott, "Isaac Newton as Phonetician," *Modern Language Review, 49* [1954], 5–12).

[31] Water was regarded with grave suspicion in sixteenth- and seventeenth-century Cambridge. During the Elizabethan age, Whitgift issued a decree when he was vice-chancellor that any student who went into a river or pool to wash or to swim should, on his first offense, be whipped publicly twice, before his college and before the whole university, and, on second offense, be expelled (Ball, *Trinity,* p. 68). James Duport's "Rules" of about Newton's time repeated the injunction without the penalty. "Goe not into the water at all or very wareily once or twice in a Summer at most, but better it were I thinke if you could quite forbeare." Duport had other reservations about water as well: "I am no great friend to going downe the water, because I have observed oftentimes it hath occasioned the going downe of the wind too much, and some under colour of going a fishing, drop into a blind house and there drink like fishes" (Trinity College, MS 0.10A.33, p. 15).

Thy day and at other times"; "Carlessly hearing and committing many sermons." He had not loved the Lord his God with all his heart and with all his soul and with all his mind: "Setting my heart on money learning pleasure more than Thee"; "Not turning nearer to Thee for my affections"; "Not living according to my belief"; "Not loving Thee for Thy self"; "Not desiring Thy ordinances"; "Not fearing Thee so as not to offend Thee"; "Fearing man above Thee"; "Neglecting to pray."[32] Relying upon this confession and upon his interpretation of the lists of words in the Morgan notebook, Professor Manuel concludes that Newton was borne down "by a sense of guilt and by doubt and self-denigration. The scrupulosity, punitiveness, austerity, discipline, and industriousness of a morality that may be called puritanical for lack of a better word were early stamped upon his character. He had a built-in censor and lived ever under the Taskmaster's eye."[33] Newton's undergraduate expenditures appear to bear out Manuel's judgment. If he treated himself now and then to cherries, "marmolet," custards, and even a little wine on occasion, he felt obliged to enter them under *Otiosi et frustra expensa* as opposed to *Impensa propria,* which included clothes, books, and academic supplies. He even considered beer and ale as *otiosi,* though we might judge them *propria* as we reflect on the water available.[34]

The Puritanical style of Newton's life would have set him apart from the ordinary pensioners even if his status of sizar had not. His conduct largely repeated the rules and ideals that had dominated the university during its heyday as the Puritan institution. The "Rules to be observed by . . . schollers in the University" composed by James Duport, a prominent tutor in Trinity still active when Newton arrived, sound a tone remarkably like Newton's confession. "Be diligent & constant at Chappell every morning." "Rise earlier on the Lords day . . . & be more carefull to trimme your soules then bodyes." "If you be with company on the Lords day, let your discourse be of the sermon or of some other point of Religion." "Think every day to be your last, and spend it accordingly."[35] Such was the public stance of the university. Chapel twice a day, at seven in the morning and at five in the evening, continued to be prescribed for all under forty. Even when Puritanism had been in its prime, such rules and ideals had been hard to impose on high-spirited youths. With the Restoration it became quite impossible. Exhortations by Duport and others were one thing; the orders and admonitions by university and college officials make it abundantly clear that reality contrasted brutally with the ideal. In 1663, Richard

[32] Westfall, "Short-writing," pp. 13–14.
[33] Frank E. Manuel, *The Religion of Isaac Newton* (Oxford, 1974), pp. 15–16.
[34] Trinity College, MS R.4.48ᶜ.   [35] Trinity College, MS 0.10A.33, pp. 1–3.

Smith was suspended from the college "in regard of his very great & heinous miscarriages, & misdemeanors (wherein he hath long continued disordering himself by intemperate drinking & lying out of ye College . . .)"; in January 1665, Young the scholar was expelled "for his foule & scandalous offence"; in 1667, the master and senior fellows agreed that the widow Powell "be not any further permitted to attend at any Chamber of any Fellow or scholar or any other within this College"; in 1676, Harry Luppincott confessed his "heinous crimes" and craved the mercy of God that "in a booth at Sturbridge fair [he] did sweare wicked oaths, and sing obscene songs . . ."[36]

While the college punished specific offenses, the university dealt with general problems. Coffeehouses had begun to appear. In 1664, the vice-chancellor and heads ordered that all *in pupillari statu* who frequented coffeehouses without their tutors should be punished "according to the statute for haunters of taverns and alehouses."[37] Since even Newton recorded several visits to a tavern, though only after he had attained the status of bachelor, we can imagine how well both orders were enforced. There were other dangers still more attractive, houses in Barnwell and even in Cambridge "infamous for harbouring lewd Women, drawing loose schollars to resort thither." Scholars were forbidden to enter such houses, which were identified very explicitly, thereby advertising their location to those who had lacked the enterprise to discover them on their own.[38] On 12 July 1675, the chancellor of the university, the Duke of Monmouth, not otherwise well known for his encouragement of moral uplift, inquired of the vice-chancellor and heads about the enforcement of the various statutes of the university. They replied in sorrow that despite their every endeavor the statute forbidding students to frequent taverns was "too frequently transgressed," that chapel services were duly celebrated but attendance at them was poor, and that coffeehouses were frequented by large numbers of all sorts – heads of houses excepted, of course.[39] Cambridge, that is, was the scene of those activities in which young men have engaged since the world began. When we compare the reality of life in Cambridge with Newton's examination of his conscience, we are perhaps not wholly surprised that Wickins found its author in the Trinity walks solitary and dejected.

Though a semblance of the old ideal remained, increasingly

---

[36] Conclusion Book, 1646–1811, pp. 78, 94, 107; Master's Old Conclusion Book, 1607–1673, new pagination at rear of book, p. 6. By and large, enforcement of discipline fell upon the sizars, who copied the style of the pensioners but did not have the connections to escape the consequences.   [37] Cooper, *Annals, 3,* 515.

[38] Cambridge University Library, MS *Mm* 1.53, f. 98. Printed in Cooper, *Annals, 3,* 571.

[39] *Ibid., 3,* 568–9.

empty in the context of the Restoration, the university itself adapted to the new order and encouraged activities forbidden a generation earlier. The stage returned to Cambridge. Like other colleges, Trinity appropriated funds in 1662 "toward ye charges of a Stage, & properties & a supper & for encouragement of ye Actors of a Comedy out of ye Commencemt moneys."[40] The following March, they found occasion to use it again when the Duke of Monmouth, Charles's illegitimate son, then all of fourteen years, received a Master of Arts degree from the university, along with thirty-four courtiers accompanying him, and was entertained by a banquet and comedy in the great hall at Trinity.[41] The city furnished its share of excitement of another order. In a period of five days in 1664, a man who stood mute and refused to answer to a charge of robbery was sentenced to be pressed to death and took an hour to die while he confessed wildly to the robbery and to all the other crimes he could recall, one Nelson was hanged for cutting his wife's throat, and a local attorney stood in the pillory on Peas Hill, all less than a quarter of a mile from Trinity.[42] No record survives of Newton's attendance at plays in Cambridge. It is hard to imagine him missing the spectacle provided by the Duke of Monmouth's ceremonial visit. It is equally hard to imagine him among the crowds who witnessed the executions and jeered the unlucky man in the pillory, though his fellow students undoubtedly made up a large proportion of them.

Meanwhile, along with the taverns and bawdy houses and comedies and executions there were also studies. By 1661 the official curriculum of Cambridge, prescribed by statute nearly a century before, was in an advanced state of decomposition. The system of tutors within the colleges, which had largely replaced university lectures, had followed its own peculiar development. Only a small minority of the fellows now engaged actively in tutoring, largely to augment their income. The others, who accepted their stipends and dividends in return for nothing at all, referred derisively to those who tutored as "pupil mongers." Whereas stories of affectionate and enduring relations between pupils and tutors, who were expected literally to stand *in loco parentis* to the boys in their charge, were common during the first half of the century, they became rare in the second half.[43] The records of Trinity College make it clear that

---

[40] Conclusion Book, 1646–1811, p. 61.   [41] Cooper, *Annals, 3,* 509.

[42] Foster, ed., *Samuel Newton's Diary,* p. 10.

[43] Cf. Matthew Robinson's relations with his tutor, Zachary Cawdrey, in the 1640s; "Autobiography of Matthew Robinson" in J. E. B. Mayor, *Cambridge in the Seventeenth Century,* 3 vols. (Cambridge, 1855–71), *2,* 16, 22, 67. Nicholas Ferrar's tutor exercised an immense influence on his pupil. Benjamin Whichcote carefully supervised his pupils' reading in order to raise up nobler thoughts in them. Henry More read a chapter of the Bible to his

the tutors, who looked after their own interests as well, functioned primarily as financial agents for the college; they dispersed payments to exhibitioners and scholars, and they oversaw payments due to the college.[44] Increasingly, as the entire structure of required studies and exercises lost its claim to legitimacy, tutors allowed students to go their own way.[45] Newton's tutor, Benjamin Pulleyn, was the champion pupil monger of Trinity during the period Newton was an undergraduate. In the five years, 1660 to 1664, which embraced Newton's undergraduate career, he took on a total of fifty-seven pupils. Pulleyn, who was then a young fellow, remained on at the college twenty-five more years. No single piece of evidence suggests even friendship, much less intimacy, between the two. The surviving record of Newton's reading indicates that Pulleyn guided him initially down the accustomed path. When Newton found a new road for himself, he was able to follow it without perceptible restraint.

The effect of the successive upheavals the university had been through in the two previous decades, first the Puritan revolution and then the Restoration, had been limited to changes of personnel almost exclusively. The curriculum of the university had remained untouched. Influences dating back to the sixteenth century had been altering aspects of the medieval course of study, however. Under the guise of rhetoric, extensive reading in the *litterae humaniores,* which continued the thrust of grammar-school education including the mastery of Latin and Greek, had been inserted in the programs recommended by tutors though the legally prescribed curriculum remained untouched.[46] Even with these innovations, study at Cam-

---

pupils in his chamber every night and lectured them on piety. Joseph Mede heard a report from each of his pupils in his chamber every night and commended them to God by prayer before he dismissed them. Alexander Akehurst watched carefully over his pupils' lives and friends and prayed with them every night in his chamber (James Bass Mullinger, *Cambridge Characteristics in the Seventeenth Century* [London, 1867], pp. 49–50, 89, 181).

[44] See various entries in Bailie's & Chamberlayne's Day Book, 1664–1673, where tutors sign for the "wages" of exhibitioners, scholars, choristers, library keepers, chapel clerks and the like under their tutelage (e.g., pp. 18, 34, 35, 80). On 25 April 1664, the steward was ordered to take special care to call on tutors for the dues that their pupils owed the college (Conclusion Book, 1646–1811, p. 83).

[45] In the early 1660s, John North's tutor gave him no direction, and he "soon fell to shift for himself, as a bird that had learned to pick alone . . ." (North, *Lives, 2,* 283). A few years later, the same John North, acting as tutor to his younger brother, did not want to be bothered, so that Roger North read whatever he wanted to (*ibid., 3,* 15). The second case, with two brothers involved, was hardly a normal tutor-pupil relation, of course.

[46] Richard Holdsworth, "Directions for a Student in the University," Emmanuel College, MS I.2.27; Duport, "Rules." As Holdsworth's "Directions" make clear, some of the classics could pose problems in a Puritan setting. "In Martial take those Epigrams wch are marked for good ones by Farnaby, passing over the obscene & scurrilous." Newton's reading notes show that he studied Gerard Vossius's *Rhetorices contractae sive, partitionum*

bridge had not broken the mold in which it had been cast four
centuries before, with its focus centered on Aristotle. It began
with a heavy dose of logic (Aristotelian logic) which together
with ethics (Aristotelian ethics) and rhetoric provided the foundation for the study of Aristotelian philosophy, and it reached its
culmination in the formal disputations, conducted with Aristotelian syllogisms, which were the standard academic exercises and
examinations. Both Holdsworth's "Directions" and Duport's
"Rules," the most important surviving records of tutorial practice
in the seventeenth century, take the traditional curriculum as an
unquestioned assumption. Student notebooks, which survive in
considerable numbers, confirm their evidence.[47] "The reading of
Aristotle will not only conduce much to your study of controversy, being read with a Commentator, but allso help you in
Greeke, & indeed crown all your other learning," Holdsworth
admonished his students, "for he can hardly deserve the name of
a Scholar, that is not in some mesure acquainted with his
works." "If at any time in your disputation you use the Authority of Aristotle," Duport exhorted in a similar vein, "be sure you
bring his owne words, & in his owne language. In your answering reject not lightly the authority of Aristotle, if his owne
words will permit of a favorable, and a sure interpretation."[48]
Certainly Duport's and probably Holdsworth's advice were written after the publication of Descartes's philosophy had inaugurated a new chapter in European thought. Duport included explicit instructions on disputations; Holdsworth assumed that the
student would spend much of his last two years in disputations,
preparing for the public Acts that would climax and conclude his
course of study, and he also justified the study of rhetoric, under
which reading in the classics was gathered, by its contribution to
effective disputation.[49] If this account suggests that the universities were conservative, perhaps reactionary, Newton's first experience at Cambridge confirmed as much. Before he left home, his
uncle William Ayscough, calling upon his own experience at
Trinity thirty years before, gave him a copy of Sanderson's logic
and told him it was the first book his tutor would read to him.
When Newton got to Trinity, his uncle's prediction was proved

   oratoriarum libri V, one of the standard university texts. There are no undergraduate notes
   from the *litterae humaniores,* but he did purchase at least two history books, Hall's *Chroni-*
   cles and Sleidan's *Four Monarchies* (as he called them), perhaps the beginning of his long
   study of chronology.

[47] William T. Costello, *The Scholastic Curriculum at Early Seventeenth-Century Cambridge*
   (Cambridge, Mass., 1958), *passim.*

[48] Holdsworth, "Directions"; Duport, "Rules," p. 11.

[49] Duport, "Rules," pp. 8–11; Holdsworth, "Directions."

perfectly correct.[50] During those thirty years, the intellectual life of Europe had been turned inside out. As far as Cambridge was concerned, nothing had happened. The story ends with a typically Newtonian flourish. He read the logic before he got to Cambridge and found that he knew more about it than the tutor—which may help to explain why no affection developed between them.[51]

One of Newton's first purchases in Cambridge was a notebook, and probably it was in this one that he entered the fruits of his reading in the established curriculum.[52] As was his custom, he entered notes from both ends, starting at one end with Aristotle's logic and at the other end with Aristotle's Nichomachean ethics. In

[50] The story appears not to be entirely accurate since Newton's copy of Sanderson has an inscription indicating that he bought it in Cambridge in 1661. The purchase at that time does tend to confirm the essence of the anecdote. There is a nice personal touch connected with it. Robert Sanderson, the author of the logic, had been the rector of Boothby Pagnell, not far from Colsterworth, when Newton was a boy. Humphrey Babington, the brother of Mrs. Clark and possibly Newton's patron in Trinity, had succeeded him when he was elevated to the bishopric of Lincoln with the Restoration. Newton spent part of his time at Boothby Pagnell during the plague years. It was there that he calculated an area under a hyperbola to fifty-two places, crowning his discovery of the binomial expansion.

Seven other books with the inscription "Isaac Newton. Trin: Coll: Cant: 1661" survive, four of them theological works. I do not know of any notes that Newton took from them at this time (John Harrison, *The Library of Isaac Newton* [Cambridge, 1978]).

[51] Conduitt's memorandum of a conversation with Newton on 31 Aug. 1726; *Keynes MS* 130.10, f. 2. The story did not end with Sanderson's logic. When Newton's tutor found him so forward, he told him that he was going to read Kepler's optics to some gentlemen commoners and that he should come to those lectures. Newton immediately bought the book, and when his tutor sent for him for the lectures he told him that he had read the book already. Frankly, this part of the story is unbelievable. What we know about the state of mathematics and natural philosophy in Cambridge makes it impossible to conceive that ordinary students there, not to mention ordinary tutors, could handle Kepler's optics, whether this refers to the *Paralipomena* or to the *Dioptrice*. Gentlemen commoners were the least likely to be able to comprehend it, and the least likely to want to. As for Newton, he never mentioned Kepler's optics elsewhere.

[52] Holdsworth's "Directions" had recommended that students not waste their time with the huge clumsy commonplace books of an earlier age but that they get some "paperbooks" of portable size; although he was not explicit, he seemed to suggest a separate one for each subject. Newton inherited the huge, predominantly blank commonplace book that his stepfather had once started. He renamed it his "Waste Book" (in reflection on his stepfather?), and he used its nearly unlimited supply of paper for important essays in mechanics and mathematics. He also brought along the two tiny notebooks from his grammar-school days (the Morgan Library and the Trinity College notebooks). Meanwhile, he had several separate octavo notebooks for different subjects. *Add MS 3996* was his basic notebook in which he originally set down notes from his reading in the established curriculum and later notes from his reading in the mechanical philosophy. I call it his philosophy notebook. *Add MS 3975* started as an extension of the "Quaestiones" in 3996 but gradually converted itself into a chemistry notebook. *Add MS 4000* was his mathematics notebook, begun probably in 1664. The notebook in the Fitzwilliam Museum never took on a distinct character but it received more mathematics than anything else. He also had a large quarto notebook that he devoted to theology; in the 1670s he collected a great deal of material in it, but he used it very little during his student days.

both cases the notes were taken in Greek, mere transcriptions of sentences, usually the first sentences of chapters and their subdivisions; they give the appearance of being as much exercises in Greek as reading notes.[53] In neither case did he proceed all the way through the work in question. Somewhat later, if we may judge by the hand, he followed the notes from Aristotle's ethics with others from the ethics of Eustachius of St. Paul, a popular seventeenth-century textbook which suitably Christianized Aristotle. Fairly early in his undergraduate career, he was introduced to Daniel Stahl's *Regulae philosophicae,* a recently published compendium of Aristotelian philosophy laid out in the form of disputations with objections and replies handily provided. Stahl must have been a popular textbook since it quickly went through seven editions. Newton also failed to finish this work, though he proceeded farther than he did with any of the others. For rhetoric, he read Gerard Vossius's *Rhetorices contractae,* proceeding only part way through the second of five books.[54]

He was introduced as well to Aristotelian physics via the seventeenth-century peripatetic, Johannes Magirus (*Physiologiae peripateticae*). Here he found an exposition of the Aristotelian cosmology, on which he took rather full notes. Something of the future Newton revealed itself when he departed entirely from Magirus's order of presentation to collect together information on the periods of the celestial spheres. "Sphaera nona & octava 4900 annis: Orbis ♄, 30: ♃, 12: ♂, 2 annis. ♀, 348 ☿, 339. ☽, 27 diebus & 8 fere horis circumferter." From Magirus he also learned of various conceptions of light, and he copied down the argument against the corporeality of light, namely that the sun would be exhausted. Again Newton stopped before he finished Magirus; later he returned to him, read two more chapters, then stopped for good and drew a line under his notes. He had been reading about phenomena such as the rainbow, which were classified as "Apparent Meteors" in the Aristotelian system. Under the line he entered a note which is symbolic of his discovery of a different school of natural philosophy: "Galilaeus sayth y^t y^e apparent diameter of starrs of y^e first magnitude is $5'' = 300''' = 0{,}083333^{\text{degr:}}$ &c. & of y^e sixt magnitude to bee $50''' = 0'{,}83333$ &c $= 0^{\text{degr:}}$, 0138888 &c. Mounsieur Auzout esteems y^e great dogs diameter to bee no more y^n $2''$ & those of y^e 6^t magnitude to bee $20'''$."[55] He never returned to Magirus.

---

[53] *Add MS* 3996, ff. 3–10, 34–6. Cambridge University Library has started foliation from the end with the notes on logic; notes, not all on logic, extend through f. 30. Foliation then shifts to the other end so that f. 34 starts the notes on ethics.

[54] *Add MS* 3996, ff. 38–40, 43–71, 77–81. Again the hand indicates that the notes on Vossius were made at a later time, after those on Stahl and Magirus.

[55] *Add MS* 3996, ff. 16–26, 20, 20^v, 26–6^v, 26^v.

It may be significant that Newton did not finish any of the books from the established curriculum that he started. Nevertheless we should be mistaken to underestimate the importance of Aristotelian philosophy in his life. He came to Cambridge in 1661 a provincial young man well grounded in Latin but entirely innocent of systematic natural philosophy. One must not confuse his interest in things mechanical with natural philosophy; while it may conceivably have predisposed him to favor the mechanical philosophy of nature once he met it, it had nothing whatever to do with natural philosophy as an attempt to account rigorously for the phenomena of nature. Although the Aristotelian philosophy was passing out of favor, it was anything but nonsense. It was the first sophisticated system Newton met, and there is no reason to think that it failed to impress him initially as he emerged from his intellectual provinciality, even though it was no longer presented with conviction. From it he learned the canons of rigorous thought, and it provided him with a system that organized the overwhelming diversity of nature into a coherent pattern. As it is impossible to imagine the scientific revolution without the background of medieval philosophy, so it is impossible to imagine Newton's achievement without his prior exposure to Aristotelianism. Nevertheless, the scientific revolution was under way. As the note on Galileo informs us, Newton discovered it before he completed his undergraduate career. Because the university started his career as a natural philosopher within the Aristotelian system, he had to recapitulate the prior history of the scientific revolution and have his own private rebellion against the orthodoxy established around him. And if he was never able to isolate limited problems from the total context of nature, if his concern with nature as an organized system never left him, part of the reason was his recollection of a completely different system to which initially he had owed allegiance. It never ceased to be necessary for him to justify the new system he had embraced.

Although Newton's notes make it clear that his initial studies at the university were directed toward the traditional curriculum, a variety of evidence from elsewhere makes it equally clear that the traditional curriculum was itself rapidly approaching a state of crisis. When it had been formulated initially, it had embodied the most advanced position of European philosophy. By 1661, European philosophy had moved on, and academic Aristotelianism represented an intellectual backwater maintained in part by the legal mandate of a curriculum enacted as law and in part by men who had a vested interest in continuing a system to which they had bound their lives. Intellectual vigor had departed long since. It was becoming an exercise performed by rote without enthusiasm. In the latter half of the seventeenth century, nearly all of the university lectureships became sinecures. The disputations in the public schools, the climactic exercises of the curriculum, gradually became mechanical performances without meaning.[56] According to the statutes, the student who had completed four years of study was required then to devote the whole of the Lent term to disputations in the public schools, where university as opposed to college exercises took place. After a series of oral examinations and ceremonies, he was admitted as a determiner to "stand *in quadragesima,*" appearing in the public schools every afternoon from one to five from Ash Wednesday to the Thursday before Palm Sunday to undergo examination by bachelors from other colleges and to defend a number of theses in disputations conducted according to the canons of Aristotelian syllogisms.[57] By Newton's day, the attendance requirements of the university had been relaxed to the point that anyone who was admitted before the end of the Easter term was allowed to commence B.A. in the Lent term four years hence. Thus Newton was admitted to Trinity on 5 June 1661 and commenced B.A. in 1665. The exercise of standing *in quadragesima,* together with the examinations preceding it, was already apparently a mere formality. The formal decision (or "grace") of the university senate to grant Bachelor of Arts degrees in 1665, which merely accepted the lists submitted by the colleges, was passed on 14 January, before any of the examinations or disputations had taken place.[58] In the summer

---

[56] Disputations had not disappeared in Newton's time, however. Roger North reported that he observed them both in his college (Jesus) and in the public schools in 1667–8. In the early 1670s, disputations bulked rather large in the undergraduate career of William Taswell in Oxford (*Autobiography and Anecdotes,* ed. George P. Elliott, in *Camden Miscellany, 2* [Camden Society, 1852], 17–18; see p. 28 for an Act in 1679, when he was a Master of Arts).

[57] George Peacock, *Observations on the Statutes of the University of Cambridge* (London, 1841), pp. 8 (the exercises as prescribed by the statutes of 1570, which remained in effect until the middle of the nineteenth century), v-xiii (the exercises as described by Mathew Stockys Esquire Bedel his Book in the second half of the sixteenth century), lxv-lxxii (the exercises as described by Beadle John Buck's Book of 1665).

[58] Cambridge University Library, Grace Book H, p. 339. Certainly the College considered its action decisive. Two days earlier Trinity had entered the list to whom Bachelor's degrees were granted in its Conclusion Book (Conclusion Book, 1646–1811, p. 93). Cf. the letter of William Sancroft, newly elected Master of Emmanuel, to his old tutor Ezechiel Wright in 1663. Sancroft told Wright that it would grieve him to hear a public examination in the college. In the whole university, learning in Greek and Hebrew was out of fashion. The old philosophy was not studied and the new was not followed systematically and seriously. "In fine, though I must do the present society right and say that divers of them are very good scholars and orthodox (I believe) and dutiful both to king and church, yet, methinks I find not that old genius and spirit of learning generally in the college that made it once so deservedly famous . . ." (cited in Curtis, *Oxford and Cambridge,* pp. 278–9). Much of the letter, of course, is about Emmanuel College, the old Puritan stronghold, which fell on desperate times in the Restoration.

of 1665, the university dispersed for most of two years because of the plague; it is impossible to discern from the statistics for degrees granted that the plague ever occurred.[59]

With the colleges completing their conquest of the university, one might perhaps look for the prosecution of serious exercises and examinations in the colleges, with the university ratifying their decisions. What we know about the colleges hardly supports such a conclusion. Most of the fellows treated their fellowships as freeholds to be enjoyed without corresponding duties. Only a small number tutored, and they did so primarily for the extra income they gained. In 1659, John Wilkins, the master of Trinity, drew up new regulations because of laxities in the examinations.[60] There were still examinations of some sort in Trinity, at least in connection with scholarships and fellowships, when Newton was there. Nevertheless, the general outlines of the picture are clear. The established curriculum had ceased to command the allegiance even of those who maintained it.[61] A serious and perceptive student could not have failed to observe as much. Small wonder that Newton never finished the Aristotelian texts he was assigned.

Small wonder also that he found other reading. Perhaps one should not regard history as alternative reading; it figured strongly in Holdsworth's program. At any rate, two history books, Hall's

[59] James Bass Mullinger, *A History of the University of Cambridge* (London, 1888), between pp. 212 and 213. Masters of Arts degrees, which were given in July, did fall in 1666; the university dispersed in June of that year. In February 1667, when the university had been vacated, with one intermission of about three months, for a year and a half, it got royal sanction to enact ordinances that such questionists as might be prevented by the plague from being present in Cambridge on Ash Wednesday to receive degrees should not forfeit their seniority in their colleges (Mullinger, *The University of Cambridge*, 3, 621). This action by the university demonstrates in two ways that the exercises had become formalities, first by dispensing with them, and second by designating Ash Wednesday, when by statute they were supposed to begin, as the date on which degrees were received. Cf. the account of Thomas Fuller of the earlier plague of 1630, which also forced the university to disperse: "But this *corruption* of the aire proved the *generation* of many Doctours, graduated in a clandestine way, without keeping any Acts, to the great disgust of those who had fairly gotten their degrees with publick pains and expence. Yea, Dr. Collins, being afterwards to admit an able man Doctour, did (according to the pleasantnesse of his fancy) distinguish *inter Cathedram pestilentiae, & Cathedram eminentiae*, leaving it to his Auditours easily to apprehend his meaning therein" (*The History of the University of Cambridge and of Waltham Abbey*, new ed., notes by James Nichols [London, 1840], p. 231.) No one would have gotten the joke in 1667. Indeed, there would have been no joke. His story is about doctors, not bachelors, but part of the change in Cambridge was the decline almost to disappearance of advanced degrees.

[60] Mullinger, *The University of Cambridge*, 3, 547.

[61] Roger North indicated that his brother John's account of Aristotelian logic was "with more freedom than the humour of the university, among the seniors at least at that time, would have allowed." John North thought that it was not only useless but also pernicious (*Lives*, 2, 313).

Chronicles and Sleidan's *Four Monarchies,* were among his early purchases in Cambridge.[62] Although he left nothing among his undergraduate notes from his reading in these works, chronology remained, in close association with his study of the prophecies, one of
his abiding interests. For a brief time, about 1663, he examined
judicial astrology, according to a conversation he had with Conduitt
near the end of his life.[63] Astrology was never part of the curriculum.
Phonetics and a universal philosophical language also had nothing to
do with established studies, although they, or at least the idea of a
universal language, were live centers of intellectual interest at the
time. There had been a number of schemes for a universal language
based, as Newton expressed it, on "ye natures of things themselves
wch is ye same to all Nations . . ."[64] Sometime during his undergraduate career, Newton came across this literature; he drew especially upon George Dalgarno's *Ars signorum* (1661). To it he attached
an interest in phonetics which may have derived from his study of
Shelton's system of short-writing.[65] The basic feature of such a language was a classification of things and concepts into categories. All
words within one category would begin with the same letter; subsequent letters (and phonetic symbols) would indicate the subdivisions
within the category. Because it would express the nature of things
themselves, such a language would transcend the barriers of nation-
alities and be universal. Newton proposed a number of modifications
to Dalgarno's scheme.[66] Other interests soon pushed the universal
language aside, and he never returned to it.

Frequently, as in the case of John Wilkins's *Essay Toward a Real
Character and a Philosophic Language* (published in 1668, after Newton's venture in this field), the concept of a universal language was
coupled with criticism of Aristotelian philosophy, which was held
not to express the "real" nature of things. Newton's youthful exercise was not so. Couched in Aristotelian terms, it reflected the sole
philosophy to which he had been introduced. Such was not long
the case, however. In the notebook in which he had entered the
fruits of his study advancing from each end, about a hundred pages
remained empty in the center. Two pages devoted to Descartes's

62   Trinity College, MS R.4.48^c· The two books referred to were Edward Halle, *The Union
     of the Two Noble and Illustrate Famelies of Lancastre & Yorke* . . . (London, 1548) and
     Joannes Philippson (Sleidanus), *The Key of Historie. Or, A Most Methodicall Abridgement of
     the Four Chiefe Monarchies* . . . (London, 1631), a translation of Sleidanus's *De quatuor
     summis imperiis* . . . (Geneva, 1559).
63   Conduitt's memorandum of the conversation of 31 Aug. 1726; *Keynes MS 130.10,* f. 2.
64   Ralph W. V. Elliott, "Isaac Newton's 'Of An Universall Language'," *Modern Language
     Review, 52* (1957), 9. Elliott's article prints most of the manuscript, which is located now
     in the University of Chicago Library, MS 1073.
65   *Ibid.,* p. 2. Cf. the phonetics in the Morgan Library notebook, which belong to the same
     period.   66 *Ibid.,* p. 5.

metaphysics bluntly interrupted the Aristotelianism of the texts he had been reading.[67] A few pages further on he entered the title, "Questiones quaedam Philosophcae," and laid out a set of headings under which to collect the notes from a new course of readings.[68] Somewhat later, he wrote a slogan over the title, "Amicus Plato amicus Aristoteles magis amica veritas." Whatever there may be of truth in the pages that follow, certainly there is nothing from Plato or Aristotle. Notes from Descartes, whose works Newton thoroughly digested in a way that he never had Aristotle, appear throughout the "Quaestiones." Nor had he confined himself to Descartes. He had also read Walter Charleton's English epitome and translation of Gassendi, and perhaps some of Gassendi as well. He had read Galileo's *Dialogue,* though apparently not his *Discourses.* He had read Robert Boyle, Thomas Hobbes, Kenelm Digby, Joseph Glanville, Henry More, and no doubt others as well. *Veritas,* Newton's new friend, was none other than *philosophia mechanica.*[69]

There is no way conclusively to date the beginning of the "Quaestiones" although various considerations suggest some time not too late in 1664.[70] Equally there is no way to state with assur-

---

[67] *Add MS* 3996, ff. 83–3ᵛ.

[68] *Add MS* 3996, ff. 88–135. I shall amend Newton's title, the first word of which is archaic in spelling and the third word, in haste, simply incorrect, and refer to the passage as "Quaestiones quaedam Philosophicae," "Certain Philosophical Questions." Fifty years later Newton told Conti that originally–by which I take him to refer to the period I am now discussing–he had been a Cartesian (Antonio-Schinella Conti, *Prose e poesi,* 2 vols. [Venice, 1739–56], *2,* 26).

[69] Cf. A. R. Hall, "Sir Isaac Newton's Note-book, 1661–65," *Cambridge Historical Journal, 9* (1948), 239–50, and Richard S. Westfall, "The Foundations of Newton's Philosophy of Nature," *British Journal for the History of Science, 1* (1962), 171–82.

[70] It had to be before 9 Dec. 1664. On that day and on the following one he made and entered observations of a comet (*Add MS* 3996, ff. 115, 93ᵛ). Many years later he mentioned his extended observations of this comet in a conversation with Conduitt on 31 Aug. 1726 (*Keynes MS* 130.10, f.4ᵛ). Already his hand had altered from that of the initial entries and had assumed the tiny perpendicular form that characterized it in 1665 and 1666. The earliest entries in the "Quaestiones" were made in a hand transitional between his grammar-school hand, in which he took his early undergraduate notes, and the hand of 1665–6. I know of no way to date the transitional hand with assurance. On f. 93ᵛ there is a passage with a reference to Descartes in it. Unfortunately, the passage to which it apparently refers does not fall on that page in any edition of Descartes. It comes close in two editions of 1664, however: the Weyerstaeten edition (from the Netherlands) and the Hart edition (from London). Since very little writing in the transitional hand survives, it is reasonable to assume that the transition went rather quickly from the grammar-school hand to that of 1665–6; the latter furnished the permanent model of his mature hand, being modified only slowly over the years ahead. Intellectual continuity suggests the same date, and so do accounts of his introduction to mathematics. Once the new program of study seized him, it seized him completely. Although entries were made in the "Quaestiones" over a period of time perhaps as long as two years, the main body was not composed of notes made at leisure. The initial "Quaestiones" were put together in hot haste by a man who had found his calling.

ance the agency involved, but everything we know about Cambridge suggests it had little to do as an institution with leading Newton to the new philosophy. One piece of testimony indicates that Descartes was very much in the air at the time, so that the advice of a tutor would scarcely have been required. Roger North, an undergraduate in Cambridge in 1667–8, whose tutor, his brother, did not wish to be bothered and left him to follow his own inclinations, "found such a stir about Descartes, some railing at him and forbidding the reading him as if he had impugned the very Gospel. And yet there was a general inclination, especially of the brisk part of the University, to use him . . ."[71] Newton's notes imply that he also found a stir about Descartes and decided to investigate him. Beyond Descartes, we are left wholly to speculation, but it is not hard to imagine the process whereby Newton was led on from one author to another into a totally new world of thought. At last he had found what he came seeking in Cambridge. Without hesitation, he embraced it as his own. The very laxity of the university now worked to his advantage. Pulleyn was probably happy enough not to be bothered, and Newton could pursue his interest unhindered.

He set down forty-five headings under which to organize the fruits of his reading, beginning with general topics on the nature of matter, place, time, and motion, proceeding to the cosmic order, then to a large number of tactile qualities (such as rarity, fluidity, softness), followed by questions on violent motion, occult qualities, light, colors, vision, sensation in general, and finally concluding with a set of miscellaneous topics not all of which appear to have been in the initial list. Under some of the headings he never entered anything; under others he found so much that he had to continue the entries elsewhere. The title "Quaestiones" adequately describes the whole in that the tone was one of constant questioning. The questions were posed within certain limits, however. They probed details of the mechanical philosophy; they did not question the philosophy as a whole. Newton had left the world of Aristotle forever.

One product of his new world view was a temporary interest in perpetual motion. The mechanical philosophy pictured a world in constant flux. Newton, the tinkerer from Grantham, thought of various devices, in effect windmills and water wheels, to tap the currents of invisible matter. For example, he adopted the view that

---

[71] North, *Lives, 3,* 15. Hugh Kearney has found five student notebooks from roughly this time with notes from Descartes. Henry More's letters to Lady Conway indicate that he was introducing students to Descartes early in the 1670s; from what we know of More's intellectual life, he was probably doing so earlier as well (Hugh Kearney, *Scholars and Gentlemen. Universities and Society in Pre-industrial Britain* [London, 1970], p. 151).

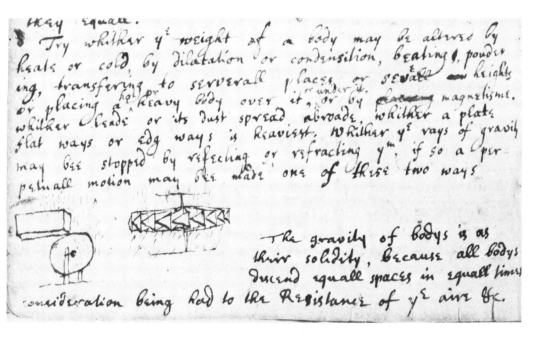

Figure 3.1. Newton's sketches of perpetual-motion machines powered by the flux of the gravitational stream. (Courtesy of the Syndics of Cambridge University Library.)

gravity (heaviness) is caused by the descent of a subtle invisible matter which strikes all bodies and carries them down. "Whither ye rays of gravity may bee stopped by reflecting or refracting y^m, if so a perpetuall motion may bee made one of these two ways." (Figures 3.1 and 3.2). Under the heading of magnetism, he proposed analogous devices.[72]

Most of the entries in the "Quaestiones" were derivative, notes from Newton's reading. Nevertheless, the whole carried the unmistakable imprint of its author. To a remarkable degree the "Quaestiones" foreshadowed the problems on which his career in science would focus and the method by which he would attack them. As to the latter, the title "Quaestiones," which describes not only the set of headings but their content, suggests the active questioning that lay behind Newton's procedure of experimental enquiry. Many of the questions were directed to the authors he was reading, whose opinions he did not merely register passively. Descartes's theory of light raised a number of objections.

Light cannot be by pression &[c] for y^n wee should see in the night a wel or better y^n in y^e day we should se a bright light above us becaus

[72] *Add MS* 3996, ff. 121^v, 102.

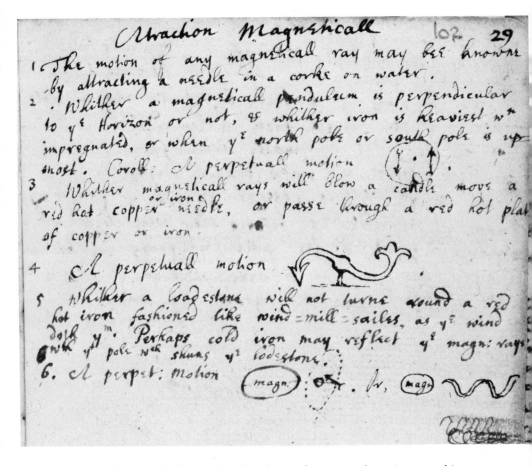

Figure 3.2 Newton's sketches of perpetual-motion machines powered by the flux of the magnetic stream. (Courtesy of the Syndics of Cambridge University Library.)

we are pressed downewards . . . there could be no refraction since ye same matter cannot presse 2 ways. a little body interposed could not hinder us from seing pression could not render shapes so distinct. ye sun could not be quite eclipsed ye Moone & planetts would shine like sunns. A man goeing or running would see in ye night. When a fire or candle is extinguish we lookeing another way should see a light. The whole East would shine in ye day time & ye west in ye night by reason of ye flood wch carrys o^r Vortex a light would shine from ye Earth since ye subtill matter tends from ye center. There is ye greatest pression on y^t side of ye earth from y^e ☉ or else it would not move about in equilibrio but from y^e ☉, therefore ye nights should be lightest.[73]

[73] *Add MS 3996*, 103^v.

These were very searching questions indeed directed to the Cartesian explanation of light. Under the heading "Of yᵉ Celestiall matter & orbes" he added a few more, pointing out that eclipses would be impossible according to the Cartesian theory since solid bodies could transmit the pressure in the vortex as well as the fluid matter of the heavens.[74] Every statement in these passages was an implicit experiment, an observation of a critical phenomenon that ought to appear if the theory were true. When he considered theories of colors, he proceeded in the same way. Do colors arise from mixtures of darkness and light? If they do, a printed page, black letters on a white sheet, ought to appear colored at a distance–another implicit experiment. Some of the experiments were posed explicitly. Descartes had referred the tides to the pressure of the moon on the fluid matter of the tiny vortex surrounding the earth. In a work by Boyle, Newton found a proposal to test the theory by correlating tides with the readings of barometers, which ought to register the same pressure. Immediately he began to think of other consequences the theory should entail.

> Observe if yᵉ sea water rise not in days & fall at nights by reason of yᵉ earth pressing from ☉ upon yᵉ night water &c. Try also whither yᵉ water is higher in mornings or evenings to know whither ⊖ or its vortex press forward most in its annuall motion . . . Try whither yᵉ seas flux & reflux bee greater in Spring or Autume in winter or Sommer by reason of yᵉ ⊖s Aphelion & perihelion. Whither yᵉ Earth moved out of its Vortexes center bye Moones pression cause not a monethly Parallax in Mars. &c.[75]

There was no suggestion that Newton had made any of these observations. Nevertheless, if the essence of experimental procedure is active questioning whereby consequences that ought to follow from a theory are put to the test, Newton the experimental scientist was born with the "Quaestiones." In 1664, such a method of inquiry had been little used. Newton's example was to be a powerful factor in helping experimental procedure convert natural philosophy into natural science.[76]

As he became interested in light and vision, for which some forms of experimentation required no equipment beyond his own eyes, Newton plunged forward with little thought of the consequences. To test the power of fantasy, he looked at the sun with

---

[74] *Add MS* 3996, f. 93.   [75] *Add MS* 3996, ff. 111–12.

[76] Two comments from later periods suggest something of the extent of Newton's eager embrace of experimentation at this time. In the kitchen at Trinity, he remembered, he had cut the heart of an eel into three pieces and observed how they went on beating in unison (13 Nov. 1712; *JBC 10,* 428). For Dr. William Briggs, who was an undergraduate at Corpus Christi while Newton was at Trinity, he recalled an occasion when they had dissected an eye together (Newton to Briggs, 25 April 1685; *Corres 2,* 417–18).

one eye until all pale bodies seen with it appeared red and dark ones
blue. After "yᵉ motion of yᵉ spirits in my eye were almost decayed" so that things were beginning to appear normal, he closed
his eye and "heightned [his] fantasie" of seeing the sun. Spots of
various hues appeared to his eye, and when he opened it again pale
bodies appeared red and dark ones blue as though he had been
looking at the sun. He concluded that his fantasy was able to excite
the spirits in his optic nerve quite as well as the sun.[77] He also came
close to ruining his eyes, and had to shut himself up in the dark for
several days before he could rid himself of the fantasies of color.
Newton left the sun alone after that but not his eyes. A year or so
later when he was developing his theory of colors he slipped a
bodkin "betwixt my eye & yᵉ bone as neare to yᵉ backside of my
eye as I could" in order to alter the curvature of the retina and to
observe the colored circles that appeared as he pressed (Figure
3.3).[78] How did he fail to blind himself? In the grip of discovery,
Newton did not pause to reckon the cost.

The content of the "Quaestiones" is equally redolent of the future Newton. The passages "Of Motion" and especially "Of violent Motion" mark his introduction to the science of mechanics.
The latter passage, really an essay, attacked the Aristotelian explanation of projectile motion and concluded that the continued motion of a projectile after it separates from the projector is due to its
"naturall gravity."[79] By "gravity" he referred in this instance to an
atomistic doctrine that every atom has an inherent motility, called
gravity, by which it moves. The doctrine was similar, though by
no means identical, to the medieval theory of impetus, which struggled with the principle of inertia for Newton's allegiance for twenty
years. He considered the cosmic order and Descartes's system of
vortices.[80] Elsewhere in the notebook, in a hand that corresponds
with the later entries in the "Quaestiones," Newton also took notes
from Thomas Streete, *Astronomia carolina,* which effectively introduced him to Keplerian astronomy. He pondered the cause of gravity (that is, heaviness) and pointed out that the "matter" that causes
bodies to fall must act on their innermost particles and not merely
on their surfaces. As I have already indicated, light and colors occupied a considerable portion of the "Quaestiones"; in their pages
Newton recorded the central insight to the demonstration of which
his entire work in optics was directed, that ordinary light from the
sun is heterogeneous, and that phenomena of colors arise, not from

---

[77] *Add MS* 3996, ff. 109, 125–5ᵛ. The temporary impairment of his sight must have impressed him. He mentioned the experiment in his *Lectiones opticae* about five years later
(*Add MS* 4002, p. 21) and as late as 1691 in a letter to Locke (*Corres 3,* 152–4).

[78] *Add MS* 3975, p. 15.   [79] *Add MS* 3996, ff. 98–8ᵛ, 113–14.

[80] *Add MS* 3996, in several places but especially ff. 93–3ᵛ.

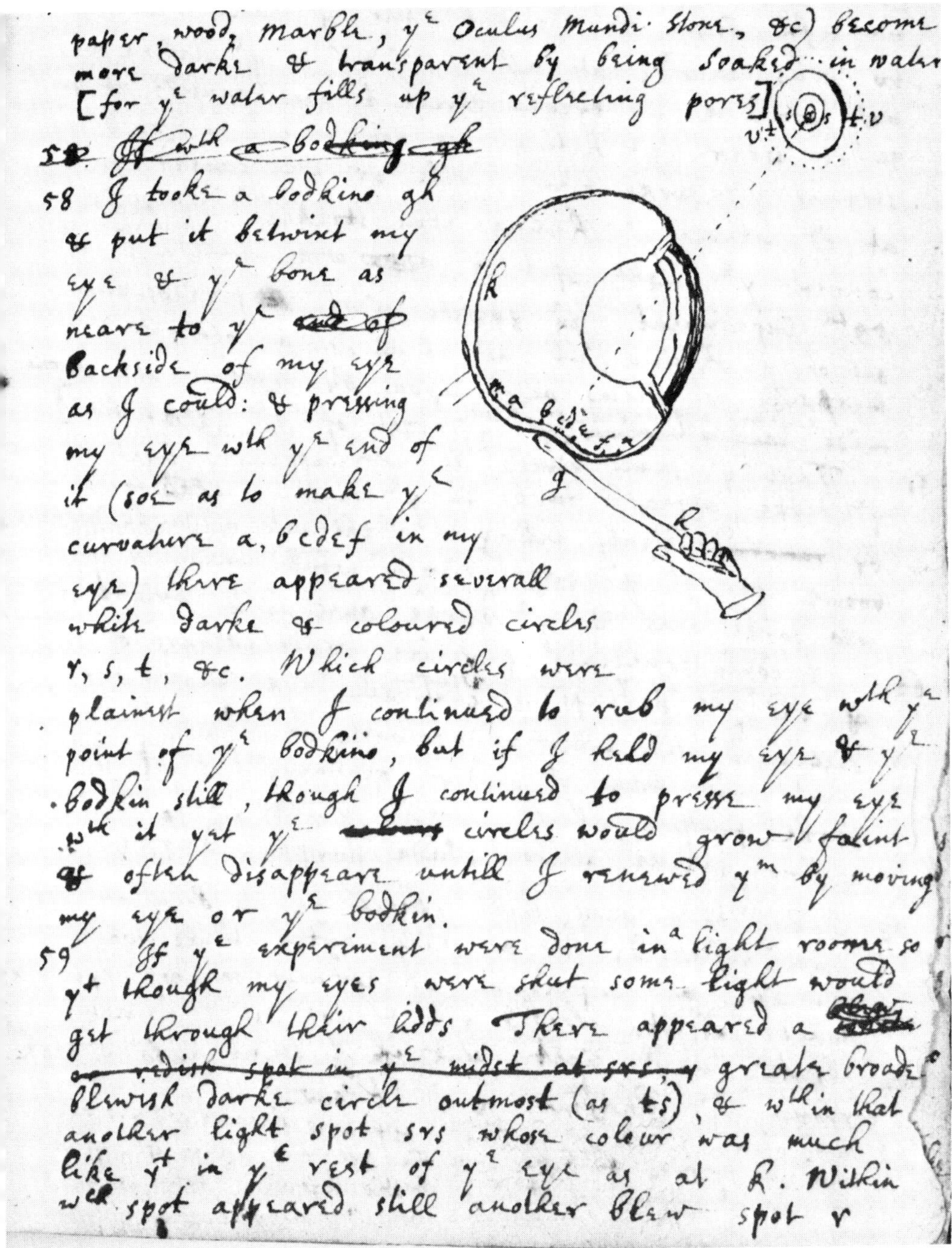

paper, wood, marble, y^e Oculus Mundi stone, &c) become
more darke & transparent by being soaked in water
[for y^e water fills up y^e reflecting pores]

58 If with a bodking gh

58 I tooke a bodkine gh
& put it betwixt my
eye & ye bone as
neare to y^e ~~cut of~~
backside of my eye
as I could: & pressing
my eye with ye end of
it (soe as to make y^e
curvature a, b c d e f in my
eye) there appeared severall
white darke & coloured circles
r, s, t, &c. Which circles were
plainest when I continued to rub my eye with y^e
point of ye bodkine, but if I held my eye & y^e
bodkin still, though I continued to presse my eye
wth it yet y^e ~~severall~~ circles would grow faint
& often disappeare untill I renewed y^m by moving
my eye or ye bodkin.

59 If ye experiment were done in a light roome so
y^t though my eyes were shut some light would
get through their lidds There appeared a ~~great~~
~~redish spot in ye midst at srs,~~ greate broad
blewish darke circle outmost (as ts), & w^{th}in that
another light spot srs whose colour was much
like y^t in ye rest of ye eye as at R. Within
wch spot appeared still another blew spot r

Figure 3.3 Newton's sketch of an experiment to produce sensations
of color by distorting his eyeball. (Courtesy of the Syndics of Cambridge University Library.)

the modification of homogeneous light as prevailing theory had it, but from the separation or analysis of the heterogeneous mixture into its components.[81]

The topics above ultimately expanded into the content of the two great works, the *Principia* and the *Opticks,* on which Newton's enduring reputation in science rests. He was never merely an empirical scientist, however. In his own eyes, he was a philosopher, intent on understanding the nature of things in the fullest sense of the phrase. From the perspective of natural philosophy, the "Quaestiones" were the first of the series of speculations that form the warp on which he wove the fabric of his scientific career. Throughout his life, his speculations turned on a limited set of crucial phenomena which seem to have functioned in his eyes as keys to the understanding of nature. Nearly all of them appear in the "Quaestiones." Chemical phenomena do not, it is true. Even though he was reading Boyle, Newton did not discover chemistry – and alchemy – until sometime after the composition of the "Quaestiones"; in the end, they furnished the most critical information of all. Beyond chemical phenomena, there were a number of others that showed up in every speculative scheme he composed from the *Hypothesis of Light* of 1675 (and its precursors) to the final version of the *Queries* in the second and third English editions of the *Opticks* near the end of his life. The cohesion of bodies, capillary action, surface tension of fluids, the expansion and pressure of air – these four (though not necessarily under these rubrics) all seized his attention in his initial reading in the new philosophy and never released it.[82]

Although the stance of questioning remained predominant in the "Quaestiones," an inchoate natural philosophy beginning to take shape can be dimly perceived. If Descartes was cited most frequently, his influence did not in the end dominate the "Quaestiones." Two other systems challenged his authority. On the one hand, Gassendi's atomistic philosophy, known to Newton at this time primarily through Charleton's *Physiologia,* offered a rival mechanical system. More than anything else, the "Quaestiones" were a dialogue in which Newton weighed the virtues of the two systems. Although he appeared to reach no final verdict, it is clear that he inclined already toward atomism. After deploying the standard arguments against a plenum, Newton opted for atoms, though not, or at least not initially, Gassendi's atoms.[83] I have already quoted

---

[81] *Add MS* 3996, ff. 27–30, 97, 122ᵛ.

[82] *Add MS* 3996, "Conjunction of bodys," ff. 90ᵛ −1; "Attraction Electricall & Filtration [capillary action]," f. 103; surface tension appears under "Of Water & Salt," ff. 100ᵛ, 111. "Mr. Boyle's receiver" was mentioned several times and the "utmost naturall" expansion of air was discussed (very briefly) under "Of Aer," f. 100.

[83] *Add MS* 3996, "Off yᵉ first mater," ff. 88–8ᵛ, 89ᵛ; "Of Attomes," ff. 89, 119–20.

Newton's objections to Descartes's conception of light and his explanation of the tides, and I have indicated that he accepted a different view of the cause of gravity (heaviness). Matter and light were the most important; to reject Descartes's opinions on these two issues was to shatter the cohesion of his natural philosophy beyond hope of repair. In his discussions of light and color, Newton left no doubt that he held the corpuscular conception.[84] Descartes may have introduced him to the mechanical philosophy, but Newton quickly transferred his allegiance to atomism.

There is also the possibility that the writings of Henry More guided Newton into the mechanical philosophy.[85] Descartes's name appeared so frequently in them that he could not have failed to notice it. Whichever he came to first, More represented the other current of thought that tempered Newton's enthusiasm for Descartes. More's views exerted a strong influence on the original essay on atoms that Newton wrote in the "Quaestiones." Newton later crossed the essay out, however, and it was not here that More's position was vital. Like the other Cambridge Platonists, Henry More was concerned by the mechanical philosophy's possible exclusion of God and spirit from the operation of physical nature. Whereas initially he welcomed Descartes as an ally of religion, the more he contemplated his sytem of nature, the more its implications alarmed him. In Hobbes he saw the dangers spelled out explicitly. More was concerned to reinstall spirit in the continuing operation of nature, all of nature. Especially in the last four entries of the "Quaestiones," "Of God," "Of ye Creation," "Of ye soule," and "Of Sleepe and Dreams &c," which appear by their position to be later additions to the original set of headings, similar concerns made a tentative appearance in the "Quaestiones."[86] Their role in Newton's thought was destined to grow, diluting and modifying his initial mechanistic views.

[84] "Of light," ff. 103^v, 128^v; "Of Species visible," f. 104^v; "Of Colours," ff. 105^v, 122–4.

[85] Although Henry More had attended the same grammar school, and although the two men would later become acquainted, I think it is out of the question that More, one of the foremost intellectual figures in Cambridge, perhaps the foremost, and a fellow of a different college, would have known Newton the undergraduate. Given More's eminence, with the added inducement of his connection with Grantham, there is no problem at all in imagining how Newton could have picked up his works. The enduring influence of Henry More and Cambridge Platonism on Newton has been the central theme of J. E. McGuire in a number of articles: – see especially "Force, Active Principles, and Newton's Invisible Realm," *Ambix*, *15* (1968), 154–208; "Atoms and the 'Analogy of Nature': Newton's Third Rule of Philosophizing," *Studies in History and Philosophy of Science, 1* (1970), 3–58; "Body and Void and Newton's De Mundi Systemate: Some New Sources," *Archive for History of Exact Sciences, 3* (1966), 206–48; and "Neoplatonism and Active Principles: Newton and the *Corpus Hermeticum*," in Robert S. Westman and J. E. McGuire, *Hermeticism and the Scientific Revolution*, (Los Angeles, 1977).

[86] *Add MS* 3996, ff. 128, 129, 130–30^v, 132.

Meanwhile, natural philosophy was not the only new study Newton discovered. He found mathematics as well. As with natural philosophy, we have Newton's original notes that chart his course. We also have a number of accounts, several in Newton's own words of which one from 1699 is the most important, one in Conduitt's memorandum of a conversation with Newton on 31 August 1726, and another in a memorandum of November 1727, soon after Newton's death, by Abraham DeMoivre. The earliest of them dated from thirty-five years after the events it described. Nevertheless, a reasonably consistent account, which is also reasonably consistent with Newton's reading notes, emerges from them.

> July 4th 1699. By consulting an accompt of my expenses at Cambridge in the years 1663 and 1664 [Newton wrote as he looked over some early notes] I find that in ye year 1664 a little before Christmas I being then senior Sophister, I bought Schooten's Miscellanies & Cartes's Geometry (having read this Geometry & Oughtred's Clavis above half a year before) & borrowed Wallis's works & by consequence made these Annotations out of Schooten & Wallis in winter between the years 1664 & 1665. At wch time I found the method of Infinite series. And in summer 1665 being forced from Cambridge by the Plague I computed ye area of ye Hyperbola at Boothby in Lincolnshire to two & fifty figures by the same method.
>
> Is. Newton[87]

In Conduitt's memorandum, it all began when Newton lit upon some books on judicial astrology (an event which DeMoivre placed at Sturbridge Fair in 1663). Being unable to cast a figure, he bought a copy of Euclid and used the index to locate the two or three theorems he needed; when he found them obvious, "he despised that as a trifling book . . ." DeMoivre's account agreed with Conduitt's except that he had Newton go on in Euclid to more difficult propositions such as the Pythagorean theorem, whereupon he changed his opinion, and read all of Euclid through twice. Such early study of Euclid does not agree either with Newton's notes or with other parts of what he told Conduitt. Pemberton also recorded Newton's regret that he had not given more attention to Euclid before he applied himself to Descartes.[88]

> He bought Descartes's Geometry & read it by himself [Conduitt continued in language very similar to that in the DeMoivre account] when he was got over 2 or 3 pages he could understand no farther

[87] *Add MS* 4000, f. 14^v. See a similar chronology in Newton to Wallis, late 1692 (*Corres 7*, 394). Newton composed a number of other résumés of his mathematical development some years later at the height of the controversy with Leibniz. They all agree essentially with the earlier one quoted here. Cf. *Add MS* 3968.5, f. 21; *Add MS* 3968.41, ff. 76, 85, 86^v.

[88] Henry Pemberton, *A View of Sir Isaac Newton's Philosophy* (London, 1728), preface.

than he began again & got 3 or 4 pages farther till he came to another
difficult place, than he began again & advanced farther & continued
so doing till he made himself Master of the whole without having
the least light or instruction from any body.[89]

Both accounts agree in making Newton an autodidact in mathematics, as he was in natural philosophy. Nearly twenty years later,
when he was recommending Edward Paget for the position of
mathematical master at Christ's Hospital, Newton probably had his
own experience in mind as he specified Paget's qualifications. Paget
understood the several branches of mathematics, he said, "& wch is
ye surest character of a true Mathematicall Genius, learned these of
his owne inclination, & by his owne industry without a Teacher."[90]
There had been even less mathematics in the university than natural
philosophy; not surprisingly, no stories survive of undergraduates
being stirred by Descartes's *Geometry*. Nevertheless, there is a curious coincidence of time that has generally been ignored. The Lucasian chair of mathematics, which Newton himself would soon occupy, was established in 1663, and the first professor, Isaac Barrow,
delivered his inaugural series of lectures in 1664, beginning on 14
March. Contrary to frequent assertions, Barrow was not Newton's
tutor, and there is no evidence of any familiarity between them at
this time. At least twice, Newton implied that he had attended the
lectures, however; and though they would probably not have directed him to Descartes, given Barrow's mathematical predilec-
tions, and though Barrow was not a major influence on him, they
could have stimulated his interest in mathematics.[91] One wonders
as well who in Cambridge could have loaned him a copy of Wallis
if it was not Barrow. In any event, the coincidence in time is so
close that it strains credulity to deny any connection between the
lectures and Newton's sudden interest.

Newton's own notes agree with the accounts of Conduitt and

---

[89] *Keynes MS* 130.10, f. 2^v. "Memorandums relating to Sr Isaac Newton given me by M^r
Abraham Demoivre in Novr 1727"; University of Chicago Library, MS 1075–7. The
account in the DeMoivre memorandum of Newton's struggle with Descartes is so similar
to Conduitt's memorandum of the conversation of 31 Aug. 1726 that it is hard to believe
they were wholly independent. Since the DeMoivre memorandum is in Conduitt's hand,
it is entirely possible that it was Conduitt's account of a conversation with DeMoivre
instead of a copy of a paper written by DeMoivre. Conduitt did not make copies of other
items he received as he collected information about Newton. In the period 1715–16
Newton gave a very similar account of his early mathematical studies to Abbé Conti,
who was then in England (Conti, *Prose e poesi*, *2*, 24).

[90] Newton to the Governors of Christ's Hospital, 3 April 1682; *Corres 2*, 375. I wish to
thank Professor Gale Christianson for pointing out the significance of this passage to me.

[91] Both references from around 1715, in statements connected with the controversy with
Leibniz (*Add MS* 3968.5, f. 21 *Add MS* 3968.41, f. 86^v). He also told Conti that he
attended Barrow's lectures (*Prose e poesi*, *2*, 25).

DeMoivre that he plunged straight into modern analysis with no appreciable background in classical geometry. They agree as well with the centrality given Descartes. Schooten's second Latin edition of the *Geometry,* with its wealth of additional commentaries, was his basic text, supplemented by Schooten's *Miscellanies,* Viète's works, Oughtred's algebra (the *Clavis* Newton mentioned), and Wallis's *Arithmetica infinitorum.* In roughly a year, without benefit of instruction, he mastered the entire achievement of seventeenth-century analysis and began to break new ground.

Newton's surrender to his new studies was not without danger. In order to pursue them to a fruitful conclusion, he had to win a permanent position in Cambridge, but rewards in Cambridge were not being passed out for excellence in mathematics and mechanical philosophy. Fellowships in Trinity went only to those who had first been elected as an undergraduate to one of the sixty-two scholarships supported by the college. In his first three years, Newton had not distinguished himself in any way. Trinity had twenty-one exhibitions which carried annual stipends of about four pounds each. The college records give no indication of the criteria of selection. It is difficult to imagine that academic promise did not figure, though need may have been the decisive factor. Suffice it to say that Newton did not appear among the ten, nearly all Pulleyn's pupils, who received exhibitions in 1662 and 1663.[92]

Many features of the college worked to lessen his chances of a scholarship. Statistics indicate that sizars had less chance than pensioners, especially when enrollment was up and demand high, as they were in the 1660s. Influence and connections were essential characteristics of the system of patronage that impinged on the whole university to the injury of those, sizars above all, who lacked sponsors in high places.[93] Newton's chances were further dimin-

[92] A complete list of the twenty-one exhibitions and their stipends can be found in the Senior Bursar's Accounts of Trinity College for 1664. Five of them endowed by Mr. (i.e., Magister) Hylord, were added only in 1663. Awards of exhibitions in 1662 and 1663 are listed in the Conclusion Book, 1646–1811, pp. 72, 76, 77.

[93] Perhaps it is a mistake to project the system of patronage back to the level of scholarships, which were pretty small pickings although they were essential steppingstones to fellowships. Fellowships were significant; they provided a permanent income on which a man (such as a younger son or the son of a client) could live. Royal mandates, which were prompted by the intercession of one courtier or another and were evidence then of a courtier's access to the king's ear, were fairly common for fellowships, although the number I have been able to locate is not proportionate to the indignation against them in some quarters of the university. (Mandates for masterships, important preferments, were nearly the invariable rule.) Trinity received a letter mandate dated 7 Sept. 1664 ordering the college to elect John Howarth to a fellowship "in consideracion of his owne merits, & of his relation to one of Our principall Servants . . ." (Trinity College, Box 29.D). He was elected. On 23 Nov. 1666, there was a similar one for Henry Cary, on account of his

ished by the privileged group of Westminster scholars who automatically received at least a third of the scholarships year after year, and with the scholarships the top rungs on the ladder of seniority for their year. During the entire century, a good half of Trinity's fellows came from Westminster School, and roughly that proportion held for the large group elected scholars in 1664.[94] Indeed, with 1664 Newton faced a crisis. Trinity held elections to scholarships only every three or four years. The election in 1664 was the only one during his career as a student. If Newton were not elected then, all hope of permanent residence in Cambridge would vanish forever. He chose exactly that time to throw over the recognized studies and pursue a course which had no standing whatever in the college's scheme of values.

Perhaps the approaching elections, to be held in April, with their attendant examinations explain an otherwise anomalous feature of Newton's notes on the established curriculum. Having dropped Magirus's peripatetic *Physics,* he took it up again and plowed his way through two more chapters. Likewise, he started Vossius's *Rhetoric* and the *Ethics* of Eustacius of St. Paul about this time – and likewise failed to finish both works. In the three cases, the notes suggest last-minute boning for an examination.[95] Newton's own

own merit and the service and suffering of his father in the royal cause in the Civil War (*CSPD: Charles II, 6,* 280). For whatever reason, this mandate was not obeyed. On 24 July and 30 Oct. 1667, there were mandates for the election of Valentine Pettit and John Goodwin to fellowships *CSPD, 7,* 322, 553). Both of these mandates were obeyed. There were also petitions to the Crown in 1665 and in 1668 by two Westminster scholars, who had been disappointed in their expectation of fellowships, asking for letters mandate *CSPD, 5,* 141, *8,* 597). They were further disappointed in their expectation of letters mandate, and neither man ever obtained a fellowship.

94 Trinity held elections to scholarships in 1661, 1664, and 1668. Both in 1662 and in 1663, the Westminster group admitted the previous year (eight in 1662, seven in 1663) took up scholarships by right without election. The Trinity records also show that one other student became a scholar in 1662. (No explanation is given; it may be simply a typographical error.) The election held in 1664 also pre-elected for 1665; in all, forty-four students were elected to scholarships at that time, including the Westminster group (five) admitted in 1663. In 1665, five more Westminster students took up scholarships; no other students became scholars in that year. The total number elected (or, in the case of the Westminster scholars, admitted) to scholarships over the four-year span was sixty-five, of which twenty-five were from Westminster. Of the sixty-five, twenty-five went on to fellowships; nine of them were from Westminster. In addition, another Westminster scholar became a chaplain in the college. Only five sizars and subsizars admitted in 1661, Newton's year, were elected to scholarships; two of them went on to fellowships (Ball and Venn, *Trinity Admissions*).

95 There is reason to think that there was an examination. There is record of the preceding one. "11 April 1661. Ordered that there be an election of scholars on April 22 & that they stand for places & be examined on the Friday & Saturday before" (Master's Old Conclusion Book, 1607–1673, p. 268). The method of examination in Caius was an extension of the practice of disputations; the candidates in Caius sat in the chapel for three days, being examined by scholars on the first day and by the Dean and fellows on the second and

account, as related to Conduitt, implies that his tutor Pulleyn may have recognized his pupil's brilliance and tried to help him by enlisting Isaac Barrow, the one man in Trinity fit to judge his competence in the unorthodox studies he had undertaken. The gesture nearly capped the debacle, since Newton had been unorthodox even in his unorthodoxy.

> When he stood to be scholar of the house his tutour sent him to Dr Barrow then Mathematical professor to be examined, the D^r examined him in Euclid wch Sr I. had neglected & knew little or nothing of, & never asked him about Descartes's Geometry wch he was master of Sr I. was too modest to mention it himself & Dr Barrow could not imagine that any one could have read that book without being first master of Euclid, so that Dr Barrow conceived then but an indifferent opinion of him but however he was made scholar of the house.[96]

The final clause is true; on 28 April 1664 Newton was elected to a scholarship. It also poses a quandary: What can explain the decision? Perhaps the explanation is the obvious one that springs immediately to mind. Newton's genius readily outshone the mediocrity around him even in studies he had abandoned. Such an explanation seems to conflict, however, with Newton's account of the impression he made on Barrow, the leading intellect of the college. Moreover, the realities of Cambridge in 1664 suggest another explanation, that Newton had a powerful advocate within the college. There is good reason to think he had such an advocate. In 1669, as a new fellow, he was appointed tutor of a fellow commoner. Tutoring fellow commoners was lucrative business usually reserved for important fellows. Two candidates for Newton's patron present themselves. One is Barrow himself despite the story. It is not impossible that Newton was misled as to the impression he made. Indeed it may be that Barrow told Pulleyn to send him; if attendance at his lectures paralleled that of Newton's later ones, he may have been curious about a familiar face. That is all speculation, however. It is not speculation that in 1668–9, Barrow was familiar enough with Newton's work to send him Mercator's *Logarithmotechnia* when he saw that it seemed to forestall some of his work. In 1669, he obtained the Lucasian chair for Newton when he himself

third (John Venn, *Caius College*, [London, 1901], p.201). In 1692, William Lynnet, a senior fellow, described the procedure then used in Trinity for the election of fellows and scholars. He said that they sat three days in the chapel from seven to ten and from one to four. The account assumes that fellows, or at least senior fellows, examined them orally. In the case of candidates for fellowships, at least, they wrote an essay on the fourth day. (*Edleston*, pp.xlii–xliii).

96 *Keynes MS* 130.10, f. 2^v. Newton also told Conti a version of this story (*Prose e poesi*, 2, 24).

resigned, and in 1675, Barrow appears to have been decisive in obtaining a royal dispensation for Newton. The other and more likely candidate is Humphrey Babington. Recall Newton's accounts which show that he employed "Mr Babingtons Woman." Recall his statement that during the plague he was, at least part of the time, at Boothby, not far from Woolsthorpe, where the same Mr. Babington was rector. Mr. Babington was also the brother of Mrs. Clark, with whom Newton had lodged in Grantham. Most important of all, he was approaching the status of a senior fellow, one of the eight fellows at the top of the ladder of seniority, who ran the college in conjunction with the master. Moreover, the college would not have forgotten that he stood well with the king; twice he had obtained letters mandate in his favor in recent years. When Babington was later bursar of the college, Newton drew up tables to aid him in renewing college leases, and the two continued to be associated in various academic affairs until Babington's death. Since Babington was resident only four or five weeks of the year at this time, however, his opportunity to influence the election may have been small. Four years earlier, the Reverend William Ayscough and Mr. Stokes had rescued Newton from rural oblivion. Someone performed that service again in April 1664, and on the whole Humphrey Babington appears most likely to have been the one.

With his election, Newton ceased to be a sizar. He now received commons from the college, a livery allowance of 13s/ 4d per year, and a stipend of the same amount.[97] Far more important, he received the assurance of at least four more years of unconstrained study, until 1668 when he would incept M.A., with the possibility of indefinite extension should he obtain a fellowship. The threat had lifted. He could abandon himself completely to the studies he had found. The capacity Newton had shown as a schoolboy for ecstasy, total surrender to a commanding interest, now found in his early manhood its mature intellectual manifestation.[98] The tentativeness suggested by the earlier unfinished notes vanished, to be replaced by the passionate study of a man possessed. Such was the characteristic that his chamber-fellow Wickins remembered, having observed it no doubt at the time with the total incomprehension of the Woolsthorpe servants. Once at work on a problem, he would forget his meals. His cat grew very fat on the food he left standing

---

[97] Junior Bursar's Accounts and Senior Bursar's Accounts.

[98] Recall his confession of 1662: "Setting my heart on . . . learning . . . more than Thee." Newton learned to rationalize his pursuit of learning, the commanding passion of his life, in religious terms. He was studying God's word and God's work, and it seems clear enough where his heart was set. It had nothing to do with rational decision – he was helpless in the hands of his own genius.

on his tray.[99] (No peculiarity of Newton's amazed his contemporaries more consistently; clearly food was not something they trifled with.[100]) He would forget to sleep, and Wickins would find him the next morning, satisfied with having discovered some proposition and wholly unconcerned with the night's sleep he had lost. "He sate up so often long in the year 1664 to observe a comet that appeared then," he told Conduitt, "that he found himself much disordered and learned from thence to go to bed betimes." Part of the story is true; he entered his observations of the comet into the "Quaestiones."[101] The rest of it is patently false as Conduitt knew from personal experience. Newton never learned to go to bed betimes once a problem seized him. Even when he was an old man the servants had to call him to dinner half an hour before it was ready, and when he came down, if he chanced to see a book or a paper, he would let his dinner stand for hours. He ate the gruel or milk with eggs prepared for his supper cold for breakfast.[102] Conduitt observed Newton long after his years of creativity. The tension of the quest that consumed him in 1664 and the years that followed stretched whatever neuroses he had brought from Woolsthorpe to their utmost limits. He was "much disordered" more than once, and not only from observing comets.

His discovery of the new analysis and the new natural philosophy in 1664 marked the beginning of Newton's scientific career. He considered the "Quaestiones" important enough that he later composed an index to them to supplement their initial organization under topics.[103] Sailing away from the old world of academic Aristotelianism, Newton launched his voyage toward the new. The passage was swift.

---

99 Apparently Newton told this to his niece. Conduitt credited the story to "C.C." It was specifically his cat at the university. (*Keynes MS* 130.6, Book 2).

100 In addition to those already cited, Humphrey Newton mentioned it in both of his letters about his period with Newton in the 1680s. (*Keynes MS* 135). Stukeley heard stories about it in Cambridge (*Stukeley,* p. 48) and Conduitt himself observed it in Newton's old age (*Keynes MS* 130.6, Book 1). As late as 1742, Dr. George Cheyne reported hearing a similar story. (*Natural Method of Curing Diseases of the Body and Disorders of Mind* [London, 1742], p. 81).

101 Nicholas Wickins to Robert Smith, 16 Jan. 1728; *Keynes MS* 137. Conduitt's memorandum of 31 Aug. 1726; *Keynes MS* 130.10, f. 4ᵛ. *Add MS* 3996, ff. 93, 115–16.

102 *Keynes MS* 130.6, Book 1; *Keynes MS* 130.5, sheet 1. Since Conduitt assigned this set of anecdotes to "C.C." (Catherine Conduitt), they may refer to the early years of Newton's residence in London. Newton may have become sensitive about the stories of his eccentricities. They were apparently widespread; Nicholas Wickins spoke of them as much repeated. At any rate, in addition to telling Conduitt that he learned to go to bed betimes, he said the same thing to Stukeley as though he were trying to deny the stories (Stukeley to Conduitt, 15 July 1727; *Keynes MS* 136, p. 11).

103 *Add MS* 3996, ff. 87–7ᵛ.

4

Resolving problems by motion

I N his age of celebrity, Newton was asked how he had discovered the law of universal gravitation. "By thinking on it continually," was the reply.[1] No better characterization of the man can be given, not only in its delineation of a life whose central adventure lay in the world of thought rather than action, but also in its description of his mode of work. Seen from afar, Newton's intellectual life appears unimaginably rich. He embraced nothing less than the whole of natural philosophy, which he explored from several vantage points which ranged all the way from mathematical physics to alchemy. Within natural philosophy, he gave new direction to optics, mechanics, and celestial dynamics, and he invented the mathematical tool which has enabled modern science further to explore the paths he first blazed. He sought as well to plumb the mind of God and His eternal plan for the world and mankind as it was presented in the biblical prophecies. When we examine Newton's grandiose adventure minutely, it turns out to be a mixture of discrete pieces rather than a homogeneous melange. His career was episodic. What he thought on, he thought on continually, which is to say exclusively, or nearly exclusively. What seized his attention in 1664, to the virtual exclusion of everything else, was mathematics.

John Conduitt, the husband of Newton's niece and his intended biographer, almost invariably smothered whatever insight he had in a froth of grandiloquence. One of his figures, applied to Newton's early career, bears repetition, however: "he began with the most crabbed studies (like a high spirited horse who must be first broke in plowed grounds & the roughest & steepest ways or could other-

---

[1] The anecdote rests on somewhat shaky authority. The first instance of it that I have found is in a note to Voltaire's *Éléments de la philosophie de Newton*, Part III, chapter 3 (in French, "en y pensant sans cesse"). It did not appear in any of the editions published during Voltaire's life, but first in the so-called Kehl edition of the *Oeuvres complètes de Voltaire*, 70 vols. (Paris, 1785–9), *31,* 175. Presumably, such notes were based on annotations that Voltaire made in an earlier edition, but the relevant volume of the annotated edition has been lost. There is reason then to think that the anecdote rests on Voltaire's authority and derives from his stay in England, where he knew Newton's niece, but its authenticity can hardly be called secure. It may be a corruption of Conduitt's note: "Bentley said – Sr I. told him all his merit was patient thought" (*Keynes MS* 130.5, sheet 1). Conduitt's note, in turn, may have derived, through Bentley, from Newton's letter of 10 Dec. 1692: "But if I have done ye publick any service this way 'tis due to nothing but industry & a patient thought" (*Corres 3,* 233).

wise be kept within no bounds."[2] Newton was to voyage over many strange seas of thought, speculative adventures from which more than one explorer of the seventeenth century never returned. The discipline that mathematics imposed on his fertile imagination marked the difference between wild flights of fancy and fruitful discovery. It was supremely important that, almost first, mathematics commanded his attention.

The surviving notes of his initial studies in mathematics bear out the various anecdotes that he plunged straightforward into Descartes's *Geometry* and modern analysis. The time was almost certainly 1664, probably in the spring or summer. His primary vehicle was Schooten's pivotal second Latin edition of Descartes's *Geometry* with its wealth of additional commentaries, supported by reading in algebra, especially from the works of Viète. He also made early contact with the mathematics of infinitesimals as represented by John Wallis. It is quite impossible to determine from the notes which came first. It is equally impossible to see that anything important hinges on their chronological order. What matters is the voracity with which he devoured whatever mathematics he found. Whiston later remarked that in mathematics Newton "could sometimes see almost by Intuition, even without Demonstration . . ."[3] Whiston had a proposition in the *Principia* in mind, but an examination of Newton's self-education in mathematics compels one to a similar judgment. Within six months of his initiation into mathematics, some of his reading notes were changing imperceptibly into original investigations. Within a year, he had digested the achievement of seventeenth-century analysis and had begun to pursue his own independent course into higher analysis.[4]

---

[2] *Keynes MS* 130.4, p. 4.

[3] William Whiston, *Memoirs of the Life and Writings of Mr. William Whiston* (London, 1749), p. 39.

[4] D. T. Whiteside, "Introduction", *Math 1*, 14. I revised this chapter after spending much time with two students, George Anastaplo and Richard Ferrier, on volume 1 of Whiteside's edition. I owe many points in what follows to their insights, which they worked out with extraordinary tenacity.

In addition to the editorial apparatus in volume 1 of *Math*, Whiteside has written a number of excellent articles on Newton's discoveries in mathematics: "Isaac Newton: Birth of a Mathematician," *Notes and Records of the Royal Society, 19* (1964), 53–62; "Newton's Discovery of the General Binomial Theorem," *Mathematical Gazette, 45* (1961), 175–80; "Sources and Strengths of Newton's Early Mathematical Thought," in Robert Palter, ed., *The 'Annus Mirabilis' of Sir Isaac Newton* (Cambridge, Mass., 1970), pp. 69–85; and "Newton's Mathematical Method," *Bulletin of the Institute of Mathematics and Its Applications, 8* (1972), 173–8. See also H. W. Turnbull, *The Mathematical Discoveries of Newton* (London, 1945), and "Newton: the Algebraist and Geometer," in The Royal Society, *Newton Tercentenary Celebrations* (Cambridge, 1947), pp. 62–72; Joseph E. Hofmann, "Studien zur Vorgeschichte des Prioritätstreites zwischen Leibniz und Newton um die Entdeckung der höheren Analysis," *Abhandlungen der Preussischen Akademie der Wissenschaften,*

Newton's early notes on analysis investigated the relation of particular axes to the equation of a curve. He tried various transformations of axes in order to simplify equations, seeking a regularized procedure whereby one of the new axes would also be an axis of the curve.[5] He was not yet beyond fairly simple mistakes. Like the early analysts, he did not at first comprehend the significance of negative roots. He drew the cubic parabola, $x^3 = a^2y$, as though it were symmetrical with respect to the y axis, and somewhat later he confined his diagram of Descartes's folium, the curve $x^3 - axy + y^3 = 0$, to the first quadrant, tacitly assuming that its shape is identical in all four quadrants.[6] He learned quickly, however. By October, he comprehended negative roots clearly enough to set down the rule that when the x axis is the diameter of a curve such that it bisects all the ordinates, then y cannot appear in the equation in odd powers because there is a negative root of equal absolute value corresponding to every positive one.[7] Above all, he seized and made his own the central insight of analytic geometry. Already in September he proposed a problem to himself in the following terms: "Haveing ye nature of a crooked line expressed in Algebr: termes to find its axes, to determin it & describe it geometrically &c."[8] Perhaps because he came to it without a background in classical geometry, Newton had comprehended the thrust of analysis more clearly than its originators. The algebraic equation is not merely a device to aid geometric constructions. The equation is more basic than the curve; the equation defines, or as Newton put it, expresses the nature of the curve. Hence the equation itself should become the focus of attention.[9]

Not that he was uninterested in the construction of curves. Very early, he collected from Schooten's works various ways in which the conics can be constructed. He found nine constructions of the ellipse, six of the parabola, and ten of the hyperbola. For each of the three curves, one construction was the basic section through a cone. Most of the others treated the curve as the path of a moving point and described some device, such as the trammel for the ellipse, which constrains the moving point such that it traces the curve (Figure 4.1). It is misleading, however, merely to say that he found

Mathematisch-naturwissenschaftliche Klasse, 1943; Carl B. Boyer, *The Concepts of the Calculus. A Critical and Historical Discussion of the Derivative and the Integral* (New York, 1939); and J. O. Fleckenstein, *Der Prioritätstreit zwischen Leibniz und Newton* (Basel and Stuttgart, 1956).   [5] *Math 1*, 155–65.

[6] *Math 1*, 160, 184.   [7] *Math 1*, 169.   [8] *Math 1*, 236.

[9] Timothy LeNoir, "The Social and Intellectual Roots of Discovery in 17th-Century Mathematics" (dissertation, Indiana University, pp. 396–525). It is impossible wholly to document the extent to which I have profited from Mr. LeNoir's work.

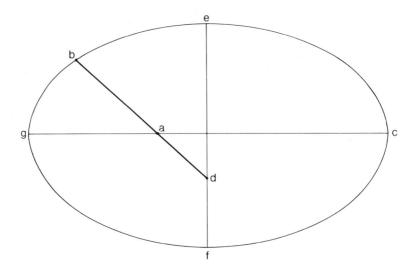

Figure 4.1   The trammel construction of the ellipse. The trammel *dab*
moves so that *a* follows track *gc* and *d* follows track *ef*.
The length *bd* of the trammel equals half the major axis of
the ellipse, the length *ab* half the minor axis.

the constructions in Schooten, since he ventured to discover
some constructions of his own and to effect improvements in
some of Schooten's.[10] He was also interested in a device he
found in Viète, the mesolabum, by which one could lay off a
given distance between two intersecting lines or between a circle
and a line intersecting it (Figure 4.2).[11] Since the instrument opened the way to the construction of two mean proportionals
between given lines, if its use were granted, that is, in effect, to
the construction of cube roots, the mesolabum was Newton's
introduction to cubics.

The collection of constructions of conics compiled from various
locations in Schooten was a typical Newtonian exercise, the ana-
logue in mathematics to the "Quaestiones quaedam Philosophicae"
put together a few months earlier from the mechanical philosophers
he had been reading. During the autumn, Newton sought further
to order his growing knowledge of the conics and of analytic geometry in a passage that went through five successive stages as it
gradually transformed itself in November into his own original
essay: "To find y^e Axis or Diameter of any Crooked Line supposeing it hath y^m." In its final version, it presents a general procedure
by which to determine axes or, as in cubics, the nonexistence of

[10] *Math 1*, 29–39.   [11] *Math 1*, 72–7.

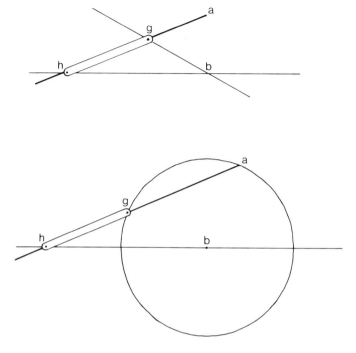

Figure 4.2 The mesolabum is a physical instrument with arm *gh* of
fixed length which slides along the beam *agh* while the
beam in turn rotates around *a*. With it one can lay off
along a line through *a* a given length *gh* between two
intersecting lines, *bh* and *gb*, or a distance equal to the
radius of a circle between the circle and its extended diameter *bh*.

axes.[12] He also investigated vertices and asymptotes. Scarcely six
months into his career as a mathematician, he was already consistently concerned to develop general procedures which would apply
to all curves and the equations that define them, and not merely to
solve particular problems. At the same time, he was discovering the
simplifying advantage of usually reducing equations and curves to
standard coordinates, *x* and *y*, set perpendicular to each other, what
we call Cartesian coordinates.

Inevitably, one of the problems he undertook as he explored the
world of analytic geometry was the problem of drawing tangents to
curves. He began with Descartes's procedure of determining the
normal. Quickly he discovered Hudde's rule in Schooten's edition,
which he was using, and as quickly he mastered it. His innate urge
to systematize and to generalize asserted itself again. Once he had

[12] *Math 1*, 167–201. Further notes relevant to the same general topic continue to p. 212.

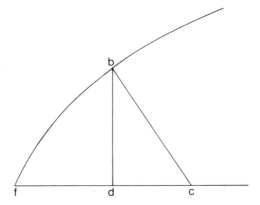

Figure 4.3. Newton's method to determine the length of the subnormal *dc*.

grasped the method in individual problems, he extended it into an equation for the length of the subnormal (which would determine the normal by locating the point at which it cuts the axis) to any curve (Figure 4.3).

> In finding $dc = v$ observe this rule. Multiply each terme of y^e equat: by so many units as x hath dimensions in y^t terme, divide it by x & multiply it by y for a Numerator. Againe multiply each terme of the equation by soe many units as y hath dimensions in each terme & divide it by $-y$ for a denom: in the valor of v.[13]

Among other things, he proceeded through ordered series of equations, especially the series

$$y = \frac{a^2}{x}; \qquad y = \frac{a^3}{x^2}; \qquad y = \frac{a^4}{x^3} \cdots$$

in which he found a pattern. In such patterns, Hudde's rule began to reveal deeper levels of meaning. Newton did not invariably use the exponents of the variable as Huddeian multipliers. Hudde's rule was a method of simplifying equations with double roots; one chose the sequence of multipliers in order to "blot out" the term multiplied by zero.[14] Thus, in one problem, Newton faced an equation with three terms in x. Using multipliers one less than the exponents of x, he blotted x out and was left with an equation in which v (the subnormal plus x) appeared in only one term and hence could be readily expressed.[15] Such problems remained par-

---

[13] *Math 1*, 236. In his editorial notes, Whiteside summarizes this in modern notation as follows: the subnormal $= y(dy/dx) = -y(f_x/f_y)$, where f_x and f_y are the implicit derivatives, dy/dx and dx/dy. Cf. earlier work on the same problem, pp. 216–233.

[14] *Math 1*, 258.   [15] *Math 1*, 219.

ticular cases; when exponents were used as multipliers, allowing the equation to be reduced by one degree since the constant was blotted out, general patterns began to emerge. Newton did not forget them.

Other new horizons also began to open before Newton. Why not carry Descartes's procedure a step farther? Descartes had determined the normal to a curve at a given point by locating on its axis the center of a circle tangent to the curve at that point. His method depended on the realization that in the circle tangent to the curve the radii to the two points where larger circles on the same center cut the curve merge into one. Newton imagined two normals to the curve at neighboring points. He began to use the symbol o for a small increment, so that the x coordinates of the two points were x and $x + o$.[16] Now let the two normals merge into one. Newton reasoned that the two coincident normals must lie along the radius of the circle tangent to the curve and equal to its curvature, or "crookednesse," at that point. Hence an extension of Descartes's method one step farther would enable him to locate the center of curvature, usually not on the axis, for a given point and to determine the curvature as the reciprocal of the radius of curvature (Figure 4.4).[17] In his usual fashion, he immediately set about ordering and generalizing his procedure to all the conics.[18] He extended the procedure yet another step by developing a general equation for the circle of curvature from which, via Hudde's rule, he could determine points of greatest and least curvature.[19] The investigation of the crookedness of lines carried him beyond anything he found in his readings. He achieved success by December 1664, probably before his twenty-second birthday. By the following May, he was

---

[16] *Math 1*, 246–8. He had employed the notation earlier, in September, when he was working on Descartes's ovals, the curve of the surface that would refract rays emanating from a point to a second point or focus (*Math 1*, 557).

[17] *Math 1*, 245–51. For the parabola $rx = y^2$, $ab = x$, $cb = y$, $bd = v$, $af = c$, and $fe = d$. The problem is to determine c and d, the coordinates of e, the point where the two normals, ce and me, intersect. First Newton determined the subnormal $v = r/2$, a constant for the parabola. By similar triangles, he expressed c and d in terms of x, y, and v, and arrived at the equation $x + v + cv/y = d$, or $-dy + cv + vy + xy = 0$. He removed v by substituting its value, $r/2$. He removed x by substituting its value, y^2/r, from the original equation. Thus $2y^3 + r^2y - 2dry + cr^2 = 0$. "Now tis evident y^t when y^e lines *em* & *ce* are coincident y^t *ce* is y^e radius of a circle wch hath y^e same quantity of crookednesse wch y^e Parabola *mca* hath at y^e point c. Wherefore I suppose *cb* & *nm*, 2 of y^e rootes of y^e equation $2y^3 + rry - 2dry + crr = 0$, to be equall to one another." Applying Hudde's rule, he got $d = (6y^2 + r^2)/2r$ and $c = 4y^3/r^2$. Cf. a later reworking of the same problem which solved for c and d in terms of x (*Math 1*, 256), and a brilliant extension of the analysis in which he clarified the geometric significance of the steps in his procedure in terms of various possible circles cutting the parabola and tangent to it (*Math 1*, 259–61).   [18] *Math 1*, 252–9.

[19] *Math 1*, 265–71. Whiteside points out an error in Newton's solution; the program itself was not mistaken, however.

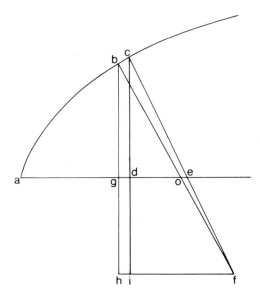

Figure 4.4. Newton's method to determine the location of the center of curvature f.

using his method of determining crookedness to rectify the evolutes, the loci of centers of curvature, of certain curves.[20]

Other aspects of seventeenth-century mathematics also caught his interest. I have mentioned his notes on Viète's mesolabum, a mechanical device by which cube roots and trisections of angles can be constructed. An extension of this passage introduced methods to develop tables of trigonometric functions from propositions on the functions of multiples of angles.[21] In fact, he set out to construct a complete table of sines to every minute. In the Fitzwilliam notebook he assigned one page to each degree from 0 to 90. Using methods from Viète, he computed a total of nine sines to fifteen places, before the sheer drudgery of the project exhausted his patience.[22] In the computations, however, and in other notes on methods of extracting roots and on Viète's calculation of π, Newton imbibed the concept of continued decimals as approximations which can never reach the exact value sought but can approach it as closely as one chooses through the calculation of additional places. Thus he computed sines to fifteen places and found devices by which he could compute roots and the value of π to fifteen places, or to fifty or five hundred if he wished.

[20] Specifically the semicubical parabola, $ky^2 = x^3$, which is the evolute of the simple parabola. The length of a segment of the evolute is equal to the difference in the two radii of curvature drawn from its ends (*Math 1*, 263–4).   [21] *Math 1*, 78–88.
[22] *Math 1*, 486–8.

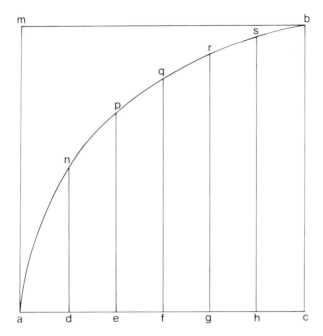

Figure 4.5. The use of infinitesimals and infinite series to find areas.

In Wallis's *Arithmetica infinitorum*, Newton also met the method of indivisibles. According to his own memorandum written in 1699 beside his early notes, this occurred in the winter of 1664–5. He mastered Wallis's method of squaring (or, as we now say, integrating) surfaces under second-degree equations by comparing the infinitesimals that exhaust them with the squared infinitesimals of a triangle which are stacked into a pyramid. From the parabola $ry = bx - x^2$, which he squared satisfactorily, he tried to extend the method to the deceptively similar hyperbola, $x^2 = dy + xy$.[23] Inevitably he failed, and in the process he made the elementary error of treating a curve without a term in y^2 as though it were symmetrical about the x axis.

Further notes from Wallis took up the method of infinite series with which Wallis had pushed back some of the limits that confined the method of infinitesimals. If the line *ac* is divided into an infinite number of equal parts, *ad, de,* . . . , from each of which the parallel lines *dn, ep,* . . . , which increase in proportion to one of the defined series, are drawn, all the lines may be taken for the surface *bqnac* (Figure 4.5). The proportion of the square *acbm* to the surface under the curve is as the ratio of the index increased by one to one.[24] If the curve is the parabola $y^2 = rx$, the lines increase according to the subsecundanary series (in Wallis's terminology), the index of which is ½. Hence the area of the square is to the area under the curve as $(\frac{1}{2} + 1)/1 = 3/2$. Still following Wallis, he recorded Wallis's sequence of compound series, $y = a^2 - x^2$, $y = (a^2 - x^2)^2$, $y = (a^2 - x^2)^3$, . . . , and his interpolation of the series $y = (a^2 - x^2)^{1/2}$ as a means of approximating π. The approximation he also recorded; $4/\pi$ is greater than $3/2 \times 3/4 \times 5/4 \times 5/6 \times 7/6 \times 7/8 \times 9/8 \times 9/10$ and less than $3/2 \times 3/4 \times 5/4 \times 5/6 \times 7/6 \times 7/8 \times 9/8 \times 9/10 \times 11/10$, a result analogous to roots and sines computed to however many figures one chooses.

In the winter of 1664–5, or sometime near then, Newton's urge continually to organize his learning led him to draw up a list of "Problems." Initially he put down twelve, one of which he later canceled. He added further problems on several occasions, as different inks show, until he had listed twenty-two in five distinct groups. The first group included most of the problems in analytic geometry to which he had addressed himself–to find the axes, diameters, centers, asymptotes, and vertices of lines, to compare their crookedness with that of a circle, to find their greatest and least crookedness, to find the tangents to crooked lines, and so on. The third group looked mostly toward the problems of quadratures (another seventeenth-century term for what we now call integration) to which Wallis had introduced him–to find such lines whose areas, lengths, and centers of gravity may be found, to compare the areas, lengths and gravities of lines when it can be done, to do the same with the areas, volumes, and gravities of solids, and so on. Several of the problems were frankly mechanical, and one treated a curve as the path traced by the end of the line y, perpendicular to x, as the line moves along x. In both respects, the problems looked forward toward distinctive features of his mathematics and of his mechanics. In all, the "Problems" laid out much of the program that would occupy Newton during 1665.[25]

His first important step beyond his mentors, which he dated on several occasions to the winter of 1664–5, was his extension of Wallis's use of series to evaluate areas. Wallis's quadratures had all been computed within fixed limits; his series had been series of numbers which he evaluated by comparing them with other series equal to their largest terms, and by inducing the limit toward which the ratio converges from a few instances. As he studied it, Newton realized that Wallis's method was more flexible than Wallis himself

[24] *Math 1*, 96–8.   [25] *Math 1*, 453–5.

had realized. It is not necessary always to compare the area under a curve with the area of the same fixed square. In the case of the simple power functions ($y = x$, x^2, x^3, ...), for example, any value of x provides a base line that can be divided into an infinite number of segments, and with the corresponding value of y it implicitly defines a rectangle with which the area under the curve can be compared. In the case of the curve $y = x^3$, for any positive value of x there is a rectangle with sides x and $x^3(= y)$. Newton accepted Wallis's demonstration that the area under the curve between the origin and x equals one-fourth the area of the rectangle, or $x^4/4$, an expression not confined to one fixed numerical value but valid for every value of x. Since a ratio was implied in the result, it did not have to be expressed explicitly.

From Wallis he had also learned that the areas under curves defined by polynomial equations, such as $y = x + 3x^2 + 2x^3$, can be treated as the sums of the areas under the individual terms. It followed that the quadrature of the polynomial above would be $x^2/2 + x^3 + x^4/2$, another polynomial stated in terms of the variable x and subject to evaluation for any value of x. The quadrature of each term in the polynomial subsumed Wallis's infinite series. Hence, by accepting Wallis's results, Newton could extend Wallis's method to construct series of a wholly different type. By making the upper limit of quadrature a variable, he could compose series made up of powers of the variable.

As Wallis had taught, polynomials are simple; finite series in powers of x express their areas. Could he use the same terms as elements in infinite series to calculate areas that could not be expressed as finite polynomials? To be specific, could he use them to evaluate Wallis's problem, the area under $y = (1 - x^2)^{1/2}$, the circle? Newton had learned more than infinite series from Wallis. He had also learned an inductive method. With no more regard for rigor than Wallis had shown, he set off confidently down the same road. Consider the set of curves $y = (1 - x^2)^0$, $(1 - x^2)^{1/2}$, $(1 - x^2)^1$, $(1 - x^2)^{3/2}$, ... (Figure 4.6). Squaring the curves with integral coefficients presents no problems. For any value of x (implicitly confined by the diagram to the range between zero and one, with the y axis, or $x = 0$, as the other boundary of the segment), the area under $y = (1 - x^2)^0$ equals x, under $y = (1 - x^2)^1$ equals $x - x^3/3$, under $y = (1 - x^2)^2$ equals $x - 2x^3/3 + x^5/5$, ... Newton drew up the results in a table as Wallis had done. Opposite each row he placed a power of x with a fractional coefficient; each column corresponds to the index of the term $(1 - x^2)$. The table, manifestly a Pascalian array, gives the coefficients by which the term for that row must be multiplied in the quadrature of the curve.

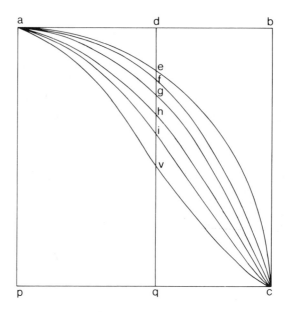

Figure 4.6. Newton's quadrature of the circle by interpolation, using a family of related curves.

		[0]	[1]	[2]	[3]	[4]	[5]	. . .
x	×	1	1	1	1	1	1	. . .
$-x^3/3$	×	0	1	2	3	4	5	. . .
$x^5/5$	×	0	0	1	3	6	10	. . .
$-x^7/7$	×	0	0	0	1	4	10	. . .
.		.	.	.	0	1	5	. . .
.		.	.	.	.	0	1	. . .
.		.	.	.	.	.	0	. . .

Now the problem was to interpolate the columns for 1/2, 3/2, 5/2, . . . , exactly Wallis's problem, though couched now in different terms. The first two rows were evident: 1, 1, 1, . . . and 1/2, 3/2, 5/2, . . . both filling in the obvious sequences. How could he proceed farther down the table to the remaining rows?

To intercalate the rest [he explained later] I began to reflect that the denominators 1, 3, 5, 7, etc. were in arithmetical progression, so that the numerical coefficients of the numerators only were still in need of investigation. But in the alternately given areas [the columns under integral powers] these were the figures of powers of the number 11, namely of these, 11^0, 11^1, 11^2, 11^3, 11^4, that is, first 1; then 1, 1; thirdly 1, 2, 1; fourthly 1, 3, 3, 1; fifthly 1, 4, 6, 4, 1, etc. And so I began to inquire how the remaining figures in these series could be derived from the first two given figures, and I found that when the second figure m was given, the rest would be produced by continual multiplication of the terms of this series,

$$\frac{m-0}{1} \times \frac{m-1}{2} \times \frac{m-2}{3} \times \frac{m-3}{4} \times \frac{m-4}{5}, \quad \text{etc.}$$

For example, let $m = 4$, and $4 \times \frac{1}{2}(m - 1)$, that is 6 will be the third term, and $6 \times \frac{1}{3}(m - 2)$, that is 4 the fourth, and $4 \times \frac{1}{4}(m - 3)$, that is 1 the fifth, and $1 \times \frac{1}{5}(m - 4)$, that is 0 the sixth, at which term in this case the series stops. Accordingly, I applied this rule for interposing series among series, and since, for the circle, the second term was $(\frac{1}{2}x^3)/3$, I put $m = \frac{1}{2}$, and the terms arising were:

$$\frac{1}{2} \times \frac{1/2 - 1}{2} \quad \text{or} \quad -\frac{1}{8}, \quad -\frac{1}{8} \times \frac{1/2 - 2}{3} \quad \text{or} \quad +\frac{1}{16},$$

$$\frac{1}{16} \times \frac{1/2 - 3}{4} \quad \text{or} \quad -\frac{5}{128},$$

and so to infinity. Whence I knew that the area of the circular segment which I wanted was

$$x - \frac{(1/2)x^3}{3} - \frac{(1/8)x^5}{5} - \frac{(1/16)x^7}{7} - \frac{(5/128)x^9}{9} \quad \text{etc.}$$

And by the same reasoning the areas of the remaining curves, which were to be inserted, were likewise obtained: as also the area of the hyperbola and of the other alternate curves in this series

$$(1 + x^2)^{0/2}, (1 + x^2)^{1/2}, (1 + x^2)^{2/2}, \quad (1 + x^2)^{3/2} \text{ etc.}$$

. . . This was my first entry into these considerations . . .[26] His papers show that the initial insight did not immediately appear with the clarity this statement implies, and that the process by which he arrived at it reflected the influence of Wallis, which the statement obscures.[27] Nevertheless, he did perceive the rule by which the column that corresponds to the circle is generated. Whereas integral powers expand into finite series when multiplied out, fractional powers expand into infinite series in ascending

---

[26] Newton to Oldenburg, 24 Oct. 1676 (the *Epistola posterior* intended for Leibniz); *Corres 2*, 130–1. The original Latin is on pp. 111–12.

[27] Thus his original statement of the continued product by which successive coefficients are found repeated a similar sequence that Wallis had used in his quadrature of the circle (and which Newton had copied into his notes), except that in Newton's sequence the figures in the numerator are all displaced two positions to the left (*Math 1*, 101, 108).

powers of x. There is no indication that Newton was unnerved by the prospect. He must have anticipated it. As he knew from Wallis, these columns involved π, and π can only be expressed by an infinite extended decimal. You should treat such problems, he later explained, "as if you were resolving ye equation in Decimall numbers either by division or extraction of rootes or Vieta's Analyticall resolution of powers; This operation may bee continued at pleasure, ye farther the better. & from each terme ariseing from this operation may bee deduced a parte of y^e valor of y."[28] Without this additional insight, the binomial theorem would have been impossible.

With the pattern now clear, he filled in all the intercalated columns of his initial table, corresponding to the fractional powers 1/2, 3/2, 5/2, . . . of the binomial $(1 - x^2)$.[29] Could a more Wallisian procedure be imagined? Perhaps he might wish to check it empirically by computing the value of π that his series would yield?[30] Not at all! He was perfectly confident. Calculating the well-known value of π was no challenge. It did not seriously occur to him that the principle of continuity on which he relied might betray him. Later he would recognize how flimsy the foundation was, and he would place the binomial expansion on a firmer footing. For the moment, untroubled by doubts and filled with the thrill of discovery, he rushed forward to exploit his powerful new tool.

Now he could attack the curve that had thwarted his effort to square it with the method of indivisibles. As he noted, $y = (1 + x^2)^{1/2}$ was virtually the same problem as the circle. The rectangular hyperbola, $xy = 1$, was more interesting for several reasons. "In y^e Hyperbola," he had written on an early page of the *Waste Book,* referring implicitly to the hyperbola $xy = 1$, "y^e area of it beares y^e same respect to its Asymptote which a logarithme doth its number."[31] The same relation applied to problems of compound

---

[28] The statement comes from his tract of October 1666, about eighteen months after the original discovery (*Math 1,* 413).

[29] He noted that with one column given, one could construct the other columns by the same rule used in the basic Pascalian array – any number (in an intercalated column) plus the number above it equals the number (in the intercalated column) to its right. The specific numbers in one of his early tables, in which the denominators under the 1/2 column correspond to the series, whereas the other columns have been reduced, suggest that he filled in the other columns by this procedure (*Math 1,* 123).

[30] For example, if we set $x = \frac{1}{2}$, the area under the section of the circle minus the area of the right triangle with legs .5 and $\sqrt{.75}$ equals $\pi/12$. Indeed Newton developed a series for the segment of a circle (*Math 1,* 108), but he did not calculate any value.

[31] *Math 1,* 457.

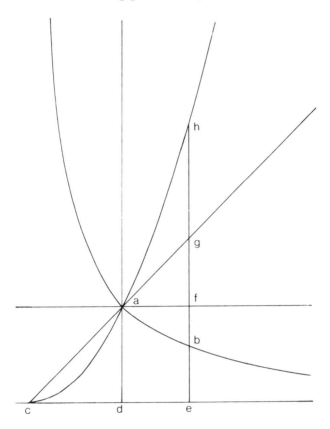

Figure 4.7. Newton's quadrature of the hyperbola by extrapolation,
using a family of related curves.

interest.[32] Since the area under $y = 1/x$ is infinite when it is computed from the y axis, Newton used the identical hyperbola to a
different axis, (defined by the equation $y = 1/(1 + x)$ (Figure 4.7).
As before, he used a sequence of curves $y = (1 + x)^0$, $y = (1 + x)^1$,
$y = (1 + x)^2$, . . . , and again he constructed a table of coefficients
for successive terms. With a different binomial, the sequence of
powers that correspond to each row is different: x, $x^2/2$, $x^3/3$, $x^4/4$,
. . . ; the Pascalian array of coefficients is the same. His problem in
this case was not to interpolate fractional powers but to extrapolate
back to the column that corresponds to $(1 + x)^{-1}$. This he could do
by the pattern of the array even without the new rule for expanding
binomials. Since the coefficient in the top row must be 1, the coeffi-

[32] *Math 1*, 461. About ten years later, when Humphrey Babington was bursar of Trinity,
the finances of the college benefited from a table Newton drew up to aid in the renewal of
leases (*Edleston*, p. lvi).

cient below it must be -1 to yield the 0 in the column to the right. Hence the extended table appeared as shown here.

		[−1]	[0]	[1]	[2]	[3]	[4]	
x	×	1	1	1	1	1	1	. . .
$x^2/2$	×	−1	0	1	2	3	4	. . .
$x^3/3$	×	1	0	0	1	3	6	. . .
$x^4/4$	×	−1	0	0	0	1	4	. . .

The series for a value of x – in effect, the logarithm for $(1 + x)$ – is expressed by $x - x^2/2 + x^3/3 - x^4/4 \ldots$ [33] Computing π once more had been no challenge. A new method quickly to compute logarithms, the new device recently bestowed on seventeenth-century calculators, was another matter. Filled with enthusiasm, he determined the values of $x = 1/10$ and $1/100$ (logs 1.1 and 1.01) to 46 places. [34] He was fascinated with his new device to the extent that he later recalculated the same logs and a few more to 55 places [35] (Figure 4.8). He had not yet graduated B.A. Scarcely a year had passed since he began to study mathematics under his own untutored tutelage.

Somewhat later, possibly in the summer or autumn of 1665, Newton systematized his work on quadratures in four propositions. He had not been able to find anything general in quadratures, he later told Wallis, "till he had reduc'd the Business to the sole Consideration of Ordinates." [36] Such had been the meaning of his two sequences of curves, $y = (1 - x^2)^{m/m}$ and $y = (1 + x)^m$, and the diagrams which accompanied them, which had furnished the keys to unlock the binomial expansion. In the paper on quadratures, he made this approach systematic and general. Let $ab = x$ and $bd = y$ so that the equation of y as a function of x defines the curve *addc* (Figure 4.9). When y is expressed as some power of x, positive or negative, to find the area *abd* multiply the value of y by x and divide by the number which expresses the power of x (after the multiplication). To illustrate, he went through two sequences: $y = 1, x, x^2, x^3 \ldots$, and $y = a/b, (a/b)x, (a/b)x^2, (a/b)x^3, \ldots$ The areas are expressed by $x, x^2/2, x^3/3, x^4/4, \ldots$, and $(a/b)x, (a/2b)x^2, (a/3b)x^3, (a/4b)x^4, \ldots$, regular patterns which did not escape his perception.

[33] *Math 1*, 112. [34] *Math 1*, 113–14. [35] *Math 1*, 134–41.
[36] Joseph Raphson, *The History of Fluxions* (London, 1715), p. 33.

Figure 4.8. The calculation of several logarithms to 55 places, ca. 1665. (Courtesy of the Syndics of Cambridge University Library.)

If $ca \| vb \| de \perp ac \| cv = vb = a$. & $bc = x$. & $dc = y = \dfrac{aa}{a+x}$. Then is vdd an Hyperbola &c: And If $\dfrac{aa}{a+x}$ bee divided as in decimalls fractions y^e production

$$\dfrac{aa}{a+x} = y = a - x + \dfrac{xx}{a} - \dfrac{x^3}{a^3} + \dfrac{x^4}{a^3} - \dfrac{x^5}{a^4} + \dfrac{x^6}{a^5} - \dfrac{x^7}{a^6} + \dfrac{x^8}{a^7} - \dfrac{x^9}{a^8} + \dfrac{x^{10}}{a^9} - \dfrac{x^{11}}{a^{10}} + \dfrac{x^{12}}{a^{11}} - \dfrac{x^{13}}{a^{12} + a^{13}x} \&c$$

Which valor of y being each terme thereof multiplyed by x & divided by y^e number of its dimensions: The product will bee y^e area $vbcd$. viz

$$vbcd = ax - \dfrac{xx}{2} + \dfrac{x^3}{3a} - \dfrac{x^4}{4a^3} + \dfrac{x^5}{5a^3} - \dfrac{x^6}{6a^4} + \dfrac{x^7}{7a^5} - \dfrac{x^8}{8a^6} + \dfrac{x^9}{9a^7} - \dfrac{x^{10}}{10a^8} + \dfrac{x^{11}}{11a^9} - \dfrac{x^{12}}{12a^{10}} \&c$$

As for example. If $a = 1$. & $x = 0,1$. The calculation is as ffollowth.

[Large block of handwritten numerical computations — columns of repeating digits 0,10000000000,00000000000... 0,3333... 1,42857,14285,7... etc., with right-hand labels $= ax$, $= \frac{xx}{2a}$, $= \frac{x^3}{3a}$, $= \frac{x^4}{4a^3}$, etc.]

... $9,31031$ ...
... $5,1612$ ...
$9,112$ ... $3,48637$ ...
$9,57446,$
$9,16326,$
$9,07843$
$4,21,68679$
$378,03 = $ summe.

[Second large block of numerical computations, with right-hand labels $= \frac{1}{2}xx$, $\frac{x^4}{4aa}$, $\frac{x^6}{6a^4}$, $\frac{x^8}{8a^6}$, $\frac{x^{10}}{10a^8}$, $\frac{x^{12}}{12a^{10}}$, $\frac{x^{14}}{14a^{12}}$, $\frac{x^{16}}{16a^{14}}$, $\frac{x^{18}}{18a^{16}}$ \&c]

... $9 \ 3 \ 6 \ 8 = $ summe.

The Summe of these two summes is squall to y^e area $bcfv$, supposing $ac = 0,9$.

And their Difference is equall to y^e area $bcdv$, supposing $ac = 1,1$. viz:

$bcfv = 0,10536051565782630122750098083939279830612037298327,4079$ &c If $ac = 0,9$. or $cb = 0,1$.
$bcdv = 0,09531017980432486004,39521,23280,84509,22206,05365,30864,4199$ &c If $ac = 1,1$. or $bc = 0,1$.

In like manner if $x = 0,01$. & $a = 1$. The calculation is as followth.

[Third block of numerical computations with right-hand labels $= ax + \frac{x^3}{3a} + \frac{x^5}{5a^3}$, $\frac{x^7}{7a^5}$, $\frac{x^9}{9a^7} + \frac{x^{11}}{11a^9}$, $\frac{x^{13}}{13a^{11}}$ \&c]

... $= $ summe.

The summe of these two summes is equall to y^e area $bcfv$, supposing $ac = 0,09$. And their Difference to $bcdv$, if $ac = 1,01$. viz:

$bcfv = 0,0100503358535014411835488575585477960855110076746298738$ &c If $ac = 0,99$.
$bcdv = 0,00995033085316808284821535754426074168872960994095,87984$ &c If $ac = 1,01$.

So if $x = 0,001$. Then $0,00100000000000003333333,5333347619047619047619047619$ &c $= ax + \frac{x^3}{3a} + \frac{x^5}{5a^3}$ &c.

the Calculation will bee.

... $= $ summe.

Therefore $\begin{cases} bcfv = 0,0010005000033335000,2500016,66667,91666,76666675000071,42863,3,3928 \text{ summe of } \frac{xx}{2} + \frac{x^4}{4} \&c \\ bcdv = 0,0009,9500,33083,6330,0142,82252,460 \end{cases}$ If $ac = 0,999$.

And if $x = 0,0001$. Then ...

Therefore $\begin{cases} bcfv = \\ bcdv = \end{cases}$ ... If $ac = 1,001$.

$ac = 0,9999$. $ac = 1,0001$.

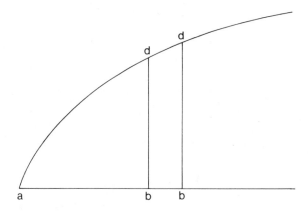

Figure 4.9. Newton's general approach to quadratures.

He noted that the formula applies to negative exponents as well as to positive ones, and that if the equation defining the curve expresses y as a polynomial in x, the area is the sum of the areas for each term. Terms with x to the first power in the denominator obviously would not fit into the pattern. For the general equation $y = a^2/(b + x)$ he gave the expansion $a^2/b - (a^2/b^2)x + (a^2/b^3)x^2 - (a^2/b^4)x^3 \ldots$, which can be squared term by term according to the stated formula. Finally, for binomials raised to fractional powers, he stated his rule to expand the binomial into an infinite series in a more elegant form than he had devised before. For any binomial of the form $(b + x)^{m/n}$, the coefficients of successive terms can be found by the progression

$$\frac{1 \times m \times (m - n) \times (m - 2n) \times (m - 3n) \times (m - 4n) \ldots}{1 \times n \times 2n \times 3n \times 4n \times 5n \ldots}$$

Again each term is squared (or integrated) by the general formula.[37]

Fifty years later, in his anonymous *Account of the Commercium Epistolicum*, Newton said that his method of quadratures rested on three rules: first, the quadrature of $x^{m/n} = nx^{(m+n)/m}/(m + n)$; second, the addition and subtraction of areas when the equation for y has two or more terms; and third, the reduction of fractions and radicals and affected equations into infinite series.[38] Although he had not yet applied his method to affected equations (that is, equations in which x and y appear in the same term such that it is impossible to state one as an explicit function of the other), the three rules had been implicit from the time of his initial steps with infinite series. They became explicit in the systematic paper of 1665. In varying degrees,

[37] *Math 1*, 126–33.
[38] "An Account of the Book entituled *Commercium Epistolicum* . . . ," *Philosophical Transactions of the Royal Society*, 29 (1714–16), 176–7.

all three owed a debt to Wallis. The first was Newton's acceptance of the results of Wallis's cumbersome procedure with infinite series. In Newton's hands, the first rule became more flexible and usable, but he did not initially supply it with a foundation that was independent of Wallis. Though later he would do so by treating quadrature (or integration) as the inverse of differentiation, it was Wallis's result which allowed him to arrive at this insight. The second rule also came directly from Wallis. The evaluation of the terms to be added and subtracted depended, of course, on the first rule. The third rule was Newton's own discovery inspired by the suggestion in Wallis's method. The third rule was also critical. By giving the quadrature of virtually all the algebraic equations then known to mathematicians that could not be squared by the first two rules, it made the method general. As was usually the case with his most signal inspirations, Newton appeared to resent his dependence on Wallis. He was not overly forward in acknowledging his debt.

His operations continually displayed patterns. The quadrature of $y = x^n$ is $[1/(n + 1)]x^{n+1}$. Could one not then use this pattern to "shewe ye nature of another crooked line y^t may be squared?"[39] In the spring of 1665, Newton began seriously to explore the possibilities to which this avenue could lead. By now surely he had become a stranger to his bed. More than one morning Wickins must have discovered a taut figure bent over his incomprehensible symbols, unaware that a night had passed and for that matter unconcerned. He was repaid by the discovery of the fundamental theorem of the calculus.[40]

Consider a parabola defined by the equation $rx = y^2$ (Figure 4.10): $ac = r/4 = ad/2$. $ap = dt = a$. Now define a new variable z, which traces the curve po. The subnormal appeared to be involved in the variable z, but by itself it was not the expression which,

[39] *Math 1*, 225.

[40] D. T. Whiteside, the editor of the *Mathematical Papers*, has separated a few theorems from ff. 92^v – 116 of *Add MS* 4000 (the mathematical notebook) and dated them to the summer of 1665 (*Math 1*, 299–302), while he has dated most of the passage to autumn 1664 (*Math 1*, 221–33). His notes indicate that the separation is based on the writing alone, and he agrees that the dates are uncertain. What he places in summer 1665 comes at and near the beginning of the passage and is not consecutive, though most of it appears to have been written (on verso sides and in blank spaces) after the material near it. I cannot find any adequate reason to suppose there was a long interruption, and I find his division misleading for two reasons. First, the entire passage assumes Newton's work on quadratures. Second, it seems incredible to me that he would have set aside what was obviously an important insight for nine months, as Whiteside's arrangement demands. Finally, in its reliance on Descartes's method of tangents, it appears to me to precede an investigation of tangents explicitly dated to May 1665 which broke free of Descartes's method. Moreover, the new approach to tangents contains strong suggestions that it emerged from the expanded investigation of quadratures.

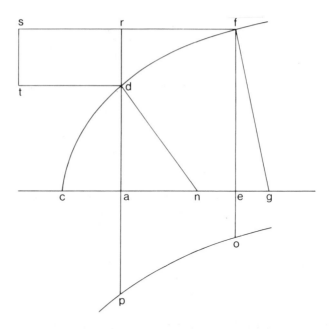

Figure 4.10. The fundamental theorem of the calculus.

upon the application of the formula for quadratures, would yield the original equation. If it were divided by y, however, then $gelef = r/2(rx)^{1/2} = z/a$. $z = (a/2)(r/x)^{1/2}$, "wch shews y^e nature of y^e crooked line *po*. now if $dt = ap$, y^n *drst* = *eoap*. for supposeing *eo* moves uniformly from *ap*, & *rs* moves from *dt* wth motion decreaseing in y^e same proportion y^t y^e line *eo* doth shorten." [sic][41] Quickly he went through a number of other examples, of which the hyperbola $y = a^2/x$ is perhaps the most revealing (Figure 4.11). $ak = a = kh$. $oa = x$. $od = y$. $ca = v$. $og = z$. By his usual method, Newton found v and then the subnormal oc $(= x - v) = a^4/x^3$. He defined z by the ratio

$$\frac{od}{oc} = \frac{kh(= a)}{og(= z)} \qquad \frac{a^2/x}{a^4/x^3} = \frac{a}{z}$$

$z = a^3/x^2$, "wch equation conteines y^e nature of y^e crooked line *gh*. Now supposeing y^e line *og* always moves over y^e same superficies in y^e same time, it will increase in motion from *kh* in y^e same proportion y^t it decreaseth in lenght & y^e line *ne* will move uniformly from *(mq)*, so y^t y^e space *mqen* = *gokh*."[42] In the operation of defining the new curve (*gh* in this case), Newton's method of tangents via subnormals began to alter itself into differentiation.

[41] *Math 1*, 299.   [42] *Math 1*, 228–9.

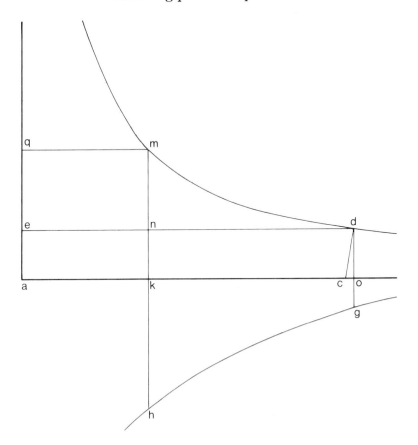

Figure 4.11. The fundamental theorem of the calculus.

Since *dc* is normal to the curve, the ratio *oc*/*od* used to define *z*, is exactly equivalent to the derivative of *y*. The slope of the parent curve in turn controls the motions by which the two areas are described. Two equal lines begin to move and in the first instant describe equal areas. In the case of the parabola, the line *ap* moves uniformly with a decreasing length while the constant length *dt* moves at a decreasing rate. For any value of *x* between *a* and *e*, the slope of the parent curve determines the length of *ap* and the rate at which *dt* moves. In the example with the hyperbola, uniform motion is attributed to the line *mq*. The ordinate of the curve *gh* moves with a velocity which varies inversely as its length, that is with a velocity inversely as the slope of the parent curve *dm*. Hence the area under the derivative curve is proportional in each case to the difference in the two corresponding ordinates of the original curve. Always systematizing, he ran through the sequence of curves $y = a^2/x$, a^3/x^2, a^4/x^3, a^5/x^4 defining *z* in the same manner and finding a

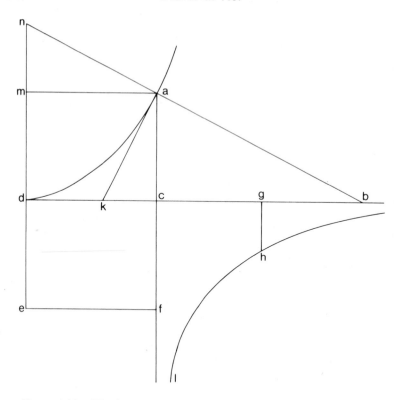

Figure 4.12. The kinetic approach to quadratures.

pattern, which he tabulated, for the quadratures of the newly defined curves between the ordinates *a* and *b*.[43]

A whole new approach to quadratures was opening itself to
Newton's advance. Before, with Wallis, he had considered areas as
static summations of infinitesimals. Now he was treating them ki-
netically, the areas swept out by a moving line. In a separate passage, he had explored this idea in terms, not of areas alone, but of
areas in equilibrium about an axis. In the diagram, $gc = ac = y$. The
area *ghlc* generated by the motion of *gh*, is in equilibrium about the
axis *acf* with the area *cdef* generated by the constant *de*, which moves
along *cd* ($= x$) as it increases (Figure 4.12):"*ck*: *ca*: : motion of y^e
point *a* from *c*: motion of y^e point *a* from *m*."[44] In this context the
problems of tangents and of quadratures suddenly were seen to
have an inverse relation to each other. If the calculus had not been

[43] *Math 1*, 228–33.
[44] *Math 1*, 240. Cf. the entire passage, pp. 238–41. As Whiteside notes, it appears in the
     mathematics notebook shortly before the passage I have been discussing, separated only
     by an earlier set of notes.

born, certainly it had been conceived. Newton had received his Bachelor of Arts degree, if he had yet received it at all, less than a month before. In mathematics, he had passed further beyond the status of student than a single month can conceivably imply. He had absorbed by now what books could teach him. Henceforth he would be an independent investigator exploring realms never before seen by human eye.

The significance of the new insight was not lost on him. He set out at once to systematize it and to set it in the context of his growing mastery in a paper, which went through two drafts, called "A Method whereby to square those crooked lines wch may bee squared."[45] Whereas initially his insight rested on the patterns of coefficients and exponents which revealed the inverse nature of the two operations separately grounded in procedures he had adopted from others, he now offered a direct demonstration of it. For the two curves $z = x^3/a$ (above the x axis in this case, the roles of y and z being reversed) and $y = 3x^2/a$, he showed as before that the area under y (the derivative curve) is proportional to the difference between corresponding ordinates of z. Draw tangents to the curve of z at points f, m, and π, and from the points of tangency and the intersections of tangents draw two perpendicular sets of parallel lines (Figure 4.13).

$$\frac{mb}{nb} = \frac{bt}{bm} = \frac{\Omega\beta}{\beta m} = \frac{kl}{bg} \qquad \frac{kl}{bg} = \frac{a}{y}$$

When $\Omega\beta$ and βm are infinitely small, which he explicitly allowed, they are equivalent to the infinitesimal increments Δx and Δz. From the proportions, $\Omega\beta \cdot bg = \beta m \cdot kl$, and $kl\nu\mu = bpsg$. We can extend the same reasoning to similar rectangles, "so y^t y^e rectangle $\rho\sigma hd$ is equall to any number of such-like squares inscribed twixt y^e line $n\psi$ & y^e point d, wch squares if they bee infinite in number, they will be equall to y^e superficies $dn\psi wg\xi$."[46] Together with a second draft which included another example, this was the only demonstration of the fundamental theorem of the calculus that Newton offered.[47] It was confined to the two related curves $z = x^3/a$ and y

[45] First draft, *Math 1*, 302–13; second draft, with a slightly different title, *Math 1*, 318–21. The second draft does not contain a long table of integrals, which makes the first version much longer.   [46] *Math 1*, 304.

[47] *Math 1*, 315. In a third draft, he radically altered the demonstration. Proposition 1 gives a general equation for the subnormal v. When $z = ax^{mm}/b$, then $v = mazx^{(m-n)m}/nb$. Proposition 2 states that when r is a constant, if $rv = zy$, the area $akhi$ [$= rz$] $= abef$ [$=\int ydx$]. Proposition 3 states that if $y = ax^{mm}/b$, the area $abef = [n/(m + n)](a/b)x^{(m+n)m}$. "Demonstracon. For Suppose $akhi$ is a parallelogram & equall to $nax^{(m+n)m}/(nb + mb)$. y^n is $nax^{(m+n)m}/(nbr + mbr) = ai = z$. & (prop 1) $az^2x^{m+n)m}/brxz = azx^{mm}/br = v$. & (prop 2^d) $rv = zy$; y^t is $ax^{mm}/b = y$" (*Math 1*, 319). That is, the demonstration now rests entirely on the inverse nature of the two operations.

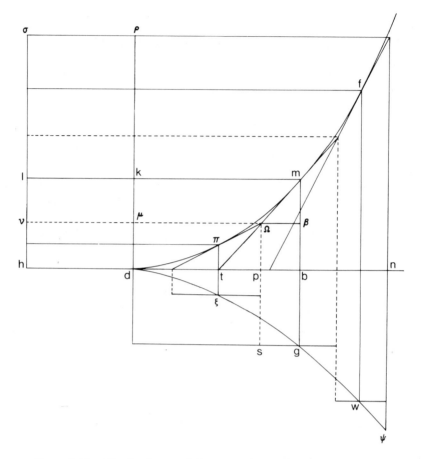

Figure 4.13. The fundamental theorem of the calculus.

$= 3x^2/a$, (and, in the second draft, $z = a^3/x$ and $y = -a^3/x^2$) though the procedure as such was general. For the most part, he preferred to define the integral as the inverse of the derivative, resting his assurance on his earlier work with quadratures. Using his fresh insight, he included an extended table of integrals and their corresponding derivatives (to use our rubrics) in the paper.[48]

Now the problem of tangents had taken on fresh interest, and in May 1665 Newton returned to it. Perhaps his exposure to infinitesimals in Wallis's method of quadratures encouraged him to apply a similar device to Descartes's method of tangents. At any rate, he now chose two points, *e* and *f*, on the curve *aef* that were removed from each other by an infinitesimal distance (Figure 4.14). He employed the symbol *o*, with which he had experimented before, for

[48] Later in 1665, he drew up two more tables of integrals (*Math 1*, 348–54, 354–63).

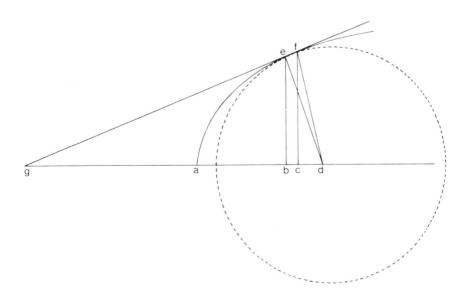

Figure 4.14. A new approach to tangents.

the increment of x. Hence if $ab = x$, $ac = x + o$, and $cf = z = f(x + o)$. He ended up with $v = o + a/2 + x$. When the two points e and f conjoin, "wch will hapen when $bc = o$, vanisheth into nothing," the terms with o are "blotted out" and (in this example, $ax + x^2 = y^2$) $v = x + a/2$.[49] The individual example interested him less than the procedure which, in typical fashion, he immediately set about generalizing. He noticed the familiar pattern of binomial coefficients. When y is expressed in several terms containing x, z has the same terms and also, since $z = (x + o)^n$, the terms "multiplied by so many units as x hath dimensions in y^t terme & againe multiplied by o & divided by x."[50] He had already indicated that terms with o to higher powers can be dropped since they will in the end be "blotted out" when o vanishes. "Blotting out" was also an operation familiar from Hudde's rule. We may assume that obvious analogies prompted his use of the term here. He set up general expressions for v in equations in y^2, y^3, and y^4, which he promptly generalized further to what he called "An universall theorem for tangents to crooked lines . . . "[51] With the device of infinitesimal increments in hand, he set about revising Descartes's method. The method in-

[49] *Math 1*, 273.
[50] *Math 1*, 273–4. That is, if y has the term x^4, z has $x^4 + 4x^3o$ (or $4x^4o/x$)).
[51] *Math 1*, 276. In fact, he had arrived at the identical expression the previous autumn by a different method (*Math 1*, 236; cf. n. 13).

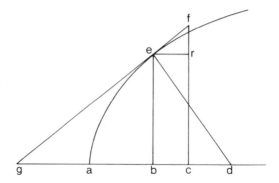

Figure 4.15. Finding tangents from infinitesimal increments.

volved the use of two radii to get an equation, $z^2 = y^2 + 2yo$, which he used to substitute for z in terms without an o. When the equation had a term in z to an odd power, the radical complicated the algebra horrendously, yielding an equation which even a seventeenth-century mathematician must have faced with a sinking heart. Why should he bother with radii? The infinitesimal increment o suggested an "infinitely little" triangle *efr* (Figure 4.15). The increment of y is vo/y, and $z = y + vo/y$, which introduced a ray of hope into the bleak algebraic scene.

The simplification was purchased at the price of a questionable device, however. These operations cannot be allowed, he remarked "unlesse infinite littlenesse may bee considered geometrically."[52] Nevertheless, he persevered. If he could dispense with Descartes's circle, why not dispense with the subnormal as well, and deal with the tangent directly by finding the subtangent *bg* ($= t$) which determines the point where the tangent cuts the axis? After all, his method of finding derivative curves that could be squared had revealed the crucial significance of the slope which the tangent expressed directly. By similar triangles, $t/y = (t + o)/z$, $z = y + yo/t$. Where before he solved for v, the subnormal, now he solved for t and got an equation exactly the inverse of the other except for a factor y.[53] He rewrote his earlier conclusions on centers of curvature in terms of his new method of infinitesimals. The entire investigation had striven consistently, in typical Newtonian style, for generality. Now he proposed a new notation suitable to "a general Theoreme whereby the crookedness of any line may be readily determined." Let X represent the equation that expresses the nature of the line set equal to zero. Then ·X represents the same terms ordered according to the power of x and multiplied by any arithmetical progression. Interestingly, in striving for generality he returned to Hudde's multipliers as more general than exponents. X·

[52] *Math 1*, 282–3.   [53] *Math 1*, 279–80.

represents the results of the same operation on the equation ordered by y, and $\cdot\cdot X$, $X\cdot\cdot$, and $\cdot X$ the equation when the operations are repeated a second time. With these symbols he constructed indigestible general equations, which only eyes dazzled by the excitement of discovery could have contemplated with equanimity, to determine the location of the center of curvature.[54]

Apparently Newton was uncomfortable with the infinitesimal basis on which his method of tangents now rested. In the fall of 1665, he began to extend his kinematic approach to areas to the generation of curves as well and to treat them as the locus of a point moving under defined conditions. Years later, during the controversy with Leibniz, Newton said that Barrow's lectures may have led him to consider the generation of figures by motion.[55] It is necessary to remark that the idea was not unique to Barrow; it was part of the mathematical culture of the day. What he now proposed was a new approach to tangents which cast away once and for all the Cartesian scaffolding which had supported him during his mathematical apprenticeship.

1. If two bodys c, d describe y^e streight lines ac, bd, in y^e same time, (calling $ac = x$, $bd = y$, $p = $ motion of c, $q = $ motion of d) & if I have an equation expressing y^e relation of $ac = x$ & $bd = y$ whose termes are all put equall to nothing. I multiply each terme of y^t equation by so many times py or p/x as x hath dimensions in it. & also by soe many times qx or q/y as y hath dimensions in it. the sume of these products is an equation expresing y^e relation of y^e motions of c & d. Example if $ax^3 + a^2yx - y^3x + y^4 = 0$. y^n $3apxx + a^2py - py^3 + aaqx - 3qyyx + 4qy^3 = 0$.[56]

---

[54]
$$c = \frac{\cdot X \; \cdot X \; \; X \cdot yy + \; \; X \cdot \; \; X \cdot \; \; X \cdot xx}{- \; \cdot X \; \cdot X \; \; X \cdot\cdot \; x + 2 \cdot X \; \; X \cdot\cdot \cdot \; \; x - \; X \cdot \; X \cdot\cdot X \; x}$$

$$d = \frac{\cdot X \; \; X \; \cdot X \; yy + \; \; X \; \; X \cdot \; \; X \cdot xx}{\cdot X \; \; X \; \; X \cdot\cdot \; x - 2 \cdot X \; \; X \cdot \; \; X \cdot \; \; x - \; X \cdot \; X \cdot\cdot X \; x}$$

c and d are the coordinates of the center of curvature. In modern notation, as Whiteside shows, $\cdot X = xf_x$, the homogenized first-order partial derivative with respect to x; $X\cdot = yf_y$; $\cdot\cdot X = x^2 f_{xx}$; $X\cdot\cdot = y^2 f_{yy}$; and $\cdot X\cdot = xyf_{xy}$ (*Math 1*, 289–90).

[55] About 1714, in connection with the controversy with Leibniz, Newton wrote that "its probable that Dr Barrows Lectures might put me upon considering the generation of figures by motion, tho I not now remember it." (*Add MS* 3968.41, f. 86^v). Newton made a similar comment about Barrow's influence to Conti (*Prose e poesi, 2*, 25); see also Newton to [?] Sloane, ca. February 1712 (*Corres 5*, 213). In fact, there are not many references to Barrow among Newton's papers. I have found one very favorable mention of Barrow's work in a draft of a paper connected with the calculus controversy. He suggested that the reader compare his letters from the early 1670s with *De analysi* to find if his method of fluxions is not there. "And then let him compare the differential method of Tangents of Dr Barrow published 1670 & that of Mr Leibnits in his Letter of 21 June 1677 & see if they be not the same, & if Mr Leibnitz hath added any thing more to the differentia of Mr Barrow then what Mr Newton in his Letters above mentioned gave him notice of." This passage lends support to Barrow's influence on Newton's mathematical development (*Mint* 19.2, f. 353^v). [56] *Math 1*, 344.

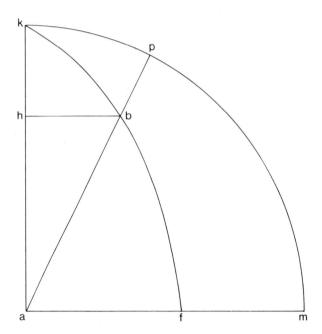

Figure 4.16.   The quadratrix *kbf* generated by the rotation of the radius *ap* and the motion of the line *hb*.

In our language, which becomes increasingly applicable with the
progress of Newton's investigation, q/p is the derivative of y with
respect to x.

About the same time, he began to apply analogous considerations
to the construction of tangents to mechanical curves. He examined
the spiral (generated by the uniform motion of a point along a line
that is rotating uniformly around one end), the quadratrix (the
point of intersection between a radius *ap* that rotates through 90
degrees and a line *hb*, perpendicular to the initial position of the
radius, *ak*, that moves uniformly from *k* to *a*) (Figure 4.16), the
ellipse (treated as the point of intersection of the two lines rotating
around the two foci such that the increase in one equals the decrease
in the other), and the cycloid (traced by a point on a rolling
wheel).[57] The method had its perils, which Newton initially failed
to avoid entirely.[58] Excited by his new approach, he immediately
undertook its revision and perfection. In doing so, he located and
corrected his own errors.

By now, no doubt, Newton was neglecting his meals. Working
in feverish haste, he was ready by 13 November to systematize the

[57] The initial paper on this topic probably dates from 30 Oct. 1665 (*Math 1*, 369–76).
Already on 8 Nov., Newton revised it (*Math 1*, 377–80).
[58] *Math 1*, 371–2. Whiteside's notes explicate Newton's errors.

new method in a paper entitled "To find ye velocitys of bodys by ye lines they describe."[59] When an equation is given expressing the relation of two or more lines x, y, and z described in the same time by two or more bodies A, B, and C, we seek the relation of their velocities p, q, and r at any point. Set the equation equal to zero. Multiply each term by p/x times the exponent of x in that term, then each by q/y times the exponent of y, finally by r/z times the exponent of z. Add all the terms together and set the sum equal to zero. The new equation expresses the relation of p, q, and r. In a lemma, Newton offered what he explicitly labeled a demonstration. As before, the lack of a generalized symbolism forced him to couch it in terms of a specific problem.

> Lemma. If two bodys A/B move uniformly ye one/other from a/b to c/d, e/f, g/h, &c in ye same time. y^n are ye lines ac/bd & ce/df & eg/fh &c as their velocitys p/q. And though they move not uniformly yet are y^e infinitely little lines wch each moment they describe as their velocitys are wch they have while they describe them. As if y^e body A wth y^e velocity p describe y^e infinitely little line o in one moment. In y^t moment the body B wth y^e velocity q will describe the line oq/p. For $p{:}q{::}o{:}oq/p$. Soe y^t if y^e described lines be x & y in one moment, they will bee $x + o$ and $y + oq/p$ in y^e next.

Let the equation expressing the line be $rx + x^2 - y^2 = 0$. Newton substituted $x + o$ and $y + oq/p$ for x and y, getting the equation, $rx + ro + x^2 + 2ox + o^2 - y^2 - 2qoy/p - q^2o^2/p^2 = 0$. By the original equation, $rx + x^2 - y^2 = 0$. Hence $ro + 2ox + o^2 - 2qoy/p - q^2o^2/p^2 = 0$. Divide by o: $r + 2x + o - 2qy/p - oq^2/p^2 = 0$. Since those terms with o are infinitely smaller than those without o, they can be blotted out. $r + 2x - 2qy/p = 0$, or $pr + 2px = 2qy$.

> Hence may bee observed: First, y^t those termes ever vanish in wch o is not because they are y^e propounded equation [$= 0$]. Secondly y^e remaining Equation being divided by o those termes also vanish in wch o still remaines because they are infinitely little. Thirdly y^t y^e still remaining termes will ever have y^t forme wch by y^e first preceding rule [the algorithm given above, with which the paper began] they should have.[60]

Clearly, Newton thought the concept of generating curves by motion provided a new, more solid foundation for the results he had already obtained. Later in the paper, after he had stated and demonstrated its basic propositions, he added, "Hence may bee pronounced those theorems in Fol. 47 [of the *Waste Book*, the paper of May]."[61] The new procedure did not produce new results. Rather it offered a new basis on which to establish old results more firmly. From the idea of motion he derived the term "fluxional,"

[59] *Math 1*, 382–9. [60] *Math 1*, 385–6. [61] *Math 1*, 387.

which became his permanent descriptive word for his method. In fact, he had not escaped from infinitesimals, however. His definitions of p and q assume a third, invisible variable, time. The concept of absolute time entered inextricably into his mathematics at this time and found here its permanent rationale in his thought. If we represent a moment by t, then $p = o/t$ and $q = oq/pt$.[62] In the intuitive idea of continuous motion or flow, infinitesimals have been replaced by instantaneous velocities. The only device that Newton had available to express instantaneous velocity, however, was the ratio of distance traveled to the infinitesimal unit of time in which it took place. The idea of motion concealed its infinitesimal ingredient by transferring it to the unexpressed variable time. For the unit of time he had only one term, a "moment." The "infinitely little lines" which bodies describe in each moment are the velocities with which they describe them.[63] His equations treated the increments as algebraic entities to be handled like other entities. He divided the equations by o and then "blotted out" terms in which o still appeared because they were infinitely little in relation to the others. The concept of continuously varying motion, which appears intuitively to overcome the discontinuity of indivisibles, never ceased to appeal to Newton's imagination. Nevertheless, he would in the end seek for another, more rigorous foundation for his calculus.

The autumn of 1665 passed in incandescent intensity. Then, with the completion of the paper of 13 November, the light went out, as suddenly and totally as if Newton had extinguished a candle. Six

---

[62] Whiteside offers a rather different interpretation of this method in his notes to the *Mathematical Papers,* in which he develops a concept of what he calls "limit-motions," and in his "Patterns of Mathematical Thought," p. 361. See Philip Kitcher, "Fluxions, Limits, and Infinite Littlenesse. A Study of Newton's Presentation of the Calculus," *Isis,* 64 (1973), 33–49; and Tyrone Lai, "Did Newton Renounce Infinitesimals?" *Historia mathematica,* 2 (1975), 127–36.

[63] Hence I find Whiteside's phrase "limit-motions," and his rephrasing of Newton's process as the modern definition of differentiation when o [dx] approaches zero, misleading. The language of limits does not appear at all in Newton's papers from this time. o is an infinitesimal increment of x in the moment t. In the final step, terms containing o are "blotted out" because they are "infinitely less" than those without o. Fifty years later, in his anonymous "Account of the *Commercium Epistolicum,*" Newton said that he used o for a finite moment of time or of any quantity that flows uniformly; he performed the whole calculation in finite figures without any approximation, and when the calculation was completed, he supposed the moment o to decrease *in infinitum* and to vanish. When he was not demonstrating, but only investigating, he supposed o to be infinitely little in order to proceed more swiftly (*Philosophical Transactions, 29* [1714–16], 179). Whatever the accuracy of the description for later periods, only the last statement appears to apply to the early documents. I agree that the concept of motion implies an intuitive approach to a limit, but in 1665–6 Newton (and the mathematics of the age) did not have the tools with which to express it.

months passed in which, if we can trust the surviving record, he did not turn a finger toward mathematics. In May, something stirred his interest anew, and he devoted three days to further elaboration of the idea of motion in two separate papers composed on 14 and 16 May. Again the light went out, and once more something stirred him in October, when he drew his thoughts together in a more definitive essay. A third time the light went out. It was as though the successful resolution of the problems that had been set for him had exhausted his interest in mathematics. There was no lack of other enthralling enquiries to command his attention. As far as we can tell, he scarcely looked at mathematics for the following two years.

The three papers of 1666 all explore the method based on motion. The second two carry similar titles, finally phrased "To resolve Problems by Motion these following Propositions are sufficient."[64] The first, theoretical, section of the paper is surprising in its close approximation to mechanics. The initial propositions choose to speak of bodies in motion, rather than points, and they lay down rules for the analysis of motion into its components. To rectilinear motion Newton added rotation, and noted that all motions in a plane can be reduced to one of the two or a compound of them. The principle would figure in his mechanics. He also defined what was meant by the center of gravity of a body and devoted the final section of the paper to locating centers of gravity of planes and to determining figures that are in equilibrium with each other about a given axis. The solutions to such problems are exercises in mathematics, of course. The content of the problems suggests some of the other interests which, in 1666, began to distract his attention from mathematics.

Much the greatest part of the paper, however, cannot be mistaken for anything but what it is, the exposition of a sophisticated mathematical method which Newton would later call his method of fluxions. With Proposition 6, bodies are forgotten, and the instantaneous motions of a point in reference to two intersecting and moving lines are shown as vectors, a geometrical foundation for his method of tangents. Proposition 7 states the method, that is, the algorithm for finding the velocities p and q of bodies A and B when an equation in x and y, which defines the relation between the lines they describe in equal times, is given. Unlike the paper of November 1665, the tract of October 1666 uses o to represent a moment of

---

[64] 14 May: "To resolve these & such like Problems these following propositions may bee very usefull" (*Math 1*, 390–2). 16 May: "To resolve Problems by motion yᵉ 6 following prop: are necessary & sufficient" (in fact, he added a seventh before he was done). (*Math 1*, 392–9). "October 1666. To resolve Problems by Motion these following Propositions are sufficient" (*Math 1*, 400–48).

time, so that *op* and *oq* represent the infinitesimal lines described in a moment. Interestingly, in composing what was meant as a general method, Newton returned to Huddeian multipliers. First, he called for multiplication by exponents. Then he added that when the equation is ordered first by powers of *x* and then by powers of *y*, any arithmetic progression, 3, 2, 1, 0, −1, −2, will do. "Or more Generally y^e Equation may bee multiplyed by y^e termes of these progressions $(ap + 4bp)/x. (ap + 3bp)/x. (ap + 2bp)/x. (ap + bp)/x. ap/x. (ap − bp)/x. (ap − 2bp)/x$, &c. And $(aq + 2bq)/y. (aq + bq)/y. aq/y. (aq − bq)/y$ &c. (*a* and *b* signifying any two numbers whither rationall or irrationall)."[65] Important lessons learned early were not easily surrendered even in the face of demonstrations that revealed the peculiar relevance of the series of exponents.

Proposition 8 proceeded to the inverse. Given the same two bodies *A* and *B* with velocities *p* and *q* and an equation expressing the relation between one of the lines *x* and the ratio *q/p*, find the line *y*. "Could this ever [i.e., always] bee done all problems whatever might bee resolved", he stated. "But by y^e following rules it may bee very often done."[66] Newton proceeded then to present his method of quadratures and offered suggestions for simplifying equations into integrable forms. "But this eighth Proposition may bee ever thus resolved mechanichally," he added, and he introduced an expansion of $a/(b + cx)$ by continued division and of $(a^2 − x^2)^{1/2}$ by continued root extraction.[67] With such procedures, which implied generality even though they were confined to specific examples, he placed the binomial expansion, which had rested heretofore on nothing more substantial than interpolations in patterns of exponents, on a new and more solid foundation. By expansion into infinite series, those problems which cannot be resolved can be approximated to whatever accuracy one chooses.

Section 1 concludes with demonstrations of Propositions 7 and 8, the two crucial elements of the method. The demonstration of Proposition 7 repeats closely the steps of his demonstration of the previous November, differing only in the equation it employs. "Prop 8th is ye Converse of this 7th Prop.," he asserted succinctly, "& may bee therefore Analytically demonstrated by it."[68] Such was Newton's final concept of quadrature; the operation of squaring an area under a curve is the inverse of the operation of finding the ratios of velocities. By the latter method, he suggested, "may bee gathered a Catalogue of all those lines wch can bee squared."[69]

The introductory, theoretical section of the tract comprises roughly a third of it. The rest is given over to the solution of

---

[65] *Math 1*, 402. Of course the factors *p/x* and *q/y* were in the other series also.
[66] *Math 1*, 403.   [67] *Math 1*, 413.   [68] *Math 1*, 415.   [69] *Math 1*, 428.

problems. Newton found tangents to curves. He reworked his earlier investigation of curvature to express it in terms of the new method, and went on to establish points of maximum and minimum curvature and points of inflection, where the radius of curvature becomes infinite. As he had done before, he rectified certain curves, which are evolutes of others, by means of his solution of curvatures. With the equation of one curve given, he showed how to find other curves the areas of which have a given relation to that of the given curve. What appears as an arcane problem is a method to transform equations into integrable form. At the end, as I have mentioned, the paper deals with centers of gravity and the equilibrium of planes about given axes.

Along the way, Newton dropped a number of clues that provide insights into the nature of his mathematical genius. He had a highly developed sense of continuity which helped him to understand the relations of figures even without demonstrations. When he developed equations for the center curvature in terms of perpendicular coordinates, he remarked that with oblique coordinates the circle of curvature would transform into an ellipse.[70] He understood the generality of procedures applied to particular cases. His algorithm for differentiation was never demonstrated for a general case, for which he had no adequate symbolism; rather he illustrated its validity for individual curves and intuited the general statement from the generality of his procedures. So also with his general equations for the center of curvature. After finding the center of curvature for one curve, he used the result to induce equations valid for all curves.[71] Wallis's method of induction found in Newton a higher application. Always interested in general solutions, he was impatient of mere computation. Thus he started to compute the points of maximum and minimum curvature of a conchoid, $x^2y^2 - (c^2 - x^2)(b + x)^2 = 0$. "The computation is too tedious," he decided, as he gave up a task which no rational human being could contemplate quietly.[72]

Taken all in all, the tract of October 1666 on resolving problems by motion was a virtuoso performance that would have left the mathematicians of Europe breathless in admiration, envy, and awe. As it happened, only one other mathematician in Europe, Isaac Barrow, even knew that Newton existed, and it is unlikely that in 1666 Barrow had any inkling of his accomplishment. The fact that he was unknown does not alter the other fact that the young man not yet twenty-four, without benefit of formal instruction, had become the leading mathematician of Europe. And the only one who really mattered, Newton himself, understood his position

[70] *Math 1*, 424.   [71] *Math 1*, 421–3.   [72] *Math 1*, 426n.

clearly enough. He had studied the acknowledged masters. He knew the limits they could not surpass. He had outstripped them all, and by far. Ten years later, Newton wrote to John Collins that

> there is no curve line exprest by any aequation of three terms, though the unknown quantities affect one another in it, or ye indices of their dignities be surd quantities . . . but I can in less then half a quarter of an hower tell whether it may be squared or what are ye simplest figures it may be compared w^{th}, be those figures Conic sections or others. And then by a direct & short way (I dare say ye shortest ye nature of ye thing admits of for a general one) I can compare them. . . . This may seem a bold assertion because it's hard to say a figure may or may not be squared or compared wth another, but it's plain to me by ye fountain I draw it from . . . [73]

Though the fountain had acquired a few additional jets by 1676, it remained essentially the instrument of 1666.

The instrument of 1666 derived in turn from the insights of 1665. As I understand him, the year 1665 was crucial to Newton's self-awareness. Almost from his first dawning of consciousness, he had experienced his difference from others. Neither in Grantham nor in Cambridge had he been able to mingle successfully with his fellow students. The servants at Woolsthorpe had despised him. Always his insatiable lust to know had set him apart. Now, finally, he had objective proof that his quest for learning was not a delusion. In 1665, as he realized the full extent of his achievement in mathematics, Newton must have felt the burden of genius settle upon him, the terrible burden which he would have to carry in the isolation it imposed for more than sixty years. From this time on, there is little evidence of the futile efforts to ingratiate himself with his peers that appeared intermittently during his grammar-school and undergraduate days. Accepting as sufficient his one close relationship with his chamber-fellow Wickins, he abandoned himself, as he had always longed to do, to the imperious demands of Truth.

Though he turned aside from mathematics at the end of 1666, Newton was not done with the method he had created by any means. Significantly, he never tried to publish the tract of October 1666.[74] As he returned to it intermittently in the years ahead, he paid primary attention to improving the foundation of the method; his descriptions of his method at the time of the controversy with Leibniz indicate how far he moved in that respect during about forty years of periodic revision. What he wished to be known for

---

[73] Newton to Collins, 8 Nov. 1676; *Corres 2,* 179–80.

[74] Nor did he allow manuscript copies of it to circulate nearly as much as copies of later versions of the method did. Some copies did get abroad, apparently toward the end of his life and in connection with the calculus controversy; see Whiteside's discussion, *Math 1,* 400n.

was not what he wrote in 1666, though its inspiration derived directly from the early tract. He also extended the method to recalcitrant problems, such as affected equations, that he could not handle in 1666, and he tackled other areas of mathematics as well. Nevertheless, as he said, he never minded mathematics so intensely again. His great period of mathematical creativity had come to a close. For the most part, his future activities as a mathematician would draw upon the insights of 1665. Years later he told Whiston "that no old Men (excepting Dr. Wallis) love Mathematicks . . ."[75] True, he was not yet an old man. Other fascinating subjects clamored for the attention of the genius in which he was now confident, however.

[75] Whiston, *Memoirs,* pp. 315–16.

5

Anni mirabiles

LOOKING back from the beginning of 1666, one finds it difficult to believe that Newton touched anything but mathematics during the preceding eighteen months. Clearly mathematics did dominate his attention during the period, but it did not completely obliterate other interests. Sometime during this period he also found time to compose the "Quaestiones," in which he digested current natural philosophy as efficiently as he did mathematics. Other natural philosophers were as ignorant of his existence as mathematicians were. To those who knew of him, his fellow students in Trinity, he was an enigma. The first blossoms of his genius flowered in private, observed silently by his own eyes alone in the years 1664 to 1666, his *anni mirabiles*.

In addition to mathematics and natural philosophy, the university also made certain demands on his time and attention. He was scheduled to commence Bachelor of Arts in 1665, and regulations demanded that he devote the Lent term to the practice of standing *in quadragesima*. Pictured in our imagination, the scene has a surrealistic quality, medieval disputations juxtaposed with the birth pains of the calculus. An investigation of curvature was dated 20 February 1665, in the middle of the quadragesimal exercises, and in his various accounts of his mathematical development he assigned the binomial expansion to the winter between 1664 and 1665.[1] While Stukeley was a student in Cambridge over thirty years later, he heard that when Newton stood for his B.A. degree, "he was put to second posing, or lost his groats as they term it, which is look'd upon as disgraceful."[2] The story raises several problems. The senate had already passed the grace granting his degree before the exercises were held, and Newton signed for this degree with the other candidates.[3] If the story has any substance, it would have to apply to prior examinations in the college. Nevertheless, as Stukeley re-

---

[1] *Math 1*, 259–63. Cf. Newton's memorandum quoted in chapter 3, (*Add MS* 4000, f. 14^v) and various statements composed at the time of the calculus controversy (*Add MS* 3968.5, f. 21, and *Add MS* 3968.41, ff. 76, 85, 86^v).

[2] *Stukeley*, p. 53. "GROATS. To save his groats; to come off handsomely: at the universities, nine groats are deposited in the hands of an academic officer, by every person standing for a degree: which if the depositor obtains with honour, the groats are returned to him" (Francis Grose, *A Classical Dictionary of the Vulgar Tongue*, ed. Eric Partridge [London, 1963], p. 172). See Allan Ferguson, "A Note on a Passage in Stukeley's 'Memoirs of Sir Isaac Newton's Life'," *Philosophical Magazine*, 7th ser., *34* (1943), 71.

[3] Cambridge University Library, Subscriptiones II, p. 163.

marked, it does not seem strange since Newton was not much concerned with the standard curriculum. Once more the laxity of the university worked to his advantage. Newton commenced B.A. largely because the university no longer believed in its own curriculum with enough conviction to enforce it.

The one surviving letter to Newton from his mother arrived in Cambridge at this time. Because the sheet has been torn, a few words are missing.

> Isack
> received your leter and I perceive you
> letter from mee with your cloth but
> none to you your sisters present thai
> love to you with my motherly lov
> you and prayers to god for you I
> your loving mother
> hanah
> wollstrup may the 6. 1665[4]

Obviously Hannah Ayscough Newton Smith was a barely literate woman; there is no reason to think that there was ever an extensive and revealing correspondence between them. Years later Stukeley heard in Grantham that Newton had written a number of letters to a friend near Colsterworth when he was a student.[5] As far as we know, nothing of them survives.

In the summer of 1665, a disaster descended on many parts of England including Cambridge. It had "pleased Almighty God in his just severity," as Emmanuel College put it, "to visit this towne of Cambridge with the plague of pestilence."[6] Although Cambridge could not know it and did little in the following years to appease divine severity, the two-year visitation was the last time God would choose to chastise them in this manner. On 1 September, the city government canceled Sturbridge Fair and prohibited all public meetings. On 10 October, the senate of the university discontinued sermons at Great St. Mary's and exercises in the public schools.[7] In fact, the colleges had packed up and dispersed long before. Trinity recorded a conclusion on 7 August that "all Fellows & Scholars which now go into the Country upon occasion of the Pestilence shall be allowed ye usuall Rates for their Commons for ye space of

[4] *Corres 1*, 2.
[5] Stukeley to Conduitt, 29 Feb. 1728; *Keynes MS* 136. Among the notes and anecdotes he collected, Conduitt noted that Ralph Clark, an apothecary in Grantham, had seen several letters of Newton when he was a student at the university. (*Keynes MS* 130.6, Book 4). The note appears to derive from Stukeley's letter.
[6] E. S. Shuckburgh, *Emmanuel College* (London, 1904) p. 114.
[7] Charles Henry Cooper, *Annals of Cambridge*, 5 vols. (Cambridge, 1842–1908), *3*, 517.

ye month following."[8] The records of the steward make it clear that the college, though ahead of the university, was behind many of its residents who had fled already and therefore collected no allowance for the last month of the summer quarter. For eight months the university was nearly deserted. At Corpus Christi College, only one fellow, two scholars, and a few servants inhabited the entire structure. They took a "preservative powder" in their wine, burned charcoal, pitch, and brimstone in the gatehouse, and somehow managed to survive both the plague and their own precautions.[9] To control panic, the vice-chancellor of the university joined with the mayor of the town in issuing bulletins of mortality every fortnight, separating deaths due to the plague from others. Each issue carried the announcement, "All the Colledges (God be praised) are and have continued without any Infection of the Plague."[10] Perhaps the flight of their residents contributed as much as the mercy of God; the inhabitants of the town, who appear at the least equally worthy of mercy but had nowhere to go, fared less well. In the middle of March when no deaths had been reported for six weeks, the university invited its fellows and students to return. By June it was evident that the visitation was not concluded. A second exodus occurred, and the university was able to resume in earnest only in the spring of 1667.

Many of the students attempted to continue organized study by moving with their tutors to some neighboring village.[11] Since Newton was entirely independent in his studies and had had his independence confirmed with a recent B.A., he found no occasion to follow Benjamin Pulleyn. He returned instead to Woolsthorpe. He must have left before 7 August in 1665 because he did not receive the extra allowance granted on that date.[12] His accounts show that he returned on 20 March 1666. He received the standard extra commons in 1666 and hence probably left for home in June. His accounts show again that he returned in 1667 late in April.[13]

[8] Conclusion Book, 1646–1811, p. 97.   [9] Cooper, *Annals*, 3, 518.

[10] James Bass Mullinger, *The University of Cambridge*, 3 vols. (Cambridge, 1873–1911), 3, 620.

[11] For example, John Sharp, who was later an archbishop, removed to Sawston, a few miles south of Cambridge, where he boarded with John Covell of his own college and other tutors who stayed there with their pupils (J. E. B. Mayor, *Cambridge under Queen Anne* [Cambridge, 1870], p. 470).

[12] Steward's Book, 1665. Newton missed a payment of £1 1s 8d. He began to draw commons for fiscal year 1666, which began with the Michaelmas term in 1665, receiving full commons for the three terms when the college was not in session, in all £6 6s 8d. In fiscal year 1667 he received the standard allowance for the full Michaelmas term and five weeks of the Lent term, £2 16s 8d in all, although he himself did not return until considerably after the college had officially resumed.

[13] His accounts in the Fitzwilliam notebook list his receipt of £10 on 22 April 1667 followed by his expenses for the return to Cambridge.

Much has been made of the plague years in Newton's life. He mentioned them in his account of his mathematics. The story of the apple, set in the country, implies the stay in Woolsthorpe. In another much-quoted statement written in connection with the calculus controversy about fifty years later, Newton mentioned the plague years again.

In the beginning of the year 1665 I found the Method of approximating series & the Rule for reducing any dignity of any Binomial into such a series. The same year in May I found the method of Tangents of Gregory & Slusius, & in November had the direct method of fluxions & the next year in January had the Theory of Colours & in May following I had entrance into ye inverse method of fluxions. And the same year I began to think of gravity extending to ye orb of the Moon & (having found out how to estimate the force with w^{ch} [a] globe revolving within a sphere presses the surface of the sphere) from Keplers rule of the periodical times of the Planets being in sesquialterate proportion of their distances from the center of their Orbs, I deduced that the forces wch keep the Planets in their Orbs must [be] reciprocally as the squares of their distances from the centers about wch they revolve: & thereby compared the force requisite to keep the Moon in her Orb with the force of gravity at the surface of the earth, & found them answer pretty nearly. All this was in the two plague years of 1665–1666. For in those days I was in the prime of my age for invention & minded Mathematicks & Philosophy more then at any time since.[14]

From this statement, combined with the other statements about his mathematics and the story of the apple, has come the myth of an *annus mirabilis* associated with Woolsthorpe. From one point of view, the leisure of his forced vacation from academic requirements gave him time to reflect. From another point of view, his return to the maternal bosom provided a crucial psychological stimulus.[15] Either theory is impossible to prove or to disprove. We may be moderately skeptical of the second as we recall the less than total bliss of his year at home in 1660. It may be relevant as well that his last act before he returned to Cambridge was to pry an extra £10 from the tight fist of his mother. In any event, exclusive attention to the plague years and Woolsthorpe disregards the continuity of his development.[16] Intellectually, he departed from Cambridge

---

[14] *Add MS* 3968.41, f. 85.

[15] "If 1666 is the *annus mirabilis,* most of it was spent in the protective bosom of his mother while the plague raged without . . ." (Frank E. Manuel, *A Portrait of Isaac Newton* [Cambridge, Mass., 1968], p. 80).

[16] Cf. D. T. Whiteside, "Newton's Marvellous Year: 1666 and All That," *Notes and Records of the Royal Society, 21* (1966), 32–41. See also J. W. Herivel, "Newton at Cambridge; the Development of a Genius," *Times Educational Supplement,* 9 June 1961, p. 1194.

more than a year before the plague drove him away physically. He took important steps toward the calculus in the spring of 1665, before the plague struck, and he wrote two important papers during May in 1666 while he was back. Similarly, his development as a physicist flowed without break from the "Quaestiones quaedam Philosophicae." If we focus our attention on the record of his studies, the plague and Woolsthorpe fade in importance in comparison to the continuity of his growth. 1666 was no more *mirabilis* than 1665 and 1664. The miracle lay in the incredible program of study undertaken in private and prosecuted alone by a young man who thereby assimilated the achievement of a century and placed himself at the forefront of European mathematics and science.

About the beginning of 1666, Newton's nearly exclusive concentration on mathematics stopped as suddenly as it had begun a year and a half before.[17] As we have seen, he returned to it briefly in May to write two papers which carried earlier work forward, and again in October to compose the important tract on fluxions. During the following two years he apparently did not touch mathematics at all. Indeed, as he said, he never minded mathematics again with the same exclusive concentration.

Newton was not a man of halfhearted pursuits. When he thought on something, he thought on it continually. By thinking continually on mathematics for a year and a half, he arrived at a new method that allowed him to solve the initial problems, set for him by earlier mathematicians, with which he began. Now other interests represented by the "Quaestiones" could claim his attention. Once they had claimed it, he thought on them as hard as he had on mathematics.

One of these was the science of mechanics. The essay "On violent Motion" in the "Quaestiones" had introduced him to mechanics. There he espoused the doctrine that a force internal to bodies keeps them in motion. In Descartes's *Principles* and in Galileo's *Dialogue,* he confronted the radically different conception of motion that we call today, using language which Newton himself later made common, the principle of inertia. In Descartes he also found two problems posed and imperfectly answered, the mechanics of impact and of circular motion. They became the focus of his investigation. It is incorrect to imply that Newton turned his attention seriously to mechanics only after his initial absorption in mathematics was broken. To some extent, the two overlapped. He dated his first important investigation in mechanics 20 January 1665, about

---

[17] In this paragraph I am relying on the dating of papers in Whiteside's edition of the *Mathematical Papers, 1* and *2.*

the time he began to stand *in quadragesima*. It does appear that his concern with mechanics was intermittent at first and that it continued into 1666 more vigorously than his work in mathematics.

The investigation of January 1665, recorded in the *Waste Book*, carried the title, "Of Reflections," by which Newton meant impact.[18] A tone of confidence not present in the "Quaestiones" infused the passage. No longer the questioning student, he began to propound alternative solutions. To be sure, he based his treatment of impact squarely on Descartes's conception of motion.

> Ax:100 Every thing doth naturally persevere in y^t state in wch it is unlesse it bee interrupted by some externall cause, hence . . . [a] body once moved will always keepe ye same celerity, quantity & determination of its motion.[19]

Of Descartes's law of impact, which completed his discussion of motion, however, Newton did not say a word. He did not even bother to refute it. Instead he launched directly into his own analysis of impact. Descartes's law of impact had led to conclusions which contradicted common experience. Newton saw that only a different approach could correct its faults. Like other early students of impact, Descartes had been mesmerized by the force with which a body strikes, what he called the "force of a body's motion."[20] The concept appeared obvious. Everyone had experienced the force of a moving body; everyone knew that it increased both with the size of the body and with its speed. Behind the obvious, however, lay unexpected difficulties on which attempts to analyze impact had repeatedly suffered shipwreck. Only Christiaan Huygens had understood like Newton the necessity of a new approach; he had avoided the difficulties by converting impact into a problem in kinematics. In 1665, his work remained unpublished. Newton's solution differed as radically from Huygens's as from Descartes's. Instead of eliminating the treacherous concept of force, he set out to modify it into a usable form. As far as its cause was concerned, he did not question Descartes. "Force is ye pressure or crouding of one

[18] *Add MS 4004 (Waste Book)*, ff. 10–15, 38^v. Except for canceled passages, the entire investigation has been published in *Herivel*, pp. 132–82. See J. W. Herivel, "Sur les premières recherches de Newton en dynamique," *Revue d'histoire des sciences, 15* (1962), 105–40, and "Newton on Rotating Bodies," *Isis, 53* (1962), 212–18. D. T. Whiteside, "Before the *Principia*: The Maturing of Newton's Thoughts on Dynamical Astronomy, 1664–84," *Journal for the History of Astronomy, 1* (1970), 5–19; Ole Knudsen, "Newton's Earliest Formulation of the Laws of Motion," *Actes du XIe congrès internationales d'histoire des sciences, 3,* 344–8; Philip E. B. Jourdain, "The Principles of Mechanics with Newton from 1666 to 1679, from 1679 to 1687," *The Monist, 24* (1914), 188–224, 515–64.

[19] *Herivel*, p. 153.

[20] Cf. Richard S. Westfall, *Force in Newton's Physics* (London, 1971), especially chaps. 2, 3, and 5, and Alan Gabbey, "Force and Inertia in Seventeenth-Century Dynamics," *Studies in History and Philosophy of Science, 2* (1971), 1–67.

body upon another."[21] This was a metaphysical proposition, however. How was one to shape it to the uses of a quantitative dynamics? Newton set out toward that goal by examining the implications of the very conception of motion proposed by Descartes. If a body perseveres in its state unless some external cause acts upon it, there must be a rigorous correlation between the external cause and the change it produces. "Soe much force as is required to destroy any quantity of motion in a body soe much is required to generate it; & soe much as is required to generate it soe much is alsoe required to destroy it."[22] "Tis knowne by ye light of nature," he asserted later as he revised his initial effort, "y^t equall forces shall effect an equall change in equall bodys. . . . For in loosing or to [*sic*] getting ye same quanty of motion a body suffers ye same quantity of mutaion in its state, & in ye same body equall forces will effect a equall change."[23] Here was a new definition of force in which a body was treated as the passive subject of external forces impressed upon it instead of the active vehicle of force impinging on others. More than twenty years of patient if intermittent thought would in the end elicit his whole dynamics from this initial insight.

All of the possibilities inherent in the insight did not appear immediately to the young man who was being introduced to the science of mechanics and was grappling with the new conception of motion for the first time. Earlier, in the "Quaestiones," he had accepted the idea of a force internal to bodies which keeps them in motion. It is clear to us that the earlier notion was incompatible with the principle of inertia and his new conception of force. It was not at first equally clear to Newton. Instead of rejecting the idea of internal force outright, he attempted to reconcile it with the new concept he was developing.

> The force wch ye body (*a*) hath to preserve it selfe in its state shall bee equall to the force w^{ch} [pu]t it into y^t state; not greater for there can be nothing in ye effect wch was not in ye cause nor lesse for since ye cause only looseth its force onely by communicating it to its effect there is no reason why it should not be in ye effect w^n tis lost in ye cause.[24]

The more he worked on impact, however, the more the incompatibility of internal force and the principle of inertia revealed itself. In reworking his initial analysis, he insisted more and more on changes of motion rather than motion itself. "A Body is saide to have more or lesse motion as it is moved wth more or lesse force, y^t is as there is more or lesse force required to generate or destroy its whole motion."[25] In such statements, the force with which a body

---

[21] *Herivel*, p. 138.   [22] *Herivel*, p. 141.

[23] *Herivel*, pp. 157–8.

[24] *Add MS* 4004, f. 12^v. Newton canceled the passage.   [25] *Herivel*, p. 157.

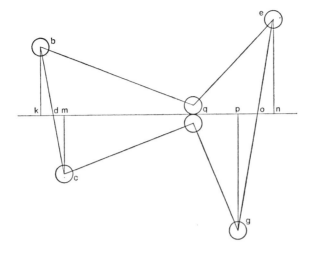

Figure 5.1. Oblique impact of two perfectly elastic bodies. (From Westfall, *Force in Newton's Physics.*)

moves approached our concept of momentum. Its quantity was relative to the inertial frame of reference. Force was measured by the generation (or destruction) of a given quantity of motion in any inertial frame of reference.

The climax of the investigation of impact came in Newton's recognition that any two bodies isolated from external influences constitute a single system whose common center of gravity moves inertially whether or not they impinge on each other. Huygens's analysis of impact had arrived at the same conclusion by exploring the implications of the principle of inertia.[26] Newton arrived there by a dynamic analysis which thereby revealed the implications of the principle of inertia to him more fully. Examining what he called "ye mutuall force in reflected bodys," he concluded that each one acts equally on the other and produces an equal change of motion in it.[27] To those who had tried to understand impact through the force of a body's motion, this conclusion had appeared patently false except for the special case of equal bodies with equal motions. Newton appears to have reached it by realizing that every impact can be viewed from the special frame of reference of the common center of gravity of the two bodies. First he demonstrated that two bodies in uniform motion have equal motions in relation to their common center of gravity and that their common center of gravity is either at rest or in uniform motion, whether the two bodies are in the same plane or in different planes. Now let the two strike each other and rebound (Figure 5.1). From the earlier propositions it

[26] Cf. Westfall, *Force*, chap. 4.   [27] *Herivel*, p. 159.

follows that *b* and *c* are in equal motion toward the line *kp,* the line of motion of their common center of gravity. When they meet at *q,* "so much as (*c*) presseth (*b*) from ye line *kp;* so much (*b*) presseth (*c*) from it . . ." Therefore, when the two bodies are in *e* and *g* some time after the impact, they will have equal motions away from their common center of gravity, which remains in uniform motion on the line *kp.*[28] Newton never forgot this conclusion. It contained the first adumbration of his third law of motion, and the conclusion itself appeared as Corollaries III and IV to the laws, where it was treated as a necessary consequence of Law 1 (the principle of inertia) and Law 3.

In 1665, the *Principia* was more than twenty years away. The path toward it proved not to be direct. Impressive as were the strides Newton had made, complexities remained that tended to reinforce his original idea of a force internal to bodies which keeps them in motion. These complexities were associated with the mechanics of circular motion, the second problem posed by Descartes. Following both Descartes and common experience, Newton agreed that a body in circular motion strives constantly to recede from the center, like a stone pulling on its string as it is whirled about.[29] The endeavor to recede appeared to be a tendency internal to a moving body, the manifestation in circular motion of the internal force which keeps a body in motion. Unlike the internal force of rectilinear motion, it could not be made to disappear by shifting inertial frames of reference. Every indication from Newton's papers suggests that the problems of circular motion, together with considerations external to mechanics, soon led him to reject the principle of inertia. Twenty years later, the same problems of circular motion viewed from a new perspective would be decisive in his final conversion to inertia.

In seizing on circular motion, Newton again paralleled the path taken earlier by Christiaan Huygens, who had also been dissatisfied with Descartes's treatment of it. Although their approaches differed, the discipline that mathematics exerted on both was evident; by instinct, as it were, they sought to reduce the mechanics of circular motion to quantitative terms. In Newton's case, the analysis grew out of his treatment of impact. When he first considered it, he decided that the "whole force" by which a body endeavors to recede from the center in half a revolution is double the force able to generate or destroy its motion.[30] That is, half a revolution is like a perfectly elastic rebound from an immovable obstacle. In both

---

[28] *Herivel,* pp. 168–9.
[29] Newton himself used this illustration; *Herivel,* p. 147. See J. W. Herivel, "Newton's Discovery of the Law of Centrifugal Force," *Isis, 51* (1960), 546–53.
[30] *Herivel,* p. 147.

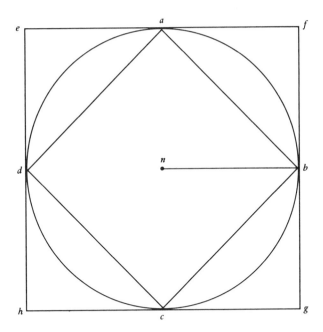

Figure 5.2. The force of a body moving in a circle derived from impact.

cases the original motion is exactly reversed, which requires a force twice that necessary to generate an equal motion. Later, after further consideration revealed the differences in the two cases, he added the words "more y^n" before the phrase "double to the force . . ." Why did he speak of "whole force"? Undoubtedly because he saw a difference between the force from the center exerted continuously through an entire revolution and the force of an impact which acts in an instant. It did not occur to him to distinguish the two by employing different words for them.

To correct his mistaken measure of the force from the center, Newton employed the model of impact with more sophistication. He imagined that a square circumscribes the circular path and that a body follows a square path inside the circle rebounding at the four points where the circle touches the outer square (Figure 5.2). From the geometry of the square he was able to compare the force of one impact, in which the component of the body's motion perpendicular to the side it strikes is reversed, to the force of the body's motion. "*2fa:ab::ab:fa::* force or pression of *b* upon *fg* at its reflecting: force of *b*'s motion." In one complete circuit of four reflections, the total force is to the force of the body's motion as *4ab:fa*, that is, as the length of the path to the radius of the circle. Newton

was mathematician enough to see without pausing to demonstrate
that the same relation of path to radius would hold if the number of
sides and impacts were doubled, then doubled again and again and
again.

> And soe if body were reflected by the sides of an equilaterall circum-
> scribed polygon of an infinite number of sides (i.e. by y^e circle it
> selfe) y^e force of all y^e reflections are to y^e force of y^e bodys motion
> as all those sides (i.e. y^e perimeter) to y^e radius.[31]

For one complete revolution the total force F is to the body's motion mv as $2\pi r/r$. Or $F = 2\pi mv$. To see the significance of the result,
convert the total force of the body in one revolution to the "force
by wch it endeavours from y^e center" at each instant by dividing
each side of the equation by the time of one revolution, $2\pi r/v$. The
division yields $f = mv^2/r$, the formula we still use in the mechanics
of circular motion.[32]

The formula for a body's endeavor to recede from the center, for
which Huygens coined the name "centrifugal force," gave Newton
the means to attack a problem that he found in Galileo's *Dialogue*. It
was an effort to answer one argument against the Copernican system by showing that the earth's rotation does not fling bodies into
the air because the force of gravity, measured by the acceleration of
falling bodies, is greater than the centrifugal force arising from the
rotation. Newton's solution, chaotically recorded on a piece of
parchment, the front side of which had been used by his mother for
a lease, was closely associated with the investigations of mechanics
in the *Waste Book*. In his first attempt, he used his first mistaken
idea of the total force in half a revolution; and when he returned for
a second attempt, he employed his later, correct formula. All he
needed in addition to his new formula were the size of the earth and
the acceleration of gravity. For both, he used the figures he found
with Galileo's solution of the problem in the Salusbury translation
of the *Dialogue*, which appeared in 1665. He arrived at the conclusion "y^t ye force of ye Earth from its center is to ye force of Gravity
as one to 144 or there abouts." But why accept Galileo's figure for
the acceleration of gravity? He suddenly realized that his measure of
centrifugal force opened a further possibility; he could use it to
measure g indirectly via a conical pendulum. The measurement,
with a conical pendulum 81 inches long inclined at an angle of 45
degrees, revealed that a body starting from rest falls 200 inches in a
second, a figure very close to the one we accept but roughly twice
as large as the one he had found in Galileo's *Dialogue*. Hence he

---

[31] *Herivel*, pp. 129–30.

[32] The use of mass here is anachronistic. Newton did not clearly define mass until early in
1685. I have been willing to insert it since he did realize the significance of the body's size
or quantity.

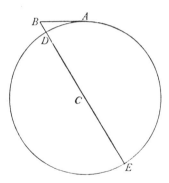

Figure 5.3. The force of a body moving in a circle derived from its deviation from the tangent.

returned to his calculation and doubled the ratio of gravity to centrifugal force.[33]

Somewhat later, in a paper which appears to date from the years immediately following his undergraduate career, Newton returned to the same problems. On this occasion, he calculated centrifugal *conatus* more elegantly by utilizing the geometry of the circle instead of impact. When a body moves in uniform circular motion, time is proportional to length of arc. Since a body will move in a straight line if it is not constrained to move in a circle, Newton set the centrifugal tendency for the motion *AD* equal to *BD,* the distance the tangent diverges from the circle at *D* (Figure 5.3). When the arc *AD* is "very small," Newton could then apply the known ratio of *BD* to *AD* to calculate the instantaneous force, and with the force he could calculate the distance it would impel a body in a straight line, starting from rest, during the time of one revolution. He employed Galileo's conclusion that distances traversed in uniformly accelerated motion from rest vary as the squares of the times, implicitly interpreting Galileo's kinematics in dynamic terms. His answer, that in the time of one revolution the centrifugal force would move a body through a distance equal to $2\pi^2 r$, is mathematically equivalent to the earlier formula derived from impact.[34] Again he compared centrifugal force at the earth's surface to gravity; and since he did not round off his more accurate measurement of *g,* he arrived this time at a slightly higher ratio, 1:350.

[33] *Add MS* 3958.2, f. 45. The manuscript is published and analyzed both in *Herivel,* pp. 183–91, and in *Corres 3,* 46–54.

[34] In the *Waste Book,* he concluded that in the time of uniform motion through one radian (or a distance along the circle equal to its radius), the centrifugal force would generate a motion equal to the body's motion. Hence the distance traversed by an equal body starting from rest would be ½r. To compare the results of the present paper, divide the distance, $2\pi^2 r$, by the squared ratio of times, $4\pi^{2.}$

So far he had come before. He was ready now to take a further step. He compared the "endeavour of the Moon to recede from the centre of the Earth" with the force of gravity at the surface of the earth. He found that gravity is somewhat more than 4,000 times as great. He also substituted Kepler's third law (that the cubes of the mean radii of the planets vary as the squares of their periods) into his formula for centrifugal force: "the endeavours of receding from the Sun [he discovered] will be reciprocally as the squares of the distances from the Sun."[35] Here was the inverse-square relation resting squarely on Kepler's third law and the mechanics of circular motion. To catch the full significance of the statement one must reflect on the earlier ratio of gravity to the moon's tendency to recede from the earth. He had found a ratio of about 4,000:1. Since he was using 60 earth radii as the moon's distance, the exact ratio according to the inverse square relation should have been 3,600:1. It is difficult to believe that this paper was not what Newton referred to when he said that he found the comparison of the force holding the moon in its orbit to gravity to "answer pretty nearly."[36]

Newton attempted to summarize his early work in mechanics in a paper from about the same time which he called "The lawes of Motion."[37] It is not an unfamiliar title to students of Newton or to students of physics. The laws of motion as Newton understood them in the 1660s differed sharply, however, from the laws of motion he pronounced in the *Principia*. In the early paper he was concerned primarily with impact. The paper attempted to arrive at a general solution of the impact of two bodies with any arbitrary linear and rotational motions. The inclusion of the latter marked an advance over the treatment of impact in the *Waste Book*. In "The lawes of Motion," Newton proposed a definition of quantity of circular motion—the product of a body's bulk and the velocity of a point on what he called its "Equator of circulation."[38] He asserted

---

[35] *Add MS* 3958.5, f. 87. The paper is also printed and analyzed in *Herivel*, pp. 195–7, in *Corres 1*, 297–303, and in A. R. Hall, "Newton on the Calculation of Central Forces," *Annals of Science, 13* (1957), 62–71. See also Léon Rosenfeld, "Newton and the Law of Gravitation," *Archive for History of Exact Sciences, 2* (1965), 365–86; and Ole Knudsen, "A Note on Newton's Concept of Force," *Centaurus, 9* (1963–4), 266–71.

[36] In 1694, about twenty years before he wrote this famous line, Newton showed David Gregory a paper written before 1669 that contained "all the foundations of his philosophy . . . : namely the gravity of the Moon to the Earth, and of the planets to the Sun: and in fact all this even then is subjected to calculation" (*Corres 1*, 301). It is also difficult to believe that the paper in question here was not the one Gregory saw.

[37] *Add MS* 3958.5, ff. 81–3ᵛ (the first version, called "The laws of Reflection"), 85–6ᵛ. The final version is printed in *Herivel*, pp. 208–18.

[38] Herivel has shown that Newton's "radius of circulation" (which describes the equator of circulation) is equivalent to what mechanics now calls the radius of gyration (k). He has also shown that Newton's quantity of circular motion ($mk\omega$) differs from what we call angular momentum ($mk^2\omega$) by a factor k.

the principle of the conservation of angular momentum for the first time in the history of mechanics: "Every body keepes the same reall quantity of circular motion and velocity so long as tis not opposed by other bodys."[39] Using what he called the "smallnesse of resistance" to change of motion, he developed a formula which parcelled out the total change of velocity of the two points that come into contact among four factors, the progressive motions of the two bodies and their rotational motions.[40]

For all its sophistication in the treatment of impact, "The lawes of Motion" contained internal problems which help to illuminate the limits of Newton's early achievement in mechanics. On the one hand, the paper employed the principle of inertia. On the other hand, it began with an assertion of absolute space which conflicted with the relativity of motion, a corollary of inertia, but which conformed to the earlier conception of motion he had held.[41] Probably factors external to mechanics influenced Newton most strongly, but within mechanics the questions associated with rotation were the leading advocates of absolute space, as his use of the adjective "reall" with "quantity of circular motion" implies. He noted that a perfectly balanced rotating body will maintain the absolute inclination of its axis, but if it is not perfectly balanced the "endeavours from the axis" will cause it to wobble with a motion that is not merely relative but absolute.[42] The most basic issue in his mechanics had not been settled after all. Newton had not adopted the new conception of motion with finality. Within three years his hesitation would end for the time being with a passionate rejection of relativism and inertia and a return to the concept of an internal force generating absolute motion in absolute space.

The fact that a paper entitled "The lawes of Motion" could have as its goal a general solution of impact indicates how far from completed Newton's dynamics was in the 1660s. His investigation of impact had led to a concept of force fully suited to impact alone. By good fortune, the concept of force worked equally well with either conception of motion, as long as rectilinear motions alone were involved. When he applied it to problems other than impact, such as circular motion, he immediately confronted ambiguities that he had only begun to learn to control. Newton had caught sight of the dynamics that would crown and complete Galileo's kinematics; he had scarcely begun to examine its depths.

Above all, he had only begun to peer into the mysteries of circular motion. The man who would coin the phrase "centripetal force" had found the quantitative formula for its illusory mirror image,

[39] *Herivel*, p. 211.   [40] *Herivel*, pp. 212–13.
[41] *Herivel*, p. 208.   [42] *Herivel*, p. 211.

the endeavor to recede from the center, or centrifugal force. Since he believed that bodies continue to move in a straight line unless something diverts them, it followed that any body moving in a circle must be constrained to do so by something else. It followed also that Newton in the 1660s, like Descartes before him, treated circular motion as a state of equilibrium between opposing forces. Thus, in his original calculation of their quantity, the ball in circular motion presses against the confining cylindrical shell, and the shell confines the ball to its circular path by pressing back. The circular motion manifests the equilibrium of the two forces. As long as he conceived of circular motion in such terms, the internal endeavor of a body to recede from the center would appear to him as an argument against the principle of inertia.

What then is one to make of the story of the apple? It is too well attested to be thrown out of court. In Conduitt's version, one of four independent ones, it ran as follows:

> In the year 1666 he retired again from Cambridge . . . to his mother in Lincolnshire & whilst he was musing in a garden it came into his thought that the power of gravity (wch brought an apple from the tree to the ground) was not limited to a certain distance from the earth but that this power must extend much farther than was usually thought. Why not as high as the moon said he to himself & if so that must influence her motion & perhaps retain her in her orbit, whereupon he fell a calculating what would be the effect of that supposition but being absent from books & taking the common estimate in use among Geographers & our seamen before Norwood had measured the earth, that 60 English miles were contained in one degree of latitude on the surface of the Earth his computation did not agree with his theory & inclined him then to entertain a notion that together with the force of gravity there might be a mixture of that force wch the moon would have if it was carried along in a vortex . . . [43]

[43] *Keynes MS* 130.4, pp. 10–12. The account in the finished memorandum on Newton that he sent to Fontenelle differed slightly. Conduitt's account is in many respects almost identical to that in the DeMoivre memorandum on Newton (Joseph Halle Schaffner Collection, University of Chicago Library, MS 1075.7) except for the apple, which does not appear in DeMoivre's version. However, DeMoivre does place him in a garden. Perhaps Conduitt heard that detail from his wife, for it was his wife whom Voltaire cited as his authority for the apple (*Lettres philosophiques,* Lettre XV; *Oeuvres de Voltaire,* ed. M. Beuchot, 72 vols. [Paris, 1829–40], *37,* 198–9). William Stukeley also wrote that Newton told him the story on 15 April 1726 (*Stukeley,* p. 20). Robert Greene repeated substantially the same story in 1727 on the authority of Martin Folkes, vice-president of the Royal Society while Newton was president, and later president himself (*The Principles of the Philosophy of the Expansive and Contractive Forces* [Cambridge, 1727], p. 972). It is surely significant that all of these accounts surfaced at about the same time and appear to date from the final year of Newton's life. In my opinion, the date does not seriously compromise acceptance of the incident itself, a concrete event that would have readily been recalled. On the other hand, Newton's age does not generate much confidence in his

Small wonder that such an anecdote, redolent of the Judaeo-Christian association of the apple with knowledge, continues to be repeated. Together with the myth of the *annus mirabilis* and with Newton's memorandum that said he found the calculation to answer pretty nearly, it has contributed to the notion that universal gravitation appeared to Newton in a flash of insight in 1666 and that he carried the *Principia* about with him essentially complete for twenty years until Halley pried it loose and gave it to the world. Put in this form, the story does not survive comparison with the record of his early work in mechanics. The story vulgarizes universal gravitation by treating it as a bright idea. A bright idea cannot shape a scientific tradition. Lagrange did not call Newton the most fortunate man in history because he had a flash of insight. Universal gravitation did not yield to Newton at his first effort. He hesitated and floundered, baffled for the moment by overwhelming complexities, which were great enough in mechanics alone and were multiplied sevenfold by the total context. What after all was in the paper that revealed the inverse-square relation? Certainly not the idea of universal gravitation. The paper spoke only of tendencies to recede, and to Newton the mechanical philosopher an attraction at a distance was inadmissible in any case. It was no accident that he placed impact at the center of the laws of motion. Revealingly, Conduitt (or DeMoivre) brought in the vortex.[44] Nevertheless, Newton must have had something in mind when he compared the moon's centrifugal force with gravity, and there is every reason to believe that the fall of an apple gave rise to it. Though he did not name the force explicitly, something had to press back on the moon if it remained in orbit. Something had to press back on the planets. Moreover, Newton remembered both the occasion and the calculation, so that fifty and more years later they seemed to constitute an important event in his development. Some idea floated at the border of his consciousness, not yet fully formulated, not perfectly focused, but solid enough not to disappear. He was a young man. He had time to think on it as matters of great moment require.

recollection of the conclusions he drew at that time, especially when his own papers tell a rather different story. See Jean Pelseneer, "La Pomme de Newton," *Ciel et terre, 53* (1937), 190–3; and Douglas McKie and G. R. de Beer, "Newton's Apple," *Notes and Records of the Royal Society, 9* (1951–2), 46–54, 333–5.

[44] So did William Whiston in what may or may not be an independent account (*Memoirs*, p. 37). Henry Pemberton, whose independence from DeMoivre and Whiston again can not be demonstrated, did not mention vortices specifically but did say that Newton long suspected the operation of some other cause beyond gravity (*View of Newton's Philosophy,* [London, 1728] preface). Newton's notes on the endpapers of his copy of Vincent Wing, *Astronomia Britannica* (London, 1669), now in the Trinity College Library, support the story that he believed some action of the vortex on the moon upset the perfect inverse-square relation in its case.

Motion and mechanics were not the only topics in natural philosophy that commanded Newton's interest. As important in his own eyes were what he later called the "celebrated Phaenomena of Colours."[45] The phenomena of colors had become a celebrated topic in optics for at least two reasons. What we call chromatic aberration appeared in every telescopic observation, coloring the images and confusing their focus. In contrast, colors sharply focused the different stances of the Aristotelian and the mechanical philosophies of nature. It is scarcely surprising that colors were among the "Quaestiones quaedam Philosophicae" compiled by the young mechanical philosopher in Cambridge.[46] He had found the issue in Descartes, in Boyle's *Experiments and Considerations Touching Colors* (1664), and in Hooke's *Micrographia* (1665). Dissatisfied with their explanations of colors, as his notes show, he turned his hand to his own.

Like the chronology of his early mechanics, the chronology of his optical research is confused. In the memorandum on the plague years, he stated bluntly that he "had the Theory of Colours" in January 1666. He had said substantially the same thing not long after the discovery when he sent his first paper to the Royal Society early in 1672: "in the beginning of the Year 1666 (at which time I applyed my self to the grinding of Optick glasses of other figures than *Spherical*,) I procured me a Triangular glass-Prisme, to try therewith the celebrated *Phaenomena of Colours*."[47] He proceeded then to describe his theory before adding that the plague intervened, driving him from Cambridge, and it was more than two

[45] Newton to Oldenburg, 6 Feb. 1672; *Corres 1*, 92.

[46] Johannes Lohne, "Newton's 'Proof' of the Sine Law and His Mathematical Principles of Colours," *Archive for History of Exact Sciences*, 1, (1961), 389–405, has suggested that we trace Newton's interest in colors to his grammar-school days and the recipes for mixing colors that he copied from Bate's book. Frank Manuel, in contrast, sees all of Newton's optical experimentation as the manifestation of his unconscious desire for intimate visual exchange with his mother (*Portrait*, pp. 78–9). My presentation is based on the premise that his work in optics is best understood in terms of the current context of natural philosophy.

[47] Newton to Oldenburg, 6 Feb. 1672; *Corres 1*, 92. Descartes had made the prism the primary instrument of experimentation with colors (*Les Météores*, Discours 8; *Oeuvres de Descartes*, ed. Charles Adam and Paul Tannery, 12 vols. [Paris, 1897–1910], 6, 330). In his *Experiments and Considerations Touching Colours* (1664), which was a major source of information to Newton, Boyle said the prism was the "usefullest instrument" men had ever employed for the investigation of colors (*The Works of the Honourable Robert Boyle*, ed. Thomas Birch, new ed., 6 vols. [London, 1772], 1, 738). Newton's notes on the grinding of lenses are found in his mathematical notebook, *Add MS 4000*, ff. 26–33ᵛ. They have no date, but they follow his annotations on Wallis's *Arithmetica infinitorum*, which Newton's own memorandum written in 1699 placed in the winter between 1664 and 1665. See Lloyd W. Taylor, "Newton's Prism in the British Museum," *Nature, 138* (1936), 585; Rudolf Laemmel, "Die Prismen von 1665," in *Miszellen um Isaac Newton* (Zürich, 1953), pp. 2–4; I. Bernard Cohen, "I prismi del Newton e i prismi dell'Algarotti," *Atti della Fondazione Giorgio Ronchi, 12* (1957), 213–23; and Vasco Ronchi, "I 'prismi del Newton,' del Museo Civico di Treviso," *ibid., 12* (1957), 224–40.

years before he returned to the question.[48] There are at least three problems in this narrative: Newton's location, the height of the sun, and the procurement of the prism. If the beginning of 1666 meant January which the later memorandum explicitly asserted, Newton was not then in Cambridge as he implied. Moreover, in order to project a spectrum more or less parallel to the floor, as he had to do if he was to view it on a wall, he needed a sun about 40 degrees above the horizon, whereas the sun does not climb above 20 degrees in January either in Cambridge or in Woolsthorpe. Finally, in January he would have had to get the prism in Grantham, and prisms seem an unlikely item of commerce for a small market town. Newton's conversation with Conduitt on 31 August 1726 indicated that they were hard enough to come by in Cambridge.

> In August 1665 Sr I. who was then not 24 bought at Sturbridge fair a prism to try some experiments upon Descartes's book of colours & when he came home he made a hole in his shutter & darkened the room & put his prism between that & the wall found instead of a circle the light made ⊂⊃ with strait sides & circular ends &c. wch convinced him immediately that Descartes was wrong & he then found out his own Hypothesis of colours tho^u he could not demonstrate it for want of another prism for wch he staid till next Sturbridge fair & then proved what he had before found out.[49]

Conduitt's chronology does agree with the height of the sun. Unfortunately, Newton had left Cambridge before the truncated Sturbridge Fair of 1665 took place, and there was none at all in 1666. Conduitt seems to have realized as much, for he altered his original 1665 to 1663. Perhaps we should not place much weight on the recollections of a very old man about events that occurred sixty years earlier. The indicated interruption does correspond to the paper of 1672, however, although the length of the interruption differs in the two accounts.[50] No resolution of the discrepancies is entirely satisfactory. Newton's recollection of Sturbridge Fair may have been mistaken; there was also an annual Midsummer Fair,

48 Newton to Oldenburg, 6 Feb. 1672; *Corres 1*, 95–6.
49 *Keynes MS* 130.10, ff. 2^v–3.
50 Newton's accounts show that he purchased three prisms sometime after 12 Feb. 1668 (Fitzwilliam notebook). The chronology of his discovery is further muddied by a sentence in a letter to Oldenburg, 10 Jan. 1676, replying to an intimation by Gascoines that the directions for his basic prismatic projection that he sent to Linus were different from those he printed. Newton denied they differed in any way from those he had followed "above these seven years" (*Corres 1*, 410). This seems to place the crucial experimentation in 1668, although the word "above" leaves it ambiguous. His initial lecture said the projection of the prism was the experiment that led him to develop his theory of colors. If his later assertions about 1666 have any meaning, the statement in 1670 places the definitive investigation, which gave the projection central importance, at that time (*Add MS 4002*, p. 3).

which managed to escape both plagues. If he purchased a prism there in 1665, he could have taken it home with him and performed there basic, though perhaps crude, experiments connected with his initial insight. His more elaborate experiments to confirm it demanded an assistant.[51] Perhaps he could have dragooned a servant at Woolsthorpe, but one thinks instead of Cambridge and Wickins. The extra prism probably demanded Cambridge, where the Midsummer Fair of 1666 might have furnished it, and the sun was then high enough in the heavens. No more than two or three years hinges on the chronology in any case. He had the theory fully elaborated before January 1670 when he lectured on it. Three years are not sufficient cause to drown the excitement of discovery in a sink of erudition.

Suffice it then to say that a passage on colors, in which Newton sharply questioned current theories, was among the earliest entries in the "Quaestiones."[52] Later, probably in 1665, he returned to colors again, utilizing empty pages at the end of his original set of headings.[53] There is a good likelihood that the theory of colors in Robert Hooke's *Micrographia* (1665) stimulated him. His immediate negative reaction to Hooke's account inaugurated forty years of antipathy between two incompatible men. As in mechanics, he was not content any longer simply to question. An alternative theory sprang to mind. Hooke proposed that "Blue is an impression on the Retina of an oblique and confus'd pulse of light, whose weakest

---

[51] Few of us measure time by the movement of shadows as Newton did. He would not have failed to realize what his first attempt would have taught him in any case – that his spectrum moved rapidly across the wall. On a wall twenty-two feet away, it moved more than an inch every minute. Without an assistant and some practice, no measurements could have been taken and no experiments involving more than one prism set up. One needs to recall as well that on any one day he had less than two hours when the sun was properly located; and in Cambridge the sun does not shine every day. Lord Adrian has combined the measurements of rooms in Trinity with the height of the sun to support the story that Newton lived at this time in a room on the north side of the great court (and hence with a window facing the noonday sun) between the master's lodge and the chapel. He suggests that the experiments were performed there in early April, when the sun was at the right elevation. The room in question is wide enough to allow the twenty-two feet between prism and spectrum that Newton mentions (The Lord Adrian, "Newton's Rooms in Trinity," *Notes and Records of the Royal Society, 18* (1963), 17–24). As it happens, the dimensions of his room at Woolsthorpe, which measures exactly twenty-two feet from the shutter of the window facing south to the wall opposite (if we ignore the partition that now walls off a corner but does not appear to have been there in the seventeenth century), fit with Newton's description of the experiment and make it possible that he performed it there. (I owe this information to Mr. K. A. Baird of Colsterworth, who recently measured the room, and to Professor Gale Christianson, who communicated Baird's letter to me.)

[52] *Add MS* 3996, f. 105^v.

[53] *Add MS* 3996, ff. 122–24^v, 133–5. See A. R. Hall, "Sir Isaac Newton's Note-book, 1661–65," *Cambridge Historical Journal, 9* (1948), 239–50.

part precedes, and whose strongest fellows." Red is the impression of "an oblique and confus'd pulse" of reversed order.[54] On the first page of his new set of notes, Newton contradicted the two fundamental assertions of Hooke's theory, that light consists of pulses and that colors arise from confused impressions. "The more uniformely the globuli move ye optick nerves ye more bodys seme to be coloured red yellow blue greene &c but ye more variously they move them the more bodys appear white black or Greys."[55] If Hooke was the immediate target of the assertion, much more than Hooke's theory was involved. Like other mechanical philosophers, Hooke had merely provided a mechanism for the existing theory of apparent colors, phenomena such as the rainbow and the colored fringes seen through telescopes and prisms. The theory had been fatally easy to mechanize. It had employed a scale of colors, which was also a scale of strength, running from brilliant red, considered to be pure white light with the least admixture of darkness, to dull blue, the last step before black, which was the complete extinction of light by darkness.

The proposal of the freshly minted Bachelor of Arts implied a completely different relation of light and color. White light, ordinary sunlight, is a confused mixture. Individual components of the mixture, which he considered to be corpuscles rather than pulses, cause sensations of individual colors when they are separated from the mixture and fall on the retina alone. Already he had drawn a picture of an eye looking through a prism at the colored fringes along a border between black and white. From the two sides of the border two rays were shown following different paths through the prism as they were refracted at different angles and emerged along the same line incident on the eye (Figure 5.4). " 1 Note y^t slowly moved rays are refracted more then swift ones."[56] Though he would modify details as he clarified his understanding of its implications, point one above contains the insight on which Newton built his work in optics. The insight fundamental to his dynamics had happened less than a year before, both of them less than two years after he turned seriously to natural philosophy. He had a keen eye for the critical point at which to seize a problem.

He started with an idea rather than an observation. Under the diagram of the prism and the eye was a table in which he tried to

---

[54] Robert Hooke, *Micrographia* (London, 1665), p. 64.

[55] *Add MS* 3996, f. 122. Newton's notes on *Micrographia* explicitly included Hooke's theory of colors. The word "confused" did not appear in them. Newton did question Hooke's theory, which was inextricably bound up with his conception of light. If light is a pulse, why does it not deflect from a straight line as sound does? How can the weaker pulse move as fast as the stronger? (*Halls*, p. 403).   [56] *Add MS* 3996, f. 122[v].

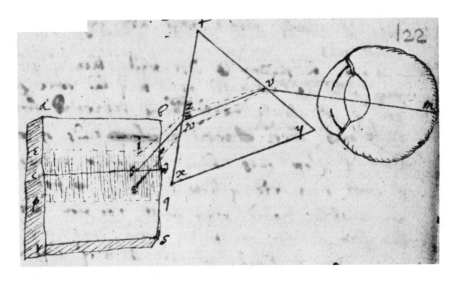

Figure 5.4. Newton's first suggestion of the differential refrangibility of light. (Courtesy of the Syndics of Cambridge University Library.)

reason out the colors that would appear along the borders between various combinations other than black and white. Quickly the complexities of mentally sorting out slowly and swiftly moving rays reflected from various bands along the border became more confused than Hooke's pulses. As suddenly as the original insight, a simplifying experiment presented itself. An experimental orientation pervaded the "Quaestiones," but until the investigation of colors experimentation had been implicit rather than explicit, questions put but not experiments performed. At this point the period of adolescence was fulfilled, and Newton the experimental scientist reached maturity.

That ye rays wch make blew are refracted more y^n ye rays wch make red appears from this experim^{nt}. If one hafe [one end] of ye thred *abc* be blew & ye other red & a shade or black body be put behind it y^n lookeing on ye thred through a prism one halfe of ye thred shall appear higher y^n ye other. & not both in one direct line, by reason of unequall refractions in ye 2 differing colours.[57]

The idea had been provisionally confirmed. Newton never forgot this experiment; he continued to cite it as one of the basic supports of his theory of color.

At the time he made the experiment, however, the theory hardly existed. It was only a promising idea supported by a single experi-

[57] *Add MS 3996, f. 122^v.*

ment. Its implications are obvious to us who profit from three hundred years of digesting them. Newton had to grope his way forward as he denied a tradition two thousand years old which seemed to embody the dictates of common sense. The concept of slow and swift rays was formulated within the context of a mechanical philosophy, and it carried the usual connotations of weak and strong. It inclined him to think in terms of a two-color system, blue and red. It inclined him as well to imagine mechanisms by which the "elastick power" of a body's particles determined how much of the motion of a ray was reflected; "then y^t body may be lighter or darker colored according as ye elastick virtue of that bodys parts is more or lesse."[58] Such ideas returned to the assumption that colors arise from the modification of light, against which his central insight was directed.

Perhaps it was here that his consideration of devices to grind elliptical and hyperbolic lenses, which he mentioned in his paper of 1672, intervened. The background of the investigation was Descartes's announcement of the sine law of refraction in his *Dioptrique*. As use of the telescope had spread in the early seventeenth century, experience had shown that spherical lenses do not refract parallel rays, such as those from celestial bodies, to a perfect focus. In *La Dioptrique*, Descartes had shown that hyperbolic and elliptical lenses would do so, given the sine law of refraction. Grinding them was another question. Spherical surfaces present no problem. Since they are symmetrical in all directions, constant turning and shifting of a lens adjust the lens and the form against which it is being ground to each other so that a spherical surface is bound to result. On the other hand, the grinding of an elliptical or hyperbolic surface is complicated indeed, exactly the problem to challenge the model-builder from Grantham, who was now equipped with a thorough knowledge of the conics. He sketched out several devices by which to produce them (Figure 5.5).[59] And as he did so, he possibly reflected on the meaning of his earlier experiment with the prism and the red and blue thread. Descartes's demonstration had assumed the homogeneity of light. What if he did succeed in grinding elliptical and hyperbolic lenses? He still would not obtain a perfect focus because light is not homogeneous; the blue rays are refracted more than the red. At this moment, it appears to me, Newton began to realize the significance of his experiment and of the idea behind it. He stopped working on nonspherical lenses and never

---

[58] *Add MS* 3996, f. 123.

[59] *Add MS* 4000, ff. 26–33^v. All would have been subject to the basic difficulty, which does not arise with spherical lenses; the mutual grinding of lens and tool would gradually make the tool inexact.

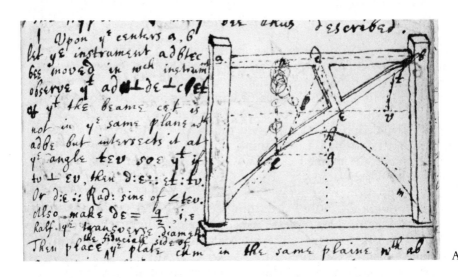

A

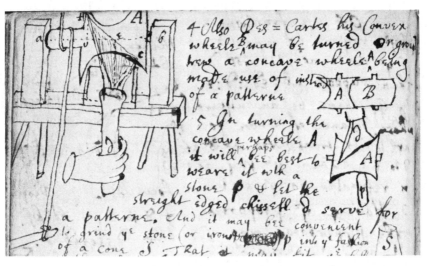

B

Figure 5.5. Devices to produce hyperbolas. In drawing *A*, the apparatus turns about the axis *ab*. The arm *cet*, which is not in the same plane with *ab*, touches the hyperbola *chm*. By moving the instrument to and fro, the operator can wear the plate into a hyperbola. Or, as Newton suggested, he can file the plate and use the instrument to check its accuracy. In drawing *B*, on the left, the straight edge *cet* of a chisel, held at any angle so as not to be in the same plane as the axis *ab* of the stock, cuts a concave hyperboloid in the stock mounted on a lathe. In drawing *B*, on the right below, a similar device is used to check the accuracy of the hyperboloid which the grinding stone *p*, beneath the stock, shapes. On the right above, the concave hyperboloid is used as a pattern to grind a convex one, *B*, the tool to be used to grind a concave hyperbolic lens. (Courtesy of the Syndics of Cambridge University Library.)

returned to them.[60] Later he showed that chromatic aberration introduces much larger errors in lenses than spherical aberration. Instead of lenses, he turned his attention to an experimental investigation of the heterogeneity of light and its role in the production of colors. I assume that this investigation came in 1666 and that he referred to it both in 1672 and in 1726. Only with it did he "have the Theory of Colours" in any legitimate sense of the phrase.

Newton recorded the investigation as the essay "Of Colours" in a new notebook in which he extended several of the topics of the "Quaestiones."[61] With his purpose now more clearly focused, he marshaled known phenomena of colors, which he had found in Boyle and Hooke, that exhibited the analysis of light into its components. Thus thin leaves of gold appear yellow from one side in reflected light but blue in transmitted light from the other; with a solution of *lignum nephriticum* (nephritic wood, infusions of which were used medicinally at the time) the colors are the reverse. In both cases, the transmission of some rays and reflection of others

---

[60] Newton did consider the possibility of an achromatic lens composed of two different media on the two folios of the notebook immediately following the passage on grinding lenses. These two folios were later torn out by an unknown person and are now in Shirburn Castle. Cf. Zev Bechler, " 'A less agreeable matter': The Disagreeable Case of Newton and Achromatic Refraction," *British Journal for the History of Science, 8* (1975), 101–26.

[61] *Add MS* 3975, pp. 1–20. Later he added another note on p. 22. "Of Colours" is not a set of notes compiled from his reading and experimentation; it is a consistent exposition of his theory of colors. With its general paucity of corrections, it may be the distillation of other papers now lost. It has no date, and it may have been entered into the notebook later than 1666. If so, it should be seen as a polished statement of Newton's investigation of 1666. The assertion in the paper of 1672 especially appears decisive in placing the investigation in 1666. Other headings reminiscent of the "Quaestiones" that he set down initially in the new notebook included "Of Cold, & Heate," and "Rarity, Density, Elasticity, Compression &c."

On Newton's theory of colors, see A. R. Hall, "Further Optical Experiments of Isaac Newton," *Annals of Science, 11* (1955), 27–43; Michael Roberts and E. R. Thomas, *Newton and the Origin of Colours. A Study of One of the Earliest Examples of Scientific Method* (London, 1934); J. A. Lohne, "Isaac Newton: The Rise of a Scientist, 1661–1671," *Notes and Records of the Royal Society, 20* (1965), 125–39; J. A. Lohne and Bernard Sticker, *Newtons Theorie der Prismenfarben* (Munich, 1969); Thomas S. Kuhn, "Newton's Optical Papers," in *Cohen,* pp. 27–45; Zev Bechler, "Newton's Search for a Mechanistic Model of Colour Dispersion: A Suggested Interpretation," *Archive for History of Exact Sciences, 11* (1973), 1–37; Alan E. Shapiro, "The Evolving Structure of Newton's Theory of White Light and Color: 1670–1704," forthcoming in *Isis;* R. Furth, "Newton's 'Opticks' and Quantum Theory," *The Lodestone, 41* (1948–9), 23–30; Lord Rayleigh, "Newton as an Experimenter," *Nature, 150* (1942), 706–9; Léon Rosenfeld, "La Théorie des couleurs de Newton et ses adversaires," *Isis, 9* (1926), 44–65; Vasco Ronchi, "L'Ottica del Keplero e quella di Newton," *Atti della Fondazione Giorgio Ronchi, 11* (1956), 189–202; Augusto Guzzo, "Ottica e atomistica newtoniane," *Filosofia, 5* (1954), 383–419; and Maurizio Mamiani, *Isaac Newton filosofo della natura. Le lezioni giovanili di ottica e la genesi del metodo newtoniano* (Florence, 1976).

analyzes white light into its components. Newton was convinced that all solid bodies would behave like gold if pieces thin enough could be obtained, and that the solution of *lignum nephriticum* would appear blue from all sides if it were made thick enough so that no light could pass through.[62] If Newton turned available observations to the advantage of his theory, however, he relied primarily on his own experiments with the prism. Guided by his ingenuity, the prism became an instrument of precision with which he dissected light into its elementary components. No other investigation of the seventeenth century better reveals the power of experimental enquiry animated by a powerful imagination and controlled by rigorous logic.

Among the authors Newton had read, both Boyle and Hooke had employed variations of Descartes's projection of a prismatic spectrum to examine colors. Newton saw that he could bend the same experiment to test his own theory by imposing carefully prescribed conditions on it. If in fact light is heterogeneous and different rays are refracted at different angles, a round beam should be projected by a prism into an elongated spectrum (Figure 5.6). It would require enough distance to spread out, however. Rays are ideal entities; in actual experimentation a physical beam had to be used, and one big enough to give visible effects. If the screen were placed close to the prism as it had been in earlier experiments, the expected elongation would not appear. Descartes had received his spectrum on a screen only a few inches from the prism. Hooke, who employed a deep beaker filled with water instead of a prism, had about two feet between refraction and screen. Boyle apparently used the floor, and hence had a distance of perhaps four feet. Newton projected his spectrum on a wall twenty-two feet away. Where earlier investigators saw a spot of light colored at its two edges, Newton saw a spectrum five times as long as it was wide.[63] His exposition in 1672 contrived to suggest an element of chance and surprise when he saw it; it was as accidental as the observation of a barometer by Pascal's brother-in-law on the summit of the Puy de Dôme. Newton had constructed his experiment to test what he wanted to test. Had the spectrum not been elongated, his promising idea would have been refuted at its second step, and he could not have elaborated it into a theory.

If he had not refuted himself, he was far from having proved anything, as he knew very well. What he proposed was a radical re-

---

[62] *Add MS* 3975, p. 1.

[63] Such was his report in 1672. In his first recorded performance of the experiment, he set down the dimensions of the spectrum as 2 1/3 inches by 7 or 8 inches (*Add MS* 3975, p. 2). It is true that Hooke's diagram of a similar experiment indicated dispersion, but coloration was the object of Hooke's attention.

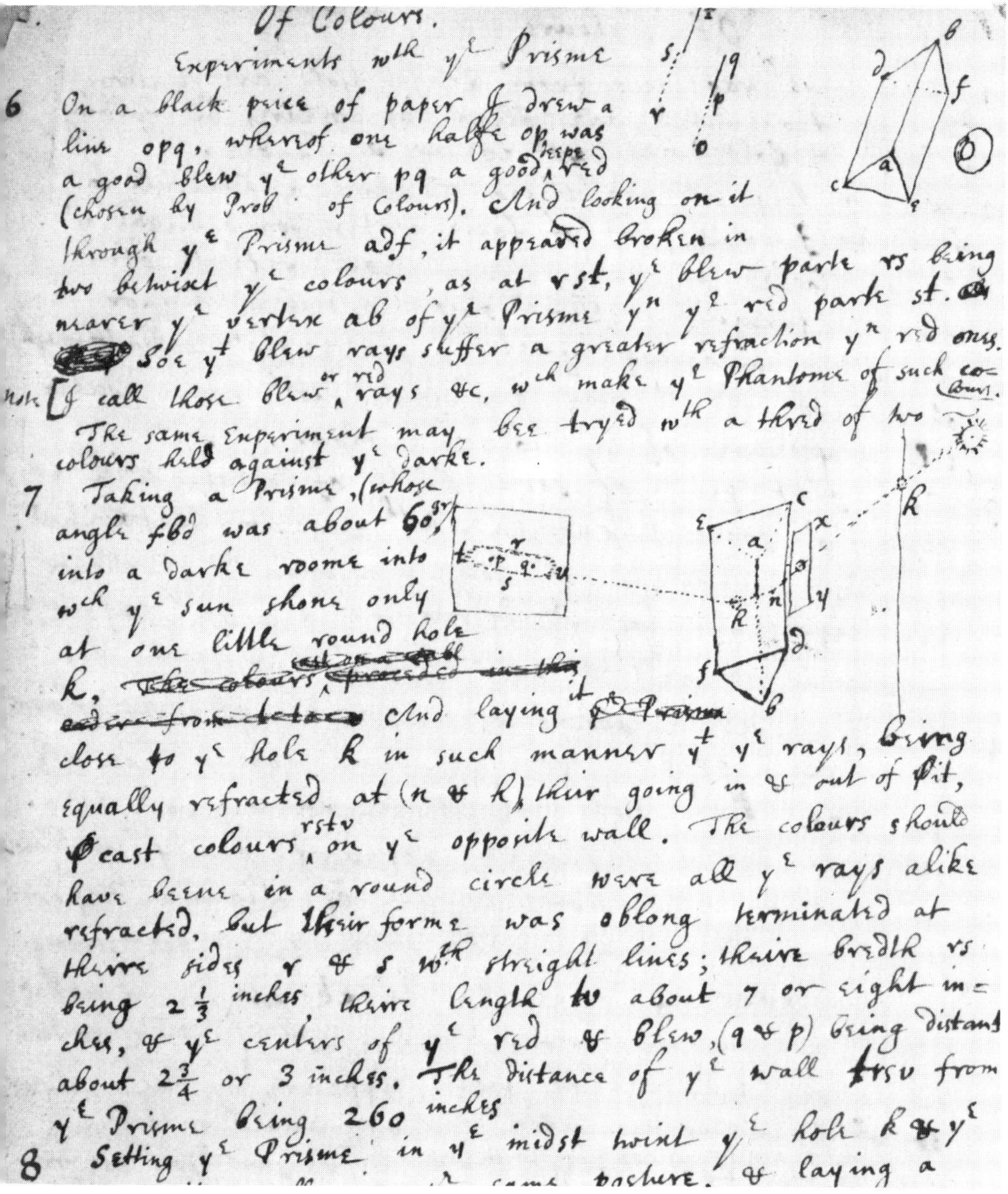

Of Colours.

Experiments with y^e Prisme s.

6 On a black piece of paper I drew a
line opq, whereof one halfe op was
a good blew y^e other pq a good red
(chosen by Prob: of Colours). And looking on it
through y^e Prisme adf, it appeared broken in
two betwixt y^e colours, as at rst, y^e blew parte rs being
nearer y^e vertex ab of y^e Prisme y^n y^e red parte st.
Soe y^t blew rays suffer a greater refraction y^n red ones.
note [I call those blew or red rays &c, wch make y^e Phantome of such co=
lours.
The same Experiment may bee tryed wth a thred of two
colours held against y^e darke.

7 Taking a Prisme (whose
angle fbd was about 60^{gr},
into a darke roome into
wch y^e sun shone only
at one little round hole
k, ~~the colours spreaded in the~~
~~red is from b to c~~ And laying it close Prisme b
close to ye hole k in such manner y^t y^e rays, being
equally refracted at (n & k) their going in & out of fit,
cast colours, rstv on y^e opposite wall. The colours should
have beene in a round circle were all y^e rays alike
refracted. But their forme was oblong terminated at
their sides r & s wth streight lines; their bredth rs
being $2\frac{1}{3}$ inches, their length to about 7 or eight in=
ches, & y^e centers of y^e red & blew, (q & p) being distant
about $2\frac{3}{4}$ or 3 inches. The distance of y^e wall rsv from
y^e Prisme being 260 inches.

8 Setting y^e Prisme in y^e midst point y^e hole k & y^e
m   r same posture. & laying a

Figure 5.6. Newton's drawing of the prismatic spectrum with the
board xy used further to narrow the pencil of light that
enters the room through the hole k. (Courtesy of the Syndics of Cambridge University Library.)

ordering of the relation of light and color. Whereas the received opinion considered white light as simple and colors as modifications of it, Newton asserted that the light which provokes the sensations of individual colors is simple and white light a complex mixture. Long-established views are not easily surrendered. Possible objections were many; they would need to be answered. Perhaps the elongation was caused in some manner by limiting the beam incident on the prism – "the termination with shadow or darkness," which had long played a role in accounts of color.[64] When he placed the prism outside the shutter so that unconfined light fell on it, however, it projected an identical spectrum through the hole. Perhaps the globuli of light acquired a spinning motion and followed a curved trajectory. When he followed the spectrum with a movable screen, he found that it spread rectilinearly. Perhaps it resulted from imperfections in the glass. He projected the beam through various parts of the prism, always with the same result when the prism was set at the same angle. When a second prism was available, he placed it immediately behind the first with its vertex reversed to make it refract in the opposite direction. If the elongation were due to imperfections in the prism, a second prism (presumably with similar imperfections) ought to increase them. Instead it restored a round beam. He was not dealing with contingent phenomena; if two opposite refractions neutralized each other, they must be governed by law.

The most important objection of all still remained. Because the sun fills a visual angle of 31 minutes, the beam incident on the prism was not composed of parallel rays. By the sine law of refraction, rays incident at different angles are refracted at different angles. Could the elongated spectrum be an unexpected product of the sine law? Newton employed various devices to get a beam composed of rays more nearly parallel. By using a board with a hole in it placed some twelve feet from the shutter he narrowed the incident beam to about 6 minutes; all the dimensions of the spectrum decreased by an equal amount, further accentuating the elongation.[65] Later he even managed to perform the experiment with the light from Venus, which he collected with a lens. When he placed the prism between the lens and its focus, he found the focus "drawn out into a long splendid line . . . "[66] Such experiments displayed his imagination and his dexterity, but Newton knew well

---

[64] I am citing from his account in 1672 (Newton to Oldenburg, 6 Feb. 1672; *Corres 1*, 92–4).
[65] *Add MS* 3975, p. 2.
[66] Newton to Oldenburg, 13 Apr. 1672; *Corres 1*, 137. For the most part, little esthetic response crept into Newton's accounts of his experiments with colors, but it can hardly be ignored in this case. He cited the experiment with Venus in his *Lectiones opticae* (*Add MS* 4002, pp. 15–16).

that only a theoretical demonstration could finally meet the objection. It was not a difficult exercise for a mathematician of his accomplishment. When the central ray of an incident pencil of homogeneous light contained within an angle of 31 minutes is refracted equally at both faces of a prism, it emerges as a pencil contained within an angle of 31 minutes.[67] As it happens, equal refraction at each face is also the condition for minimal refraction, so that Newton had only to turn the prism until the spectrum reached its lowest position on the wall to obtain it. His first recorded projection of a spectrum noted that the rays were equally refracted by both faces of the prism.[68] Along with the distance of projection, equal refraction at the two faces was a planned condition of the initial experiment. So far was the elongated spectrum from a chance observation.

Although the mathematical demonstration supplied necessary rigor to the evidence of the spectrum, Newton ultimately found a further experiment that seemed to confirm it no less strongly. Like the final theory, the experiment did not appear in one burst of inspiration. It evolved through several stages until, as he realized its force, he called it the *experimentum crucis*.[69] In its initial form, it was ill-defined and hardly compelling. He simply held a second prism in the spreading spectrum five or six yards from the first. The blue rays suffered a greater refraction than the red. In neither case did the second refraction produce further coloration; blue remained blue and red remained red.[70] In 1666 he went no farther. Only later did he realize the demonstrative potential he could gain by refining the experiment.

Once Newton had fully seized the concept of analysis, he was able readily to generate other experiments to illustrate it. He could analyze sunlight into its components by tilting a prism to the critical angle, where blue rays, which are the most refrangible, began to be reflected from the second face while red rays were still transmitted through it (Figure 5.7). He obtained an analogous separation with a thin film of air trapped between two prisms bound together.[71] Realizing that it was necessary to demonstrate that he could reconstitute white, he cast the spectra of three prisms onto each other so that they overlapped without coinciding. In the center, where all the colors fell, the combined spectrum was white. He attached a paper to the face of a prism with several slits parallel to the edges. On a screen held near the prism a line of color appeared for each slit. As he moved the screen away, the center of the spectrum became white, but the full spectrum appeared again,

[67] *Add MS* 4002, pp. 4–7.   [68] *Add MS* 3975, p.2.
[69] The phrase appeared only in 1672 (Newton to Oldenburg, 6 Feb. 1672; *Corres 1*, 94). Actually it was Hooke's phrase, an embroidery on Bacon's *instantia crucis*.
[70] *Add MS* 3975, p. 12.   [71] *Add MS* 3975, p. 6.

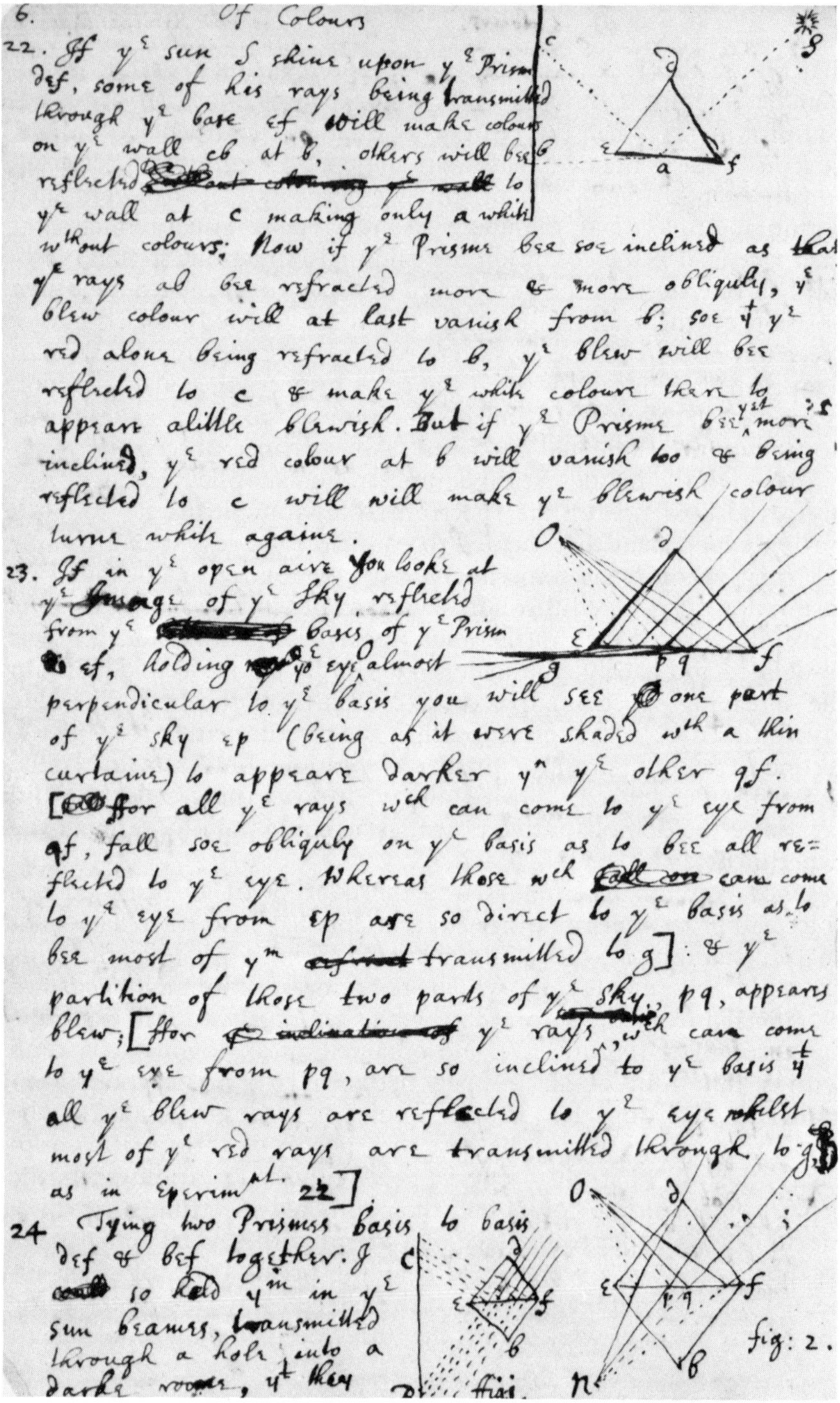

6. Of Colours

22. If y^e^ sun S shine upon y^e^ Prism def, some of his rays being transmitted through y^e^ base ef will make colours on y^e^ wall cb at b, others will bee reflected ~~without colouring ye wall~~ to ye wall at c making only a white without colours; Now if ye Prisme bee soe inclined as ~~that~~ ye rays ab bee refracted more & more obliquely, ye blew colour will at last vanish from b; soe y^t^ ye red alone being refracted to b, ye blew will bee reflected to c & make ye white colour there to appeare alittle blewish. But if ye Prisme bee y^et^ more inclined, ye red colour at b will vanish too & being reflected to c will will make ye blewish colour turne white againe.

23. If in ye open aire you looke at y^e^ ~~image~~ of ye sky reflected from y^e^ [illegible] basis of ye Prism [illegible] ef, holding [illegible] ye eye O almost perpendicular to ye basis you will see [illegible] one part of ye sky ep (being as it were shaded wth a thin curtaine) to appeare darker y^n^ ye other qf. [For all ye rays wch can come to ye eye from qf, fall soe obliquely on ye basis as to bee all reflected to ye eye. Whereas those w^ch^ ~~fall on~~ can come to ye eye from ep are so direct to ye basis as to bee most of y^m^ ~~refract~~ transmitted to g]: & ye partition of those two parts of ye sky, pq, appeares blew; [For ~~ye inclination of~~ ye rays, wch can come to ye eye from pq, are so inclined to ye basis y^t^ all ye blew rays are reflected to ye eye whilst most of ye red rays are transmitted through to g as in Experim^t^. 22]

24 Tying two Prismes basis to basis def & bef together. I ~~could~~ so held y^m^ in ye sun beames, transmitted through a hole into a darke roome, y^t^ they

fig: 2.

Figure 5.7. Three means of analyzing light with prisms. At the top, a beam falls at *a* on the base *ef* of the prism. At the critical without further experimental manipulation, as he moved the screen still farther.[72]

Some seven years later, after the publication of his first paper in the *Philosophical Transactions,* Newton replied to a critique by Huygens with a methodological homily.

> It seems to me that M. Hugens takes an improper way of examining the nature of colours whilst he proceeds upon compounding those that are already compounded, as he doth in the former part of his letter. Perhaps he would sooner satisfy himself by resolving light into colours as far as may be done by Art, and then by examining the properties of those colours apart, and afterwards by trying the effects of reconjoyning two or more or all of those, & lastly by separating them again to examin w^t changes that reconjunction had wrought in them. This will prove a tedious & difficult task to do it as it ought to be done but I could not be satisfied till I had gone through it.[73]

No doubt Huygens, the doyen of European science, did not relish the lecture from an unknown professor in Cambridge. Nevertheless it is a reasonable description of Newton's procedure as he unraveled the implications of his central idea.

When we follow his papers, we realize what hard work it was to cast off the dictates of common sense embodied in a tradition two thousand years old. His ultimate theory did not reveal itself all at once. He started with a new idea, a different solution to the problem of colors. It took Newton several years, however, to fully realize that in pursuing his idea, that colors appear from the analysis of light instead of its modification, he had demonstrated a new and

---

[72] *Add MS* 3975, pp. 12–13.
[73] Newton to Oldenburg, 3 April 1673; *Corres 1*, 264.

angle, the more refrangible rays at the purple end of the spectrum are reflected to *c*, while the less refrangible rays at the red end pass through the base and are refracted to *b*. In the middle, an eye at O (close to the prism) views the sky as it is reflected from the bottom of the prism, *ef*. Because of the different angles between the eye and the two halves of the base, *qf* and *ep*, light is reflected to the eye from *qf*, but less from *ep*, where much of it is transmitted through the base and refracted toward *g*. Newton reported that the sky seen at *ep* looked shaded. The partition between the two halves *pq* appears blue because the red rays, being less refrangible, are transmitted and refracted toward *g*. At the bottom, the thin film of air between two prisms effects similar analyses. In all three diagrams, Newton did not attempt to indicate any refraction at the other faces of the prism, since only the face *ef* effects the analysis. (Courtesy of the Syndics of Cambridge University Library.)

unexpected property of light, its heterogeneity. "Of Colours" appears to embody the results of his investigation of 1666 as he first began seriously to explore his insight. Without stating them explicitly, it assumes many of the propositions of his ultimate theory. Rays of light differ both in the colors they exhibit and in their refrangibility. Though he frequently referred to blue rays and red rays, Newton never understood the rays themselves to be colored. A mechanical philosopher would never have entertained that notion. Rays exhibit colors. "I call those blew or red rays &c wch make ye Phantome of such colours."[74] Because they differ in refrangibility as well, the heterogeneous rays in sunlight are separated by the prism. Allied to the concept of analysis was the further proposition that individual rays are immutable in their properties. Newton had not comprehended this at the time of his original idea when he had toyed with mechanisms of reflection which attempted, in effect, to combine modification with analysis. By 1666, immutability had become clear, and experiments seemed to support it. He painted a red patch and a blue patch on a sheet of paper. When he cast "Prismaticall blew" on them, they both appeared blue, though one was much fainter than the other. When he cast "Prismaticall red" on them, both appeared red; the other one was now fainter.[75] Nothing he did, neither refraction nor reflection, could alter the inherent properties of a ray of light. As he finally realized, analysis entails immutability. Colors are never generated; they are only made apparent by some process that separates them from the heterogeneous mixture of white light.

Heterogeneity, the new property of light on which the concept of analysis depended, challenged the assumption of two thousand years of optical research. It reversed the relation of colors and whiteness. Whereas white had been associated universally with simplicity and purity, both within optics and beyond it, Newton's theory made individual colors simple and pure and demoted white to the status of a secondary appearance, the sensation of no single ray but of a heterogeneous mixture. "I perswade my selfe," he later remarked, "that this assertion above the rest appeares *Paradoxicall,* & is with most difficulty admitted."[76] Apparently in 1666 he did not yet understand how paradoxical the assertion was. Only a couple of minor experiments attempted to illustrate the recomposition of white. This too would await further elaboration in the future.

The implications of his insight, only dimly perceived at first, began to emerge more clearly in his investigation of 1666. Nev-

---

[74] *Add MS* 3975, p. 2.   [75] *Add MS* 3975, p. 3.
[76] Newton to Oldenburg, 11 June 1672; *Corres 1,* 183.

ertheless, a comparison of the essay "Of Colours" with the *Lectiones opticae,* the polished product of the investigation of 1668 and 1669, indicates that Newton groped forward a step at a time. In 1666, several important points remained obscure. The essay "Of Colours" assumed a two-color system, red and blue. Twice, early in the essay he started to write down the order of colors, first in the spectrum and then in the film of air between two prisms. Both times he left a blank space that he did not bother to fill in, as though the order of colors were of little interest.[77] When finally he did list them for the colored circles in a film between a lens and a flat sheet of glass, he named five: red, yellow, green, blue, purple.[78] "As white was made by a mixture of all sorts of colours . . . [he explained] Greene is made by a mixture of blew & yellow, purple by a mixture of red & blew &c."[79] He had indicated already that yellow is merely dilute red. Perhaps no aspect of the essay of 1666 better illustrates the gradual unfolding of the full meaning of his idea.

More significant was the continuing confusion between his theory of colors and his corpuscular conception of light. Already with the "Quaestiones" Newton accepted the corpuscular conception of light as one aspect of his atomistic philosophy of nature.[80] When he met an embryonic wave conception of light in Hooke's *Micrographia,* he reacted against it with one of his enduring objections – that light ought then to stray from its rectilinear path as sound, which was agreed to be a wave motion, does.[81] Moreover, it seems clear that the corpuscular conception functioned as a propaedeutic to his theory of colors. The notion of strong and weak rays in which he couched his original insight readily clothed itself in the images of large and small corpuscles. His geometrization of colors found its counterpart in the paths of single corpuscles, whereas the wave theory of light had always to think of a physical beam of finite dimensions. As the immutability of rays impressed itself on his understanding, it became a further argument for the corpuscular conception, finding its physical basis in the definition of an atom. Thus there were many filiations binding the corpuscular conception of light to the new theory of color. His initial experiments with thin films in 1666 served to reinforce them. He observed that the circles of colors between a lens and a flat piece of glass appeared to grow in diameter the more obliquely he observed them. He immediately interpreted their size

---

[77] *Add MS* 3975, pp. 3, 7.
[78] *Add MS* 3975, p. 10. He cited the same five as the components of white under somewhat different circumstances on p. 12.   [79] *Add MS* 3975. p. 13.
[80] *Add MS* 3996, ff. 104ᵛ, 122–3ᵛ.   [81] *Halls,* p. 403.

in terms of the strength of the ray's blow on the film. A stronger blow allowed the rays to pass through more readily in a circle of smaller diameter, while a weaker blow caused it to pass through less readily and hence to appear as a larger circle. He worked out a formula, which he later crossed out, in which the diameter of the circles varied inversely as the component of motion perpendicular to the film.[82] Newton never entirely separated his theory of colors from his conception of light. When he wrote Query 28 in the *Opticks* in 1706, he asserted that wave conceptions of light always embodied modification theories of color.[83] Nevertheless, the two were separate issues. He had created powerful experimental evidence to support the heterogeneity of light, whereas the corpuscular conception remained a matter of speculation, as he acknowledged forty years later when he argued for it only in the Queries of the *Opticks*. In 1666, the distinction between the two had not become clear in his mind.

One other issue remained: the colors of solid bodies. Newton built his theory of colors from experiments with prisms. The vast majority of the colors we see, however, are associated with solid bodies. Unless he accounted for their colors, his theory would be extremely limited. From the moment of his initial insight, of course, he had a general account of the colors of solid bodies. Reflection can also analyze white light into its components. A body is disposed to reflect some rays more than others and appears to be the color it reflects best. His account of colors never deviated from this position. In the beginning, however, the statement expressed an idea without empirical foundation and without quantitative content. The essay "Of Colours" provided some empirical foundation. When he painted red and blue patches on a piece of paper and viewed them in "Prismaticall blew" and "Prismaticall red," both patches appeared to have the color of the incident light, but the blue patch was fainter in red light and the red fainter in blue. "Note y^t ye purer ye Red/Blew is ye lesse tis visible wth blew/Red rays."[84] Later he would add further empirical evidence as the immutability of rays became clearer to him.

Quantitative content was a more difficult matter. It was abso-

---

[82] In fact, the formula was more complicated in two ways. It employed the components of motion (or, as he put it, the sines of the angle of obliquity) both in the film of air and in the glass that confined the film. It further multiplied both components by the motion of the ray in the medium; in effect this made the formula a comparison of the squares of the motions because the index of refraction between the two media was understood to be the ratio of their motions. Since Newton crossed the formula out, its details are not important for my purposes. Its importance for me here lies in its implicit conception of light, expressed in its assumption of corpuscles moving with greater or lesser strength in relation to a surface.   [83] *Opticks,* p. 362.   [84] *Add MS* 3975, p. 3.

lutely essential. After his experiments with prismatic spectra, analysis by refraction could be expressed in rigorous quantitative terms. Color ceased to be a wholly subjective phenomenon since it was attached immutably to a given degree of refrangibility.[85] With reflected colors, in contrast, he had attained no similar quantitative treatment, and reflected colors constitute the overwhelming bulk of color phenomena in the world. He had picked up a suggestion, however. In Hooke's *Micrographia* he found descriptions of colors in a variety of thin transparent bodies – in Muscovy glass (or mica), in soap bubbles, in the scoria of metals, in the air between two pieces of glass. Newton himself observed the colors in a film of air between two prisms, both in the transmitted and in the reflected light. The "plate of air (*ef*) is a very reflecting body," he noted, and later he indicated that the colors of solid bodies are related to the colors of thin transparent films.[86] He even realized a means of doing what Hooke had confessed himself unable to do, measuring the thickness of the films in which colors appear. When a lens of known curvature was pressed on a flat piece of glass, a thin film of air was constituted between them. Circles of color appeared around the point of contact. Using the geometry of the circle, the same proposition indeed which he used in calculating centrifugal force, he computed the thickness of the film from the curvature of the lens and the measured diameter of the circles.

"Of Colours" recorded Newton's first observation of "Newton's rings." Hooke had asserted that the colors in thin films are periodic, that is, that successive additions of a given increment of thickness cause the same colors to reappear. Newton observed that successive circles around the center did not increase linearly in diameter; without measuring, he inferred that their diameters corresponded to a linear increase in thickness of the film. "Let cd = the radius of curvature of the glass; $efghik$ circles of colors; & $el = fm/2 = gn/3 = hp/4 = ig/5 = kr/6$ = the thicknesses of air."[87] With a lens 25 inches in radius, he measured the diameter of the fifth circle and calculated that the thickness of el, the first circle, was 1/64,000-inch, a figure which he altered several years later to 1/83,000-inch after more refined measurements. Thus he had the outlines of a theory of reflected colors and an experimental method whereby to elaborate it. It appears, however, that he did not yet grasp the full connection between the two. The circles were merely "circles of colours," not

---

[85] Or so Newton finally believed. At one point, early in his career, he at least entertained the idea that dispersion might vary with media so that an achromatic lens was theoretically possible (cf. Bechler, " ' A less agreeable matter' ").

[86] *Add MS* 3975, pp. 8, 13.

[87] *Add MS* 3975, p. 9. On his diagram, *el*, *fm*, *gn*, *hp*, *iq*, and *kr* are the thicknesses of the thin film where successive circles appear at *e*, *f*, *g*, *h*, *i*, and *k* on the flat surface.

circles of particular colors. He did nothing to measure the different thicknesses for different colors. This aspect of his theory was in 1666 the least developed.

On close examination, the *anni mirabiles* turn out to be less miraculous than the *annus mirabilis* of Newtonian myth. When 1666 closed, Newton was not in command of the results that have made his reputation deathless, not in mathematics, not in mechanics, not in optics. What he had done in all three was to lay foundations, some more extensive than others, on which he could build with assurance, but nothing was complete at the end of 1666, and most were not even close to complete. Far from diminishing Newton's stature, such a judgment enhances it by treating his achievement as a human drama of toil and struggle rather than a tale of divine revelation. "I keep the subject constantly before me," he said, "and wait 'till the first dawnings open slowly, by little and little, into a full and clear light."[88] In 1666, by dint of keeping subjects constantly before him, he saw the first dawnings open slowly. Years of thinking on them continuously had yet to pass before he gazed on a full and clear light.

By any other standard than Newtonian myth, the accomplishment of the *anni mirabiles* was astonishing. In 1660, a provincial boy ate his heart out for the world of learning which he was apparently being denied. By good fortune it had been spread before him. Six years later, with no help beyond the books he had found for himself, he had made himself the foremost mathematician in Europe and the equal of the foremost natural philosopher. What is equally important for Newton, he recognized his own capacity because he understood the significance of his achievements. He did not merely measure himself against the standard of Restoration Cambridge; he measured himself against the leaders of European science whose books he read. In full confidence he could tell the Royal Society early in 1672 that he had made "the oddest if not the most considerable detection wch hath hitherto beene made in the operations of Nature."[89]

The parallel between Newton and Huygens in natural philosophy is remarkable. Working within the same tradition, they saw the same problems in many cases and pursued them to similar conclusions. Beyond mechanics, there were also parallel investigations in optics. At nearly the same time and stimulated by the same book, Hooke's *Micrographia*, they thought of identical methods to measure

[88] *Biographia Britannica* (London, 1760), *5*, 3241. No source for the quotation is given.
[89] *Corres 1*, 82–3.

the thickness of thin colored films.[90] No other natural philosopher even approached their level. In the very year 1666, Huygens with all his acclaim was being wooed by Louis XIV to confirm the renown of his Académie royale des sciences. There was no occasion for a young man recently elevated to the dignity of Bachelor of Arts and working in isolation to be ashamed of his achievement, even if the Sun King, in his presumption, had not placed a crown of laurel on his brow.

[90] Huygens dated his experiments November 1665; *Oeuvres complètes,* pub. Société hollandaise des sciences, 22 vols. (The Hague, 1888–1950), *17,* 341–8.

6

Lucasian professor

POSSIBLY the fact that the university closed down for the better part of two years, freeing Newton from control and supervision, aided his unorthodox program of study. Such was the progressive decomposition of the established curriculum, however, that it is far from clear that his intellectual development would have differed in any respect had the plague not intervened. While the requirements prescribed an impressive array of lectures, disputations, and acts over a three-year period before a Bachelor of Arts might be admitted as an inceptor for the Master of Arts degree, most of them had ceased to be operative by the Restoration, as repeated orders from Westminster that they be observed testify. Even residence had ceased to be required. Though Newton himself was not frequently absent from Cambridge once he returned after the plague, the college Exit and Redit Book reveals that other scholars of the house felt free to come and go as they pleased, even in term time. Newton's friend Wickins, for example, who was a scholar drawing a stipend, left the college on 15 October 1667 and did not return until the middle of January.[1] Unorthodox study was not manifestly worse than no study at all. The one discipline exerted was financial; Newton received his stipends through his tutor.[2] Although his papers do not indicate that he again made any gestures whatever toward the established curriculum, his academic career nevertheless marched through the remaining stages to full membership in the college and university without hesitation.

Shortly after his return from Woolsthorpe late in April 1667, the magnificent funeral of Matthew Wren, bishop of Ely, escorted by the entire academic community in full regalia according to their ranks and degrees, must have reminded him that he stood then only on the first step of the university hierarchy and that others loomed immediately ahead.[3] In only a few months he would face the first and by far the most important of these, the fellowship election. As with the scholarship three years earlier, Newton's whole future hung in the balance of this election. It would determine whether he would stay on at Cambridge and be free to pursue his studies or whether he would return to Lincolnshire, probably to the village

[1] Exit and Redit Book, 1667–1703.
[2] Junior Bursar's *Accounts,* and Bailie's & Chamberlayne's *Day* Book, 1664–1673.
[3] See the account of the funeral in J. E. Foster, ed., *The Diary of Samuel Newton, Alderman of Cambridge (1662–1717)* (Cambridge, 1890), pp. 18–20.

vicarage that his family connections could have supplied, where he might well have withered and decayed in the absence of books and the distraction of petty obligations. On the face of it, his chances were slim. There had been no elections in Trinity for three years, and as it turned out, there were only nine places to fill. The phalanx of Westminster scholars exercised their usual advantage. The growing role of political influence, whereby those with access to the court won letters mandate from the king commanding their election, was notorious.[4] For the rest, all depended on the choice of the master and eight senior fellows, and stories of influence peddling filled the air.[5] The candidates had to sit in the chapel four days in the last week of September to be examined *viva voce* by the senior fellows, the dying embodiment of the curriculum Newton had systemically ignored for nearly four years. How could an erstwhile subsizar of whatever capacity hope to prevail against such odds?

If he too had a patron, he might do more than hope. In 1667, Humphrey Babington joined the ranks of the senior fellows. Neither in Newton's papers nor in the surviving anecdotes does a hint of tension over the outcome appear. His accounts present a picture of relaxation which almost belies our other evidence of unremitting, introverted study. Soon after his return, he spent 17s 6d to celebrate his Bachelor's Act and on subsequent occasions tossed away another of the £10 he had pried loose from Hannah

[4] There were two mandates at Newton's election, one for John Goodwin, who was elected fellow, and one for Valentine Pettit (*CSPD:* Charles II, 7, 322, 553). Since the college was in that year establishing the practice of electing only third-year bachelors, and since Pettit was only in his second year, the college deferred his election until 1668. On the role of mandates, see E. F. Churchill, "The Dispensing Power of the Crown in Ecclesiastical Affairs," *Law Quarterly Review, 38* (1922), 297–316, 420–34; James Bass Mullinger, *The University of Cambridge,* 3 vols. (Cambridge, 1873–1911), *3,* 626–9; John Venn, *Caius College* (London, 1901), pp. 105, 111, 144. Contemporary comments and accounts are found in Roger North, *The Lives of the Right Hon. Francis North, Baron Guilford; the Hon. Sir Dudley North; and the Hon. and Rev. Dr. John North. Together with the Autobiography of the Author,* ed. Augustus Jessopp, 3 vols. (London, 1890), 3. (London, 1890), and Nathaniel Johnston, *The King's Visitorial Power Asserted* (London, 1688), pp. 274–9. Johnston's book was an apology for the intrusions of James II into the universities that made effective use of established practices to defend James's actions. Thomas Baker, *History of the College of St. John the Evangelist, Cambridge,* ed. J. E. B. Mayor, 2 vols. (Cambridge, 1869), 1, pp. 298–9, 543, contains a nearly contemporary account.

[5] Such is a major theme of North's life of his brother John, master of Trinity (*Lives, 2*). Inevitably there is much about it in Johnston, *Visitorial Power.* Baker accused Gunning, the master of St. John's at the time when Newton was a student, and a man whom Baker generally admired, of misusing his influence in the election of fellows. (*History of St. John's* p. 237); Cf. J. Beeby of Oxford to Secretary Williamson, 1668, in which Beeby deplored the destruction of free elections by mandates, but nevertheless thought it better that the King should command an election "than that any particular fellow or head of a house may sell it for £200 or £300 contrary to law or reason" (quoted in Churchill, "Dispensing Power," p. 315).

Figure 6.1. The gowns of seventeenth-century Cambridge as David Loggan presented them. 3. An undergraduate of Trinity College. 5. A Bachelor of Arts. 7, 8, 9, 10. Masters of Arts. (From David Loggan, *Cantabrigia illustrata,* n.d., late seventeenth century.)

Smith and then some with "acquaintances" at taverns. He cheerfully confessed to a loss of 15s at cards, compensated perhaps by a purchase of oranges for his sister. The accounts radiate confidence as well. He invested £1 10s in tools, real tools including a lathe, such as he must have longed for in Grantham – not the purchase of a man seriously expecting to move on a year hence. His manifest intent to convert their chamber into a factory loft may have encouraged Wickins in his extended absences. And Newton invested handsomely in cloth for a bachelor's gown, which could be converted later into a master's, eight and a half yards of "Woosted Prunella" plus four yards of lining, for which he paid nearly £2 in all.[6] On 1 October a bell tolled at eight in the morning to summon the seniors to the election. The bell tolled again the following day at one to call those chosen to be sworn in: It tolled for Newton.

Now at last the way was clear. The election promised permanent membership in the academic community with freedom to continue

[6] Fitzwilliam notebook.

the studies so auspiciously, as he at least understood, begun. True, two more steps remained for him to mount. In October 1667, he became only a minor fellow of the college, but advancement to the status of major fellow would follow automatically when he was created Master of Arts nine months hence. The exercises for the degree had become wholly *pro forma;* no one was known to have been rejected. The final step could come at any time in the following seven years. The incumbents of two specific fellowships excepted, the sixty fellows of the college were required to take holy orders in the Anglican church within seven years of incepting M.A. Shortly after one o'clock on 2 October 1667, Newton became a fellow of the College of the Holy and Undivided Trinity when he swore "that I will embrace the true religion of Christ with all my soul . . . and also that I will either set Theology as the object of my studies and will take holy orders when the time prescribed by these statutes arrives, or I will resign from the college."[7] The final requirement was not likely to pose more of an obstacle to a pious and earnest young man than the Master's degree.

As a minor fellow, Newton received a stipend (or "wages" in the blunter language of the day) of £2 a year, a livery allowance of £1 6s 8d, and a dividend of £10. Not yet admitted to the high table, he continued to dine with the scholars. The college assigned a room to him. On a list dated 5 October 1667, his name appears on the last line: "to Sr Newton—Spirituall Chamber."[8] Edleston has speculated that this referred to the ground-floor room next to the chapel in the northeast corner of the great court. Wherever the room, the mere fact that the college assigned it to Newton tells us nothing about where he lived. The room was one of his perquisites. He might rent it out if he chose and pocket the income, and his accounts do show that he received £1 11s 0d chamber rent in the summer of 1668. Stukeley later heard that he lived on the north side of the court between the master's lodge and the chapel. In any event, he continued to share a chamber with Wickins, who would also have one at his disposal in another year. Undoubtedly the two split the rent from the room they did not use.[9] Two puzzling accounts of payments for a chamber to or for Humphrey Babington also survive, further testimony of a connection between the two whatever it means about Newton's residence.[10] For one room or

---

[7]   Trinity Statutes, *Fourth Report from the Select Committee on Education* (London, 1818), p. 373.   [8] *Edleston,* p. xliii.
[9]   Fitzwilliam notebook. *Stukeley,* p. 56. Nicholas Wickins to Robert Smith, 16 Jan. 1728; *Keynes MS* 137. Wickins's son reported that Newton collected room rent for his father and forwarded it.
[10]   *Add MS* 3970.3, f. 469^v. *Yahuda MS* 34. Both appear to be related to Newton's final chamber next to the great gate.

another, he spent 2s 6d during 1667 to hire a glazier and to point the windows and fireplace.

Newton spent Christmas of 1667 at home, leaving the college on 4 December and returning on 12 February. There must have been some blunt talk about money and the style a regent master of the university expected to keep. He returned this time with £30 of Hannah Smith's money in his pocket. He spent a significant part of it on clothes.[11] On 1 April 1668, the senate passed the grace granting M.A. degrees to 148 inceptors including Newton. He paid the Proctor £2 and the college £2 10s and he paid out 15s further, "Expenses caused by my degre," which undoubtedly went into the pocket of a tavern keeper.[12] He was created Master of Arts on 7 July and became therewith a major fellow of Trinity College.

As a major fellow, Newton's stipend increased to £2 13s 4d per year and his livery allowance to £1 13s 4d. On behalf of each fellow, the college paid its steward 3s 4d commons for every week he was in residence, though in fact the high table expended twice that sum to feed the fellows. The Elizabethan statutes had set the official stipend and allowances. Inflation since then had rendered the sums insignificant, but as income from endowments kept pace with inflation, the college devised a new system to support the fellows, an annual division of the surplus, or dividend. All the colleges of Oxford and Cambridge adopted this device during the middle years of the seventeenth century. By the definitive plan of division which Trinity adopted in 1661, the ordinary fellow of low seniority received a dividend of £25. Until the latter years of the century, the college paid out a dividend annually with only one exception. In addition, fellows resident more than half the year, as Newton invariably was, shared the profits of the college bakery and brewery. The pandoxator's dividend, as it was called, brought the ordinary fellow £5 per year. In sum, Newton found himself the recipient of

[11] In his accounts immediately after his return in April 1667 Fitzwilliam notebook:

Two paire of shoos	0	–	8	–	0
Later in 1667:					
Shoos	0.		4.		0
Cloth 2 yards & buckles for a Vest.	2.		0.		0
To the Taylor Octob 29. 1667.	2.		13.		0
To the Taylor. June 10. 1667	1.		3.		10.
Shoos & mending	0.		4.		10.
In 1668:					
Shoemaker	0.		5.		8.
Making &c of my last suit	1.		11.		9
In 1669:					
16 yards of Stuffe for a Suit	2.		8.		0
For making &c.	1.		13.		0
For turning a Cloth suit	1.		3.		3.

[12] *Ibid.*

approximately £60 per year, between £20 and £25 of which he received in the form of food and lodgings. He received in addition perquisites which cannot be reduced to measure, such as the silver tankard reserved for him in the great hall and access to the bowling green. Since he was assessed ten shillings for the latter, it may have appeared to him more as an expense than a perquisite; twenty years later Humphrey Newton observed that he never used it.[13] Though his income derived from the landed endowment of the college, it was by law exempt from taxation. To put his financial position into perspective, recall that the skilled workmen who labored on the Trinity library a decade later received between 20 and 25d per day, which would have amounted to £26 to £31 per year if they worked six days a week fifty-two weeks of the year; they more probably earned between £15 and £20.[14] Unlike the enforced celibates of the college, they supported families on their income. Unskilled workmen received half that sum. In 1658, Samuel Pepys lived with a wife in London on a salary of £50 as a clerk at the Exchequer.[15] Though one would not call £60 per annum handsome, certainly it was comfortable, especially for one whose aspirations were intellectual. In addition, Newton owned a small property near Woolsthorpe, though income from it never appears in his accounts, and in due time he would inherit much more.

Newton received an added dividend during his first year as fellow. On 2 April 1669, the college assigned St. Leger Scroope, a fellow commoner, to his tuition. Since substantial fees accompanied fellow commoners, they were usually reserved for important members of the college; for a number of years before he migrated to the master's lodge in Magdalene, for example, James Duport had made it his business to engross Trinity's supply of fellow commoners. The assignment of such a plum to a new fellow without seniority or influence of his own can only be seen, I believe, as further evidence that he had a patron in high place in the college. No mention of the relationship with Scroope made its way into Newton's papers. Scroope neither matriculated in the university nor graduated from it. He left no trace on Newton's life. He also left no plate to the college, which expected such from its fellow commoners. Until Newton resigned his fellowship, the junior bursar's accounts annually carried the entry under the heading "Plate not Received": "From Mr Newton, Mr Scroope's."

Thomas Fuller referred to Trinity College as "the stateliest and most uniform College in Christendom, out of which may be carved

---

[13] Humphrey Newton to Conduitt, 17 Jan. 1728; *Keynes MS* 135. *Keynes MS* 135 also contains a second letter from Humphrey, 14 Feb. 1728.

[14] The accounts for the library are in Trinity College, 0.4.47.

[15] Arthur Bryant, *Samuel Pepys,* 3 vols. (London, 1933–8), *1,* 46.

three Dutch Universities."[16] There is reason to believe that Newton savored the dignity to which he had risen. He spent time and money appointing his chamber. After the college had it plastered early in 1668, Newton paid £1 6s 0d for putty and the services of a joiner and painter. He purchased a leather carpet and joined with Wickins to purchase a couch. Ticking and feathers for his bed cost nearly £2, and he acquired a tablecloth and six napkins.[17] Somewhat later he added another rug and refurbished his bedroom.[18] He took care at once to prepare himself for his role in the colorful university processions. He purchased no less than eighteen yards of tammy (£1 13s 0d) for his master's gown, and lining besides (3s 6d), and he paid a tailor £1 0s 6d to turn his bachelor's gown and to make his new one. A hat cost 19s 0d and a hood £1 3s 6d.[19]

To celebrate his new dignity, Newton treated himself to his first visit to London. His accounts record that he left the college on 5 August and that he spent nearly £10 in the city. A year earlier he had purchased Sprat's *History of the Royal Society* and had begun to purchase and read the *Philosophical Transactions*. He had read as well the works of prominent fellows of the Royal Society such as Robert Boyle and Robert Hooke. He could not have failed to appreciate the affinity of the society with the studies in which he had immersed himself. Nevertheless, while he was in London Newton did not attempt to approach the society, with which his future was to be inextricably woven. Membership in it would have to wait for their overture. We do not know what he did in London during the month he was there. On his return, he detoured to Woolsthorpe, possibly to show his mother the new finery her money had purchased. She must have been impressed. She came across with £4 6s to cover part of his London expenses and later in the year sent £11 more. Newton returned to the college on 28 September according to his account, on 29 September according to the college's.[20]

He did not have long to wait before Cambridge provided an occasion to display his new grandeur. On 1 May 1669, Cosimo de'Medici, Prince of Tuscany, visited the town and university. As he was conducted from the Rose Tavern to the public schools, all the students and masters in their robes lined the walk. The university entertained the prince, whose uncle had organized the famous

[16] Thomas Fuller, *History of the University of Cambridge,* ed. James Nichols (London, 1840), p. 174.

[17] Fitzwilliam notebook. The Junior Bursar's Accounts for 1668 in Trinity show an expenditure in the second quarter for plastering Newton's chamber, though it need not refer to the chamber in which he lived.

[18] *Yahuda MS* 34, f. 1. This is an account from 10 March 1670 in a nearly indecipherable hand. It definitely mentions a rug and a bed for which Newton paid in part with his old bed.   [19] Fitzwilliam notebook.   [20] *Ibid.* Edleston, p. lxxxv.

Accademia del Cimento (Academy of Experiment) in Florence, with a philosophic act, "De methodi philosophandi in experimentis fundata, et contra systema Copernicanum." We are not informed that Newton was invited to participate. The party moved on to the show places of the university, King's College chapel, St. John's, and Trinity, "where schollers, Bachelors and Master of Arts of that Coll. were orderly placed all along the first walke on both sides to yᵉ Lodge . . ." In the evening students performed a comedy in the hall.[21] The next year the Prince of Orange, the future monarch William III, visited, and in 1671 Charles himself. As he was conducted in the usual way from the town to the university through the ordered ranks of scholars and masters, the "Conduit run claret wine when his Majestie passed by who was well pleased with it." The King was entertained at dinner in Trinity, where the Masters of Arts in their robes served; Newton probably did not join them since he had by then assumed the dignity of university professor. The college presented the inevitable comedy with which his majesty "expressed himself to be well pleased, as also with the good Order of the whole University . . ." The only thing with which he did not express himself well pleased was the Bible the university presented to him.[22] The visit cost the university £1,039 5s 1d and the town £268 11s 2d, including 6d paid to John Fox to sweep the butchers' stalls.

Behind the pomp of public display, however, mortal illness ate at the life both of the university and of Trinity. On the surface they seemed to flourish. After the upheavals and contraction of the Puritan revolution, they were expanding again and approaching the peak in numbers that both had reached in the 1620s. The expansion proved deceptive; a precipitous decline, which reduced the university in the space of two decades nearly to half its former size, was about to set in early in the 1670s.[23] University institutions were increasingly shams. Ostensibly the senate still functioned as the

---

[21] Foster, *Samuel Newton's Diary*, pp. 44–5. Charles Henry Cooper, *Annals of Cambridge*, 5 vols. (Cambridge, 1842–1908), *3*, 536.

[22] Cooper, *Annals 3*, 548–9. Foster, *Samuel Newton's Diary*, pp. 64–6.

[23] A census of the university was published in 1622, which was very near the peak in the early part of the century. It showed a total population of 3,050 resident in the university, a number which included fellows, students, and servants. In that year about 285 students graduated B.A. Matriculations were running above 450 per year; with the normal discrepancy between admissions and matriculations, about 550 new students were probably admitted. Another census was taken in 1672, the peak of the Restoration surge. It showed a total population of 2,522. About 265 graduated B.A. that year, and about 350 matriculated. No census was published in the early 1690s. B.A.s had dropped to about 145, however, and matriculations below 200. By extrapolation, the size of the university must have been about 1,500, of whom well less than 1,000 would have been undergraduates. A chart showing B.A.s is found in James Bass Mullinger, *A History of the University of*

supreme governing body; in fact, a small oligarchy of college masters (or heads) appointed directly or covertly by the Crown dominated it. The university officers, once chosen by the regent masters from their ranks, were now imposed on them from outside or selected by rotation from senior fellows of the colleges. In the fall of 1669, scarcely a year after Newton became a member, the senate tried to retain its power to elect the Esquire Bedell. It not only failed but received for its pains a rebuke from the king, who empowered the vice-chancellor to suspend any regent master who challenged his authority.[24] Four years later, the heads seized effective nomination of the university orator. Since the act deprived Isaac Cravens, a fellow of Trinity, of the position, members of Trinity, led by senior fellows including Humphrey Babington, lodged a protest. Newton signed it along with others. As usual, the heads won.[25] In case there were any doubt, Charles informed the senate of its status with abundant clarity in 1674. On 11 July, he wrote to inform them that he was removing the Duke of Buckingham from his service. To curry favor with the Crown, the senate had elected Buckingham chancellor of the university only three years earlier. Charles now declared his election void, ordered the university to proceed at once to a new election, and recommended to their consideration his bastard son, the Duke of Monmouth. Without a pause, the senate elected Monmouth, unanimously. They also thanked Charles for his goodness in giving them the liberty of such a choice.[26] Only a major crisis in which the heads and the teaching masters might see their interests coincide would suffice to reanimate the senate. Passing graces, which granted degrees to those whom the colleges named, had become the senate's primary function. Thus were the teaching masters disinherited from the institution their predecessors had created and governed.

It is perhaps not surprising that during the restoration the teaching masters for the most part chose not to teach. Acts and disputations ceased to be performed. The statutes required all Masters of Arts of four years' standing to perform divinity acts, the most dignified exercise of a curriculum which saw theology as its culmination. In fact, they neglected divinity acts to the point that the

Cambridge (London, 1888) between pp. 212 and 213. Matriculations are listed in the *Historical Register of the University of Cambridge*, ed. J. R. Tanner (Cambridge, 1917), pp. 988–9. There are a number of revealing statistics in David Arthur Cressy, "Education and Literacy in London and East Anglia, 1580–1700" (dissertation, Cambridge University, 1972), especially a table on p. 237. Three different interpretations of the decline of Cambridge can be found in Hugh Kearney, *Scholars and Gentlemen. Universities and Society in Pre-Industrial Britain* (London, 1970), pp. 141–73; Mark H. Curtis, *Oxford and Cambridge in Transition, 1558–1642* (Oxford, 1959), pp. 272–81, and Richard S. Westfall, "Isaac Newton in Cambridge," forthcoming.
24 Cooper, *Annals, 3,* 537–9. Cambridge University Library, Baker MSS, Mm.1.53, ff. 62–3.
25 *Edleston*, pp. xlvii–viii.   26 Cooper, *Annals, 3,* 559–60.

court admonished the university to see to their performance, with the usual null effect. Despite his solemn oath at Trinity, Newton, for one, never kept a divinity act. University professors converted their positions into sinecures, even those founded during the Restoration. As Lucasian professor of mathematics, Newton resisted this trend only during the first half of his tenure.

To some extent, the decline of the university proceeded in step with the rise of the colleges. The colleges were the younger institutions, most of them like Trinity creations of the sixteenth century which the Tudor government fostered as instruments of discipline in the two universities. The apparent prosperity of the colleges had to do with externals alone. Their inner malaise was no less profound than that of the university, and Trinity did not differ from the rest. In government, the tyranny of the heads over the senate repeated itself in microcosm in the tyranny of each head over his house.[27] The neglect of exercises and obligations repeated itself as well.[28] The foundation of Trinity provided for eight college lecturers; like university professorships, the positions became sinecures. The great majority of fellows evaded tutoring, which concentrated itself in the hands of a small number, like Newton's tutor Pulleyn, who undertook the duty for the extra income it provided. The others called them derisively "pupil mongers."[29] Other facets of the old Cambridge discipline collapsed as well. It was symptomatic of the new atmosphere that the master and seniors of Trinity agreed on 11 August 1668 that "no other Saturdaies be fish-daies in ye Hall but those which in ye Rubrick are appointed to be observed as daies of fasting and abstinence."[30]

---

[27] Cf. a story about Lazarus Seaman, master of Peterhouse, in the 1650s (Mullinger, *The University of Cambridge, 3*, 392–416); William Taswell's relations with Dr. Fell, dean of Christ Church a generation later (William Taswell, *Autobiography and Anecdotes*, ed. George P. Elliott, in *Camden Miscellany, 2* [Camden Society, 1852] pp. 24–5); Richard Bentley's use of his powers as master of Trinity early in the eighteenth century (James Henry Monk, *The Life of Richard Bentley, D. D.*, 2nd ed., 2 vols. [London, 1833], *1*, 231–2).

[28] In Trinity a conclusion recorded in 1683 noted the "very great failure and neglect" of college disputations and established healthy (indeed prohibitive) fines for nonperformance (Conclusion Book, 1646–1811, p. 162). Signed by the master, John North, alone without any of the senior fellows, this incident reflected the tension between North and the rest of the college, but their bad relations do not call the facts asserted into question. Late in 1684, after North's death, the vice-master and seniors fined three fellows for "their great contempt" in being present in the college but not opposing in their turn (*ibid.*, p. 167). In 1693, fines had to be imposed for nonattendance at chapel (*ibid.*, p. 187). In 1698, fines were levied anew (at a much lower rate than North had tried to collect) for neglecting to oppose at divinity disputations (*ibid.*, p. 197).

[29] Cf. William Taswell's comments on his income from tutoring (*Autobiography*, pp. 28–9).

[30] Master's Old Conclusion Book, 1607–1673, p. 276. Cf. Taswell's comments on the observance of days of abstinence at Oxford: "These nights were so far from being kept as they should be, that we commonly lived more sumptuously than usual, at inns or coffeehouses." He spoke of "cramming myself with meat and drink . . ." (*Autobiography*, pp. 31–2).

No small share of responsibility for the deterioration of tone belonged to the system of seniority. Seniority had always played a role in the life of the college, but in the statutes it vied with another governing principle, academic achievement. For example, the college bestowed stipends according to degree; a Doctor of Divinity drew more than twice as much as a Master of Arts. By the Restoration, however, the dividend furnished the main substance of a fellow's support, and according to the definitive formula of division adopted by the college in 1661, the year of Newton's admission, seniority alone determined the level of the dividend.[31] Seniority also determined virtually all the other substantial rewards of college life: assignment of chambers, college offices, university offices in Trinity's turn in the rotation, especially presentation to benefices in the college's gift. The Steward's Book for 1666 listed Newton (who was then a scholar), below another scholar, John Herring, who stood above him solely because he entered Trinity on 3 June 1661, two days before Newton. As it happened, Herring did not attain a fellowship, and his name dropped out after 1668. Herring would otherwise have continued one step ahead of Newton in college preferment until death or resignation, and any amount of activity in tutoring, any signal service to the college, or twenty *Principias* could not have moved Newton ahead of him. Nor could any record of sloth have dropped Herring back.

If the dead hand of seniority lay heavily on Trinity, as it did on every college, the throttling hand of patronage clutched at its throat. In 1664, as alarm over the rising tide of mandated fellowships spread through the university, St. John's protested to the king that mandates would injure the college by "causing deserving persons . . . to seek interest at court rather than proficiency in learning."[32] Already the system of patronage had pervaded the university beyond the possibility of elimination, however. And nowhere did it entrench itself more deeply than at the pinnacle of the university hierarchy. Almost by definition, to be a master of a college was to be the object of patronage. John North, master of Trinity from 1677 to 1683, inveighed against the corruption of patronage but accepted that tendered to him as a matter of right.[33] Benjamin Laney, restored to the mastership of Pembroke in 1660, was rewarded for his loyalty during the Interregnum with further preferments as dean of Rochester, bishop of Peterborough, and canon of Westminster, all of which appointments he held at the same time. The mastership of St. John's, the Regius professorship of divinity, a prebend in Canterbury, and two parsonages likewise rewarded Peter Gunning for his

[31] Master's Old Conclusion Book, 1607–1673, p. 265.
[32] Mullinger, *The University of Cambridge, 3*, 626–7.   [33] North, *Lives, 2,* passim.

loyalty.[34] He was master when St. John's deplored the encouragement that mandates gave to seek interest at court. Small wonder such protests bore little fruit. In lesser amounts, patronage filtered down through the entire university. In Trinity, in 1661, Nathaniel Willis obtained a dispensation to hold his fellowship along with a rectory, which exceeded the statutory limit. A year later, Humphrey Babington, whom a letter mandate had recently restored to his fellowship, had the same statute dispensed with so that he might hold the rectory in Boothby Pagnell. Two years more and Robert Boreman received the same privilege by dispensation.[35] Cambridge might deplore letters mandate for others, but men who wanted letters mandate for themselves filled Cambridge. A good three-quarters of the students proceeded to careers in the church; they knew full well whence came livings, as they were frankly called.

Nothing is more revealing of the university than the ritual volumes of Latin verse published on every occasion that touched the Crown. To fill these productions, every ambitious man in Cambridge periodically tortured his muse for his quota of lines in order that his name not be missed. Masters and professors seemed particularly anxious to appear. None labored more diligently in the production of verse than Isaac Barrow, recently Regius Professor of Greek and presently Lucasian Professor of Mathematics, but clearly hungry for more. *Lacrymae Cantabrigienses* proclaimed the title of the 1670 edition, which mourned the death of Charles's sister by poison, a hideous compilation of sycophancy unmoistened by any suggestion of a real tear.

Andrew Marvell sketched the whole university when he characterized Francis Turner, master of St. John's, as "this close youth who treads always upon the heels of Ecclesiastical Preferment."[36] To savor the full ambiguity of the situation, realize that Newton, who never contributed to a valedictory volume, and stood resolutely aside from the scramble for place, would in less than ten years owe his survival in Cambridge to a manipulation of the statutes by royal dispensation. There was no one in Cambridge who did not have a favor to seek.

The colleges themselves held extensive patronage, which they dispensed, inevitably, according to seniority. Trinity alone con-

[34] Mullinger, *The University of Cambridge, 3*, 566–7. Baker, *History of St. John's*, p. 236.

[35] Churchill, *Dispensing Power*, pp. 310–11. Conclusion Book, 1646–1811, p. 68. Trinity College, Box 29, D. Cf. D.A. Winstanley, *The University of Cambridge in the Eighteenth Century* (Cambridge, 1922), which focuses primarily on the role of patronage in the university, exercised now, not by the monarch, but by the Duke of Newcastle, a parliamentary magnate who was chancellor of the university for twenty years in the middle of the century.

[36] Andrew Marvell, *Mr. Smirke: Or the Divine in Mode*, in *The Complete Works of Andrew Marvell*, ed. Alexander B. Grosart, 4 vols. (private circulation, 1872–5), *4*, 11.

trolled nearly fifty advowsons and added more continually.[37] It took a fellow nearly twenty years to achieve the seniority requisite for a college living. Once he had waited it out, he could if he chose resign his fellowship, belatedly marry, and with income secure rear a family. In addition, masters bestirred themselves in recommending fellows as chaplains, tutors, and vicars. Or if, for whatever reason, marriage held no charms, a fellow with seniority could become a college preacher, which bestowed the right to hold a living in conjunction with a fellowship.

In either case, the fellowship rapidly came to appear primarily as a mode of support, an end in itself rather than a means to an academic end. Everybody treated it as a freehold which imposed no obligations in return for its benefits, which were considerable. Its income did not even require one's presence. The Steward's Books of Trinity indicate that year in and year out more than a quarter and often more than a third of the fellows resided in Cambridge less than half the year; more than 10 percent never showed up at all. The fate of the statute *de mora* in Emmanuel and Sidney Sussex colleges, the two preeminent Puritan institutions, is revealing. Sir Walter Mildmay founded Emmanuel to enrich the church with an educated clergy; hence the statute *de mora,* which limited tenure to ten years beyond the M.A. degree. A fellowship, Mildmay insisted, was not meant to provide a permanent abode. Under Laud, the statute was revoked. The revolutionary Puritan government resurrected it. With the restoration, it disappeared forever.

By the late seventeenth century, stories of the effects of these changes on the fellows themselves filled the air both of Cambridge and of Oxford. Freed from obligations, shorn of useful functions, the fellows surrendered meekly to a life of indolence and boredom relieved primarily by the solace of the table and the tavern. In the splendid phrase of Roger North, they became "wet epicures," succumbing to a corpulent lethargy liberally flavored with alcohol.[38] William Whiston told of a student at Clare with him in the 1690s who decided that a reputation as a heavy drinker was the surest path to a fellowship.[39] At much the same time in Oxford,

---

[37] They are listed in the Trinity Register, volumes for 1664–72, 1673–80, 1680–8, and 1688–1702. Four grants of advowsons to the college in 1667, 1673, 1678, and 1681 are in *Miscellaneous Papers Relating to Trinity College,* vol. *3,* Nos. 21, 27, and 39. In a manuscript from about 1736, Thomas Baker listed the number of livings held by the various colleges; by that time Trinity held seventy (more than twice as many as any other college), ten of which had incomes above £100 per annum (Cambridge University Library, Baker MSS Mm.1.48).   [38] North, *Lives, 2,* 272.

[39] William Whiston, *Memoirs of the Life and Writings of Mr. William Whiston* (London, 1749) p. 129. In 1727, Louis de Jaucourt, a collaborator of Diderot's in the *Encyclopedia,* visited Cambridge; "whoever is ignorant of the art of drinking a lot and smoking a lot," he wrote in a letter, "is very unwelcome in this University . . ." (quoted in Arthur M. Wilson, *Diderot* [New York, 1972], p. 481).

Humphrey Prideaux, when asked to recommend a college for his nephew, replied that most of the men in authority in the colleges there were "such as I could scarce commit a dog to their charge."[40] In its original concept, a fellowship was meant to support learning. "When any person is chosen fellow of a college," Nicholas Amherst declared early in the eighteenth century, "he immediately becomes a freeholder, and is settled for life in ease and plenty . . . He wastes the rest of his days in luxury and idleness: he enjoys himself, and is dead to the world: for a senior fellow of a college lives and moulders away in a supine and regular course of eating, drinking, sleeping, and cheating the juniors."[41] Amherst was a satirist whose words should not be mistaken for disinterested description, but Dr. Johnson was not playing the satirist when he visited his old Oxford college, Pembroke, in 1754 and met the Reverend Mr. Meeke, a student with Johnson who had envied his learning, and now a fellow of the college. "About the same time of life," Johnson said to his friend Wharton as they left, "Meeke was left behind at Oxford to feed on a Fellowship, and I went to London to get my living: now, Sir, see the difference of our literary characters."[42]

Early in the seventeenth century, Trinity had been the leading academic institution in England. It had nurtured six of the translators of the Authorized Version, more than any other college in Oxford or Cambridge; it had furnished more bishops to the Church of England than any other foundation. Likewise, Cambridge had functioned as the point of ferment in English intellectual life. Traditions do not die at once. John Pearson, master of Trinity from 1662 to 1673 (during Newton's early years there), was one of the most esteemed clergymen of his age, and his successor, Isaac Barrow, was an outstanding scholar of many facets. Elsewhere in the university the Cambridge Platonists survived for a time from the earlier age. As they died off, no one replaced them. Early in the eighteenth century, the German traveler Zacharius von Uffenbach visited Cambridge. His recital of the wretched state of the university and colleges is a depressing chronicle of the level to which they had fallen. Though he did not meet the university librarian, Dr. Laugh-

---

[40] Quoted in Charles Edward Mallet, *A History of the University of Oxford*, 3 vols. (London 1924–7), *3*, 56. Cf. the diatribe of Dr. John Edwards written about 1715, against the "laziness and debauchery" of the fellows of the various colleges (quoted in James Bass Mullinger, *St. John's College* [London, 1901], pp. 203–5). Edwards was then a very old man, but it is not possible to pass off what he says entirely as the cluckings of the old against the young. See also the descriptions of life in Restoration Oxford (Mallet, *Oxford*, *2*, 422–3), in Jacobite Oxford (*ibid.*, *3*, 1–55), and in eighteenth-century Cambridge (D. A. Winstanley, *Unreformed Cambridge* [Cambridge, 1935], pp. 256–67).

[41] Quoted in A. D. Godley, *Oxford in the Eighteenth Century* (London, 1908), p. 77.

[42] James Boswell, *The Life of Samuel Johnson L.L.D.* (New York, n.d.), p. 161.

ton, who was out of town, he heard Laughton extolled as a man of great learning. "Rara avis in his terris," Uffenbach remarked tartly to his diary.[43] To be sure, Uffenbach rather made a career of denigrating all things English. Ample evidence elsewhere confirms his judgment, however. Cambridge was fast approaching the status of an intellectual wasteland.

Consider the forty-one men who became fellows of Trinity in the three elections of 1664, 1667, and 1668. One of the forty-one was Newton, of course. Of the others, Robert Uvedale became a prominent educator and horticulturist; he pursued his career entirely outside Cambridge though he held onto his fellowship for fifteen years. Edward Pelling, who resigned his fellowship after one year, went on to become an Anglican polemicist of heroic proportions though his writings are, I believe, virtually unknown today. Samuel Scattergood, who held his fellowship for sixteen years, published many sermons; Henry Dove, George Seignior, and William Baldwin all published a small number. John Batteley gained some prominence as an antiquarian after his tenure of seventeen years in his fellowship. John Allen, a fellow for thirty years, mostly in absentia, published one sermon with the intriguing title, in view of the oaths he had taken, "Of Perjury." Newton aside, they do not form an imposing group of intellectuals by any standard. Nor were they more impressive as tutors. Four chose the role of pupil monger, in the pejorative phrase of the day. Of the other thirty-seven, only ten ever tutored a pupil, and those ten tutored a total of sixteen. Newton with three and Wickins with two accounted for five of the sixteen. The average tenure of the forty-one was seventeen and a half years; eleven stayed more than twenty years; and four fulfilled Mildmay's fears by making the college their permanent abode. George Modd, Patrick Cock, William Mayor, and Nicholas Spencer all stayed on at Trinity over forty years. After more than thirty years, Spencer did take a degree as Doctor of Divinity; the other three took no further degrees. None of the four ever tutored a pupil. None of the four ever published a word. All survived to become senior fellows of the college and to reap its ripest rewards.

After his creation as Master of Arts, Newton lived in Trinity for twenty-eight years. Those years coincided roughly with the most disastrous period in the history both of the college and of the university. Whatever his initial expectations may have been, he did not find a congenial circle of fellow scholars. A philosopher in search of

[43] J. E. B. Mayor, *Cambridge under Queen Anne* (Cambridge, 1870) p. 140. According to Uffenbach, Jean LeClerc, the Huguenot scholar, shared his opinion. LeClerc complained to him "of the great laziness of Englishmen, and justly too; enjoying such large *beneficia* and noble libraries, they produced very little in the way of learning; which is only too true, with a few bright exceptions" (*ibid.,* p. 427).

truth, he found himself among placemen in search of a place. This fundamental fact colored the scene in which virtually the whole of his creative life was set.

Against this background, we can read the various anecdotes that have survived about his life in the college. Three sources furnished most of the stories: William Stukeley, a student at Cambridge early in the eighteenth century and later a friend of Newton, who made it his business to collect information about him after his death; Humphrey Newton, who served as Newton's amanuensis in Cambridge for five years in the 1680s and wrote two letters about the experience after Newton's death; and Nicholas Wickins, the son of John Wickins, who wrote about his father's recollections to Robert Smith soon after the death of Newton and eight years after the death of his father. The first two were not entirely independent since both men lived in Grantham in 1627–8 and conferred together. Some of Stukeley's anecdotes sound enough like Humphrey Newton's that they seem to derive from him instead of from Cambridge. Humphrey Newton lived with Isaac Newton during a unique period while he was composing the *Principia*. Perhaps we should exercise some caution in treating his recollections as typical, though Newton's capacity to be dominated by a problem did not confine itself to the *Principia*. Stukeley reported that stories of Newton's absentmindedness were rife in Cambridge.

> As when he has been in the hall at dinner, he has quite neglected to help himself, and the cloth has been taken away before he has eaten anything. That sometime, when on surplice days, he would goe toward S. Mary's church, insted of college chapel, or perhaps has gone in his surplice to dinner in the hall. That when he had friends to entertain at his chamber, if he stept in to his study for a bottle of wine, and a thought came into his head, he would sit down to paper and forget his friends.[44]

Humphrey Newton's chaotic stream of consciousness contained similar recollections.

> He always kept close to his studyes, very rarely went a visiting, & had as few Visiters, excepting 2 or 3 Persons, Mr Ellis of Keys, Mr Lougham [called Laughton in his other letter] of Trinity, & Mr Vi-

---

[44] *Stukeley*, p. 61. Nicholas Wickins reported that his father also called the stories of Newton forgetting meals "what ye world has so often heard of Sr Isaac . . ." (*Keynes MS* 137). The story about the surplice was heard and recorded by Thomas Parne, among papers he assembled early in the eighteenth century for a history of Trinity. He attributed it to a Mr. Burwell, who Edleston speculates may have been Alexander Burrell, eleven years Parne's senior in Trinity and possibly related to a chaplain of that name in Trinity from 1673 to 1681. "Newton hath come into the Hall without his Band, and went toward St. Maries in his surplice" (*Edleston*, p. lxxx). On college surplice days, fellows were required to wear their surplices to the college chapel, not to St. Mary's, which was the university church, and obviously not to dinner in the hall.

gani, a Chymist, in whose Company he took much Delight and Pleasure at an Evening, when he came to wait upon Him. I never knew him take any Recreation or Pastime, either in Riding out to take ye Air, Walking, Bowling, or any other Exercise whatever, Thinking all Hours lost, y^t was not spent in his studyes, to wch he kept so close, y^t he seldom left his Chamber, unless at Term Time, when he read in ye schools, as being Lucasianus Professor . . . He very rarely went to Dine in ye Hall unless upon some Publick Dayes, & then, if He has not been minded, would go very carelesly, wth Shooes down at Heels, Stockins unty'd, surplice on, & his Head scarcely comb'd.[45]

"He would with great acutness answer a Question," Humphrey added in his second letter, "but would very seldom start one."[46] During five years, Humphrey saw Newton laugh only once. He had loaned an acquaintance a copy of Euclid. The acquaintance asked what use its study would be to him. "Upon which Sir Isaac was very merry."[47]

It is not hard to recognize in these anecdotes the man who unconsciously sketched his own portrait in his papers, a man ravished by the desire to know. Equally, it is not hard to recognize his status in Trinity – isolation, indeed alienation. True, Stukeley mentioned friends being entertained in his chamber, and Humphrey Newton named three of them. The references scarcely suffice to erase the impression left by the rest. Newton seldom leaves his chamber. He prefers to eat there alone. When he does dine in the hall, he is hardly a genial companion; rather he sits silently, never initiating a conversation, as isolated in his private world as though he had not come. He does not join the fellows on the bowling green. He rarely visits others. None of those who visit him are fellows of Trinity. Of the three, we know that Newton later broke with Vigani because he "told a loose story about a Nun . . ."[48] Whatever his

---

45 *Keynes MS* 135. John Laughton, admitted pensioner to Trinity in 1665, was elected scholar but not fellow. He did become a chaplain in the college, however, and college librarian. Later he was university librarian from 1686 to 1712, the one referred to by Uffenbach. John Ellis, admitted sizar to Caius in 1648, was a fellow there from 1659 to 1703 and an eminent tutor. Elected master in 1703, he was serving as vice-chancellor of the university in 1705 when he was knighted on the same day as Newton. John Francis Vigani came to Cambridge about 1682 and taught chemistry there informally for twenty years before the university conferred the title of professor of chemistry on him; it did not confer any stipend to go with the title. At least part of the time, he experimented and taught in Trinity, where his laboratory was apparently in the room (still called Vigani's room) which Newton himself had used for that purpose earlier.

46 *Keynes MS* 135.

47 *Stukeley*, p. 57. Stukeley ascribed the story to Humphrey, who alluded to it without giving the whole since Stukeley had already relayed it to Conduitt.

48 *Keynes MS* 130.6, Book 2. Conduitt attributed this story to his wife, Catherine, Newton's niece.

friendships with Laughton and Ellis, they were not close enough to elicit correspondence from either side after Newton left Cambridge.

On 18 May 1669, Newton wrote a letter to Francis Aston, a fellow of Trinity who had received leave to travel abroad and was then departing. According to the letter, Aston had asked his advice about traveling, and Newton complied rather fully. The bulk of the letter was worldly advice cribbed from a discourse still found among the Newton papers: "An Abridgement of a Manuscript of Sr Robert Southwell's concerning travelling."[49] Aston should adapt his behavior to the company he is in. He should ask questions but not dispute. He should praise what he sees rather than criticize. He should realize that it is dangerous to take offense too readily abroad. He should observe various things about the economy, society, and government of the countries he visits. A final paragraph added a number of particular enquiries that Newton wished Aston to make, mostly about alchemy and mostly based on Michael Maier, *Symbola aureae mensae duodecim nationum* (Frankfort, 1617).[50] The letter to Aston is among the most eloquent in Newton's correspondence. Not for its content—with its borrowed air of worldliness, the letter itself is more ludicrous than eloquent. It is found today among Newton's own papers, which suggests that he recognized he was cutting a ridiculous figure as he assumed a worldly posture on the basis of one month in London and an essay by Southwell, and decided not to send it. The eloquence of the letter lies in its uniqueness. It is the only personal letter to or from a peer in Cambridge in the whole corpus of Newton's correspondence. In its uniqueness, it adds color to the portrait of isolation in Stukeley's and Humphrey Newton's anecdotes.

So do surviving accounts of the college that were collected in the second decade of the eighteenth century. Thomas Parne, B.A. 1718, collected materials for a history of the college, including the recollections of elderly fellows such as George Modd. He recorded particulars about Ray, Pearson, Barrow, Thorndike, and Duport. Newton was a famous man when Parne was drawing his materials together, far more famous than the men above, but only three references to him appear in the collection: his name (without any comment whatever) at the head of the list of writers; the dates of his elections to Parliament and of his later unsuccessful bid for election; and one brief anecdote about his absence of mind.[51] At much the same time, James Paine, who was elected to a fellowship in 1721, set down a conversation with Robert Creighton, who had been a fellow from 1659 to 1672. Creighton recalled Pearson,

---

[49] *Keynes MS* 152.
[50] *Corres 1*, 9–11. Newton's early notes on Maier are in *Keynes MS* 29.
[51] *Edleston*, pp. lxxix–lxxx.

Dryden, Gale, Wilkins, and Barrow; he did not mention Newton.[52] Neither did Samuel Newton, who as registrar and auditor was employed by the college during the whole of Newton's tenure as fellow, enter his name in the diary he kept until Newton's election to Parliament in 1689.[53] To be sure, Samuel Newton filled his diary more with events than with references to individuals. Nevertheless, it seems evident that Newton did not loom prominently on the college scene.

Two stories do suggest that for their part the other fellows, whatever their amusement at his absentmindedness, regarded him with awe. In 1667, he seemed to have prophetic powers when the Dutch fleet invaded the Thames.

> Their guns were heard as far as Cambridg, and the cause was well known; but the event was only cognisable to Sir Isaac's sagacity, who boldly pronounc'd that they had beaten us. The news soon confirm'd it, and the curious would not be easy whilst Sir Isaac satisfy'd them of the mode of his intelligence, which was this; by carefully attending to the sound, he found it grew louder and louder, consequently came nearer; from whence he rightly infer'd that the Dutch were victors.[54]

When he walked in the fellows' garden, "if some new gravel happen'd to be laid on the walks, it was sure to be drawn over and over with a bit of stick, in Sir Isaac's diagrams; which the Fellows would cautiously spare by walking beside them, and there they would sometime remain for a good while."[55]

As far as we know, Newton formed only three close connections in Trinity, all of them rather elusive. With John Wickins, the young pensioner he met on a solitary walk in the college, he continued to share a chamber until Wickins resigned his fellowship in 1683 for the vicarage of Stoke Edith. Wickins was frequently absent for extended periods, and during his final five years he was hardly there at all. When Robert Smith, Plumian professor of natural philosophy

---

[52] Quoted in W. W. Rouse Ball, *Notes on the History of Trinity College Cambridge* (London, 1899), pp. 97–9.

[53] Foster, *Samuel Newton's Diary*, pp. 97–8. He merely gave his name and title. Newton appeared only one other time, when William III proposed to appoint him provost of King's late in 1689 (*ibid.*, p. 102). Samuel Newton did not even mention his knighting, though he did mention that of his kinsman, John Ellis, on the same day. Apparently the two Newtons were acquainted since Newton loaned "Mr Newton" 18s in 1665 (Fitzwilliam notebook).

[54] *Stukeley*, pp. 58–9. This must have been a common story, because it survived in a second account as well, though placed on this occasion in 1672 with the battle of Southwold Bay. Edleston quotes it from John Nichols, *The History and Antiquities of Hinckley* (London, 1780), p. 61n (pp. xlvi–vii).

[55] *Stukeley*, p. 61. Stukeley referred this story to Humphrey Newton; hence it belonged to the time when Newton was composing the *Principia*.

wrote to Wickins's son Nicholas after Newton's death, he could find only three short letters to his father, which he thought were not even worth transcribing, plus four or five other very short letters in which Newton forwarded rent and dividends. Although it is not clear when the first three letters were written, it appears that Newton severed connections with his chamber fellow of twenty years. Nicholas Wickins did say that Newton paid for Bibles to be distributed among the poor in the parish.[56] The relation with Wickins was special and the break with him probably more significant than the break with Vigani. In addition to Wickins, there were relationships with Humphrey Babington and Isaac Barrow about which we know even less.

The alienation from college society worked to Newton's advantage. As Dr. Johnson's friend, the Reverend Mr. Meeke, was to learn, the increasing triviality of the fellows' lives could entangle a promising man and destroy him. Passionately inclined to study in any case, Newton turned away from his peers and in upon himself and surrendered completely to the pursuit of knowledge. The Steward's Books of the college show that he seldom left. In 1669 (which means, in the college books, the twelve months that ended with Michaelmas, 29 September 1669) he was there all fifty-two weeks; in 1670, forty-nine and a half; in 1671, forty-eight; in 1672, forty-eight and a half. When he did leave, it was usually for a trip home. A decade later, Humphrey Newton found that he seldom went to morning chapel since he studied until two or three every morning. For that matter, he seldom interrupted his studies to attend evening chapel either, though he did go to church in St. Mary's on Sundays.[57] "I believe he grudg'd y^t short Time he spent in eating & sleeping," Humphrey Newton observed.[58] The Reverend John North, master of the college from 1677 to 1683 and resident in it for a time before that, who rather fancied himself a scholar, "believed if Sir Isaac Newton had not wrought with his hands in making experiments, he had killed himself with study."[59]

The laxity of the system, which had helped him already as an undergraduate, continued to favor him. If it demanded nothing of the George Modds and Patrick Cocks, likewise it demanded nothing of him. His use of his leisure may have unsettled the others, but the essence of the system was tolerance. Unrelenting study on a fellowship intended to support study was not demonstrably more subversive than drawing dividends in absentia. Supported comfortably, Newton was free to devote himself wholly to whatever he chose. To remain on, he had only to avoid the three unforgivable

---

[56] *Keynes MS* 137.
[57] *Stukeley*, p. 60. Stukeley referred the information to Humphrey Newton.
[58] *Keynes MS* 135.   [59] North, *Lives*, 2, 284.

sins: crime, heresy, and marriage. Safely ensconced with Wickins in the orthodox fastness of Trinity College, he was not likely to sacrifice his security for one of them.

Besides the three topics of the *anni mirabiles*, a new subject began now to engross him. His accounts show that in 1669 he spent 14s for "Glasses" in Cambridge and 15s more for the same in London.[60] He made some other purchases in London as well.

For Aqua Fortis, sublimate, oyle perle [*sic* – per se?] fine Silver, Antimony, vinegar Spirit of Wine, White lead, Allome Nitre, Salt of Tartar, ☿	2.	0.	0
A Furnace	0.	8.	0
A tin Furnace	0.	7.	0
Joyner	0.	6.	0
Theatrum Chemicum	1.	8.	0[61]

He also paid 2s to have the oil transported to Cambridge. *Theatrum Chemicum* referred to the huge compilation of alchemical treatises published by Lazarus Zetzner in 1602 and recently expanded to six volumes. More than woodworking was going on in the chamber shared by the long-suffering Wickins. Years later Newton remarked to Conduitt that Wickins, who was stronger than he, used to help him with his kettle, "for he had several furnaces in his own chambers for chymical experiments."[62] When Newton's hair turned gray early in the 1670s, Wickins told him it was the effect of his concentration. Newton, whom Humphrey Newton saw laugh only once, would jest that it was "ye Experim^{ts} he made so often wth Quick Silver, as if from Hence he took so soon that Colour."[63]

Chemical experimentation had medical implications also, and Newton had more worries than gray hair. He suspected he had tuberculosis, for which, Wickins recalled, he treated himself with Lucatello's Balsam. As it happens, Newton's formula for "Lucatello's Balsome" has survived, a heady mixture of turpentine, the best damask rosewater, beeswax, olive oil, and sack, flavored with a pinch of red sandalwood and a dash of oil of St. John's wort. According to Newton's recipe, it was good for measles, plague, and smallpox, for all of which one was to mix a quarter of an ounce in a little broth, take it warm, and sweat afterwards (which should have been no problem). For the bite of a mad dog, one applied it to

60 The Trinity Exit and Redit Book shows that he was absent from 26 November to 8 December (*Edleston*, p. lxxxv). This is the absence shown in the Steward's Book for 1670.

61 Fitzwilliam notebook.

62 Conduitt's memorandum of 31 Aug. 1726; *Keynes MS* 130.10, f. 3^v.

63 Nicholas Wickins to Smith; *Keynes MS* 137. Humphrey Newton also reported that Newton was gray by thirty (*Keynes MS* 135).

the wound as well, while wind colic, green wounds, sore breasts, burns and bruises were treated by external application alone.[64] Although "Lucatello's Balsome" did not claim to be effective against consumption, Newton apparently concluded that such a sovereign remedy would not likely fail in this, either. As Wickins remembered it, "when he had compos'd Himself, He would now and then melt in Quantity ab[t] a Q[r] of a Pint & so drink it."[65]

Meanwhile, chemistry was not his only study. In 1669, he took up mathematics and optics again. Probably Newton devoted at most a bare modicum of attention to mathematics during 1667. Perhaps he composed a paper on analytic geometry and translated a short fluxional essay into Latin at that time; it is impossible to date them with assurance.[66] Two pieces related to analytic geometry rather than the fluxional calculus possibly stem from 1668. One, on the organic construction of conics, generalized individual methods of generating conics, which he had met in Schooten's work, into a device that could describe the general conic with a continuous, uninterrupted motion. Two angles fixed in size rotate about poles *d* and *e*; their legs intersect at points *b* and *p* (Figure 6.2). When intersection *p* traces a given curve, the directrix, intersection *b* describes a new curve, the describend, which is either of equal degree or, when the directrix embodies certain conditions, of higher degree. With certain rectilinear directrices, *b* can describe a conic through *d* and *e*. Similarly Newton argued that the device can trace a cubic and a quartic from a conic, and a quartic, a quintic and a sextic from a cubic directrix of suitable properties.[67] Twenty years later, he incorporated the burden of the paper into Section V, Book I, of the *Principia*.

The other paper that may date from 1668, though it may also belong to 1670 when John Collins was thrusting related questions before Newton, is more significant.[68] *Enumeratio curvarum trium dimensionum* (*The Enumeration of Cubics*) was a sophisticated exercise

---

[64] Stanford University MS 538.   [65] *Keynes MS* 137.

[66] "Problems of Curves"; *Math 2*, 175–84. *De solutione problematum per motum*; *Math 2*, 194–200.

[67] The paper exists in four successive forms; *Math 2*, 106–50. The final version (pp. 134–50) bears the title *De modo describendi conicas sectiones et curvas trium dimensionum quando sint primi gradus. &c.* Whiteside dates it provisionally to late 1667 or 1668.

[68] Dating papers by the hand in which they are written is chancy business at best. On that evidence alone, and with explicit reservations about it, Whiteside places this paper in late 1667 or 1668. From such a time, the paper springs before us without motive and without intimate connection to the main themes of Newton's mathematical work. 1670 at least provides a motive, the questions Collins persistently raised about cubics. See W. W. Rouse Ball, "On Newton's Classification of Cubic Curves," *Proceedings of the London Mathematical Society*, 22 (1890–1), 104–43; and H. Hilton, "Newton on Plane Cubic Curves," in W. J. Greenstreet, ed., *Isaac Newton, 1642–1727* (London, 1927), pp. 115–16.

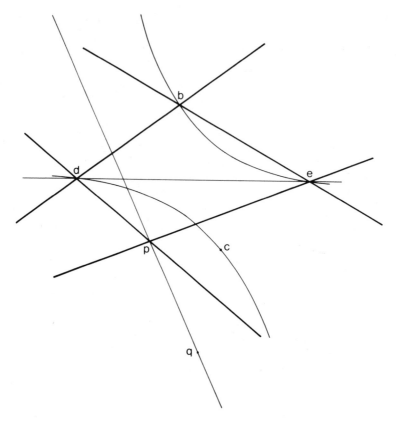

Figure 6.2. Newton's organic construction of curves. The two fixed
angles *bdp* and *bep* pivot on the points *d* and *e*. As the
intersection *p* traces the directrix *pq*, intersection *b* traces
the describend *be*, *dc*, a conic (in this case, a hyperbola)
through *d* and *e*.

in analytic geometry which burst the swaddling clothes in which
the infant science had been nourished and extended it successfully to
curves of higher order than the conics. To be sure, analysis had
confronted cubics before, but in 1668 only five individual cubics
had been described, and they imperfectly. When Newton had tried
to describe two cubics in 1664, he had failed to do so adequately.
What he now undertook was a systematic taxonomy of cubics
which classified them into cases, species, and (within species) forms
and the grades they can take on.

He began with a kinematic description of curves. The line *BC*
(the describer) moves on the base *AB*, with which it maintains a
fixed angle, while the point *C* moving on *BC* traces the curve
CE (Figure 6.3). That is, *AB* and *BC* are the two coordinates of
the curve.

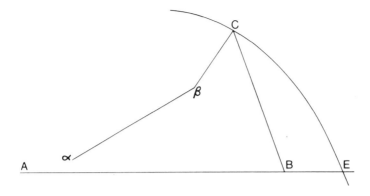

Figure 6.3. The reduction of the cubic with coordinates AB, BC to the different coordinates $\alpha\beta$, βC.

And since any fixed line $\alpha\beta$ may be the base and the describer βC may be ordinate to it in any given angle, it is evident that the nature of the same single curve may be expressed in an infinity of ways. It is our purpose to show in what manner one may arrive at the simplest and thereby enumerate and determine the species of curves.[69]

Newton then proceeded to show that by a suitable choice of coordinates the general cubic

$$ay^3 + bxy^2 + cx^2y + dx^3 + ey^2 + fxy + gx^2 + hy + kx + l = 0$$

can always be reduced to eliminate four terms. The conversion to new coordinates involved a herculean exercise in algebra which passed through an equation with eighty-four terms. He emerged successfully with the canonical equation of the general cubic

$$bxy^2 + dx^3 + gx^2 + hy + kx + l = 0$$

He went on to illustrate nine cases of cubics, and within the cases sixteen species. With the further divisions of species into forms and grades, he arrived in all at fifty-eight distinct types of cubics, all of which he plotted with some care (Figure 6.4). As an aid in classifying and plotting, Newton defined the concept of the diametral hyperbola, a conic hyperbola drawn to the axis and one of the asymptotes of the cubic. Like the diameter of a conic, the diametral hyperbola bisects chords (including those between two branches of the cubic) and cuts the curve at extrema. A quartic equation, obtained from the cubic and diametral hyperbola solved simultaneously to yield the extrema, bore much of the burden of Newton's taxonomy.[70]

---

[69] *Math 2*, 11.
[70] The *Enumeratio* went through three versions: a first incomplete one (*Math 2*, pp. 10–16); a second incomplete one (*ibid.*, pp. 18–36); and the final one for that time (*ibid.*, pp. 36–84).

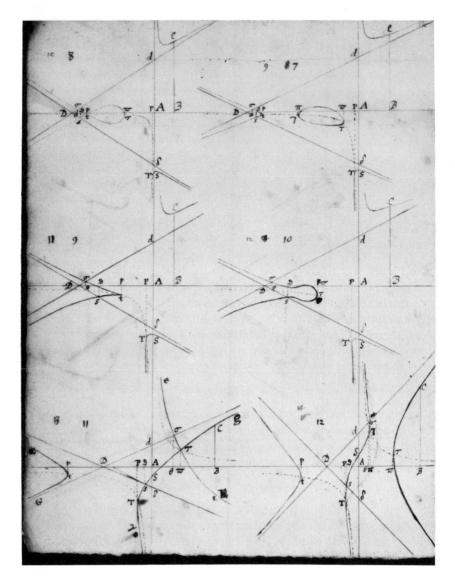

Figure 6.4. Newton's drawing of some of the cubics. (Courtesy of the Syndics of Cambridge University Library.)

In his treatment of cubics, Newton amply displayed the power of his geometric imagination. In analogy to the continuity of conics, whereby the hyperbola passes into the parabola, the ellipse, and finally the circle, so in the cubics "the first [case] passes into the second via the third, that is, as d passes into $-d$ via 0; . . . the third passes into the fourth and fifth, the fourth into the fifth via the sixth,

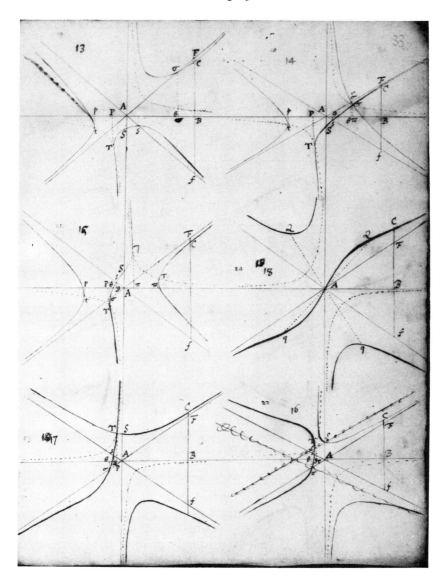

and the first three into the seventh, the seventh into the eighth and the eighth into the ninth."[71] A separate paper composed at much the same time, perhaps as an introduction to the enumeration, similarly set out to generalize such characteristics and properties of conics as ordinates, diameters, axes, vertices, and asymptotes to make them applicable to curves in general.[72] In all, Newton's work on cubics

[71] *Math 2*, 99.
[72] Early drafts, *Math 2*, 90–2; finished version, *ibid.*, pp. 94–104.

raised the study of analytic geometry to a new plane of generality. As with his method of fluxions, he thrust it back in his desk, and there it remained until he dusted it off twenty-five years later and finally published it after still another decade.

In 1669, events focused Newton's attention once more on his fluxional method and forced him to take it from his desk. Though he did not then publish it, at least he made it known. Toward the end of 1668, Nicholas Mercator published a book, *Logarithmotechnia*, in which he gave the series for log (1 + *x*), which he had derived by simply dividing 1 by (1 + *x*) and squaring the series term by term. As the title suggests, he realized that the series offered a simplified means to calculate logarithms. Some months later – the exact time is unknown, but it appears from the dates of following events to have been in the early months of 1669 – John Collins sent a copy of the book to Isaac Barrow in Cambridge. Collins was a mathematical impresario who had made it his business to foster his favorite study. To that end, he functioned as a clearinghouse for information, attempting by his correspondence to keep the growing mathematical community of England and Europe abreast of the latest developments. No doubt Collins was playing this role when he sent a copy of Mercator's work to the Lucasian professor of mathematics. Late in July, Collins received in reply a letter which informed him that a friend of Barrow's in Cambridge, "that hath a very excellent genius to those things, brought me the other day some papers, wherein he hath sett downe methods of calculating the dimensions of magnitudes like that of Mr Mercator concerning the hyperbola, but very generall . . ."[73] Barrow was not mistaken in thinking the paper would please Collins, and he promised to send it with his next letter. About ten days later, Collins did receive a paper with the title *De analysi per aequationes numero terminorum infinitas(On Analysis by Infinite Series)*. Late in August, he learned who the author was. "His name is Mr Newton; a fellow of our College, & very young (being but the second yeest Master of Arts) but of an extraordinary genius & proficiency in these things."[74]

Among other things, the episode informs us that Barrow and Newton were now acquainted. Apparently they had been for some time; Collins would later note that Newton contrived a general method of infinite series "above two yeares before Mercator pub-

---

[73] Barrow to Collins, 20 July 1669; *Corres 1*, 13.

[74] Barrow to Collins, 20 Aug. 1669; *Corres 1*, 14–15. The editor of the *Correspondence* suggests that "yeest" is a dialect form of "youngest." In August 1669, Newton was far from the second youngest Master of Arts in Trinity. I have not seen the original letter, but the phrase would make sense if "the second yeest" said, or was meant to say, "a second year."

lished any thing, and communicated the same to Dr Barrow, who accordingly hath attested the same."[75] Hence when Barrow received Mercator's book, he realized its implication for Newton's work and showed it to him.

The episode served also to confront Newton with the enormous anxieties that prospective publication aroused. Describing the events a few years later, he said that upon the appearance of Mercator's book, "I began to pay less attention to these things, suspecting that either he knew the extraction of roots as well as division of fractions, or at least that others upon the discovery of division would find out the rest [of the binomial expansion] before I could reach a ripe age for writing."[76] Forget the main and the final clauses; they imposed later reflections on his initial reaction. What he found in Mercator's book was half of the discovery that had set him on his way four years earlier. If Mercator had done it for the hyperbola, would he not do it for the circle as well (i.e., the series for $(1 - x^2)^{1/2}$, "the extraction of roots")? Moreover, Mercator had applied series expansion to quadratures. To Newton at least, the whole of his proud advance stretched out directly beyond the door Mercator had opened. We know from Collins's correspondence that others caught the published hint. Lord Brouncker claimed to have found a series for the area of a circle. James Gregory was working toward one. More than once, Mercator himself claimed to have one.[77] It is unlikely that Newton heard of these claims, but his imagination would have filled them in, for he knew that infinite series were in the air and that other mathematicians were at work. In haste, he composed a treatise, drawn from his earlier papers, which by its generality (in contrast to Mercator's single series) would assert his priority. Still in haste, he took it to Barrow, who

---

[75] Collins to James Gregory, 21 Jan. 1671; *Corres 1*, 60. Nearly fifty years later, during the priority dispute, Newton wrote a paper stating that Barrow's letter to Collins was grounded on what Newton "had communicated to him from time to time before the Logarithmotechnia came abroad . . ." (*Add MS* 3968.27, f. 390^v).

[76] Newton to Oldenburg, 24 Oct. 1676 (the *Epistola posterior* for Leibniz); *Corres 2*, 133; original Latin, p. 114. If William Derham's memory served him well, Newton gave him a somewhat different account of the effect of Mercator's book upon him. "That when he was a Junior [*sic* – Junior Sophister?] at the University, he had thoughts of these things, but brought them to no perfection, by reason he was forced to leave the University in the Plague-year 1665, & 1666. But at his return to Cambridge, Mercator's Logarithmotechia, & what Dr Wallis published about y^t time, revived his thoughts of these matters, and then he brought them to more perfection, & communicated them to Dr Barrow, at y^t time of the same college; who acquainted Mr Collins & others wth them" (Derham to Conduitt, 18 July 1733; *Keynes MS* 133).

[77] I am following Whiteside's convincing argument here; *Math 2*, 165–8. In fact, others did pick up Mercator's hint and proceed to the derivation of other series. In 1671 James Gregory, and in 1673 Leibniz used analogous methods to derive the inverse tan series (*Math 2*, 34n).

proposed to do the obvious thing and send it to Collins. As he faced the implications of that move, Newton's haste melted suddenly away. Hitherto he had communed with himself alone, aware of his own achievement but secure from profane criticism. Lately he had communicated something to Barrow, but Barrow was a member of the closed society of Trinity, the only one in it able even to read his paper. What now opened before Newton was something much more, and apparently he shrank back in fright, all too aware of his unripe age and much else besides. When Barrow wrote on 20 July, he had the paper in his possession but was not allowed to send it. We can only imagine what went on during the following days, though his letter of 31 July gives more than a hint.

> I send you the papers of my friend I promised . . . I pray having perused them as much as you thinke good, remand them to me; according to his desire, when I asked him the liberty to impart them to you. and I pray give me notice of your receiving them with your soonest convenience; that I may be satisfyed of their reception; because I am afraid of them; venturing them by the post, that I may not longer delay to corrispond with your desire.[78]

Only when Collins's enthusiastic response calmed his fears did Newton allow Barrow to divulge his name and give permission for Brouncker to see the paper. In all, it was a fair show of apprehension by a man who knew himself to be the leading mathematician of Europe.

As both its title and the circumstances of its composition suggest, *De analysi* concerned itself primarily with infinite series in their application to quadratures, though it did contrive to indicate something of the scope of the general method of fluxions. Drawn from the October tract, it added a few features to the earlier treatment. It began with three rules: the quadrature of $y = ax^{m/n}$, the quadrature of curves composed of several such terms, and the reduction of more compounded terms to simpler ones like $ax^{m/n}$ by division and the extraction of roots. The third rule, of course, bore the burden of the paper. Instead of stating the general formula of the binomial expansion, Newton chose to develop series by division – of $a^2/(b + x)$ for example – and root extraction – of $(a^2 + x^2)^{1/2}$ for example. Any equation, he asserted, "however complicated with roots and denominators it may be," can be reduced to an infinite series of simple terms, each of which can be squared.[79]

What he now added to his method for the first time was an iterative procedure by which to reduce affected equations, such as $y^3 - a^2y - 2a^3 + axy - x^3 = 0$, to an infinite series which expressed y as a function of increasing or decreasing powers of x. "Indeed,"

[78] *Corres 1*, 14.   [79] *Math 2*, 219.

he concluded, "since every problem on the length of curves, the quantity and surface of solids and the centre of gravity may ultimately be reduced to an inquiry into the quantity of a plane surface bounded by a curve line, there is no necessity to adjoin anything about them here."[80] Nevertheless, he did adjoin a problem in rectification and a general exposition of how the problems listed can be treated as problems of areas. And by means of his iterative procedure for affected equations, he developed from the series expressing the area of an hyperbola and the arc of a circle new series to express their bases – in effect, the series for e^x, sin x, and cos x, here expressed for the first time in European mathematics.

With a touch of virtuosity, Newton went on to develop series for the areas of two mechanical curves, the cycloid and the quadratrix. Even the length of the quadratrix's arc can be determined, he added, though the computation is difficult.

> Nor do I know anything of this kind to which this method does not extend itself, and then in various ways. . . . And whatever common analysis performs by equations made up of a finite number of terms (whenever it may be possible), this method can always perform by infinite equations: in consequence, I have never hesitated to bestow on it also the name of analysis. Indeed, deductions in the latter are no less certain than in the other, nor its equations less exact . . .[81]

De analysi did not confine itself to the method of calculating areas. It also expounded Newton's concept of the generation of areas by the motion of lines, whereby infinitesimal moments are continuously added to the finite area already generated. At the end of the paper, to assert his priority, he briefly exposed his method of tangents as the inverse of the method of quadratures. A demonstration of Rule 1 used the method of tangents to obtain, from an equation for the area z as a function of x, the equation for y as a function of x, the curve of which z represents the area.[82] Thus, however briefly, *De analysi* did indicate the full extent and power of the fluxional method.

With the transmission of *De analysi* to John Collins in London, Newton's anonymity began to dissolve. Though a mediocre mathematician at best, Collins could recognize genius when he saw it. He received *De analysi* with the enthusiasm it deserved. Before he fulfilled Barrow's request and returned the paper, he took a copy. He showed the copy to others, and he wrote about the contents of the tract to a number of his correspondents: James Gregory in Scotland, René de Sluse in the low countries, Jean Bertet and the Englishman Francis Vernon in France, G. A. Borelli in Italy, Rich-

---

[80] *Math 2*, 233.
[81] *Math 2*, 241–3. Translation altered slightly.   [82] *Math 2*, 243–5.

ard Towneley and Thomas Strode in England. Years later, when Newton went through Collins papers, he was surprised to learn just how widely the paper had circulated. "Mr. Collins was very free in communicating to able Mathematicians what he had receiv'd from Mr. Newton . . . ," he wrote anonymously in the supposedly impartial *Commercium epistolicum*.[83] Meanwhile, Collins and Barrow wanted to publish it as an appendix to Barrow's forthcoming lectures on optics. This was more than Newton could contemplate. He drew back. A letter of Collins indicates they applied more suasion than a mere suggestion in passing; he thought Newton would "give way" eventually.[84] He was mistaken. Newton withheld the publication of his method, the first episode in a long history of similar withdrawals. Thus quietly did Newton's apprehensions sow the seeds of vicious conflicts.

De analysi was not, however, devoid of effect on Newton's life. At the very time he communicated it to Barrow, Barrow was contemplating resignation from the Lucasian professorship of mathematics. The professorship, established scarcely five years earlier by the bequest of Henry Lucas, was the first new chair founded in Cambridge since Henry VIII had created the five Regius professorships in 1540. With the Adams professorship of Arabic, established in 1666, it brought the number of similar positions in the university to eight. It was the only one concerned in any way with mathematics and natural philosophy, which were otherwise hardly touched upon by the curriculum. By existing standards, Lucas endowed it magnificently; with its stipend of £100, more or less, from the income of lands purchased in Bedfordshire, it ranked behind the masterships of the great colleges and the two chairs in divinity (which were usually occupied by college masters) as the ripest plum of patronage in an institution much concerned with patronage. On 29 October 1669, this plum fell into the lap of an obscure young fellow of peculiar habits, apparently without connections, in Trinity College – to wit, Isaac Newton.

There are various stories about Barrow's resignation and Newton's appointment. One version holds that Barrow recognized his master in mathematics and resigned in his favor.[85] Frankly, it is quite impossible to square this account with the features of life in the Restoration university as we know them. Another more recent

[83] *Comm epist*, p. 120.
[84] Collins to James Gregory, 25 Nov. 1669 and 12 Feb. 1670; *Corres 1*, 15, 26.
[85] Newton himself was responsible for this story. He told Conti that Barrow had worked out a rather long solution to a problem about the cycloid and was struck with amazement when Newton gave a solution in six lines. Barrow then resigned his chair to Newton, confessing "that he was more learned than he" (Antonio-Schinella Conti, *Prose e poesi*, 2 vols. [Venice, 1739–56], 2, 25–6).

one, more in harmony with the times, suggests that Barrow was angling for a higher position.[86] It is known, I think, beyond doubt that Barrow was a man ambitious for preferment. One has only to recall his invariable contributions to the valedictory volumes published by the university – not to mention their length! – to realize as much. Within a year of his resignation, he was appointed chaplain to the king, and within three years master of the college. Nevertheless, there was no rule that demanded he resign the Lucasian chair before he courted further preferment, or for that matter that he forgo the ubiquitous royal dispensation to enjoy both at once, and it is hard to leave the third account of his resignation wholly out of the picture. In his own eyes, Barrow was a divine, not a mathematician; he resigned to devote himself to his true calling. Seventeenth-century society being what is was (that is, not wholly incommensurable with twentieth-century society), this motive was in no way incompatible with the other.

Contemporary comments agreed in the assertion that Barrow effectively appointed Newton. Collins understood as much, and a generation later Conduitt did also.[87] There is every reason for us to accept the account as well. By the Lucasian statutes, Lucas's two executors, Robert Raworth, an attorney, and Thomas Buck, the university printer, appointed the professor while they were alive, as they were in 1669. The mere fact of Barrow's appointment as the first Lucasian professor suggests that he stood in well with them or had means to influence them. So do the Lucasian statutes, which they had drafted in 1663, and which embodied modifications of the statutes for the Regius chairs of Greek and Hebrew that Barrow had obtained in 1661 when he held the Greek professorship. The letter patent with the modifications specifically mentioned Barrow and thus seems to have been inspired by him. It allowed the two professors, who were paid by Trinity College, to retain the incomes from their fellowships with the chairs; it also allowed them to hold ecclesiastical appointments that did not entail cure of souls. The Lucasian statutes extended the same rights and restrictions to the new professor. Barrow's anticipation of his own appointment may perhaps be evident in one omission from the statutes, which did not forbid the incumbent to hold office in his college. Others were not prepared to grant Barrow a coup of this dimension, however, and the letter patent of 1664, confirming the statutes, reinstalled the prohibition. In all, the two documents made the terms of the professorship

[86] *Math 3*, xivn; and Whiteside's life of Barrow in *Dictionary of Scientific Biography, 1*, 473.

[87] Collins to James Gregory, 25 Nov. 1669; *Corres 1*, 15. Collins to David Gregory *père*; 11 Aug. 1676; *Comm epist*, p. 48. Collins to John Wallis, probably 1678; *Corres 2*, 241. Conduitt's memoir; *Keynes MS* 129A, p. 11.

nearly identical to those of the Regius chairs. Barrow wrote his hand all over the statutes. It is not surprising that he was able to effect the nomination of his successor. The publication of Mercator's book had been almost an act of Providence. *De analysi*, which it provoked, raised Newton to the Lucasian chair.

According to the statutes, the Lucasian professor was required to read and expound "some part of Geometry, Astronomy, Geography, Optics, Statics, or some other Mathematical discipline," each week during the three academic terms. For each lecture that he missed, he would be fined forty shillings. Each year, he had to deposit in the university library copies of ten of the lectures he had read; here too the statutes established a fine for failure to comply. In addition to the lecture, the professor was to make himself available two hours each week to respond to questions and to clarify difficulties. He had to reside in the university continuously during term time. He could leave for more than six days only for serious reasons approved by the vice-chancellor; and if he absented himself for as long as half a term, the vice-chancellor was instructed to appoint a substitute who would receive all the professorial income. The letter patent that confirmed the statutes defined the audience of the professor; it required all undergraduates beyond their second year and all Bachelors of Arts until their third year to attend under pain of the usual penalties. Although they allowed him to hold his fellowship, the statutes, as I have indicated, forbade the professor to accept any ecclesiastical promotion that involved cure of souls and required residence elsewhere. To this restriction, the letter patent added the prohibition from college or university office. In compensation, it confined his tutorial activity to the wealthy fellow commoners. Besides the considerable income, the professor also received a considerable promotion in status; his scarlet gown set him off from the mere teaching masters. Though we do not have Newton's accounts, we may presume that he did not stint in assuming his new dignity. The statutes did not fail to require that the professor be a man of learning and of good repute. He was subject to removal for conviction "of any serious crime (such as Lese-Majesty, heresy, schism, voluntary homicide, notable theft, adultery, fornication, perjury)," or for being "intolerably negligent" in his duties.[88]

Newton did not so immerse himself in his studies as to be indifferent to his material welfare or incompetent to attend to it. In the mid-1670s, the tax commissioners of Bedfordshire tried to collect taxes on the lands from which his income came. He wrote one of

---

[88] The Lucasian statutes and the letter patent confirming them are printed in full in *Math 3*, xx–xxvii.

them indignantly, explaining that the properties were not college lands but were attached rather to his professorship. Citing the words of the Act, he concluded that they plainly included him, "& by consequence excuse me expresly from paying for any of the profits of my Professorship."[89]

In Restoration Cambridge, performance tended to diverge, often wildly, from statutory requirements. As far as the students were concerned, the fresh burden of lectures imposed upon them was only another item on a list now universally ignored. By 1660, college tutoring had virtually completed its conquest of university instruction. Barrow had complained of the neglect of his lectures when he was professor of Greek. "Sophocles and I acted in an empty theatre; . . . there was no chorus, not even of boys . . ."[90] Edmund Castell, the first Adams professor of Arabic, met the same indifference. Without further ado, he posted a sign on the door, "Tomorrow the Professor of Arabic goes into the wilderness," and converted the position into a sinecure, which it remained for a century and a half.[91] While we have no information about Newton's early experience, we do know what Humphrey Newton found upon his arrival fifteen years later. When Newton lectured, he recalled, "so few went to hear Him, & fewer y^t understood him, y^t oftimes he did in a manner, for want of Hearers, read to ye Walls." He usually lectured for half an hour, though he returned in less than a quarter of an hour when he had no audience. Humphrey added that Newton had no pupils at that time.[92] In this he was mistaken; the college records indicate that Newton took on his third and last fellow commoner during Humphrey's stay with him. Perhaps the discrepancy calls the account into question; more probably it is a comment on the closeness of Newton's relation with the pupils he tutored. As for the two hours a week for consultations, no evidence whatever of such survives.

---

[89] *Corres* 7, 371. The hand is about 1675, and the draft of the letter appears on a sheet with experiments on diffraction that confirm that time.

[90] Quoted in James Bass Mullinger, *Cambridge Characteristics in the Seventeenth Century* (London, 1867), p. 55.

[91] Quoted in Winstanley, *Unreformed Cambridge*, pp. 132–3. In 1710, Uffenbach was amazed to find that no lectures were being given in the university. It was summertime; in the winter three or four were given, "to the bare walls, for no one comes in" (Mayor, *Cambridge under Queen Anne*, p. 124). When Conyers Middleton did battle with Richard Bentley in 1727, he employed Bentley's neglect of the Regius professorship of divinity as one of his weapons. Earlier professors of divinity, he said, had put up notices of lectures, showed up, "and actually read a theological lecture whenever they found an audience ready to attend them, which was sometimes the case" (quoted in Winstanley, *Unreformed Cambridge*, p. 98). Cf. Winstanley's and Godley's descriptions of the increasing neglect of professorial duties in both universities (*ibid.*, pp. 103–38. Godley, *Oxford*, pp. 43–7).

[92] *Keynes MS* 135.

One of the fundamental facts of Newton's tenure as Lucasian professor of mathematics is the paucity of reference to his teaching. For forty years after 1687, he was the most famous intellectual in England, and there was every incentive for former students at the university to recall their connections with him. Even William Whiston, who became his disciple and successor, could barely remember having heard him.[93] As far as we know, only two others ever claimed to have been instructed by him.[94] The Lucasian chair was not created as part of a consciously formulated revision of the curriculum. It came into being because one man, Henry Lucas, thought Cambridge should have its equivalent to the Savilian chairs in Oxford. Mathematics was further removed from the interests of the average Restoration student than Arabic, and Newton pitched his lectures on a plane apt to make them incomprehensible. Moreover, no man, no matter what his genius, could have reversed the decay of the university lecture system.

The prohibition against college office did operate, though probably only because Newton chose that it should. During the same period, Regius professors of Greek and Hebrew held college offices.[95] At a time when all the life in Cambridge was concentrating itself in the colleges, exclusion from college office tended further to isolate Newton within the university.

[93] Whiston, *Memoirs*, p. 36.

[94] Both of them used his fame to their full advantage, and in so doing emphasized the lack of other such recollections. Henry Wharton, admitted to Caius in 1680, was the pupil there of John Ellis, whom Humphrey Newton mentioned as a friend of Newton. According to the "Life" of Wharton, prefaced to the second edition of his *Fourteen Sermons* (London, 1700), he attained considerable skill in mathematics. "Which last was much encreased by the kindness of Mr. *Isaac Newton*, Fellow of *Trinity College,* the incomparable *Lucas-Professor* of *Mathematicks* in the University, who was pleased to give him further instruction in that noble Science, amongst a select Company in his own private Chamber" (Quoted in *Math 4*, 11). Among Wharton's papers is a copy in his hand of Newton's "Epitome of Trigonometry" with the note that Newton gave it to him (i.e., to copy) in 1683. I cannot forebear to wonder why none of the others in that "select company" (which had dispersed by the time Humphrey Newton arrived shortly thereafter) came forward similarly to pat himself on the back in public. Sir Thomas Parkyns, admitted to Trinity in 1680, and later author of *Progymnasmata. The Inn-Play: or Cornish-Hugg Wrestler* (London, 1727), asserted that he owed his application of mathematics to wrestling to his tutor, Dr. Bathurst, and to Sir Isaac Newton, mathematics professor. "The latter, seeing my Inclinations that Way, invited me to his publick Lectures, for which I thank him, tho' I was Fellow Commoner, and seldom, if ever, any such were call'd to them . . ." (p. 12). Whatever we can make out of this bizarre reference, it does not appear from Parkyns's account that he belonged to Wharton's select group. Possibly the time was right; Parkyns left for Gray's Inn in 1682.

[95] Cf. "The Case of the Hebrew & Greek Professors in regard to Preacherships &c"; Trinity College, Box 29, C.III,a. The paper, a plea for the right to hold church livings with the professorships, asserted (correctly) that Benjamin Pulleyn, Greek professor from 1674 to 1686, and Wolfran Stubbe, Hebrew professor from 1688 to 1699, both held college offices with the chairs.

Meanwhile, there were the weekly lectures. Or perhaps there were the weekly lectures. As far as the record informs us, Barrow had already reduced the requirement to lectures during one of the three terms; Newton acquiesced in that schedule. He delivered a course of lectures in the Lent term of 1670, soon after his appointment. Thereafter he gave a series during the Michaelmas term (or at any rate he deposited manuscripts with such dates on them) each year through 1687. After 1687, he succumbed to the prevailing mode and held the position as a sinecure for fourteen years, during five of which he was not even resident in Cambridge. The record of his absences from Trinity during the earlier period supports the conclusion that he lectured only one term per year. Although he did not leave frequently, when he did do so, he was as apt to go during term time as during vacations. Indeed he left for two weeks in London less than a month after his appointment.[96] Nor did he concern himself excessively with the requirement that he deposit copies of ten lectures each year. In all, he eventually deposited four manuscripts that purported to contain annual courses of lectures through 1687.[97] Questions have been raised about all four, so that it is quite impossible to know for sure on what he lectured. It appears highly probable that the lectures through 1683 corresponded roughly to those in the deposited manuscripts.[98] Beginning in 1684, he may have lectured on the *Principia,* but the deposited manuscripts were merely drafts of that work which he sent in as the easiest way to fulfill the requirement.

As the topic for his first course of lectures, Newton chose, not the subject of *De analysi,* indeed not mathematics at all, but optics. He had required external stimulus to compose *De analysi.* During the following two years, he devoted a fair amount of time to mathematics, but again under external stimulus. What he turned to of his

[96] *Edleston,* p. lxxxv.
[97] Cambridge University Library, Dd.4.18; Dd.9.46; Dd.9.67; Dd.9.68.
[98] Although Dd.9.67 purports to be the optical lectures delivered in four courses, there is an earlier version among Newton's papers (*Add MS* 4002) that was organized differently, did not give all of the same dates, and presented as two courses of lectures virtually all that the deposited lectures presented as four. There is good reason to think the deposited manuscript was originally a revision of *Add MS* 4002 prepared for publication and deposited (with suitable dates inserted) when he abandoned the project. Dd.9.68 claimed to be lectures on algebra delivered between 1673 and 1683. Newton put the manuscripts together only in 1683–4. When John Flamsteed visited Cambridge in July 1674, Newton gave him a paper of notes for a lecture supposedly delivered in midsummer of that year. The deposited manuscript contains virtually the identical materials split between two lectures in October 1674. The manuscript further pretended that a course of lectures was given in the autumn of 1679, when we know that Newton was in Woolsthorpe after the death of his mother (*Math 5,* vii–xii, 3–6).

own accord was optics and the theory of colors.[99] His accounts recorded the purchase of three prisms some time after February 1668, probably during one of the summer fairs. His earliest surviving letter, dated 23 February 1669, described his first reflecting telescope and referred obliquely to his theory of colors. Thus there is reason to think that Newton had resumed his investigation of colors before his appointment, and that he chose to lecture on the topic then foremost in his mind. Two or three years earlier, he had sketched out his theory of colors. Now the problem seized him in earnest and would not release him until he had conquered it. Years later Dr. Cheyne was "credibly informed" that when Newton was carrying out the investigation, "to quicken his faculties and fix his attention, [he] confined himself to a small quantity of bread, during all the time, with a little sack and water, of which, without any regulation, he took as he found a craving or failure of spirits."[100] Wickins probably had to contend with more than neglected meals and a room plunged into midnight at noon. The new experiments, which were more sophisticated, required an assistant, and Wickins undoubtedly had to endure forced labor. Newton besieged his friends to supplement his supply of equipment. A paper composed about this time, which gave a method to determine the curvature of lenses from a measurement of focal distance, employed the object glass of Dr. Babington's telescope as a concrete example.[101] Primarily, he drew upon his own earlier work. Reaching back to his incomplete investigation of colors, he now worked out the full implications of his central insight and brought his theory of colors virtually to the form that he published more than thirty years later as his *Opticks*.

Aside from specifically mathematical sections, Newton's *Lectiones opticae* corresponded to Book I of the ultimate *Opticks*. Hence the lectures concerned themselves primarily with prismatic phenomena as they related to the heterogeneity of light.

> The late Invention of Telescopes [he began, for the benefit of whatever audience was present] has so exercised most of the Geometers, that they seem to have left nothing unattempted in Opticks, no room for further Improvements. . . . But since I observe the Geometers hitherto mistaken in a particular Property of Light, that belongs to its Refractions, tacitly founding their Demonstrations on a certain Physical Hypothesis not well established; I judge it will not be unacceptable if I bring the Principles of this Science to a more strict

---

99 The *Lectiones opticae*, published in the eighteenth century from his desposited manuscript, began with an extensive mathematical section. In the manuscript that seems clearly to represent the lectures given (*Add MS* 4002), the experimental investigation of colors came first.

100 Quoted in *Edleston*, pp. xli-xlii.   101 *Math 3*, 525.

Examination, and subjoin, what I have discovered in these Matters, and found to be true by manifold Experience, to what my reverend Predecessor has last delivered from this Place.[102]

He launched forthwith into an exposition of his theory of colors and the heterogeneity of light.

The renewed research and thought on which the lectures rested clarified aspects of his theory that had remained relatively crude in 1666. Earlier he had expressed his theory in terms of a traditional two-color system. Now the meaning of the continuous spectrum forced itself upon him. If there were only two colors, he should have seen two separate circles. Instead, of course, he always obtained a continuously illuminated oblong. Though he frequently spoke of seven colors, both in the lectures and the years ahead, and though he even compared their positions in the spectrum to the divisions of the musical octave, he understood that such divisions were wholly arbitrary. Instead of two colors, or seven, there were an infinite number corresponding to the infinite angles of refraction between the extremes for purple (which he now used, in place of blue) and red.

As he clarified the theory in 1669, so also he strengthened its experimental foundation. In 1666, he had begun, in a crude way, to employ a second prism to refract separate parts of the spreading spectrum. Now he refined the experiment into a form which could rigorously refute the theory of modification. He placed the second prism half-way across the room with its axis perpendicular to the first so that the full spectrum fell upon it. If, as the theory of modification might argue, dispersion as well as coloration were a modification introduced by the prism, the second prism ought to spread the spectrum into a square. Quite the contrary, it produced a spectrum inclined at an angle of 45 degrees.[103] Further improvements suggested themselves. With the second prism set parallel to the first, he covered its face except for a small hole which admitted individual colors isolated from the rest of the spectrum, and he compared the quantities of their refractions. Finally he realized the

---

[102] *Lectiones Opticae* (London, 1729), pp.1–2; *Add MS* 4002, p.1. The final clause referred to Barrow's optical lectures that had been delivered the year before.

[103] *Lectiones*, p. 32; *Add MS* 4002, p.25. Johannes Lohne has pointed out an error in Newton's description of his results ("Newton's 'Proof' of the Sine Law and His Mathematical Principles of Colors," *Archive for History of Exact Sciences, 1* (1961), 389–405). What appears to me as a simple error arising from the geometry of the room in which Newton had to perform the experiment appears to Lohne as a crisis in the foundation of Newton's mathematical science of light. I am unable to accept this interpretation of the basic thrust of Newton's optics. See also Lohne, "Experimentum crucis," *Notes and Records of the Royal Society, 23* (1968), 169–99; and Ronald Laymon, "Newton's *Experimentum crucis* and the Logic of Idealization and Theory Refutation," *Studies in History and Philosophy of Science, 9* (1978), 51–77.

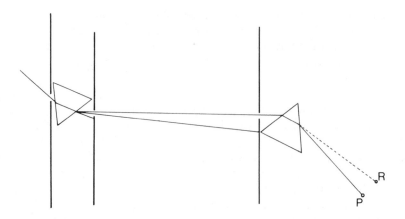

Figure 6.5. The *experimentum crucis*.

importance of a demonstrably fixed angle of incidence on the second prism. To achieve this, he used two boards with small holes, one placed immediately beyond the first prism, the other immediately in front of the second (Figure 6.5). Since the boards were fixed in position, the two holes defined the path of the beam which fell on the second prism, also fixed in position beyond the second hole. By turning the first prism slightly on its axis, Newton could transmit either end of the spectrum, pretty well if not perfectly isolated from the rest, into the second prism. There, as he expected, the blue rays were refracted more than the red. Neither beam suffered further dispersion. It was this experiment which Newton called his *experimentum crucis* in 1672, though he did not employ that phrase either in his lectures or in his *Opticks*. No doubt an experiment that is crucial in the full sense of the word for the confirmation of a theory is impossible. If we consider only the two alternatives under consideration at the time, it does appear that Newton's *experimentum crucis* refuted his competitor, the theory of modification. When the science of optics had fully digested the meaning of the experiment, it never returned to the concept of modification again.

In 1669, Newton also greatly expanded his experimental demonstration that white is merely the sensation caused by a heterogeneous mixture of rays. No single part of his investigation underwent greater expansion, almost as though Newton found this consequence of his theory difficult to accept and needed to convince himself. To the brief experiments of 1666 in which he cast overlapping spectra on each other, he added one in which a lens collected a diverging spectrum and restored it to whiteness. If he intercepted the converging rays before the focus, he obtained an elongated spectrum reduced in size. At the focus, the spectrum disappeared into a white spot. Be-

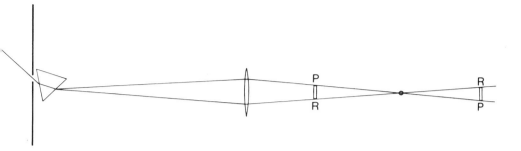

Figure 6.6. The reconstitution of white light with a lens.

yond the focus, the spectrum reappeared with its order reversed (Figure 6.6). Beyond the lens, no operation was performed on the light. When the spectrum merged into the focus, colors merged into whiteness. Since the individual rays retained their identity, colors reappeared as they separated anew beyond the focus. Newton was aware that impressions on the retina endure for about a second. The further thought occurred that all the elements of the heterogenous mixture that produces the sensation of white need not be present at once. He mounted a wheel beyond the lens so that cogs intercepted individual colors of the converging spectrum. When he turned the wheel slowly, a succession of colors appeared at the focus. When he turned it fast enough that the eye could no longer distinguish the succession, white appeared once more.[104]

One consequence of the heterogeneity of light was the distinction between colors on the one hand and the scale from white to black on the other. The scale remained as a scale of intensity alone. Its intermediate steps are the shades of gray. As far as color is concerned, the entire scale is identical, the sensation of a heterogeneous mixture more or less intense.[105] Newton's experimental imagination rose to the challenge of illustrating this further paradox. When colored bodies reflected a beam of sunlight onto a white sheet of paper, the paper appeared in the color of the body. If his theory were correct, the paper ought to appear white in the light reflected from a shiny black body. It did.[106]

In addition to the theory of colors, the lectures also contained

[104] *Lectiones*, pp. 171–213; *Add MS* 4002, pp. 30–57. These pages contain the entire discussion of white, not just the experiments described in my text. See Alan E. Shapiro, "The Evolving Structure of Newton's Theory of White Light and Color: 1670–1704," forthcoming in *Isis*.

[105] He associated white, black, and the grays already at the time of his initial insight (*Add MS* 3996, f. 122).

[106] I take this experiment from a slightly later period. Newton to Oldenburg, 11 June 1672 (his reply to Hooke's criticism of his paper); *Corres 1*, 183–4.

extensive mathematical demonstrations which further revealed
Newton's mathematical powers (if that were needed) but did not
materially advance the argument central to his contribution to optics. One demonstration did compare the errors in lenses arising
from spherical and chromatic aberration. On the assumption that
the diameter of a lens is small in relation to its circle of curvature,
he demonstrated that the error introduced by chromatic aberration
is "far greater" than that by spherical aberration. To be specific,
when the diameter of the lens is 2 inches and its radius of curvature
120 inches, the minimum circle at the focus within which all the
differentially refracted rays of different colors pass is 1,500 times
greater than the minimum circle of homogeneous rays.[107]

As far as the theory of colors is concerned, the *Lectiones opticae,*
which Newton probably composed late in 1669 and in 1670, concentrated on prismatic phenomena. They did not elaborate the explanation of the colors of solid bodies that Newton had sketched in
his essay, "Of Colours." I assume, therefore, that the investigation
which became the foundation for such an elaboration, contained in
a paper with the title, "Of ye coloured circles twixt two contiguous
glasses," dated from 1670 at the earliest.[108]

Newton employed the technique suggested in 1666. He placed a
lens of known curvature on a flat piece of glass, causing a series of
colored rings to appear to him as he looked down on the apparatus.
In 1666, he used a lens with a radius of curvature of 25 inches. Now
he employed a lens with a radius of 50 feet, increasing the diameters
of the rings nearly five times. Beyond the curvature of the lens, the
paper gave very little information about the technique he employed.
He indicated that he had to press the glasses hard to bring them
together, and he spoke of tying them together. Apparently he built a
frame of some sort, which pressed the glasses together when he tied
it tightly and held them in position for measurements. Since one
experiment mentioned that the only force used in it was the weight
of the uppermost glass, he seems to have placed them flat on a table
and viewed them from above. In the *Opticks* he indicated that his eye

---

[107] *Math 3,* 511, 510n. An error by Newton, whereby he forgot to square one ratio, made
his result roughly three-halves as large as it should have been.

[108] *Add MS* 3970.3, ff. 350–53^v. The hand in which the paper was written also suggests a
date around 1670. Since Newton composed the first draft of the so-called "Discourse of
Observations," which he finally sent to the Royal Society in 1675, and which is mostly
identical to Book II of the *Opticks,* in 1672, the paper in question here cannot have been
composed later than 1672 (see Richard S. Westfall, "Newton's Reply to Hooke and the
Theory of Colors," *Isis, 54* [1963], 82–96). I have published the paper in its entirety:
"Isaac Newton's Coloured Circles twixt Two Contiguous Glasses," *Archive for History of
Exact Sciences, 2* (1965), 181–96. See also Vasco Ronchi, "Un Grande uomo fronte a un
grande mistero. Newton e le interferenze luminose," *Bolletino dell'Associazione Ottica
Italiana occazione, 3* (1938), 154–64.

was 8 to 9 inches above the lens and that he used a pair of compasses to take the diameters from the surface of the upper glass.

As with his prismatic observations, the powerful experimental imagination which conceived the means to reduce thin films to measurement cannot fail to impress. Once again, the course of the unknown young man in Cambridge ran parallel to that of Huygens, the acknowledged leader of physical science in Europe. Stimulated by Hooke's observations, Huygens thought of the same device to measure thin colored films at almost exactly the same time as Newton's original observations in 1666.[109] Newton's skill in performance, the legacy of his years in Grantham, outran that of Huygens from the beginning, and in the more sophisticated experiments of 1670 he simply eclipsed his unsuspected rival.[110] What he demanded of his measurements tells us much about the man. Measuring with a compass and the unaided naked eye, he expected accuracy of less than one-hundredth of an inch. With no apparent hesitation, he recorded one circle at 23 1/2 hundredths in diameter and the next at 34 1/3. When a small divergence appeared in his results, he refused to ignore it but stalked it relentlessly until he found that the two faces of his lens differed in curvature. The difference corresponded to a measurement of less than one-hundredth of an inch in the diameter of the inner circle and about two-hundredths in the diameter of the sixth. "Yet many times they imposed upon mee," he added grimly to his successful elimination of the error.[111] No one else in the seventeenth century would have paused for an error twice that size. Newton was confident enough in his technique that he used his results to correct the radius of curvature of the lens; in the "Discourse of Observations" of 1672 (and in the *Opticks*) he put it at about fifty-one feet.

In addition to the error from the curvature, he found that the diameter of the first circle was consistently too large. Huygens had discovered the same thing. For him, the first circle appeared at a

[109] Huygens, *Oeuvres complètes,* pub. Société hollandaise des Sciences, 22 vols. (The Hague, 1888–1950), *17,* 341–5. Huygens dated his investigation November 1665, a few months before the probable time of Newton's first performance of the experiment, early in 1666. Initially Huygens used two lenses, both of ten-foot radius, though on a second try he saw the advantage of one lens with a greater radius (forty-five feet) pressed on a flat plate of glass. Like Newton in his first experiment, he simply assumed the periodicity of the rings; he measured only the innermost and outermost.

[110] Huygens divided the difference between the calculated thicknesses of the innermost and outermost circles by the number of circles minus one to get the increment of thickness that causes a circle to appear. His calculated thickness of the innermost circle was three times too large, that of the outermost two times. The combination of the two errors reduced his error on the increment to only 50 percent too large. In comparison, Newton's calculated thickness of the innermost circle in 1666 was about 50 percent too large, and he improved his accuracy greatly in 1670.

[111] *Add MS* 3970.3 f. 352ᵛ; Westfall, "Newton's Coloured Circles," p. 195, cf. n. 10.

calculated thickness of .000034 inch; seven additional circles appeared with increments of .000014 inch, and no more appeared beyond the thickness of .000134 inch. The bizarre nature of the result probably contributed to Huygens's abandonment of his observations. In contrast, Newton refused to accept the anomalous first circle. The fact that only a finite number of circles appeared never bothered him; their gradual fading into white as successive circles overlapped followed directly from his theory of colors. The thickness for the first circle, however, should have equaled the increment for successive ones. He had been impressed by the amount of pressure necessary to bring the glasses into contact. Was it possible that he had distorted the shape of the lens by his pressure, thus artificially enlarging the first circle? Various modifications of the experiment allowed him to conclude legitimately that he had. Again, in his measurements comparing circles in water with those in air, he modified his procedure until he obtained results that agreed with the accepted index of refraction between the two media to two significant figures.

It is impossible to compare Newton's results against modern measurements with finality. For the most part he measured dark circles. Since the lower end of the third spectrum overlaps the upper end of the second, there is only one circle that is truly dark; subsequent ones are not really analogous. The figure Newton obtained at this time for the thickness of the first dark circle is between 10 and 20 percent too large. Other comparisons involve arbitrary choices of wavelengths; insofar as they can be done, they indicate about the same degree of error, with the thickness always too high.[112] Significantly, he reduced the sizes a bit when he composed the "Discourse of Observations" in 1672, and still more for the *Opticks*. As far as the comparison is meaningful, his final figures do not appear to diverge from modern measurements, an extraordinary achievement for a pioneering investigation.

The paper on colored circles fit into many of Newton's interests. The difficulty of bringing a convex lens into contact with a flat sheet of glass impressed him; it appeared in all of his subsequent speculations as one of the key phenomena for understanding the nature of things. The experiment with water between the glasses gave him first-hand experience with capillary action, though earlier he had made notes in the "Quaestiones" about it from his reading. He made a film of water by letting a drop "creepe" between the glasses. From the varying results of two experiments, in one of which the glasses were pressed much harder, he concluded that the creeping in of the water altered the curvature of the glass because

---

[112] Cf. Westfall, "Newton's Coloured Circles," pp. 185–6.

water has less "incongruity" with glass than air. Congruity and incongruity were concepts he had met in Hooke's *Micrographia;* they too had a long history ahead of them in Newton's speculations. So did the aether to which he alluded several times, as he had done in the essay of 1666. Already at that time he had implied the principal features of his mechanical account of optical phenomena. It is not, he asserted, "ye superficies of Glasse or any smoth pellucid body y^t reflects light but rather ye cause is ye diversity of Aether in Glasse & aire or in any contiguous bodys." He spoke of pulses in the aether in connection with thin films.[113] References to pulses which implied his mechanical explanation of the periodic rings, also filled the paper on colored circles. The pulses were not light. Rather they were vibrations in the aether, set up by the blow of a corpuscle of light on the first surface of a film, which determined whether or not the corpuscle would be able to penetrate the second surface and thus be transmitted, or would be reflected.

Beyond its revelation of Newton's experimental acumen and its incidental information on his speculative system, the paper tells us much about the progress of his theory of colors. As late as 1670 (if my proposed dating is correct), Newton had still not separated his theory of colors from his conception of light. At least a good half of his interest in the circles seemed to stem from his conviction that they supported the corpuscular conception of light. Thus he devoted much of his effort to measuring circles at various obliquities. The more obliquely he viewed them, the larger they appeared. In one experiment, he measured them from five different angles, in another from four, in four others from two. He concluded that the diameter of a circle is proportional to the cosecant of the ray's obliquity (= secant of the angle of incidence in our terminology) "or reciprocally as y^e sines of its obliquity; that is, reciprocally as y^t part of the motion of y^e ray in y^e said filme of aire wch is perpendicular to it, or reciprocally as y^e force it strikes y^e refracting surface wth all." Hence the thicknesses of the films vary as the squares of the cosecant "or reciprocally as ye quares of ye sines, motion, or percussion."[114] A stronger blow allowed the corpuscle to pass more easily and corresponded then to a thinner film. Similar considerations stood behind his comparison of circles in water and in air. The thickness here varied reciprocally as the subtlety of the medium. He was satisfied to see his measurements of films in air and water yield a ratio equal to the index of refraction between the two media, since

---

[113] *Add MS* 3975, pp. 14, 10.

[114] *Add MS* 3970.3, f. 350; Westfall, "Newton's Coloured Circles," p. 191. I cannot explain his use of the square here, It had appeared earlier in his attempted formula for the same phenomena in the essay "Of Colours." In fact, his data showed that the thickness varied inversely as the first power of the perpendicular component of motion, not as the square.

he considered that indices of refraction were another expression of the resistances of media to light. In the end, however, these speculations led into contradictions. When he compared colors, he found that red rings were larger than blue. That is red rays cause bigger pulses. He was already convinced from refractions that red rays are stronger, and he was not prepared to set a higher priority on his speculations about obliquity, in which the weaker blow caused the bigger pulse. By the time he drew up the "Discourse of Observations," the critiques of his paper of 1672 had taught him to separate the theory of colors from the conception of light. He allowed the concern with obliquity to recede into the background, and the speculations about the strength of blows to disappear.

Similarly, when he began to measure rings, he had not yet firmly fixed their relation to colors in his mind. Only at the end of the paper did the observations come to the aid of the theory of colors. Perhaps this was a stroke of good fortune, for it allowed Newton to concentrate his attention on an aspect of the circles, their periodicity, which had no necessary connection to his theory of colors. Hooke had asserted the periodicity of colors in thin films. Newton and Huygens had both assumed the periodicity of the circles they saw in 1666, but neither had demonstrated it. The paper "Of coloured circles" first established the periodicity of some optical phenomena by careful measurement. From the geometry of the circle, the thickness of the film between the lens and the flat sheet is proportional to the square of the diameter of the colored circles. Newton measured dark circles between colored circles, squared their diameters, and found a simple arithmetic progression. If the thickness of the first dark circle were set at 2 units, successive circles appeared at thicknesses of 4, 6, 8, 10 and 12 units; colored circles appeared between them at thicknesses of 1, 3, 5, 7, 9, and 11 units, becoming less distinct until they merged completely into whiteness. No matter what the obliquity, no matter what the medium, air or water, the same progression held.[115] There could be no question about the periodicity of the circles. Whereas Huygens simply forgot his measurements when periodicity became an embarrassment to his treatment of light, Newton's measurements etched its reality so deeply on his consciousness that he could not forget it even though it eventually became an even greater embarrassment to his treatment.

Nevertheless, in the "Discourse of Observations," he pushed periodicity away from the center of attention almost as far as the effects of obliquity, and he pointed the whole investigation firmly toward the explication of colors in solid bodies, that is toward the analysis of heterogeneous light by reflection. As irony would have it, the details

---

[115] *Add MS* 3970.3, ff. 350–2ᵛ; Westfall, "Newton's Coloured Circles," pp. 191–4.

of his explication have not survived, whereas the general statement made already in 1666, that bodies reflect some rays more than others, has survived. On the other hand, when periodicity was found to be a property of light itself, it constituted as important an addition to optics as heterogeneity. To cap the irony, periodicity played the central role in the overthrow of Newton's corpuscular conception of light in the nineteenth century, though he had seen his observations initially as a support for corpuscularity.

On its final page, the paper on colored circles turned directly to the theory of colors when Newton tried to measure the difference in the thicknesses of films in which red and purple appeared. The fact of different thicknesses was evident, of course, in the succession of colors in the inner rings. Measuring them proved to be more difficult. Instead of attempting to measure the colors on the circles themselves, where only the innermost and least reliable circle was free from overlap, he tried to cast colors separated by the prism on his apparatus. Limitations of space dictated his method. There was no place where he could project a spectrum down vertically far enough to let the colors separate. Hence he projected the colors onto a white sheet of paper which reflected the light down on his lens. Apparently the intensity was not great enough to allow measurements of circles, but Newton's ingenuity rose to the challenge. Simply laying the lens on the sheet of glass, he cast purple on the paper. As he rotated the prism through the spectrum, circles expanded and new ones appeared in the center. The number of circles from purple to red corresponded to the number of extra half-pulses for purple at that thickness. When he then pressed the lens down on the glass, more circles appeared in the center and expanded until the glasses came into contact. The number of red circles corresponded to the number of half-pulses for red at that thickness. The ratio of the total number of circles to the number of red circles gave the ratio between the half-pulses for purple and red. Newton set it at 14:9 or 20:13.[116] This ratio remained the empirical foundation of Newton's quantitative treatment of colors in solid bodies. Bodies are composed of transparent particles the thickness of which determines the colors they reflect. He had demonstrated with the prism that ordinary sunlight is a heterogeneous mixture of rays, each with its own immutable degree of refrangibility. "And what is said of their refrangibility may be understood of their reflexibility; that is, of their dispositions to be reflected, some at a greater, and others at a less thickness of thin plates or bubbles, namely, that those dispositions are also connate with the rays, and immutable . . ."[117] Hence

[116] *Add MS* 3970.3, f. 353ᵛ; Westfall, "Newton's Coloured Circles," p. 196.
[117] From the "Discourse of Observations"; *Cohen*, p. 224.

all the phenomena of colors derive from processes of analysis, whether refraction or reflection, which separate individual rays from the mixture. In 1666, Newton laid out the program and carried it through for refractions. Only about in 1670 did he fully work out the details for the colors of solid bodies.

With 1670, Newton's creative work in optics virtually came to an end. He had worked out the implications of his initial insight, answering to his own satisfaction the questions he had set himself. Though he would devote considerable time to the exposition of his theory, first in 1672, later in the 1690s, and carry out some minor experimentation, he had effectively exhausted his interest in the subject. Never again was it able to command his undivided attention.

During this same period, willy nilly, Newton also did some work on mathematics. Two enthusiastic and persuasive men, Isaac Barrow and John Collins, now knew his power and refused to let it rest. Barrow involved him in the publication of his two sets of lectures. To the *Lectiones XVIII* (1669) on optics, Newton added two small improvements for which Barrow thanked him in the preface. He did not name him, however, probably at Newton's request. At one point, Barrow's lectures employed the theory of colors that Newton had disproved. Its use suggests at least that Newton had not communicated this discovery to his new patron. Newton did advise Barrow to include a certain passage in his *Lectiones geometricae* (1670). In July, Barrow presented an inscribed copy of the work to him.[118] In his own optical lectures, Newton went out of his way several times to defer to his predecessor.[119] Nor was the relation one-sided. Barrow allowed him to use his extensive mathematical library. Barrow also set him mathematical tasks. In the fall of 1669, he suggested that Newton revise and annotate the *Algebra* of Gerard Kinckhuysen, which had recently been translated from Dutch into Latin. It was also Barrow who set him at work a year later on a revision and expansion of *De analysi*. The episodes from 1669 to 1671 constitute most of the known relationship between the two men. It was just as well for Newton's career that he chose to please the older scholar. Already Barrow had shown himself to be a powerful patron. Newton was to need his aid one more time.

John Collins proved a more pertinacious gadfly. He had been the ultimate source of the Kinckhuysen *Algebra,* which he had had translated from Dutch to supply the lack of a good introduction to the subject. Late in November 1669, when Newton made his sec-

[118] *Math 1,* xv; *Math 3,* 70–1n, 440n, 479n, 490n.
[119] *Math 3,* 441, 453, 455–65, 461.

ond trip to London, he met Collins. Newton's accounts suggest that alchemy rather than mathematics was his primary purpose on the trip. Either he or Barrow informed Collins of the visit, however, and Collins could not let the opportunity pass. A year later, he described the meeting to James Gregory.

> I never saw Mr Isaac Newton (who is younger then yourselfe) but twice viz somewhat late upon a Saturday night at his Inne, I then proposed to him the adding of a Musicall [i.e., harmonic] Progression, the which he promised to consider and send up. . . . And againe I saw him the next day having invited him to Dinner.[120]

It was out of the question that Collins should allow his new discovery to escape from his net of communication. Newton's exchange with Collins effectively introduces his surviving correspondence.

Collins had planted the question about the harmonic series artfully. For his part, Newton glowed in the warmth of appreciation, which Collins reinforced by the gift of a copy of Wallis's *Mechanics* sent via Barrow. In January, Newton initiated the correspondence with a lengthy and difficult letter, on which he must have spent much of the intervening time, responding to the problems Collins had proposed at their meeting. He offered various devices to sum up a finite number of terms in any harmonic series and a method to approximate the answer.[121] Collins never did comprehend what Newton had written, but he sent in return his own thoughts on the harmonic series and a new problem. How could one compute the rate of interest N on an annuity of B pounds for thirty-one years purchased for A pounds? On 6 February 1670, Newton sent a formula by which to compute N when A and B are given, and he described to Collins the sort of annotations he was preparing on the Kinckhuysen *Algebra*.[122] By 18 February, Collins had added the questions of Michael Dary, a computer and gauger, and Newton enclosed a letter for Dary with a series for the area under a circle.[123] In July, Collins sent the London bookseller Moses Pitts, who was planning to publish the Kinckhuysen, to see Newton in Cambridge, although ultimately the two men never met.[124]

Newton completed his annotations to Kinckhuysen by the summer of 1670, though he later added to them at Collins's behest. This paper does not approach his fluxional papers in significance. Nevertheless, it was the work of a man who had probed the depths of his topic, and it revealed his mastery of basic algebra. His comments strove to simplify Kinckhuysen's more cumbersome procedures and to propose general methods where Kinckhuysen dealt in

---

[120] Collins to James Gregory, 24 Dec. 1970; *Corres 1*, 53.   [121] *Corres 1*, 16–20.
[122] *Corres 1*, 23–5.   [123] *Corres 1*, 27. The letter to Dary is lost.
[124] Newton to Collins, 11 July 1670; *Corres 1*, 31. Collins to Newton, 13 July 1670; *Corres 1*, 32.

particular ones. At Collins's request, he added significant portions on extracting the roots of cubic equations, including the identification of imaginary roots, and he wrote a masterful exposition of how to reduce problems to equations, in which he treated algebra as a language akin to other languages and the construction of equations as an exercise in translation.[125] Newton's "Observations on Kinckhuysen" served further to increase his fame within a limited circle of mathematicians. John Wallis, to whom Collins had not shown *De analysi* because of his reputation for plagiary, did hear about the annotations; he offered the opinion that Newton could bring them out as an independent treatise of his own.[126] Towneley longed to see the Kinckhuysen volume "with those wonderfull additions of Mr Newton."[127] James Gregory, a mathematician who approached Newton's stature, continued to correspond with Collins about Newton's method of expanding binomials into infinite series.[128]

Whether he understood the full extent of the publicity or not, Newton no sooner sensed the consequences of Collins's adulation than his anxiety, lulled initially by the pleasure of recognition, began to mount anew. It was evident already in his letter of 18 February 1670. Collins had asked to publish the formula for annuities. Newton agreed, "soe it bee wthout my name to it. For I see not what there is desirable in publick esteeme, were I able to acquire & maintaine it. It would perhaps increase my acquaintance, ye thing wch I cheifly study to decline." Already he had begun to fend off the suggestions that he publish *De analysi*. Now he began to withdraw from Collins's overeager embrace as well. He informed Collins in the letter of 18 February that he had found a way to compute the harmonic series with logarithms, but he did not include it since the calculations were "troublesom."[129] Collins did not hear from him again until July.

When he finally sent Collins the "Observations on Kinckhuysen" in July, Newton accompanied them with a letter filled with defensive diffidence. He hoped he had done what Collins wanted. He left it entirely to Collins whether to print any or all of it. "For I assure you I writ w^t I send you not so much wth a designe y^t they should bee printed as y^t your desires should bee satisfied to have me revise

[125] "In algebram Gerardi Kinckhuysen observationes"; *Math 2*, 364–444. See also the related paper, "Problems for Construing Aequations," *Math 2*, 450–516. See Christoph J. Scriba, "Mercator's Kinckhuysen-Translation in the Bodleian Library at Oxford," *British Journal for the History of Science*, 2 (1964), 145–58.

[126] Wallis to Collins, 25 Jan. 1672; Stephen Peter Rigaud, ed., *Correspondence of Scientific Men of the Seventeenth Century*, 2 vols. (Oxford, 1841), 2, 529.

[127] Towneley to Collins, 4 Jan. 1672; *Corres 1*, 78.

[128] James Gregory to Collins, 23 Nov. 1670; *Corres 1*, 45–8.   [129] *Corres 1*, 27.

ye booke. And so soone as you have read ye papers I have my end of writing them."

> There remains [he added] but one thing more & thats about the Title page if you print these alterations wch I have made in the Author: For it may bee esteemed unhandsom & injurious to Kinck huysen to father a booke wholly upon him wch is soe much alter'd from what hee had made it. But I think all will bee safe if after ye words [nunc e Belgico Latine versa] bee added [et ab alio Authore locupletata.] or some other such note.[130]

Not "enriched by Isaac Newton," but "enriched by another author"! Others might long to see his wonderful additions. Newton himself was primarily concerned that his name not appear.

Encouraged by receiving the annotations, Collins hastily wrote back that he noted Newton's agreement with him on the insufficiency of Kinckhuysen's treatment of surds. He sent along three books and asked Newton to pick out the best discussion of surds to insert in the volume. Rather wearily, Newton asked to have the manuscript back.[131] It came at once with another letter filled with further questions and the promise of further publicity; "your paines herein," Collins assured him, "will be acceptable to some very eminent Grandees of the R Societie who must be made accquainted therewith . . ."[132] It was a clumsy thing to say to a man who had recently told him he studied chiefly to diminish his acquaintance. Over two months passed before Newton replied. On 27 September he informed Collins that he had thought some of composing a completely new introduction to algebra.

> But considering that by reason of severall divertisements I should bee so long in doing it as to tire you patience wth expectation, & also that there being severall Introductions to Algebra already published I might thereby gain ye esteeme of one ambitious among ye croud to have my scribbles printed, I have chosen rather to let it passe wthout much altering what I sent you before.[133]

Collins never saw the manuscript again. He also heard no more from Newton for ten months.

Although he had misjudged Newton initially, Collins now realized that he was dealing with a man extraordinary in more ways than mathematical genius. He responded to Newton's silence with his own, and in December he described his relations with Newton to James Gregory. Gregory was eager to learn about Newton's general method of infinite series. Collins told him how Newton communicated individual series but not the general method, though

[130] *Corres 1*, 30–1.
[131] Collins to Newton, 13 July 1670; *Corres 1*, 32–3. Newton to Collins, 16 July 1670; *Corres 1*, 34–5.
[132] Collins to Newton, 19 July 1970; *Corres 1*, 36.   [133] *Corres 1*, 43–4.

he understood that he had written a treatise about it. Collins had sent him the annuity problem, hoping thereby to learn the general method. Newton had sent back only the formula; "hence observing a warinesse in him to impart, or at least an unwillingness to be at the paines of so doing, I desist, and doe not trouble him any more . . ."[134]

But in the end Collins could not deny his self-imposed mission and desist forever. In July 1671, he wrote a chatty letter about mathematics and the Kinckhuysen edition, which would have a better sale, he remarked, if it carried Newton's name. He also sent Newton a copy of Borelli's new book. Newton responded with deliberate incivility by suggesting that Collins not send him any more books. It would be sufficient if he merely informed him what was published. He did mention that he had intended to visit Collins on the occasion of the recent induction of the Duke of Buckingham as chancellor of the university, but a bout of sickness had prevented his making the trip to London. Somewhat grudgingly, it appears, he also added that he had reviewed his introduction to Kinckhuysen during the winter.

> And partly upon Dr Barrows instigation, I began to new methodiz ye discourse of infinite series, designing to illustrate it wth such problems as may (some of them perhaps) be more acceptable then ye invention it selfe of working by such series. But being suddainly diverted by some buisinesse in the Country, I have not yet had leisure to return to those thoughts, & I feare I shall not before winter. But since you informe me there needs no hast, I hope I may get into ye humour of completing them before ye impression of the introduction, because if I must helpe to fill up its title page, I had rather annex something wch I may call my owne, & wch may bee acceptable to Artists as well as ye other to Tyros.[135]

The new methodized discourse, known as the *Tractatus de methodis serierum et fluxionum* (*A Treatise of the Methods of Series and Fluxions*), though Newton himself did not give it a title, was the most ambitious exposition of his fluxional calculus that Newton had yet undertaken.[136] Drawing on both *De analysi* and the tract of October 1666, he produced an exposition of his method directed to that circle of mathematical artists with whom he had communed so far only passively in the reading of their works. Although the treatise

---

134 *Corres 1*, 53–5. I am unsure how to reconcile Collins's reception of *De analysi* with this letter. It is true that *De analysi* gave methods of expanding by division and root extraction without stating the general binomial theorem. Nevertheless, it outlined Newton's entire fluxional calculus as well as quadratures by infinite series. Probably Collins did not understand what he had.

135 Collins to Newton, 5 July 1671; *Corres 1*, 65–6. Newton to Collins, 20 July 1671; *Corres 1*, 67–9.   136 *Math 3*, 32–328.

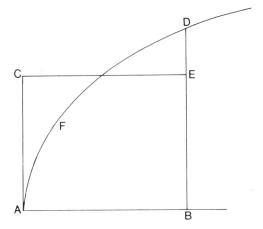

Figure 6.7. Areas treated as fluent quantities with ordinates as their fluxions.

was immensely important, there is no need to repeat here what I have described at some length earlier. Suffice it to say that in virtually every respect, Newton both deepened and expanded his treatment and made his exposition more systematic. He gave special attention to the two basic operations, posed here as problems 1 and 2, to which all difficulties in the analysis of curves reduce.

1. Given the length of the space continuously (that is, at every time), to find the speed of motion at any time proposed.
2. Given the speed of motion continuously, to find the length of the space described at any time proposed.[137]

Dissatisfied with his exposition of integration, Newton later added a new problem: To determine the area of any proposed curve. "The resolution of the problem is based on that of establishing the relationship between fluent quantities from one between their fluxions (by Problem 2)." If the line BD by the motion of which the area $AFDB$ is described advances along AB, conceive that the rectangle $ACEB$ is described by part of the line, BE, one unit in length (Figure 6.7). Take BE as the fluxion of the rectangle; then BD will be the fluxion of the area. $AB = x$. Hence $ABEC = 1 \cdot x = x$. $BE = \dot{x}$ $(= 1)$. $AFDB = x$. $BD = \dot{z} = \dot{z}/\dot{x}$ (since $\dot{x} = 1$). "Consequently, by the equation defining BD is at once defined the fluxional ratio $\dot{z}/\dot{x}$, and from this (by Problem 2, Case 1) will be elicited the relationship of the fluent quantities x and z."[138] That is, from

[137] *Math 3*, 71.
[138] *Math 3*, 211. I have followed Whiteside's use of $\dot{x}$ and $\dot{z}$ in his translation. Actually Newton used *m* and *r* at this time and developed the dot notation only later.

the formula of the curve expressing y ($= \dot{z}$) as a function of x, we can, via problem 2, the method of quadratures, arrive at a new formula expressing the area z as a function of x. From this introduction, Newton proceeded to the quadrature of polynomials, of equations that can be expanded into infinite series, and of affected equations, and he included an extensive table of integrals.

> By means of the preceding catalogues [he concluded] not merely the areas of curves but also other quantities of any kind generated at an analogous rate of flow may be derived from their fluxions, and that through the medium of this theorem: A quantity of any kind is to the unity of its own class as the area of a curve to the surface unity if only the fluxion generating that quantity shall be to the unity of its own kind as the fluxion generating the area to its own unity – that is, as the line, moving normally upon the base, by which that area is described, to the linear unity. In consequence, if a fluxion of whatever kind be expressed by an ordinate line of this sort, the quantity generated by that fluxion will be expressed by the area described by that ordinate. Or if the fluxion be expressed by the same algebraic terms as the ordinate line, the generated quantity will be expressed by the same ones as the area described.[139]

One of the notable features of *De methodis* is Newton's growing attention to the foundation of his fluxional calculus. From the beginning of his use of the concept of motion, its intuitive idea of continuous flow had clashed with the infinitesimal devices by which Newton expressed it. *De analysi* offered some tentative steps toward the resolution of the conflict.[140] *De methodis* extended them. Significantly, the treatise began with an exposition of infinite series in which Newton enunciated a concept of convergence. As the quotients of affected equations are extended, they "ever more closely approach the root till finally they differ from it by less than any given quantity and so, when they are infinitely extended, differ from it not at all . . ."[141] Although *De methodis* continued for the most part to use the language of infinitesimals, the concept of limit values which can be approached more closely than any defined difference probably influenced Newton's understanding of his terms.[142] When he dealt with problems such as the area of the cissoid, in which he appealed directly to comparable quantities being generated in equal moments, the ideas implicitly present in his concept of motion of flux emerged most clearly. In such demonstrations, he stated,

[139] *Math 3*, 285.
[140] For example, in his derivation of the algorithm for finding a fluxion; *Math 2*, 243–5.
[141] *Math 3*, 67. Cf. *De Analysi*; *Math 2*, 245–7.
[142] For examples of usage that implies such, see *Math 3*, 79–83, 123, 133, 299, 305, 311.

I take quantities as equal whose ratio is one of equality. And a ratio of equality is to be regarded as one which differs less from equality than any ratio of inequality which can possibly be assigned. . . . I have here used this method . . . since it has an affinity to the ones usually employed in these cases. However, that based on the genesis of surfaces by their motion of flow appears a more natural approach.[143]

So also Newton's willingness to use the fluxions of fluxions and his treatment of angles of contact indicate that his fluxional method rested on something far more subtle than a simpleminded appeal to infinitesimals. When Newton introduced the concept of the quality (or variation) of curvature, he had in effect to employ a third derivative. It is true that he tended to mask the issue by setting up a new equation of which he found the fluxion. Thus, in this case, he defined the radius of curvature as a variable (stated as a function of x), the fluxion of which expressed the variation of curvature. Nevertheless, he was handling the equivalent of a third derivative, and he was not one to deceive himself about his own procedures.[144] The problem of curvature raised the related issue of degrees of infinity. His investigation of the cycloid led to the conclusion that its radius of curvature at its end is zero. The cycloid, he asserted, is "more curved at its cusp F than any circle, forming with its tangent βF produced a contact angle infinitely greater than a circle can with a straight line. There are also contact angles infinitely greater than cycloidal ones and others in turn infinitely greater than these, and so on infinitely, but still the greatest are infinitely less than rectilinear ones."[145]

In a separate paper associated with *De methodis* and perhaps intended as an addendum to it, Newton broke through the limitations of available language and placed his method on a new foundation more adequate to its central inspiration. It began with four axioms.

Axiom 1. Magnitudes generated simultaneously by equal fluxions are equal.
Axiom 2. Magnitudes generated simultaneously by fluxions in given ratio are in the ratio of the fluxions.
Note: by simultaneous generation I understand that the wholes are generated in the same time.
Axiom 3. The fluxion of a whole is equal to the fluxions of its parts taken together.
Here note that increasing fluxions are to be set positive, decreasing ones negative.
Axiom 4. Contemporaneous moments are as their fluxions.[146]

---

[143] *Math 3*, 283. Cf. his rectification of the Archimedean spiral by transforming it into a parabola of equal length; *Math 3*, 313.
[144] *Math 3*, 187–91. [145] *Math 3*, 165–7.
[146] *Math 3*, 331. By a moment, Newton meant the incremental change of a fluent magnitude. A fluxion was the velocity of change, the moment divided by the "moment" of time.

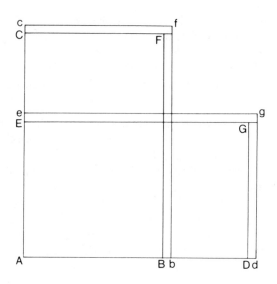

Figure 6.8. The concept of nascent or ultimate ratios.

Newton then proposed as theorem 1 that when there are four perpetually proportional fluent quantities, the sum of the products of each extreme with the other's fluxion is equal to the sum of the products of each middle term with the other's fluxion. That is, if $A/B = C/D$ (so that $AD = BC$), then $A \cdot \text{fl}D + D \cdot \text{fl}A = B \cdot \text{fl}C + C \cdot \text{fl}B$. To prove this theorem, he set out with a purely analytic approach which depended on eliminating terms in which a moment M of "infinite smallness" appeared. He stopped in mid-sentence and set out anew with a geometrical model in which four lines, AB, AC, AD, and AE, became the fluent quantities. $AB/AD = AE/AC$; hence the rectangle $ACFB$ equals the rectangle $AEGD$ (Figure 6.8). Let the lines "increase fluxionally by their respective moments Bb, Dd, Ee, and Cc . . . " Since the rectangles remain equal by assumption, their increments must also be equal. $Ab \cdot Cc + AC \cdot Bb = Ad \cdot Ee + AE \cdot Dd$. By axiom 4, moments are as their fluxions; hence we can replace Cc, Bb . . . with $\text{fl}(AC)$, $\text{fl}(AB)$ . . . Therefore

$$Ab \cdot \text{fl}(AC) + AC \cdot \text{fl}(AB) = Ad \cdot \text{fl}(AE) + AE \cdot \text{fl}(AD)$$

With his geometric model, Newton initially treated the moments once again as static, infinitesimal increments. Such was not his goal, however. He crossed the passage out and pursued the intuitive concept of motion to a different end.

Now let the rectangles *Af* and *Ag* diminish till they go back into the primary rectangles *AF* and *AG*: *Ab* will then come to be *AB* while *Ad* becomes *AD*. Hence at the last moment of that infinitely decreasing fluxion – that is, at the first moment of flux of the rectangles *AF* and *AG* when they start to increase or diminish – there will be

$$AB \cdot \text{fl} (AC) + AC \cdot \text{fl}(AB) = AD \cdot \text{fl}(AE) + AE \cdot \text{fl}(AD).$$

As was to be proved.[147]

Thus was born the concept of nascent or ultimate ratios which brought the notion of limit values, present in his method of infinite series, to his method of fluxions and enabled Newton more adequately to express the intuitions present in the idea of fluxional change.

As corollaries to theorem 1, Newton derived the basic rules for the fluxions of products, quotients, roots, and powers. He added eleven more theorems as "foundations for demonstrations"; and he illustrated the power of the method he was proposing by quickly doing the areas under three of the more intractable curves known to contemporary mathematics, the Gutschoven quartic $(x^2(a^2 - y^2) = y^4)$, the cissoid, and the conchoid.[148]

For all its brilliance, the most remarkable thing about *De methodis* with its associated papers is the fact that Newton never completed it. From his letters, it appears that he started the treatise in the winter of 1670–1. A trip home in the spring interrupted him, and when he wrote to Collins on 20 July 1671, he said he had not returned to the papers and did not expect to before winter. "I hope I may get into ye humour of completing them . . . " Since Newton was prone to similar comments, which were defensive maneuvers to ward off criticism by pretending lack of interest, we should pause before we take the comment seriously. The manuscript itself does seem to bear him out, however. It reveals an initial effort which came to a halt, a renewed attempt which advanced a bit further, and ultimate abandonment.[149] In May 1672, Newton informed Collins that he had written the better half of the treatise the

---

[147] *Math 3*, 331–5.

[148] The paper, which has no title, is found in *Math 3*, 328–52.

[149] *Add MS* 3960.14. Newton folded fifty small sheets into a rough pamphlet containing two hundred pages which he sewed together at some point (including at least one other half-sheet with an early emendation). I find in the pamphlet's size the suggestion that he expected to fill about two hundred pages. He never did get to the second half, which remained empty except for a few emendations entered there. When he returned to the treatise, he added a few more sheets which he simply folded into the center without sewing, and (perhaps at this time) added some corrections and amplifications to earlier parts. There is a list of problems from ca. 1670 (*Math 3*, 28–30). which appears to be an outline of the projected treatise. What he completed does not contain a number of the problems on the list, though it contains well over half of them.

previous winter, but it had proved larger than he expected. It was not done; he might "possibly" complete it. By July he did not know "when I shall proceed to finish it."[150] In fact, he never did.

No doubt the reluctance of London booksellers to publish mathematical books, which usually lost money, played a role in Newton's dilatoriness.[151] It is impossible to assign the determining role to this factor, however. The Royal Society subsidized the publication of Horrox's *Opera* in 1672. Edmund Gunter's *Workes* were republished the following year. In 1674, Barrow brought out a new edition of his lectures, and in the years ahead proceeded with publications of Euclid's *Data* and *Elements* and of Archimedes and Apollonius. Other mathematical works, mostly elementary ones to be sure, appeared continually. Had Collins ever gotten his hands on Newton's *De methodis*, he would have moved heaven and earth to put it into print as Halley did with another treatise fifteen years later. It was not the depression of the publishing trade, rather it was Newton who aborted the publication of a work which would have transformed mathematics. He never returned his annotations on Kinckhuysen to Collins, and he finally killed the edition by buying out the interest of the bookseller Pitts for four pounds. Collins never saw anything of the major treatise beyond the barest of tantalizing hints in a couple of letters. The unresolved, unresolvable tension that pulled Newton to and fro, as he responded to the warmth of praise, then fled in anxiety at the scent of criticism, worked now to suppress his masterful treatise.

For that matter, he was not terribly interested in it in any case. As he had told Collins, he was not in the humor to complete it. Nearly all of Newton's burst of mathematical activity in the period 1669–71 can be traced to external stimuli, to Barrow (armed initially with Mercator's work) and to Collins. His own interests had moved on. By 1675, Collins, who confessed that he had not heard from him for nearly a year, reported to Gregory that Newton was "intent upon Chimicall Studies and practises, and both he and Dr Barrow &c [were] beginning to thinke mathcall Speculations to grow at least nice and dry, if not somewhat barren . . ."[152]

However, it proved to be impossible for Newton to retire again to the anonymity of his sanctuary. An irresistible current that would not let his gifts be hidden bore him forward. If not mathematics, then something else. Fittingly, it was a product of his hands rather than a child of his brain which brought the craftsman from Gran-

---

[150] *Corres 1*, 161, 215.   [151] *Cf. Math 3*, 6–7n.
[152] Collins to James Gregory, 19 Oct. 1675; *Corres 1*, 356.

tham fully into the view of the European scientific community. Though, contrary to the account long accepted, we now know that Newton's theory of colors did not lead him wholly to despair of refracting telescopes,[153] he did build a reflecting telescope nevertheless. He cast and ground the mirror from an alloy of his own invention. He built the tube and the mount. And he was proud of his handiwork. He was still proud when he recalled it for Conduitt nearly sixty years later: "I asked him," Conduitt recorded, "where he had it made, he said he made it himself, & when I asked him where he got his tools said he made them himself & laughing added if I had staid for other people to make my tools & things for me, I had never made anything of it . . ."[154] The telescope was about six inches long, but it magnified nearly forty times in diameter, which, as Newton could be brought to admit, was more than a six-foot refractor could do.

Later he made a second telescope. "When I made these," he confessed in the *Opticks*, "an Artist in London undertook to imitate it; but using another way to polishing them than I did, he fell much short of what I had attained to . . ." The Lucasian professor was unable to restrain himself from proceeding to lecture the artisans of London on the secrets of their craft.

The Polish I used was in this manner. I had two round Copper Plates, each six Inches in Diameter, the one convex, the other concave, ground very true to one another. On the convex I ground the Object-Metal or Concave which was to be polish'd, 'till it had taken the Figure of the Convex and was ready for a Polish. Then I pitched over the convex very thinly, by dropping melted Pitch upon it, and warming it to keep the Pitch soft, whilst I ground it with the concave Copper wetted to make it spread eavenly all over the convex. Thus by working it well I made it as thin as a Groat, and after the convex was cold I ground it again to give it as true a Figure as I could. Then I took Putty which I had made very fine by washing it from all its grosser Particles, and laying a little of this upon the Pitch, I ground it upon the Pitch with the concave Copper, till it had done making a Noise; and then upon the Pitch I ground the Object-Metal with a brisk motion, for about two or three Minutes of time, leaning hard upon it. Then I put fresh Putty upon the Pitch, and ground it again till it had done making a noise, and afterwards ground the Object-Metal upon it as before. And this Work I repeated till the Metal was polished, grinding it the last time with all my strength for

[153] *Math 3*, 467–9, 512–13n. Cf. Zev Bechler, " 'A less agreeable matter': The Disagreeable Case of Newton and Achromatic Refraction," *British Journal for the History of Science, 8* (1975), 101–26.
[154] Conduitt's memorandum of 31 Aug. 1726; *Keynes MS* 130.10, ff. 3–3^v.

a good while together, and frequently breathing upon the Pitch, to keep it moist without laying on any more fresh Putty.[155]

Whatever the success of the reflector, the telescope Humphrey Newton found fifteen years later, stationed at the head of the stairs down to the garden, where Newton used it to observe comets and planets, was a refractor.[156]

Meanwhile he found it quite impossible not to show off his creation. His earliest surviving letter, of February 1669, is a description of it to an unknown correspondent written as a result of a promise to Mr. Ent, to whom he had presumably shown or mentioned the telescope.[157] When he met Collins in London at the end of 1669, he told him about the telescope, allowing it in his account to magnify 150 times.[158] He must have been showing it off in Cambridge. In December 1671, Collins repeated to Francis Vernon what Mr. Gale (a fellow of Trinity) had written of it from Cambridge.[159] By January, Towneley was asking about it excitedly, and Flamsteed had heard of it both from London and from a relative who had recently been in Cambridge.[160] Perhaps Collins had never informed the eminent grandees of the Royal Society about Newton's mathematical achievements, but they heard about the telescope all right and asked to see it late in 1671. At the very end of the year, Barrow delivered it to them (Figure 6.9).

When it arrived, the telescope caused a sensation. Early in January, Newton received a letter from Henry Oldenburg, the secretary of the society.

S^r

Your Ingenuity is the occasion of this addresse by a hand unknowne to you. You have been so generous, as to impart to the Philosophers here, your Invention of contracting Telescopes. It having been considered, and examined here by some of ye most eminent in Opticall Science and practise, and applauded by them, they think it necessary to use some meanes to secure this Invention from ye Usurpation of forreiners; And therefore have taken care to represent by a scheme that first Specimen, sent hither by you, and to describe all ye parts of ye Instrument, together wth its effect, compared wth an ordinary, but much larger, Glasse; and to send this figure, and description by y^e

[155] *Opticks*, pp. 104–5.
[156] *Keynes MS* 135. See A. A. Mills and P. J. Turvey, "Newton's Telescope. An examination of the Reflecting Telescope Attributed to Sir Isaac Newton in the Possession of the Royal Society," *Notes and Records of the Royal Society, 33* (1979), 133–55.
[157] *Corres 1*, 3–4.
[158] Collins to James Gregory, 24 Dec. 1970; *Corres 1*, 59.   [159] *Corres 1*, 5.
[160] Towneley to Collins, 4 Jan. 1672; *Corres 1*, 78. Flamsteed to Collins, 31 Jan. 1672; *Corres 1*, 88.

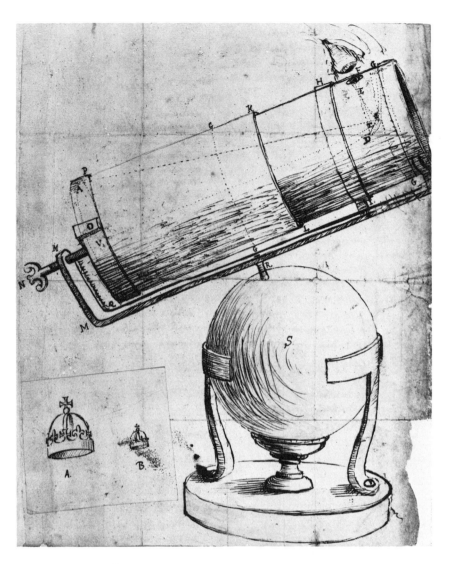

Figure 6.9. The Royal Society's drawing of Newton's telescope. The two crowns are the same object, an ornament on a weathercock 300 feet away, as seen through Newton's telescope (*A*) and as seen through a refracting telescope 25 inches long (*B*). (Courtesy of the Royal Society.)

Secretary of ye R. Soc. (where you were lately by ye Ld Bp. of Sarum [Seth Ward] proposed Candidat) in a solemne letter to Paris to M. Hugens, thereby to prevent the arrogation of such strangers, as may perhaps have seen it here, or even wth you at Cambridge; it being too frequent, y^t new Invention and contrivances are snatched away from their true Authors by pretending bystanders; But yet it was not thought fit to send this away wth out first giving you notice of it, and sending to you ye very figure and description, as it was here drawne up; y^t so you might adde, & alter, as you shall see cause; wch being done here w^th, I shall desire your favour of returning it to me wth all convenient speed, together wth such alterations, as you shall think fit to make therein . . .

> S^r
>
> your humble servant
> Oldenburg[161]

True to their word, the Royal Society sent a description of the instrument to Huygens; they even sent a general account written on 1 January ahead, so concerned were they to secure the credit to Newton.[162] Huygens was no less pleased than they; he called it the "marvellous telescope of Mr. Newton . . ."[163] The Society employed Christopher Cock, an instrument maker in London, to construct a relecting telescope four feet in length, and later one of six feet, though both failed for want of satisfactory mirrors.[164] Having nominated Newton, they proceeded to his full election to the society on 11 January.

The ritual dance performed with Collins now commenced anew. Newton fairly beamed as the warm glow of praise fell about him.

> At the reading of your letter [he replied to Oldenburg] I was surprised to see so much care taken about securing an invention to mee, of wch I have hitherto had so little value. [*sic!*] And therefore since the R. Society is pleased to think it worth the patronizing, I must acknowledg it deserves much more of them for that, then of mee, who, had not the communication of it been desired, might have let it still remained in private as it hath already done some yeares.[165]

Despite his pretense of indifference, Newton picked up their intention to send a description to Huygens and suggested that they be sure he realized the telescope eliminated colors from the image. He volunteered instructions about its maintenance, and in his next two letters he sent information about alloys he had tried for mirrors. He readily agreed to the publication of the description without even suggesting that his name be withheld.

---

[161] *Corres 1*, 73. [162] *Corres 1*, 72. [163] *Corres 1*, 89. [164] *Corres 1*, 83, 104.
[165] Newton to Oldenburg, 6 Jan. 1672; *Corres 1*, 79.

I am very sensible of the honour done me by ye B^p of Sarum in proposing mee Candidate [he concluded his initial reply to Oldenburg] & wch I hope will bee further conferred upon mee by my Election into the Society. And if so, I shall endeavour to testify my gratitude by communicating what my poore & solitary endeavours can effect towards ye promoting your Philosophicall designes.[166]

The Royal Society could not have guessed that the final sentence contained a hidden promise. Newton unveiled it on 18 January. He informed the society that "I am purposing them, to be considered of & examined, an accompt of a Philosophicall discovery wch induced mee to the making of the said Telescope, & wch I doubt not but will prove much more gratefull then the communication of that instrument, being in my Judgment the oddest if not the most considerable detection wch hath hitherto beene made in the operations of Nature."[167] The ritual dance had further figures, however. As he had found already with Collins, it proved to be not quite that simple to divulge his discovery. A week and a half later, he had not yet sent it, and he felt compelled to go through the remaining steps. He wrote that he hoped he could "get some spare howers" to send off the account.[168] Wickins needed the spare hours as much as Newton, who set him to work copying the paper. The die had been cast, however. It was too late to withdraw. On 6 February 1672, Newton finally mailed an account of his theory of colors to London. For the moment the positive pole prevailed. Swept along by the success of his telescope, Newton stepped publicly into the community of natural philosophers to which he had hitherto belonged in secret.

[166] *Corres 1*, 80.   [167] *Corres 1*, 82.   [168] *Corres 1*, 84.

7

Publication and crisis

THE paper on colors that Newton sent to the Royal Society
early in 1672 in the form of a letter addressed to Henry Oldenburg did not contain anything new from Newton's point of view.
The occasion provided by the telescope had come at an opportune
time. At Barrow's behest, Newton had been revising his lectures
for publication during the winter.[1] He had not found it a great
chore to produce a succinct statement of his theory buttressed by
three prismatic experiments that he took to be most compelling. He
thought it relevant to include a special discussion of how the discovery had led him to devise the reflecting telescope. The continuing correspondence provoked by the initial paper, which intruded
intermittently on his time and consciousness during the following
six years, also involved only one addition to his optics, his introduction to diffraction and brief investigation of it. Aside from diffraction, the entire thrust of his concern with optics during the
period was the exposition of a theory already elaborated.

The continuing discussion forced Newton to clarify some issues.
When he wrote in 1672, he had not yet fully separated the issue of
heterogeneity from his corpuscular conception of light, and he allowed himself to assert that, because of his discovery, it could "be
no longer disputed . . . whether Light be a Body."[2] He could
hardly have been more mistaken. Within a week of the paper's
presentation, Robert Hooke produced a critique that mistook corpuscularity for its central argument and proceeded to dispute it with
some asperity. The lesson was not wasted. Though he continued to
believe in the corpuscular conception, Newton learned to insist that
the essence of his theory of colors lay in heterogeneity alone.[3] This

---

[1] Cf. Collins to Gregory, 23 Feb. 1672 (H. W. Turnbull, ed., *James Gregory Tercentenary Memorial Volume* [London, 1939], p. 218); Collins to Gregory, 14 March 1672 (*Corres 1,* 119); Collins to Newton, 30 April 1672 (*Corres 1,* 146). *Add MS* 4002, the original manu-script of the lectures, contained eighteen lectures. Dd.9.67, the manuscript that Newton deposited as his lectures, contained thirty-one. Although the numbers exaggerate the ex-tent of expansion, there was certainly a revision, and it appears to me that Dd.9.67 was the product of the winter of 1671–2. See George Sarton, "Discovery of the Dispersion of Light and of the Nature of Color (1672)," *Isis, 14* (1930), 326–41.   [2] *Corres 1,* 100.

[3] Cf. his statement in February 1676 against objections based on hypotheses about the nature of light. "*That in any Hypothesis whence ye rays may be supposed to have any originall diversities, whether as to size or figure or motion or force or quality or any thing els imaginable wch may suffice to difference those rays in colour & refrangibility, there is no need to seek for other causes of these effects* [colors and different refractions] *then those original diversities.* This rule being laid down, I argue thus. In any Hypothesis whatever, light as it comes from ye Sun must be

was a matter of clarification and exposition, however, not an alteration of his theory. The very fact that six years of discussion effected no change, such that his *Opticks,* finally published in 1704, merely restated conclusions worked out in the late 1660s, testifies to the intensity and rigor of the early investigation.

The discussion that followed on the paper of 1672 tells us less about optics than about Newton. For eight years he had locked himself in a remorseless struggle with Truth. Genius of Newton's order exacts a toll. Eight years of uneaten meals and sleepless nights, eight years of continued ecstasy as he faced Truth directly on grounds hitherto unknown to the human spirit, took its further toll. And exasperation that dullness and stupidity should distract him from the further battles in which he was already engaged on new fields added the final straw. By 1672, Newton had lived with his theory for six years, and it now seemed obvious to him. For everyone else, however, it still embodied a denial of common sense that made it difficult to accept. Their inability to recognize the force of his demonstrations quickly drove Newton to distraction. He was unprepared for anything except immediate acceptance of his theory. The continuing need to defend and explain what he took to be settled plunged him into a personal crisis.

To be sure, the initial response gave no hint of the crisis to follow. Almost before the ink had dried on his paper of 6 February, Newton received a letter from Oldenburg. Filled with lavish praise, it informed him that his paper had been read to the Royal Society, where it "mett both with a singular attention and an uncommon applause . . ." The Society had ordered it to be printed forthwith in the *Philosophical Transactions* if Newton would agree.[4] The tension caused by the decision to send the paper can be heard in Newton's relief as he read Oldenburg's letter.

> I before thought it a great favour to have beene made a member of that honourable body; but I am now more sensible of the advantage. For beleive me S^r I doe not onely esteem it a duty to concurre wth them in ye promotion of reall knowledg, but a great privelege that instead of exposing discourses to a prejudic't & censorious multitude (by wch means many truths have been bafled & lost) I may wth freedom apply my self to so judicious & impartiall an Assembly.[5]

He assented to the publication of the paper with only a slight – for him, compulsory – demur.

---

supposed either homogeneal or heterogeneal. If ye last, then is that Hypothesis comprehended in this general rule & so cannot be against me: if the first then must refractions have a power to modify light so as to change it's colorifick qualification & refrangibility; wch is against experience." Newton to Oldenburg, 15 Feb. 1676; *Corres 1,* 419–20.

[4] Oldenburg to Newton, 8 Feb. 1672; *Corres 1,* 107–8.

[5] Newton to Oldenburg, 10 Feb. 1672; *Corres 1,* 108–9.

Accordingly, the paper appeared in the *Philosophical Transactions* for 19 February 1672. Together with the description of his telescope, which the following issue carried, it established Newton's reputation in the world of natural philosophy. Oldenburg took care to publicize both items in his extensive correspondence with natural philosophers throughout Europe. The replies he received indicate that both were noticed.[6] The telescope caught the eye of leading astronomers everywhere – Cassini, Auzout, and Denis in Paris, and Hevelius in Danzig. In the summer, Oldenburg received a report from Florence that a reflector built on Newton's plan had already been constructed there. A Frenchman, Cassegrain, did Newton the ultimate honor of claiming priority for the invention. In the same letter that reported Cassegrain's assertion, Oldenburg felt compelled to mention that James Gregory had published a plan for a reflecting telescope in his *Optica promota,* 1663.[7] As far as Cassegrain was concerned, no one seems to have taken his claim, advanced after the publication of Newton's telescope, seriously. Gregory never pressed his; and ever generous, he acknowledged the importance of Newton's paper on colors in establishing the theoretical significance of the reflecting telescope.[8] The Gregorian and the Cassegranian models differed from the Newtonian and from each other in the shape and placement of the secondary mirror. Newton argued, not without some heat, for the superiority of his arrangement.[9] No one challenged the fact that Newton had been the first to produce a reflecting telescope that worked.

The paper on colors also received its share of attention. Oldenburg specifically called it to Huygens's attention when he mailed the *Philosophical Transactions* to him. Huygens replied that "the new Theory . . . appears very ingenious to me." To be sure, Huygens later expressed reservations about the theory; meanwhile, in April, Newton received what can only have appeared as praise from the recognized leader of European science.[10] A young English astronomer, John Flamsteed, who would soon become the first Astronomer Royal, commented on the paper, though without much comprehension.[11] A young German savant resident in Paris, Gottfried Wilhelm Leibniz, then unknown but as determined to make his way in natural philosophy as he was destined to, indicated that he

6   A. Rupert and Marie Boas Hall, eds., *The Correspondence of Henry Oldenburg,* 12 vols. (Madison, 1965–), *9, passim.*
7   Oldenburg to Newton, 2 May 1672; *Corres 1,* 150.
8   Gregory to Collins, 23 Sept. 1672; *Corres 1,* 240.
9   Newton to Oldenburg, 4 May 1672; *Corres 1,* 153–5.
10 Oldenburg sent the comment in a letter on 9 April 1672; *Corres 1,* 135.
11 Flamsteed to Collins, 17 April 1672; *Corres 1,* 145.

had seen it.[12] Towneley reported to Oldenburg that Sluse had asked him to translate it into French so that he might read it. For himself, Towneley found the paper "so admirable" that he urged the publication of a Latin translation for the benefit of philosophers across Europe.[13] As a result of the telescope and the paper on colors, Newton soon found himself the recipient of presentation copies of books by Huygens and Boyle. Never again could he return to the anonymity of the early years in Cambridge. Once and for all, he had installed himself in the community of European natural philosophers, and among its leaders.

Newton did not see every comment on his theory of colors that Oldenburg and others received. He did see enough that he should have been gratified by its overwhelmingly favorable reception. The praise was not unanimous, however. Newton had concluded the paper with a seeming invitation to comment and criticism: "That, if any thing seem to be defective, or to thwart this relation, I may have an opportunity of giving further direction about it, or of acknowledging my errors, if I have committed any."[14] Alas, within two weeks he received a lengthy critique from Robert Hooke, the established master of the subject in England, a condescending commentary that contrived to imply that Hooke had performed all of Newton's experiments himself while it denied the conclusions Newton drew from them. Initially, Newton chose to ignore Hooke's tone.

> I received your Feb 19th. And having considered Mr Hooks observations on my discourse, am glad that so acute an objecter hath said nothing that can enervate any part of it. For I am still of the same judgment & doubt not but that upon severer examinations it will bee found as certain a truth as I have asserted it. You shall very suddenly have my answer.[15]

The critique must have rankled more than he let on, however. Instead of receiving the answer suddenly, Oldenburg had to wait three months; and when it arrived, its tone was rather less unruffled.

Meanwhile other comments and critiques arrived. Sir Robert Moray, the first president of the Royal Society, proposed four experiments (which betrayed no understanding of the question) to test

[12] Leibniz to Oldenburg, 26 Feb. 1673; *Oldenburg Correspondence, 9,* 491.
[13] Towneley to Oldenburg, 24 April 1673; *ibid., 9,* 622.   [14] *Corres 1,* 102.
[15] Newton to Oldenburg, 20 Feb. 1672; *Corres 1,* 116. See Léon Rosenfeld, "La Théorie des couleurs de Newton et ses adversaires," *Isis, 9* (1926), 44–65, and "La Premier conflit entre la théorie ondulatoire et la théorie corpusculaire de la lumière," *Isis, 11* (1928), 111–22; A. R. Hall and Marie Boas, "Why Blame Oldenburg?" *Isis, 53* (1962), 482–91; Zev Bechler, "Newton's 1672 Optical Controversies: A Study in the Grammar of Scientific Dissent," in Y. Elkana, ed., *The Interaction between Science and Philosophy* (Atlantic Highlands, N.J., 1974), pp. 115–42; and Alan E. Shapiro, "Newton's Definition of a Light Ray and the Diffusion Theories of Chromatic Dispersion," *Isis, 66* (1975), 194–210.

the theory.[16] More significant were the objections of the French
Jesuit, Ignace Gaston Pardies, professor at the Collège de Louis-
le-Grand and a respected member of the Parisian scientific community. He pointed out that for certain positions of the prism the
sine law of refraction could account for the diverging spectrum
since all the sun's rays were not incident on the prism's face at the
same angle, and he questioned the *experimentum crucis* on the same
grounds of unequal incidence.[17] In fact, Newton's initial paper had
adequately answered both objections. Nevertheless, Pardies's letter
was an intelligent comment by a man obviously knowledgeable in
optics. It was also respectful in tone, though Pardies made the
mistake of opening with a reference to Newton's "very ingenious
Hypothesis . . ." Hooke had also called the theory of colors Newton's "hypothesis" several times. Now he began to bridle.

> I am content [he concluded his reply to Pardies, manifestly
> discontent] that the Reverend Father calls my theory an hypothesis if
> it has not yet been proved to his satisfaction. But my design was
> quite different, and it seems to contain nothing else than certain
> properties of light which, now discovered, I think are not difficult to
> prove, and which if I did not know to be true, I should prefer to
> reject as vain and empty speculation, than acknowledge them as my
> hypothesis.[18]

Pardies did not propose to start a quarrel. He apologized handsomely and accepted Newton's explanation of why the unequal
incidence of the sun's rays on the prism could not explain the
divergence of the spectrum. He raised a further question, however: Could not Grimaldi's recent discovery, diffraction, explain
the divergency?

> In answer to this [Newton replied], it is to be observed that the
> doctrine which I explained concerning refraction and colours, con-
> sists only in certain properties of light, without regarding any hypo-
> theses by which those properties might be explained. For the best
> and safest method of philosophizing seems to be, first to enquire
> diligently into the properties of things, and to establish those prop-
> erties by experiments and then to proceed more slowly to hypotheses
> for the explanation of them. For hypotheses should be employed
> only in explaining the properties of things, but not assumed in deter-
> mining them; unless so far as they may furnish experiments. For if
> the possibility of hypotheses is to be the test of the truth and reality
> of things, I see not how certainty can be obtained in any science;
> since numerous hypotheses may be devised, which shall seem to

[16] *Cohen*, pp. 75–6. Newton responded in his letter to Oldenburg of 13 April 1672; *Corres 1*, 136–9.

[17] Pardies to Oldenburg, 30 March 1672; *Corres 1*, 130–3.

[18] Newton to Oldenburg, 13 April 1672; *Corres 1*, 144. Original Latin, p. 142.

overcome new difficulties. Hence it has been here thought necessary to lay aside all hypotheses, as foreign to the purpose . . .[19]

As the rest of the correspondence would further demonstrate, the discussion of colors provided Newton with his first serious occasion to explore questions of scientific method. Pardies expressed himself satisfied with the additional explanations that Newton offered, though there is no evidence that he accepted the theory.

During all this time, Hooke's critique of the February paper and the need to reply hung over Newton's head. Hooke and Newton were probably fated to clash. Newton had conceived his theory of colors in reaction to Hooke's. For his part, Hooke considered himself the authority on optics and resented the appearance of an interloper. When Newton's telescope set the Royal Society agog, he submitted a memorandum about a discovery using refractions that would perfect optical instruments of all sorts to the limit anyone could desire, far beyond Newton's invention. Unfortunately, he concealed the discovery itself in a cipher.[20] He approached the paper on colors in much the same way, with a magisterial tone of authority which would have been galling to a person less sensitive than Newton. Two scientists more different are hard to imagine. Though highly gifted, Hooke was more plausible than brilliant. He had ideas on every subject and was ready to put them into print without much hesitation. Newton in contrast was obsessed with the ideal of rigor and could hardly convince himself that anything was ready for publication. Hooke later confessed that he spent all of three or four hours composing his observations on Newton's paper.[21] He had cause to regret his haste. Newton spent three months on his response. It may be relevant as well that Hooke was sick enough with consumption that later in the year he was not expected to survive.[22]

Hooke submitted his critique to the Royal Society on 15 February, one week after Newton's paper was read. Newton had a copy by 20 February. Hooke granted Newton's experiments, "as having by many hundreds of tryalls found them soe," but not the hypothesis by which he explained them. "For all the expts & obss: I have hitherto made, nay and even those very expts which he alledged, doe seem to me to prove that light is nothing but a pulse or motion

---

[19] *Cohen*, p. 106. I have made some slight alterations in this eighteenth-century translation. Original Latin in *Corres 1*, 164.

[20] Collins's memorandum, written in early 1672, on Newton's letter of 23 Feb. 1669 to a friend; *Corres 1*, 4. Cf. the similar report in the minutes of the Royal Society for 18 Jan. 1672; Thomas Birch, *The History of the Royal Society of London*, 4 vols (London, 1756–7), *3*, 4. Cf. also Oldenburg to Huygens, 12 Feb. 1672; *Corres 1*, 110.

[21] In a paper about Newton's reply; *Corres 1*, 198.

[22] Cf. Collins to Gregory, 26 Dec. 1672; *Corres 1*, 255.

propagated through an homogeneous, uniform and transparent medium: And that Colour is nothing but the Disturbance of y^t light . . . by the refraction thereof . . . ." The burden of Hooke's critique was the reassertion of his own version of the modification theory as he had published it in *Micrographia*. He protested as well against Newton's abandonment of refracting telescopes. "The truth is, the Difficulty of Removing that inconvenience of the splitting of the Ray and consequently of the effect of colours, is very great, but not yet insuperable." He had already overcome it in microscopes, he asserted, but had been too busy to apply his discovery to telescopes. Like a true mechanical philosopher, Hooke kept returning to picturable images such as split rays to express his theory of colors. He saw Newton's theory in similar terms, as primarily an exposition of the corpuscular hypothesis, and he assured Newton that he could solve the phenomena of light and colors not only by his own hypothesis, but by two or three others as well, all different from Newton's. He failed entirely to come to grips with Newton's experimental demonstration of the fact of heterogeneity.[23]

Although Newton initially promised to reply at once, he planned an answer that required more time. Whatever else Hooke's critique contained, it reasserted the modification theory of colors without compromise. Newton decided to seize the opportunity it offered for a fully elaborated exposition of his own theory of analysis. He drew heavily on his *Lectiones opticae* for experimental support that he had omitted in his brief initial paper. Nor did he stop there. He composed as well an exposition of the phenomena of thin films as they pertained to the colors of bodies and the heterogeneity of light, more than a first draft of the "Discourse of Observations" of 1675 and Book II of the *Opticks*, since extensive passages appeared verbatim as they would be published thirty years hence. What Newton drafted in the early months of 1672 was a treatise on optics which contained a sometimes briefer exposition of all the elements of his ultimate work except Book II, Part IV (the phenomena of thick plates), Book III (his brief exposition of diffraction), and the Queries. Since it included the first sketch of his "Hypothesis of Light" (1675), it did contain material analogous to some of the Queries.[24] Published in 1672, the small treatise would have advanced the science of optics by thirty years.

[23] *Corres 1*, 110–14. For a more favorable assessment of Hooke's critique, see A. I. Sabra, *Theories of Light from Descartes to Newton* (London, 1967), pp. 251–64.

[24] *Add MS* 3970.3, ff. 433–44, 519–28. Although separated, these two batches go together; f. 519 follows the introduction to it at the bottom of f. 442^v. For a fuller discussion see Richard S. Westfall, "Newton's Reply to Hooke and the Theory of Colors," *Isis, 54* (1963), 82–96.

It was not published, however. In March, he told Oldenburg he had not yet completed it, and in April he delayed again.[25] Perhaps by now he was looking for an excuse not to send it. Two years earlier, he had refused to have his name attached to a formula for annuities lest it increase his acquaintance, which he studied chiefly to diminish. The telescope and the paper on colors had shown how right he had been. By early May, four months after he sent the telescope to London, he had received twelve letters and written eleven answers about the telescope and colors – hardly a crushing burden, but not a decrease in acquaintance either.[26] Discussing the events of the spring four years later, Newton told Oldenburg that "frequent interruptions that immediately arose from the letters of various persons (full of objections and of other matters) quite deterred me from the design [of publishing the *Optical Lectures*] and caused me to accuse myself of imprudence, because, in hunting for a shadow hitherto, I had sacrificed my peace, a matter of real substance."[27] If Newton was looking for an excuse, Oldenburg's letter of 2 May offered one. Oldenburg urged him to omit Hooke's and Pardies's names from his answers and to deal with their objections alone, "since those of the R. Society ought to aime at nothing, but the discovery of truth, and ye improvem^t of knowledge, and not at the prostituting of persons for their mis-apprehensions or mistakes." Another statement in the letter, that "some begin to lay more weight upon [your theory of light] now, than at first," may have served to increase his unhappiness with the suggestion.[28] Had Oldenburg misled him about the reception of his paper? Initially, Newton acquiesced in the request, though it clearly irritated him. He was more than irritated after he had brooded over it for two weeks.

> I understood not your desire of leaving out Mr Hooks name, because the contents would discover their Author unless the greatest part of them should be omitted & the rest put into a new Method wthout having any respect to ye Hypothesis of colours described in his *Micrographia*. And then they would in effect become new objections & require another Answer then what I have written. And I know not whether I should dissatisfy them that expect my answer to these that are already sent to me.

He had decided, he continued, not to send all that he had prepared, though he still intended to include a discourse on "the Phaenomena of Plated Bodies," in which he showed that rays differ in reflexibil-

---

[25] Newton to Oldenburg, 19 March 1672, 13 April 1672; *Corres 1,* 122, 137.

[26] I am counting all of the letters for which solid evidence exists, not all of the letters that survive.

[27] Newton to Oldenburg, 24 Oct. 1676 (the *Epistola posterior* for Leibniz); *Corres 2,* 133. Original Latin, p. 114.   [28] *Corres 1,* 151.

ity as well as refrangibility and related the colors of bodies to the thickness of their particles.[29]

A few days later, in a state of great agitation, he wrote to Collins, thanking him for his offer to undertake the publication of his optical lectures.

But I have now determined otherwise of them; finding already by that little use I have made of the Presse, that I shall not enjoy my former serene liberty till I have done with it; wch I hope will be so soon as I have made good what is already extant on my account.

He could not put the subject out of his mind, and after a paragraph about his mathematical work he returned to it.

I take much satisfaction in being a Member of that honourable body the R. Society; & could be glad of doing any thing wch might deserve it: Which makes me a little troubled to find my selfe cut short of that fredome of communication wch I hoped to enjoy, but cannot any longer without giving offense to some persons whome I have ever respected. But tis no matter, since it was not for my own sake or advantage y^t I should have used that fredome.[30]

When he finally sent the reply on 11 June, Newton had eliminated the discourse on thin films along with most of the material from his *Lectiones*. What he did send was an argument pointed to the issue of analysis versus modification. Though not so well known as the paper of February, the answer to Hooke supplemented it brilliantly in the use of prismatic phenomena to support the theory of colors.[31] Equally, it presented an argument *ad hominem*. Far from omitting Hooke's name, Newton inserted it in the first sentence of the reply, in the last, and in more than twenty-five others in between. He virtually composed a refrain on the name Hooke. Successive drafts of various passages progressed through three and four stages, each one more offensive than the last.[32]

I must confesse [he said in the final version of the opening paragraph] at ye first receipt of those Considerations I was a little troubled to find a person so much concerned for an *Hypothesis*, from whome in particular I most expected an unconcerned & indifferent examination of what I propounded. . . . The first thing that offers itselfe is lesse agreable to me, & I begin with it because it is so. Mr Hook thinks himselfe concerned to reprehend me for laying aside the thoughts of improving Optiques by *Refractions*. But he knows well y^t it is not for

[29] Newton to Oldenburg, 21 May 1672; *Corres 1*, 159–60.
[30] Newton to Collins, 25 May 1672; *Corres 1*, 161.
[31] *Corres 1*, 171–88. The faithful Wickins made a copy of the letter that is now in *Add MS* 3970.3, ff. 489–500.
[32] Successive drafts in *Add MS* 3970.3, ff. 433–4, 445, 447. Cf. Westfall, "Reply to Hooke."

one man to prescribe Rules to ye studies of another, especially not without understanding the grounds on wch he proceeds.

After setting Hooke straight on that score, Newton turned to Hooke's considerations of his theory.

And those consist in ascribing an Hypothesis to me wch is not mine; in asserting an Hypothesis wch as to ye principall parts of it is not against me; in granting the greatest part of my discourse if explicated by that Hypothesis; & in denying some things the truth of wch would have appeared by an experimentall examination.

Newton showed Hooke how to reconcile the wave hypothesis to his theory of colors, before he offered the opinion that Hooke's theory was "not onely *insufficient*, but in some respects *unintelligible*." He even instructed Hooke, who had boasted that he could perfect optical instruments in general, how to improve microscopical observations, Hooke's special province, by the use of monochromatic light. So much for Hooke's pretended perfection of refracting instruments!

In a covering letter of the same date to Oldenburg, Newton assumed that Hooke would find nothing objectionable in the reply since he had avoided "oblique & glancing expressions . . ."[33] That is, he employed the broadsword instead of the rapier. Where Hooke's observations had been irritatingly patronizing, Newton's reply was viciously insulting – a paper filled with hatred and rage. It established a pattern for his relations with Hooke that was never broken. The Royal Society forebore to print Hooke's critique lest it appear disrespectful to Newton.[34] It did allow Hooke to endure the humiliation, first of hearing the response read at a meeting, then of seeing it in print in the *Philosophical Transactions*.

Hooke was not the sole cause of Newton's exasperation. The entire exchange, the need to amplify and explicate what seemed perfectly obvious, annoyed him. On 19 June, he asked Oldenburg that he "not yet print any thing more concerning the Theory of light before it hath been more fully weighed."[35] On 6 July, he requested again that Pardies's second letter not be published, though he relented when Oldenburg told him it was already at the printers.[36] In the letter of 6 July, he tried to restate the question in a form that would terminate discussion.

I cannot think it effectuall for determining truth to examin the severall ways by wch Phaenomena may be explained, unless there can be a perfect enumeration of all those ways. You know the proper

[33] *Corres 1*, 193.
[34] Hooke was allowed to read a brief reply to Newton at the meeting of 19 June (*Corres 1*, 195–7). He did not even get to read his longer reply (*Corres 1*, 198–203).
[35] *Corres 1*, 194.
[36] Newton to Oldenburg, 6, 13 July 1672; *Corres 1*, 210, 217.

Method for inquiring after the properties of things is to deduce them from Experiments. And I told you that the Theory wch I propounded was evinced to me, *not by inferring tis thus because not otherwise,* that is not by deducing it onely from a confutation of contrary suppositions, but *by deriving it from Experiments concluding positively & directly.* The way therefore to examin it is by considering whether the experiments wch I propound do prove those parts of the Theory to wch they are applyed, or by prosecuting other experiments wch the Theory may suggest for its examination.

He proceeded then to reduce his theory to eight queries which could be answered by experiments. Let all objections from hypotheses be withheld. Either show the insufficiency of his experiments or produce other experiments that contradict him. "For if the Experiments, wch I urge be defective it cannot be difficult to show the defects, but if valid, then by proving the Theory they must render all other Objections invalid."[37]

Newton obviously meant that the experiments he had sent already answered his eight queries. Alas, the Royal Society directed that experiments be performed to test them, and Oldenburg, with all the delicacy of an uncoordinated cow, asked Newton to suggest some. It was 21 September before he brought himself to reply, and then only to say he was busy with other things. Oldenburg heard no more that autumn. Neither did Newton's other correspondent, Collins, until he received a lengthy commentary on Gregory's remarks about telescopes in December. Newton explained to Collins that he had written such a "long scribble . . . because Mr Gregory's discours looks as if intended for the Press."[38] Finally, in January, Oldenburg succeeded in eliciting a reply from Newton to a query on the improbable subject of cider ("wch liquor I wish, wth you, propagated far and near in England . . .")[39] Indeed, cider later became one of their staples of correspondence, a subject free of emotional investment whatever its content of spirit.

Having not understood the silence, Oldenburg immediately followed Newton's response by forwarding a new critique, from no less a figure than Huygens. This was the fourth comment Newton had received from Huygens, each one less enthusiastic than the one before. When the paper appeared, Huygens found it "very ingenious." In the summer, it still seemed "very probable" to him, though he doubted what Newton said about the magnitude of chromatic aberration. Newton sent a brief explication.[40] By

---

[37] *Corres 1,* 209–10.   [38] Newton to Collins, 10 Dec. 1672; *Corres 1,* 252.

[39] The initial exchange has been lost. It is known from Oldenburg to Newton, 18 Jan. 1673; *Corres 1,* 255.

[40] Oldenburg to Newton (containing Huygens's comment), 2 July 1672; *Corres 1,* 206–7. Newton to Oldenburg, 8 July 1672; *Corres 1,* 212–13.

autumn, Huygens thought things could be otherwise than the theory held and suggested that Newton be content to let it pass as a very probable hypothesis. "Moreover, if it were true that from their origin some rays of light are red, others blue, etc., there would remain the great difficulty of explaining by the mechanical philosophy in what this diversity of colors consists."[41] Oldenburg forwarded the comment to Newton; Newton did not answer. Of all the natural philosophers in Europe, Huygens was subjecting Newton's theory to its most searching scrutiny. In January 1673, he sent his fourth and fullest comment. It was also his most critical.

> I have seen, how Mr. Newton endeavours to maintain his new Theory concerning Colours. Me thinks, that the most important Objection, which is made against him by way of Quaere, is that, Whether there be more than two sorts of Colours. For my part, I believe, that an Hypothesis, that should explain mechanically and by the nature of motion the Colors Yellow and Blew, would be sufficient for all the rest, in regard that those others, being only more deeply charged (as appears by the Prismes of Mr. Hook) do produce the dark or deep-Red and Blew; and that of these four all the other colors may be compounded. Neither do I see, why Mr. Newton doth not content himself with the two Colors, Yellow and Blew; for it will be much more easy to find an Hypothesis by Motion, that may explicate these two differences, than for so many diversities as there are of other Colors. And till he hath found this Hypothesis, he hath not taught us, what it is wherein consists the nature and difference of Colours, but only this accident (which certainly is very considerable,) of their different Refrangibility.[42]

Once again, the mechanical philosophy with its demand for picturable explanatory images obstructed the understanding of Newton's discovery that light is heterogeneous.

Newton waited two more months to respond, only to indicate then that Huygens's private letter to Oldenburg did not call for an answer from him. If Huygens expected an answer, however, and intended "y^t they should be made publick," he would do so if he had Huygens' agreement "y^t I may have liberty to publish what passeth between us, if occasion be." In case that were not curt enough, he added something else for Oldenburg.

> S^r I desire that you will procure that I may be put out from being any longer fellow of ye R. Society. For though I honour that body, yet since I see I shall neither profit them, nor (by reason of this

---

[41] Huygens to Oldenburg, 17 Sept. 1672; *Corres 1*, 235–6.

[42] Oldenburg to Newton (containing Huygens's comment), *Corres 1*, 255–6. I quote Oldenburg's translation; *Cohen*, 136. The "Prismes of Mr. Hook" referred to experiments in *Micrographia* with glass containers of prismatic shape filled with yellow and blue liquids.

distance) can partake of the advantage of their Assemblies, I desire to withdraw.[43]

In regard to the threat of withdrawal, Oldenburg expostulated with Newton briefly and offered to have him excused from "ye trouble of sending hither his qterly payments & without any reflection."[44] Newton, who was seeking to avoid complications, not to multiply them, did not pursue the issue, and it simply passed. In April, he sent a reply to Huygens which returned again to the issue of explanatory hypotheses. He could not rest satisfied with two colors because experiments showed that other colors are equally primary and cannot be derived from red and blue. Nor was it easier to frame a hypothesis for only two "unless it be easier to suppose that there are but two figures sizes & degrees of velocity or force of the aetherial corpuscles or pulses rather then an indefinite variety, wch certainly would be a very harsh supposition." No one is surprised that the waves of the sea and the sand on the shore reveal infinite variety. Why should the corpuscles of shining bodies produce only two sorts of rays?

> But to examin how colours may be thus explained Hypothetically is besides my purpose. I never intended to show wherein consists the nature and difference of colours, but onely to show that *de facto* they are originall & immutable qualities of the rays wch exhibit them, & to leave it to others to explicate by Mechanicall Hypotheses the nature & difference of those qualities; wch I take to be no very difficult matter.[45]

He went on to discuss the other issues Huygens had raised; and though he avoided the deliberately insulting tone of his reply to Hooke, he could not conceal his vehemence. Certainly Oldenburg did not miss it. "I can assure you, " he wrote to Huygens, "that Mr. Newton is a man of great candor, as also one who does not lightly put forward the things he has to say."[46] Huygens, who was not used to being addressed as a delinquent schoolboy, did not miss it either; "seeing that he maintains his doctrine with some warmth," he replied, "I do not care to dispute."[47] He did permit himself a few pointed comments in a tone of icy hauteur. With one more letter from a somewhat chastened Newton, the exchange came to an end. Though Huygens had ample excuse to take offense, he recognized the quality of his opponent and chose not to.

[43] Newton to Oldenburg, 8 March 1673; *Corres 1,* 262.
[44] Oldenburg to Newton, 13 March 1673; *Corres 1,* 263.
[45] Newton to Oldenburg, 3 April 1673; *Corres 1,* 264.
[46] Oldenburg to Huygens, 7 April 1673; *Oldenburg Correspondence, 9,* 571. Original French in *Corres 1,* 268.
[47] Huygens to Oldenburg, 10 June 1673; *Corres 1,* 285.

The very letter that carried his response enclosed a list of English scientists to whom Oldenburg should present copies of his newly published *Horologium oscillatorium;* Newton was among them. What is more to the point, Huygens allowed himself to be convinced, even though the heterogeneity of light posed difficulties, which he never surmounted, to the specific form in which he couched his wave theory of light.[48]

Oldenburg had mentioned Newton's threat to withdraw from the Royal Society to Collins, who in turn commented on it to Newton.

> I suppose there hath been done me no unkindness [Newton wrote him in May], for I met wth nothing in y^t kind besides my expectations. But I could wish I had met with no rudeness in some other things. And therefore I hope you will not think it strange if to prevent accidents of that nature for ye future I decline that conversation wch hath occasioned what is past.[49]

When Collins showed him this, Oldenburg asked Newton to "passe by the incongruities" committed against him by members of the Royal Society. After all, every assembly had members who lacked discretion.

> The incongruities you speak of, I pass by [Newton told him]. But I must, as formerly, signify to you, y^t I intend to be no further sollicitous about matters of Philosophy. And therefore I hope you will not take it ill if you find me ever refusing doing any thing more in y^t kind, or rather y^t you will favour me in my determination by preventing so far as you can conveniently any objections or other philosophicall letters that may concern me.[50]

Oldenburg did not receive another letter from Newton for eighteen months.

Collins also found his correspondence interrupted. In the summer of 1674, Newton acknowledged the receipt of a book on gunnery and even commented on its content. "If you should have occasion to speak of this to ye Author," he added, "I desire you would not mention me becaus I have no mind to concern my self further about

---

[48] In paragraph 21, chap. 5 of the 1678 version of his *Treatise of Light,* which he read to the Académie, Huygens appeared to accept Newton's results though he still hoped to find a mechanical explanation of them (*Oeuvres complètes,* pub. Société hollandaise des Sciences, 22 vols. (The Hague, 1888–1950), *19,* 385–6). In letters to Leibniz (11 Jan. 1680) and to Fullenius (12 Dec. 1683 and 31 Aug. 1684), he accepted Newton's demonstration that chromatic aberration is much greater than spherical (*ibid., 8,* 257, 478, 534). He accepted that result also in his *Dioptrica,* which belonged to the period 1685–92 (*ibid., 13,* 483–7; cf. pp. 551–7, 621–73). He did not accept Newton's corpuscular conception of light, of course, and he remained dissatisfied that Newton offered no explanation in mechanical terms of what color is (Huygens to Leibniz, 29 May 1694; *ibid., 10,* 610–11).

[49] Newton to Collins, 20 May 1673; *Corres 1,* 282.

[50] Newton to Oldenburg, 23 June 1673; *Corres 1,* 294–5.

it."[51] Late in 1675, Collins told Gregory that he had neither seen nor written to Newton for a year, "not troubling him as being intent upon Chimicall Studies and practices, and both he and Dr Barrow &c beginning to thinke mathcall Speculations to grow at least nice and dry, if not somewhat barren."[52] Collins's correspondence with Newton never revived.

Oldenburg and Collins had functioned as Newton's contact with the learned world outside Cambridge. Although ample opportunities to correspond directly with men of the caliber of Gregory and Huygens had presented themselves, Newton had refused to grasp them. He communicated with others through the two intermediaries, who virtually monopolized his correspondence. In cutting their access to him, Newton attempted to regain his former solitude. After the publications of 1672, however, that was impossible. Huygens's presentation copy of the *Horologium* demonstrated as much, and in September 1673, Boyle confirmed the point by presenting him a copy of his book on effluvia. Nevertheless, for the moment, a modicum of criticism had sufficed, first to incite him to rage, and then to drive him into isolation.

Meanwhile, life in Cambridge continued its course. As far as we know, Newton completed his lectures on optics in the autumn of 1672 and the following autumn began a series on algebra (ultimately published as *Arithmetica universalis,* 1707) which continued for eleven years. In the summer of 1672, he appeared in print as the editor of Bernard Varenius, *Geographia universalis* (originally published in the Netherlands in 1650). Although we know nothing about the occasion of this publication, it is reasonable to see the hand of Isaac Barrow behind it. Newton later confessed that all he did was supply the schemes which were referred to in the original edition but not present.[53] Early in 1673, his erstwhile patron, who had been resident in London for a period, returned to Trinity as master of the college. The news had preceded the fact; already in December, Newton wrote to Collins about it as common knowledge, adding that no one rejoiced over the appointment more than he.[54] At that time, he could not know how important it would be to him two years hence.

Newton was absent from Cambridge for a month in the summer of 1672. After a visit to Bedfordshire, undoubtedly in connection

---

[51] Newton to Collins, 20 June 1674; *Corres 1,* 309.

[52] Collins to Gregory, 19 Oct. 1675; *Corres 1,* 356. Earlier, on 29 June 1675, Collins told Gregory that Newton's main attention was centered on chemical studies and experiments; *Corres 1,* 345.

[53] *Keynes MS* 130.5, sheet 1, and *Keynes MS* 130.6, Book 1.

[54] Newton to Collins, 10 Dec. 1672; *Corres 1,* 252.

with the lands that supported his professorship, he spent a couple of weeks at home. On his return, he paid a mysterious call to Stoke Park, Northamptonshire, where he stayed with unidentified friends for nearly two weeks.[55] His three weeks' absence in the spring of 1673 probably indicated another trip home. He did not return to Woolsthorpe again for at least two and a half years, when a ten-day absence in October 1675 may have been spent there.

Newton probably moved into his permanent chamber in Trinity on the first floor beside the great gate near the end of 1673 (Figure 7.1). An account headed "The Income of that Chamber in wch Mr Isaac Newton now inhabits" began with an entry for a payment to Thomas Coppinger attached to "Mr Thorndicks income".[56] Herbert Thorndike had inhabited the chamber until he resigned his fellowship in 1667. His lease from the college of the tithes and parsonage of Trumpington that year had forced his resignation, since the college statutes forbade a fellow to hold a lease from the

---

[55] He wrote three letters to Oldenburg (6, 8, 13 July) and one to Collins (13 July) from there; *Corres 1,* 208–11, 212–13, 217–18, 215–16.

[56] The Income of that Chamber in wch Mr Isaac Newton now inhabits.

Mr Thorndicks Income	Imprimis paid by the said Isaac Newton to Mr Tho Coppinger for Mr Thorndicks income upon Dr Babingtons account	2.^{lb} 18.^{s} 6^d
	Paid more by Dr Babington to Mr Tho. Coppinger upon the same account	3. 0. 0

Paid to Silk for a new door out of the Chamber
Portal into the Garden   0.   8. 6
To the Porter for a Pump & setting of it down   2.   6. 8

Add MS. 3970.3, f.469^v. Newton crossed the final item out. The account itself dated from the 1690s. On the same sheet, he referred to an observation Halley made about light and colors from a diving bell, an observation he included in *Opticks,* Book I, Part II, Proposition X. Halley did not begin to dive with a bell until 1691. I do not pretend to know why Newton should have been drawing up the account at that time; possibly Babington's death at the beginning of 1692 provided the occasion. The terms of the account baffle me. Babington may have appeared in it because of his office as bursar of the college in 1674. I do not understand what the "income" of the chamber could refer to or in what sense a former resident of a chamber could command its income. There is another account, apparently related to the chamber, which I date by its hand alone to the late 1670s. It also included references to income and to Dr. Babington.

Paid in part of Income due to Dr Babington   8.   0. 0
Paid more for Income to Dr Bab.   3.   8. 0
Paid for making ye Oven mouthed Chimney in
ye Chamber   0.   7. 8
Paid for ye fire irons there   0. 14. 6
Paid for a stone roll in ye Garden besides
ye frame   0. 17. 0
Paid for making a new Cellar behind y^e
Chappel   3.   5. 2
For a door bolts & a lock to ye Cellar   0. 10. 0

college. In his will of 1672, Thorndike bequeathed the lease to the bishop of Rochester, the dean of Christ Church, and the master of Trinity, and to their successors, and decreed that they should allow the profits of the lease to the resident vicar, Thomas Coppinger, M.A. 1626, a graduate of Trinity and brother of Thorndike's sister-in-law.[57] Coppinger died on 25 April 1674. The records in the Junior Bursar's Accounts of expenditures on fellows' gardens indicate the possibility that Thomas Gale succeeded Thorndike in the chamber. Gale resigned his fellowship sometime between Michaelmas and Christmas 1673, freeing the chamber.[58] The same college

Paid for altering & hanging ye Chamber

To ye Joyner		1.	6.	0
Smith		0.	19.	0
Painter		1.	0.	0
Upholster		14.	13.	9
Glasier		1.	15.	0
		36.	16.	1
Deduct Oven Chimney & Glasier		2.	02.	8
Remains		34.	13.	5
Thirds deducted		23.	2.	3

Two Chairs wth arms 1^{lb}6^s   ⎫
Eight Chairs wthout arms 4^{lb}   ⎬   8.^{lb} 0.^s 0.^d
Ten Cushions 5^{lb}   ⎭

Scritore 3^{lb}	3.	0.	0
6 Russia Leather chairs	0.	15.	0
A Chest	0.	2.	6
Two Spanish Tables wth neats leather			
Carpets	1		

Yahuda MS 34, f. 2. In accordance with his own instructions, Newton crossed out the entry for making the oven-mouthed chimney, though he did not cross out the entry for the glazier. The sum for the chairs and cushions was originally (and correctly) £10 6s 0d. I assume this was a joint account, and that he understood the special chimney, clearly for his alchemical experiments, to have been his private expense. One possible explanation for the deduction of a third instead of a half is Wickins's nonresidence by the late 1670s. I have no explanation for Babington's presence in the account beyond the speculation offered above for the other account.

[57] Herbert Thorndike, *Theological Works*, 6 vols. (Oxford, 1844–56), 6, 144–5.

[58] The Senior Bursar's Book recorded a payment to Gale of one quarter's stipend for the year that began with Michaelmas 1673.

Figure 7.1. Loggan's engraving of Trinity College in the late seventeenth century. The chamber in which Newton first lived as a fellow may have been on the north side of the court (the right-hand side in this print), more or less opposite the figures on the path there. His final chamber was on the first floor (i.e., above the ground floor) immediately north of the great gate. The garden beside the chapel was attached to that chamber. (From David Loggan, *Cantabrigia illustrata*, n.d., late seventeenth century.)

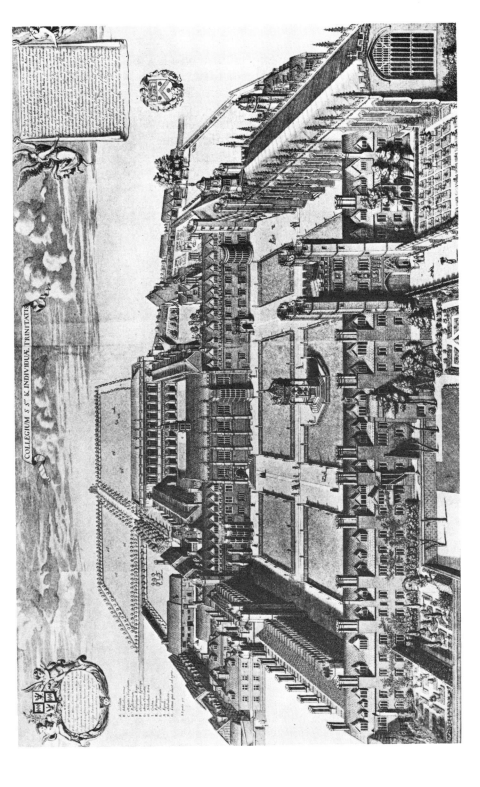

COLLEGIUM S.S. K INDIVIDUÆ TRINITATIS

accounts do not show any expenditure on Newton's garden before 1683; the recorded expenditure in that year has been taken as the earliest evidence of Newton's habitation of the chamber. The entry in his account, with the payment to Coppinger, seems unequivocal, however. Newton continued to reside in the chamber, first with Wickins when he was resident and then alone, until he left Trinity in 1696.

Newton traveled to London at the end of August 1674 to participate in the installation of the Duke of Monmouth as chancellor of the university. The vice-chancellor and heads had decided that at least six Masters of Arts from the three great colleges and three from all the others should attend, and that most of them should be regents "whose ornaments (as twas thought) would give the greatest grace & beauty to the Procession." The scarlet robe of a university professor manifestly filled that category. On 3 September they gathered at Derby House, some 480 in all, including nobles and knights graduated from the university, and they set out at four o'clock when word arrived that Monmouth was ready to receive them. Preceded by mounted members of the King's Life Guards to clear a way through the crowd, the junior bedell led the way with the bedell's staff, followed by the regents, non-regents, university officers and doctors of the various faculties, "all habited in the Ornaments agreeable to their respective Orders & Degrees." Monmouth received them at the door of Worcester House, where musketeers kept the throng from entering with the procession. The usual round of fulsome orations accompanied the administration of the oath before they all sat down to the obligatory feast. In "a few but very full & affectionate words," Monmouth declared his pleasure at his election, which he owed of course to the explicit order of his father, Charles II. For his part, Charles declared his pleasure at the university's docility by distributing £300 among its officers and granting them what was called a "large Concession" to admit such as they thought fit to the degree of Master of Arts. They made large use of the large concession, undoubtedly with no small profit to themselves to supplement the £300.[59] Though he was in London for a week, Newton made no recorded effort to meet with members of the Royal Society.

Willy nilly, Newton devoted some time to mathematics during these years. Every indication suggests that he did not do so spontaneously. He did not pursue a coherent program of mathematical

[59] Charles Henry Cooper, *Annals of Cambridge*, 5 vols. (Cambridge, 1842–1908), *3*, 560–3. The Fellows Exit & Redit Book in Trinity shows that Newton left the college on 28 August and returned on 5 September. Two other fellows left earlier in August, one other on 28 August, and four on 1 September. I assume that most of them represent the Trinity delegation at the installation.

investigation. He frequently protested his lack of interest. Nevertheless, his mathematical genius was known now, and mathematical discourse was forced upon him whether he wanted it or not, much of it through the agency of John Collins. In December 1672, for example, Collins wrote him about a method of drawing tangents developed by René de Sluse, a mathematician in the low countries, which was later published in the *Philosophical Transactions*. Newton replied that the method was apparently identical to his own. As though determined to tantalize Collins beyond endurance, he worked one specific problem without saying a word to explain his procedure.

> This S^r [he concluded maddeningly] is one particular, or rather a Corollary of a Generall Method wch extends it selfe wthout any troublesome calculation, not onely to the drawing tangents to all curve lines whether Geometrick or mechanick or how ever related to streight lines or to other curve lines but also to the resolving other abstruser kinds of Problems about the crookedness, areas, lengths, centers of gravity of curves &c. Nor is it (as Huddens method de maximis et minimis & consequently Slusius his new method of Tangents as I presume) limited to aequations wch are free from surd quantities. This method I have intervowen wth that other of working in aequations by reducing them to infinite series. I remember I once occassionally told Dr Barrow when he was about to publish his Lectures that I had such a method of drawing Tangents but some divertisement or other hindered me from describing it to him.[60]

So much for Collins. Six months later, after Oldenburg had relayed this hint of a hint to Sluse, who immediately wanted to learn more, Newton refused to discuss it in a manner calculated to make Collins realize he was not to repeat what Newton said.[61] By this time, if not before, Newton's exasperation with the correspondence about colors was interacting with his mathematical correspondence further to enhance his already considerable reluctance to communicate. The very next sentence of his letter to Collins in May 1673 took up his threat to resign from the Royal Society, which Collins had heard about from Oldenburg. Years later, when he cited the December letter to Leibniz as evidence of his early discovery of the calculus, Newton wished he had said more.

Probably it was the receipt of Huygens's *Horologium oscillatorium* in the summer of 1673 which prompted Newton to develop a demonstration of the isochronous property of the cycloid.[62] However briefly, Huygens's work recalled Newton to the topics in mechan-

---

[60] Newton to Collins, 10 Dec. 1672; *Corres 1*, 247–8.
[61] Newton to Collins, 20 May 1673, and Collins to Newton, 18 June 1673; *Corres 1*, 282, 288.   [62] *Math 3*, 420–30.

ics he had pursued in the 1660s. Huygens's investigation of centrifugal force, he wrote to Oldenburg, "may prove of good use in naturall Philosophy & Astronomy as well as mechanicks."[63] He would find occasion to recall this letter also, to cite against Hooke rather than Leibniz.

Barrow continued to press him into service. Newton went over both Barrow's edition of Euclid and his edition of Archimedes to find corrections for the list of errata. Thomas Horne, a fellow of King's, applied to him for help in understanding Descartes's *Geometry*. From outside Cambridge, John Lacy, one of the King's surveyors, put a question in mathematics to him.[64]

A note dated 1673 in the appendix to Edward Sherburne, *The Sphere of Manilius* (1675), stated that Newton had a treatise on dioptrics ready for the press, "and divers *Astronomical Exercises,* which are to be subjoyned to Mr. *Nicholas Mercator's Epitome of Astronomy,* and to be printed at *Cambridge.*" He added that Newton also planned to publish a general analytic method based on infinite series for the quadrature of figures, centers of gravity, volumes, surfaces, and rectifications.[65] Whatever the source of Sherburne's information, we know nothing more about the astronomical exercises. They did not appear in Mercator's book when he published it in 1676, though a reference to Newton, who had shown the author a very elegant hypothesis on the moon's libration, establishes that the two had met.[66]

Though Newton was holding Collins at arm's length now, Collins was still the primary agent who broadcast the news of his mathematical ability. Late in 1673, he set Michael Dary, a computer and gauger who, at the ripe age of sixty, had become Collins's protégé, on Newton. For more than a year, Dary forced a correspondence, mostly about algebra, on him.[67]

Of more significance was the correspondence in 1675 with another computer, John Smith, whom Collins also directed to Newton. Smith planned to prepare a table of square, cube, and fourth roots of all the numbers from 1 to 10,000. He applied to Newton for help in reducing the appalling burden of computation. Newton advised a method of interpolation. In the beginning, Smith

---

63 Newton to Oldenburg, 23 June 1673; *Corres 1,* 290.
64 Collins to Newton, ca. July 1675; *Corres 1,* 346. Horne to Newton, 22 Aug. [?1676]; *Corres 2,* 86–7. Newton to Lacy, n.d. but hand of early 1670s; *Corres 7,* 361–2.
65 Edward Sherburne, *The Sphere of Manilius* (London, 1675), p. 116.
66 Cited in *Edleston,* p. li. In his letter to Oldenburg of 23 June 1673, commenting on Huygens's *Horologium,* Newton mentioned the moon's motion (*Corres 1,* 290).
67 Newton to Dary, 6 Oct. 1674; Dary to Newton, 15 Oct. 1674; Newton to Dary, 22 Jan. 1675; *Corres 1,* 319–20, 326, 332–3. There is evidence of quite a few other letters. It is worth noting that Newton corresponded directly with Dary, possibly because he was only a computer and in no way a peer.

would need to extract a hundred roots to ten or eleven places for each table, one for each hundredth number. Newton sent him some instructions based on the binomial theorem on how to minimize that task. Working from the hundred roots, he could compute all the intervening ones correct to eight places by a method of interpolation, first every tenth root, and from them the nine in between. Once he computed the initial set of roots, nearly all the additional computation would consist only of addition and subtraction.[68] Unlike the elementary problems set him by Dary, the problem of interpolation presented a challenge which seized Newton's interest, and he devoted some time to it. Nothing came of Smith's project, but Newton's interest produced the foundations of modern interpolation theory. Late in 1676, he started a brief treatise on interpolation under the title *Regula differentiarum*.[69] In typical fashion, he strove to transcend particular methods in order to "embrace everything in one single general rule . . ."[70] He abandoned the *Regula* unfinished to start anew on what became a systematic exposition of interpolation by means of central differences. In the latter paper, the foundation of his *Methodus differentialis* published in 1711, Newton derived what are now referred to as the Newton-Stirling and Newton-Bessel formulas.[71] Early in the twentieth century, a commentator on Newton's method of interpolation compared his work, an occasional piece flung off hastily in the midst of other concerns, with the account of interpolation that was then considered authoritative. "Modern workers," he concluded, "have struggled up to the level reached by Newton . . ."[72] In a letter for Leibniz written in October 1676, Newton referred to his work on interpolation and stated the basic theorem on which it rested, that interpolation is equivalent to finding, at the desired point, the ordinate of a curve that passes through the given values between which one is interpolating.[73] He did not, however, include in the letter the method of interpolation, which he may not yet have completed. Like the rest of his mathematics, it remained a private possession shared as yet by no one.

In 1676, a mathematical correspondence that Oldenburg carried on with the aid of Collins as part of his program of philosophical

---

[68] Newton to Smith, 8 May, 24 July, 27 Aug. 1675; *Corres 1*, 342–4, 348–9, 350–1. Note that Newton also corresponded directly with Smith.

[69] *Math 4*, 36–50. See Duncan Fraser, "Newton and Interpolation," in *Isaac Newton, 1642–1727, a Memorial Volume*, ed. W. J. Greenstreet (London, 1927), pp. 45–69, and *Newton's Interpolation Formulas* (London, n.d.). [70] *Math 4*, 47. [71] *Math 4*, 54–68.

[72] Fraser, *Newton's Interpolation Formulas*, p. 71.

[73] Newton to Oldenburg, 24 Oct. 1676 (the *Epistola posterior* for Leibniz); *Corres 2*, 137. Original Latin, p. 119.

communication spilled over to include Newton. The correspondence dated back beyond the early months of 1673, when a young German philosopher, Gottfried Wilhelm Leibniz, visited the Royal Society. Leibniz had arrived in Paris the previous year in the diplomatic service of the Elector of Mainz. Already marked as a man of genius, he readily made himself at home in the intellectual circle of the French capital and among the members of the Académie des Sciences. He convinced himself that he must become proficient in mathematics, for which task he recruited the aid of no less a tutor than Christiaan Huygens. Pressing financial necessities also played a role in Leibniz's activities. Without means of his own, he had only his intellect on which to live, though it was not an inconsiderable asset. When his patron in Mainz died, leaving him virtually stranded, his practical need became acute. He hoped desperately for an appointment to the Académie, or for some other position which would enable him to stay in a center of learning where he could thrive amidst an intellectual community not to be found in his native Germany, still shattered by the trauma of the Thirty Years' War. Both for intellectual and for practical reasons, Leibniz cultivated learned circles in western Europe. The Royal Society in London rivaled the Académie in Paris. Already in 1671, he had thrust himself before it with his dedication of an essay on motion. He visited London and the Royal Society in January and February 1673 and was elected to membership; after he returned to Paris, he took care to maintain a steady correspondence with Oldenburg, who needed little encouragement in any case. At the beginning of 1673, Leibniz was still very much a tyro in mathematics, but he was advancing with giant strides toward its leading ranks. He made mathematics the focus of the correspondence with Oldenburg. Oldenburg, who was not a mathematician, pressed Collins into service to support his end of the exchange. Collins, of course, had made it his business to stay in touch with the leaders of British mathematics, especially Gregory and Newton.

In response to Leibniz's questions, Oldenburg mailed him on 6 April 1673 a long report, which Collins had drawn up for him, on the status of British mathematics. Though Newton's name was by no means the only one mentioned, it did figure prominently; one paragraph especially gave an intriguing résumé of the problems he could solve with his method of infinite series, though it did not give any suggestion of what the method was.[74] During the following years, the correspondence continued while Leibniz's own grasp of mathematics broadened rapidly. He tended to ask more than he revealed, but in this he did not differ from the established mode of

---

[74] *Oldenburg Correspondence, 9,* 563–7.

the day. He had powerful practical motives not to reveal his own progress. His inventions constituted his sole capital. The possibility of a position in Paris, membership in the Académie or the Ramus chair of mathematics (vacated by Roberval's death), depended on his achievement. Besides, the replies from London did not communicate anything fundamental. In April 1675, in response to specific questions, Leibniz received the most informative letter yet, a résumé (ultimately from Collins) of progress with infinite series, which seemed to Leibniz to be the focus of British mathematics. He was not unacquainted with the topic himself. Gregory and Newton dominated the report, which included a number of series expansions.[75] About the same time, in expectation of a visit to England and a discussion with Collins, Leibniz used Oldenburg's letter of 1673 as the basis of a memorandum in which he listed questions he wanted to ask.[76] The memorandum is of interest for its unselfconscious revelation of what Leibniz thought about Newton in 1675. Newton's name appeared in it, of course. Leibniz wanted to learn more about his method of infinite series, the one thing he had heard about his mathematics. But Newton's was only one name among others. By no means did Leibniz see him in 1675 as the commanding figure of English mathematics.

Building on his own earlier progress, Leibniz achieved the fundamental insights of his differential calculus, which was virtually identical to Newton's fluxional method, during the autumn of 1675. He developed his distinctive notation, in which the calculus still expresses itself, at that time. All of this has been established, not from Leibniz's assertions, but from his manuscripts, just as Newton's invention of the fluxional method has been.[77] At the end of 1675, it is doubtful that Newton was aware of Leibniz's existence, though he may have heard his name spoken at the Royal Society earlier that year in connection with a bitter exchange between Oldenburg and Hooke about Huygens's spring-driven watch. To the best of our knowledge, Newton was also unaware that reports on his mathematical achievements, with materials from his letters and from *De analysi,* were being sent to Leibniz. For this, he alone was to blame. He had consistently discouraged communication and cut himself off when others were eager to discuss and learn. Years later, after a bitter priority dispute had broken out, when Newton learned what Collins had sent, he drew his own sinister conclusions. What is clear from the correspondence, how-

---

[75] Oldenburg to Leibniz, 12 April 1675; *Der Briefwechsel von Gottfried Wilhelm Leibniz mit Mathematikern,* ed. C. I. Gerhardt (Berlin, 1899), pp. 113–22.

[76] *Corres 2,* 235–6. See also Joseph E. Hoffman, *Leibniz in Paris, 1672–1676. His Growth to Mathematical Maturity* (Cambridge, 1974), p. 291. Hoffman was the first to identify this memorandum.   [77] *Ibid.,* pp. 187–201.

ever, is that by the end of 1675, the critical period in Leibniz's own development, he had received only some of Newton's results without demonstrations, and that these results had been confined to infinite series. To be sure, infinite series formed an integral part of the fluxional method, but Leibniz had not heard of their broader ramifications.

In 1676, Newton both learned who Leibniz was and entered the correspondence himself. Leibniz wrote to Oldenburg in May about two series he had recently received from Collins via a Danish mathematician, Georg Mohr, the series expressing the sine of an angle given the arc, and the inverse series which expressed the arc or angle given the sine. Actually, the letter from Oldenburg a year earlier had also included both series; only now did their elegance impress Leibniz. He asked for demonstrations of them.[78] Both Oldenburg and Collins urged Newton to respond. The request came at an unwelcome time. A new correspondence that challenged his theory of colors had opened, and Newton was allowing it to agitate him excessively. Nevertheless, he acceded to the request, and on 13 June 1676, he completed a letter for Leibniz and Ehrenfried von Tschirnhaus, another German mathematician then resident in Paris and a friend of Leibniz, who was also involved in the correspondence tangentially. Once again, Newton chose not to enter into direct communication. He addressed the letter to Oldenburg, who forwarded a copy, together with another résumé of the work of Gregory and others that Collins prepared, to Leibniz and Tschirnhaus on 26 July. Not wishing to trust a letter he recognized to be important to the mail, Oldenburg had waited until he found a carrier who would deliver it to Leibniz.

Newton wrote two letters for Leibniz in 1676. Nearly forty years later, he cited them as evidence against Leibniz in the priority dispute and labeled them the earlier letter and the later letter, the *Epistola prior* and the *Epistola posterior*. Replying in the first to Leibniz's question about the foundation of the two series, he drew upon the combined reservoir of *De analysi* and *De methodis* to present a general exposition of series. The foundation of the reduction of functions to infinite series lies in division and root extraction, he explained, carried out in symbols just as they are carried out with decimals, but the operations are shortened by the use of the binomial theorem. He stated the theorem and illustrated its use with nine examples. He proceeded then to his method of extracting the roots (expressed as infinite series, of course) of affected equations. Nine further examples derived the two series about which Leibniz

[78] Leibniz to Oldenburg, 2 May 1676; *Corres 2*, 3–4. See Christoph J. Scriba, "The Inverse Method of Tangents: A Dialogue between Leibniz and Newton (1675–1677)," *Archive for History of Exact Sciences, 2* (1962–6), 113–37.

had asked, suggested the use of series to determine areas, volumes, and so on, and showed the reduction of mechanical curves, such as the quadratrix, to series. In these examples, Newton assumed and employed his algorithms for differentiation and integration without expounding them. With all the letter offered, Newton could not refrain from tantalizing Leibniz as he had tantalized Collins by suggesting that he held more in reserve.

> From all this it is to be seen how much the limits of analysis are enlarged by such infinite equations; in fact by their help analysis reaches, I might almost say, to all problems, the numerical problems of Diophantus and the like excepted. Yet the result is not altogether universal unless rendered so by certain further methods of developing infinite series. For there are some problems in which one cannot arrive at infinite series by division or by the extraction of roots either simple or affected. But how to proceed in those cases there is now no time to explain . . .[79]

If Leibniz had hitherto considered Newton merely as one among a number of English mathematicians, the *Epistola prior* disabused him. Nor was he reluctant to express his admiration. "Your letter," he wrote to Oldenburg immediately upon its receipt, "contains more numerous and more remarkable ideas about analysis than many thick volumes published on these matters . . . Newton's discoveries are worthy of his genius, which is so abundantly made manifest by his optical experiments and by his catadioptrical tube [the reflecting telescope]." He went on to show Newton that he knew a thing or two about infinite series himself, to expound his general method of transformations, as he called it (withholding somewhat as Newton had done), and to put some specific questions to Newton.[80] Tschirnhaus also responded enthusiastically, but with his departure from Paris he ceased to participate in the exchange.

Leibniz's new questions provided the occasion of the *Epistola posterior*. Before Newton could write it, however, Leibniz himself visited London for ten days in October. Despite all his hopes, his search for a position in Paris had failed, and he had finally accepted an appointment at the court of the Duke of Brunswick-Lüneburg. He stopped in London on his way to Hanover. While he was there, he conversed with Collins. It was hardly a meeting of equals, Collins scarcely more than a computer, Leibniz a mathematical genius of the first order, who swept Collins quite off his feet and left him convinced that English mathematicians lagged far behind.[81] Daz-

---

[79] *Corres 2*, 39. Original Latin, p. 29. The whole *Epistola prior* is found on pp. 20–32, and the English translation on pp. 32–41.

[80] Leibniz to Oldenburg, 26 July 1676; *Corres 2*, 65–71. Original Latin, pp. 57–64.

[81] Collins to Strode, 24 Oct. 1676; *Corres 2*, 109.

zled by the visitor, Collins opened his files to him. Leibniz read *De
analysi* and a fuller exposition of Gregory's work than that sent to
him, a piece called the *Historiola* which included Newton's letter on
tangents. Though he took notes on the last, he did not take notes
on Newton's fluxional propositions at the end of *De analysi* nor on
Gregory's method of maxima and minima. His notes concentrated
on infinite series, which he saw as the subject in which British
mathematics could instruct him. The absence of notes on the fluxional calculus implies that he saw nothing there he did not know
already.[82] After Collins regained his equilibrium following Leibniz's departure, he realized the extent of his indiscretion. He did not
tell Newton what he had shown to Leibniz. Apparently, from the
content of the *Commercium epistolicum,* Newton learned only later
that Leibniz saw *De analysi.* For his part, Leibniz chose not to
mention it.

Even before Leibniz's visit, Collins had been impressed enough
by the reply to the *Epistola prior* to urge Newton anew to publish
his method. Obsessed with the latest exchange on colors, Newton
thought otherwise.

> I look upon your advice as an act of singular friendship [he wrote],
> being I beleive censured by divers for my scattered letters in y^e
> *Transactions* about such things as no body els would have let come
> out w^{th}out a substantial discours. I could wish I could retract what
> has been done, but by that, I have learnt what's to my convenience,
> wch is to let what I write ly by till I am out of ye way.

Collins's expressed fear that Leibniz's method would prove more
general left him wholly unmoved. With serene confidence he described what his method could accomplish in a passage quoted
above at the end of Chapter 4. "This may seem a bold
assertion . . . ," he continued, "but it's plain to me by ye fountain I
draw it from, though I will not undertake to prove it to others."[83]

Meanwhile he completed the second response to Leibniz's questions, the *Epistola posterior,* one week after Leibniz left London for
Hanover. Wickins transcribed the copy sent to London, probably
the last episode in his career as Newton's amanuensis. The letter
began with an autobiographical passage on Newton's discovery of
the binomial theorem and the various aborted plans for publication,
a precious passage from a man not much given to self-revelation.
Inexorably, the pattern of the letter drew him into *De methodis* and
his fluxional method, which he discussed again in tantalizing in-
completeness. More even than the *Epistola prior,* the second letter
was a veritable treatise on infinite series, but twice, as he ap-

---

[82] See the discussion of these notes in Hoffman, *Leibniz in Paris,* pp. 278–87.
[83] Newton to Collins, 8 Nov. 1676; *Corres 2,* 179–80.

proached the fluxional method, he drew back and concealed critical passages in anagrams. Thus he mentioned that *De methodis* contained a method of tangents similar to Sluse's though more general, and that it dealt with questions such as maxima and minima. "The foundation of these operations is evident enough, in fact; but because I cannot proceed with the explanation of it now, I have preferred to conceal it thus: 6accdæ13eff7i3l9n4o4qrr4s8t12ux. On this foundation I have also tried to simplify the theories which concern the squaring of curves, and I have arrived at certain general Theorems."[84] He went on to state what he called the first theorem of his method of squaring curves, illustrated by three examples, before he turned the letter back to infinite series. Indeed, this was to reveal a good deal and to hint at much more. What the anagram concealed was a general statement in Newton's terminology of the fundamental theorem of the calculus: "Data aequatione quotcunque fluentes quantitates involvente, fluxiones invenire; et vice versa" (given an equation involving any number of fluent quantities to find the fluxions, and vice versa). Toward the end of the letter a similar, longer anagram concealed statements of two methods to solve equations containing fluxions. Leibniz recognized in general terms what the anagrams concealed; after all, he was reading *De analysi* about the time Newton was composing them.

Leibniz did not receive the *Epistola posterior* until the following June. As with the earlier one, Oldenburg had recognized its importance and refused to send it until he heard Leibniz was settled in Hanover and until he had a reliable carrier. On 11 June 1677, immediately upon receiving it, Leibniz penned a response filled with praise. In it, he communicated the essence of his differential calculus, asked some probing questions such as only an expert could have formulated, and virtually implored further exchange. A month later, when he had had time to digest the letter, he wrote again. Oldenburg warned him in August that Newton was preoccupied with other affairs.[85] And in September Oldenburg died. Both of Leibniz's letters were forwarded to Newton. It is impossible to imagine that he did not recognize the significance of their contents. Perhaps the long delay had aroused his suspicions, though there is no evidence to warrant a projection of later attitudes back onto 1677. The fact is that Newton had made his decision five years earlier. There is no good reason to think that he would have communicated to a German mathematician he had never met what he had been unwilling to let Collins publish five years earlier. With Oldenburg dead, he did not reply, and the correspondence lapsed.

---

[84] *Corres 2*, 134. Original Latin, p. 115.
[85] Leibniz to Oldenburg, 11 June and 12 July 1677; Oldenburg to Leibniz, 9 Aug. 1977; *Corres 2*, 212–19, 231–2, 235.

An unpleasant paranoia pervaded the *Epistola posterior*. The autobiographical passage insisted on the pressure of Collins and Oldenburg to publish, and he concealed two vital passages in anagrams.
Two days after he sent it, he wrote to Oldenburg again: "Pray let
none of my mathematical papers be printed w^{th}out my special licence."[86] Leibniz was probably not the object of the paranoia at this
time. Rather Newton was obsessed with his correspondence on
colors and allowed his frustration with it to influence his response
to Leibniz. In so doing, he sowed the seeds of unlimited turmoil. In
1676, Leibniz had not published his calculus. He had not communicated it. A free and open communication from Newton would have
plunged him, undeservedly, into a cruel dilemma. Before he had
established any claims of his own, he would have learned that
another mathematician had invented substantially the same method
before him. Since the correspondence passed through Oldenburg,
he would have learned it publicly. One can only speculate what the
outcome would have been – and hope it would have been less dis-
creditable to both than what did finally happen. As far as Newton is
concerned, such a letter would have secured what his futile conceal-
ments gave away, an unassailable claim to prior invention of the
calculus.

In 1676, Leibniz and Newton stood in different personal positions. In effect, Leibniz had arrived where Newton had been ten
years earlier. He had just invented the calculus. He understood its
significance. The excitement in his letters expressed in his fashion
what Newton had acted out a decade earlier by staying up all night
and ignoring his meals. Meanwhile, Newton had moved on to
other things. On the one hand, there was the torment (as he perceived it) of the correspondence on colors. He concluded the *Epistola prior* with a paragraph about it; he wrote letters about it immediately before and immediately after the *Epistola posterior*. Optics
was not where his heart lay in 1676, however; it too was another
diversion. Both of the letters for Leibniz protested his lack of interest in the mathematical exchange. He was pressed for time, he
insisted, and could not explain things fully. "For I write rather
shortly because these theories long ago began to be distasteful to
me, to such an extent that I have now refrained from them for
nearly five years."[87] In the *Epistola posterior*, he noted (correctly)
that he had not completed the treatise, *De methodis*, which he began
in 1671, "nor has my mind to this day returned to the task of
adding the rest."[88] In the covering letter that he sent to Oldenburg
with the *Epistola posterior*, he gave more blunt vent to his distaste.

[86] *Corres 2*, 163.   [87] *Corres 2*, 39. Original Latin, p. 29.
[88] *Corres 2*, 133–4. Original Latin, p. 114.

"I hope this will so far satisfy M. Leibnitz that it will not be necessary for me to write any more about this subject. For having other things in my head, it proves an unwelcome interruption to me to be at this time put upon considering these things."[89] The manuscript remains from the 1670s support Newton's assertions. Oldenburg's death had little to do with the break in the correspondence with Leibniz. Newton participated under protest from the beginning. Collins and Wallis continued to press him to publish. Wallis, who was always suspicious of foreign designs to steal English inventions, inserted passages from the two *Epistolae* in his *Algebra* (1685). Newton only wanted to be left alone. Surrendering to his exasperation with the need to explain and discuss, he withdrew within his shell. Leibniz went on to publish the calculus. Newton eventually harvested the bitter fruit his own neuroses had planted. He forced Leibniz to share it with him.

Meanwhile, optics had refused to leave him alone. In the autumn of 1674, Oldenburg received a letter criticizing Newton's original paper from Francis Hall (or Linus, as he latinized his name), an English Jesuit who was a professor at the English college in Liège. It inaugurated an extended exchange with Linus and his pupils which lasted into 1678 and proved to be for Newton the most trying yet. Ten years earlier, Linus had conferred immortality on Robert Boyle by challenging his concept of air pressure and provoking the experiments that led to Boyle's law. He was seventy-nine years old in 1674. The tone of his letter suggests that senility had set in. Citing experiments he claimed to have done thirty years before, he confidently denied that the projected spectrum could have appeared on a clear day as Newton described it and opined that clouds near the sun had misled him.[90] He misunderstood the description of the spectrum and thought it was parallel to the length of the prism. Newton took two months to bring himself even to acknowledge the letter, and then he refused to answer it on the grounds that it deserved no answer and that he had "long since determined to concern my self no further about ye promotion of Philosophy." As an afterthought, he suggested that Linus be told "(but not from me)" that the experiment was performed on a clear day and the spectrum was transverse to the axis of the prism.[91]

The matter did not come to rest there. Early in 1675, Newton visited London in connection with the royal dispensation he needed to remain in his fellowship without taking orders. On 18 February,

89 *Corres 2*, 110.
90 Linus to Oldenburg, 6 Oct. 1674; *Corres 1*, 317–19. See Conor Reilly, "Francis Line, Peripatetic (1595–1675)," *Osiris, 14* (1962), 222–53.
91 Newton to Oldenburg, 5 Dec. 1674; *Corres 1*, 328–9.

he attended his first meeting of the Royal Society and became a full-fledged member by signing the register. He attended two other meetings during his stay. It is evident from later remarks that the visit made a deep impression. Far from finding himself the object of criticism, he was covered with attention. While he was there, a second letter arrived from Linus, as vehement as the first in denying the possibility of a projected spectrum as Newton described it. The Society ordered the experiment to be performed before them to testify to the matter of fact, though none other than Robert Hooke assured them that the experiment was beyond question. Indeed Newton heard, or thought he heard, Hooke accept his theory of colors.[92] He met Robert Boyle, whose works he had read with care, and conversed with him. Though the immediate aftermath of the visit to London was six months of further silence, Newton's realization of the respect in which he was held prepared the way for something else.

Linus provided the proximate cause. In the autumn he wrote anew demanding that his second letter to Oldenburg be printed lest people conclude he had been mistaken. When Newton had seen the letter in London, he had deemed it not to be worth an answer. Upon the receipt of Linus's demand in November, however, he wrote out explicit instructions on how the experiment should be performed, cited all the others who had confirmed his description, and asked the Royal Society to try it at a meeting if they had not yet done so. Emboldened by his recollection of the spring, he added something more.

> I had some thoughts of writing a further discours about colours to be read at one of your Assemblies, but find it yet against ye grain to put pen to paper any more on y^t subject. But however I have one discourse by me of y^t subject written when I sent my first letters to you about colours & of wch I then gave you notice. This you may command w^n you think it will be convenient if ye custome of reading weekly discourses still continue.[93]

Although the letter does not survive, Oldenburg must have informed him it was then convenient. The familiar routine had to be performed. Two and a half weeks later, on 30 November, he had not yet sent the papers because, when he reviewed them, "it came into my mind to write another little scrible to accompany them."[94] Perhaps the convenience of Wickins, who was pressed into service as amanuensis, contributed to the delay. What he finally sent on 7 December contained two items, a "Discourse of Observations," which was virtually identical to Parts I, II, and III of Book II of the

---

[92] Newton to Oldenburg, 7 Dec. 1675 (the "Hypothesis of Light"); *Corres 1,* 362–3.
[93] Newton to Oldenburg, 13 Nov. 1675; *Corres 1,* 358. [94] *Corres 1,* 359.

Opticks published nearly thirty years later, and "An Hypothesis explaining the Properties of Light discoursed of in my severall Papers."

The first of the two dated from 1672, though Newton may have revised the early version into its final form in 1675.[95] I shall not repeat what I have said before about its content. In many respects, the "Hypothesis of Light" was also not new. He had begun to draft it in 1672 as part of his reply to Hooke, and things like it had appeared already in his essay "Of Colours" in 1666. We need to read Newton's comment about it in a covering letter to Oldenburg against this background.

> S[r]. I had formerly purposed never to write any Hypothesis of light & colours, fearing it might be a means to ingage me in vain disputes: but I hope a declar'd resolution to answer nothing that looks like a controversy (unles possibly at my own time upon some other by occasion) may defend me from y[t] fear. And therefore considering that such an Hypothesis would much illustrate y[e] papers I promis'd to send you, & having a little time this last week to spare: I have not scrupled to describe one so far as I could on a sudden recollect my thoughts about it, not concerning my self whether it shall be thought probable or improbable so it do but render y[e] papers I send you, and others sent formerly, more intelligible. You may see by the scratching & interlining 'twas done in hast, & I have not had time to get it transcrib'd . . .[96]

For the first time, Newton undertook to reveal his thoughts about the ultimate constitution of nature; he did not find it a task that he could do lightly.

In the introduction to the "Hypothesis," as in the covering letter,

---

[95] In addition to the draft that definitely dated from 1672 (*Add MS* 3970.3, ff. 519–28) there is a second autograph (ff. 501–17) and a transcript by Wickins with several corrections and additions in Newton's hand, which was the copy sent to London and later returned (ff. 549–68). I do not know of any way to date the last two with assurance. Newton's letter of 30 Nov. 1675 certainly asserted that he was merely sending a paper completed earlier. Since he was prone to similar remarks, such as his comment on the "little scrible" that accompanied it, which fended off possible criticism by implying a distance between him and his work, I am not inclined to accept his word without further evidence.

[96] *Corres 1*, 361. In fact, in addition to the copy sent to London and later returned (*Add MS* 3970.3, ff. 538–47), there is the inevitable transcript in Wickins's hand with two sheets missing (ff. 573–81), which could have been made after the other was returned, of course, but probably was not. There is also a draft (ff. 475, 476, 534, 533, in that order) and drafts for it (ff. 535–6). The draft even included a version, which did not hedge as much as the final statement, of the disclaimer sent to Oldenburg in the covering letter. On various issues associated with the "Hypothesis of Light" see Philip E. B. Jourdain, "Newton's Hypothesis of Ether and of Gravitation from 1672 to 1679, from 1679 to 1693, from 1693 to 1726," *The Monist, 25* (1915), 79–106, 234–54, 418–40; A. I. Sabra, "Newton and the 'Bigness' of Vibrations," *Isis, 54* (1963), 267–8; and Roger H. Stuewer, "Was Newton's 'Wave-Particle Duality' Consistent with Newton's Observations?" *Isis, 60* (1969), 392–4.

Newton insisted that he sent it merely to illustrate his optical papers. He did not assume it; he did not concern himself whether the properties of light he had discovered could be explained by this hypothesis or by Hooke's or by another. "This I thought fitt to Expresse, that no man may confound this with my other discourses, or measure the certainty of one by the other, or think me oblig'd to answer objections against this script. For I desire to decline being involved in such troublesome & insignificant Disputes."[97] It is quite impossible to reconcile the actual "Hypothesis" with Newton's deprecations of it, however. For one thing, it presented far more than an explanation of optical phenomena. For another, a feeling of intensity pervaded it. In it, Newton presented himself in his preferred role, not of positive scientist, but of natural philosopher confronting the entire sweep of nature. For ten years he had contemplated the order of things in solitude. Now he was disclosing, partially, to a limited audience, where ten years of speculation had carried him. No amount of feigned indifference and hard words about insignificant disputes could obscure the significance the enterprise held for him.

As far as light was concerned, the "Hypothesis" presented with one exception an orthodox mechanical philosophy. The one exception was a principle of motion that he ascribed to the light corpuscles themselves. He assigned reflections and refractions to the causation of a universal aether which stands rarer in the pores of bodies than in free space and causes corpuscles of light to change directions by its pressure. A mechanism of vibrations in the aether explained the periodic phenomena of thin films. The "Hypothesis of Light" contained much more than an explanation of optical phenomena, however. The first half of it presented a general system of nature based on the same aether. All of the crucial phenomena that appeared in his "Quaestiones" a decade earlier appeared now in the "Hypothesis" either to be explained by aetherial mechanisms or to offer illustrative analogies. For example, the pressure of the aether explained the cohesion of bodies, and surface tension illuminated an aethereal mechanism. In passing, Newton described an experiment with static electricity which played a major role in the early history of that science. He set a disk of glass in a brass ring which held it about an eighth or a sixth of an inch from the table. Placing the glass over some tiny bits of paper, he rubbed it briskly with cloth until the papers began to move;

> after I had done rubbing the Glass, the papers would continue a pretty while in various motions, sometimes leaping up to the Glass & resting there a while, then leaping downe & resting there, then leaping up &

97 *Corres 1*, 364.

perhaps downe & up againe, & this sometimes in lines seeming perpendicular to the Table, Sometimes in oblique ones, Sometimes also
they would leap up in one Arch & downe in another, divers times
together, without Sensible resting between; Somtimes Skip in a bow
from one part of the Glasse to another without touching the table, &
Sometimes hang by a corner & turn often about very nimbly as if they
had been carried about in the midst of a whirlwind, & be otherwise
variously moved, every paper with a divers motion.[98]

Newton saw no way to explain these motions except by an aether
condensed in the glass which was vaporized and set in motion by
the rubbing.

Because the aether condensed continually in bodies such as the
earth, there is, according to the "Hypothesis," a constant downward stream of it which impinges on gross bodies and carries them
along. Newton explicitly extended this explanation of gravity to
the sun and suggested that the resulting movement of aether holds
the planets in closed orbits. The passage contains the first known
hint of the concept of universal gravitation in Newton's papers; he
did not fail to refer to it when Hooke cried plagiary in 1686.

The "Hypothesis of Light" refuses to be presented solely as a
mechanical system of nature, however. If it showed the enduring
influence of the mechanical philosophy, it was an ambiguous document which contained vestiges of other influences that had begun to
bear on Newton's conception of nature. One of the features that
distinguished it was the prominent role of chemical phenomena,
which had played no part in the "Quaestiones" ten years earlier.
They epitomized the new influences that would, in the years ahead,
carry him on beyond his position in 1675. I shall return to them in a
different context.

Whatever his announced intention to avoid disputes, the new
papers plunged Newton directly into a new round of correspondence, of explication, and very quickly of controversy. The papers
were read at the Royal Society immediately, the "Hypothesis" on 9
and 16 December, and after a Christmas recess and two meetings
monopolized by discussions stemming from the "Hypothesis," the
"Discourse of Observations" from 20 January to 10 February. Like
the paper of 1672, they caused a sensation. The Royal Society requested the immediate publication of the "Discourse," which Newton declined. Questions also arose. The static-electric experiment in
the "Hypothesis" caught the Society's eye, but their first effort to
reproduce it failed. They applied to Newton for instructions, of
course, and he had to write two letters about it.[99]

---

[98] *Corres 1*, 364–5.
[99] Newton to Oldenburg, 21 Dec. 1675 and 10 Jan. 1676; *Corres 1*, 404, 407–8.

More significant was Hooke. Whether deliberately or through
inadvertence, Newton had introduced him into the "Hypothesis"
rather prominently, both in the introduction, which justified the
whole enterprise by a reference back to Hooke's critique of 1672,
and in the discussion of diffraction at the conclusion. Newton owed
his very knowledge of diffraction to Hooke's discourse at a meeting
of the Royal Society early in 1675, where Hooke presented it as his
own new discovery. Newton had remarked then that diffraction
was only a new kind of refraction – which corresponded exactly to
the way he later explained it in the "Hypothesis." Seeing the in-
vader from Cambridge demolish yet another of his prizes, Hooke
had bridled and replied "that though it should be but a new kind of
refraction, yet it was a *new one*."[100] In the "Hypothesis," Newton
was unkind enough to point out, in connection with a few experiments on diffraction, that it was not a new one after all. Faber's
book on light had mentioned it, and Faber had learned about it
from Grimaldi.[101] Small wonder that Hooke rose when the reading
of the "Hypothesis" was completed to assert "that the main of it
was contained in his *Micrographia,* which Mr. Newton had only
carried farther in some particulars."[102]

Without much thought about the provocation he had offered,
Newton erupted in anger at the charge. Since Oldenburg's letter
reporting the incident does not survive, we do not know exactly
what Newton heard, or for that matter exactly what happened.
Indeed, none of Oldenburg's letters to Newton from this period
survive, possibly a suspicious circumstance because Hooke believed
that Oldenburg, with whom he was at swords' points, had deliberately fomented trouble. Even minutes of the Royal Society do not
offer an independent account; Oldenburg kept them. It is not hard
to believe that an incident occurred, however. Hooke was a prickly
personality in his own right, and he had reason to feel aggrieved
with Newton. What the "Hypothesis" poured on his wounds was
more embalming fluid than balm. For his pains he now received a
further dose of gall. Hooke's hypothesis of light, Newton asserted,
was merely an embroidery on Descartes's. His own was entirely
different, to the extent that the experiments, which were new to
Hooke, on which Newton based his treatment of thin films undermined everything Hooke had said about the subject. The more he
wrote, the hotter Newton became. True, he had learned about
colors in thin films from Hooke. Hooke had confessed that he did

---

[100] Newton to Oldenburg, 7 Dec. 1675 (the "Hypothesis of Light"); *Corres 1,* 384. This is,
of course, Newton's description of the episode. See Roger H. Stuewer, "A Critical
Analysis of Newton's Work on Diffraction," *Isis, 61* (1970), 188–205.
[101] *Corres 1,* 384.   [102] Birch, *History, 3,* 269.

not know how to measure the thickness of the films, however, "&
therefore seing I was left to measure it my self I suppose he will
allow me to make use of what I tooke ye pains to find out."[103]
Three weeks of thinking about it left Newton even more incensed.
Initially, he was inclined to grant that he got the idea of vibrations
in the aether from Hooke. Now he retracted that as well; it was a
common idea. "I desire Mr Hooke to shew me therefore, I say not
only ye summ of ye Hypothesis I wrote, wch is his insinuation, but
any part of it taken out of his *Micrographia:* but then I expect too
that he instance in what's his own."[104]

The handling of Newton's first letter tends to confirm Hooke's
suspicions that Oldenburg egged him on. Though Oldenburg read
a passage from it about the electrical experiment to the Royal Society on 30 December, he neither read the comment on Hooke nor
told Hooke about it. Hooke heard it with surprise at the meeting on
20 January. At that point, he took matters into his own hands and
wrote directly to Newton the same day. He feared Newton had
been misinformed about him, a "sinister practice" which had been
used against him before. He protested his disapproval of contention, his desire to embrace truth by whomever discovered, and the
value he placed on Newton's "excellent Disquisitions," which proceeded farther than anything he had done. Finally, he proposed a
correspondence in which the two could discuss philosophical matters privately. "This way of contending I believe to be the more
philosophicall of the two, for though I confess the collision of two
hard-to-yield contenders may produce light yet if they be put together by the ears of other's hands and incentives, it will produce
rather ill concomitant heat which serves for no other use but . . .
kindle cole [*sic*]."[105]

Newton replied in kind, calling Hooke "a true Philosophical
spirit." "There is nothing wch I desire to avoyde in matters of
Philosophy more then contention," he agreed, "nor any kind of

---

[103] Newton to Oldenburg, 21 Dec. 1675; *Corres 1,* 406.

[104] Newton to Oldenburg, 10 Jan. 1676; *Corres 1,* 408. By the time he finished the second
letter, he had cooled down a good bit; he asked to have his service presented to Hooke,
"for I suppose there is nothing but misapprehension in w^t has lately happend" (*Corres 1,*
411).

[105] *Corres 1,* 412–13. Hooke made an entry in his diary the same day in which he referred to
Newton's letter "seeming to quarrell from Oldenburg fals suggestions. . . . Wrot letter
to Mr. Newton about Oldenburg kindle Cole" (*The Diary of Robert Hooke M.A., M.D.,
F.R.S. 1672–1680,* ed. Henry W. Robinson and Walter Adams [London, 1935], p. 213).
The phrase, "kindle cole," which appeared both in the letter and in the diary, is elucidated by a use that Cromwell made of it in a speech to Parliament on 25 Jan. 1658; "I
speak of men going about that cannot tell *what* they would have, yet are willing to kindle
coals to disturb others" (*The Letters and Speeches of Oliver Cromwell with Elucidations by
Thomas Carlyle,* 3 vols., ed. S. C. Lomas [London, 1904], *3,* 181).

contention more then one in print . . . ." Accepting the offer of a
private correspondence, he went on to praise Hooke's contribution
to optics. "What Des-Cartes did was a good step. You have added
much several ways, & especially in taking ye colours of thin plates
into philosophical consideration. If I have seen further it is by
standing on ye sholders of Giants."[106] Sentiments too lofty drift
away from human reality. A lack of warmth was evident on both
sides. Neither man endeavored to institute the philosophic correspondence both professed to want, and their basic antagonism remained undissolved.

Another correspondence refused to go away, the one instituted by
Linus, which provided the occasion for the two papers of December. That same month, a letter from Liège written by John Gascoines, a pupil of Linus, informed Oldenburg and Newton that
Linus was dead but that Gascoines intended to defend his professor's honor. As Linus had done, he denied the basic experiment.
Newton replied with further instructions on how to perform the
experiment and with a heated rejoinder to what he took as an
implication that he had dealt underhandedly with Linus.[107] A
month later, when he saw his own letter in the *Philosophical Transactions* with Linus's second letter (written a year before), he recalled
the whole episode and whipped himself into a fine fury, first at
Linus for failing to comprehend the experiment, and then at Gascoines for impugning his honor.[108] Finally on 27 April 1676, the
Royal Society had the basic experiment of the prismatic spectrum
performed before them to confirm its outcome – a mere four years
and three months after they had begun to discuss the questions it
raised.

The success of the experiment did not settle the matter with
Liège. In June, at the time of the *Epistola prior,* a new letter arrived
from a third correspondent, Anthony Lucas, another English Jesuit,
whom Gascoines had recruited to take up cudgels too heavy for
him. Lucas began by conceding the sole point that hitherto had
been in contention: a prism does project a spectrum elongated per-
pendicularly to the axis of the prism. His spectrum did not have the
same proportions as Newton's, however. Its length was not five
times its breadth, but only three or three and a half times. Further-

---

[106] Newton to Hooke, 5 Feb. 1676; *Corres 1*, 416. I do not accept the interpretation that the
last phrase was a deliberate, oblique reference to Hooke's twisted physique. As Newton
said once before in regard to Hooke, he avoided oblique thrusts. When he attacked, he
lowered his head and charged.
[107] Gascoines to Oldenburg, 15 Dec. 1675 (N.S.), *Corres 1*, 393–5. Newton to Oldenburg,
10 Jan. 1676; *Corres 1*, 409–11.
[108] Newton to Oldenburg, 29 Feb. 1676; *Corres 1*, 421–5.

more, Lucas proceeded to relate nine other experiments he had performed to test Newton's theory. Far from confirming it, their outcome seemed to negate it.[109] Through four years of discussion, Newton had challenged opponents to bring forward experiments instead of hypotheses. To be sure, Lucas's letters betrayed no particular acumen; the experiments he presented were not well designed.[110] In regard to them, as Newton insisted on the careful examination of the import of his own experiments, he made another penetrating pronouncement which modified the seeming empiricism of other assertions in order to cover the other flank of the methodological debate. "For it is not number of Exp^{ts}, but weight to be regarded; & where one will do, what need of many?"[111] Nevertheless, what Lucas sent was a reasoned letter which deserved a reasoned response. Newton greeted it quite otherwise. As the correspondence continued, he became increasingly agitated and irrational. He convinced himself that the Liègois (papists, of course) had formed a conspiracy to engage him in perpetual disputation and to undermine his credit. He refused to discuss Lucas's experiments but insisted that Lucas discuss his, in particular the *experimentum crucis*. When Lucas obliged and did not find the *experimentum crucis* to be conclusive or for that matter to work as Newton described it, Newton grew angrier yet. Nor was he mollified when Lucas suggested that the real question was not the length of the spectrum but the theory of colors.

> The Question in hand is this [he stormed]. Whether ye Image in ye Experiment set down in my first letter about colours . . . could be five times longer then broad as I have there exprest it, or but three or at most three and a half as Mr Lucas has represented it. To this I desire a direct answer: which I hope will be so free as (wthout putting me to any further wayes of justifying myself) may take off all suspicion of my misrepresenting matter of fact.

And again:

> Tis ye truth of my experiments which is ye business in hand. On this my Theory depends, & which is of more consequence, ye credit of my being wary, accurate and faithfull in ye reports I have made . . .[112]

"I see I have made my self a slave to Philosophy," he exclaimed to

---

[109] Lucas to Oldenburg, 27 May 1676 (N.S.); *Corres 2*, 8–12. See S. M. Gruner, "Defending Father Lucas: A Consideration of the Newton-Lucas Dispute on the Nature of the Spectrum," *Centaurus*, 17 (1973), 315–29.

[110] Cf. Rosenfeld, "La Théorie des couleurs de Newton."

[111] Newton to Oldenburg, 18 Aug. 1676; *Corres 2*, 79.

[112] Newton to Oldenburg, 28 Nov. 1676; *Corres 2*, 183–5. Newton replied to Lucas's letter of 27 May on 18 August (*Corres 2*, 76–81). Lucas's second letter, to which Newton's present one replied, was dated 23 October (N.S.) (*Corres 2*, 104–8). His letters had an unfortunate way of arriving just as Newton was engaged in writing to Leibniz.

Oldenburg in desperation, "but if I get free of Mr Linus's buisiness I will resolutely bid adew to it eternally, excepting what I do for my privat satisfaction or leave to come out after me. For I see a man must either resolve to put out nothing new or to become a slave to defend it."[113] Recall that at the time he wrote, Newton's "slavery" consisted of five replies to Liège, totaling fourteen printed pages, over a period of a year. Recall also that he had completed the *Epistola posterior* less than a month before.

The length of the spectrum, which became for Newton the symbol of his honor, has been the object of considerable historical comment. Lucas might have used a prism made of glass with a different dispersive power than Newton's. Had Newton not been so dogmatic, he could have pursued the source of the discrepancy and discovered the possibility of the achromatic lens. The case is by no means proved, however. Newton had to push hard to get Lucas to measure distances and angles with a rigor that even approached his own. He always suspected that Lucas projected his spectrum with a prismatic angle less than 60 degrees, which could have accounted for the difference. In the end, Lucas did concede the question of the length of the spectrum.[114] Along the way, he also mentioned that the faces of his prism were concave (which would have tended to decrease the proportions of the spectrum), and he indicated that he usually performed the experiment between six and seven in the morning. The last admission was particularly damning. It called everything Lucas had said into question. Lucas had claimed to project his spectrum eighteen feet. With a horizontal sun, he would have had to project it upward at an angle of about 45 degrees. While a room that would both allow such a trajectory and enable an observer to receive the spectrum on a screen perpendicular to its trajectory in order to measure it properly is not impossible, it is also not common. Newton's dogmatism in the exchange is beyond challenge, but it probably obscured Lucas's sloppiness instead of a new discovery.[115]

When a third letter from Lucas arrived in February 1677, Newton decided on a different mode of answer. During the previous year, he had mentioned the possibility of a major publication on optics designed to confirm his theory and hence to settle all disputes. At least once, he had considered publishing all the correspondence his papers had provoked. Several pieces of evidence indicate that Newton was planning such a volume during 1677. In March, Collins mentioned that David Loggan, who was resident in Trinity College

---

[113] Newton to Oldenburg, 18 Nov. 1676; *Corres 2*, 182–3.

[114] Lucas to Oldenburg, 2 Feb. 1677 (N.S.); *Corres 2*, 191.

[115] In regard to Newton's dogmatism in the exchange, see Johannes Lohne, "Newton's 'Proof' of the Sine Law and His Mathematical Principles of Colours," *Archive for History of Exact Sciences*, 1 (1961), 389–405.

as he prepared the engravings for his *Cantabrigia illustrata* (1690), had told him he had drawn Newton's picture for an engraving to be included in a book on light and colors.[116] In December, after Lucas had enquired about the printing of his letters, Newton mentioned his plan to Robert Hooke, who had succeeded Oldenburg as secretary of the Royal Society. "Mr Oldenburg being dead I intend God willing to take care y^t they be printed according to his mind, amongst some other things wch are going into ye Press."[117] A few printed sheets with part of his paper of February 1672 and notes which explicate certain passages have recently been discovered.[118] There is every reason to believe they constituted the beginning of the proposed volume. Then fire struck his chamber destroying part of the collection of papers.[119] Though he tried briefly to get new copies, he finally abandoned the project.

Fourteen years later, Abraham de la Pryme, a student in Johns, recorded in his diary a story he had heard.

Febr: [1692] What I heard to-day I must relate. There is one Mr. Newton . . . fellow of Trinity College, that is mighty famous for his learning, being a most excellent mathematician, philosopher, divine, etc. . . . but of all the books that he ever writt there was one of colours and light, established upon thousands of experiments, which he had been twenty years of making, and which had cost him many

---

[116] Collins to Newton, 5 March 1677; *Corres 2*, 200–1. Oldenburg appeared to refer to an intended publication in his letters to Newton on 2 Jan. 1677 (*Corres 2*, 188) and to Leibniz on 2 May 1677 (*Corres 2*, 209).

[117] Newton to Hooke, 18 Dec. 1677; *Corres 2*, 239.

[118] Derek Price found the sheets, which had been used as stuffing in the binding of a book ("Newton in a Church Tower: the Discovery of an Unknown Book by Isaac Newton," *The Yale University Library Gazette, 34* [1960], 124–6). They are both reproduced and discussed fully in I. Bernard Cohen, "Versions of Isaac Newton's First Published Paper," *Archives internationales d'histoire des sciences, 11* (1958), 357–75. The evidence for Newton's plans to publish at this time is fully laid out in A. Rupert Hall, "Newton's First Book (I)," *ibid., 13* (1960), 39–54.

[119] The evidence for a fire at this time is strong. Newton wrote to Lucas on 2 Feb. 1678 for copies of Lucas's letters. The letter has disappeared, but in reply Lucas referred to "your losse" (Lucas to Newton, 14 March 1678 [N.S.]; *Corres 2*, 251). Newton also wrote to Hooke on 5 March and to Aubrey about June, mentioning the loss of papers (*Corres 2*, 253, 266). In his memorandum of a conversation with Newton on 31 Aug. 1726, Conduitt recorded Newton's recollection of a fire. "When he was in the midst of his discoveries he left a candle on his table amongst his papers & went down to the Bowling green & meeting somebody that diverted him from returning as he intended the candle sett fire to his papers & he could never recover them I asked him wether they related to his opticks or to the method of fluxions & he said he believed there were some relating to both" (*Keynes MS* 130.10, ff. 4–4^v; *Keynes MS* 130.4, pp. 14–15). Humphrey Newton said that he had heard of "his Opticks being burnt" before he came to Cambridge in 1683. Writing in 1683 for a book published in 1685, Wallis mentioned the loss of papers by fire (*Opera mathematica*, 3 vols. [Oxford, 1693–9], *2*, 390). Stukeley also mentioned that Humphrey Newton had told him of several sheets of optics being set on fire by a candle, and he added a story about the burning of a work on chemistry, apparently at another time (Stukeley to Conduitt, 15 July 1727; *Keynes MS 136*, p. 10).

a hundred of pounds. This book which he valued so much, and which was so much talk'd off, had the ill luck to perish and be utterly lost just when the learned author was almost at putting a conclusion at the same, after this manner. In a winter morning, leaving it amongst his other papers on his studdy table, whilst he went to chappel, the candle which he had unfortunately left burning there too cachd hold by some means or other of some other papers, and they fired the aforesayd book, and utterly consumed it and several other valuable writings, and that which is most wonderful did no further mischief. But when Mr. Newton came from chappel and had seen what was done, every one thought he would have run mad, he was so troubled thereat that he was not himself for a month after. A large account of this his system of light and colours you may find in the transactions of the Royal Society, which he had sent up to them long before this sad mischance happened to them.[120]

De la Pryme's story is usually attached to Newton's established breakdown in the autumn of 1693, which a story Huygens heard, also involving a fire, tends to support. However, the date of de la Pryme's entry antedates the breakdown of 1693 by more than eighteen months, and the past perfect tense in the final sentence does not seem to place the fire in the recent past. It could refer to the well-authenticated fire that halted an optical publication in the winter of 1677–8.[121] There is a hiatus in Newton's correspondence from 18 December to February, though his correspondence in this period was very light in any case. Twice at least in earlier correspondence, with Hooke and with Huygens, he had lost partial control of himself as the heat of his own vehemence swept him away, and the tone of his letters to Lucas implies a complete loss of control that is compatible with a breakdown. As later in 1693, Newton was in a state of acute intellectual tension throughout the 1670s, not just from answering objections to his optics, but even more from other studies then foremost in his mind, studies which excited him keenly. Another parallel with 1693 may also be relevant. The crisis in his relations with Fatio de Duillier at that time had its counterpart in Wickins's decision to leave Trinity.[122]

When he discarded the intended publication, Newton wrote two further letters to Lucas on the same day, 5 March 1678, one answering Lucas's first two and the other answering his third (of February

---

[120] *The Diary of Abraham de la Pryme,* ed. Charles Jackson (Durham, 1870), p. 23.

[121] Edleston concluded that Pryme's story referred to a fire between 1676 and 1683 (*Edleston,* pp. lx–lxiii).

[122] The Steward's Books show that in 1676 (i.e., the year that ended on Michaelmas 1676) Wickins, who had hitherto resided in the college most of each year, was in Cambridge only 25 1/2 weeks. In 1677, he resided in college 13 1/2 weeks, in 1678 6 weeks. He was never there that long again before his final resignation in 1684.

1677). Even the earlier letters, furious as they were, could not have prepared Lucas for the flood of paranoia that now burst over him.

> Do men use to press one another into Disputes? Or am I bound to satisfy you? It seems you thought it not enough to propound Objections unless you might insult over me for my inability to answer them all, or durst not trust your own judgement in choosing ye best. But how know you y^t I did not think them too weak to require an answer & only to gratify your importunity complied to answer one or two of ye best? How know you but y^t other prudential reasons might make me averse from contending wth you? But I forbeare to explain these things further for I do not think this a fit Subject to dispute about, & therefore have given these hints only in a private Letter. of wch kind you are also to esteem my former answer to your second. I hope you will consider how little I desire to explain your proceedings in public & make this use of it to deal candidly wth me for ye future.[123]

Arrogant and brutal, the two letters made it clear that nothing less than the public humiliation of his antagonists could satisfy Newton – hateful letters, did not our knowledge of the circumstances incline us to sympathy with the author's anguish.

The correspondence gave one last spasm before it died. In May, Newton acknowledged a letter from Lucas, though he probably did not answer it.[124] Later that month, he heard there was another waiting for him in London.

> Mr Aubrey
> I understand you have a letter from Mr Lucas for me. Pray forbear to send me anything more of that nature.[125]

With that he brought his correspondence about colors to an end. Oldenburg was dead. Newton had ceased to correspond with Collins. As far as he could, he isolated himself. To the best of our knowledge he wrote only two letters, one to Arthur Storer (which has not survived) and one to Robert Boyle, between June 1678 and December 1679.[126] To the end of his days Newton remembered this

---

[123] *Corres 2,* 254–60, 262–3.

[124] Newton to Hooke, 18 May 1678; *Corres 2,* 264.

[125] Newton to Aubrey, ca. June 1678; *Corres 2,* 269. I believe that this letter was not sent. It exists only as an autograph draft, and it looks very much like a sentence in a letter sent to Aubrey about June (*Corres 2,* 266–8). Though not so succinct – or rather, because it is not so succinct – the one sent offers a further exhibit in Newton's display of paranoia, if another is needed.

[126] I do not believe in the authenticity of the supposed letter to Maddock on 7 Feb. 1679 (*Corres 2,* 287), which surfaced in a funeral sermon in the middle of the eighteenth century. Its tone of light irony resembles Newton's style about as much as a gazelle resembles a tiger.

withdrawal as a conscious decision and considered that it had marked an epoch in his life. "Its now about fifty years," he wrote to Mencke in 1724, "since I began for the sake of a quiet life to decline correspondencies by Letters about Mathematical & Philosophical matters finding them tend to disputes and controversies . . ."[127]

---

[127] A draft of a letter; *Keynes MS* 111. The letter apparently sent (*Corres 7*, 255) did not include this passage.

8

Rebellion

NEWTON'S repeated protestation that he was engaged in other studies supplied an ever-present theme to his correspondence of the 1670s. Already in July 1672, only six months after the Royal Society discovered him as a man supremely skilled in optics, he wrote to Oldenburg that he doubted he would make further trials with telescopes, "being desirous to prosecute some other subjects."[1] Three and a half years later, he put off the composition of a general treatise on colors because of unspecified obligations and some "buisines of my own wch at present almost take up my time & thoughts."[2] Apparently the other business was not mathematics, because later in 1676 he hoped the second letter for Leibniz would be the last. "For having other things in my head, it proves an unwelcome interruption to me to be at this time put upon considering these things." He was not only preoccupied, he was almost frantic in his impatience. "S^r," he concluded the letter, "I am in great hast, Yours . . ."[3] In great haste because of what? Surely not because of ten lectures on algebra that he purportedly delivered in 1676. And not because of pupils or collegial duties, for he had none of either. Only the pursuit of Truth could so drive Newton to distraction that he resented the interruption a letter offered. Newton was in a state of ecstasy again. If mathematics and optics had lost the capacity to dominate him, it was because other studies had supplanted them.

One of the studies was chemistry. Collins mentioned his absorption in it twice in letters to Gregory.[4] Years later, when he chatted with Conduitt about his early life in Cambridge, Newton himself mentioned that Wickins helped in his "chymical experi-

---

[1] Newton to Oldenburg, 8 July 1672; *Corres 1*, 212. On 21 September he wrote to Oldenburg that, in response to Oldenburg's request, he had begun to draw up a list of experiments to answer the eight queries he had posed in July; "before it was finished falling upon some other business, of wch I have my hands full, I was obliged to lay it aside, & now know not when I shall take it again into Consideration" (*Corres 1*, 237–8).

[2] Newton to Oldenburg, 25 Jan. 1676; *Corres 1*, 414. Ten days later, in accepting Hooke's offer of a philosophical correspondence, he added that he had grown tired of optics and doubted that he would ever again become interested enough to spend much time on it (*Corres 1*, 416). In fact, he never did. On 11 May 1676, he told Oldenburg that he hoped to use the two papers he had sent in December, if he could "get some time" to write the other discourse, about the colors revealed by the prism, to accompany them (*Corres 2*, 6).

[3] Newton to Oldenburg, 24 Oct. 1676; *Corres 2*, 110.

[4] Collins to Gregory, 29 June, 19 Oct. 1675; *Corres 1*, 345, 356.

ments."[5] His interest in it developed somewhat later than his interest in natural philosophy. When he composed the "Quaestiones quaedam philosophicae" in the mid-1660s, he entered almost nothing that one would call chemistry, even though Robert Boyle was one of the major sources of his new mechanical philosophy. When he extended his notes on a number of the headings under "Quaestiones" in a new notebook, however, chemistry did begin to appear, and the notes indicate that Boyle supplied his introduction to the subject.[6] The format of the new notebook with its series of headings repeated Newton's practice of recording newly acquired knowledge in an ordered fashion. A decade and more later, he was still entering a few items under these headings from further reading.[7] At about the time of his earliest chemical notes, around 1666, he also composed a chemical glossary based largely on Boyle, whom alone he cited in it. In the glossary, under a number of headings, he set down basic information needed by a chemist. He began with "Abstraction," the evaporation or distillation of a solution to obtain the salt in it, and proceeded through quite a number of terms such as "Amalgam," "Crucible," "Extraction," "Sublimation," and so on, giving a brief summary of operative information that he had culled.[8] Under "Testing," he described a means to refine gold and silver by heating them with lead. Newton's ability to organize what he learned so that he could retrieve it was a significant aspect of his genius. Years later, he described exactly the same process in a paper he prepared at the Mint and used some of the same language he had entered fifty years earlier under "Testing."[9]

As an example of the level of detail Newton demanded and of the sophistication his early knowledge of chemistry attained, consider his entry under "Furnace" (Figure 8.1).

---

[5] Conduitt's memorandum of 31 Aug. 1726; *Keynes MS* 130.10, f. 3ᵛ. One of the anecdotes that Conduitt collected indicated that Newton's "furnace at Cambridge" was an item of interest shown to visitors (*Keynes MS* 130.5, sheet 1; *Keynes MS* 130.6, book 1).

[6] *Add MS* 3975. The early sections of this notebook, on colors, cold and heat, compression and elasticity, and fire and flame, did not contain any chemistry. With the exception of the first, they drew primarily on Boyle, and their frequent citation of his *Origine of Formes and Qualities* placed them in 1666 at the earliest. Later sections, "Of Formes & Transmutations wrought in them" (pp. 61–6) and "Of Salts, & Sulphureous bodys, & Mercury. & Mettalls" (pp. 71–80, 88–100), which also cited Boyle's *Origine,* were definitely chemical in content.

[7] *Add MS* 3975, pp. 38–41, experiments on freezing, January 1670; p. 46, an experiment on degrees of heat and expansion of air and water by heat, 10 March 1693; p. 51, a formula for phosphorus, early 1680s; pp. 65–6, notes from the Earl of Sandwich's translation of *The Art of Metals* (1674 edition); pp. 88–100, notes from Boyle, *Certain Physiological Essays* (1669 edition); p. 162, under a later heading, "The medicall virtues of Saline & other Praeparations," notes from Boyle's *Essay of . . . Effluviums* (1673).

[8] Bodleian Library, Oxford, MS Don. b. 15.   [9] *Mint* 19.2, f. 293.

Figure 8.1. Newton's drawing of various chemical furnaces. 1, 2, 3, and 4. Wind furnaces. 5. Reverberatory furnace. 6. Athanor, Piger Henricus, or Furnace Acediae. (Courtesy of the Joseph Halle Schaffner Collection, University of Chicago Library.)

Furnace. As 1 ye Wind furnace (for calcination, fusion, cementation &c) wch blows it selfe by attracting ye aire through a narrow passage 2 ye distilling furnace by naked fire, for things y^t require a strong fire for distillation. & it differs not much from ye Wind furnace only ye glasse rests on a crosse barr of iron under wch bar is a hole to put in the fire, wch in ye wind furnace is put in at ye top. 3 The Reverberatory furnace where ye flame only circulating under an arched roof acts upon ye body. 4 ye Sand furnace when ye vessel is set in Sand or sifted ashes heated by a fire made underneath. 5 Balneum or Balneum Mariae when ye body is set to distill or digest in hot water. 6 Balneum Roris or Vaporosum ye glasse hanging in the steame of boyling water Instead of this may bee used ye heat of hors dung (cald venter Equinus) i:e: brewsters grains wheat bran, Saw dust, chopt hay or straw, a little moistened close pressed & covered. Or it may in an egg shell bee set under a hen. 7 Athanor, Piger Henricus, or Furnace Acediae for long digestions [y^e] vessel being set in sand heated wth a Turret full of Charcoale wch is contrived to burne only at the [botto]m the upper coales continually sinking downe for a supply. Or the sand may be heated [by a] Lamp & it is called the Lamp Furnace. These are made of fire stones, or bricks.[10]

Putting the skills developed in Grantham to use, Newton usually built his own furnaces. Manifestly, he knew what he was about.

Not all of the entries in the glossary confined themselves to straightforward prosaic chemistry – or "rational chemistry," as those call it who wish to pretend that Newton did not leave behind a vast collection of alchemical manuscripts. He included quite a few entries on mercury, including mercury sublimate which "opens" copper, tin, and silver, but not gold. "Yet perhaps," he added, "there may bee Sublimates made (as by subliming common sublimate & Sal Armoniack well powdered together) wch besides notable operations on other metalls, may act upon Gold too."[11] One entry described Boyle's *menstruum peracutum,* which dissolved gold and even carried some gold with it in distillation. Boyle invested the *menstruum peracutum* with alchemical significance; Newton's entry implied that he did too. Antimony and its power to purify gold appeared. As with the refining of gold by lead, Newton later employed his knowledge of refining by antimony in an emotionally charged memorandum when the standard of his coinage was impugned at the trial of

[10] MS Don. b. 15, f. 3. The brackets fill in a damaged corner of the manuscript. Cf. Joseph Halle Schaffner Collection, University of Chicago Library, MS 1075-2.
[11] MS Don. b. 15, f. 4.

the pyx in 1710.[12] His early glossary also included instructions to make regulus of antimony, regulus of Mars, and "Regulus Martis Stellatus," the star regulus of Mars, which would soon figure prominently in an explicitly alchemical setting.

In similar fashion, the chemical notebook changed its character. Notes from George Starkey's *Pyrotechny Asserted* succeeded those from Boyle. Starkey was probably the pseudonymous Eirenaeus Philalethes, whose numerous treatises on alchemy exercised enormous influence on Newton. Under "Medical Observations," he entered the recipe for *primum ens* of Baulm, which had the power to restore youth. The instructions called for a dose in wine every morning "till ye nailes hair & teeth fall of & lastly the skin be dryed & exchanged for a new one . . ." The hesitant are assured that the *ens* started a woman of seventy years menstruating again.[13] Its alchemical connotation rather than its effect is the point at issue here, however, and a pointing hand drawn in the margin, a device familiar to students of Newton's manuscripts of every sort, asserted its importance. One of the last sections of the notebook, added perhaps a decade after the initial set, carried the heading, "Of ye work wth common ⊙ [gold]." He drew the content of the entry from Philalethes' commentary on Ripley.[14]

No solid evidence allows us to date Newton's plunge into alchemy with precision. A number of items suggest 1669. The completion of his optical research before his appointment to the Lucasian chair may have cleared the way for a new intellectual passion. His impatience with questions about the theory of colors in the 1670s sprang in part from his total absorption in a new investigation.

The order of development of Newton's chemical notebook was significant. He did not stumble into alchemy, discover its absurdity, and make his way to sober, "rational," chemistry. Rather he started with sober chemistry and gave it up rather quickly for what he took to be the greater profundity of alchemy. The latest notes attributed to Boyle referred to his *Essay of . . . Effluviums* of 1673. An unattributed recipe for making phosphorus (which began with the heroic instruction, "Take of Urin one Barrel") undoubtedly stemmed from Boyle's investigation of phosphorus in the early 1680s, but an isolated recipe for a new and unusual substance is a different matter from notes on sustained reading.[15] Boyle himself

---

[12] Ibid., f. 4^v. In the Mint memorandum, Newton said the Goldsmiths Company were of the opinion that gold could not be made finer than 24 carats. First he told them how to do it with aqua fortis. "Chymists also tell us," he continued, "that gold may be made finer by Antimony then by Aqua fortis & by consequence then by the Assay [which used aqua fortis]; but the Goldsmiths know not how to refine Gold by Antimony" (*Mint* 19.1, f. 250^v). [13] *Add MS* 3975, p. 189.

[14] *Add MS* 3975, pp. 243–4. [15] *Add MS* 3975, p. 51.

was deeply involved in alchemy in any case, and once they became acquainted, the two men corresponded on the subject until Boyle's death in 1691. Meanwhile, reading that began with Boyle in the 1660s turned overwhelmingly to explicitly alchemical authors about 1669. His accounts show that on his trip to London that year he purchased the great collection of alchemical writings, *Theatrum chemicum,* in six heavy quarto volumes. He also purchased two furnaces, glass equipment, and chemicals.[16] Possibly some practitioner of the Art introduced Newton to it. Evidence exists that Cambridge had its adepts while Newton was there.[17] We are not obliged to look for an alchemical father, however. Newton had already found his way alone to a number of studies. With collections such as *Theatrum chemicum* available, his independent discovery of alchemy would have been easy enough.

Solid evidence further shows that however it began, Newton's alchemical activity included his personal introduction into the largely clandestine society of English alchemists. His reading in alchemy was not confined to the printed word. Among his manuscripts is a thick sheaf of alchemical treatises, most of them unpublished, written in at least four different hands.[18] Since Newton copied out five of the treatises plus some recipes, the collection appears to have been loaned to him for study but then, for whatever reason, not returned.[19] In the late 1660s, he copied Philalethes' "Exposition upon Sir George Ripley's Epistle to King Edward IV" from a version which differed from published ones though it agreed with two manuscripts now in the British Library.[20] He took extensive notes from a manuscript of Philalethes' "Ripley Reviv'd" about

[16] Fitzwilliam notebook.

[17] Betty Jo Teeter Dobbs, *The Foundations of Newton's Alchemy. The Hunting of the Greene Lyon* (Cambridge, 1975), pp. 95–121. Mrs. Dobbs argues that Barrow and Henry More may have introduced Newton to alchemy. On Newton and alchemy see also A. R. and Marie Boas Hall, "Newton's Chemical Experiments," *Archives internationales d'histoire des sciences,* 11 (1958), 113–52; Karin Figala, "Newton as Alchemist," *History of Science,* 15 (1977), 102–37; R. J. Forbes, "Was Newton an Alchemist?" *Chymia,* 2 (1949), 27–36; Douglas McKie, "Newton and Chemistry," *Endeavour,* 1 (1942), 141–4, and "Some Notes on Newton's Chemical Philosophy Written upon the Occasion of the Tercentenary of his Birth," *The London, Edinburgh & Dublin Philosophical Magazine and Journal of Science,* ser. 7, *33* (1942), 847–70; and Mary S. Churchill, "*The Seven Chapters,* with Explanatory Notes," *Chymia,* 12 (1967), 29–57.

[18] *Keynes MS 67.* On f. 68ᵛ Newton wrote a paragraph, of a nonalchemical nature, in a hand that appears to belong to the mid-1660s. Elsewhere, he corrected some of the manuscripts against Ashmole's *Theatrum* and numbered some recipes. That is, he studied the collection intensely.

[19] *Keynes MS 62* contains the material he copied from *Keynes MS 67.*

[20] *Keynes MS 52.* Ronald S. Wilkinson, "Some Bibliographical Puzzles concerning George Starkey," *Ambix, 20* (1973), 235, has collated Newton's copy with Sloan MSS 633 and 3633 (which contains excerpts).

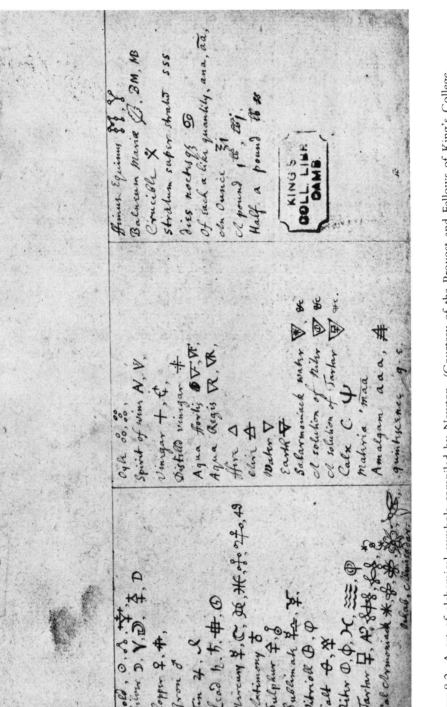

Figure 8.2. A set of alchemical symbols compiled by Newton. (Courtesy of the Provost and Fellows of King's College, Cambridge.)

ten years before it was published.[21] During the following twenty-five years, Newton continued to receive a flow of alchemical manuscripts which he himself copied.[22]

These manuscripts offer one of the most intriguing aspects of his career in alchemy. Where did they come from? The Philalethes manuscripts circulated initially among the group of alchemists associated with Samuel Hartlib in London. Hartlib had died well before Newton took up alchemy, but he may have been in touch with remnants of the group. Since William Cooper, who kept a shop at the sign of the Pelican in Little Britain, later published "Ripley Reviv'd" and at least two other treatises that Newton copied, the contact may have been through him. Robert Boyle had known the Hartlib circle as well as Philalethes-Starkey, though it seems clear that Newton first met Boyle in 1675. One of the copied manuscripts concluded with letters dated 1673 and 1674 from A. C. Faber to Dr. John Twisden with Twisden's notes on them and on the manuscript.[23] Faber (A. D. rather than A. C.) was a physician to Charles II who published a treatise on potable gold. Twisden, also a physician in London known for his defense of Galenic medicine, does not seem a likely clandestine alchemist, but the notes attributed to him are those of a serious practitioner. There was at least the possibility of personal contact and direct transmission in connection with this paper. On another manuscript, "Manna," which is not in his hand, Newton entered two pages of notes and variant readings "collected out of a M.S. communicated to Mr F. by W. S. 1670, & by Mr F. to me 1675."[24] Mrs. Dobbs has argued plausibly that "Mr F." was Ezekiel Foxcroft, a fellow of King's who died in that same year of 1675.[25] Foxcroft, the nephew of Benjamin Whichcote, relative by marriage of John Worthington, and friend of Henry More (all Cambridge Platonists), translated the Rosicrucian tract, "The Chymical Wedding," which was published fifteen years after his death. Newton read it and made notes on it at that time. Whether or not "Mr F." was Ezekiel Foxcroft, the essential mystery of the alchemical manuscripts remains unclarified. The

---

[21] *Keynes MS* 51. This manuscript is perplexing since the notes have page references, though they do not correspond to the pages in the published version. The hand is unmistakably that of the late 1660s.

[22] *Keynes MSS* 22, 24, 31, 33, 39, 50, 51, 52, 55, 58 (part only), 62, 65, 66, and a manuscript in the Yale Medical Library are such. *Keynes MSS* 65 and 66, differing copies of a treatise by the English alchemist of Newton's day, William Y-Worth, a treatise of which he had two other versions in other hands, may imply that Newton was in direct contact with Y-Worth.   [23] *Keynes MS* 50.   [24] *Keynes MS* 33.

[25] Dobbs, *Foundations*, p. 112. Figala, "Newton as Alchemist," pp. 103–4, questions this identification on the basis of the Eton College register, which shows that Foxcroft died in 1674. Since the Eton register used Old Style dates, however, Foxcroft could have lived into 1675, the year that Venn's *Alumni cantabrigiensis* gives for his death.

man who isolated himself from his colleagues in Trinity and discouraged correspondence from philosophical peers in London apparently remained in touch with alchemists from whom he received manuscripts.

The mystery refuses to be ignored. The manuscripts survive—unpublished alchemical treatises, copied by Newton, the originals of which are unknown. The not very illuminating references to Twisden, Faber, "W. S." and "Mr F." excepted, the network of acquaintance that brought them to him left virtually no tangible evidence behind. In March 1683, one Fran. Meheux wrote to Newton from London about the success of a third alchemist, identified only as "hee," in extracting three earths from the first water.[26] Meheux's letter mentioned a continuing correspondence, but the letters have disappeared. Meheux and "hee" have all the substance of shadows. In 1696, an unnamed and equally shadowy figure, a Londoner acquainted with Boyle and Edmund Dickinson (a well-known alchemist whom Charles II had patronized), visited Newton in Cambridge to discourse on alchemy. They did not meet by chance; the man came to find him. Newton recorded the conversation in a memorandum.[27] Alchemy formed the initial subject of a correspondence with Robert Boyle which commenced in 1676. His friendships with John Locke and Fatio de Duillier involved alchemy, but both of them began only in the late 1680s. Otherwise nothing. One of the major passions of his life, as testified by a vast body of papers which stretched over thirty years, a pursuit which included contact with alchemical circles as attested by his copies of unpublished treatises, remained largely hidden from public view and remains so today.

The experience of another collector, Elias Ashmole, helps in assessing Newton's manuscripts. In the preface of *Theatrum chemicum britannicum,* Ashmole declined to name the source of his treatises because they preferred not to see their names in print.[28] His diary recorded a visit, not wholly unlike that which Newton received in 1696, when an unknown and mysterious man appeared at his door ready to reveal the Art.[29] One remembers as well the elusive Eirenaeus Philalethes, who cloaked his identity in a pseudonym so effectively that only in this present generation have we learned with reasonable assurance he was George Starkey. We know that Newton also composed an alchemical pseudonym—Jeova sanctus unus, an anagram of Isaacus Neuutonus—and as knowledge of his al-

---

[26] *Corres 2,* 386.
[27] *Keynes MS* 26; published in *Corres 4,* 196–8. Indeed Newton wrote two versions of the memorandum; the second is University of Chicago MS 1075–3.
[28] Elias Ashmole, *Theatrum chemicum britannicum* (London, 1652), Prolegomena.
[29] C. H. Josten, *Elias Ashmole (1617–1692),* 5 vols. (Oxford, 1966), 4, 1747–8.

chemical activity becomes known, we may learn that Newton fed treatises into the same network from which he received them.

Meanwhile, against the background of deliberate secrecy, we can at least speculate that otherwise unexplained events in his life were alchemically motivated. In the summer of 1672, he spent two weeks with "friends" at Stoke Park near Towcester, Northampton-shire. There is no other known reference in his life to Stoke Park or these friends. Between 1668 and 1677, Newton went to London at least five times. We know very little about the visits, where he stayed and whom he saw, except that in 1669 he returned loaded with alchemical gear. In February and March 1675, it is true, he attended the Royal Society three times and he received visits from members at his inn. Later in the year, in justifying his conduct vis-à-vis Linus, he recalled an occasion when Oldenburg arrived and found an unidentified "Gentleman" there, from the context clearly not a member of the Royal Society.[30] Perhaps this gentle-man rather than Ezekiel Foxcroft was the "Mr F." who gave New-ton a manuscript to copy in 1675; perhaps he was the "W. S." who gave it to "Mr F." This is only speculation, of course. It is not speculation that Newton had alchemical manuscripts which he must have received from someone since they did not, I believe, material-ize out of air.

Nor is it speculation that about 1669 Newton began to read exten-sively in alchemical literature. His notes on the reading survive, the hand not datable with precision but unmistakably from the general period of the late 1660s and perhaps 1670–1. In her recent study of Newton's early alchemy, Mrs. Dobbs asserts that Newton probed "the whole vast literature of the older [i.e., pre-seventeenth century] alchemy as it has never been probed before or since."[31] He also studied seventeenth-century alchemists, especially Sendivogius, d'Espagnet, and Eirenaeus Philalethes, with equal intensity. Much of Newton's attention to alchemy came later. I have carried out a rather careful quantitative study of the alchemical manuscripts he left behind, in which I divide them into three chronological groups. Of the total, which I estimate to include well over a million words devoted to alchemy, about one-sixth appear to stem from the pe-riod before 1675.[32] In his usual fashion, Newton purchased a note-

[30] Newton to Oldenburg, 13 Nov. 1675; *Corres 1*, 357.   [31] Dobbs, *Foundations*, p. 88.

[32] I have had to rely mostly on Newton's handwriting in dating his alchemical manuscripts. There is some internal evidence, such as the dates in *Keynes MSS* 33 and 50, and Mint business on *Keynes MSS* 13 and 56, to supplement its evidence. Especially important are the publication dates of books cited, though one must collate page references with an actual work to be sure Newton was using it instead of a manuscript. The work of Theodore Mundanus, *De quintessentia philosophorum* (1686), offers major assistance.

book in which he entered twelve general headings and a number of subheadings under which to organize the fruits of his reading – headings such as "Conjunctio et liquefactio," "Regimen per ascensum in Caelum & descensum in terram," and "Multiplicatio." In this case, he did not carry the plan out beyond a small number of entries.[33] His later Index chemicus would supply its lack on a heroic scale. Meanwhile, the reading proceeded apace.

I have already indicated the copies he made from the borrowed sheaf of alchemical treatises, the copy of a treatise by Philalethes, and notes on another.[34] He also studied Philalethes' *Secrets Reveal'd* (1669), his heavily annotated copy of which survives.[35] Very early, he read Sendivogius' *Novum lumen chymicum,* on which he took at least two sets of notes. The second set also included notes on d'Espagnet's *Arcanum hermeticae philosophiae opus.*[36] Already he was more than merely a reader. He divided the pages of the second set of notes with a vertical line. On one side he entered notes from Sendivogius; on the other he made a running commentary, which frequently translated Sendivogius' imagery into the language of laboratory chemicals, and which also pointed out Sendivogius' deficiencies on occasion. At much the same time, he read Michael Maier's *Symbola aureae mensae duodecim* (on the notes of which he drew for his letter to Aston in May 1669), the *Opera* of the medieval English alchemist, George Ripley, and Basil Valentine's *Trium-*

Newton cited it extensively; in light of Newton's other activities in the period 1686–9, I have taken a reference to Mundanus to place a paper in the 1690s. I tried to introduce a rough quantitative scale. As a unit of measurement, I used a single page of standard size written in a small hand. Newton wrote the great bulk of his alchemical papers on sheets folded twice to make eight pages about seven inches by nine. Four such pages in a small, early hand, chosen at random, averaged 750 words. Since I realized how rough the measure was in any case, I did not spend further time counting. When Newton wrote in a hand of medium size, I converted pages to units by reducing the number by one-third; when he used a large hand, I reduced by one-half. When he wrote on folios, I multiplied by two. I think I understand the limitations of the figures I obtained as well as the reader will. Among other things, the measurement treats every page of writing, whether it be the record of experiments, the comparison of alchemical authors, the copy of a manuscript, or Newton's own composition, as equal. Manifestly, they involved greatly differing amounts of time. In order to get a rough measurement of his alchemical activity, I divided the papers into three chronological groups: early (to the middle of the 1670s), middle (from the late 1670s to 1687), and late (the early 1690s). By my count, there are 1,443 units in the papers I have been able to see. Using the indications of length in the Sotheby catalogue, I estimate that there are about 85 units I have not seen, or a total of 1,528 units, or nearly 1,200,000 words. I have not included the closely allied material in drafts of some Queries for the *Opticks,* which would swell the total considerably. Of those I counted, a little under 20 percent belonged to the early period, roughly 30 percent to the middle, and roughly 50 percent to the late.

33 *Yahuda MS* Var. 260.   34 *Keynes MSS* 62, 52, 51.
35 University of Wisconsin Library.
36 *Yahuda MS.* Var. 259, *Keynes MS* 19. Cf. also *Babson MS* 925.

phal Chariot of Antimony.[37] Notes on van Helmont, who did not exercise much influence over Newton, date from a few years later, perhaps 1674 or 1675; additions to these notes about ten years later suggest again how reading notes did not become dead repositories but functioned as living parts of continuing investigations.[38] Notes on individual authors fail to convey the extent of his early reading, however. At one point, he made a list of the most important items in the six-volume *Theatrum chemicum*.[39] On a folded sheet with the title, "Of proportions," that is, of the proportions of ingredients in the Work, he cited the opinions of nineteen separate authorities such as Morienus, Hermes, Thomas Aquinas, Bacon, Rosinus, the *Turba philosophorum,* the *Scala,* and the *Rosary*. The list drew, not only on the *Theatrum chemicum,* but also on the other two great collections of alchemical literature then available, the three-volume *Ars auriferae* and the *Musaeum hermeticum*.[40] Already he had begun what would become one of the great private collections of alchemical literature. When he died, his library contained one hundred and seventy-five alchemical books plus numerous pamphlets not listed separately. Alchemical works made up about one-tenth of his total library.[41] Rather quickly, as one would expect, he developed standards of judgment. He crossed out one short passage among his early notes from the *Theatrum chemicum* and dismissed it summarily: "I believe that this author is in no way adept."[42]

Always striving to extract the general from the specific, Newton set out at once to reduce the multitudinous testimony of his reading to the one true process. He drew up a set of forty-seven axioms that summarized the Work with references to the authorities on which he based them.[43] He also began to look for common referents behind the manifold extravagant imagery of his sources.

> Concerning Magnesia or the green Lion [he wrote in a list of *Notae,* which included similar explications of other alchemical terms]. It is called prometheus & the Chameleon. Also Androgyne, and virgin verdant earth in which the Sun has never cast its rays although he is its father and the moon its mother: Also common mercury, dew of heaven which makes the earth fertile, nitre of the wise. Instruct de Arb. sol. [*Instructio de arbore solari*] Chap. 3. It is the Saturnine stone.[44]

---

[37] Notes on Maier: *Keynes MS* 29 and a manuscript in the library of St. Andrew's University. There are later notes on Maier in *Burndy MS* 14, and from the 1690s very extensive notes in *Keynes MS* 32. Notes on Ripley: *Keynes MS* 17 and in the third, fifth, sixth, and seventh items of the manuscript in the Countway Medical Library of Harvard University. Notes on Basil Valentine: *Keynes MS* 64; British Library, *Add MS* 44888.

[38] Eighth item in the Countway MS; the notes are not solely from van Helmont. Later notes on ff. 21, 24ᵛ.   [39] *Ibid.,* third item, f. 10ᵛ.   [40] *Ibid.,* second item, ff. 5–6ᵛ.

[41] See the list of the library in John Harrison, *The Library of Isaac Newton* (Cambridge, 1978).   [42] *Yahuda MS* 259, no. 9.   [43] Countway MS, fourth item.

[44] *Ibid.,* third item, f. 7. Cf. *Keynes MS* 19.

Similar but much longer passages, in which he listed as many as fifty different images for the same ingredient or product of the Work, were to be common features of Newton's later alchemical manuscripts.[45]

Whatever else alchemy meant to him, Newton was always convinced that the treatises he read referred to changes that material substances undergo. It was his goal to penetrate the jungle of luxuriant imagery in order to find the process common to all the great expositions of the Art. To assert as much is not to say that the chemistry he pursued would have been acceptable in the scientific academies of his day, or that scientists of the twentieth century would be willing even to recognize it as chemistry. Nevertheless, he did understand that chemical processes, not mystical experience expressed in the idiom of chemical processes, formed the content of the Art. Thus his reading in the literature of alchemy proceeded hand in hand with laboratory experimentation. The progress made of late in penetrating the maze of his alchemical endeavor has rested on the correlation of his surviving experimental notes with the alchemical manuscripts.[46]

Most of the experimental notes derive from 1678 and after. There are some undated ones in his chemical notebook, however, which seem definitely to have belonged to the late 1660s and early 1670s. His earliest experiments, based on Boyle and showing perhaps the influence of Michael Maier as well, attempted to extract the mercury from various metals. In the intellectual world of alchemy, mercury – not common quicksilver, but the mercury of the philosophers – was the common first matter from which all metals were formed. Liberating it from its fixed form in metals, cleansing

---

[45] "Now this green earth is the Green Ladies of B. Valentine ye beautifully green Venus & the green Venereal Emrauld & green earth of Snyders wth wch he fed his lunary ☿ & by vertue of wch Diana was to bring forth children & out of wch saith Ripley the blood of ye green Lyon is drawn in ye beginning of ye work. Nam viridis et vegetabilis nostri argenti vivi substantia est Basilisci philosophici pabulum saith Mundanus p 180. The spirit of this earth is ye fire in wch Pontanus digests his feculent matter, the blood of infants in wch y^e ☉ & ☽ bath themselves, the unclean green Lion w^{ch}, saith Ripley, is ye mean of joyning ye tinctures of ☉ & ☽, the broth wch Medea poured on ye two serpents, the Venus by mediation of w^{ch} ☉ vulgar & the ☿ of 7 eagles saith Philalethes must be decocted, & in wch ye same ☿ digested alone will give you the Philosophic Lune & ☉, & the Spirit of Grasseus where he saith y^t in the via humida the 4 Elem^{ts} become by steps one ☿ & this ☿ is divided into ☉ & ☿ wch must be reconjoyned by mediation of ye Spirit to give a third thing" (*Keynes MS* 46, f. 1). This passage was composed in the 1690s; cf. *Keynes MSS* 46, f. 1^v; 38, ff. 4–4^v, 9^v; 40, f. 23^v; 56, ff. 7, 12, 12^v; and *Burndy MS* 15. I think I could find hundreds of others. Especially see numerous entries in the Index chemicus (*Keynes MS* 30). All of these passages dated from the 1690s; they were the fruit of more than twenty years of alchemical study. The early one, cited in the text, indicates that Newton's study aimed in this direction from the beginning.

[46] Dobbs, *Foundations,* and Figala, "Newton as Alchemist."

it of contaminating feces, was equivalent to vivifying it and making it fit for the Work. The two images found here, images of purification and of vivification, which included generation by male and female, pervaded the alchemical literature that Newton read. Newton's first method of extracting mercury involved the dissolution of quicksilver in aqua fortis (nitric acid). When he added lead filings to the solution, a running mercury, which he took to be the mercury of lead, was released and a white powder precipitated. He did it with tin and copper as well, but he did not entirely like what he saw. "I know not whither y^t ☿ come out of ye liquor or of ♀ [copper] for ye liquor dissolves ♀. Also ♀ will draw ☿ out of ye limus [the white precipitate] wch falls down in dissolving ♃ [tin] or ♄ [lead] & also out of ye liquor both during ye solution & afterward."[47] Hence he turned to a different method, one that used sublimate (mercuric chloride) to "open" the metals, as he had put it in his chemical glossary, and release the philosophic mercury.[48]

More powerful alchemical methods presented themselves, especially one based on the regulus of antimony. What the seventeenth century called antimony was not the metal but its ore, stibnite (Sb_2S_3). From the ore, they produced regulus of antimony (our metallic element) by reducing it with a variety of agents. Because of the efficacy of the regulus of antimony in refining gold, alchemists of the seventeenth century seized on it as a powerful agent of the Art. Regulus per se resulted from reduction with charcoal. Metals could also act as reducing agents, and with them one got regulus of Saturn (lead), regulus of Jupiter (tin), regulus of Venus (copper), and especially regulus of Mars (iron). In fact they were all identical, pure antimony, except insofar as improper proportions left contaminants. However, alchemists thought the different reguli contained the seeds of the metals with which they had been produced.

About 1669 or 1670, Newton composed an experimental essay on the production of the reguli of antimony. Together with his other alchemical experiments, the essay serves to remind us that Newton came to the Art with unique intellectual equipment such as no other alchemist ever possessed, and though it appears to me and to others that he entered into the fantastic world of alchemy, he did not leave his special gifts at the door. Indeed, he could not. He brought with him standards of intellectual rigor born of mathematics, which he applied to his own experimentation. We have already seen them at work above. He brought with him the mechanical philosopher's sense that nature is quantitative; both now and later

---

[47] *Add MS* 3975, p. 80.

[48] In connection with his early experiments, cf. *Keynes MS* 31, *Liber mercuriorum corporum* (The Book of the Mercuries of Bodies), an early collection of recipes to obtain the mercuries of metals.

his alchemical experimentation revealed an overriding concern for quantitative measurement such as cannot be found in the many tomes he read. He brought with him as well experimental skill, nourished originally in Grantham and recently honed to a fine edge in his optical work. All of these characteristics marked the essay on reguli. Manifestly the result of prolonged experimentation, it set down instructions for the successful production of the various reguli. He started with the proper proportions to use and the procedure to follow

> If ye scoria of ♄ bee full of small eaven rays there is two little ♄ in proportion. If any reg swell much in the midst of the upper surface it argues two much ♁ [antimony] if it bee flat it argues two little. The better yor proportions are the brighter and britler will ye Reg bee & ye darker ye scoria & the easier will they part . . . The work succeeds best in least quantitys. If there bee stuff like pitch long in cooling tis noe good signe & often argues too much Antimony. . . .
>
> These rules in generall should bee observed. 1st y^t ye fire bee quick. 2dly y^t ye crucible bee throughly heated . . . 4tly That they stand some time after fusion before they bee poured of . . .
>
> Also these signes may bee observed in generall. That if ye scoria & Regulus part not well there is two much metall; that if they doe part well & yet yeild not a dew quantity of Regulus there is too little metall (unlesse ye fire hath not been quick enough or the regulus not had time to sattle) That if the reg bee tough it argues too much metall unlesse in tin wch is therby made ye brittler.

The regulus of Mars especially interested Newton. He worked out its ideal proportions. "Thus wth a good quick & smart fire 4 of ♂ to 9 of ♁ gave a most black & filthy scoria & ye Reg after a purgation or two starred very well. But in a lesse heat a greater proportion of ♁ gave ye blackest scoria."[49] A black and filthy scoria meant a successful purification of the regulus. As for the regulus of Mars, Newton believed it contained the sulfuric seed of iron planted in the mercuric matrix of antimony. It was the alchemical hermaphrodite, the union of the male and female principles, which formed the necessary first matter of the Art. As Philalethes told him, "Art first impregnates ☿ [mercury] (ye soule) w^{th} ♁^{r} [sulfur] (the spirituall seed) by wch it becomes powerfull in the dissolution of metals, & then adds mature ♀ (the body) to make

---

49 *Add MS* 3975, pp. 81–2. Mrs. Dobbs publishes the entire essay as an appendix to her book (pp. 249–50). See her excellent discussion of the chemistry of the regulus in terms of twentieth-century chemistry (pp. 146–8) and of its role in Newton's alchemy (pp. 148–60). My account both of this process and of all Newton's experiments draws heavily on hers. See also Figala, "Newton as Alchemist."

it elixir."[50] Properly made, the regulus revealed the crystalline structure of antimony, in which alchemists saw a star. The star regulus was obviously an alchemical agent of preeminent virtue. By early 1672 at the latest, Newton was familiar with its production, for at that time he told Oldenburg he had tried the "stellate Regulus of Mars" as a reflector in his telescope.[51] It did not last long as a mirror, but it had a long career ahead of it in Newton's alchemy.

With the star regulus and copper, Newton proceeded to the production of another hermaphrodite, called the net, which combined the male seed of Mars with the female principle of Venus. The concept of the net derived from the classical myth that Vulcan fashioned a golden net about his bed and with it trapped his wife Venus *in flagrante delicto* with Mars. One might search some time to find an alchemical image more evocative. It may or may not be significant that Newton employed it far more than alchemical literature in general. About 1670, Newton drew up a list of topics he needed to investigate and understand. It included the entry, "De concubitu Martis et Veneris, et rete Vulcani" (Concerning the copulation of Mars and Venus and the net of Vulcan). Not long thereafter, he satisfied his demand and was able to record the proportions of copper and of regulus of Mars.

R ♂ [regulus of Mars] 9 1/4, ♀ 4 gave a substance wth a pit hemisphericall & wrought like a net wth hollow work as twere cut in.

R ♂ 8 1/2, ♀ 4 gave noe pit but a net work forme spread all over ye top, yet more impressed in ye middle

R ♂ 2 ♀ 1 gave net worke but not so notable as ye former, & so did R ♂ 5 ♀ 2

The best proportion is about 4, 8 1/2 or 9.[52]

As with the star regulus, the physical appearance of the net matched its name, but its alchemical significance lay in its fertile union of female and male. Like the star regulus again, it looked forward to a long career both in Newton's laboratory and in his natural philosophy. So did another substance, sal ammoniac, which appeared briefly though it did not yet hold the prominent place in his early experiments that it would later occupy.[53]

[50] *Keynes MS* 51, f. 1^v. Cf. the commentary on Philalethes' *Exposition of the Epistle to King Edward IV* in the British Library, Sloane MS 646: "But as to why Philalethes wanted to have the iron regulus of antimony, he gives the following reasons: 1. So that ordinary mercury in so far as it is cold and damp may be made hot by the volatile sulfur of gold that is in iron, may be dried, and may be impregnated by the very goldmaking sulfur. 2. So that the male should be produced by the impregnation of this volatile sulfur but remain female because of its material and hence be an hermaphrodite" (cited by Figala, "Newton as Alchemist," p. 131).

[51] Newton to Oldenburg, 18 Jan. 1672; *Corres 1,* 82.

[52] *Add MS* 3975, p. 83. Cf. Dobbs, *Foundations,* pp. 161–3.

[53] *Add MS* 3975, p. 84.

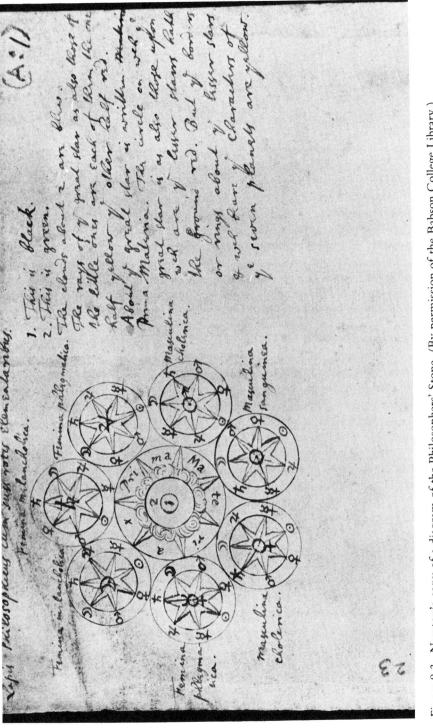

Figure 8.3. Newton's copy of a diagram of the Philosophers' Stone. (By permission of the Babson College Library.)

From the testimony of others, from his own reading notes, and from his experiments with substances alchemically significant, it is clear beyond doubt that Newton devoted great attention to alchemy in the late 1660s and early 1670s. We are left to decide for ourselves what his purpose may have been. As everyone knows, alchemy sought to make gold. Nothing whatever in the vast corpus of Newton's alchemical manuscripts even hints that gold making, in the vulgar sense of the phrase, ever dominated Newton's concern. While Newton was not indifferent to his material welfare, it was never money that kept him from his meals and drove him to distraction. Truth and Truth alone held that power over him. To the great figures and monuments of the alchemical tradition, the men and works that Newton studied, Truth was also the goal of the Art. As Elias Ashmole insisted in the preface to his *Theatrum chemicum britannicum,* gold making was the lowest use to which the adepts applied their knowledge.

> For they being lovers of Wisdome more then Worldly Wealth, drove at higher and more Excellent Operations: And certainly He to whom the whole Course of Nature lyes open, rejoyceth not so much that he can make Gold and Silver, or the Divells to become Subject to him, as that he sees the Heavens open, the Angells of God Ascending and Descending, and that his own Name is fairely written in the Book of life.[54]

Newton copied much the same sentiment in the preface to "Manna" that "Mr F." showed him in 1675.

> For Alchemy tradeth not wth metalls as ignorant vulgars think . . . This Philosophy is not of that kind wch tendeth to vanity & deceipt but rather to profit & to edification inducing first ye knowledg of God & secondly ye way to find out true medicines in ye creatures. . . . so y^t ye scope is to glorify God in his wonderful works, to teach a man how to live well, & to be charitably affected helping o^r neighbours.[55]

The philosophical tradition of alchemy had always regarded its knowledge as the secret possession of a select few who were set off from the vulgar herd both by their wisdom and by the purity of their hearts. By the time he turned seriously to alchemy about 1669, Newton had carried two investigations of capital importance to completion; he could not have doubted his right to claim membership in an intellectual elite. We know less of what he thought about the purity of his heart, but convictions on that score are endemic throughout mankind.

The concept of a secret knowledge for a select few aside, all the above characteristics applied as well to the mechanical philosophy,

[54] Ashmole, *Theatrum,* Prolegomena.   [55] *Keynes MS* 33, f. 5^v.

which Newton had recently embraced. In the nature of the truth they offered, however, the two philosophies differed profoundly. In the mechanical philosophy, Newton had found an approach to nature which radically separated body and spirit, eliminated spirit from the operations of nature, and explained those operations solely by the mechanical necessity of particles of matter in motion. Alchemy, in contrast, offered the quintessential embodiment of all the mechanical philosophy rejected. It looked upon nature as life instead of machine, explained phenomena by the activating agency of spirit, and claimed that all things are generated by the copulation of male and female principles. Among his "Notable Opinions" that he collected some ten years later Newton included the argument of Effararius the Monk that the stone is composed of body, soul, and spirit, that is, imperfect body, ferment, and water.

> For a heavy and dead body is imperfect body per se. The Spirit that purges, lightens, and purifies body is water. The soul that gives life to imperfect body when it does not have it, or raises it to a higher plane, is ferment. Body is Venus and feminine; spirit is Mercury and masculine; soul is the Sun and the Moon.[56]

And in a later collection of "Enlightening Opinions and Notable Conclusions," Newton included an unattributed expression of the concept of sexual generation to which Effararius alluded in his final sentence.

> A double mercury is the sole first and proximate matter of all metals, and these two mercuries are the masculine and feminine semens, sulfur and mercury, fixed and volatile, the Serpents around the caduceus, the Dragons of Flammel. Nothing is produced from masculine or feminine semen alone. For generation and for the first matter the two must be joined.[57]

Newton also met another idea in alchemy that refused to be reconciled with the mechanical philosophy. Where that philosophy insisted on the inertness of matter, such that mechanical necessity alone determines its motion, alchemy asserted the existence of active principles in matter as the primary agents of natural phenomena. Especially it asserted the existence of one active agent, the philosophers' stone, the object of the Art. Images of every sort were applied to the stone, all expressing a concept of activity utterly at odds with the inertness of mechanical matter characterized by extension alone. Flammel called it "a most puissant & invincible king," Philalethes, the "miracle of the world" and "the subject of wonders." The author of *Elucidarius* insisted that "it is impossible to express [its] infinite virtues . . ."[58] In Sendivogius and Philalethes,

---

[56] *Keynes MS 38*, f. 12.   [57] *Keynes MS 56*, f. 2ᵛ.
[58] *Keynes MSS 40*, ff. 20, 19ᵛ; 41, f. 15ᵛ. *Babson MS 417*, p. 35.

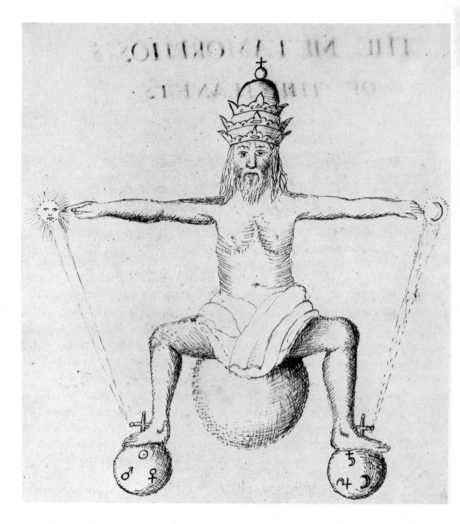

Figure 8.4. Jupiter enthroned. Newton's drawing with his copy of
John de Monte-Snyder's *Metamorphosis of the Planets.*
(Courtesy of the Yale Medical Library.)

the activity sometimes took on the specific form of an attraction,
and they called it a magnet. Philosophic reformers such as Descartes
had explicitly worked to eliminate "occult" concepts, such as attractions, from natural philosophy; they had invented whirlpools of
various invisible particles to explain away the apparent fact of magnetism. Not Sendivogius and Philalethes. To them, the magnet
offered an image of the operation of nature. "They call lead a
magnet," Newton recorded in his early notes on Sendivogius, "because its mercury attracts the seed of Antimony as the magnet

attracts the Chalybs." And again he noted that "our water" is drawn out of lead "by the force of our Chalybs which is found in the belly of Ares." In a note, Newton explained that this meant "the force of our sulfur which lies hidden in Antimony."[59]

It is necessary, I believe, to see Newton's interest in alchemy as a manifestation of rebellion against the confining limits that mechanistic thought imposed on natural philosophy. If the pursuit of Truth expressed the essence of his life, there is no reason to expect that he should have remained satisfied forever with his first love. Mechanical philosophy had surrendered to his desire, perhaps too readily. Unfulfilled, he continued the quest and found in alchemy, and in allied philosophies, a new mistress of infinite variety who never seemed fully to yield. Where others cloyed she only whet the appetite she fed. Newton wooed her in earnest for thirty years.

Perhaps "rebellion" is too strong a word, and I should speak rather of partial rebellion. Newton never wholly abandoned his first love. He never ceased to be a mechanical philosopher in a fundamental and important sense. He always believed that particles of matter in motion constitute physical reality. Where mechanical philosophers of strict persuasion insisted that particles of matter in motion alone constitute physical reality, however, Newton came rather early to find those categories too confining to express the reality of nature. The significance of alchemy in his intellectual odyssey lay in the broader vistas it opened to him, additional categories to supplement and complete the narrow mechanistic ones. His enduring fame derived from his seizing the possibilities thus spread before him.[60]

From the beginning, Newton had felt some reservations about the mechanical philosophy. Henry More, the Cambridge Platonist, was among the early authors he read, and in More he found anxiety, which he recorded in the "Quaestiones," about the religious implications of the exclusion of spirit from nature. By the late 1660s, Newton's anxiety had grown to the point of alarm, which expressed itself in an essay, "De gravitatione et equipondio fluidorum" (On the Gravity and Equilibrium of Fluids).[61] Internal evidence places it in 1668 at the earliest. As the title indicated, "De

---

59 *Keynes MS* 19, ff. 1, 3.
60 Cf. P. M. Rattansi, "Newton's Alchemical Studies," in Allen G. Debus, ed., *Science, Medicine and Society in the Renaissaince*, 2 vols. (New York, 1972), *2*, 167–82, and "Some Evaluations of Reason in Sixteenth- and Seventeenth-Century Natural Philosophy," in Mikulas Teich and Robert Young, eds., *Changing Perspectives in the History of Science* (London, 1973), pp. 148–66.
61 *Add MS* 4003. Published in *Halls*, pp. 90–121; English translation pp. 121–56. For the bearing of the fluid mechanics in the piece on optics, see Alan E. Shapiro, "Light, Pressure, and Rectilinear Propagation: Descartes' Celestial Optics and Newton's Hydrostatics," *Studies in History and Philosophy of Science, 5* (1974), 239–96.

gravitatione" set out to be a treatise on fluid mechanics. Aside from some definitions and two propositions (which eventually found their way into the *Principia,* where they were compressed into Proposition XIX, Book II), what he completed was a prefatory essay which attacked Descartes's philosophy. Newton was a man of strong preferences. Even in the privacy of his study, he worked himself into a passionate fury against the philosopher who, scarcely five years earlier, had introduced him to a new world of thought. The gravamen of the charge was atheism. By his separation of body and spirit, Descartes denied the dependence of the material world on God. The ultimate cause of atheism, Newton asserted, is "this notion of bodies having, as it were, a complete, absolute and independent reality in themselves . . ."[62] To refute that notion he had to attack the Cartesian equation of matter and extension, and in consequence Descartes's relativistic conceptions of place and motion. Lest his disagreement be overlooked, he described the latter as "absurd . . . confused and incongruous with reason . . ."[63]

The argument on this point held considerable importance for the development of Newton's mechanics. It was in "De gravitatione" that he first pronounced his doctrines of absolute space and absolute time and propounded the phenomena of circular motion with which he supported the first. The argument led him on ineluctably to the conclusion that "physical and absolute motion is to be defined from other considerations than translation, such translation being designated as merely external."[64] What considerations defined motion then? Newton couched his refutation of relative motion in terms of causal considerations which made force central. Hence the importance of Definition 5:

> Force is the causal principle of motion and rest. And it is either an external one that generates or destroys or otherwise changes impressed motion in some body; or it is an internal principle by which existing motion or rest is conserved in a body, and by which any being endeavours to continue in its state and opposes resistance.

Definition 8 went on to assert that "Inertia is force within a body, lest its state should be easily changed by an external exciting force."[65] His use of the word "inertia" here is apt to be misleading. We should not give it a modern meaning. In his assault on Cartesian philosophy and Cartesian conceptions of motion, Newton went all the way and effectively denied something else he had earlier found in Descartes under a different name – what we call today the principle of inertia.

[62] *Halls,* p. 144. Original Latin, p. 110.
[63] *Halls,* p. 124. Original Latin, p. 92.
[64] *Halls,* p. 128. Original Latin, p. 96.
[65] *Halls,* p. 148. Original Latin, p. 114.

Though a passionate rejection of Descartes, "De gravitatione" did not reject the mechanical philosophy. It drew its four opening definitions (of place, body, rest, and motion) from Gassendi and its doctrines of absolute space and time reflected similar doctrines in that philosopher's *Syntagma philosophicum.*[66] Likewise it advanced a particulate conception of matter and implied that all bodies are composed of such particles. Nevertheless, it would be difficult to reconcile the assertion that force, including the force internal to bodies, is the defining characteristic of motion with orthodox mechanical philosophies. The central feature of Newton's conception of matter would present equal difficulty. Newton's goal was to demonstrate the dependence of matter on God. He denied the possibility of knowing the essence of matter, the existence of which depends (Descartes to the contrary) on the arbitrary will of God. What he could do was to describe something that would have all the characteristics of bodies though he would not assert that God had in fact made matter in this form. Imagine then that God chose to prevent bodies from entering a certain volume of absolute space. This hole in space (or "determined quantity of extension," in Newton's phrase) would be tangible because impenetrable, and visible because opaque. In short, it would have all the properties of a particle. Imagine further that God moves it about according to certain laws. "In the same way if several spaces of this kind should be impervious to bodies and to each other, they would all sustain the vicissitudes of corpuscles and exhibit the same phenomena. And so if all this world were constituted of this kind of being, it would seem hardly any different."[67] For the existence of such bodies, Newton declared, nothing is required but "extension and action [*actus*] of the divine will . . ." Such a conception emphasized the distinction between extension and body: "Because extension is eternal, infinite, uncreated, uniform throughout, not in the least mobile, nor capable of inducing change of motion in bodies or change of thought in the mind; whereas body is opposite in every respect . . ." In addition to extension, bodies also possess "faculties by which they can stimulate perceptions in the mind and move other bodies." Corporeal nature belongs, not to extension, but to their faculties; "spaces are not the very bodies themselves but are only the places in which bodies exist and move . . ."[68] Here was a view of matter intimately related to alchemical views – a passive mass animated by an active principle. The affinity of outlook was further emphasized by the parallel he drew between God, with His capacity to move matter, and the human soul, with its capacity to

[66] Pierre Gassendi, *Opera Omnia,* 6 vols. (Lyons, 1658), *1,* 179–228.
[67] *Halls,* p. 139. Original Latin, p. 106.
[68] *Halls,* pp. 140–8. Original Latin, pp. 107–14.

move its body. Even created mind is "of a far more noble a nature than body so that perhaps it may eminently contain it in itself." Whereas Descartes's distinction separated mind and body, his allowed them to combine.[69]

"De gravitatione" is not an alchemical essay. If Newton's conception of absolute space drew upon Gassendi, so also it drew upon the Neoplatonism of Henry More, especially in its assertion that extension is a disposition of being *qua* being. Equally, More had pointed out the limitations of mechanical science and the necessary role of spirit in nature. The influence of More and the Cambridge Platonists on Newton, both in "De gravitatione" and later, has been effectively demonstrated of late, and the question has been raised why there is need to call upon the influence of alchemy.[70] The concepts that expanded the narrow categories of his early mechanical philosophy were present in the works of the Cambridge Platonists. Why raise the perplexed issue of alchemy? The answer to this must be empirical. The records of Newton's work in alchemy exist. For every page of notes from a Cambridge Platonist, there are at least a hundred pages on alchemy. To bring up the role of alchemy is not to deny the influence of More, and later of Cudworth, on Newton. It is not necessary to trace the modification of his natural philosophy to a single factor. Poring over his books until dawn, night after night, Newton laid many sources under tribute. Extensive evidence exists that alchemy was among them, and prominently so.

At much the same time that he wrote "De gravitatione," Newton composed two other papers which cannot be mistaken for anything but alchemy. One of them was a short list of "prepositions" which he supported by citations of alchemical authorities. Since many of the citations came from the *Theatrum chemicum*, he probably drew up the list after his purchase of that work in the fall of 1669. It concluded with a briefly stated philosophy of nature.

> The vital agent diffused through everything in the earth is one and the same
>
> And it is a mercurial spirit, extremely subtle and supremely volatile, which is dispersed through every place.
>
> The general method of operation of this agent is the same in all things; that is, it is excited to action by a gentle heat, but driven away by a great one, and when it is introduced into a mass of

---

[69] *Halls,* pp. 141–2. Original Latin, pp. 108–9.

[70] J. E. McGuire and P. M. Rattansi, "Newton and the 'Pipes of Pan'," *Notes and Records of the Royal Society, 21* (1966), 108–43. J. E. McGuire, "Force, Active Principles, and Newton's Invisible Realm," *Ambix, 15* (1968), 154–208, and "Neoplatonism and Active Principles: Newton and the *Corpus Hermeticum,*" in Robert S. Westman and J. E. McGuire, *Hermeticism and the Scientific Revolution* (Los Angeles, 1977).

substances its first action is to putrefy and confound into chaos; then it proceeds to generation. . . .

In a metallic form it is found most abundantly in Magnesia [antimony]

And all species of metals derive from this single root

And in this order ☿, ♄, ♃, ☽, ♀, ♂, ☉ [mercury, lead, tin, silver, copper, iron, gold][71]

The second paper, which is much longer, has no title but is usually called, by an extension of phrases in its first lines, "The Vegetation of Metals." A reference to the spring in Hungary that he had read about in Michael Maier suggests that it was at least as late as the letter to Aston in the spring of 1669. It was not a single connected exposition, though all its parts were related. Not only did it expand on the theme of the "Prepositions," but the central concepts of "De gravitatione" reappeared here in an explicitly alchemical context. "Vegetation" opened with a set of twelve statements which, with the brief summaries of evidence that accompanied some of them, had all the appearance of chapter headings for an intended treatise.

1 Of Natures obvious laws & processes in vegetation.
2 That metalls vegetate after the same laws. . . .

6 That vegetation is ye sole effect of a latent spt & that this spt is ye same in all things only discriminated by its degrees of maturity & the rude matter.[72]

A later set of general statements that insisted on the necessity of putrefaction in every process of change concluded with the following assertions:

All things are corruptible
All things are generable
Nature only works in moyst substances
And wth a gentle heat[73]

This list too spoke of a "powerfull agent," undoubtedly the same as the "spirit" above.

Much of the paper devoted itself to describing a system of nature in which the spirit, variously called "metallin fumes," "vapors," and "minerall spt," played the central role. Much of it resembled his later aethereal hypothesis, the first draft of which dated from 1672, scarcely three years after the "Vegetation of Metals." His focus in "Vegetation," however, was the earth and its mineral products. He described how the metalline spirit ascends from the bowels of the earth, is fixed into salts and minerals when it meets water,

[71] *Keynes MS* 12, ff. 1^v–2.   [72] *Burndy MS* 16, f. 1.   [73] *Burndy MS* 16, f. 5.

and is thereby alienated from its metallic nature. The alienation arises because concretion is not a process of vegetation but only "a gros mechanicall transposition of parts." If the spirits can be freed from their fixed compositions, they can again "receive metallick life & by degrees recover their pristine metalline forme."[74]

Newton described how some metallic vapors ascend into the air, become aether, and by their pressure force other aether to descend: "Thus this Earth resembles a great animall or rather inanimate vegetable, draws in aethereall breath for its dayly refreshment & vitall ferment & transpires again wth grosse exhaltations." "This is the subtil spirit," he said of the aether, "this is Natures universall agent, her secret fire, ye onely ferment & principle of all vegetation. The material soule of all matter wch being constantly inspired from above pervades & concretes wth it into one form & then if incited by a gentle heat actuates & enlivens it . . ." All sensible matter may be nothing but aether condensed and interwoven into various textures, the life of which depends on that part of the aether not yet condensed.

> Note that tis more probable ye aether is but a vehicle to some more active spt. & ye bodys may bee concreted of both together, they may imbibe aether as well as air in gen[er]ation & in y^t aether ye spt is intangled. This spt perhaps is ye body of light becaus both have a prodigious active principle both are perpetuall workers.[75]

Ordinary aether carrying the active spirit, a replication of passive extension animated by God, repeated the pattern of the alchemical hermaphrodite which Newton produced in the star regulus and the net.

In a discussion of putrefaction included in "Vegetation," Newton directly faced the relation of alchemy to the mechanical philosophy. "Natures actions," he asserted, "are either vegetable . . . or purely mechanicall." The principles of vegetable actions are "the seeds or seminall virtues of things those are her only agents, her fire, her soule, her life." The seed in anything is that part of it which has attained the fullest maturity. "Vegetation is nothing else but ye acting of w^t is most maturated or specificate upon that wch is less specificate or mature to make it as mature as it selfe. And in that degree of maturity nature ever rests." Only a tiny portion of things, never seen alone but only "inclothed wth watry humidity" is seed, "The maine bulk being but a watry insipid substance in wch rather then upon wch the action is performed." Again, he described the matter in which the seed operates as "dead earth & insipid water," mere vehicles for the active seed, the "invisible inhabitant." As the particles of the grosser substances are separated and recom-

[74] *Burndy MS 16, f. 3.*   [75] *Burndy MS 16, ff. 3^v–4.*

bined, they may take on different appearances. Both the operations of vulgar chemistry and some natural processes can produce such changes, which appear on occasion to be profound transformations though they are nothing but "mechanicall coalitions or seperations of particles . . ."

> But so far as by vegetation such changes are wrought as cannot bee done wthout it wee must have recourse to som further cause And this difference is vast & fundamentall because nothing could ever yet bee made wthout vegetation wch nature useth to produce by it . . . There is therefore besides ye sensible changes wrough in ye textures of ye grosser matter a more subtile secret & noble way of working in all vegetation which makes its products distinct from all others & ye immediate seate of thes operations is not ye whole bulk of matter, but rather an exceeding subtile & inimaginably small portion of matter diffused through the masse wch if it were seperated there would remain but a dead & inactive earth.[76]

With unmistakable clarity, "The Vegetation of Metals" proclaimed Newton's conviction that mechanical science had to be completed by a more profound natural philosophy which probed the active principles behind particles in motion.

The same conviction repeated itself in the "Hypothesis of Light" of 1675, though he disguised it considerably, perhaps because of the audience. On the surface, the "Hypothesis" presented a mechanical cosmology based on a universal aether, and for three hundred years it has been read as a representative expression of seventeenth-century mechanical philosophy. It contained strange elements, however, though they look less strange after one has read the "Vegetation of Metals." Several times the "Hypothesis" referred to a "secret principle of unsociablenes" by which fluids and spirits do not mix with some things but do with others. Mechanical philosophers explained such phenomena by sizes of particles and shapes of pores. Newton said explicitly that "liquors & Spirits are disposed to pervade or not pervade things on other accounts then their Subtility . . . So some fluids (as Oyle and water) though their pores are in freedome enough to mix with one another, yet by some secret principle of unsociablenes they keep asunder . . ."[77] An analogous principle without the name had appeared in the "Vegetation"; and in the "Hypothesis," Newton justified it by evidence drawn solely from chemistry. Active principles appeared as well. Though the "Hypothesis" treated reflection and refraction in straightforward mechanical terms, it ascribed to the corpuscles of light a "Principle of motion" which accelerates them until the resistance of the aether equals its force. "God who gave Animals self motion beyond our

---

[76] *Burndy MS* 16, ff. 5^v–6^v.   [77] *Corres 1,* 368.

understanding is without doubt able to implant other principles of motion in bodies wch we may understand as little. Some would readily grant this may be a Spiritual one; yet a mechanical one might be showne, did not I think it better to passe it by."[78]

As in the "Vegetation," he distinguished the aether into the "maine flegmatic body of aether" and "other various aethereall Spirits," such as electrical and magnetic effluvia and the gravitating principle.

> Perhaps the whole frame of Nature may be nothing but various Contextures of some certaine aethereall Spirits or vapours condens'd as it were by praecipitation . . . and after condensation wrought into various formes, at first by the immediate hand of the Creator, and ever since by the power of Nature, wch by vertue of the command Increase & Multiply, became a complete Imitator of the copies sett her by the Protoplast.[79]

The gravitating attraction of the earth may be caused by the condensation of such a spirit, which bears the same relation to the main phlegmatic body of aether as the "vitall aereall Spirit" that conserves flame and life does to the air. If such an aethereal spirit is condensed in fermenting & burning bodies, "the vast body of the Earth, wch may be every where to the very center in perpetuall working," may cause a continual descent of the spirit, which would carry bodies down with it. Other matter would have to rise in the form of vapors and exhalations to replenish the aether.

> For nature is a perpetuall circulatory worker, generating fluids out of solids, and solids out of fluids, fixed things out of volatile, & volatile out of fixed, subtile out of gross, & gross out of subtile, Some things to ascend & make the upper terrestriall juices, Rivers and the Atmosphere; & by consequence others to descend for a Requitall to the former.[80]

Small wonder that the "Hypothesis" has recently been called an alchemical cosmology.

Not long after he composed the "Hypothesis," Newton read in the *Philosophical Transactions* an account by "B.R." of a special mercury that heated gold when mixed with it. B.R. asked for advice as to whether he should publish the recipe for the mercury.[81] As far as we know, Newton was the only one who offered an opinion. Without asking, he knew that B.R. stood for Robert Boyle. He expressed doubt that anything of importance was concealed in the mercury, the action of which he explained by his principle of sociability.

[78] *Corres 1*, 370.
[79] *Corres 1*, 364. Cf. number 11 in the list of twelve statements that introduced "Vegetation": "Of protoplasts yt nature can onely nourish, not form them, Thats Gods mechanism in these natures" (*Burndy MS* 16, f. 1).   [80] *Corres 1*, 365–6.
[81] "Of the Incalescence of *Quicksilver* with *Gold*, generously imparted by B. R.," *Philosophical Transactions, 10* (1676), 515–33.

But yet because ye way by w^{ch} ☿ may be so impregnated [with metallic particles], has been thought fit to be concealed by others that have known it, & therefore may possibly be an inlet to something more noble, not be to communicated wthout immense dammage to ye world if there should be any verity in ye Hermetick writers, therefore I question not but that ye great wisdom of ye noble Authour will sway him to high silence till he shall be resolved of what consequence ye thing may be either by his own experience, or ye judgm^t of some other that throughly understands what he speakes about, that is of a true Hermetic Philosopher, whose judgm^t (if there be any such) would be more to be regarded in this point then that of all ye world beside to ye contrary, there being other things beside ye transmutation of metalls (if those great pretenders bragg not) wch none but they understand.[82]

Before the extent of Newton's involvement in alchemy was understood, this letter was generally interpreted as a forthright expression of his skepticism about the whole enterprise. It may have expressed a temporary disenchantment. As far as one can tell from the uncertain evidence of handwriting, his vigorous application to alchemy at the beginning of the decade did slacken in its middle years before it flamed up again more brightly at the decade's end. However, Boyle's paper was about a specific approach to alchemy with which Newton had already experimented briefly, and his letter can be read without forcing it as merely a rejection of amalgamation as the way to the Work. The "Hypothesis," written less than half a year earlier, does not support the conclusion that he had bidden farewell to alchemy, nor does the letter for Boyle itself. The most interesting thing about the letter is the fact that Newton wrote it. At the very time when he was frantically trying to terminate his correspondence on optics and mathematics, he volunteered a letter on alchemy which looks like an effort to initiate a correspondence. Later evidence confirms that one ensued, the only sustained one we know of during his years of silence, and for that matter, after the initial letter, a direct correspondence without intermediary.

If, as I indicated, Newton's active involvement in alchemy slackened for a time after the early years of the decade, he did not lack for other interests. His manuscripts show that about this time he turned to a new field of study, theology. Perhaps it is wrong to call it "new." Speculation about his reading in his stepfather's library aside, there is solid evidence of early theological interest. Four of the ten books known by his accounts and by his dated signature to have been purchased soon after his arrival in Cambridge were theo-

[82] Newton to Oldenburg, 26 April 1676; *Corres 2*, 1–2.

logical.[83] Nevertheless, no body of theological manuscripts survive from earlier than about 1672. At that time, Newton was completing his fourth year as a Master of Arts and fellow of Trinity. Within the next three years, he would need to be ordained to the Anglican clergy or face expulsion from the college. The beginning of serious theological study may have stemmed from the approaching deadline. Whatever the cause, the fact itself cannot be denied. Nor should we imagine that he devoted himself to theology reluctantly, for the subject quickly seized him as others had before. His notes reveal a massive commitment to it. There are very few secure dates internal to the manuscripts, and in locating them chronologically, one is thrown back primarily on the uncertain evidence of handwriting. Nevertheless, there can be no reasonable question that at least part of the time, when Newton expressed impatience at the interruptions caused by optical and mathematical correspondence during the 1670s, it was theology that preoccupied him. He jotted down a number of theological references on a draft of his letter of 4 December 1674, in which he told Oldenburg that he intended "to concern my self no further about promotion of Philosophy."[84]

If it is impossible to date most of the manuscripts with precision, so also it is impossible to be certain of their order. Surely the standard Newtonian exercise in organization was among the earliest, however. In a notebook he entered a number of headings that summarized Christian theology: "Attributa Dei," "Deus Pater," "Deus Filius," "Incarnatio," "Christi Satisfactio, & Redemptio," "Spiritus Sanctus Deus," and the like.[85] Apparently he intended to

---

[83] In one of his accounts, Newton listed four purchases: a logic, Hall's *Chronicle* (Edward Halle, *The Union of the Two Noble and Illustrate Families of Lancaster & Yorke* . . . [London, 1548]), Sleiden's *Four Monarchies* (Joannes Sleidanus, *The Key of Historie. Or, a Most Methodical Abridgement of the Four Chiefe Monarchies* . . . [London, 1631], an English translation of *De quatuor summis imperiis*), and "Schrepelius his lexicon" (Cornelius Schrevelius, *Lexicon manuale Graeco-Latinum et Latino-Graecum* . . . [Leyden, 1654]). Eight books survive with Newton's signature and the inscription, "Trin: Coll: Cant: 1661." One of them is Robert Sanderson, *Logicae artis compendium* (Oxford, 1631), probably the "logic" in his account. The other six are Theodore Beza, *Annotationes maiores in novum . . . testamentum . . .* (Geneva, 1594); John Calvin, *Institutio christianae religionis . . .* (Geneva, 1561); Isaac Feguernekinus, *Enchiridii locorum communium theologicorum . . .* (Basle, 1604); Sebastian Fox Morcilla, *De naturae philosophia . . .* (Paris, 1560); Lucas Trelcatius, *Locorum communium s. theologiae institutio . . .* (London, 1608); Homer, *Poemata duo, Ilias et Odyssea . . .* (Paris, 1589); and Robert Stephan, *Dictionarium nominum propriorum . . .* (Cologne, 1576) (Harrison, *Library of Isaac Newton*).   [84] *Add MS* 3970.3, f. 456.

[85] *Keynes MS* 2. As he frequently did, Newton used both ends of the notebook, entering a set of headings in each. The end I am now considering, which I take to have come first, he foliated with Roman numerals; he paginated the other end with Arabic numerals. Although he knew the Bible thoroughly, not many other notes from Biblical studies survive. I have met two other sheets of such among his theological papers: *Yahuda MSS* 8.3, f. 3, and 14, f. 11.

use the notebook to systematize his study of the Bible; the references that he entered, the foundation of his extensive knowledge of the holy writings, came almost entirely from the Scriptures. While the list of headings appear unexceptionably orthodox, Newton's entries under them suggest that certain doctrines, which had the inherent capacity to draw him away from orthodoxy, had begun to fascinate him. In his original list, he alloted one folio to "Christi Vita" and the following one to "Christi Miraculi." When an earlier entry spilled over onto the first, he joined it to the second, and he entered nothing at all under the combined heading. He left five full folios, or ten pages, for the heading "Christi Passio, Descensus, et Resurrectio," and two folios, or four pages, for "Christi Satisfactio, & Redemptio." He filled fewer than two of the ten pages intended for the first and under one of the four pages intended for the second.[86] The heading that had spilled over its allotted two pages was "Deus Filius." Under it, Newton collected passages from the Bible that defined the relation of the Son to God the Father. From Hebrews 1 he quoted verses 8–9, which say God set Christ on His right hand, called him God, and told him that because he had loved righteousness "therefore God, even *thy God,* hath annointed thee with the oil of gladness above thy fellows." Opposite the two words he had underlined, Newton inserted a marginal note: "Therefore the Father is God of the Son [when the Son is considered] as God."[87] A later entry reinforced the implication of the note.

Concerning the subordination of Christ see Acts 2.33.36. Phil 2.9.10. 1 Pet 1.21. John 12.44. Rom 1.8 & 16. 27. Acts 10.38 & 2.22. 1 Cor 3.23, & 15.24, 28. & 11.3. 2 Cor. 22, 23.[88]

Under "Deus Pater," he had already entered half a page of references on the same topic, including three that began to sound rather pointed.

There is one God & one Mediator between God & Man ye Man Christ Jesus. 1 Tim. 2.5.

The head of every man is Christ, & ye head of ye woman is ye man, & the head of Christ is God. 1 Cor. 11.3.

He shall be great and shall be called ye son of y^e *most high.* Luke 1.32.[89]

It was Newton who underlined "most high." The reiterated implication in the two headings of a real distinction between God the Father and God the Son suggests that almost the first fruit of Newton's theological study was doubt about the status of Christ and the doctrine of the Trinity. If the approaching need for ordination had started Newton's theological reading, the reading itself started to threaten ordination.

[86] *Keynes MS* 2, ff. XIV, XV, XVII–XXI, XXII–XXIII. [87] *Keynes MS* 2, f. XII.
[88] *Keynes MS* 2, f. XIII^v. [89] *Keynes MS* 2, f. XI.

In the other end of the notebook, Newton entered a new set of headings under which he recorded notes from other theological readings, mostly from early fathers of the church. The nature of the headings (e.g., "De Trinitate," "De Athanasio," "De Arrianis et Eunomianis et Macedonianis," "De Haerisibus et Haereticis") together with a couple of citations of Fathers at the end of headings in the other end, strongly implies that this end of the notebook involved new reading undertaken to explore the questions already raised. The content of the notes exercised lasting influence over Newton's life. He drew up an index to facilitate access to them, and a number of entries in later hands demonstrate that he did return to them.[90] The convictions that solidified as he collected the notes remained unaltered until his death.

The longest entry, "De Trinitate" (Concerning the Trinity), filled nine pages. The passage was studious rather than contentious. Newton returned to the works of the men who formulated trinitarianism – Athanasius, Gregory Nazianzen, Jerome, Augustine, and others – to inform himself correctly about the doctrine. Other "Observations upon Athanasius's works," together with notes elsewhere, contributed to the same goal.[91] More than the doctrine interested him. He became fascinated with the man Athanasius and with the history of the church in the fourth century, when a passionate and bloody conflict raged between Athanasius and his followers, the founders of what became Christian orthodoxy, on the one hand, and Arius and his followers, who denied the Trinity and the status of Christ in the Godhead, on the other; and he read extensively about them.[92] Indeed, once started, Newton set himself the task of mastering the whole corpus of patristic literature. To a quotation from Justin Martyr on the names of God he added a comment: "Many others say the same thing."[93] He was in a position to know. In addition to those mentioned above, Newton cited in the notebook Irenaeus, Tertullian, Cyprian, Eusebius, Eutychius, Sulpitius Severus, Clement, Origen, Basil, John Chrysostom, Alexander of Alexandria, Epiphanius, Hilary, Theodoret, Gregory of Nyssa, Cyril of Alexandria, Leo I, Victorinus Afer, Rufinus, Manentius, Prudentius, and others. He seemed to know all the works of prolific theologians such as Augustine, Athanasius, and

---

[90] *Keynes MS 2*, pp. 18, 93, 101, and 103. Cf. a similar, later entry in the first end of the notebook, f. XIV.

[91] *Keynes MS 2*. "De Trinitate", pp. 33–6, 79–82, 89. "Observations upon Athanasius's works", pp. 13–14. Cf. *Babson MS 436*, a folded sheet, four pages in all and all filled, on Athanasius' conception of *homoousia*.

[92] *Keynes MS 2*. "De Athanasio, & Antonio", pp. 49–50. "De Arrianis et Eunomianis et Macedonianis", p. 67. Notes on the history of the church in the fourth century, pp. 51–62. A long note on the Council of Constantinople from the *Annals* of Eutychius, p. 30.

[93] *Keynes MS 2*, p. 83.

Origen.[94] In addition to the early Fathers, he read some later theology as well; he especially cited the work of the seventeenth-century French Jesuit, Denis Petau (or Petavius, as Newton knew him). He even cited one book by Herbert Thorndike, who had preceded him in his chamber in Trinity. The bulk of his citations came from the early Fathers, however. There was no single one of importance whose works he did not devour. And always, his eye was on the allied problems of the nature of Christ and the nature of God.

The conviction began to possess him that a massive fraud, which began in the fourth and fifth centuries, had perverted the legacy of the early church. Central to the fraud were the Scriptures, which Newton began to believe had been corrupted to support trinitarianism. In the notebook, he recorded doubts about a number of passages, not only 1 John 5:7 and 1 Timothy 3:16, on which he later wrote an essay called "Two Notable Corruptions of Scripture," but also a number of other passages that appeared in a further essay.[95] Since differences in ink, especially in marginal additions, indicate that Newton returned to his original notes with further thoughts, it is impossible to say exactly when the conviction fastened upon him. The original notes themselves testify to early doubts. Far from silencing the doubts, he let them possess him. "For there are three that bear record in heaven, the Father, the Word, and the Holy Ghost: and these three are one." Such is the wording of 1 John 5:7, which he read in his Bible. "It is not read thus in the Syrian Bible," Newton discovered. "Not by Ignatius, Justin, Irenaeus, Tertull. Origen, Athanas. Nazianzen Didym Chrysostom, Hilarius, Augustine, Beda, and others. Perhaps Jerome is the first who reads it thus."[96] "And without controversy great is the mystery of godliness: God was manifest in the flesh . . ." Thus 1 Timothy 3:16, in the orthodox version. The word "God" is obviously critical to the usefulness of the verse to support trinitarianism. Newton found that early versions did not contain the word but read only, "great is

[94] Among the other Fathers on whom Newton made notes, some of which came from secondary sources such as Petau, were Athenagoras, Theophilus, Tatian, Dionysius of Alexandria, Paul of Samosata, Arnobius, Lactantius, Hermas, Ignatius, Hilary, Sabellius, and Apollinarius. Cf. *Yahuda MSS* 2.5b, ff. 40–7 (a brief history of the early church); 13.1, ff. 1–4; 14, *passim;* 15.7, f. 218; and *Keynes MS 4.*

[95] "An Historical Account of Two Notable Corruptions of Scriptures" was written in the form of a letter, really two letters, one for each passage, to John Locke, 14 Nov. 1690 (*Corres 3,* 83–122). A third letter, undoubtedly also to Locke and probably composed about the same time, dealt with twenty-five other corruptions (*Corres 3,* 129–42). Newton argued that all of the corruptions inserted trinitarian sentiments into the Scriptures where the originals did not contain such. For early notes that look forward to these essays, cf. *Keynes MS 2,* ff. XII^v, XIII^v, and pp. 19–20, 34, 93, 99.

[96] *Yahuda MS 14,* f. 57^v. For early notes on this verse, cf. *Keynes MS 2,* pp. 19–20. Collations of versions of the Bible, including manuscripts, are found in *Keynes MS 2,* p. 99, and *Yahuda MS 14,* ff. 201–2.

the mystery of godliness which was manifested in the flesh." "Furthermore in the fourth and fifth centuries," he noted, "this place was not cited against the Arians."[97]

The corruptions of Scripture came relatively late. The earlier corruption of doctrine, which called for the corruption of Scripture to support it, occurred in the fourth century, when the triumph of Athanasius over Arius imposed the false doctrine of the trinity on Christianity. Central to trinitarianism was the adjective *homoousios,* which was used to assert that the Son is consubstantial *(homoousios)* with the Father. Newton tended to call the Athanasians "homousians." In an early sketch of the history of the church in the fourth century, he described how the opponents of Arius in the Council of Nicaea wanted to base their argument solely on scriptural citations as they rejected Arianism and affirmed their own convictions that the Son is the eternal uncreated *logos.* However, the debate drove them to assert that the Son is *homoousios* with the Father even though that word is not in Scripture. "That is, when the Fathers were not able to assert the position of Alexander [the Bishop of Alexandria who had charged Arius with heresy] from the scriptures, they preferred to desert the scriptures than not to condemn Arius." Eusebius of Nicomedia had introduced the word *homoousios* into the debate as a clearly heretical, intolerable consequence of the anti-Arian position.

> Thus you see these fathers took ye word not from tradition but from Eusebius's letter, in wch though he urged it as a consequence from Alexander's doctrin which he thought so far from ye sense of ye Church y^t even they themselves would not admit of it, yet they chose it for it's being opposite to Arius.[98]

Athanasius claimed that the orthodox use of the term *homoousios* did not begin with the Council of Nicaea but could be found, for example, in the writing of the third-century Father, Dionysius of Alexandria. Careful study revealed to Newton than Athanasius had deliberately distorted Dionysius to make it appear he accepted a term which in fact he considered heretical.[99] Other early Fathers had been tampered with as well. Words were "foisted in" the epistles of Ignatius, of the second-century, for example, to give them a trinitarian flavor.[100] Athanasius had also distorted the proclamation of the Council of Serdica for the same purpose.[101]

In Newton's eyes, worshipping Christ as God was idolatry, to him the fundamental sin. "Idolatria" had appeared among the original list of headings in his theological notebook.[102] The special hor-

[97] *Keynes MS* 2, f. XIII^v.   [98] *Yahuda MS* 2.5b, ff. 40^v–41.
[99] *Yahuda MS* 14, f. 83^v.   [100] *Yahuda MS* 14, f. 61^v.
[101] *Keynes MS* 2, p. 77.   [102] *Keynes MS* 2, ff. V–VI.

ror of the perversion that triumphed in the fourth century was the reversion of Christianity to idolatry after the early church had established proper worship of the one true God. "If there be no transubstantiation," he wrote in the early 1670s, "never was Pagan Idolatry so bad as the Roman, as even Jesuits sometimes confess."[103] Newton held that the pope in Rome had aided and abetted Athanasius and that the idolatrous Roman church was the direct product of Athanasius' corruption of doctrine.

In the end – and the end did not wait long – Newton convinced himself that a universal corruption of Christianity had followed the central corruption of doctrine. Concentration of ecclesiastical power in the hands of the hierarchy had replaced the polity of the early church. The perverse institution of monasticism sprang from the same source. Athanasius had patronized Anthony, and the "homousians" had introduced monks into ecclesiastical government. In the fourth century, trinitarianism fouled every element of Christianity. Though he did not say so, he obviously believed that the Protestant Reformation had not touched the seat of infection. In Cambridge of the 1670s this was strong meat indeed. It is not hard to understand why Newton became impatient with interruptions from minor diversions such as optics and mathematics. He had committed himself to a reinterpretation of the tradition central to the whole of European civilization.

Well before 1675, Newton had become an Arian in the original sense of the term. He recognized Christ as a divine mediator between God and man, who was subordinate to the Father Who created him. Christ had earned the right to be worshiped (though not with the worship suitable to the Father) by humbling himself and being obedient unto death. The man Jesus was to Newton, not a hypostatical union of divinity with human nature in one person, but the created *logos* incarnate in a human body so that he, and not man, might suffer in the flesh. For his obedience, God exalted him and raised him to sit at his right hand. Newton summarized his Arian christology in twelve points which appear to date from the period 1672–5.

1. The [word] God is no where in ye scriptures used to signify more then one of the thre persons at once.
2. The word God put absolutely without particular restriction to ye Son or Holy ghost doth always signify the Father from one end of the scriptures to ye other.
3. When ever it is said in the scriptures that there is but one God, it is meant of ye Father
4. When, after some heretiques had taken Christ for a meare

---

103 *Yahuda MS* 14, f. 9^v. Cf. other early notes on idolatry, f. 9.

man & others for the supreme God, St John in his Gospel indeavoured to state his nature so y^t men might have from thence a right apprehension of him & avoyd those haeresies & to that end calls him ye word or λογος: we must suppose that he intended that terme in ye same sence that it was taken in ye world before he used it when in like manner applied to an intelligent being. For if the Apostles had not used words as they found them how could they expect to have been rightly understood. Now the term λογος before St John wrote, was generally used in ye sense of the Platonists, when applied to an intelligent being, & ye Arrians understood it in ye same sence, & therefore theirs is the true sense of St John.

5. The son in several places confesseth his dependance on the will of the father.

6. The son confesseth ye father greater then him calls him his God, &c

7. The Son acknowledgeth the original praescience of all future things to be in ye father onely.

8. There is no where made mention of a humane soul in o^r saviour besides the word, by the mediation of wch ye word should be incarnate. But ye word it self was made flesh & took upon him ye form of a servant.

9. It was ye son of God wch he sent into ye world & not a humane soul y^t suffered for us. If there had been such a human soul in o^r Saviour, it would have been a thing of too great consequence to have been wholly omitted by the Apostles.

10. It is a proper epithete of ye father to be called almighty. For by God almighty we always understand ye Father. Yet this is not to limit the power of ye Son, For he doth what soever he seeth ye Father do; but to acknowledg y^t all power is originally in ye Father & & that ye son hath no power in him but w^t he derives from ye father for he professes that of himself he can do nothing.

11. The son in all things submits his will to ye will of the father. wch could be unreasonable if he were equall to ye father.

12. The union between him & the father he interprets to be like y^t of ye saints one wth another. That is in agreement of will & counsil. [104]

---

[104] *Yahuda MS* 14, f. 25. The bracket in the first line fills in a space obliterated by physical damage to the top of the sheet. Newton entered the number for a thirteenth point, but he did not write anything after it. Cf. Newton's comments following notes on Lactantius' opinion of the relation of Christ to God the Father:

In another manuscript from the same time, Newton spoke of Christ's "glory and exhaltation/dominion . . . wch he acquired by his Death . . . He was a son before his incarnation but now he was made heir . . . for Deity & worship are relative terms. . . . ἴσα Θεῷ [as a God] I understand . . . not of his congenit divinity but of this exhaltation to honour & dominion after his death . . ."[105] Nearly a century later, in support of the truth of Christianity, Dr. Johnson argued that "Sir Isaac Newton set out an infidel, and came to be a very firm believer."[106] Dr. Johnson was misinformed about Newton's starting point. He was not mistaken when he said that Newton became a very firm believer. He would have been shocked to learn what Newton believed.

An incidental acquisition by Newton, which came with his Arianism, was a particular conception of God. For Arius, the need for a mediator between God and man, a mediator different in essence from God, arose from the ineffable nature of God, the absolute Lord of creation, who was too far removed from His creatures to

"Note 1 That to say there are two Gods, would supose them collaterall & univocall 2 That as in a family ye tile of Master is to be understood of ye supreme master unless when by some circumstance it is limited to ye son or other subordinate master soe ye title of God is to be understood of ye supreme God, unless when it is limited to ye son or holy ghost. 3 That to say there is but one God, ye father of all things, excludes not the son & Holy ghost from ye Godhead becaus they are virtually conteined & implied in the father. 4 To apply ye name of God to ye Son or holy ghost as distinct persons from the father makes them not divers Gods from ye Father becaus the divinity of ye son & holy ghost is derived from y^t of ye father. To make this plainer suppose a, b & c are 3 bodies of wch a hath gravity originally in it self by wch it presseth upon b & c wch are w^thout any originall gravity but yet by ye pressure of a communicated to y^m do presse downwards as much as A doth. Then there would be force in a, force in b & force in c, & yet they are not thre forces but one force wch is originally in a & by communication/descent in b & c Soe there is divinity in ye Father, divinity in ye Son, & divinity in ye holy ghost, & yet there are not 3 divinities but one divinity wch is originally in ye father & by descent or communication in ye son & holy ghost. And as in saying there is but one force, that in body a I deprive not ye body b & c of that force wch they derive from a so by saying there is but one god, the father of all things, I deprive not ye son & holy ghost of the divinity wch they derive from ye father &c." (*Yahuda MS* 14, ff. 173–3^v).

Cf. also Newton's early sketch of a history of the church, that is, primarily a history of the Arian controversy in the fourth century (*Yahuda MS* 2.5b, ff. 40–7). The "Homousians," he said, deny a true and full generation of the Son, "both in denying that the substance of the Son was generated and in asserting that his existence is by necessity and not an effect that originated from and depends on the will of the father" (f. 43^v.) For further discussion of Newton's theological stance see Herbert McLachlan, *The Religious Opinions of Milton, Locke, and Newton* (Manchester, 1941).

105 *Keynes MS* 2, p. 15. Cf. *Yahuda MS* 14, f. 84^v, and places in his early writings on Revelation such as *Yahuda MSS* 1.4, ff. 156–63, and 10.3, f. 5. In the first of these last two, Newton cited Hebrews 2:10, underlining the words, that Christ was made "perfect through suffering" (f. 159).

106 James Boswell, *The Life of Samuel Johnson, L.L.D.* (New York, n.d.), p. 274.

be approached by them. In his reading about Platonic theologians of the third century, on whom Arius drew, Newton met expositions of such a God. God, said Clement of Alexandria, is "not divided, not disjointed, not moving from place to place, nor in any way circumscribed but existing everywhere always; all mind, all paternal light, all eye, seeing everything, hearing everything, knowing everything, examining powers by his power."[107] Novitian considered God to be "immense and without limit, not one who is enclosed in a place but one who encloses every place, not one who is in a place but rather in whom every place is, one containing everything and clasping it together, so that accordingly he neither ascends or descends since he himself contains and fills everything."[108] Newton may have been influenced by the Cambridge Platonists, but he also went beyond them directly to the sources on which they drew.

It is useful to set Newton's behavior in the early 1670s against the background of his Arianism. He identified himself with Arius, both intellectually and emotionally. He relived the terrible struggles of the fourth century, when doctrine counted for more than charity, came to see Athanasius as his personal nemesis, and learned to hate him fiercely. When questions, which look legitimate to us, about his theory of colors seemed to drive him frantic, the pattern that disagreement took in the fourth century may have determined his conduct. He wished to avoid controversy at all costs. On the most important question of all, he had to avoid controversy.

His new convictions probably influenced his relation to Cambridge as well. Whatever the factors in his personality and position that made for isolation, his heretical convictions in a society of pliant orthodoxy operated far more powerfully to the same end. Cambridge was tolerance itself in regard to performance; it did not extend tolerance to belief. In 1669, a fellow of Caius, Daniel Scargill, was expelled from the university "for asserting impious and atheistical tenets"; he had to undergo a humiliating public recantation to regain admission. In 1675, Samuel Rolls was admitted to the degree of Doctor of Physic only when he similarly disowned and disavowed everything against the Church of England in a book he had published.[109] One needs to remember that Newton's patron, Barrow, composed a *Defense of the Blessed Trinity,* and that Barrow's successor as master of the college let it be known he intended "to batter the atheists and then the Arians and Socinians."[110] Since

[107] *Keynes MS* 4, f. 14.   [108] *Keynes MS* 4, f. 41.

[109] Charles Henry Cooper, *Annals of Cambridge,* 5 vols. (Cambridge, 1842–1908), *3,* 532, 570.

[110] Roger North, *The Lives of the Right Hon. Francis North, Baron Guilford; the Hon. Sir Dudley North; and the Hon. and Rev. Dr. John North. Together with the Autobiography of the Author,* ed. Augustus Jessopp, 3 vols (London, 1890), *2,* 310, 312.

any discussion was fraught with the danger of ruin, Newton chose silence. Significantly, with one exception, none of his theological papers appear in the hand of Wickins. The one exception, an anti-Catholic interpretation of Revelation that Wickins copied for him, was not such as to raise doubts about his orthodoxy.[111] There is no evidence to imply that Wickins ever suspected the transformation taking place under his eyes. Newton concealed his views so effectively that only in our day has full knowledge of them become available.

The seventh point in his Arian credo, that only the Father has foreknowledge of future events, indicated another dimension of Newton's early theological studies, the interpretation of the prophecies. Newton's interest in the prophecies, Daniel and the Revelation of Saint John the Divine, has been known since the publication of his *Observations upon the Prophecies* shortly after his death. It has generally been assumed that the work was a product of his old age, as the treatise published was. Nevertheless, references to the prophecies filled his early theological notebook.[112] Already in the 1670s, he believed that the essence of the Bible was the prophecy of human history rather than the revelation of truths beyond human reason unto life eternal. Already at that time he believed what he asserted later about Revelation: "There [is] no book in all the scriptures so much recommended & guarded by providence as this."[113] He put that belief into practice by composing his first interpretation of Revelation while engaged in his earliest theological study.[114] It proved to be more than a passing interest. His first full discourse contains many insertions in later hands, showing that he referred

---

[111] *Yahuda MS* 23.

[112] *Keynes MS* 2, ff. I, XIV–XVI, XXVII–XXVIII; pp. 17, 21–2, 41–2, 43–4, 45, 65. The last page contains notes under the heading, "Ecclesiastical History after the Time of Theodosius"; the notes are about an invasion by Goths, an episode intimately related to Newton's interpretation of the trumpets. In his *Observations upon the Prophecies*, Newton insisted that no other book of the New Testament was commented on so much in the early period of Christianity, and he cited a large number of Fathers to support his assertion (*Observations upon the Prophecies of Daniel, and the Apocalypse of St. John* [London, 1733], pp. 246–9). Among his papers are many notes on these early commentaries: *Yahuda MSS* 2.4b, f. 33; 13.1, ff. 4–4^v; 14, f. 17^v, 85, 173^v–4.

[113] *Yahuda MS* 7.2i, f. 4.

[114] What appears to me to have been the earliest sketch of his interpretation, as opposed to notes such as those in *Keynes MS* 2, is found in *Yahuda MS* 10.3, ff. 14–29. It has no date, but the hand appears to belong to the mid-1670s. The draft of a letter to Oldenburg of January 1675, on his expected loss of his fellowship, has material on the interpretation of Revelation that is very similar to his early full-scale treatise in *Yahuda MS* 1. Since Oldenburg brought the letter up at the 28 Jan. 1675 meeting of the Council of the Royal Society (Thomas Birch, *The History of the Royal Society of London*, 4 vols. (London, 1756–7), *3*, 178), the draft can be dated with assurance. It provides the only solid date for his early treatise of which I know.

back to it frequently.[115] He composed numerous revisions of it, one of which was probably the last thing on which he was at work when he died more than fifty years later.

An introduction, which insisted on the critical importance of the prophecies, opened the original treatise.

> Having searched after knowledg in ye prophetique scriptures [he began], I have thought my self bound to communicate it for the benefit of others, remembering ye judgment of him who hid his talent in a napkin. For I am perswaded that this will prove of great benefit to those who think it not enough for a sincere Christian to sit down contented with ye principles of ye doctrin of Christ such as ye Apostel accounts the doctrin of Baptisms & of laying on of hands & of the resurrection of ye dead & of eternall judgment, but leaving these & the like principles desire to go on unto perfection until they become of full age & by reason of use have their senses exercised to discern both good & evil. Hebr. 5.12

People should not be discouraged by past failures to understand these writings. God vouchsafed the prophecies for the edification of the church. They are not about the past. They were written for future ages. When the time comes, their meaning will be revealed in their fulfillment. Let people be warned by the example of the Jews, who have paid dearly for failing to recognize the promised Messiah. If God was angry with the Jews, He will be more angry with Christians who fail to recognize Antichrist. "Thou seest therefore that this is no idle speculation, no matter of indifferency but a duty of the greatest moment." One cannot be too careful. Antichrist comes to seduce Christians, and in a world of many religions, of which only one can be true "& perhaps none of those that thou art acquainted with," one must be circumspect in finding the truth.[116] Already we perceive that Newton's interpretation of the prophecies was not unconnected with his Arianism.

In fact, Arianism – or perhaps its victorious opponent, trinitarianism – provided the key to Newton's interpretation. Interest in the prophecies had been rife in Puritan England, and a standard Protestant interpretation of Revelation, in which the Roman church inevitably played the role of the Beast, had emerged. The appearance among Newton's papers of a manuscript treatise, copied by Wickins, on the mystery of iniquity (the papacy, of course) testifies to the vitality of prophetic studies in the England of his day.[117] New-

[115] E.g., *Yahuda MS* 1.1, ff. 2, 28^v, 34, 46, 52.

[116] *Yahuda MS* 1.1, ff. 1–10. The introduction is published in full in Frank E. Manuel, *The Religion of Isaac Newton* (Oxford, 1974), pp. 107–116.

[117] *Yahuda MS* 23. In his preface to Joseph Mede's *Apostacy of the Latter Times*, William Twisse mentioned that he had received a number of manuscript treatises on the subject (second edition, London, 1644).

ton met the Protestant interpretation primarily in the writings of Joseph Mede and Henry More, whose contribution to the understanding of the prophecies he acknowledged. Central to their interpretation was what they called the great apostasy, in which Christendom abandoned the true faith to embrace a perversion of it. Needless to say, the great apostasy was Roman Catholicism, and they dated its triumph within the church somewhere around the reign of Theodosius and the year 400 (Figure 8.5). From this starting point, the prophecies provided an internal chronology. Different visions mentioned periods of 1,260 days, 42 months, and three and a half years (literally a time times and half a time, taken to mean a year, two years, and half a year). By Mede's principle of synchronism, these visions were all taken to refer to the same period of time, 1,260 years, the period during which the great apostasy would reign. Revelation promised the destruction of Antichrist at the end of the period. Simple arithmetic supplied the conclusion that the restoration of true apostolic Christianity would come in the mid- or late seventeenth century, the exact date depending on the year taken for the beginning of the apostasy. More especially was content to believe that the reign of the Beast had indeed ended. With the Anglican Church pure apostolic Christianity was restored and would rapidly triumph everywhere.

Newton was not prepared to identify the established Church of England with pure apostolic Christianity. Accepting the broad outlines of the Protestant interpretation, he altered the meaning of the great apostasy to fit his new perception of Christianity. Over the years, Newton continually revised his interpretation, and with each revision made its thrust more obscure. In the version ultimately published, a melding together of two different manuscripts composed during his old age, a couple of references to the great apostasy appeared without anything that even hinted at the new meaning he attached to the phrase. Indeed, it is quite impossible to find any point to the rambling chronologies that compose Newton's published *Observations upon the Prophecies*. There is no similar problem with the manuscript of the 1670s. The great apostasy was trinitarianism.

Newton's interpretation differed from most contemporary ones in another way. Most students of the prophecies looked to them for understanding of contemporary events.[118] Newton never did. Over the years he exhibited some, though not intense, interest in the date of the second coming, which he never placed in or near the late seventeenth century. Not once that I have seen did he attempt to

---

[118] See the discussion in Margaret C. Jacob, *The Newtonians and the English Revolution, 1689–1720* (Ithaca, 1976), chap. 3, The Millennium, pp. 100–42.

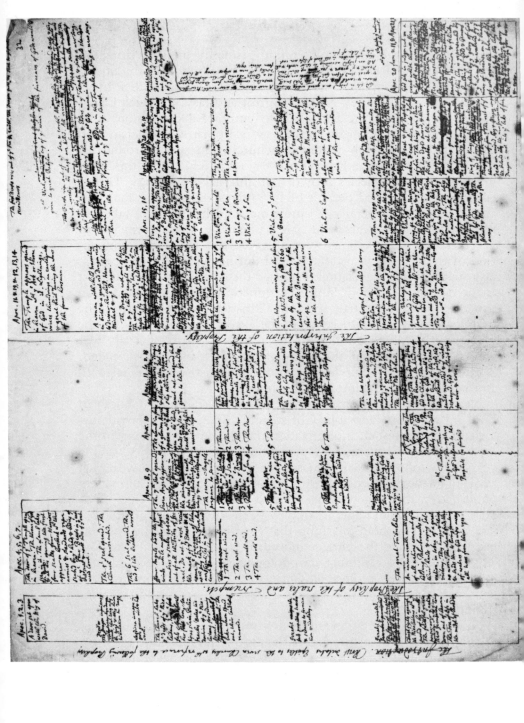

relate the political history of England in his day to the prophecies. Rather, he riveted his attention on the fourth century, the crucial century of human history, when the Great Apostasy seduced mankind from worship of the one true God.

In Revelation, as Newton understood it, the opening of the first six seals, which represent successive periods of time, concern themselves with the history of the church until its definitive establishment within the empire in the reign of Theodosius. The seventh seal, within which the seven trumpets (which also represent successive periods of time) are included, began with the year 380. Until its conclusion at the sounding of the seventh trumpet, it portrays "one & the same continued Apostasy . . . And indeed so notable are the times of this Apostasy y^t ye whole Apocalyps from ye fourth chapter seems to have been written for ye sake of it."[119] Until then, trinitarian doctrines, though formulated by Athanasius, had been professed only by a few western bishops led by the Pope. At that time, however, Theodosius became its patron and called the Council of Constantinople in 381 to ratify it.

> The year 381 is therefore wthout all controversy that in wch this strange religion of ye west wch has reigned ever since first overspread ye world, & so ye earth wth them that dwell therein began to worship ye Beast & his Image, y^t is ye church of ye western Empire & the afforesaid Constantinopolitan Counsel its representative . . .[120]

The mere thought of trinitarianism, the "fals infernal religion," was enough to fan Newton into a rage. With it had come the return of idolatry in a more degraded form, not the stately worship of dead kings and heroes in magnificent temples, but "ye sordid worship in sepulchres of ye Christian Divi . . . ye adoration of mean & despicable plebeians in their rotten reliques." Superstitions of every sort, fanned and spread by monks with feigned stories of false miracles, accompanied the new worship.[121] "Idolaters," Newton thundered at them in the isolation of his chamber, "Blasphemers & spiritual fornicators . . ." They pretended to be Christians, but the devil knew "that they were to be above all others ye most wicked wretched sort of people. . . . the worst sort of men that ever

---

[119] *Yahuda MS* 1.3, ff. 57-9.
[120] *Yahuda MS* 1.4, f. 50. *Yahuda MS* 1.3 and 1.2 are two successive drafts (in that order) of Newton's interpretation presented in broad outline in the form of ten general propositions. He filled in the details elsewhere.   [121] *Yahuda MSS* 1.2, f. 23; 1.5, f. 6; 1.4, f. 52.

Figure 8.5. Newton's schematic interpretation of the Book of Revelation. Successive repetitions of the prophecy refer to the same historical events with time measured along the vertical axis. The first full horizontal line (from the top) represents the beginning of the reign of Theodosius. The last horizontal line at the bottom is the Day of Judgment. (Courtesy of the Jewish National and University Library.)

reigned upon the face of yᵉ earth till that same time . . ."[122] The first six trumpets and the six vials of wrath corresponding to them represent successive invasions of the Empire – "like Furies sent in by the wrath of God to scourge yᵉ Romans," repeated punishments of an apostate people who whored after false gods.[123]

As the passion with which Newton expressed himself suggests, his early treatise on the prophecies was a very personal document. In his view, the triumph of trinitarianism had stretched beyond the limits of doctrine. It had won dominance by allying itself with base human motives, such as "covetousness & ambition . . ."

> It's plain therefore that not a few irregular persons, but yᵉ whole clergy began at this time to be puft up, to set their hearts upon power & greatness more then upon piety & equity, to transgress their Pastoral office & exalt themselves above yᵉ civil magistrate; not regarding how they came by praerogatives or of what ill nature or consequence they were, so they were but praerogatives, nor knowing any bounds to their ambition but impossibility & yᵉ Imperial edicts.[124]

Behind the specifically anti-Catholic element in Newton's denunciation, one can see a much broader condemnation. Restoration Cambridge supplied a more familiar example of covetousness and ambition in the church. Already Newton had separated himself in his daily life from the aspiring clerics about him. By embracing Arianism, he had expressed his scorn for their belief. Now, in his prophetic studies, he justified his rebellion by appealing to the divinely ordained course of human history. Men will call you a fanatic and heretic if you study the prophecies, he warned.

> But yᵉ world loves to be deceived, they will not understand, they never consider equally, but are wholly led by prejudice, interest, the prais of men, & authority of yᵉ Church they live in: as is plain becaus all parties keep close to yᵉ Religion they have been brought up in, & yet in all parties there are wise & learned as well as fools & ignorant. There are but few that seek to understand the religion they profess, & those that study for understanding therein, do it rather for worldly ends, or that they may defend it, then to examin whither it be true wᵗʰ a resolution to chose & profess that religion wᶜʰ in their judgment appears the truest. . . . Be not therefore scandalised at the re-

---

[122] *Yahuda MS* 1.4, ff. 67–8. Cf. the whole passage, ff. 68–84, in which Newton built a case for the special depravity of the generation following Theodosius. Not the least interesting aspect of this passage is formed by the extensive additions to it on the verso sides. I gain the impression that Newton pored over possible sources looking for particulars to add to his indictment of the age.

[123] *Yahuda MS* 1.4, f. 127. Newton devoted very extensive attention to the period of the barbarian invasions. Cf. *Yahuda MSS* 1.4, 1.5, 1.6, 1.7, 1.8, 10.3, 14 (ff. 34–46), 28.1, 28.2, 28.3, 39, and *Keynes MS* 1.   [124] *Yahuda MS* 1.4, f. 92.

proaches of ye world but rather looke upon them as a mark of ye true church.[125] By the true church, for whom the prophecies are intended, he did not mean all who call themselves Christians, "but a remnant, a few scattered persons which God hath chosen, such as without being led by interest, education, or humane authorities, can set themselves sincerely & earnestly to search after truth."[126] There is no doubt that Newton placed himself among the select few. Some of his descriptions of the remnant have the poignancy of personal experience. Consider, he said in the earliest sketch of his interpretation, "ye Apostacies of ye Jewish Church under ye Law & particularly in Ahab's reign when ye undefiled part of ye Church so far disappeared y^t ye Elijah thought himself ye only person left."[127] Isolated in his chambers from the hedonism and triviality of Restoration Cambridge, Newton may have wondered if he was another Elijah, like the first almost the only true believer left.

A tension ran through the work. On the one hand, it contained a chiliastic flavor. Now at last the meaning of the prophecies was being revealed and therefore the end had to be near, when the sealed saints who refused the mark of the Beast would resume their place at the head of the church. On the other hand, Newton was far from ready to identify anything about him as true apostolic Christianity. His deliberately chosen internal chronology set the day of the final trumpet two centuries away. About this, he was explicit. The beginning of the critical period of 1,260 years had come, not in 380 with the beginning of the seventh seal, but at the close of the fourth trumpet, when the apostasy reached its peak, in the year 607.[128] Whatever Newton's contempt for the society about him, perhaps because of his contempt, he did not think its conversion was near.

There is a further tension between his own passion and his dispassionate method. It is hardly surprising that a man whose first step in any new study was to organize his knowledge methodically should want to proceed similarly with Revelation. He complained of interpreters who, "w^{th}out any such previous methodising of ye Apocalypse . . . twist ye parts of ye Prophesy out of their natural order according to their pleasure. . ."[129] Though he admitted the validity of mystical and allegorical meanings, he insisted that they be used with caution. "Too much liberty in this kind savours of a luxuriant ungovernable fansy & borders on enthusiasm. . . . He that without better grounds then his private opinion or the opinion of any human authority whatsoever shall turn scripture from the

---

[125] *Yahuda MS* 1.1, ff. 5–6.   [126] *Yahuda MS* 1.1, f. 1.   [127] *Yahuda MS* 10.3, f. 27^v.
[128] *Yahuda MSS* 1.2, ff. 60–1; 1.3, ff. 40–8. A chronology based on an event connected with the sixth trumpet implied a slightly different date, 1844, for the seventh (*Yahuda MS* 1.7, ff. 65–6).   [129] *Yahuda MS* 1.1, f. 8.

plain meaning to an Allegory or to any other less naturall sense declares thereby that he reposes more trust in his own imaginations or in that human authority then in the Scripture."[130] In place of private fancy, Newton wanted certainty, for only then could the Bible be a clear rule of faith. His "Method" then would be the key to his approach. First, he would lay down rules of interpretation. Then he would give a key to the prophetic language which would remove the liberty of twisting passages to private meanings. Third, he would compare the parts of Revelation with each other and digest them into order by the internal characters imprinted on them by the Holy Ghost, what he called opening scripture by scripture. The harmony of different visions with each other was a principle of great importance to Newton. "For the parts of Prophesy are like the separated parts of a Watch," he later said. "They appear confused & must be compared & put together before they can be usefull, & those parts are certainly to be put together wch fit without straining."[131] One cannot miss the fact that this interpreter of the prophecies had trained his mind in the hard school of mathematics. It was not by accident that he proceeded in his exposition in a mathematical style, beginning with ten general "Propositions" (later renamed "Positions") followed by a larger number of particular ones, which drew upon them to build a demonstrative argument. Newton believed he had attained the certainty he sought. The original opening sentence of his treatise began: "Having searched & by the grace of God obtained knowledg . . ."[132]

As their introduction proclaimed, Newton's fifteen rules intended to move the interpretation of the prophecies from the world of private fantasy into the realm of public discourse: Assign only one meaning to one passage. Keep a single meaning for words within any one vision. Prefer meanings that are closest to the literal sense of the words except where an allegory is clearly demanded. Accept that meaning which follows most naturally from the use and propriety of language and context. Prefer interpretations that, without straining, reduce things to the greatest simplicity. Interpretations must apply to the most important events of an age, not to obscure ones.[133] These were not the rules of an enthusiast in direct communion with God, but the rules of a sober man who regarded the prophecies as objective revelations of a God Who does not trifle. "Truth is ever to be found in simplicity . . . ," he said in his ninth

---

[130] *Yahuda MS* 1.1, ff. 12^v–13

[131] *Yahuda MS* 7.2j, f. 114^v. This statement appears to date from the period 1705–10, but the general principle on which it rests, called synchronism by Mede, governed Newton's approach from the beginning.   [132] *Yahuda MS* 1.1, f. 1.

[133] *Yahuda MS* 1.1, f. 12–19. The rules are published in Manuel, *Religion*, pp. 116–25.

rule. Thus it is in natural philosophy. So it is with the prophecies. "It is ye perfection of God's works that they are all done wth ye greatest simplicity. He is ye God of order & not of confusion."[134] Following Mede, Newton held that there had been a prophetic language common to the ancient Near East. Hence one could draw, not only upon the other books of the Bible, but also on sources beyond it, to interpret the meaning of images. He chose, he said, "to rely rather upon ye traditions of ye ancient Interpreters then upon ye suggestions of private Phantasy . . ."[135] From Mede, he learned about an Arab, Achmet, who had gathered the interpretations of dreams from the ancient monuments of the Egyptians, Persians, and Indians. He also read the work of the Hellenistic Greek, Artemidorus, on the interpretation of dreams.[136] Above all, he combed the Bible for other instances of figures and types. From the prophecies, he drew up a catalogue of seventy "Prophetic figures," and in a following section, called "The Proof," he marshaled evidence for their meaning.[137] One example will suffice to indicate the rigor of his procedure.

> Some have supposed that ye Ballance in ye third Seale (Rev. 6) might be an emblem of famin, but wthout ground, there being no authority either in scripture or other authentic writings, that I have met wth, for such an interpretation. There are other ways of expressing famin, as by ye tearing of doggs Jer: 15. 3: & where ye Ballance is mentioned in scripture it is either wth respect to judgment as in Job. 31.6, Psal: 62.9. Dan: 5.27. or to Justice as in Hosea 12.7. Micah 6.11 Prov: 16.11. &c. And to ye same purpose is ye doctrine of ye Indian interpreters . . .[138]

A lengthy citation from Achmet established the Indian interpreters' use of the figure.

Other problems of rigor arose. It was Newton's intention to establish the exact correlation of prophecy and history. He needed to be certain of the content of prophecy, but already he had begun to doubt the authenticity of the received Scriptures. He began to compare different versions of Revelation. In the end, he collated about twenty different ones, plus at least two manuscript versions owned by John Covel of Christ's College (now in the Harleian collection in the British Library) and individual passages as he could find them in ancient commentators such as Cyprian, Irenaeus, and

134 *Yahuda MS* 1.1, f. 14.
135 *Yahuda MS* 1.1, f. 2bis. Cf. f. 28.   136 *Yahuda MS* 14, ff. 78–80.
137 *Yahuda MS* 1.1, ff. 24–7, 20–3, two successive versions of what he called first "Definitions" but retitled "Prophetic figures." Ff. 1–31bis, 28–55 contain two successive versions of "The Proof." *Yahuda MS* 2.1, ff. 1–10, are part of a later draft of "The Proof."
138 *Yahuda MS* 1.1, f. 48.

Tertullian.[139] Rather early, he composed a "Variantes Lectiones Apocalypticae" (Variant Apocalyptic Readings) which proceeded through Revelation verse by verse, indicating variant readings from his many sources.[140] The student of Newton quickly learns not to be surprised. Newton tossed off this massive piece of Biblical scholarship as a mere incident on his way to vaster projects.

The text of the Bible was not enough. To vindicate the dominion of God, he must demonstrate as well that the facts of history have corresponded to the words of prophecy. There were modern historians on whom he could draw – Cesare Baronio, Philipp Clüver, and Gilles Bouchier – and Newton did consult them. It was not his way to be satisfied with such authorities, however. Rather he had to return to the original sources and to reconstruct the order of events in the Christian era himself. What Newton did, he did with intensity. As his study of theology had led him through the whole corpus of patristic literature, so now he ransacked the ancient historians and chroniclers, pagan and Christian alike, plus orations, letters, the Theodosian Code, anything that could help him establish the order of events. For example, the second trumpet corresponded to the invasion of the Western Empire. It was essential to his chronology that it occurred in 408, and he cited Orosius, Prosper, and Marcellinus, who related the invasion to other, dated events. Actually, Prosper placed the invasion a year earlier as did Cassiodorus, but Newton cited Jornandes, Orosius, and Zosimus to prove that Stilicho, whose death (it was generally agreed) occurred soon after the invasion began, died in 408. Moreover, Zosimus related that in 407 Stilicho was preparing a military action in Illyria, something he would hardly have done had a barbarian invasion been in progress. Hence Newton felt justified in placing the invasion in 408.[141] In describing the barbarian kingdoms eventually set up in the West, he quoted a passage from Gregory of Tours that mentioned Resplendial, "Rex Alanorum" (king of the Alans). A note in the margin said that in some editions of Gregory, Resplendial was called "Rex Alemannorum" (king of the Alemans), but the edition of Bladius of 1512, based on the oldest manuscript, called him Rex Alanorum, and the testimony of others such as Sigebertus, Trithemius, the

---

[139] In a manuscript that can be dated precisely, by the draft of a letter, to the summer of 1680, Newton composed a critical commentary on the manuscript sources of Revelation and drew up three schemes of symbols to refer to the different versions. One of the lists had twenty-one symbols. What appears to be the final list had fourteen (*Yahuda MS* 2.5c, ff. 1–2). In his manuscript of variant readings, which seems to come from the early 1670s, he referred to quite a few other versions not present in any of these lists; I have not tried to go through the manuscript to enumerate them all.

[140] *Yahuda MSS* 4.1 and 4.2 contain two successive versions. The second is thirty-two folios long, with a good half of the verso sides filled. *Yahuda MS* 7.2j, ff. 91–4, have two lists of further variant readings.   [141] *Yahuda MS* 1.4, f. 140^v.

author of the *Annales Boiorum,* Vasaeus, and Bucher confirmed that title.[142] In one short passage of ten pages, in which Newton was concerned to establish that peace broke out for a short period beginning in 380, he cited Zosimus, Theodoret, Cedrenus, Baronio, Marcellinus, Ammianus Marcellinus, Socrates (the historian), Sozomen, Prudentius, Symmachus, Paulinus, Gregory Nazianzen (two works), Jornandes, Sigonius, Prosper, Mede, Rufinus, Petrus Diaconus, Platina, Idatius, Victor, Gregory the Presbyter, Claudianus, Orosius, Cassiodorus, Pacatus, Jerome's addition to Eusebius' *History of the Church* and two of his letters, the Alexandrian Chronicle, the Theodosian Code, and Gothofredus' commentaries on the code.[143]

No one, I think, would call Newton a great historian. He approached history with an *a priori* pattern of interpretation, and he produced indigestible catenae of quotations instead of readable narratives. His goal was rigor rather than *belles lettres,* however, and I suspect that no one would sneer at him on that score. He brought the standards of scientific demonstration to historical research. He pursued evidence relentlessly. I seriously doubt that any historian has ever attained a firmer grasp on the facts relating to the barbarian invasions of the fifth and sixth centuries. To Newton, the correspondence of prophecy with fact demonstrated the dominion of God, a dominion exercised over human history even as it is exercised over the natural world. For him, the dominion of God was primary, more important than compassion and love. Hence he considered the prophecies to be the very heart of divine revelation. He found symbolic meaning in the scene in Revelation in which the Lamb is given a book by him who sits on the throne in the court of heaven. The book, Newton argued, is "y^t plenary revelation wch ye great God imparted to o^r Saviour after his resurrection & to none but him. . . . this book signifies one of ye greatest treasures that he who sat upon ye Throne ever conferred upon ye Lamb, & consequently nothing less then all that fulnes of knowledg of things past & to come wch God gave him after his resurrection."[144] By placing the means of salvation below the prophecies in the scriptures, Newton found another point of contact with his Arianism.

Even the established Christian view of history underwent alteration, with a shift of emphasis from the first to the second coming of Christ. In his early treatise on Revelation, Newton did not devote much time to describing the final end. Later he would do so at some length. Nevertheless, he said enough to make it clear that for him, as for his contemporary expositors, the apocalyptic had van-

---

[142] *Yahuda MS* 1.5, f. 17.   [143] *Yahuda MS* 1.4, ff. 33–42.
[144] *Yahuda MS* 1.4, ff. 156–7.

ished from the Apocalypse. Though he used scriptural phrases, such as the "end of the world," he imposed on them a theological meaning that contrasted the world with the spirit. Thus the second coming of Christ would bring the final triumph of Christianity rather than a cataclysmic destruction of the physical world. He spoke of it in terms of "ye establishment of true religion," and "the preaching of ye everlasting gospel to every nation & tongue & kindred & people . . ."[145] The great apostasy would endure until the seventh trumpet

> at wch time ceases & ye mystery of God is finished (Apoc 10.6, 7) & ye Kingdoms of ye world become ye kingdoms of Christ for ever & ye dead are judged & saints rewarded (chap 11.15, 18) & ye Word of God comes to destroy ye wicked wth ye sword of his mouth & cast ye Beast & fals Prophet into ye Lake of fire (chap 19) wch time is ye great day of God Almighty even The day wherein o^r Saviour comes as a Thief (chap 16.14, 15.)[146]

One historian of the ancient church, in commenting on its expectation of the second coming, has spoken of the tension between already and not yet. With the institutionalization of Christianity in the church, that tension was resolved on the side of already. The historic coming of Christ, the injection of the divine into human life, was the climactic event of human history. The church embodied its meaning. To Newton, the Arian, the first coming of Christ could never assume such significance. In his version of the Christian view of history, the emphasis fell on not yet. Only with the second coming will there be a final conversion of the kingdoms of this world to the kingdom of Christ forever.

All that was two hundred years away. There was a pressing problem, however, that could not wait two hundred years. In 1675, Newton would have to be ordained in the Anglican church or he would have to resign his fellowship. In the general laxity of Trinity, the requirement of ordination was one rule which was enforced. In 1661 and 1666, the college ejected three fellows in all who had not been ordained.[147] A society of would-be clerics intent on preferment and constrained by the principle of seniority did not allow the ladder all must climb to be clogged with non-clerics who could hold their fellowships forever. Four times within the previous decade Newton had been willing to assert his orthodoxy under oath. When he signed for his degrees in 1665 and 1668, he affirmed

---

[145] *Yahuda MS* 1.3, ff. 55–6. Cf. *Yahuda MSS* 1.3, ff. 38–48; 1.4, ff. 1–4.
[146] *Yahuda MS* 1.2, f. 30^v.
[147] Master's Old Conclusion Book, 1607–1673, p. 269. Conclusion Book, 1646–1811, p. 106.

his belief in the Thirty-nine Articles of the church.[148] When he became a fellow in 1668, he swore to embrace the true religion of Christ with all his soul, in a context that clearly implied the Anglican definition of true religion.[149] Finally, in 1669, when he took up the Lucasian professorship, he swore that he would conform to the liturgy of the Church of England.[150] To remain a fellow of the College of the Holy and Undivided Trinity, he would need to affirm his orthodoxy one last time, in ordination. By 1675, however, the holy and undivided Trinity itself stood in the way. "If any shall worship the Beast and his image and receive his mark in his forehead or in his hand," Revelation told him, "the same shall drink of the wine of the wrath of God . . ." Newton did not doubt the literal truth of the Word. Accept ordination he could not.

Late in the eighteenth century, a story that had belonged to the lore of the Uvedale family appeared in *Gentleman's Magazine*. It concerned Newton, another fellow of Trinity, Robert Uvedale, and one of the two exempted fellowships in the college. The death of Robert Crane in February 1673 had vacated the law fellowship; and according to the Uvedale family tradition, both Newton and Uvedale had tried to get it. Uvedale was two years Newton's senior. The rule of seniority was adamant. Uvedale got the exempted fellowship.[151] The credentials of the story are questionable, resting as they do on nothing more than a family tradition then over a hundred years old and passed down through three generations to a great-grandson of Robert Uvedale. Nevertheless, the story is plausible, especially in the light of later events. Once he became heretical, one of the exempted fellowships appeared to be Newton's

[148] Cambridge University Library, Subscriptiones, II, pp. 163, 243. When he signed for his Bachelor's degree early in 1665, his pen must have been very full. When the page was turned, Newton's signature made a huge blot that virtually obliterated it. Lest one find the blot deliberate and symbolic, he did not repeat it three years later when he signed for his M.A.

[149] Trinity Statutes, *Fourth Report from the Select Committee on Education* (London, 1818), p. 373.

[150] The oath, imposed by the Act of Uniformity of 1662 on every master and head, fellow, chaplain, and tutor of a college, and every public professor and reader in the university, is in Cooper, *Annals, 3,* 499.

[151] *Edleston,* pp. xlviii–xlix. Edleston regarded the story as incompatible with the two men's sense of obligation. The story conforms to established practice in the college, however. A decade later, for example, Charles Montague, who would figure in Newton's life, was appointed to the other exempted fellowship, reserved for physic, although he did not study medicine. The fact is that Uvedale, who did not pursue law, did accept the fellowship in question. Though nonresident, he was thus enabled to enjoy its income for another six years until marriage looked more attractive to him than the annual dividend. The mere fact that it was a fellowship ostensibly devoted to law does not in any way imply that Newton intended to study law. Cf. his own remarks on accepted practice in regard to the two exempted fellowships in his letter to Sir Alexander Frazier a year later (*Corres 3,* 146–7).

only hope to remain in his fellowship. If the story is true, as I believe it is, it implies that he had become an Arian by February 1673.

More than the fellowship was at stake. If Newton despised the society of Trinity, the material support that the college provided, in a location which assured his access to the world of learning, was the bedrock of his existence. Perhaps he could have held his chair without the fellowship and remained in Cambridge, although I do not know of such another case. The problem was secrecy. Questions were bound to be asked. In itself, ordination did not entail duties. It did not entail ecclesiastical appointment. Uvedale's case differed. He merely wanted to retain the income while he pursued a career elsewhere. No one questioned greed as an honorable motive. But why would anyone in Newton's position, someone who intended to stay on in Cambridge, celibate, surrender a fellowship worth sixty pounds a year for no reason at all? Or rather, what was one to conclude about the true reason that led a man in such a position to refuse ordination? Questions were bound to be asked. Questions were exactly what Newton had to avoid. Heresy was grounds for ejection from his chair, as William Whiston later learned. Newton's particular heresy was grounds for ostracism from polite society, as Whiston also learned. It is impossible even to imagine what the consequences for Newton would have been had he been branded a moral leper in 1675. As the deadline approached, his career faced another crisis.

In fact, there was one other possibility besides an exempt fellowship. Any statute could be set aside by a royal dispensation. Late in 1674, Francis Aston, the fellow to whom Newton wrote in 1668, attempted to obtain a dispensation to free himself from the obligation of ordination. A letter from Barrow, who was then master of Trinity, to the Secretary of State, Joseph Williamson, on 3 December 1674, presented the college's case against a dispensation. It would destroy succession and subvert the principal end of the college, which was the breeding of clerics. He was sure the senior fellows would refuse to accept it.[152] Aston did not obtain a dispensation. The draft of a letter from Newton to "Sr Alexander," which indicates that Newton had been involved in Aston's attempt, survives. Sir Alexander was probably Sir Alexander Frazier, physician and confidant of Charles II, whose son, Charles Frazier, was elected to a fellowship in Trinity in 1673. In the letter, Newton thanked Sir Alexander for including him in the proposal for a dispensation, which, he said, had been successfully opposed by the college. The vice-master objected strongly, and the seniors followed him. They

---

[152] *CSPD*, 1673–5, pp. 443–4.

said it would hinder succession (almost exactly Barrow's phrase), and furthermore they did not wish to recede so far from fundamental statutes of the college.[153]

By early 1675, Newton had given up hope. He wrote to Oldenburg in January requesting that the Royal Society excuse him from payments, as Oldenburg had offered two years before. "For ye time draws near y^t I am to part wth my Fellowship, & as my incomes contract, I find it will be convenient that I contract my expenses."[154]

At the last moment, the clouds lifted. Less than a month after his letter to Oldenburg, Newton went to London. By 2 March, Secretary Coventry, with the rationale that His Majesty was willing "to give all just encouragement to learned men who are & shall be elected to ye said Professorship," sent the draft of a dispensation to the Attorney General for his opinion.[155] On 27 April, the dispensation became official. By its terms, the Lucasian professor was exempted from taking holy orders unless "he himself desires to . . ."[156] We know nothing of the factors behind the recorded events. The presence of Humphrey Babington among the senior fellows could not have injured Newton's prospects. Nevertheless, it appears more likely that on this occasion it was Isaac Barrow who rescued Newton from threatened oblivion. A dispensation was a royal act, and Barrow was the one who had the ear of the court. Although Newton's letter to Sir Alexander stated that he was involved in a joint endeavor to gain a dispensation, Barrow's letter to Secretary Williamson in December concerned Aston alone. In the letter to Sir Alexander about the proposal, Newton specifically stated that the master "received it kindly . . ." We can only speculate on what passed between Barrow and Newton. Barrow was deeply committed to the Church, and it is hard to believe that he would have acquiesced in a plea of Arianism. It is not hard to believe that he would have acquiesced in an argument that Newton had no vocation for the ministry. Barrow understood Newton's worth, and he valued learning. Moreover, he would have recognized that Newton, in contrast to Aston, would not set a precedent. As Lucasian professor, he was unique. The dispensation was granted to the Lucasian professorship in perpetuity, not to Isaac

---

[153] *Corres 3*, 146–7. The letter is misplaced here. The hand dates it in the early or mid-1670s, the content in late 1674 or early 1675.

[154] *Corres 7*, 387. As I mentioned above, we know that the letter was sent even though it does not survive. Since Oldenburg mentioned it at a meeting of the Council on 28 Jan. 1675 (Birch, *History, 3,* 178), we can date the draft to January with confidence.

[155] A draft of the letter patent in Newton's hand contains, also in Newton's hand, a copy of Coventry's letter (Cambridge University Library, Res.a.1893, Packet E).

[156] The letter patent is printed in full in *Edleston*, pp. xlix–1.

Newton, fellow of Trinity. It was probably Barrow's last service to his protégé.

Once more a crisis that threatened Newton's scientific career before it had reached fruition passed. Now at last, despite gross heresy sufficient to make him a pariah, he had surmounted the final obstacle and found himself secure in his sanctuary. And he had demonstrated a new facet of his genius: he could have his cake and eat it too.

9
Years of silence

BY the end of 1676, as absorbed in theology and alchemy as he was distracted by correspondence and criticism on optics and mathematics, Newton had virtually cut himself off from the scientific community. Oldenburg died in September 1677, not having heard from Newton for more than half a year. Newton terminated his exchange with Collins by the blunt expedient of not writing. It took him another year to conclude the correspondence on optics, but by the middle of 1678, he succeeded. As nearly as he could, he had reversed the policy of public communication that he began with his letter to Collins in 1670 and retreated to the quiet of his academic sanctuary. He did not emerge for nearly a decade.

Humphrey Newton sketched a few facets of Newton's life as he found it in the 1680s. Newton enjoyed taking a turn in his garden about which he was "very Curious . . . not enduring to see a weed in it . . ." His curiosity did not rise to the level of dirtying his hands, however; he hired a gardener to do the work. He was careless with money; he kept a box filled with guineas, as many as a thousand, Humphrey thought, by the window. Humphrey was not sure if it was carelessness or a deliberate ploy to test the honesty of others – primarily Humphrey. In the winter, he loved apples, and sometimes he would have a small roasted quince. Not much in the account suggested leisure, however. The Newton Humphrey found had immersed himself in unremitting study to the extent that he grudged even the time to eat and sleep.[1] During five years, Humphrey saw him laugh only once, and John North, master of Trinity from 1677 to 1683, feared that Newton would kill himself with study.

These were disastrous years for the college. In 1675, when it was still too soon to recognize the decline in numbers that had just set in, Isaac Barrow committed the college to the construction of an extravagant library. He had wanted Cambridge to build something to rival the new Sheldonian Theatre in Oxford, and when caution prevailed in the university, he resolved that Trinity would show its spirit. The Trinity library had been damaged by fire in any case and was near collapse. Early in January 1676, Barrow appealed for contributions to such a library as would yield "much Ornament to the *University,* and some honour to the Nation."[2] In that he succeeded.

[1] *Keynes MS* 135.
[2] A printed circular appealing for contributions; Trinity College Library, 0.11a.4⁹.

Christopher Wren contributed his services and drew the plans for a magnificent classical structure enclosing the west end of Neville's Court, which had also to be extended. The college broke ground on 23 February 1676, and eventually completed the library. It stands today, an ornament to the university and an honor to the nation as Barrow hoped. Not much else in his calculations worked out. He had told the university that contributors would come forward to support a grandiose project. He suffered a sad disappointment, for contributors external to the college covered well under half of the ultimate cost of £16,434.[3] In the end, the college squeezed the rest out of its own substance. No fellow and no incumbent in a college living escaped without his levy. One entry in the Conclusion Book implies that a promise to contribute became a condition of election to a fellowship.[4] Even the college barber was made to pay his dues. Newton donated £40, slightly less than the average contribution of one hundred and fifteen past and present fellows. In contrast, his friend Humphrey Babington gave over £900. Funds raised by contribution did not begin to suffice. By the end of the decade, the college had to appropriate its book funds to the structure.[5] They raised £303 by selling duplicates. They got an additional £64 from old flooring and scaffolding. They seized the income of an invalid fellow. They appropriated money from the general income of the college. They raised loans from fellows, including £100 from Newton in 1680, which the college repaid (without interest) only in 1689, and then by means of another loan from another fellow.[6] They even sold the college plate, £1479 worth. The last transaction offers a fascinating insight into the customary perquisites of college life. A second decision accompanied the one to sell the plate. Every senior fellow was allowed to choose three items from the collection for himself. The next eight in seniority received two, while a mere single item was tendered to the remaining forty-four.[7] Despite all the expedients, construction began to slow down by the end of the decade and nearly came to a halt in the mid-1680s. The limping construction of the library remained a fea-

---

[3] The accounts of the construction of the library, including contributions to it, are in 0.4.46.

[4] A conclusion on 4 Oct. 1682 ordered the senior bursar to pay £10 for the library from the dividend of Skinner, who had been elected to a fellowship in 1681, "according to ye said Mr Skinners promise before the Master & Seniors" (Conclusion Book, 1646–1811, p. 161).

[5] In response to an inquiry, Newton wrote to Aubrey on 22 Dec. 1683 that Trinity could not buy Sir Jonas Moore's library because the college was not buying books; all of the book funds were going into the building (*Corres 2*, 395). Cf. the library account, 0.4.46.

[6] Loans from Newton and Lynnet are recorded for the year 22 Dec. 1679–22 Dec. 1680 (*ibid.*). On 5 Feb. 1689, John Hawkins, a senior fellow, loaned the college £200 for finishing the library. The entire sum went to repay loans from Newton and another fellow (Conclusion Book, 1646–1811, p. 177).   [7] *Ibid.*, pp. 150, 151.

ture of life in Trinity virtually until Newton's departure. Although the roof was finally put on in 1684–5, the structure did not house the library until a year before Newton left in 1696.

Newton also contributed three books to the library between 1675 and 1680. Significantly, two of them were theological: Irenaeus' works and Bishop Huet's *Demonstratio evangelico*. In 1685, he sold the library a Silesian version of the Lord's Prayer for 10s.[8]

The burden of the library was no doubt increased by a crisis in college leadership. Barrow died in 1677. The Hon. John North, a younger son of the fourth Baron North and brother of two important political figures, succeeded him. Though not a fellow of Trinity, North had lived in the college for a time. He liked to think of himself as a scholar, and the slight indications we have suggest that he valued Newton. In 1677, not long before his nomination by the king, he submitted a book by his brother, Francis North, *A Philosophical Essay of Musick,* to Newton for comment.[9] Whatever his appreciation of Newton, his mastership was a disaster. Despite his connections, he proved unable to allay the financial crisis of the library. He was at odds with the fellows, whom he despised – in theory because of learning, in fact because of class – without bothering to hide the fact. On fellows and students alike, he tried to enforce a discipline gone beyond hope of recall, and he did it in a heavy-handed way which quickly united all the college against him. Scion of a wealthy family, North expected the mastership to be the final plum in his pudding, which in the words of his brother would complete his ease and repose. He inherited instead unceasing strife. One can read the mounting tension in the Conclusion Book. A rock thrown through the window of the master's lodge demonstrated it more concretely.[10] Finally, in 1680, North suffered a stroke during a confrontation with two students he was disciplining. He had the final gall to live on for three more years while the college drifted without leadership. His last order, as he lay dying, was "that he should be buried in the outward chapel, that the fellows might trample upon him dead as they had done living."[11] There is every reason to believe that they did so with gusto.

The Hon. John Montagu, a younger son of the Earl of Sandwich, followed North and held the mastership through the rest of Newton's residence in the college. Sheer indifference succeeded aristo-

---

[8] *Edleston*, pp. xxvi, xxviii, xxix. Junior Bursar's Accounts, 1685.

[9] Newton to North, 21 April 1677; *Corres 2,* 205–7.

[10] Roger North, *The Lives of the Right Hon. Francis North, Baron Guilford; the Hon. Sir Dudley North; and the Hon. and Rev. Dr. John North. Together with the Autobiography of the Author,* ed. Augustus Jessopp, 3 vols. (London, 1890), *2,* 298–304, 322–5.

[11] *Ibid., 2,* 328.

cratic hauteur in the life of the college. Montagu was absent from Cambridge more than he was present. When he was present, he offered no leadership; he had none to offer. During the period of North's and Montagu's masterships, the extraordinary longevity of the seniors meant general senility among those who might have supplied their wants. By the late 1680s, the financial crisis of the library had spread into financial chaos for the college as a whole, and dividends began to fail.

The problems of the college had to touch Newton. He did not withdraw completely from academic life. He voted in the senate in various university elections.[12] When Charles abruptly informed the university in 1682 that he had removed from his service the Duke of Monmouth, whom he had ordered them to elect as Chancellor eight years before, Newton probably cast his ballot obediently for the Duke of Albemarle, Charles's new choice, as everybody else did.[13] The Exit and Redit Book seems clearly to indicate that with others he represented the college at Albemarle's installation on 11 May. Nevertheless, no hint of the growing crisis within the college found its way into his papers or his correspondence. We know of his contribution and loan to the library only from the college records. He lived in Trinity. He never gave the college his heart.

While Humphrey Babington served as bursar in 1674–8, Newton drew up a set of tables of the fines (or premiums) to be paid upon the renewal of leases of college lands. A few years later, he endorsed the reliability of similar *Tables for Renewing and Purchasing of the Leases of Cathedrals Churches and Colleges, &c.,* published by an official of King's College. Later editions of the work, after Newton became famous, capitalized on the endorsement and called it Newton's tables.[14] The draft of a letter in the hand of the late 1670s, about a visit to Bedfordshire "to take some order about Mr Day's tenure," may indicate that he helped Babington in more ways than composing tables.[15]

In June 1680, Newton accepted his second pupil, George Markham, son of Sir Robert Markham. He was, of course, a fellow commoner. Immediately upon his arrival, Newton wrote to Markham's father; an incomplete draft of the letter survives.

Sr Robt

Your son is well arrived hither, by whom I received 5 guineys, An acquittance for it you will receive inclosed in his letter. He shall have

---

[12] *Edleston,* pp. xxviii, xxix.

[13] Charles Henry Cooper, *Annals of Cambridge,* 5 vols. (Cambridge, 1842–1908), *3,* 596–7.

[14] *Edleston,* p. lvi.

[15] *Corres 7,* 373. The context does not suggest that the letter was about Newton's own tenant on Lucasian land.

money of me as he has occasion for it, & his accounts you shal receive quarterly as the custome is. I suppose it may be enough to keep his french if a French M^r come to him once a week. If you think more necessary, be pleased to let me know your mind. Mr Battely has alread[16]

Such was the image of tutorial efficiency and concern. Perhaps the reality corresponded, but the lack of comment from either side suggests otherwise. Like St. Leger Scroope before him, Markham neither matriculated nor graduated. Like Scroope, he departed without presenting the expected piece of plate. Unlike Scroope, he finally came through with a silver tankard in 1696.[17] Such is the full extent of his known impact on Newton's life at this time—an admission, a letter, and a silver tankard for the college treasury. (They met again years later in the Royal Society.) The Junior Bursar's Accounts for 1678 mentioned the chamber of Newton's sizar. If he kept such, his sizar or sizars left an impression even less visible than Markham's.

A new arrival who came ultimately from abroad, John Francis Vigani (as he anglicized his name), had more impact on Newton than anyone in Cambridge. Vigani may have been the Italian who prompted one of Newton's rare letters from this period to the Royal Society, in December 1680.[18] Vigani was a chemist. Though Newton later broke with him, Humphrey Newton remembered him as the person in whose company Newton took the most delight. An incomplete draft of a letter, probably to Robert Boyle, to accompany something Newton intended to send by Vigani's hand, referred to the Italian chemist as one "who has been here performing a course of Chymistry to several of o^r University much to their satisfaction . . . "[19]

Late in the spring of 1679, Newton's mother died. As Conduitt heard the story, her son, Benjamin Smith, was seized with a malignant fever in Stamford. She went to nurse him and caught the fever. Newton in turn went to nurse his mother,

> sate up whole nights with her, gave her all her Physick himself, dressed all her blisters with his own hands & made use of that manual dexterity for wch he was so remarkable to lessen the pain wch always attends the dressing the torturing remedy usually applied in that distemper with as much readiness as he ever had employed it in the most delightfull experiments.[20]

Despite his ministrations, she died. An entry in the parish record of Colsterworth recorded her death. "Mrs Hannah Smith, wid.

[16] *Corres* 7, 390.   [17] Junior Bursar's Accounts, 1696.
[18] Newton to Hooke, 3 Dec. 1680; *Corres* 2, 241.   [19] *Yahuda MS* 2.4, f. 25^v.
[20] *Keynes MS* 130.8. Cf *Stukeley*, p. 59.

was burried in woollen June ye 4th 1679."[21] Filial love is an attractive quality for a variety of reasons, some moral, some psychoanalytical. I trust it is not too cynical to note that during the twelve years following his return to Cambridge after the plague, Newton is known to have paid three visits to Woolsthorpe. On three other absences from Cambridge, he probably returned home also.[22] More vigorous displays of filial affection have been recorded.

Death imposes practical demands, and his mother's death led to Newton's longest stay in Woolsthorpe, the plague years aside, after he was called home from grammar school twenty years before. He was both heir and executor, and it took him virtually all of the rest of 1679 to put his affairs in order. He devoted nearly as much time to his estate in 1679 as he had to his mother in all his visits in the previous twelve years combined. According to the Steward Book, in addition to the summer, he was absent for nearly four months of the year that began on Michaelmas 1679. Much of the time was spent in Woolsthorpe.[23] Part of the problem was a tenant, probably Edward Storer, one of the stepsons of the apothecary Clark, with whom Newton had lived in Grantham. At least Edward Storer and his sons were Newton's tenants eight years later, and their unsatisfactory relations then had accumulated during a tenure of unspecified length. The draft of a letter, in the hand of the late 1670s and suggestive of 1679 and 1680 in its content, survives among Newton's papers. It was probably addressed to one of Edward Storer's sons.

S^r

I wrote to you about 4 or 5 months since, but you were then from home & not having heard from you since, I have sent this to beg a line or two from you to let me know how my affairs stand & when I may find you & your brother at home. I desire you would put your

---

21 *Corres 2*, 303.

22 *Edleston*, p. lxxxv. Of the three I call known visits, one (in the autumn 1668) is not wholly certain. The visits that Newton himself mentioned in letters came in summer 1672 and spring 1678. Three absences of roughly a month each about which we know nothing but the dates in the Exit and Redit Book occurred in spring 1671, spring 1673, and spring 1677. Two other unexplained absences, ten days in October 1675 and six days in May 1676, were too short to make a trip to Woolsthorpe likely. During the other absences recorded in the Exit and Redit Book, he is known to have gone to London.

23 The Exit and Redit Book recorded a redit on 27 November, two months after Michaelmas. (*Edleston*, p. lxxxv). The following day, he wrote to Hooke that he had been concerned with family problems in Lincolnshire "this last half year . . . " (*Corres 2*, 300). That same day, 28 November, Humphrey Babington noted in his Day-Book that Newton delivered to him £11 15s 7d from Mr. Walker, rector of the Grantham School (*Edleston*, p. lxxxv).

brother in mind to have his arrears ready against ye time you appoint, & I will wait upon you. In ye meantime I rest[24]

Most of the sheet is covered with theological notes. Newton did not propose to waste a good half year just because business detained him in the country.

He also encountered trouble over a debt of £100 that one Todd owed. Like his problem with the Storers, the debt hung on to plague him, and somewhat later, probably in 1681, he took some pains in composing an intemperate letter to Todd which supplies further insight into the tenor of his dealings with home-town folk.

> Mr Todd
>
> Whether the order I sent you to pay ye money to my Sister miscarried or whether to gain more time you are unwilling to own the receipt of it I will not affirm. But Mr Drake is witness that such an order was sent you by ye Post enclosed in a bill of charges from himself, & letters by ye Post use not to miscarry. But be it as it will, to make sure that this come to you I have charged ye carrier to deliver it wth his own hands. About your pretenses of ye money's being ready long since & of a jugment wch you would have me beleive I had against you I do not think it material to expostulate. I shall only tell you in general that I understand your way & therefore sue you. And if you intend to be put to no further charges you must be quick in payment for I intend to loos no time. I desire you therefore to pay it to my Sister Mary Pilkington at Market Overton as soon as you can & take her acquittance for your discharge. Besides the 50^{lb} principall you are to pay the use for an hundred pounds from the time of ye date of ye Bond Jun 20 till the date of my receipt of ye 1st 50^{lb} & from that time ye use for ye other 50^{lb}, excepting only 40^s of use, wch is already payd. You are also to pay ye charges of ye suit an account of wch you will herewith receive from Mr Drake. And when I am satisfied that you have payd all this, your bond shall be delivered in here to any one you please to appoint. S^r I am
>
> Y^r [humble servant][25]

I suspect that Todd understood the extent to which Newton was his humble servant.

It was also at this time, in 1681, undoubtedly as part of the business of putting his estate in order, that Newton and others

---

[24] *Corres 7*, 373. Cf. another letter, undoubtedly to the same man since it also spoke of his brother, which mentioned the rent was three and half years overdue (*Corres 3*, 393). While the editor of the *Correspondence* placed the letter in the 1690s, it is on a sheet of mathematical work which Whiteside dates about 1680 (*Math 4*, 329). Yet another letter to the same brother, still dunning him for the rent, is found on a mathematical manuscript from 1684 (*Math 5*, 622. *Corres 7*, 369).

[25] *Add MS* 3965.11, f. 155^v. *Corres 5*, 253–4.

brought an action in Chancery against Dorothy Elston "pretending to be the Lady of the Manor of Woolsthorp," John Story, John Greenham, and Richard Elston. The Elstons were lords of the manor of Colsterworth. Styling himself "Lord of the Manor of Mortimer" (an ancient name of Woolsthorpe), Newton accused them of taking twenty acres of waste land at Woolsthorpe from him in the name of the manor of Colsterworth. Chancery found in his favor.[26]

As the oldest son, Newton also inherited obligations to his mother's three children by the Reverend Mr. Smith, which he shouldered for the rest of his life. The first concrete act of which we know was the shipment of a barrel of oysters, "wch I suppose at their first coming in may be a novelty wth you," to his sister Hannah and her husband, the Reverend Robert Barton, in Northampton.[27]

Not long thereafter, Newton sustained another loss. In 1683, after an extended period of nonresidence, when he came to the college only one or two weeks a year, Wickins decided to resign his fellowship. He visited the college for three weeks in March 1683; though he remained a fellow beyond Michaelmas 1684, he never returned to Trinity again. Probably Wickins had already been inducted into the rectory of Stoke Edith, Hereford. The Foley family, into which John North's sister married, controlled the living; there is every reason to think that North recommended one of his fellows, who was clearly on the prowl, to them. In this at least, North built well. A Wickins occupied the rectory of Stoke Edith under the patronage of a Foley for more than a century. Undoubtedly, it was the decision to marry and beget that progeny which led Wickins to resign his fellowship.

Newton's relation to Wickins remains a mystery. Despite a friendship of twenty years, we know nothing about it aside from the anecdote of their meeting as undergraduates and Wickins's service as amanuensis, to which numerous papers in his hand testify. The blank that followed Wickins's departure wraps the mystery in an enigma. After Newton's death, Robert Smith of Trinity wrote to Wickins's son Nicholas for information about Newton. Nicholas Wickins replied that his father, who had died earlier, had set out once to collect everything related to Newton that he possessed. He had transcribed into a notebook three short letters, so uninformative that Nicholas did not bother to send them, and he had in addition four or five very short letters in which Newton had merely forwarded dividends and rent. He said nothing about dates, but most of the second group at least had to have come before 1683. He went on to tell the story of how his father met Newton, and he

---

[26] Turnor Papers, Lincolnshire Archives.

[27] *Corres 3*, 393. With the letter mentioned in n. 24, this one appears on a sheet of mathematics which Whiteside places in about 1680 (*Math 4*, 329).

added three other superficial anecdotes he had heard. He concluded with the information that through his father and through him himself Newton had supplied Bibles to the poor in the parish of Stoke Edith, apparently their single topic of communication after 1683.[28] There is an unsatisfactory tone about the letter, as though someone, Wickins or his son, were concealing something. I am unable to imagine a close friendship of twenty years leaving so little residue, unless it ended in a breach. Newton's draft of the single letter to Wickins that we have, brief to the point of being curt, implied some barrier which he could not transcend. Written sometime between 1713 and 1719, it responded to a request for more Bibles to distribute in Stoke Edith; Newton indicated that he would send them via Wickins's patron, Thomas Foley. He brushed off Wickins's attempt at a friendly exchange. "I am glad to heare of your good health, & wish it may long continue, I remain . . ."[29] On the same day, 28 March 1683, that Wickins left Trinity for the last time, Newton also left. He returned on 3 May and left again for another week on 21 May.[30] We have no idea where he went.

Later that year, he arranged to have a young man from his own grammar school in Grantham, Humphrey Newton, come to live with him as his amanuensis.[31] According to Stukeley, who conversed with Humphrey in 1727 and 1728, he was "under Sr Isaac's tuition . . ."[32] This suggests the status of sizar, though Humphrey Newton was never admitted to Trinity. In the following five years, as he lived in Newton's chamber like Wickins before him, he copied out extensive writings, at first primarily on theology and

---

[28] *Keynes MS* 137.   [29] *Corres 7*, 368.

[30] While the date for Newton's exit in the Exit and Redit Book appears to be 27 March, it was entered after Wickins's exit (28 March) in the same hand and ink. The next two exits after Newton's (in a different hand) were also dated 28 March. No redit after the exit on 21 May was recorded; the Steward's Book showed that Newton was absent for a total of six weeks during 1683.

[31] Although Humphrey himself, forty-five years later, said that he went to Cambridge during the last year of Charles II (which would have been 1684), he also made particular mention of one cold winter. Both Samuel Newton and William Whiston recorded the winter of 1683–4 as very severe (J.E. Foster, ed., *The Diary of Samuel Newton, Alderman of Cambridge (1662–1717)* [Cambridge, 1890], pp. 86–7; William Whiston, *Memoirs of the Life and Writings of Mr. William Whiston* [London, 1749], p. 19). In 1709, when there was a very cold day in January, members of the Royal Society compared it with the "Greatest Cold of the Frost in the Year One Thousand Six Hundred and Eighty Three" (*JBC 10*, 203; cf. another reference to that winter in May, p. 213). As late as 1715, Halley mentioned the great frost of 1683 in a discussion (*JBC 10*, 575; cf. another reference the previous December, p. 528). Hence there is good cause to believe that Humphrey had that winter in mind. Moreover, he said that he stayed five years. There is reason to think he left before Newton's long absence while he attended the Convention Parliament, which Humphrey also failed to mention. This suggests that he left near the end of 1688, again implying that he arrived in 1683.

[32] Stukeley to Conduitt, 1 July 1727; *Keynes MS* 136.

mathematics. About a year after he arrived, a visit from Edmund Halley set off a new investigation which insured Humphrey's immortality; it was Humphrey who transcribed the copy from which the *Principia* was printed. Years later, when he married as an old man, he named his son Isaac after his "dear deceased Friend . . ."[33]

Theological study occupied much of Newton's time during the years of silence. In the late 1670s, he began a history of the church, concentrating on the fourth and fifth centuries, which repeated the themes of his interpretation of Revelation.[34] Athanasius played the role of the villain, of course. Some passages functioned as first drafts of his treatise from the same period, "Paradoxical Questions concerning the morals & actions of Athanasius & his followers," in which Newton virtually stood Athanasius in the dock and prosecuted him for a litany of sins too long to enumerate here.[35] In these papers, the passion evident in his earlier interpretation of Revelation rose to a new level of intensity as Newton sought to show, not only that Athanasius was the author of "the whole fornication" – that is, of trinitarianism, "the cult of three equal Gods"[36] – but also that Athanasius was a depraved man ready even to use murder to promote his ends. A shrill note of iconoclastic Puritanism which one does not associate with Newton also rings through these pages. He said a great deal about superstitions – "vehement superstition," indeed – deliberately fostered by "that crafty politician Athanasius . . . For when he found himself by means of ye Councils of Sirmium Ariminum & Seleucia baffled & deserted by all but the Monks: he contrived his religion for ye easy conversion of the heathens by bringing into it as much of ye heathen superstitions as

[33] Humphrey Newton to Conduitt, 17 Jan., 14 Feb. 1728; *Keynes MS* 135. Writing to Conduitt on 13 February, Stukeley said that he had attended the christening of "the young Sr Isaac Newton" (*Keynes MS.* 136).

[34] As I mentioned above, Newton had sketched a history of the church earlier (*Yahuda MS* 2.5b, ff. 40–7). He left very extensive manuscripts on it from the late 1670s and early 1680s: *Yahuda MSS* 2.2, ff. 1–66 (much of which was heavily emended); 2.3, ff. 1–103 (a number of separate sections which are not continuous, some heavily emended); 2.5b, ff. 1–55; 5.3, ff. 1–19; 11, ff. 1–40 (ff. 1–7 outlined an ecclesiastical history and sketched out a framework of eight books); 12, ff. 1–34; 14, ff. 52–4, 65–71, 86–92, 97–100, 104–15, 119–25, 128–31, 135–70, 175–90, 195–8, 203–11, 214–20; 18, ff. 1–4; 19, ff. 1–166 (intended as fair copy, although inevitably somewhat emended); 29, ff. 1–4. *Keynes MS* 10. Clarke Library MS of the "Paradoxical Questions."

[35] *Keynes MS* 10. This is one of the papers published among Newton's *Theological Manuscripts*, ed. H. McLachlan (Liverpool, 1950), pp. 61–118. Although it is more accurate than most of the papers in this misbegotten volume, McLachlan nevertheless took the liberty of imposing on it his own division into paragraphs. Early drafts of the same material in the context of his history of the church are found in *Yahuda MS* 19, ff. 61–8, 81–143. Cf. the Clarke Library MS, which includes some further questions not in *Keynes MS* 10.   [36] *Yahuda MSS* 2.2, f. 19, and 11, f. 7.

the name of Christianity would then bear." What Newton called superstitions were "monstrous Legends, fals miracles, veneration of reliques, charmes, ye doctrine of Ghosts or Daemons, & their intercession invocation & worship & such other heathen superstitions as were then brought in."[37] Along with the practices above, he included the introduction of images into churches.[38] The identity of the superstitions to which he pointed with practices of the Catholic church cannot be missed. Newton wrote these pages in or not long after 1679, the year of the great Popish Plot. I have never found an explicit reference to the plot among Newton's papers. The intensity of feeling in his indictment of Athanasius suggests that the passions of that year did not pass him by. And if Newton surrendered to them, perhaps we can understand better how an entire nation did.

Newton intended to devote Book I of his history of the church to the spread of monasticism. Monasticism was yet another atrocity spawned by the evil genius Athanasius, who had fostered it to promote his interests. Newton devoted considerable study to it, especially to the feigning of miracles by monks and to their stories of sexual temptations. The latter apparently fascinated him; he collected every story of that nature he could find. Rather solemnly, he lectured the monks on how avoid unchaste thoughts.

> For lust by being forcibly restrained & by struggling wth it is always inflamed. The way to be chast is not to contend & struggle with unchast thoughts but to decline them keep the mind imployed about other things: for he that's always thinking of chastity will be always thinking of weomen & every contest wth unchast thoughts will leave such impressions upon the mind as shall make those thoughts apt to return more frequently.[39]

In Newton's plan, Book VIII of the history of the church was to show how events had fulfilled the divine prophecies. The history

---

[37] Clarke Library MS. Under the question, whether Athanasius did not start the practice of false miracles for his own interest, Newton said that miracles began to make a great noise among the monks in Egypt and Syria in the middle of the fourth century, with Egypt leading the dance. "But for as ye persecution of Maximianus had afforded them the greatest plenty of martyrs so they were not content wth doing miracles at home but sent into all the Empire the reliques of their saints & martyrs to inflame ye whole Roman world with this kind of Superstition." Athanasius and his party adopted the pretense of miracles as a conscious design. "They found by experience y^t their opinions were not to be propagated by disputing & arguing, & therefore gave out that their adversaries were crafty people & cunning disputants & their own party simple well meaning men, & therefore imposed this law upon the Monks that they should not dispute about ye Trinity. Thus they left ye success of their cause to ye working of miracles & spreading of Monkery." Among feigned miracles he included "the superstitious or to speak more truly the magical use of the signe of ye Cross . . . " (*ibid.*). [38] *Yahuda MS* 19, f. 153–61.

[39] Clarke Library MS of "Paradoxical Questions." Cf. *Yahuda MS* 11, ff. 8–40, a chapter on the rise of monasticism, more than half of which was devoted to such stories.

had grown out of his interpretation of the prophecies, and during the same period, when he worked on the history, he continued also with them. At this time, his interpretation acquired an important new dimension. In his early treatise, Newton had insisted on the certainty of an approach founded on rigorous method. There is no evidence that he ceased to believe in the certainty of his interpretation. Nevertheless, he altered it by adding a new facet which had not been there before, the role of Jewish ceremony and worship as a "type" or figure to aid in understanding the prophecies. "God forecast everything of great moment first by types," he asserted, "and then explained those types by new prophecies." Jewish practices and Revelation, "like twin prophecies of the same things, mutually explicate each other and cannot be satisfactorily understood apart from one another." It followed that he must become a student of Jewish religion. Being the man he was, he plunged into an extensive program of reading in Josephus, Philo, Maimonides, and the Talmudic scholars.[40] Jewish ceremonies continued to play a prominent role in Newton's understanding of Revelation until his death.

In this context, the exact shape of the temple in Jerusalem became significant, and Newton set out to reconstruct it. In fact, the temple as such never figured prominently in Newton's interpretation, but it did become a problem in itself which took on a life of its own and for a time took command of Newton as problems did. He became obsessed with the temple's plan and dimensions (Figure 9.1). The tabernacle of Moses and the two temples, of Solomon and Ezekiel, he decided, had all been built to the same plan except that the dimensions of the two temples were twice those of the tabernacle, which also lacked some minor structures near the entrance. Ezekiel gave the best description of the plan, though Ezekiel was very difficult to comprehend. Newton undertook to reproduce the plan of the temple by a careful reconstruction and interpretation of the text of Ezekiel 40–42, 43:1–7, and 46:19–24. If he had not done so earlier, he learned Hebrew in order to read Ezekiel in the original. Using the Latin Vulgate as his starting point, he compared its text with other versions, including Hebrew ones, and with the demands of his own floor plan as he constructed it from the text. Proceeding through Ezekiel verse by verse, he produced an annotated, amended reading with a detailed floor plan to match it. Consider two examples. Ezekiel 40:20 began with three words, "Et adduxit me . . ." in the Vulgate. Equivalent words appeared in the Greek

[40] *Yahuda MS* 2.4, f. 46. *Babson MS* 434, f. 1. "Prolegomena ad Lexici Prophetici partem secundam, in quibus agitur De forma Sanctuarij Judaici"; *Babson Ms* 434. *Yahuda MSS* 13.2, ff. 1–22, and 28.5, ff. 1–3.

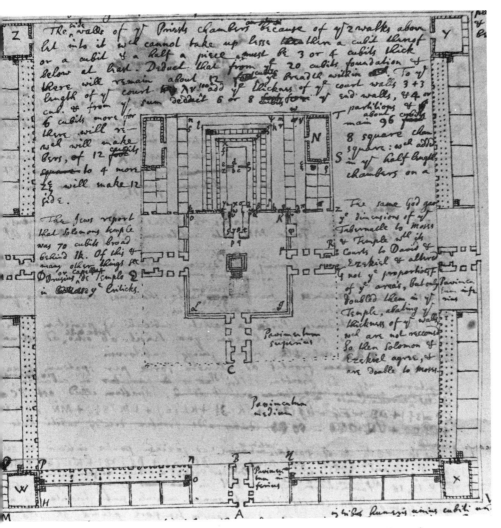

Figure 9.1. Newton's plan of the Jewish temple. (By permission of the Babson College Library.)

Septuagint. The Hebrew, however, began with the equivalent of "et ad aquilonem . . . , " which Newton accepted. One commentator had argued that these three words really belonged to the previous verse. Here Newton relied on the demands inherent in the plan. To attach the phrase to the preceeding verse, he argued, would reduce the whole structure to chaos. In the case of Ezekiel 41:10, some versions lacked an initial word, "Et," which determined what a measurement of twenty cubits referred to. Newton decided that

"Et" must stay because, with it there, the twenty cubits referred to the interior width of the atrium, a measurement consistent with the temple as a whole.[41] Newton's treatise on the temple presented itself with very little interpretation. As its title said, it was a "Prolegomena" to a *Lexicon propheticum*.[42] Here and there, phrases suggested the freight of significance that the temple's divinely ordained dimensions carried for him. "This structure," he remarked at one point, "commends itself by the utmost simplicity and harmony of all its proportions."[43] Eventually, Newton included a compressed version of his plan of the temple in his *Chronology of Ancient Kingdoms Amended*.

The plan of the temple raised a further question which seized his attention as the temple itself had done. In known units, what was its size? Ezekiel expressed it in cubits, but what was the length of a cubit? On this question, Newton took Jerome severely to task. Jerome had not understood that the Jews used two different cubits as units of measurement, a common cubit of five palms and a sacred cubit, which was used of course in descriptions of the temple, and had translated Ezekiel to say that the measuring rod consisted of six cubits and a palm. What Ezekiel had said, however, was that his measuring rod contained six cubits, each of which was one palm longer than the common cubit.[44] Having elucidated the cubit, Newton went on to collect evidence that related it to other measures and concluded that the sacred cubit was between 25 1/5 and 26 1/4 Roman inches.[45]

Newton may well have owed his perception of the role of Jewish ritual in Revelation to Henry More, who had not only employed what he called *"Israelismus"* in expounding the prophecy but had also drawn a scheme of the temple.[46] Although much has been written about the influence of More on Newton, and much specu-

---

[41] *Babson MS* 434, ff. 41–3, 47–8.

[42] A partial text of the "Prolegomena" is found in *Yahuda MS* 14, ff. 1–8. Cf. also *Yahuda MSS* 2.4, ff. 17–49; 14, ff. 32–3; and 28.4. Stukeley also reported that Newton had left a *Lexicon propheticum* as well as a discourse on the form of the tabernacle and the sacred cubit (*Stukeley*, p. 59).   [43] *Babson MS* 434, f. 58.   [44] *Yahuda MS* 14, f. 1.

[45] *Yahuda MS* 2.4 f. 40. Using a radically mistaken notion of the ratio of the English foot to the Roman foot, Newton added that the cubit was between 2.6 and 2.8 English feet (f. 43). Cf. *Babson MS* 434, f. 34. An English translation of Newton's later "Dissertation upon the Sacred Cubit of the Jews and the Cubits of the several Nations" was published in John Greaves, *Miscellaneous Works*, ed. Thomas Birch, 2 vols. (London 1737), *2*, 405–33. Here Newton set narrower limits, between 25.57 and 25.79 Roman inches, around the cubit. In his *Chronology*, he published the radically different measure of 21 1/2 or almost 22 English inches. (*Chronology*, p. 333.)

[46] Henry More, *The Theological Works* (London, 1708), pp. 530–1. More, *Opera omnia*, 3 vols., reprint ed. (Hildesheim, 1966), *1*, 26.

lated on the basis of their common origin in Grantham, concerning their mutual contact in Cambridge, we know very little that is concrete about their relations. We do know, however, that Newton cited More's contribution to understanding Revelation from the beginning of his own study, and we do know that the two men discussed the prophecy. It was in connection with such a discussion that More drew one of the most revealing sketches of Newton. Many persons repeated anecdotes about Newton neglecting his meals. Only More fully caught him in a state of ecstasy. Writing to John Sharp in 1680, he mentioned how well he and Newton agreed about the Revelation.

> For after his reading of the Exposition of the Apocalypse which I gave him, he came to my chamber, where he seem'd to me not onely to approve my Exposition as coherent and perspicuous throughout from the beginning to the end, but (by the manner of his countenance which is ordinarily melancholy and thoughtfull, but then mighty lightsome and chearfull, and by the free profession of what satisfaction he took therein) to be in a maner transported.[47]

More went on to describe their areas of disagreement, and he did not hesitate to call Newton's identification of the seven vials and the seven trumpets "very extravagant." He assumed that Newton would see his error as soon as he read More's exposition more carefully. In this, he rather underestimated Newton's tenacity. There is nothing in the letter to imply that Newton had even hinted at his private understanding of the great apostasy. The two men must have maintained some intimacy. When More died six years later, Newton was one of only five outside the fellowship of Christ's College to whom he bequeathed a funeral ring.[48]

In the middle 1680s, Newton undertook a major revision of his treatise on Revelation in which he incorporated his new perception of Jewish ritual as a type and much else besides.[49] It was one of the tasks on which he set Humphrey Newton to work as his amanuen-

---

[47] More to Sharp, 16 Aug. 1680; Marjorie Hope Nicolson, ed., *Conway Letters. The Correspondence of Anne, Viscountess Conway, Henry More, and Their Friends* (New Haven, 1930), pp. 478–9.   [48] *Ibid.*, p. 482.

[49] *Keynes MS* 5, ff. I–VI, 1–6, two successive versions of "The First Book. Concerning the Language of the Prophets." With f. 7, another manuscript on the prophecies begins, in the hand of about 1705–10. McLachlan has published part of *Keynes MS* 5 (*Theological Manuscripts*, pp. 119–26) without recognizing that the whole consists of two different manuscripts accidentally conjoined. Among the other manuscripts on the prophecies from around 1680 are a number in which Newton set his interpretation over into Latin: *Yahuda MSS* 2.5a, ff. 1–4, and 3, ff. 1–16; and *Keynes MS* 1. *Yahuda MS* 10.2 ff. 1–15, which a reference to the revocation of the Edict of Nantes together with its hand places around 1685, contains (also in Latin) the earliest extended interpretation of Daniel that I have found.

sis.[50] The principal changes were twofold, and they pose a riddle to their interpreter not much less puzzling than the riddle Revelation posed to Newton. On the one hand, the passion of the earlier period, which had reached its climax scarcely five years before, simply disappeared, to be replaced by the colorless exposition that henceforth characterized Newton's writings on the prophecies. What he composed was a dispassionate scholarly treatise which, in its effort to be demonstrative if in little else, reminds one of the *Principia*. For example, the explanation of prophetic figures that has survived from it, even in its incomplete form, was several times the earlier version in length.[51] At the same time, though he continued to assert that "To describe the times of Apostacy was the main design of the Apocalyps . . .," and devoted Section X to "The Apostacy of the latter times," he eliminated every explicit reference to trinitarianism and couched his entire description of the apostasy in deliberately ambiguous terms.[52] One of the signs of the apostate church, he said in a typical passage, was its use of sorceries and false miracles to deceive people. It taught that the sign of the cross could drive devils away and produce beneficial spiritual effects.

> So the consecrating Images Pictures Holy water, Agnus Dei's, Psalters, rings, Beads wooden crosses, & ye like by crossing them or touching y^m wth reliques or performing any other ceremonies wth invocations & prayers for endowing them with any kind of supernatural powers & vertues is a superstition of ye same kind wth ye Charmes & spells of ye old Heathen, & even wthout a figure may be truly called enchantment & sorcery: as may also ye turning a wafer into a man by ye words Hoc est corpus meum.[53]

Not one Puritan in England would have paused to identify that description with the Roman Catholic Church. Whereas earlier, Newton had assigned the cause of the apostasy to Athanasius and the "homousians," he now ascribed it to hypocrites who flocked into the church for wordly ends once it became the establishment.

No evidence at all indicates that Newton ceased to be an Arian,

---

[50] *Yahuda MSS* 9 and 13.2; several times, Newton himself took over these manuscripts so that his hand appears along with Humphrey's. Newton gave Humphrey quite a bit of theology to copy, including parts of *Yahuda MSS* 16.2 (his radically unorthodox "Theologiae gentilis origines philosophicae") and 22 (two professions of faith from the fourth and fifth centuries connected with his history of the church). He also had Humphrey copy a passage "Ex historia Ingulphi. edit. Oxonijs 1684" into his theological notebook (*Keynes MS.* 2, p. 101). I noted above that he was careful to keep his unorthodox views out of the sight of Wickins; perhaps he thought Humphrey was not sophisticated enough to understand the implications of what he was copying.   [51] *Yahuda MS* 9.1, ff. 5–51.

[52] *Yahuda MS* 9.2, ff. 96, 97–122.

[53] *Yahuda MS* 9.2, ff. 99–99^v. Cf. his harsh coments, probably directed at Grotius, on efforts to establish religious peace by identifying the fourth beast of Daniel with Antiochus Epiphanes (*Yahuda MS* 9.2, f. 105).

however. Quite the contrary. Perhaps we can correlate the development of Newton's exposition with the rising threat of Catholicism in England. Faced with James II, Newton may have decided that Anglicanism was tolerable after all and hence have couched his interpretation in new terms which would qualify as Protestant propaganda. On this view the lower level of passion could have been the result of deliberate discipline as he strove to eliminate any trace that might have compromised the struggle against popery. Newton's motives are wholly speculation, of course. It is not speculation that he revised his interpretation of the prophecies while Humphrey was with him, though he took no steps we know of to publish it as Protestant propaganda. It is also not speculation that the rebel of the 1670s made his peace with the established order, led the resistance that preserved it (in Cambridge) from change, and lived comfortably in London on the considerable material reward he accepted – all without changing his secret opinion. It is at least plausible to read the revised interpretation of the prophecies as an obscure prediction of his own imminent compromise with the world.

Once again, in a theological context, he expounded a conception of God that he would later make familiar in a different setting. When he addressed himself to the final end, he declared that the New Jerusalem refers to the whole assembly of the resurrected. Where is it? He did not know.

> Tis not ye place but ye state wch makes heaven & happiness. For God is alike in all places. He is substantially omnipresent, & as much present in ye lowest Hell as in ye highest heaven, but ye enjoyment of his blessings may be various according to ye variety of places, & according to this variety he is said to be more in one place less in another, & where he is most enjoyed & most obeyed, there is heaven & his Tabernacle & Kingdom in ye language of ye Prophets.[54]

He went further than he had a decade earlier in equating the end of the world with the final establishment of true Gospel Christianity in human society.

While Humphrey was with him, Newton also undertook a new theological venture which became henceforth the vehicle for his heterodox theological opinions. Certainly he could not have misunderstood it as part of a common anti-Catholic endeavor. In its implications, "Theologiae gentilis origines philosophicae" (The Philosophical Origins of Gentile Theology) was more radical than any Arian statement he composed during the 1670s.[55] The "Origines"

---

[54] *Yahuda MS* 9.2 ff. 139.

[55] The chaotic manuscript remains of "Theologiae gentilis origines philsophicae" indicate that Newton never put it into what he considered relatively finished form. The basic manuscript, from which I take the title, is *Yahuda MS* 16.2, a confused, missorted mass, half Latin and half English, half in Humphrey's hand and half in Newton's, with

started with the argument that all the ancient peoples worshiped the same twelve gods under different names. The gods were divinized ancestors – in fact Noah, his sons, and his grandchildren – though as this religion passed from people to people, each used it to its own ends by identifying the gods with its own early kings and heroes. Nevertheless, common characteristics distinguished the corresponding gods of all the ancient peoples. All peoples worshiped one god whom they took to be the ancestor of the rest. They described him as an old and morose man and associated him with time and with the sea. Clearly, Noah furnished the original model of the god called (among other names) Saturn and Janus. Like Noah, Saturn had three sons. Every people had a god whom they depicted as a mature man, the god they held most in honor. They had translated Ham into Zeus, Jupiter, Hammon, and others. All worshiped a voluptuous woman variously named Aphrodite, Venus, Astarte, *et alia,* originally a daughter of Ham. The histories of the gods of one people frequently became confused with those of another, and peoples invented fables which confounded the origins of the gods by claiming the gods of others for their own. "Every nation studing to honour their own ancestors they were not content to worship them them selves but sometimes pretended them to be the Gods of other nations also. & on this account ye Egyptians also pretended that Dionysius Bacchus Adonis & Pan were their Osiris."[56] Once this tendency was

successive drafts of the same material in both hands. *Yahuda MS* 16.1 belongs to the same treatise. *Yahuda MS* 17 contains rougher material pertaining to it, and 17.2, f. 14, looks like the earliest version of the basic argument. In *Yahuda MS* 41, Newton put part of the treatise into more finished form. *Yahuda MS* 7.1a, f. 1. "Chap 1 Of the original of Monarchies," is one sheet of what looks like the beginning of a finished version of the same treatise. *Keynes MS* 146, with virtually the same title, is a full version of the chapter. Rough notes relevant to it are found in *Yahuda MSS* 13.3, ff. 1–20; 25.2, ff. 26–31, 34–58; New College MS 361.3, *passim* (Bodleian Library); and the Newtonian manuscript in the American Philosophical Society Library. Humphrey's hand in *Yahuda MS* 16.2 provides the most important clue by which to date it. A number of isolated pieces of evidence demonstrate that Newton also worked on it later. *Yahuda MS* 41, f. 5, refers to a book on the ritual laws of the Jews published by John Spencer in 1685. Because of Newton's other known activities at that time, I do not think that he could have read the work before autumn 1689. *Yahuda MS* 41, f. 28^v, has the heading, "De motu Fluid," which suggests a rejected sheet from the *Principia. Yahuda MS.* 17.1, f. 14, is the back of a letter addressed to Newton in Parliament, and f. 5^v refers to a book published in 1691. *Yahuda MS* 16.2, f. 20^v, refers to the Million Act, relevant to his activities at the Mint in the late 1690s.

Although Newton never completed a full version of this work that survives, concepts from it, which did not reveal its central argument, continued to appear in his writings for the rest of his life – e.g., Paragraph 1 of the first version of the *Principia's* final book, the last paragraph of Query 31, two footnotes to the General Scholium, his much revised theological treatise, "Irenicum," and especially his *Chronology.*

[56] *Yahuda MS* 16.2, f. 19^v. See Frank Manuel, *Isaac Newton, Historian* (Cambridge, Mass., 1963), chap. 7, "The Pragmatization of Myth."

recognized, the common pattern behind the apparent chaos of gods became apparent.

The number twelve derived from the seven planets, the four elements, and the quintessence. Newton argued that peoples had identified their most eminent ancestors with the most eminent objects that nature presents. Even in the seventeenth century, Galileo named the newly discovered satellites of Jupiter after his benefactors; so in the earliest ages of human society, men had identified their most prominent ancestors with the heavenly bodies, assumed that their souls transmigrated to those bodies, and begun to ascribe divine powers to the souls.[57] Hence, as the title of the treatise proclaimed, gentile theology had its origin in natural philosophy. Chapter 1 asserted the same thing: "Gentile theology was philosophical and dependent on the astronomy and physical science of the system of the world . . ."[58] He frequently called it "astronomical theology" and similar names.

The "Origines" entailed yet another program of study. To be sure, Newton drew some of his material from contemporary sources, especially Samuel Bochart's *Geographia sacra*. Most of the notes that he took from Ralph Cudworth's *Intellectual System of the World* related to the themes of the "Origines," but the content of the notes implies that he read Cudworth only after he had worked his own treatise out.[59] I have not found a single reference to Cudworth in any version of the treatise itself. The great majority of his references came from classical authors, many of them Christian Fathers with whom he had already made himself familiar, but more of them writers of pagan antiquity: Virgil, Ovid, Macrobius, Pliny, Strabo, Lucian, Herodotus, Cicero, Plato, Homer, Plutarch, Seneca, Euripides, Horace, Xenophon, Pindar, Catullus, Aeschylus, Tacitus, and others.[60] The "Origines" helps to explain the presence of such authors in Newton's personal library.

Egypt was the original home of gentile theology. Here Noah settled after the flood, and from Egypt other lands were settled when the sons of Noah fought over the inheritance and separated. The Egyptians first developed sidereal theology, which looked back to their own ancestors. They taught it to other peoples, who had the same ancestors but remodeled the gods in order to enhance their own self-esteem.

Newton was convinced that gentile theology represented a falling away from true religion.

It cannot be believed, however, that religion began with the doctrine of the transmigration of souls and the worship of stars and elements:

[57] *Yahuda MS* 16.2, ff. 10–11.   [58] *Yahuda MS* 16.2, f. 1.
[59] "Out of Cudworth," Clarke Library MS. Also Trinity College Library, R. 16.38, f. 436.
[60] References in *Yahuda MS* 41.

for there was another religion more ancient than all of these, a religion in which a fire for offering sacrifices burned perpetually in the middle of a sacred place. For the Vestal cult was the most ancient of all.

He cited evidence to prove that a similar worship had been the most ancient in Italy, Greece, Persia, and Egypt, among other places. When Moses instituted a perpetual flame in the tabernacle, he restored the original worship "purged of the superstitions introduced by the Egyptians." Such had been the worship of Noah and his sons. In his turn, Noah had learned it from his ancestors. It was indeed the true worship instituted by God. "Now the rationale of this institution was that the God of Nature should be worshiped in a temple which imitates nature, in a temple which is, as it were, a reflection of God. Everyone agrees that a Sanctum with a fire in the middle was an emblem of the system of the world . . ."[61] In an essay about this time on the prophetic language, Newton said that "Temples were anciently contrived to represent the frame of ye Universe as the true Temple of the great God." Such a concept helps to explain his interest in the temple of Solomon, which (he later told Stukeley) was older than any other temple and the model first of the Egyptian and then of the Greek ones. Men are "ever inclined to superstitions," however.[62] The Egyptians corrupted the true worship of God by creating false gods from their ancestors; other peoples were only too ready to learn degenerate practices from them.

Another name for the false worship was idolatry. To Newton, idolatry represented the fundamental sin. Apparently he was required at some point to preach a series of sermons, probably in the Trinity chapel. He chose as his text 2 Kings 17:15, 16, in which the children of Israel, in imitation of the heathen around them, made molten images of two calves and served Baal. As Newton said, the text gave him the opportunity "to consider Idolatry in its full latitude & to discourse of its nature in general."[63] This he did in at least two discourses.[64] Their theme ran very close to the "Origines," and they probably stemmed from the same time. In passing, Newton declared that God prefers to be worshiped, not for the necessary aspects of His being, such as His omniscience and om-

---

[61] *Yahuda MS* 17.3, ff. 8–11.

[62] *Keynes MS* 5, f. 5^v; McLachlan, *Theological Manuscripts*, p. 125. *Stukeley*, pp. 17–18. *Yahuda MS* 16.2, f. 67; cf. f. 44 and *Yahuda MS* 17.3, f. 12.

[63] *Yahuda MS* 21, f. 1.

[64] *Babson MS* 437 is another draft of the discourse in *Yahuda MS* 21. Newton concluded the latter with a reference to what he would say the next time. The manuscript in the Humanities Research Center, University of Texas, of another discourse on 2 Kings 17:15, 16, appears to be the other one. There is no evidence I know of to indicate that there were more than two.

nipotence, but for what he has done: "ye wisest of beings requires of us to be celebrated not so much for his essence as for his actions, the creating preserving & governing all things according to his good will & pleasure. The wisdome power goodness & justice wch he always exerts in his actions are his glory wch he stands so much upon, & is so jealous of."[65] Since true worship took place in a sanctum that represented God's creation, it followed directly that its corruption involved a corruption of natural philosophy as well; so Newton believed. The original temple, with the fire in the center, illuminated by seven lamps representing the planets, symbolized the world.[66]

> The whole heavens they recconed to be ye true & real temple of God & therefore that a Prytanaeum [Sanctum] might deserve ye name of his Temple they framed it so as in the fittest manner to represent the whole systeme of the heavens. A point of religions then wch nothing can be more rational. . . . So then twas one designe of ye first institution of ye true religion to propose to mankind by ye frame of ye ancient Temples, the study of the frame of the world as the true Temple of ye great God they worshipped. . . . So then the first religion was the most rational of all others till the nations corrupted it. For there is no way (wthout revelation) to come to ye knowledge of a Deity but by ye frame of nature.[67]

Geocentric astronomy accompanied the spread of false religion. It was not by accident that Ptolemy was also an Egyptian.

So far, it was possible to fit the "Origines" into an orthodox Christian setting. One could have equated the universal tendency to idolatry with original sin. If he emphasized knowledge of God from his works, Newton admitted revelation as well. In taking Noah as the common ancestor of all mankind, he acknowledged the truth of the Biblical account. Nevertheless, the orthodox would have found other elements of the treatise inadmissible. Though he admitted the Mosaic account, he checked it against other ancient testimony, such as the Phoenician chronicler Sanchuniathon, and the Chaldean Berossus. He listed Berossus's set of ten kings between the creation and the flood side by side with the Patriarchs from Genesis and pronounced them identical. He argued that there was a general tradition of a deluge among the ancient peoples.[68] Though Newton never questioned that the Bible was the revelation of God, such passages treated it on a par with pagan testimony and raised the question of what Newton understood revelation to be.

More significant was the implicit deemphasis of the role of Christ, a step which came readily enough to an Arian. Instead of

65 *Yahuda MS* 21, f. 2.   66 *Yahuda MS* 17.3, f. 11.   67 *Yahuda MS* 41, ff. 6–7.
68 *Yahuda MS* 16.2, ff. 48, 50, 74.

the agent of a new dispensation, Christ was a prophet, like Moses before him, sent to recall mankind to the original true worship of God. As he revised the "Origines," Newton set down a number of chapter headings, the last of which was for chapter 11. "What was the true religion of the sons of Noah before it began to be corrupted by the worship of false Gods. And that the Christian religion was not more true and did not become less corrupt."[69] In this setting, trinitarianism with its encouragement of the worship of saints and martyrs, indeed with its worship of Christ as God, took on a new meaning. What was trinitarianism but the latest manifestation of the universal tendency of mankind to superstition and idolatry? Through Athanasius, Egypt once again played its nefarious role as the corrupter of true religion. By universalizing the Christian experience of the first four centuries, Newton denied it any unique role in human history. The Christian religion rightly understood was not more true than the religion of the sons of Noah, which was founded upon the recognition of God in His creation.

In the revision of his interpretation of the prophecies in the mid 1680s, Newton insisted "y^t ye great mystery of God to be fulfilled at ye voice of ye seventh Angel when he shal begin to sound" was the central message of the whole Bible.[70] That is, the second coming of Christ was to be the great event of history. Only with it would the kingdom of God be realized.

> For he came not ye first time as a King over men, & therefore ye kingdom of God or Heaven is never spoken of ye Church at his first coming, but only of ye invisible powers of heaven then commanded by him. This was all ye regal power & kingdom he then came w^{th} . . . Where he speaks of ye Gospel of ye kingdome, he means ye Gospel concerning ye future kingdom. . . . So then they are much mistaken who by ye kingdom of heaven in ye Gospels understand any state of men before ye second coming of Christ.[71]

Perhaps it is not altogether surprising that Humphrey Newton remembered a man who, despite what we know of his intense theological concern, did not trouble himself much with the established forms of public worship. Though he usually went to St. Mary's on Sundays, he seldom went to chapel. Morning chapel fell during the time he slept, and vespers during the time he studied, "so y^t He scarcely knew ye Hour of Prayer."[72] It is perhaps also not surprising that he kept the "Origines" to himself, so that its very existence has become known only in this day, nearly three hundred years after he conceived it.

---

[69] *Yahuda MS* 16.2, f. 45^v.   [70] *Yahuda MS* 9.2, f. 157.
[71] *Yahuda MS* 9.2, ff. 168–9.   [72] *Keynes MS* 135.

Theology was not Newton's only occupation during these years. The manuscript remains testify that alchemy vied with theology to command his attention. He continued to receive unpublished alchemical manuscripts to copy, and the letter from Francis Meheux in 1683 provides one of our most explicit pieces of evidence of his contact with alchemical circles.[73] Newton began to devote considerable time to the collation of alchemical authorities, apparently in the conviction that all true adepts described one and the same process. In one such paper, he drew up a set of seventeen chapter headings, the first two of which, "In what manner metals are generated and corrupted in the veins of the earth" and "Concerning the seed, sperm, & body of minerals," repeated themes central to the "Vegetation of Metals" composed a decade earlier.[74] The manuscript contained the texts to go with four of the headings, three of which he labeled chapters. The texts themselves were mostly and perhaps entirely compilations of passages cited from more than thirty different alchemists. Under the title, "The Regimen," another paper boiled the work down to seven aphorisms.

> Aphorism 1.
> The work consists of two parts the first of wch is called the gross work & by many imbibitions & putrefactions purges the matter from its gross feces & exalts it highly in vertue & then whitens it.
> Aph 2
> The second part also putrefies the matter by severall imbibitions & thereby purges it from ye few remaining feces & exalts it much higher in vertue & then whitens it. For the two parts of the work resemble one another & have the same linear process.
> Aph 3
> The putrefaction of ye second work lasts about five months & is done by seven imbibitions or at most by 9 or 10, & for promoting the putrefaction the spirit is drawn of at ye end of every imbibition & digestion.
> Aph 4
> In both works the Sun & Moon are joyned & bathed & putrefied in their proper menstruum & in the second work by this conjunction they beget the young king whose birth is in a white colour & ends the second work unles you shall think fit to decoct one half of it to ye red.

[73] *Corres 2*, 386.
[74] *Keynes MS* 30, f. 1. This sheet has been misplaced; it belongs with *Keynes MS* 35, which contains the texts corresponding to four of the headings. Cf. *Keynes MSS* 20, 25.

Aph. 5

The feces wch in the second work are separated by ye putrefactions when the putrefactions are over & the matter relents into a white water, fall to the bottom of the water & must be separated.

Aph 6

The young new born king is nourished in a bigger heat with milk drawn by destillation from the putrefied matter of the second work. With this milk he must be imbibed seven times to putrefy him sufficiently & then decocted to the white & red, & in passing to ye red he must be imbibed once or twice wth a little red oyle to fortify ye solary nature & make the red stone more fluxible. And this may be called the third work. The first work goes on no further then to putrefaction the second goes on to ye white & ye third to ye red.

Aph. 7.

The white & red sulphurs are multiplied by their proper mercuries (white & red) of the second or third work, wherein a little of the fixt salt is dissolved.

This Process I take to be ye work of the best Authors, Hermes, Turba, Morien, Artephius, Abraham ye Jew & Flammel, Scala, Ripley, Maier, the great Rosary, Charnock, Trevisan. Philaletha. Despagnet.[75]

In the notes, which constituted the bulk of the paper, he drew upon the authors cited to support his résumé of the Work.

Carrying the same impulse – the characteristic Newtonian urge to systematize and organize – a step further in the 1680s, Newton began what he called the Index chemicus.[76] Initially he drew up a list of 115 headings which he entered at roughly half-inch intervals on the eight pages he obtained by folding a single sheet 16 inches by 12 inches twice. Newton wrote virtually all of his alchemical papers on such sheets. Under the headings, he began to enter references from his readings.

Aes. Rosar p 143, 242, 243. Clangor p 305, 311. Scalla p 80 ℓ4, 7, 11. Turba p 5, 6. 13, 16, 17, 23, 24, 25, 33. Norton p 43.

Calcinatio Scala p 82, 83, 84 ℓ 37. Ludus puer. p 128 ℓ 11. Ripl p 126. Intr. ap. p 85.

[75] *Keynes MS* 49, f. 1.

[76] Since the first version of the Index contained references to "Ripley Reviv'd" it had to postdate 1678, the year of publication. Since the first version did not contain any reference to Mundanus (as Newton called Edmund Dickinson's publication of 1686), it probably predated 1686. Since it included *quercus cava* (the hollow oak) as one entry, it may have been begun after 1680, when that substance started to figure in Newton's experimentation. The hand appears to belong to the period around 1680.

He also squeezed new entries in between the original list. By the time the Index had grown to 251 headings, the original sheet could hold no more, and he started again with three sheets or twenty-four pages. He eliminated a few of the original headings that had gathered no references. The new Index expanded as the first had done; it had reached 714 headings when it became too full to hold anymore. By now, it contained references to well over a hundred different alchemical authors. It also began to change in character from a simple index intended to guide its author to passages about given topics. Bits of prose with brief explanations of the headings began to appear, and the Index started to become a general guide to the Art. A few references to Mundanus, added at the end of entries, suggest that Newton filled up the second version of the Index about 1686, the year Dickinson published Mundanus' letter. At that point, other affairs took control of Newton's life for a time. Later, he returned to the Index and expanded it vastly into the final form it achieved, though he continued to add to it until he gave up active alchemical practice.[77]

One sheet indicates that Newton drew upon his alchemy when he composed the "Theologiae gentilis origines philosophicae." The paper contains several tables of the twelve categories that set the number of gods – the seven planets, the four elements, and the quintessence "or chaos of elements" (Figure 9.2). Under each he entered a member of Noah's family, an Egyptian god, a Roman god, an alchemical symbol, and one or more alchemical substances. Thus, under the symbol for fire, △, he put two substances, sulfur and acid. The symbol for the quintessence was ♁, also the symbol of antimony and of the earth considered as a cosmic body, and under it he entered, on one list, "chaos," on another, both "chaos" and "magnesia." "Magnesia," he explained, "is neither fire nor air nor water nor earth, but all of them. It is fiery, airy, watry, and earthy. It is hot and dry, moist and cold. It is watry fire and fiery water. It is earthy spirit and spiritual earth. It is condensed spirit of the world, and the most noble quintessence of all things, and hence ought to be designated by the symbol of the world."[78]

Newton did more than read alchemy. He also experimented extensively. He left behind dated experimental notes beginning in 1678 and extending, with a break in the late 1680s, nearly to the

---

[77] With the exception of a partial draft of the opening pages of the final version, which is in the Yale Medical Library, all of the papers relevant to the Index chemicus are in *Keynes MS* 30. Only the final version has been foliated. In the unfoliated sheets at the end of *Keynes MS* 30, the first version immediately follows the final one, the second version follows it, and the original list of headings follows the second. Pages that Newton labeled "Supplementum Indicis Chemici," which he wrote preparatory to the final version, come at the end of *Keynes MS* 30.   [78] *Babson MS* 420, pp. 1–2.

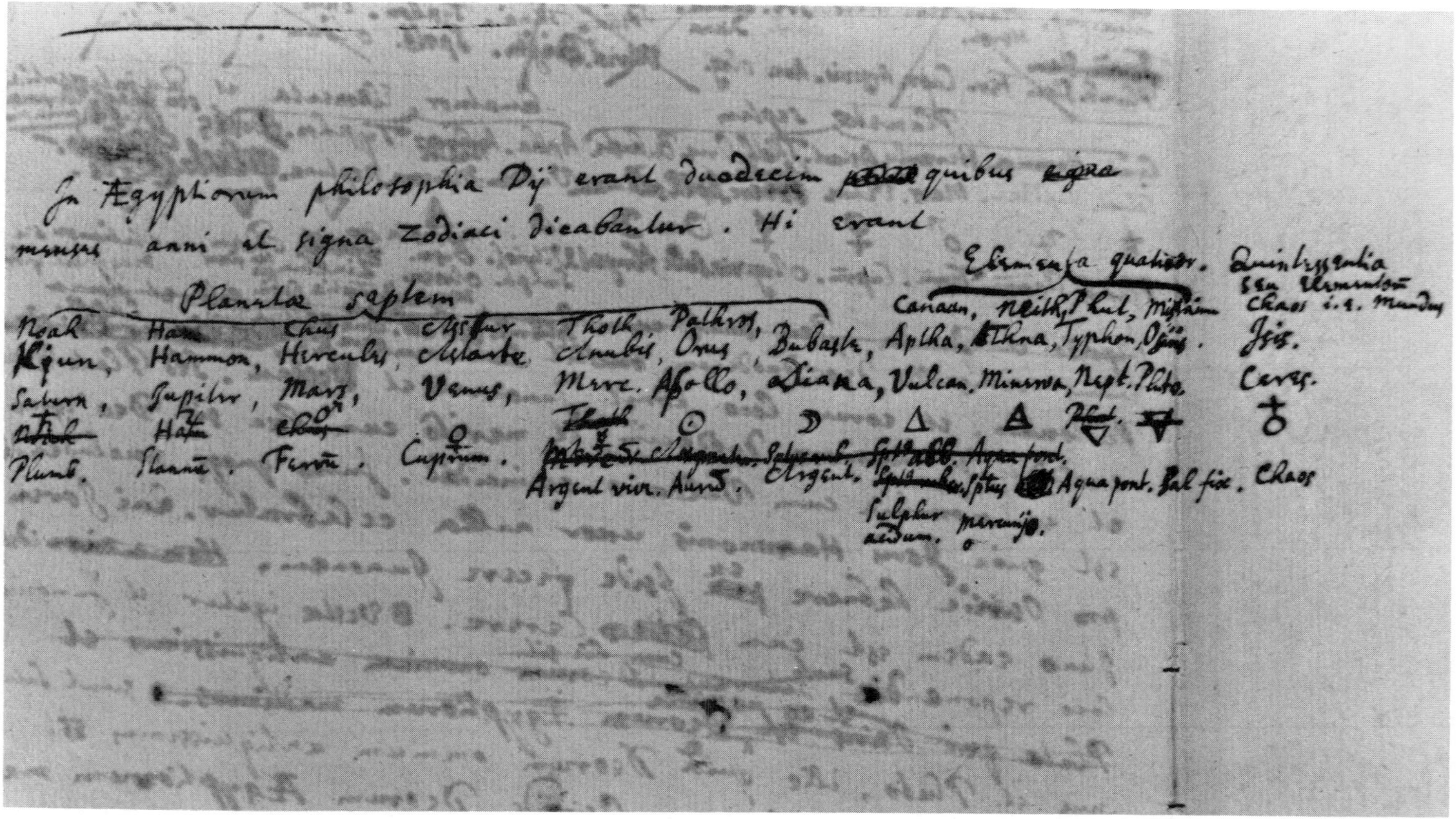

Figure 9.2. The twelve gods of the ancient peoples. (By permission of the Babson College Library.)

time of his departure from Cambridge. His experimentation left a deep impression on Humphrey Newton. We know from surviving manuscripts that Humphrey copied extensive mathematical and theological manuscripts for Newton. He also transcribed the entire *Principia,* the composition of which occupied at least half of his stay in Cambridge. Neither of the first two activities figured in Humphrey's recollections at all, and he gave only two sentences to the *Principia.* Chemical activities bulked very large in his memory, however. He recalled that Newton slept very little

> especially at Spring & Fall of ye Leaf, at wch Times he used to imploy about 6 weeks in his Elaboratory, the Fire scarcely going out either Night or Day, he siting up one Night, as I did another, till he had finished his Chymical Experiments, in ye Performances of wch he was ye most accurate, strict, exact: What his Aim might be, I was not able to penetrate into, but his Pains, his Diligence at those sett Times, made me think, he aim'd at something beyond ye Reach of humane Art & Industry.

He employed himself "with a great deal of satisfaction & Delight" in his laboratory, Humphrey added. He built and remodeled his own furnaces. The laboratory was very "well furnished with Chymical Materials, as Bodyes, Receivers, Heads, Crucibles &c, wch was made very little use of, ye Crucibles excepted, in wch he fus'd his Metals." Sometimes, though very seldom, he would look in an "old mouldy Book" which Humphrey thought was Agricola, *De metallis [De re metallica],* "The transmuting of metals, being his Chief Design, for wch Purpose Antimony was a great Ingredient."[79]

Although the sparsity of references to Agricola among Newton's papers suggests that Humphrey supplied that detail by grasping at a well-known work which he thought was appropriate, the extensive experimental notes that Newton left behind from this period do support the principal elements of his account. A manuscript in the Countway Library contains a sheet with recipes including some on the amalgamation of tin, and another sheet with experimental notes that appear to stem from the late 1670s. *Keynes MS* 58 examined the chemical meaning of imagery associated with Saturn, Jupiter, and the scepter of Jupiter. One source of this imagery to which he referred was John de Monte-Snyders' "Metamorphosis of the Planets," one of the unpublished alchemical treatises that Newton copied, apparently in the late 1670s.[80] In one of the myths to which

---

[79] *Keynes Ms* 135. He devoted considerable space in each of the two letters to describing the experimentation and the laboratory. It was one of the things in Humphrey's oral account that impressed Stukeley. Humphrey, he said, admired Newton's "nicety & constancy in making his experiments, in weighing things he would be scrupulously exact, & that his fires were almost perpetual" (Stukeley to Conduitt, 1 July 1727; *Keynes MS* 136, p. 7).

[80] Yale Medical Library.

Newton referred, Saturn attempted to eat his son Jupiter but was tricked into eating a stone which he vomited up. Jupiter then deposed him, grasped the scepter, and was carried to his throne on the back of an eagle. A substance called *aqua sicca* (dry water), one of the combinations of opposites in which alchemy delighted, also appeared in the paper. Newton posed three questions which set the imagery over into experimental laboratory terms.

Quaere 1. whether ♄ [Saturn, lead] must eat ye stone for ♃ [Jupiter, tin] so soon as ye spt has dissolved ♄ & is satiated but not yet grown to a dry black calx (as is most probable becaus otherwise ♄ will distill over wth him) or after ye sublimation of ♄? Quaere 2 Whether ye aqua sicca will not be more fixt by joyning wth some bodies as ♂ [Mars, iron] or ♀ [Venus, copper] & so let y^e ☿ [Mercury] come over alone. Quare 3 whether this stone be crude or its calx.[81]

And he concluded with three very specific experiments.

To be tried.
1. Extract ♀ from green Lyon wth Æ [aqua fortis, nitric acid] diluted & make ye menstrue of this
2. Try if y^t menstrum will dissolve Lead ore.
3. Get y^e ⊕^{s} [vitriols] & try ye ferment.[82]

Whether or not he tried these experiments, he tried many others of which he left two manuscript accounts, beginning in 1678.[83] Some of them imply a sustained program of experimentation pursued over a long period of time. Thus in August 1682, he used a lead ore that he had impregnated with salt of antimony many years before and had freed from volatile sulfur the previous year. In April 1686, he performed an experiment with spelter (zinc) that had been dissolved in distilled spirit of antimony two years earlier and then sublimed with sal ammoniac and again with aqua fortis.[84] The dates also indicate that 1682 and the winter of 1683–4 (when Humphrey was there) were periods of especially intense laboratory work.

Newton's later experiments have not been explicated as his earlier ones have been. Since he began to employ a private system of symbols, at the meaning of which we can only guess in many cases, they may never be explicated. Nevertheless, they contain enough hints that we are compelled to believe they followed the pattern of

---

[81] *Keynes MS* 58, f. 2ᵛ.

[82] *Keynes MS* 58, f. 4ᵛ. Cf. the discussion in Betty Jo Teeter Dobbs, *The Foundation of Newton's Alchemy. The Hunting of the Greene Lyon* (Cambridge, 1975) pp. 168–72, on which I have drawn heavily.

[83] *Add MSS* 3973 and 3975, pp. 101–58, 267–83. Although there is not a complete overlap between the two, much of the record in 3975 was copied over from 3973. See A. R. and Marie Boas Hall, "Newton's Chemical Experiments," *Archives internationales d'histoire des sciences, 11* (1958), 113–52.   [84] *Add MS* 3973, ff. 13, 19.

the earlier experiments and the proposed experiments in Keynes MS 58. That is, their chemical operations pursued alchemical ends. Vague correlations with themes in his alchemical papers can be perceived, leading one to hope that in time more precise correlations, such as those already established for his earlier experiments, can more fully illuminate his experimental records. For the moment, note that he experimented in the 1680s with the net, the hermaphroditic compound of Venus and the regulus of Mars that he used in the early 1670s.[85] In his compilation of the Work about this time, Newton included a passage from Basil Valentine. "The tincture of ☉ is found in none more plentiful then in ♂ & ♀ as man & wife. . . . Mercury makes active but Venus provokes giving lust & desire. Her Sulphur surpasses ye Sun. With same tincture Mars abounds in a more noble & powerful manner for ♂ is ye man & Venus ye woman."[86]

In 1680, Newton mentioned another substance, the oak (usually the hollow oak or *quercus cava*), and in 1682, he experimented with it extensively. The oak derived its name from the myth of Cadmus, whose companions were devoured by a serpent which Cadmus slew by impaling it against an oak with his spear. Newton found the image more than once in his reading.[87] Chemically, the companions of Cadmus were the inferior metals; the serpent was vinegar of antimony which would dissolve them. Newton described the oak as "Reg ♂ ♀ ☿," a description which makes it difficult to distinguish from the net.[88] Indeed, when he first mentioned it, he used the name interchangeably with powdered net. He imbibed the oak with vinegar of antimony, and in a series of experiments tried to determine the conditions under which the maximum amount of imbibed oak would sublime with sal ammoniac. Presumably the serpent was fixed to the oak when the two sublimed together. He found that 1 part of oak to 3/22 parts vinegar of antimony volatized best and when melted formed a substance grained like the oak itself. "And this I conceive to be ye right preparation of ye Oak. But I do not think it is to be volatized wth Venus because ye addition of more reg. of ☿ will volatize it better. Tis rather designed for a clean sulphur to joyn in fermentation w^{th} ☿."[89] Though the pur-

[85] *Add MS* 3975, pp. 144, 147. In connection with the hermaphrodite, note Newton's frequent use of materials "impregnated" with some liquid; e.g., *Add MS* 3973, f. 1; *Add MS* 3975, p. 109.

[86] *Keynes MS* 35, sheet 7. Cf similar passages on sheet 6 (from Snyders) and in *Keynes MS* 25 (from Philalethes). I think I could find an almost unlimited number of such passages among Newton's alchemical papers.

[87] E.g., *Keynes MS* 35, sheet 7 (from Philalethes) and the Index chemicus, where he cited Philalethes, Ripley, and Maier under the heading of Cadmus with his companions (*Keynes MS* 30, f. 19).

[88] *Add MS* 3973, f. 9, and *Add MS* 3975, p. 106.   [89] *Add MS* 3975, pp. 133–4.

pose remains obscure enough, the reference to the male and female principles in the final suggestion indicates some connection with the basic principle that all things generate by sexual union.

Much the most prominent substance used in his experiments at this time was sal ammoniac (NH_4Cl), which he designated with the symbol $*$. Both the experiments that used it and others as well investigated means of volatizing metals, and much of the time $*$ appears to represent the volatizing principle, the sal ammoniac latent in bodies, rather than the chemical sal ammoniac. Sal ammoniac did seem, in Newton's view, peculiarly to embody the volatizing principle, however.

> Sal ammoniac [he wrote later in the Index, possibly quoting some authority] is the key to the art. It does not give color but it gives entrance, it prepares and purges: Then the other spirits enter mixed bodies and it joins the bodies and spirits and itself withdraws. It is an unguent coagulated from a warm and subtle nature by the dryness of fire; penetrating from part to part it dissolves bodies. It is the unifier of contradictories and of all spirits with bodies. For it itself is a volatile spirit, a generating stone, an aid to the Elixir. . . . Therefore anyone who works without the salt accomplishes nothing, like someone who hunts with a bow that has no string.[90]

Newton proffered only a few hints about his purpose in these experiments. He appears to have been seeking to liberate the spirit or active virtue of bodies from its encumbering feces, or perhaps, as the quotation above implies, to spiritualize bodies by joining spirit to them.

In 1680, he carried out a systematic investigation of the effect of various mixtures on the volatizing virtue in antimony, the sal ammoniac, as he sometimes put it, in the belly of antimony.

> I do not yet find any way [he noted] of cleansing ye sublimate of ♂ from it's impure ♀ without destroying its volatizing virtue. If ♂ be melted wth 1/2 1/3 or 1/4th part of ☉ [nitre], the nitre does not hold down ye impure ♀ of y^e ♂ at all. But let ye whole body of y^e ♂ rise & remains it self in ye bottom without much addition of ♂.[91]

A year later, he experimented with spar and Venus volans (a sublimate of copper, probably cuprous chloride). The result was negative. Spar, he concluded, is "not to be spiritualized" immediately by Venus volans. Once, he sublimed lead antimoniate with a sublimate made from iron and antimony. The *caput mortuum* that was left behind fumed nearly all away on a piece of glass set in a hot fire. "Whence I knew it to be ye shadow of a noble exp^t." Perhaps it is indicative of his purpose that in the summer 1682 he deter-

---

[90] *Keynes MS* 30, ff. 73–4. Cf. two passages, one from Philalethes, the other apparently from Sendivogius, in *Keynes MS* 35, sheet 7.   [91] *Add MS* 3975, p. 114.

mined by experiment that unmelted antimony made the matter more fluxible and volatile than melted antimony. He had learned something about this already from Philalethes: "The fire of Vulcan is ye artificial death of metals & as many as have suffered fusion in it have lost their life." In 1684, Newton found that a salt obtained from spelter, antimony, and vinegar of antimony sublimed more successfully with a prepared sal ammoniac than salt of copper. "Is not that salt more related to mercury than the salt of copper?" he asked. "Is it not the mediator between the two for composing the caduceus?"[92] Beyond such vague hints, it is not possible to proceed at the moment.

One of the tests that Newton commonly applied to his chemicals was taste. A sharp taste indicated the presence of sal ammoniac. As earlier he had not hesitated to perform frightening experiments on his eyes, so now he did not hesitate to taste a considerable spectrum of chemicals many of which contained heavy metals and other toxic substances. He also vaporized a number of dangerous substances, including a great deal of antimony and mercury, in the cramped quarters of his laboratory in the garden. Recently, samples of Newton's hair have been tested by the latest techniques for their content of heavy metals and other chemicals. The samples of hair probably came from his old age, thirty years after his active experimentation ceased, when the concentrations in his body of the elements in question should have diminished. Nevertheless, for virtually every element tested, his hair showed concentrations several times those of average hair in the twentieth century. In the case of mercury, the concentration in one sample was more than ten times, in another sample nearly forty times the normal level of twentieth-century hair.[93] In pursuing the alchemical elixir, Newton very nearly poisoned himself.

One of the features that has struck everyone who has looked at Newton's experimental notes is their quantitative precision, which reminds us that they were composed by the same man who had established the periodicity of certain optical phenomena by the careful measurement of rings. As we have seen, the same precision left a lasting impression on Humphrey Newton. Consider but one example.

[92] *Add MS* 3975, p. 119. *Add MS* 3973, ff. 5^v–6, 14–14^v; *Keynes MS* 35, sheet 10. *Add MS* 3975, pp. 147–8.

[93] It is only fair to add that a third sample, the least well authenticated one, showed a concentration of mercury only 40 percent higher than that of average twentieth-century hair. P. E. Spargo and C.A. Pounds, "Newton's 'Derangement of the Intellect.' New Light on an Old Problem," *Notes and Records of the Royal Society, 34* (1979), 11–32. See also L.W. Johnson and M.L. Wolbarsht, "Mercury Poisoning: A Probable Cause of Isaac Newton's Physical and Mental Ills," *ibid., 34* (1979), 1–9.

Le. o. [lead ore] wch many years ago had been impregnated wth salt of ♂, & the last year freed from all ye volatile ♁ I took & from 6, 6 & 6 ^{gr} thereof I sublimed over a candle 10, 12, & 14 ^{gr} of ☉ [sublimate of antimony?] & the salts sublimed away w^{th}out fusion & left of ye matter 6, 6 & 6 ^{gr} of a light coloured calx wch did not flow upon a red hot iron but yet fumed away & shrunk the first a little the second more the third most. Afterwards in thre glass viols from 60, 60, 60, 60, 60, & 60 ^{gr} of this impregnated Le. o. I sublimed 60, 80, 100, 120, 150, 180 ^{gr} of ☉ & they all flowed & bubled in ye sublimation & sent up a white sublimate & left 56, 55 3/4, 55 2/3, 55, 54, 52 ^{gr} below of a brittle whitish mass of ye color of a grey hat: wch masses all flowed & fumed away readily & alike upon a red hot iron except ye first wch flowed not & fumed away but very slowly & difficultly. If there was any difference between the rest it was only this that ye second in flowing spread something less upon ye iron then ye 3^d & ye 3^d then ye 4th. Upon a glass over a candle the second did not grow so thin in melting as ye third nor the third as the 4th 5^t & 6^t nor spread & flowed so much over the glass but continued in a viscous & thick lump like birdlime & sooner grew into a dry unfusible mass. The 4th 5^t & sixt were almost alike fusible & the third equalled them in fusibility more nearly then the second. The sixt seemed a very little more fusible & then when melted then ye fift & the fift then ye 4th. & the sixt in fusion grew transparent, the fift had a little whitish opakeness mixt wth a transparency & ye 4 was more opake & less thin then ye fift as ye fift was then ye sixt. Yet the fift was almost totally transparent. From all wch I conclude y^t ye proportion of 2 to one or at most 2 2/3 to one is good. But since 2 to one makes 4/5 parts distill either this proportion or at most 4 to 9 is to be used. For I do not find that a greater proportion y^n 2 to one will make above 4/5 parts distill.[94]

To be sure, more simply than quantitative precision distinguishes this experimental report, but quantitative precision is perhaps its most prominent feature. Alchemy had never known anything like it before. Indeed, it appears to me that alchemy, which relied on allusions that avoided the explicit, could not long survive such treatment. Newton himself enabled us quickly to compare his experimentation with the tradition he received. In the Index, under "Pondus" he collected all the references he could find to the proportions of ingredients in the Work. It is already indicative of his outlook that "Pondus" was by far the longest entry in the Index. Even so, it offered rather thin pickings in comparison to his own experiments. The "Turba" (and a number of other treatises) told

---

[94] *Add MS* 3973, ff. 13–13^v. Cf. other experiments similar in their quantitative concern, ff. 5^v–6 and *Add MS* 3975, pp. 135 and 141–4.

him to use 1 of air to 3 of water by weight, or perhaps 2 to 7. Ripley said that the weights of gold, silver, and mercury should be 1, 2, 4, or 1, 3, 4 without imbibition. Others set them at 1, 2, 1 and 1, 2, about 1/4. Ripley also advised "equal or a little greater weight of mercury to sulfur." From the *Rosary,* supported by eleven other authorities, he learned that "the weight of gold to the weight of imperfect body in fermentation, or the weight of one of the elements to the weight of another [should be] as 1 to 3 or 4."[95] Given his proclivities as the entry "Pondus" reveals them and the paucity of information he found, it is small wonder that he frequently tried experiments with as many as six different proportions to establish the optimum conditions.

In the report cited above, it is worth noting something else that pertains to all his experiments. Newton worked with a system of weights that divided an ounce into 480 grains. In one experiment performed in 1684, he tried four different proportions of a precipitate of the net imbibed with antimonial vinegar and a salt made of prepared sal ammoniac and vitriol. The total quantity did not exceed 20 grains in any of the four trials. The first three attempts left residues of 3 1/2, 3 1/7, and 3 1/6 grains, the fraction in all cases representing the excess of the residue over the 3 grains of salt with which he started. In the fourth trial, 5 grains of precipitate and 4 grains of salt "left 1 ^{gr} exactly of a sweetish & a litle stiptick tast but not strong. This was tried wth great exactness & caution, the matters being mingled on a looking-glass that none of them might be lost. The last but one was also tryed carefully enough ye matters being mingled on a glass wth ye point of a knife."[96] In the context of the garden laboratory, confidence in weights to a fraction of a grain may seem misplaced. Newton was not one to deceive himself, however. A decade earlier he had measured colored circles to fractions of one-hundredth of an inch and arrived thereby at conclusions which remain basic to our present conception of light.

In the spring of 1681, Newton's experimentation reached a climax. In the midst of his experimental notes, which he set down in English, he entered two paragraphs in Latin that were obviously not experimental notes but appear to have been meant as interpretations of them. An exultant air pervades the paragraphs; one can almost hear the triumphant "Eureka!" ring through the garden.

> May 10 1681 I understood that the morning star is Venus and that she is the daughter of Saturn and one of the doves. May 14 I understood—☽ [the trident?]. May 15 I understood "There are indeed certain sublimations of mercury" &c as also another dove: that is a sublimate which is wholly feculent rises from its bodies white, leaves

95 *Keynes MS* 30, ff. 66–9.   96 *Keynes MS* 3975, p. 143.

a black feces in the bottom which is washed by solution, and mercury is sublimed again from the cleansed bodies until no more feces remains in the bottom. Is not this very pure sublimate —✳? [sophic sal ammoniac?][97]

The following experimental note, which intervened before the second interpretive paragraph, employed "sophic ✳," which I take as support for my interpretation of the symbol —✳.

May 18 I perfected the ideal solution. That is, two equal salts carry up Saturn. Then he carries up the stone and joined with malleable Jupiter [as much as one wants to say "tin" here, Newton did write *Jove* instead of inserting the symbol ♃] also makes ✳ [sophic sal ammoniac?] and that in such proportion that Jupiter grasps the scepter. Then the eagle carries Jupiter up. Hence Saturn can be combined without salts in the desired proportions so that fire does not predominate. At last mercury sublimate and sophic sal ammoniac shatter the helmet and the menstruum carries everything up.[98]

An entry in Newton's other set of laboratory notes, also in Latin, with the date "July 10" but no year, appears to belong to the same climactic experiments of 1681. "I saw sophic sal ammoniac. It is not precipitated by salt of tartar."[99] From this time, Newton used mostly sophic sal ammoniac, which he called a number of names such as "✳ prep" [prepared sal ammoniac] and sometimes distinguished from vulgar sal ammoniac, in his experimentation. In 1682, he contrasted "✳ prepared with ♁" to "✳ without ♁" and once he referred to "the green lion (or our ✳)."[100]

Another undated Latin note among his experimental records may also stem from the same series. At least, it suggests an interpretation of the symbol.

Dissolve volatile green lion in the central salt of Venus and distill. This spirit is the green lion the blood of the green lion Venus, the Babylonian Dragon that kills everything with its poison, but conquered by being assuaged by the Doves of Diana, it is the Bond of Mercury.

[97] "May 10 1681 intellexi Luciferam ♀ et eandem filiam ♄^{ni}, & unam columb^{rum}. May 14 intellexi ⋲. May 15, intellexi *Sunt enim quaedam* ☿^{ij} *sublimationes* &c ut & columbam alteram: nempe Sublimatum quod solum foeculentum est, a corporibus suis ascendit album, relinquitur foex nigra in fundo, quae per solutionem abluitur, rursusque sublimatur ☿^{ius} a mundatis corporibus donec foex in fundo non amplius restet. Nonne hoc sublimatum depuratissimum sit —✳?" (*Keynes MS* 3975, p. 121).

[98] "May 18 Ideam solutionis perfeci Nempe aequalia duo salia elevant ♄. Dein hic elevat lapidem, nec non cum Jove malleabili conjunctus fit ✳ idque in tali proportione ut ♃ sceptrum apprehendat. Tunc aquila ♃^{em} attollet. Potest dein ♄ sine salibus in ratione desiderata conjugi ne ignis praedominetur. Denique ☿ sublim. & ✳ praeparat feriunt cassidem, & menstruum omnia attollit" (*Keynes MS* 3975, p. 122)

[99] "July 10. Vidi ✳ Philosophicum. Hic non praecipitatur per salem tartari" (*Keynes MS* 3973, f. 17). [100] *Add MS* 3975, pp. 123, 153.

Neptune with his trident leads the Philosopher into the sophic garden. Therefore Neptune is the mineral watry menstruum and the trident is the ferment of water similar to the caduceus of mercury with which mercury is fermented, that is, the two dry Doves with the dry martial Venus.[101]

Unfortunately, the interpretive passages require interpretation; and by couching them in alchemical symbolism, Newton has made that virtually impossible. At some point, he crossed out the two triumphant paragraphs from May, probably indicating that the triumph had dissolved into failure. Nevertheless, the disillusionment was apparently not complete, for he continued from this time to employ sophic sal ammoniac in his experimentation. And ten years later, when he composed his most important alchemical treatise, he employed all the imagery that the paragraphs embodied.

In 1684, Newton squeezed in another exultant note between two lines of his experimental notes. "Friday May 23 [1684] I made Jupiter fly on his eagle."[102] By May 1684, Humphrey Newton was helping to tend the furnaces. "Nothing extraordinary, as I can Remember, happen'd in making his experiments," he recounted to Conduitt, "wch if there did, He was of so sedate & even Temper, y^t I could not in ye least discern it."[103] Newton had his own way of expressing his elation. He shouted his eurekas to himself. Nevertheless, one might think that Jupiter flying on his eagle near the east end of Trinity chapel would have attracted Humphrey's attention. Or he might have noticed that Newton got excited enough with his experiments to forget, as his notes reveal, what day of the week it was.[104]

In addition to reading, copying, and experimenting in alchemy, Newton was also composing alchemical treatises. One such, which appears by its hand to belong to the late 1670s, carried the title "*Separatio elementorum*" (The Separation of Elements). In careful if somewhat confusing detail, he described special furnaces and containers as well as the process to extract four distinct substances from

---

[101] "Dissolve ♌ vir. volat. in sale centrali veneris et destilla. hic spiritus est ♌ vir. sanguis ♌^ij vir. Venus, Draco Babelonicus omnia veneno suo interficiens, a Columbis tamen Dianae mulcendo victus, Vinculum ☿^ij.

"Neptunus cum tridente inducit Philos in hort. soph. Ergo Neptunus est menstruum aqueum minerale ac tridens fermentum aquae simile caduceo ☿^ij quocum ☿^ius fermentatur, vizt Columbae duae aridae, cum venere arida martiali" (*Add MS* 3973, f. 12).

[102] "Friday May 23 Jovem super aquilam volare feci" (*Add MS* 3975, p. 149).

[103] *Keynes MS* 135.

[104] *Keynes MS* 3975, p. 133, referred to Tuesday, July 19 [1682 or 1683]. July 19 did not fall on Tuesday in either year. p. 150 referred to Wednesday, April 26, 1686, whereas April 26 fell on Monday in 1686.

"sifted powder." Since the paper did not imply any mystery about the powder, which was the beginning of the Work rather than the end, one assumes that he described (or meant to describe) its preparation elsewhere. The "Separatio" does have the appearance of a single chapter from a more extended treatise. It is found today attached to another apparent chapter, *"Reductio et Sublimatio"* (Reduction and Sublimation); but whereas the latter was a compilation of citations from authorities on the Work, the "Separatio" contained numerous emendations, which characterized Newton's own compositions, and no citations at all. Distillations repeated six, eight, or ten times produced a "fluid celestial spirit" which separated into impure lower waters and a mercurial spirit called "our magnet," the bridegroom of a certain central water. Calcination of the *caput mortuum* separated it into "white lunar flowers . . . the foundation of the central Moon," and a salt which is "the foundation of this Treasure . . . since it is extracted from a red solar earth which is the central fire of the central sun. And so much," he concluded, "for the separation of the elements."[105]

In another paper from this period, called "Clavis" (The Key), Newton described a process by which to make a philosophic mercury. The identification of "Clavis" as Newton's own composition has been challenged.[106] Certainly its lack of corrections is a serious argument against its authenticity. Newton extensively corrected and amended nearly everything he wrote. The presence of such corrections in the "Separatio" almost certainly marks it as his own composition. Their absence from the "Clavis" is evidence against his composition. Over against this evidence one must weigh the characteristics it shares with the "Separatio." Unlike virtually all other alchemical literature, both described laboratory procedures in detailed operative terms that could be repeated today, even though we doubt the asserted results. What is most compelling, the "Clavis" employed the proportions for the star regulus that Newton had derived empirically in his own earlier experimentation. Newton also used the distinction that he made in the "Vegetation" between mere mechanical alterations and vegetative processes. And in the years ahead, he drew upon this paper without referring to it explicitly, though it was his policy to note his sources exhaustively.[107] In my estimation, the evidence weighs in favor of its authenticity.

In the "Clavis," Newton described a method that used the doves

---

[105] *Burndy MS* 10.
[106] Figala, "Newton as Alchemist," 107. D. T. Whiteside, "From his Claw the Greene Lyon," *Isis*, 68 (1977), 118.
[107] In *"Praxis," Babson MS* 420. As with the whole argument about "Clavis," I owe this point to Mrs. Dobbs, in this case in a private letter.

of Diana to mediate between the star regulus and common mercury in order to produce a philosophic mercury that would dissolve all metals including gold.

> I know whereof I write [he declared], for I have in the fire manifold glasses with gold and this mercury. They grow in these glasses in the form of a tree, and by a continued circulation the trees are dissolved again with the work into new mercury. I have such a vessel in the fire with gold thus dissolved, where the gold was visibly not dissolved by a corrosive into atoms, but extrinsically and intrinsically into a mercury as living and mobile as any mercury found in the world. For it makes gold begin to swell, to be swollen, and to putrefy, and to spring forth into sprouts and branches, changing colors daily, the appearances of which fascinate me every day. I reckon this a great secret in Alchemy . . .[108]

During the later 1670s, following his response to Robert Boyle's letter about a special mercury in the *Philosophical Transactions*, Newton entered into direct correspondence with Boyle. Late in 1676, Boyle sent two copies of his recent book via Oldenburg, one for Newton and one he was asked to deliver to Henry More.[109] Newton immediately sent his thanks through Oldenburg since he did not have Boyle's address.[110] Three months later, again in contrast to his usual reticence, he volunteered a comment on another article by Boyle in the *Philosophical Transactions*.[111] Some time before 28 February 1679, his manifest effort to establish a correspondence bore fruit. Although the letter of that date is the first that survives, it referred to earlier communication. Newton's letter has been known since it was printed in the eighteenth century, and it has often been republished as a typical example of the mechanical philosophy at work. In fact, it shows that Newton had moved still farther from the stance of orthodox mechanical philosophy, and its content suggests that alchemy had played the crucial role in moving him.

The book that Boyle presented to Newton late in 1676 was probably *Experiments, Notes, &c. about the Mechanical Origine or Production of Divers Particular Qualities*. Newton's letter of 1679 commented implicitly on some of the central arguments of the book, arguments about the dissolution, precipitation, and volatilization of substances.

---

108 *Keynes MS* 18. Printed in full in Dobbs, *Foundations*, pp. 251–3; translation (from which I quote) pp. 253–5. Although in *Foundations* Mrs. Dobbs places "Clavis" in the early 1670s, further evidence has since led her to put it in the late 1670s.

109 Mentioned in Newton to Oldenburg, 14 Nov. 1676; *Corres 2*, 182.

110 Newton to Oldenburg, 18 Nov. 1676; *Corres 2*, 183.

111 Newton to Oldenburg, 19 Feb. 1677; *Corres 2*, 193–4.

Honoured S^r [he began]

I have so long deferred to send you my thoughts about ye Physicall qualities we spake of, that did I not esteem my self obliged by promise I think I should be ashamed to send them at all. The truth is my notions about things of this kind are so indigested y^t I am not well satisfied my self in them, & what I am not satisfied in I can scarce esteem fit to be communicated to others, especially in natural Philosophy where there is no end of fansying.

He concluded the letter by saying that Boyle could "easily discern whether in these conjectures there be any degree of probability, wch is all I aim at. For my own part I have so little fansy to things of this nature that had not your encouragement moved me to it, I should never I think have thus far set pen to paper about them."[112] We can judge how seriously to take such deprecations by the fact that he composed four thousand words on the fancies he fancied so little.

In his book, Boyle had cited many chemical reactions in which heat appeared. Chemists referred the heat, he said, to violent antipathies between substances; as a mechanical philosopher, he ascribed it to the motion among the particles of the substances. By 1679, as his experimental notes testify, Newton had had first-hand experience with many such reactions. Indirectly, he raised with Boyle the question of the cause of motion among the particles of previously cold substances. He himself explained the motion by an "endeavour . . . bodies have to recede from one another . . . " When a particle is separated from a body in dissolution, it is accelerated by the endeavor to recede "so y^t ye particle shall as [it] were wth violence leap from ye body, & putting ye liquor into a brisk agitation, beget & promote y^t heat we often find to be caused in solutions of Metals."[113] He traced the phenomena of surface tension and expansion of air, which had fascinated him since his introduction to natural philosophy, to the same endeavor to recede. Under certain conditions, bodies also endeavor to approach each other, and he argued that this endeavor causes the cohesion of bodies. He derived both endeavors, in turn, from mechanisms within the aether, with the positing of which the letter began. Nevertheless the very language of "endeavors" was a step toward a radically different explanation.

The letter concerned itself primarily with the causes of solubility and volatility, two of the qualities to which Boyle had directed his treatise. Chemists, Boyle had argued, generally attributed solubility to a certain sympathy between the substance in question and its menstruum. As a mechanical philosopher, he was unable to under-

---

[112] Newton to Boyle, 28 Feb. 1679; *Corres 2*, 288–95.   [113] *Corres 2*, 291–3.

stand what such a sympathy could be unless it were solely a matter of the sizes and shapes of particles and pores. Newton did not use the word "sympathy," but he did assert that "there is a certain secret principle in nature by wch liquors are sociable to some things & unsociable to others." He expressly denied that it has anything to do with the sizes (and by implication the shapes) of pores and particles.[114] He had mentioned such a principle in his "Hypothesis of Light." Its existence and its role in nature provided the central argument of the letter to Boyle. By inference, an unsociability between aether and gross bodies caused the rarity of aether in the pores of bodies, the basis of his explanation of the endeavors to approach and recede and of the new explanation of gravity that he offered at the end of the letter. The basis of the secret principle of sociability, in turn, was entirely chemical. That is, he justified his assertion of it by chemical phenomena. Water does not mix with oil but does mix readily with spirit of wine and with salts. Water sinks into wood, but quicksilver does not. Quicksilver sinks into metals, but water does not. Aqua fortis dissolves silver but not gold; aqua regis dissolves gold but not silver. So also he illustrated by chemical phenomena the principle of mediation by which unsociable substances are brought to mix. Molten lead does not mix with copper or with regulus of Mars, but with the mediation of tin it mixes with either. With the mediation of saline spirits, water mixes with metals; that is, acids (water impregnated with saline spirits) dissolve metals. He employed the same arguments, the endeavor to recede and the principle of sociability, to explain volatization; and in passing, he dropped comments about metallic exhalations from the earth and metallic particles as the constituents of true permanent air. The letter to Boyle opened with a set of five suppositions about the aether that gave it the appearance of standard mechanical philosophy. The centrality in it of the principle of sociability, however, transformed it into an assertion of the insufficiency of the mechanical philosophy of nature for the explanation of chemical phenomena.

Although few of their letters have survived, Newton and Boyle continued to correspond and also to meet. In August 1682, Boyle thanked Newton for a transcribed book he had sent. It is difficult not to connect it with the large number of alchemical treatises copied by Newton. Boyle's letter mentioned other correspondence, including favorable comments by Newton on Boyle's book about the noctiluca (or phosphorus). He mentioned a recent visit and Newton's promise that he would come to London again soon.[115] His reference to visits in London recalls the evidence from the Steward's Books that Newton was absent from Cambridge more

---

[114] *Corres* 2, 292.   [115] Boyle to Newton, 19 Aug. 1682; *Corres* 2, 379.

during the early 1680s than he had been during the previous fifteen years. His mother's death occasioned his long absence in 1679, but he was gone even longer (fifteen and a half weeks) in the year that began on Michaelmas 1679. Since he returned from Grantham at the end of November, more than six weeks of later absence about which we know nothing remain to be accounted for. In 1682, he was gone for five and a half weeks and in 1683 for six. Boyle's letter makes it clear that he spent part of the time at least in London concerned with alchemy. His letter from Francis Meheux in 1683 implies the same.

Meanwhile, in close connection with the letter to Boyle in 1679, Newton began a treatise known, from the titles of its two chapters, as "De aere et aethere" (Concerning the Air and the Aether). From its content, it appears to have been an effort to expound the same phenomena in the form of a systematic treatise.[116] Whereas the letter to Boyle began with the postulation of an aether, "De aere" began with observed phenomena of the air, primarily its capacity to expand, one of the critical phenomena that had seized Newton's attention in 1664–5. The absence of external pressure allows air to expand. Heat causes it to expand. The presence of other bodies also causes it to expand. To justify the last assertion, Newton pointed to capillary phenomena, which arise from differences in pressure "because the air seeks to avoid the pores or intervals between the parts of these bodies . . ." Indeed, he found that in general bodies seek to avoid each other and he cited in justification, among other things, the phenomenon of surface tension. Various explanations of these phenomena might be offered, he continued.

> But as it is equally true that air avoids bodies, and bodies repel each other mutually, I seem to gather rightly from this that air is composed of the particles of bodies torn away from contact, and repelling each other with a certain large force.

Only in such terms, he argued, can we understand the immense expansions that air can undergo until it fills volumes a thousand times its normal ones,

> which would hardly seem to be possible if the particles of air were in mutual contact; but if by some principle acting at a distance [the

---

[116] The Halls, who published *De aere et aethere* (*Halls*, pp. 214–20; English translation, pp. 220–8), date it in about 1674 (p. 134). It must have postdated a book of 1673 by Boyle to which it referred, and they feel that it definitely preceded the "Hypothesis," since the paper referred to forces of attraction and repulsion, whereas the "Hypothesis" used aethereal mechanisms. The thrust of my interpretation of Newton holds that he started with aethereal mechanisms–the established philosophic orthodoxy–and that he finally rejected them in order to assert the radical concept of forces at a distance. Hence I am inclined to put the paper after the letter to Boyle. Obviously I run the danger of forcing evidence into preconceived molds. I shall not attempt to defend my interpretation further here; if it does not stand by itself, no footnote can suffice to prop it up.

particles] tend to recede mutually from each other, reason persuades us that when the distance between their centers is doubled the force of recession will be halved, when trebled the force is reduced to a third and so on . . . [117]

As in the letter to Boyle, he went on to discuss the generation of air by various processes in nature.

At one point, Newton started to speculate on the possible causes of the repulsion between bodies. He crossed the paragraph out, presumably because chapter 1 expounded phenomena only and he intended to reserve such discussion for another place. The third possible explanation he offered before he canceled the passage had a familiar look. "Or it may be in the nature of bodies not only to have a hard and impenetrable nucleus but also a certain surrounding sphere of most fluid and tenuous matter which admits other bodies into it with difficulty."[118] That is, the alchemical hermaphrodite, sulphur surrounded by its mercury, offered a model to explain the universal property that all bodies possess to act upon each other at a distance.

Chapter 2 of the treatise bore the title, "De aethere," and in it Newton started to discuss the generation of aether by further fragmentation of aerial particles into smaller pieces. Apparently he intended to use the aether, as he had in the letter to Boyle, to explain what he had there called the endeavor to recede. First, he began to list the evidence for the existence of an aether. He cited electric and magnetic effluvia and the saline spirit that passes through glass and causes metals calcined in sealed vessels to increase in weight. He also cited the fact that a pendulum in an evacuated receiver stops almost as soon as one in the open air. Already, when he composed the "Quaestiones," this experiment had seemed to Newton to demonstrate the presence of a resisting medium in the receiver, and he had cited it as evidence for the aether in the "Hypothesis of Light." Newton did not proceed very far with the chapter on the aether. After a few lines, in the middle of a sentence in the middle of a page, he stopped, and he never took it up again.[119] There are many possible reasons why Newton broke off the treatise. Mr. Laughton may have called; he may have gone to dinner and forgotten the paper when he returned; or he may have paused to reflect on the argument itself and realized that he was committing himself to an infinite regression in which a further aether would be required to explain the aether that explained the properties of air, a third aether to expain the second, and so on.

---

[117] *Halls*, pp. 221–4; original Latin, pp. 214–17. Newton crossed out the paragraph in which the specific line about repelling "with a certain large force" appeared. The concept, if not the specific words, remained present.

[118] *Halls*, p. 223; original Latin, p. 216.

[119] *Halls*, pp. 227–8; original Latin, p. 220.

He may also have begun to think more deeply about the evidence of the pendulum experiment. Newton had believed that aether resists the motion of bodies differently than air. Air encounters only the surface of a body, whereas aether penetrates its pores and strikes against all of its internal surfaces as well. Newton had accepted the experiment of the pendulum in a void from Boyle's *Spring of the Air*, one of the first books on the new natural philosophy that he read. Could the issue be more complicated than he had thought? Could he refine the pendulum experiment to make it more demonstrative? In the *Principia*, Newton described such a refined experiment, which he had to relate from memory since he had lost the paper on which he recorded it. He did not date the experiment, but it had to have been after 1675, when he cited the earlier experiment in the "Hypothesis," and if my dating of "De aere et aethere" is correct, at least as late as 1679. If he performed it as late as 1685 in connection with the *Principia*'s composition, it is difficult to imagine that he would have lost the paper. It must have impressed him profoundly, because he remembered its details clearly. Newton constructed a pendulum eleven feet long. To minimize extraneous resistance, he suspended it from a ring that hung on a hook filed to a sharp edge. The first time he tried it, the hook was not strong enough, and it introduced resistance by bending to and fro. Newton was careful enough to notice it—and of course to correct it with a stronger hook. For the bob of the pendulum, he used a hollow wooden box. He pulled it aside six feet and carefully marked the places to which it returned on the first, second, and third swings. To be certain, he did it several times. Then he filled the box with metal, and by careful weighing, in which he included the string around the box, half the length of the string, and even the calculated weight of the air inside the box, he determined that the filled box was seventy-eight times heavier than the empty one. The increased weight stretched the string, of course; he adjusted it to make the length equal to the original. He pulled it aside to the same starting point and counted how many swings it took to damp down to the marks for the empty one. The pendulum required seventy-seven swings to reach each successive mark. Since the box filled had seventy-eight times as much inertia, the resistance full apparently bore the ratio of 78/77 to the resistance empty. By calculation, he concluded that this ratio corresponded to a resistance on the internal surfaces that was 1/5,000 the resistance on the external surface.

> This reasoning [he concluded] depends upon the supposition that the greater resistance of the full box arises from the action of some subtle fluid upon the included metal. But I believe the cause is quite another. For the periods of the oscillations of the full box are less than the periods of the oscillations of the empty box, and therefore

the resistance on the external surface of the full box is greater than that of the empty box in proportion to its velocity and the length of the spaces described in oscillating. Hence, since it is so, the resistance on the internal parts of the box will be either nil or wholly insensible.[120]

In the letter to Boyle, Newton had argued that mechanical principles are inadequate to explain all phenomena. Now he had demonstrated to his own satisfaction that the aether, the deus ex machina that made mechanical philosophies run, does not exist. Driven, as it seems to me, primarily by the phenomena that he observed in alchemical experimentation, and encouraged by concepts that he encountered in alchemical study, Newton appeared to be poised in 1679 on the brink of a further break with the mechanical philosophy, which would have major impact on his future career.

If we judge by his manuscript record from the years of silence, Newton's life during this period consisted primarily of theological and alchemical study. As far as his own remains allow us to judge he would have been content, indeed would have preferred, to be left alone permanently to pursue these studies. He was not left alone, however, despite his efforts to cut himself off. The events of the early 1670s had made his extraordinary capacity in mathematics and physics known, and it tells us much about the dominant currents of European civilization that these studies continued to intrude upon him. Whatever his own desires, alchemy and theology belonged primarily to the past. Mathematics and physics belonged to the future, and the future belonged to them. Those who fostered mathematics and physics knew about Newton and refused to leave him alone, and old interests did not fail to stir when they were prodded. External forces stronger than Newton's desires frequently interrupted his solitude even during the years of silence and finally brought them to an end.

There was no point during these years when Newton divorced himself entirely from mathematics. He was the Lucasian professor, of course, and he apparently delivered a series of lectures on algebra every year with the probable exception of 1679, when his mother's death took him to Woolsthorpe. The two men who later claimed him as their teacher in mathematics, Henry Wharton and Sir Thomas Parkyns, both studied in Cambridge in the early 1680s. Mathematics was never solely a duty to Newton. It retained its capacity to excite him, and sometime in or near the year 1680, he devoted

---

[120] *Principia*, 1st ed., p. 353; *Var Prin 1*, 463. In the second and subsequent editions, Newton omitted the final two sentences, possibly because about the time of the second edition, he began to admit once more the existence of an extremely rare aether. The later form of the paragraph is found in *Prin*, p. 325.

some attention to it. For the first time, he became a serious student of classical geometry. Years later, Henry Pemberton testified to Newton's high regard for the ancient geometers.

> Of their taste, and form of demonstration Sir Isaac always professed himself a great admirer: I have heard him even censure himself for not following them yet more closely than he did; and speak with regret of his mistake at the beginning of his mathematical studies, in applying himself to the works of Des Cartes and other algebraic writers, before he had considered the elements of Euclid with that attention, which so excellent a writer deserves.[121]

What Newton had repeated to Pemberton were opinions that he first developed around 1680.

Two publications in France in 1679, the posthumous *Varia opera mathematica* of Pierre Fermat with his reconstruction of Apollonius' treatise on plane loci and of Euclid's porisms, and Phillipe de la Hire's *Nouveaux élémens des sections coniques* with a treatise on solid loci, may have provided the occasion. Books VII and VIII of Pappus' *Mathematical Collection* almost certainly provided the text. Whatever the cause, Newton undertook an investigation of the classical loci problems, the pinnacles of Greek geometry. It was the first important mathematical exercise that he had undertaken spontaneously in more than a decade. It is necessary to add that he abandoned it, like others before it, incomplete.

Newton first worked out a number of propositions on plane loci, loci that constitute either straight lines or circles. Although they were inherently less challenging than solid loci, or conics, he brought his treatment of them to a climax, in characteristic fashion, with four propositions of high generality. Given certain conditions, a considerable list of which he included, the locus in question will be a straight line or a circle or a conic (a solid locus) or a curve of higher dimension (a linear locus).[122] From the plane locus, Newton proceeded on to the solid locus. This problem had functioned as midwife at the birth of modern analysis in Descartes's *Geometry*. Newton had learned its analytic solution at the beginning of his mathematical study. Now he convinced himself both that the ancient geometers had solved the problem, Descartes's claim to the contrary notwithstanding, and also that the solution in purely geometric terms was more elegant than the modern analytic one. In seven propositions, Newton offered his own solution, in which he reduced the four-line or solid locus to his earlier organic construction of conics. Then in six additional propositions of the incomplete

121 Pemberton, *View of Newton's Philosophy*, Preface. See Markus Fierz, *Isaac Newton als Mathematiker* (Zürich, 1972).

122 *Loca plana*; *Math 4*, 230–44. Cf. related pieces, *Math 4*, 244–50, and *Quaestionum solutio geometrica*; *Math 4*, 250–68.

treatise, he began to attack the problem of demonstrating when a locus problem is a solid locus.[123]

Newton inserted a modified version of the first seven propositions into the *Principia* (Book I, Section V, Lemmas XVII–XXI, Proposition XXII), where it served no function beyond the demonstration of his mathematical prowess. Together with the geometrical format of the *Principia* and certain statements elsewhere that denigrated the rigor of modern analysis, the solution of the locus problem has contributed to the widely held conclusion that Newton's outlook in mathematics was essentially classical. The editor of his *Mathematical Papers* has argued persuasively that the classicism of Newton's work on the loci lay mostly on the surface. In the body of the treatise, he called upon techniques that appeal to limiting values as a point (or variable) approaches another point (i.e., an increment approaches zero) or tends to infinity. Such techniques, wholly seventeenth century in outlook, closely resembled those which had led him to the calculus. Newton's work on the loci made use of and further promoted a modern approach known as projective geometry. The editor of the *Mathematical Papers*, far from finding it an exercise that looked back to the classics, urges that it was a work of "pioneering modernity," one "far in advance of contemporary thought . . ."[124]

An introduction to the locus problem composed about this time opened with a forthright attack on Descartes and modern analysts. Descartes had made a great show of solving the climactic problem on which Greek geometers had failed. Newton denied that they had failed.

> Indeed their method is more elegant by far than the Cartesian one. For he achieved the result by an algebraic calculus which, when transposed into words (following the practice of the Ancients in their writings), would prove to be so tedious and entangled as to provoke nausea, nor might it be understood. But they accomplished it by certain simple proportions, judging that nothing written in a different style was worthy to be read, and in consequence concealing the analysis by which they found their constructions.[125]

At much the same time, Newton reread Descartes's *Geometry*. Sixteen years earlier, by his own account, he had struggled through it alone, three or four pages at a time, and it had introduced him to mathematics. Now he went through writing in the margins com-

---

[123] *Solutio problematis veterum de loco solido; Math 4*, 282–320.

[124] *Math 4*, 226. D. T. Whiteside, "Introduction," *The Mathematical Works of Isaac Newton*, 2 vols. (New York, 1964), 2, xix. See also J. J. Milne, "Newton's Contribution to the Geometry of Conics," in W. J. Greenstreet, ed., *Isaac Newton, 1642–1727* (London, 1927), pp. 96–114.

[125] *Veterum loca solida restituta; Math 4*, 277. I have altered one phrase in the translation.

ments such as "vix probo" (I hardly approve), "error," and "Non Geom." Probably in connection with this reading, he drew up a paper of "Errors in Descartes' Geometry." In the future, he never referred to his debt to Descartes. He often repeated his objections, however, calling Cartesian geometry "the Analysis of the Bunglers in Mathematicks."[126]

Perhaps it was the vehemence of his revulsion that led him to undertake at this time a revised statement of his fluxional calculus which would place it on a solid geometrical foundation free of any taint of modern analysis. At any rate, he introduced the piece, called *Geometria curvilinea*, with another indictment of modern analysis.

> Men of recent times, eager to add to the discoveries of the ancients, have united the arithmetic of variables with geometry. Benefiting from that, progress has been broad and far-reaching if your eye is on the profuseness of output, but the advance is less of a blessing if you look at the complexity of its conclusions. For these computations . . . often express in an intolerably roundabout way quantities which in geometry are designated by the drawing of a single line.

What he intended to do, he continued, was to demonstrate that the same problems can be solved by a geometrical synthesis. Since Euclid's elements would not suffice, he would need to formulate new ones. Euclid had established the foundations of the geometry of straight lines. Those who had tried to apply his principles to curvilinear figures had treated them in terms of infinitesimals.

> I, in fact, shall consider them as generated by growing, arguing that they are greater, equal or less according as they grow more swiftly, equally swiftly, or more slowly from their beginning. And this swiftness of growth I shall call the fluxion of a quantity. So when a line is described by the movement of a point, the speed of the point—that is, the swiftness of the line's generation—will be its fluxion. I should have believed that this is the natural source for measuring quantities generated by continous flow according to a precise law, both on account of the clarity and brevity of the reasoning involved and because of the simplicity of the conclusions and the illustrations required.[127]

Newton outlined a treatise in four books. What he completed was book 1, in which he developed the basic algorithms of differentia-

---

[126] Trinity College Library. *Math 4*, 336–44. David Gregory's diary, May 1708. *Hiscock*, p. 42.

[127] *Math 4*, 421–3. Some thirty years later Newton repeated this concept as the essence of his method in contrasting it to the Leibnizian. "This Method is derived immediately from Nature her self, that of indivisibles Leibnitian differences or infinitely small quantities not so. For there are no *quantitates primae nascentes* or *ultimae evanescentes*, there are only *rationes primae quantitatum nascentium* or *ultimae evanescentium*" (*Add MS* 3968.41, f. 83)

tion from the concept of nascent and ultimate ratios which he had originally sketched in a revision of *De methodis* eight years earlier. "Fluxions of quantities are [to each other]," he asserted in one of the axioms on which he built the work, "in the first ratio of their nascent parts or, what is exactly the same, in the last ratio of those parts as they vanish by defluxion."[128] He broke new ground by adding to his earlier formulas the fluxions of trigonometric functions derived from the concept of first and last ratios.[129] As soon as he got beyond questions of foundations and into applications, Newton lost interest in the work and abandoned it.[130] He did devote considerable time to mathematics at this time, however. During much the same period, he also revised his treatise on cubics, and in one of the *tours de force* of his geometrical endeavors, he showed that all the cubics can be generated by the optical projection of the five divergent cubical parabolas.[131]

It is impossible to ignore the coincidence in time between Newton's rebellion against Cartesian mathematics and developments in his natural philosophy. Descartes had led him into the new world of thought, both in mathematics and in philosophy. Philosophically, he had begun to repudiate what he saw as atheistical tendencies in Descartes late in the 1660s in his essay "De gravitatione." Now in the late 1670s, he stood poised to reject the fundamental tenet of Descartes's mechanical philosophy of nature, that one body can act on another only by direct contact. Newton's mathematical papers suggest that only a wholesale repudiation of his Cartesian heritage would allow him to take that step. The repudiation determined not only the content but also the form of the *Principia*.

As I have indicated, mathematics retained its power to excite Newton. So also did questions in natural philosophy. He did not turn to them spontaneously during the years of silence, but others obtruded them upon his consciousness. Despite his withdrawal, a variety of men turned to him with questions, and the questions usually stimulated more than a simple reply. Newton's letter to Boyle early in 1679 was an answer to one such question; "De aere et aethere" was

---

[128] *Math 4*, 427.   [129] *Math 4*, 440–66.

[130] *Geometria curvilinea; Math 4*, 420–84. He did revise book 1 extensively, apparently near the time of composition; pp. 484–504. Cf. further drafts, pp. 506–21. Cf. also *Geometria, Lib. I; Math 7*, 402–68. Whiteside places this in the 1690s. The hand appears to me to be that of the early 1680s; it is quite different from the hand of the Book II with which Whiteside yokes it in the 1690s. Moreover, much of the content is closely related to the *Geometria curvilinea*, while it is out of harmony with Newton's geometrical work of the 1690s.

[131] *Math 4*, 354–80, with prior drafts, pp. 382–401. *Math 7*, 410–32. Cf. his observations on asymptotes and diameters of cubics (*Math 4*, 346–51). Cf. also his work at this time on the roots of equations (*Math 5*, 34–48).

his more extended private response. Late in 1679, immediately upon his return from Woolsthorpe, another intrusion burst in upon him, this one a letter from Robert Hooke. Writing as Oldenburg's successor as secretary of the Royal Society, Hooke invited Newton to resume his earlier correspondence. He passed on a few items of information, and he specifically asked for Newton's opinion of his own hypothesis that planetary motions are compounded of a tangential motion and "an attractive motion towards the centrall body . . ."[132]

Hooke was referring to a remarkable paragraph that had concluded his *Attempt to Prove the Motion of the Earth* (1674, republished in 1679 in his *Lectiones Cutlerianae*). There he had mentioned a system of the world he intended to describe.

> This depends upon three Suppositions. First, That all Coelestial Bodies whatsoever, have an attraction or gravitating power towards their own Centers, whereby they attract not only their own parts, and keep them from flying from them, as we may observe the earth to do, but that they do also attract all other Coelestial Bodies that are within the sphere of their activity . . . The second supposition is this, That all bodies whatsoever that are put into a direct and simple motion, will so continue to move forward in a streight line, till they are by some other effectual powers deflected and bent into a Motion, describing a Circle, Ellipsis, or some other more compounded Curve Line. The third supposition is, That these attractive powers are so much the more powerful in operating, by how much the nearer the body wrought upon is to their own Centers. Now what these several degrees are I have not yet experimentally verified . . .[133]

In view of Hooke's later charge of plagiary, it is not surprising that this remarkable passage has received considerable attention. It can be shown, I believe, that Hooke did not truly hold a concept of universal gravitation, although it is obvious that he was beginning to break through the limitations of earlier ideas of particular gravities specific to each planet.[134] The most remarkable aspect of the paragraph is not its apparent statement of universal gravitation,

132 Hooke to Newton, 24 Nov. 1679; *Corres 2*, 297–8. On the ensuing correspondence with Hooke, see Alexandre Koyré, "A Documentary History of the Problem of Fall from Kepler to Newton," *Transactions of the American Philosophical Society*, n. s., *45* (1955), 329–95, and "An Unpublished Letter of Robert Hooke to Isaac Newton," *Isis, 43* (1952), 312–37; and D. T. Whiteside, "Newton's Early Thoughts on Planetary Motion: A Fresh Look," *British Journal for the History of Science*, 2 (1964), 117–37.

133 Robert Hooke, *Lectiones Cutlerianae*, in R. T. Gunther, *Early Science in Oxford*, 14 vols. (Oxford and London, 1920–45), *8*, 27–8.

134 Richard S. Westfall, "Hooke and the Law of Universal Gravitation," *The British Journal for the History of Science, 3* (1967), 245–61. Cf. Louise Diehl Patterson, "Hooke's Gravitation Theory and its Influence on Newton," *Isis, 40* (1949), 327–41, and *41* (1950), 32–45. Johannes Lohne, "Hooke *versus* Newton," *Centaurus*, 7 (1960), 6–52.

however. For the first time, it correctly defined the dynamic elements of orbital motion. Hooke said nothing about centrifugal force. Orbital motion results from the continual deflection of a body from its tangential path by a force toward some center. Newton's papers reveal no similar understanding of circular motion before this letter. Every time he had considered it, he had spoken of a tendency to recede from the center, what Huygens called centrifugal force; and like others who spoke in such terms, he had looked upon circular motion as a state of equilibrium between two equal and opposing forces, one away from the center and one toward it. Hooke's statement treated circular motion as a disequilibrium in which an unbalanced force deflects a body that would otherwise continue in a straight line. It was not an inconsiderable lesson for Newton to learn.

In his reply, written the day after his return to Cambridge, Newton began by declining the proffered correspondence. For the past six months family affairs in Lincolnshire had occupied him so fully that he had had no time for philosophical speculation.

And before that, I had for some years past been endeavouring to bend my self from Philosophy to other studies in so much y^t I have long grutched the time spent in y^t study unless it be perhaps at idle hours sometimes for a diversion . . . And having thus shook hands wth Philosophy, & being also at present taken of wth other business, I hope it will not be interpreted out of any unkindness to you or ye R. Society that I am backward in engaging my self in these matters . . .

Newton went on to tell Hooke that he had not heard of his proposed system of the world, though he cast doubt on the truth of that assertion later in the letter when, in response to a remark about an observation by Flamsteed, he referred to something else in the same lecture.[135] He also commented briefly on another item in Hooke's letter. But he was not quite able to leave it at that. Flamsteed's purported observation of parallax had to do with the motion of the earth, which was the subject and title of Hooke's lecture. Newton suggested an experiment to reveal the diurnal rotation. Later, in the heat of controversy over the charge of plagiary, he told Halley that he put it in to sweeten the answer. It appears to me rather that the idea of proving the earth's motion pricked his curiosity. The classic objection against diurnal rotation held that falling bodies would be left behind as the earth turned beneath them; hence they should appear to land to the west if the earth rotates. Newton's experiment hoped to show that they land instead to the east.

[135] In a paper from the early 1680s that recorded observations of comets, Newton also referred to Hooke's work (*Add MS* 3965.14, f. 582). And in a letter to Halley of 20 June 1686, he spoke of Hooke's *Lectures and Collections* (1678) as though he had read them when they appeared (*Corres* 2, 436).

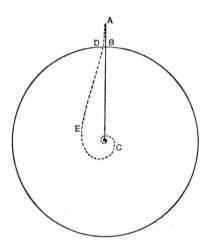

Figure 9.3. Newton's sketch of the path of a body that falls on a rotating earth. (From Westfall, *Force in Newton's Physics.*)

The tangential velocity of the top of a high tower is greater than that of the foot; therefore a falling body should slightly outrun the place directly beneath its point of release. A master experimenter, Newton carefully defined the details of the trial to insure its accuracy. He also drew a trajectory which showed the path *AD* as part of a spiral ending at the center of the earth. (Figure 9.3).

He concluded the letter by saying that he would certainly comment on Hooke's hypothesis if he had seen it and would with pleasure hear objections against any of his own notions. "But yet my affection to Philosophy being worn out, so that I am almost as little concerned about it as one tradesman uses to be about another man's trade or a country man about learning, I must acknowledge my self avers from spending that time in writing about it wch I think I can spend otherwise more to my own content & ye good of others . . ."[136]

The spiral was a gross blunder. By drawing the complete curve as though the earth were not present to offer resistance, Newton implicitly converted the problem of fall into the problem of orbital motion and showed a body with an initial tangential velocity falling into the center of attraction. Hooke can be excused for correcting the mistake. The past aside, Newton had seemed to invite objections, albeit in a rather backhanded way. Let it be said on Hooke's behalf that he couched his correction in the mildest of terms; and though he did read this letter and his next one (which also corrected Newton) before the Royal Society, he did not depart further from

[136] Newton to Hooke, 28 Nov. 1679; *Corres 2*, 300–3.

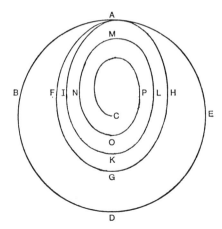

Figure 9.4. *AFGH* is the elliptical path of a body that meets no resistance. The inner spiral *AIKLMNOPC* represents its path in a resisting medium. (From Westfall, *Force in Newton's Physics*.)

his promise of 1676 to keep their correspondence confidential. Nevertheless, he did express his conviction that under conditions of no resistance a body let fall on a rotating earth would not fall to the center but would rather follow forever a path resembling an ellipse (Figure 9.4). He explicitly treated the problem as orbital motion by referring it to "my Theory of Circular motions compounded by a Direct motion and an attractive one to a Center."[137]

When he received the correction, Newton's announced pleasure in hearing objections evaporated like a dew in August. He had been caught, as Hooke had been caught in 1672, by haste. We can judge how deep an impression Hooke's correction made by the fact that six and a half years later he remembered the correspondence in detail.[138] More than thirty years later, the memory still smarted enough that he tried to explain the error away by calling it "a negligent stroke with his pen," which Hooke interpreted as a spiral.[139] At the time, he knew that the diagram was not a negligent stroke of the pen, however, and he knew as well that he had in his letter explicitly called it a spiral. His reply, as he accepted Hooke's correction, was as dry as a piece of burned bacon. Yes, Hooke was correct; the body would not descend to the center "but circulate wth an alternate ascent & descent . . ." Hooke was mistaken, however,

---

[137] Hooke to Newton, 9 Dec. 1679; *Corres 2*, 304–6.
[138] Newton to Halley, 20 June 1686; *Corres 2*, 436.
[139] De Moivre's memorandum on Newton's life; Joseph Halle Schaffner Collection, University of Chicago Library, MS 1075–7.

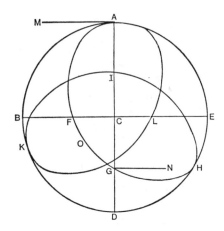

Figure 9.5. The path of a falling body under uniform gravity. (From Westfall, *Force in Newton's Physics.*)

about the ellipse. In a brief but compelling argument, he showed that under uniform gravity the body would reach its lowest point in less than 180 degrees and its original height in less than 360 degrees. Though he gave no figures, the diagram implied that the two angular positions were a little less than 120 and 240 degrees (Figure 9.5). Four brief sentences, which suggested in the most general terms how the orbit might vary if gravity were not constant, let Hooke know that Newton could handle the technical difficulties of orbital mechanics. He attempted to excuse himself by saying the question was "of no great moment," and he closed by leaving his newest offering to Hooke's correction.[140] He got it.

> Your Calculation [Hooke replied] of the Curve by a body attracted by an aequall power at all Distances from the center Such as that of a ball Rouling in an inverted Concave Cone is right and the two auges [apsides] will not unite by about a third of a Revolution. But my supposition is that the Attraction always is in a duplicate proportion to the Distance from the Center Reciprocall, and Consequently that the Velocity will be in a subduplicate proportion to the Attraction and Consequently as Kepler Supposes Reciprocall to the Distance.[141]

The passage has drawn as much attention as the paragraph from his lecture about the system of the world. On close analysis, its apparent derivation of the inverse-square law turns out to be a bastard demonstration resting on a deep confusion about dynamics and accelerated motion. Hooke employed a basic dynamic formula, $f \propto$

[140] Newton to Hooke, 13 Dec. 1679; *Corres 2*, 307–8.
[141] Hooke to Newton, 6 Jan. 1680; *Corres 2*, 309.

v^2, which he drew from a dynamic interpretation of Galileo's formula, $v^2 = 2as$, for uniformly accelerated motion from rest, and applied blindly to problems, such as the one in question, of nonuniformly accelerated motion.[142] In this case, he compounded the confusion by combining his fallacious law of force with Kepler's fallacious law of velocities. Hooke later based his charge of plagiary against Newton on the correspondence of 1679–80. It is small wonder that Newton, who had come upon the inverse-square relation more than ten years earlier by substituting the values from Kepler's third law into his formula for centrifugal force, was not impressed. In view of the confusion in Hooke's statement, it is also small wonder that Newton chose not to answer a third time.

Two weeks later, not realizing he had touched a raw nerve, Hooke wrote again to inform Newton that he had successfully tried the experiment to prove the rotation of the earth. It remained, he added, to know the properties of the path that a body would follow when attracted by an inverse-square force. "I doubt not but that by your excellent method you will easily find out what that Curve must be, and its proprietys, and suggest a physicall Reason of this proportion."[143] Whether it was Hooke's explicit proposal or the one implicit in the whole correspondence which moved him, Newton later acknowledged that he took up the challenge. He inverted the problem Hooke set, however, and instead of investigating the path in an inverse-square force field, he assumed the path and demonstrated that an elliptical orbit around an attracting body located at one focus entails an inverse-square attraction. In 1686, Newton told Halley that the correspondence with Hooke provided the occasion for the investigation, and he later said the same in memoranda connected with the calculus controversy and in a conversation with DeMoivre.[144] Recently a copy of the demonstration has been identified. In it, Newton began (as he later asserted) by demonstrating Kepler's law of areas. Using the law of areas and accepting Hooke's definition of the dynamic elements of orbital motion, he showed first that the force varies inversely as the square of the distance at the two apsides of an ellipse, and then that the same relation holds for every point on an ellipse.[145] If the inverse-square relation had

---

[142] Cf. Westfall, "Hooke and the Law of Universal Gravitation," pp. 252–9.

[143] Hooke to Newton, 17 Jan. 1680; *Corres 2*, 313. It is universally agreed that Hooke was mistaken about the experiment.

[144] Newton to Halley, 27 July 1686; *Corres 2*, 447. *Add MS* 3968.9, ff. 101, 106. Joseph Halle Schaffner Collection, University of Chicago Library, MS 1075-7.

[145] *Herivel*, pp. 247–53. Herivel originally identified the paper: "Newtonian Studies III. The Originals of the Two Propositions Discovered by Newton in December 1679?" *Archives internationales d'histoire des sciences, 14* (1961), 23–33. Cf. *Herivel*, pp. 108–17. A. Rupert and Marie Boas Hall challenged the identification: "The Date of 'On Motion in Ellipses'," *Archives, 16* (1963), 23–8. I supported Herivel: "A Note on Newton's

initially flowed from the substitution of Kepler's third law into the formula for centrifugal force under the simplifying assumption of circular orbits, the demonstration of its necessity in elliptical orbits far excelled in difficulty what had been a simple substitution. In fact, the demonstration, which probably dated from early 1680, was one of the two foundation stones on which the concept of universal gravitation rested. Newton did not even consider sending it to Hooke. Neither, in 1680, did he pursue it any further himself. He had worn out his affection for philosophy and now devoted his time to other studies more to his own content.

Nevertheless, the correspondence with Hooke had wider ramifications which bore on philosophy. Hooke posed the problem of orbital motion in terms of an attraction to the center, an action at a distance similar to the short-range attractions and repulsions Newton seemed ready to embrace in 1679. Both in his letters to Hooke and in his demonstration with the ellipse, Newton accepted the concept of attraction without even blinking. Between "De aere et aethere," which I have placed in 1679, and papers from 1686–7, Newton did not compose any essay on the system of nature that we know of. We have no way to determine if he adopted in 1679–80 opinions later expressed, and if so, whether Hooke's proposal served to crystallize opinions beginning to take form, or whether he found Hooke's concept of attraction acceptable because they had taken form already. What we do know is that Newton had recast his entire philosophy of nature by 1686–7. In papers composed with the *Principia*, first a proposed "Conclusio," then the same material in a proposed "Preface," neither of which appeared in the published work, Newton applied action at a distance to virtually all the phenomena of nature. The "Conclusio" started by noting that in addition to the observable motions in the cosmos there were innumerable unobservable ones among the particles of bodies.

If any one shall have the good fortune to discover all these, I might almost say that he will have laid bare the whole nature of bodies so

Demonstration of Motion in Ellipses," *Archives, 22* (1969), 51–60 (Cf. Richard S. Westfall, *Force in Newton's Physics*, [London, 1971], pp. 429–31). Whiteside has rejected my arguments with some vigor in a review of the *Archives* article: *Zentralblatt für Mathematik und ihre Grenzegebiete, 194* (1970), 2–3 (Cf. *Math 6*, 14, 553–4). As it appears to me, the only reason anyone has doubted the placement of this paper in 1680 is the existence of a copy, the first one generally known, among Locke's papers. It was copied by Locke's amanuensis and dated "Mar 89/90." I would think the value of this date as evidence of the date of composition, beyond the terminus it sets, is as close to zero as one could get. I shall confine myself here to stating explicitly what is implicit in my text–to wit, that nothing said so far seems to me to answer the arguments advanced for the earlier dating, or for that matter even to come close. The point has been made that corrections in existing versions imply an earlier one that is lost. The one in Newton's hand could be a revision done in 1684 after Halley's visit.

far as the mechanical causes of things are concerned. I have least of all undertaken the improvement of this part of philosophy. I may say briefly, however, that nature is exceedingly simple and conformable to herself. Whatever reasoning holds for greater motions, should hold for lesser ones as well. The former depend upon the greater attractive forces of larger bodies, and I suspect that the latter depend upon the lesser forces, as yet unobserved, of insensible particles. For, from the forces of gravity, of magnetism and of electricity it is manifest that there are various kinds of natural forces, and that there may be still more kinds is not to be rashly denied. It is very well known that greater bodies act mutually upon each other by those forces, and I do not clearly see why lesser ones should not act on one another by similar forces.[146]

In the evidence he proceeded to offer in support of his assertion, Newton composed an early draft of what later became familiar as Query 31 of the *Opticks*. All of the crucial phenomena that had caught his attention in 1664–5 and had played roles connected with aetherial mechanisms in the "Hypothesis of Light" appeared. All now furnished evidence of forces of attraction and repulsion between particles. The overwhelming burden of the argument rested, however, on chemical phenomena, as it continued to do in Query 31. Newton drew on many examples that fell into two main categories: reactions that generate heat and reactions that reveal affinities. The first repeated the argument he made to Boyle in 1679, with attractions now supplying the new motion manifest as heat. The latter converted the secret principle of sociability into specific attractions between certain substances. Without exception, all the chemical phenomena he cited had appeared in his alchemical papers, some among his own experiments, some among the other papers, most of them in both. Certain alchemical themes, such as the role of fermentation and vegetation in altering substances, the peculiar activity of sulfur, and the combination of active with passive, also made their way into the new exposition of the nature of things.[147] His alchemical compound, the net, even supplied its name to his new conception of matter, not an unorganized heap of particles like a sandpile, as mechanical philosophers pictured it, but organized crystals held in definite patterns by forces. He spoke of matter as "particulas retiformes" and "texturas retiformes," particles and textures in the form of nets.[148]

Newton understood that he was proposing a radical revision of what he referred to as "philosophy of the common sort . . . "

---

[146] *Halls*, p. 333; original Latin p. 321.
[147] *Halls*, pp. 333–47; original Latin, pp. 321–33. Cf. the partial draft of a preface for the *Principia* (*Halls*, pp. 305–8; original Latin, pp. 302–5).
[148] *Halls*, pp. 341, 306; original Latin, pp. 328, 303.

Although the very words *attractions* and *repulsions* would displease many, he wrote in one draft,

> yet what we have thus far said about these forces will appear less contrary to reason if one considers that the parts of bodies certainly do cohere, and that distant particles can be impelled towards one another by the same causes by which they cohere, and that I do not define the manner of attraction, but speaking in ordinary terms call all forces attractive by which bodies are impelled towards each other, come together and cohere, whatever the causes be.[149]

The last statement seemed to open the door again to the aetherial mechanisms of philosophy of the common sort. Recall that it was written to appear at the conclusion of a treatise which contained a pendulum experiment designed to disprove the existence of an aether and which also contained in its Book II a sustained argument against the Cartesian philosophy in particular and mechanical philosophies in general.

As it appears to me, Newton's philosophy of nature underwent a profound conversion in 1679–80 under the combined influence of alchemy and the cosmic problem of orbital mechanics, two unlikely partners which made common cause on the issue of action at a distance. Insofar as he continued to speak of particles of matter in motion, Newton remained a mechanical philosopher in some sense. Henceforth, the ultimate agent of nature would be for him a force acting between particles rather than a moving particle itself–what has been called a dynamic mechanical philosophy in contrast to a kinetic. In the realm of chemical phenomena, he never succeeded in moving the concept of force beyond the level of general speculation. His paper of 1680, however, showed the possibilities inherent in the idea when it was applied on the cosmic plane and supported by all the resources of Newton's mathematics. In 1680, the problem failed to seize his imagination, however, and he put it aside.

Hooke's letter was not the last. Late in 1680, Thomas Burnet asked him to read the manuscript for his *Telluris theoria sacra* (*Sacred Theory of the Earth*), which he published the following year. A brief correspondence ensued. Newton took exception to two arguments in Burnet's work: that Moses' seven days of creation cannot be taken literally, and that the world we see about us is not the world as it was originally created. These were topics on which Newton held opinions. He agreed that Moses spoke as a prophet, not as a philosopher; he adjusted his account of the creation to the level of the common people. Nevertheless, Moses spoke the truth, and Newton offered an interpretation of the days of creation that

---

[149] *Halls*, p. 345; original Latin, p. 331.

assumed the earth was accelerated from rest by divine force until it reached its present rate of rotation. Hence the days of creation were longer than our days. As to the form of the earth, Burnet was convinced that only a smooth earth could congeal from a uniform chaos. By way of objection, Newton gave a philosophical account of the earth's formation in the uneven shape we know–significantly, a chemical account, or rather accounts, three examples of uniform liquids congealing into solids with rough surfaces.[150]

In January 1681, the Royal Society asked him if he would give technical assistance to one John Adams, who was undertaking a survey of England. Newton agreed. He wrote a letter to his distant cousin, Sir John Newton, asking his support for Adam's survey. The enterprise did not prosper, and whatever technical aid Newton rendered disappeared with it.[151]

During the same winter, a comet appeared. Rather, as all astronomers in Europe with one exception believed, two comets appeared. The first was sighted before sunrise early in November and vanished into the morning sun at the end of the month. Two weeks later, in mid-December, another comet appeared in the early evening moving away from the sun. In late December, the second comet was immense, with a tail four times as broad as the moon and more than seventy degrees long. "I beleive scarce a larger hath ever been seen . . ." an excited Royal Astronomer, John Flamsteed, wrote to a friend in Cambridge.[152] Flamsteed was the astronomer who believed that the two comets were one and the same, a single comet which reversed its direction in the vicinity of the sun. Heretofore, astronomers had considered comets as foreign bodies which did not follow the same laws of motion as planets. Planets circled the sun in stable orbits. As far as astronomers could tell, their orbits had not changed in the two thousand years since the Greeks had begun to observe them. In contrast, comets were irregular and ephemeral, and majority opinion assigned to them rectilinear paths, the opposite of the planets' nearly circular orbits, which alone could repeat themselves forever and maintain the stability of the heavens. Flamsteed's proposal that comets reverse their direction in the vicinity of the sun marked a radical break with universally accepted opinion.

Flamsteed knew where to turn for comment. Although he had

---

[150] Newton to Burnet, Jan. 1681; *Corres 2*, 329–34. There had been at least one earlier letter, now lost, on 24 Dec. 1680.

[151] Newton to Sir John Newton, undated but probably about Jan. 1681; *Corres 2*, 335. At the meeting of the Royal Society on 19 Jan. 1681, Sir Christopher Wren mentioned the survey and Newton's promise to assist it; Thomas Birch, *The History of the Royal Society of London*, 4 vols. (London, 1756–7), *4*, 65.

[152] Flamsteed to Crompton, 15 Dec. 1680; *Corres 2*, 315.

met Newton in Cambridge in 1674, he had never corresponded with him, and he approached him now through a friend, James Crompton, a fellow of Jesus College. Newton received extracts from three letters to Crompton, written on 15 December 1680, and 3 January and 12 February 1681.[153] Sometime, probably in February, he also received a copy of Flamsteed's theory. Since Newton's first known reply, typically addressed to Crompton rather than Flamsteed, dated from 28 February, he may not have received the letters as they were written. Meanwhile, the comet had kindled his own curiosity. On 12 December, only four days after its first evening sighting, Newton observed it and recorded information about its tail (Figure 9.6).[154] He observed it almost daily from that time until it disappeared in March, keeping a log on its tail. Initially he observed it with the naked eye. Since he was shortsighted, he used a concave glass, in effect a monocle, to make it more distinct. "Such a glass, but the best, I always used," Newton explained, as though the lens made the observations suspect. He became sufficiently excited to get himself a telescope as the comet began to fade. "Jan 23 & 24. I saw the comet again with the help of a three foot telescope."[155] Within a week, he replaced the three-foot telescope with a seven-foot tube equipped with a micrometer; with the new instrument he followed the comet, its tail now reduced to one or two degrees, until 10 February, lost it for two weeks, then followed it with no visible tail until 9 March.[156]

He began to think again of his reflecting telescope and turned his hand to the construction of one 4 feet long, designed to magnify one hundred fifty times, with a glass mirror silvered on the back in place of the unsatisfactory metallic ones of his earlier reflectors.[157] He made two such mirrors ground on a sphere of 5 feet 11 inches, one ¼ inch thick, the other 5/62 inch thick.[158] Nothing came of the telescope, but the mirrors revealed new phenomena of colors to him and led him into the only significant piece of optical experimentation he carried out after the period 1669–72. It became Part IV, Book II of his later *Opticks*, "Observations concerning the Reflexions and Colours of thick transparent polish'd Plates."[159]

The majesty of the heavens had imprinted itself indelibly on

---

[153] *Corres 2*, 315, 319–20, 336.   [154] In the *Waste Book; Add MS* 4004, f. 99.

[155] *Add MS* 4004, ff. 99–101ᵛ, 101, 100ᵛ.

[156] *Add MS* 4004, f. 101. Newton described the two telescopes in a letter to Flamsteed on 16 April 1681 (*Corres 2*, 366). In the *Principia*, he referred to the second one as a six-foot tube instead of seven-foot. The table of his own observations is published in *Corres 2*, 357.

[157] *Opticks*, p. 105. From a reference to his early reflecting telescopes, this passage was written about 1687. Hence the attempt to construct the new reflector, which he placed five or six years earlier, came at the time of the comets.

[158] *Opticks*, pp. 290, 301, 305.   [159] *Opticks*, pp. 289–315.

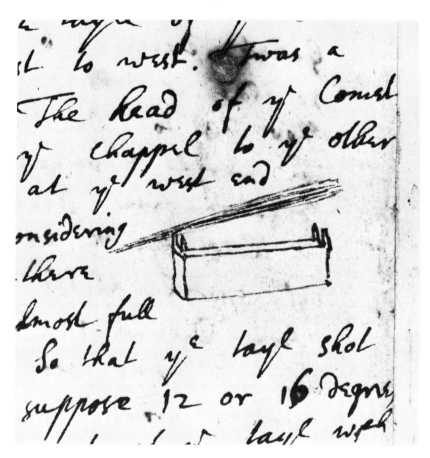

Figure 9.6. Newton's drawing (somewhat enlarged) of the great comet of 1680–1 over King's College Chapel, as seen by Humphrey Babington between 5:00 and 6:00 in the morning on 22 or 23 December. (Courtesy of the President and Fellows of Corpus Christi College, Oxford.)

Newton's imagination in Grantham and would soon dominate him again. It held him for a time in 1681. In addition to observing, he systematically collected the observations of others. He called in an undergraduate who had seen the comet in November and questioned him closely about its location. He began to read what literature he could find: Hooke, Hevelius, Gottignies, and Petit.[160] He later met Halley, probably in 1682, and questioned him about his observation of the comet on 8 December, the earliest evening sighting.[161] He worked at reducing the observations to a path through

[160] *Add MS* 4004, ff. 103–3ᵛ.   [161] *Add MS* 4004, f. 101ᵛ.

space.[162] He also wrote two long letters to Flamsteed criticizing his theory. The length of the letters, in contrast to his reticence with Hooke, and the fact that he addressed the second directly to Flamsteed, give some measure of the extent of his involvement.

True, Newton did not wholly throw off his reticence. His first letter opened with the statement that he was merely raising possible objections to Flamsteed's theory which he might wish to consider. "If hereafter he shal please to publish his Theory & think any of ye objections I propound need an answer to prevent their being objected by others, he may describe ye objections as raised by himself or his friends in general w^{th}out taking any notice of me."[163] As the statement implies, Newton refused to accept Flamsteed's new theory. Confusion between New Style and Old Style dates attached to the observations led him to argue that if the two comets were one, it must have been accelerated and retarded three times. Though Flamsteed corrected this confusion, other objections remained. Flamsteed dressed his proposal in an outlandish physical theory which treated the sun and the comet as magnets, with the sun first attracting one pole then repelling the other pole of the comet. According to the theory, the comet did not circle the sun; it turned in front of it. Manifold valid objections sprang to Newton's mind, and he laid them out at some length.[164] The primary interest of the letter lies in what it did not contain, however. Only a year before, Newton had solved the mechanics of orbital motion for a planet circling the sun. He did not now attempt to apply the same principles to the comet. Hence the letter allows us to measure crudely the progress of his thought toward the concept of universal gravitation. Most opinion held that comets were foreign bodies not related to the solar system and not governed by its laws. In his writings on comets, Hooke excluded them from the attraction between cosmic bodies that he posited. Apparently Halley held a similar view in 1680.[165] The letter to Flamsteed strongly implies that Newton did also. That is, no matter how important his demonstration was that

---

[162] What he tried at this time was a rectilinear path traversed at uniform velocity, the generally accepted theory. He inserted a problem on finding such a path from four observations into his lectures at this time (*Math 5*, 298–302). Three manuscripts survive in which he tried to find such a path for the comet of 1680–1 from four observations (*Math 5*, 524–31). Whiteside places these manuscripts in 1685 on what seem to me insufficient grounds to outweigh their contradiction then to the theory of parabolic or closed orbits, which Newton adopted a year earlier. The manuscripts appear to me to belong to 1681 (cf. James A. Ruffner, "The Background and Early Development of Newton's Theory of Comets" [dissertation, Indiana University, 1966], pp. 215–38.)

[163] Newton to Crompton for Flamsteed, 28 Feb. 1681; *Corres 2*, 340.

[164] *Corres 2*, 341–7.

[165] Flamsteed mentioned it in his letter to Crompton for Newton, 7 March 1681; *Corres 2*, 351.

elliptical orbits entail an inverse-square force, Newton apparently considered the force as specific to the solar system which contained related bodies. He had not yet formulated the idea of universal gravitation.

In a draft of his second letter to Flamsteed, Newton did begin to apply the dynamics of planetary orbits to the comet, though he couched his suggestion in terms of magnetic attraction. He described verbally an orbit around the sun in which the sun's attraction accelerated the approaching comet, reversed its direction, and retarded it as it receded.[166] In the letter he sent, however, he eliminated the passage and confined the letter to continued criticism of the idea of a single comet.[167] Two possible interpretations arise. Either he was not convinced of the single comet and hence eliminated a passage likely to support it, or he wished to terminate the correspondence and hence eliminated a passage likely to continue it. The two interpretations are not mutually exclusive. In any case, no further letter to or from Flamsteed at this time has survived.

In the spring, Newton received a table of observations of the comet from the colonies. His old antagonist of school days, Arthur Storer, sent them from Maryland, where he had settled in 1679.[168] Storer had become an amateur astronomer. A few years earlier, he had appealed to Newton for assistance with tables he had calculated of the positions of the stars and sun.[169] Newton had replied and received now his reward in the form of further data on the comet. Storer later sent him observations of the comet of 1682, together with a request for instruments and a star catalogue.[170]

Although Newton did not continue the correspondence with Flamsteed, his interest in comets did not evaporate. When the comet of 1682 appeared, which we now call Halley's comet, he made and recorded observations of its position.[171] Some time after 1680, he systematically collected information on all recorded comets and classified them under a number of headings, such as those in opposition to the sun.[172] He also revised his opinion on the paths of comets. In a set of propositions about comets, along with assertions that the sun and planets have gravitation toward their centers that decreases as the square of the distance and that the sun's gravitation is far greater than the planets', he forsook the theory of rectilinear paths for comets and accepted curved ones. The point of greatest

---

[166] *Corres 2*, 361.   [167] Newton to Flamsteed, 16 April 1681; *Corres 2*, 363–7.

[168] Storer to Babington, 18 April 1681; *Corres 2*, 368–9.

[169] In 1678; *Corres 2*, 269–85. Newton's replies are lost, but Storer mentioned one.

[170] Storer to Newton, 28 March 1683 (lost) and 26 April 1683; *Corres 2*, 387–93.

[171] *Add MS* 4004, f. 105.

[172] *Add MS* 3965.14, f. 614. Reproduced in Ruffner, "Newton's Theory of Comets," p. 314; English translation, p. 315.

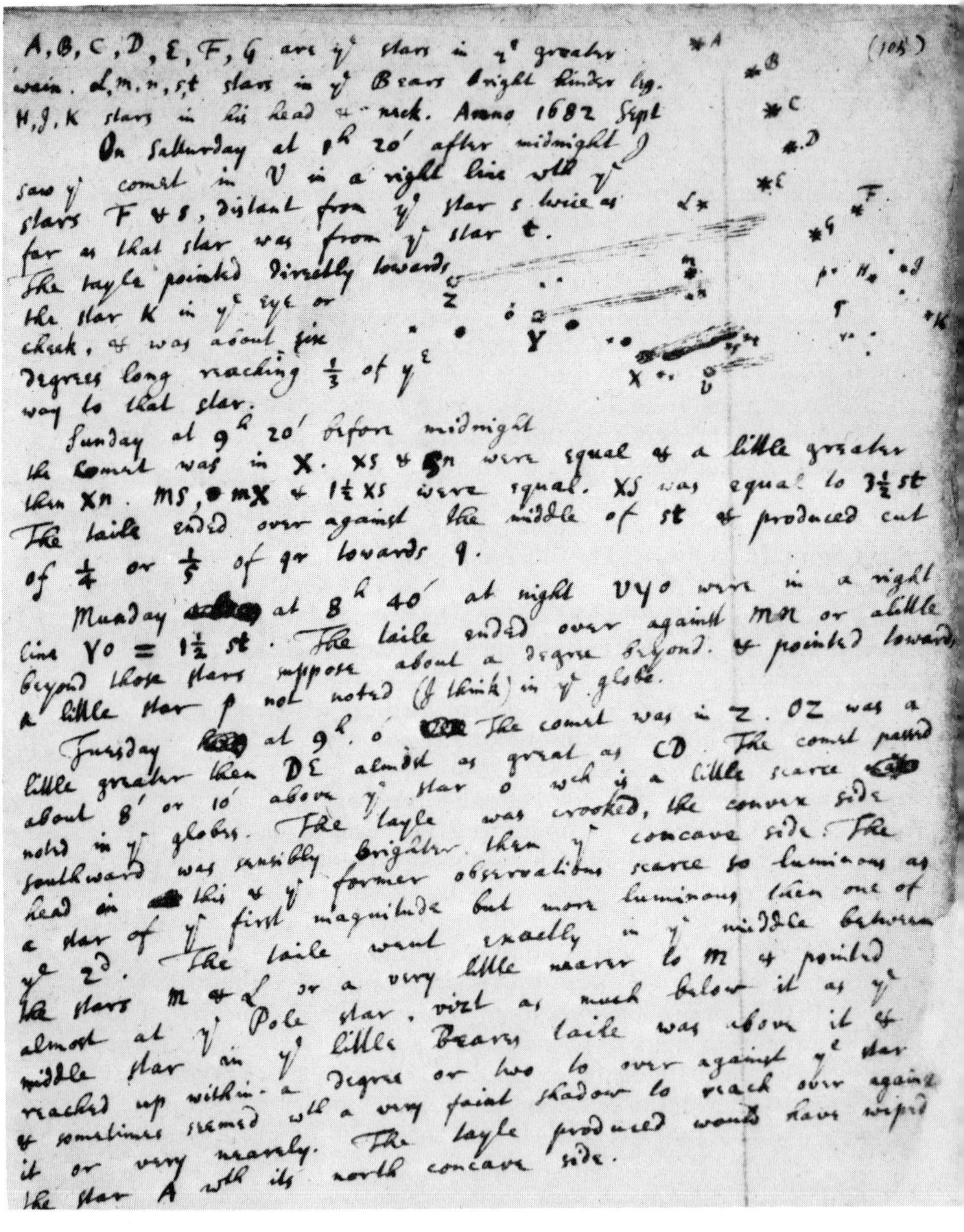

A, B, C, D, E, F, G are ye stars in ye greater wain. L, m, n, s, t stars in ye Bears bright hinder leg. H, J, K stars in his head & neck. Anno 1682 Sept

On Saturday at 9ʰ 20' after midnight I saw ye comet in V in a right line with ye stars F & s, distant from ye star s twice as far as that star was from ye star t. The tayle pointed directly towards the star K in ye eye or cheek, & was about six degrees long reaching ⅓ of ye way to that star.

Sunday at 9ʰ 20' before midnight the comet was in X. Xs & sn were equal & a little greater then Xn. ms, ● mX & 1½ Xs were equal. Xs was equal to 3½ st The taile ended over against the middle of st & produced cut of ¼ or ⅕ of qr towards q.

Munday at 8ʰ 40' at night VYo were in a right line Yo = 1½ st. The taile ended over against mn or a little beyond those stars suppose about a degree beyond. & pointed towards a little star p not noted (I think) in ye globe.

Tuesday at 9ʰ. o The comet was in Z. OZ was a little greater then DE almost as great as CD. The comet passed about 8' or 10' above ye star o wch is a little scarce noted in ye globes. The tayle was crooked, the convex side southward was sensibly brighter then ye concave side. The head in this & ye former observations scarce so luminous as a star of ye first magnitude but more luminous then one of ye 2d. The taile went exactly in ye middle between the stars m & l or a very little nearer to m & pointed almost at ye Pole star. vizt as much below it as ye middle star in ye little Bears taile was above it & reached up within a degree or two to over against ye star & sometimes seemed wth a very faint shadow to reach over against it or very nearly. The tayle produced would have wiped the star A wth its north concave side.

Figure 9.7. Newton's sketch of sightings of Halley's Comet on four successive nights in September 1682. (Courtesy of the Syndics of Cambridge University Library.)

curvature coincides with the point of perihelion. If the comet returns, the curve is an "oval"; if it does not return, it is nearly a hyperbola.[173] Whatever his reluctance in the spring of 1681, Newton now saw the application to comets of the orbital dynamics of planets.

In the spring of 1682, mathematics and optics had their turns to intrude. In the case of mathematics, his friend John Ellis asked Newton to recommend someone suitable for the vacant position of master of the Mathematical School of Christ's Hospital in London. He recommended a young fellow of Trinity, Edward Paget, for whose competence he vouched; and with considerable generosity, he wrote not only to the governors of Christ's Hospital but also to Collins and Flamsteed to enlist their support.[174] It was the first time Collins had heard from him in five and a half years and the last time he would hear. With such support, Paget got the position. Alas, it was all to no purpose. Though Newton had styled him "sober & industrious," he took to drink and neglected his duties, abandoned the post in the 1690s and went abroad, eventually showing up in India, and finally returned home to die in 1703. It says much about Cambridge that Paget, who was resident in Trinity as a fellow less than two years, continued on the college books until his death. Finally, in 1702, twenty years after he departed, the college did rescind its quarterly payments, but it did not strike him from the list of fellows.

That same spring William Briggs, a former fellow of Corpus Christi College, sent to Newton for comment a copy of his theory of vision, a paper he had just presented at the Royal Society. Sometime earlier, perhaps when they were both students, Briggs had dissected an eye for Newton's edification.[175] The present communication on vision undoubtedly stemmed from this former relation. In a noncommittal letter, Newton approved aspects of the paper, mentioned that he did not agree with other parts, but refrained from specifying his disagreement.[176] When Briggs pressed him, he wrote at greater length in the autumn.

> Though I am of all men grown ye most shy of setting pen to paper about any thing that may lead into disputes [he began], yet your friendship overcomes me so far as y^t I shall set down my suspicions

---

[173] *Add MS* 3965.14, f. 613. Reproduced in Ruffner, "Newton's Theory of Comets," pp. 310–11; English translation, pp. 312–13.

[174] Newton to Flamsteed and Newton to the governors of Christ's Hospital, 3 April 1682; *Corres 2*, 373, 375–6. Collins to the governors of Christ's Hospital, 16 May 1682, mentioned that Newton had written to him also (*Corres 2*, 376).

[175] Newton mentioned the dissection in the letter written in the spring of 1685 that Briggs prefixed to the Latin edition of the "New Theory of Vision," published that year (*Corres 2*, 417–18).

[176] Newton to Briggs, 20 June 1682; *Corres 2*, 377–8.

about your Theory, yet on this condition, that if I can write but plain
enough to make you understand me, I may leave all to your use
wthout pressing it further on. For I designe not to confute or convince you but only to present & submit my thoughts to your consideration & judgment.

He repeated similar sentiments at the end of the letter. "For having
laid Philosophical speculations aside nothing but ye gratification of a
friend would easily invite me to so large a scribble about things of
this nature."[177] Between the two disclaimers, Newton presented a
detailed rebuttal of Briggs's theory that tension in the optic nerves,
which tunes corresponding pairs to vibrate in unison, effects the
union of separate images in two eyes into a single sensation.
Though his argument seems wholly convincing to me, Briggs nevertheless proceeded to defend and republish his theory.[178]

With the arrival of Humphrey Newton in 1683 and the availabil-
ity of an amanuensis, Newton bethought himself of his obligation
to deposit lectures in the university library. Since 1673, he had
delivered lectures on algebra, but the library had seen nothing of
them. In 1683–4, Newton hastily compiled a manuscript which
William Whiston later published in 1707 under the title *Arithmetica
universalis*. In the eighteenth century, the *Universal Arithmetick* was
Newton's most popular mathematical work. It was republished a
number of times, both on the Continent and in England. Undoubtedly a large measure of its popularity arose from the fact that it was
Newton's most elementary work, and for that reason it reveals the
least about him. He compiled it in a cavalier fashion by plundering
the papers connected with his *Observations on Kinckhuysen* early in
the 1670s. Since things Newton touched had a way of turning to
gold, the work did not merely summarize algebra but advanced the
science–in its analysis of imaginary roots, for example. Leibniz
reviewed the published work anonymously in the *Acta eruditorum* in
highly laudatory terms. It contained, he said, "certain extraordinary
features which you will seek in vain in vast volumes of Analysis."[179] It is a further measure of Newton's mathematical genius
that a contribution to duty, flung off in a spare moment, could
merit such praise.

---

[177] Newton to Briggs, 12 Sept. 1682; *Corres 2*, 381–5.

[178] The initial paper, "A New Theory of Vision," had already appeared in the *Philosophical
Collections*, no. 6 (1682), 167–77. Briggs published "A Continuation of a Discourse about
Vision, with an Examination of some late Objections against it," *Philosophical Transactions*, 13 (1683), 171–82, in which he referred to Newton and responded to his objections
along with others. In 1685, he published Latin translations of both papers, which were
bound with the second edition of his *Ophthalmographia*.

[179] *Math 5*, 24. See Carl B. Boyer, "Cartesian and Newtonian Algebra in the Mid-Eigh-
teenth Century," *Actes du XIe congrès internationales d'histoire des sciences*, (Warsaw, 1968),
3, 195–202.

Beyond the extraordinary features that Leibniz found, it also contained a further expression of Newton's new taste in mathematics. In a work the heart of which was an extended exercise in analytic geometry, he lashed out at modern analysts, that is to say, at Descartes. Equations belong to arithmetical computation, he announced; they have no place in geometry except insofar as geometrical quantities such as lines or surfaces are set equal to each other. The introduction of arithmetical computations such as multiplication and division into geometry is contrary to the intent of the classical geometers, who contrived their science to escape tedious calculations by drawing lines.

> Consequently these two sciences ought not to be confused. The Ancients so assiduously distinguished them one from the other that they never introduced arithmetical terms into geometry; while recent people [read "Descartes"], by confusing both, have lost the simplicity in which all elegance in geometry consists.[180]

Forty years later he told Pemberton that he called the treatise *Universal Arithmetick* because Descartes had called a similar work *Geometry*.[181] In fact, it was Whiston who attached the title to the published work; the manuscript deposited with the university had no title.

Newton wrote dates in the margins, in his lack of concern even dividing single investigations between the lectures of two successive years, and handed the manuscript in as his lectures for eleven years. Perhaps he thought he should put the labor to more effective use by publishing it also; at least the manuscript repeatedly addressed the "reader" instead of the class. He even undertook a revision, though he lost interest in it before he had pursued it very far.[182] The work itself betrayed a similar lack of interest. It wanted cohesion, and it wandered far from the purposes announced in the opening paragraph, as though Newton had trouble keeping his mind concentrated on it. For that matter, the manuscript that he deposited, from which the book was later printed, also remained unfinished. It simply stopped in the middle of an exposition of the construction of cubic equations. Newton added one more to his catalogue of unfinished works.

If the requirements attached to his chair provided the impetus behind the *Universal Arithmetick*, Newton's mathematical renown beyond the university's walls also impinged upon his privacy. In 1683, John Wallis published *A Proposal about Printing a Treatise of Algebra*, which announced that he would expound Newton's method of infinite series. When his *Treatise of Algebra* did appear in

---

[180] *Math 5*, 429.
[181] Pemberton, *View of Newton's Philosophy*, Preface.   [182] *Math 5*, 538–610.

1685, it devoted five chapters to portions of the two *Epistolae* to Leibniz of 1676.[183] More important, both for its testimony and for its impact, was a letter from Edinburgh, which he received in June 1684, from David Gregory, the nephew of James Gregory.

> S^r
>
> Altho I have not ye honour of your acquaintance, yet ye character and place yee bear in the learned world I presume gives me a title to adress you especially in a matter of this sort, which is humbly to present you with a treatise latly published by me heir, which I am sure contains things new to the greatest parte of the geometers.

The work that the letter accompanied was *Exercitatio geometrica de dimensione figurarum* (*Geometrical Exercise on the Measurement of Figures*), an exposition of his uncle's method of infinite series with applications to quadratures, volumes, and the like. Gregory told Newton that he knew from Collins's letters to his uncle that Newton had cultivated this method and that the world had long expected to see his discoveries.

> S^r yee will exceedingly oblidge me, if yee will spare so much time from your Philosophical and Geometrical studies, as to allow me your free thoughts and character of this exercitation, which I assure you I will justly value more then that of all the rest of ye world . . .[184]

Gregory was not the only young mathematician who had heard of Newton's reputation. Early in 1685, another Scot, John Craig, came to Cambridge to meet him. We do not know much about the visit except that Newton read a paper by Craig with objections to Tschirnhaus and allowed Craig to read *De methodis* and the letters to Leibniz and to take extensive notes on them.[185]

Meanwhile, Gregory's letter stirred Newton to action. Since Gregory had handsomely acknowledged Newton in the work, no charge of plagiary could arise. In its threat of forestalling Newton, however, Gregory's book was like Mercator's earlier one, and he responded in the same way. He projected a treatise in six chapters with the title "Matheseos universalis specimina" (Specimens of a Universal System of Mathematics) in which he intended to include letters that would demonstrate his priority over James Gregory.[186] From the beginning, he also had Leibniz in mind, and he indicated his intention to publish their entire correspondence of 1676–7. In the chapters he wrote, Newton virtually forgot Gregory and devoted himself to replying to Leibniz. Hence the "Matheseos" pro-

---

[183] *Math 4*, 413–17.   [184] Gregory to Newton, 9 June 1684; *Corres 2*, 396.

[185] Craig to Campbell, 30 Jan. 1689; *Corres 3*, 8–9. The preface to Craig's *De calculo fluentium* (London, 1718); quoted in *Corres 3*, 9. David Gregory made a copy of Craig's notes, published in *Math 3*, 354–72.   [186] *Math 4*, 526–88.

vides a sharp insight into Newton. Despite his declared lack of interest, he had not been able to put the German completely out of his mind. Gregory posed no threat. Newton recognized that Leibniz did, and a defensive polemical tone characterized the treatise as he answered objections now seven years old. The projected six chapters concentrated on infinite series. The chapter 4 that he wrote expounded the method of fluxions. It began by translating the fluxional anagram in the *Epistola posterior*, and it ended by comparing the fluxional method with Leibniz's differential calculus as he had stated it in the letter of 1677. Though Leibniz's definitive publication of his calculus had not yet begun, he had published a quadrature of the circle in 1682, which Newton later cited against him. Perhaps Newton was aware of it as he composed the "Matheseos." More likely he was responding to the threat he had perceived in 1676–7.

In chapter 4, Newton also repeated his attack on modern analysts, including Leibniz among them by implication. One fascinating sentence suggests the depth his revulsion against Descartes had reached. After expounding his fluxional method, he paused to reflect. "On these matters I pondered nineteen years ago, comparing the findings of   and Hudde with each other."[187] The silence of the blank is deafening. Only one name–Descartes–could have belonged there. Newton could no longer bring himself even to acknowledge his debt.

Newton left the "Matheseos" unfinished and started a revision with the title "De computo serierum" (On the Computation of Series). He tired of it rather quickly as well, and abandoned it in the middle of the third chapter.[188] He never returned to either.

On the Continent, someone else had noted the publication of Gregory's work. In July, Otto Menke, editor of the *Acta eruditorum*, wrote to Leibniz that someone in England had attributed his quadrature of the circle to Newton.[189] Leibniz had already clashed with Tschirnhaus about papers on tangents and maxima and minima, which Leibniz called plagiary. Menke's message spurred him to compose a paper on his differential calculus, which the *Acta* published in October. In the summer of 1684, Newton and Leibniz seemed destined to collide, as indeed they were. The collision was deferred, however, largely because of yet another interruption. In August, Edmond Halley traveled up from London to put a question which only Newton could answer. No other interruption stirred him so deeply. Halley's visit changed the course of his life.

[187] *Math 4*, 571. Whiteside suggests this interpretation of the blank. Blank appears in the original.   [188] *Math 4*, 590–616.
[189] Menke to Leibniz, 16 July 1684; Gottfried Wilhelm Leibniz, *Sämtliche Schriften und Briefe* (Berlin, 1923 continuing), ser. I, 4, 475–6.

10

Principia

THE background to Halley's visit to Cambridge in August 1684 was a chance conversation of the previous January. By his own account, Halley had been contemplating celestial mechanics. From Kepler's third law, he had concluded that the centripetal force toward the sun must decrease in proportion to the square of the distance of the planets from the sun.[1] Halley wrote his account in the summer of 1686 after he had read two successive versions of what became Book I of the *Principia*. Since Newton coined the word "centripetal" after January 1684, Halley could not have framed his conclusion using that word, and he probably did not use the concept, either. The context of his statement implied that he arrived at the inverse-square relation by substituting Kepler's third law into Huygens's recently published formula for centrifugal force. He was not the only one who made the substitution. After Hooke raised the cry of plagiary in 1686, Newton recalled a conversation with Sir Christopher Wren in 1677 in which they had considered the problem "of Determining the Hevenly motions upon philosophicall principles."[2] He had realized that Wren had also arrived at the inverse-square law. When, at Newton's request, Halley asked Wren about the conversation, Wren told him that for many years he "had had his thoughts upon making out the Planets motions by a composition of a Descent towards the sun, & an imprest motion . . . ," but he had not been able to solve the problem. Although Hooke had frequently claimed that he could do so, he had never satisfied Wren with his demonstration.[3] If Hooke started with the derivation of the inverse-square relation he sent to Newton, we can understand Wren's dissatisfaction. In any case, it is clear that the problem Hooke put to Newton in the winter of 1679–80 was one which several people defined for themselves at much the same time. It was, indeed, the great unanswered question confronting natural philosophy, the derivation of Kepler's laws of planetary motion from principles of dynamics.

This same problem was discussed by Halley, Wren, and Hooke at a meeting of the Royal Society in January 1684. Hooke claimed that he could demonstrate all the laws of celestial motion from the inverse-square relation. Halley admitted that his own attempt to do

[1] Halley to Newton, 29 June 1686; *Corres 2*, 442.
[2] Newton to Halley, 27 May and 20 June 1686; *Corres 2*, 433–4, 435.
[3] Halley to Newton, 29 June 1686; *Corres 2*, 441–2.

so had failed. Wren was skeptical of Hooke's claim. To encourage the investigation, he offered a prize of a book worth forty shillings to the one who would bring him a demonstration within two months. Compared with the problem under discussion the inducement was modest enough, though Wren did observe that the winner would also gain the honor involved. Hooke again asserted that he had the demonstration, but he intended to keep it secret until others, by failing to solve the problem, learned how to value it.[4] Halley had met Newton in 1682, when they talked about comets. We can assume that he had come to know Newton's reputation as a mathematician. We do not know what took Halley to Cambridge. Since he allowed seven months to pass, we can hardly surmise that he rushed there, afire with curiosity, to lay the problem before Newton. Nevertheless, he did find himself in Cambridge in August, and he did seize the opportunity to consult an expert.

Although Halley mentioned the visit, the best account came from Newton's recollection as he told it to Abraham DeMoivre.

> In 1684 D[r] Halley came to visit him at Cambridge, after they had been some time together, the D[r] asked him what he thought the Curve would be that would be described by the Planets supposing the force of attraction towards the Sun to be reciprocal to the square of their distance from it. S[r] Isaac replied immediately that it would be an Ellipsis, the Doctor struck with joy & amazement asked him how he knew it, why saith he I have calculated it, whereupon D[r] Halley asked him for his calculation without any farther delay, S[r] Isaac looked among his papers but could not find it, but he promised him to renew it, & then to send it him . . .[5]

We can dismiss the charade of the lost paper – all the more since it survives among Newton's papers. Newton did not lightly send things abroad. The repeated faux pas in the correspondence with Hooke on this very question would have made him more wary than usual. He did commit himself to reexamining the paper, however, by promising to send the demonstration to Halley.

As Newton told the story to DeMoivre, he must have congratulated himself for restraining any impetuosity. When he tried the demonstration anew, it did not work out.[6] As he ultimately discovered, a hastily drawn diagram led him to confuse the axes of the ellipse with conjugate diameters. Not one to give up, he started anew and finally achieved his goal. In November, via the hands of

4 *Corres 2*, 442.
5 Joseph Halle Schaffner Collection, University of Chicago Library, MS 1075–7.
6 Whiteside has made the point that corrections and differences in the surviving copies of what I accept as the early demonstration imply a lost original (*Zentralblatt für Mathematik und ihre Grenzegebiete, 194* [1970], 2–3). It is possible that the copy in Newton's hand dated from the fall of 1684 and was a product of the revision prompted by Halley's visit.

Edward Paget, Halley received somewhat more than he expected, a small treatise of nine pages with the title *De motu corporum in gyrum* (*On the Motion of Bodies in an Orbit*). Not only did it demonstrate that an elliptical orbit entails an inverse-square force to one focus, but it also sketched a demonstration of the original problem: An inverse-square force entails a conic orbit, which is an ellipse for velocities below a certain limit. Starting from postulated principles of dynamics, the treatise demonstrated Kepler's second and third laws as well. It hinted at a general science of dynamics by further deriving the trajectory of a projectile through a resisting medium.[7]

When he received *De motu*, Halley did not wait another seven months. He recognized that the treatise embodied a step forward in celestial mechanics so immense as to constitute a revolution. Without delay, he made a second trip to Cambridge to confer with Newton about it; and on 10 December, he reported his activities to the Royal Society.

> Mr. Halley gave an account, that he had lately seen Mr. Newton at Cambridge, who had shewed him a curious treatise, *De motu*; which, upon Mr. Halley's desire, was, he said, promised to be sent to the Society to be entered upon their register.
>
> Mr. Halley was desired to put Mr. Newton in mind of his promise for the securing his invention to himself till such time as he could be at leisure to publish it. Mr. Paget was desired to join with Mr. Halley.[8]

When Newton persisted in reworking the treatise longer than Halley expected, he had the secretary of the society register the copy he had made with the date on which he had reported it. Letters from Newton to Flamsteed indicate that Halley was not the only one to sense the revolutionary implications of the treatise. His copy was in such demand that Flamsteed, to whom Newton offered the privilege of reading it, had to wait a month before he could see it.[9]

The cause of Halley's long wait for the amended treatise was a process at work in Cambridge, a process typically Newtonian though not less marvelous for that. The problem had seized Newton and would not let him go. There was in it that same majesty that years before had aroused the awe of a schoolboy in Grantham.

---

[7] *Herivel*, pp. 277–89; original Latin, pp. 257–74. See Stephen Peter Rigaud, *Historical Essay on the First Publication of Sir Isaac Newton's Principia* (Oxford, 1838); and W. W. Rouse Ball, *An Essay on Newton's 'Principia'* (London, 1893).

[8] Birch, *The History of the Royal Society of London*, 4 vols. (London, 1756–7), *4*, 347.

[9] On 27 December 1684, Flamsteed wrote to Newton that he would not be able to see *De motu* until Hooke and others had satisfied themselves. Newton replied in January that he was instructing Paget to send the tract to Flamsteed, and late in January, Flamsteed acknowledged that he had received it (*Corres 2*, 405, 412–13, 414).

Over the years, he had briefly heard its call several times. As a student, he had found the inverse-square relation from Kepler's third law. Under Hooke's stimulus, he had extended the inverse-square force to account for Kepler's first law. In August 1684, Halley evoked the same splendor anew, and this time Newton surrendered utterly to its allure. Halley later liked to say that he had been "the Ulysses who produced this Achilles,"[10] but Halley in London did not understand what was happening in Cambridge. Halley did not extract the *Principia* from a reluctant Newton. He merely raised a question at a time when Newton was receptive to it. It grasped Newton as nothing had before, and he was powerless in its grip. The treatise *De motu*, which Halley received in November, bore marks testifying that the challenge was at work already. Initially the treatise contained four theorems and five problems that dealt with motion in space void of resistance. As he began to glimpse broader horizons, Newton revised his initial definitions and hypotheses to allow for resistance and added two problems on motion through such a medium. At much the time when Halley was telling the Royal Society about *De motu*, Newton was writing to Flamsteed for data that would enable him to make its demonstrations more precise. "Now I am upon this subject," he told Flamsteed in January, "I would gladly know ye bottom of it before I publish my papers."[11] In getting to the bottom of it, he nearly cut himself off from human society. From August 1684 until the spring of 1686, his life is a virtual blank except for the *Principia*. In December and January 1684–5, he wrote to Flamsteed four times for information, and again in September he asked for more. In February 1685, he responded to a letter about *De motu* from the secretary of the Royal Society, Francis Aston, his old acquaintance in Trinity. Aston had also raised the question of organizing a philosophical society in Cambridge. Newton indicated his readiness to participate "so far as I can doe it without engaging the loss of my own time in those thinges."[12] With such support, the plan aborted, the *Principia*'s first victim. In April, Newton heeded the appeal of William Briggs and wrote a letter of muted praise which Briggs prefixed to the Latin translation of his *New Theory of Vision*. He allowed family obligations to draw him to Woolsthorpe briefly in the spring of 1685.[13] Beyond this meager list of known activities, most of which were related to the *Principia*, there was nothing. The investigation wholly absorbed his life. Apparently he continued to

---

[10] DeMoivre Memorandum; Joseph Halle Schaffner Collection, University of Chicago Library, MS 1075-7.   [11] *Corres 2*, 413.   [12] *Corres 2*, 415.

[13] In February, Newton told Aston that he would be in Lincolnshire for a month or six weeks (*Corres 2*, 415). The Exit and Redit Book indicates absences from 27 March to 11 April and 11 June to 20 June. The Steward's Book showed a total absence of six weeks during academic year 1685.

lecture. William Whiston, who entered Clare in 1686, remembered that he heard one or two of Newton's lectures, which he did not understand.[14] The manuscripts that Newton later deposited as lectures, however, were merely drafts of the *Principia*. He forwent alchemical experimentation, which had been a major activity since 1678.[15]

An awestruck Humphrey Newton observed the erratic behavior of a man transported outside himself.

> So intent, so serious upon his Studies, y^t he eat very sparingly, nay, ofttimes he has forget to eat at all, so y^t going into his Chamber, I have found his Mess untouch'd of wch when I have reminded him, [he] would reply, Have I; & then making to ye Table, would eat a bit or two standing . . . At some seldom Times when he design'd to dine in ye Hall, would turn to ye left hand, & go out into ye street, where making a stop, when he found his Mistake, would hastily turn back, & then sometimes instead of going into ye Hall, would return to his Chamber again. . . . When he has sometimes taken a Turn or two [in the garden], has made a sudden stand, turn'd himself about, run up ye Stairs, like another Alchimedes [*sic*], with an εὕρηκα, fall to write on his Desk standing, without giving himself the Leasure to draw a Chair to sit down in.[16]

Early in February, external events threatened to shatter his concentration. Charles II died, passing the crown on to his brother and heir, James II, an acknowledged Catholic. On the morning of 9 February the university gathered at the public schools, all attired in their robes, and proceeded to Market Hill to proclaim James. The heads, doctors, and officers retired with the vice-chancellor to drink the king's health in the master's lodge of Clare Hall. Consumed as he was with hatred of popery, Newton must have rejoiced to be spared that ritual. No sooner had he retired to his room, however, than the mayor and aldermen of the city, mounted and attired in their equally resplendent robes and accompanied by the councilmen, bailiffs and freemen of Cambridge, appeared before the gate of Trinity, immediately outside his window, to proclaim James once more. Alderman Samuel Newton, the auditor of Trinity, to whom Newton had loaned eighteen shillings nineteen years before during his career as student usurer, read the proclamation.[17] New-

[14] William Whiston, *Memoirs of the Life and Writings of Mr. William Whiston* (London, 1749) p. 36.

[15] His experiments in 1684 began in February and appear to have continued until about 23 May, a date at the end of the series (*Add MS* 3975, p. 149). On the following page, he entered experiments that began on 26 April 1686, with no intervening space for any in 1685.   [16] *Keynes MS* 135.

[17] Cambridge University Library (Baker MSS), *Mm. 1. 53.* J.E. Foster, ed., *The Diary of Samuel Newton, Alderman of Cambridge (1662–1717)* (Cambridge, 1890) p. 88. Fitzwilliam notebook.

ton left no record of what passed through his mind if indeed he brought himself to think on it at all. Two years elapsed before the crisis that the succession implied touched Cambridge. Fortunately, two years were exactly the time he required.

The *Principia* was not only Newton's monumental achievement. It was also the turning point in his life. As we know from his papers, he had performed prodigies in a number of fields. As we also know, he had completed nothing. By 1684, he had littered his study with unfinished mathematical treatises. He had not pursued his promising insights in mechanics. His alchemical investigations had produced only a chaos of unorganized notes and disconnected essays. Had Newton died in 1684 and his papers survived, we would know from them that a genius had lived. Instead of hailing him as a figure who shaped the modern intellect, however, we would at most mention him in brief paragraphs lamenting his failure to reach fulfillment. The period 1684–7 brought an end to the tentative years. Finally, he saw an undertaking through to its conclusion. True, as the goal came in sight leaving the excitement of discovery behind and the drudgery of computation ahead, he began once again to lose interest and to procrastinate. But this time the sheer grandeur of the theme carried him through to completion. The publication of the *Principia* could not reshape Newton's personality, of course, but the magnitude of its achievement thrust him into the public eye beyond the possibility of another withdrawal.

In the *Principia*, Newton's manifold activities found a focus. He picked up again the investigation of dynamics, which he had hardly looked at since he laid it down twenty years before. He brought to it now a mathematical maturity he had not possessed earlier. Alchemy had led him to consider concepts of activity and force that were susceptible to mathematical treatment in a way that aethereal mechanisms were not. To a limited degree, his work in optics fit into the new program as he saw a new, mathematically precise, derivation of the law of refraction. Even his theological studies contributed on a plane of high generality, for in the "Origins of Gentile Theology" he had just argued that true natural philosophy supports true religion.

The *Principia* redirected Newton's intellectual life, which theology and alchemy had dominated for more than a decade. It interrupted his theological studies, which he did not resume seriously for another twenty years. It did not terminate his career as an alchemist, but it diverted the thrust of alchemical concepts from the private world of arcane imagery into an unexpected and concrete realm of thought where the rigor of mathematical precision could help them reshape natural philosophy. The investigation that seized Newton's imagination late in 1684 and dominated it for the following two and a half years transformed his life as much as it transformed the course of Western science.

When he started, Newton did not realize where the pursuit would lead or the demands it would place on him. What he sent to Halley was a short treatise concerned primarily with orbital mechanics. It implied that inverse-square centripetal attractions are general in nature since it asserted both that the satellites of Jupiter and Saturn, as well as the planets about the sun, conform to Kepler's third law and that the motions of comets are governed by the same laws that determine planetary orbits. It was about these things that Newton wrote to Flamsteed. He apologized for the requests. He had not paid any attention to astronomy for some years, and he was not abreast of the latest information.[18] He wanted Flamsteed's observations of the periods of Jupiter's satellites and the dimensions of their orbits. Flamsteed indicated that they indeed also obey Kepler's third law. "Your information about ye Satellits of Jupiter gives me very much satisfaction," Newton assured him.[19] Had Flamsteed observed the new satellites of Saturn that Cassini reported? Flamsteed had not; he was convinced of the existence only of the one satellite discovered earlier by Huygens. A single satellite could not instantiate the third law or an inverse-square attraction. Newton crossed out the reference to the satellites of Saturn, and they did not reappear as support for the law of universal gravitation until the second edition of the *Principia*. Newton also asked for the precise celestial coordinates of two stars in the constellation Perseus. As seen from the earth, the great comet of 1680–1 had passed through Perseus. Flamsteed understood immediately that Newton had taken the comet under consideration again. Newton confirmed his surmise. "I do intend to determin ye lines described by ye Comets of 1664 & 1680 according to ye principles of motion observed by ye Planets . . ."[20]

De motu had already included the matter above. When Newton asked Flamsteed for observations of the velocities of Jupiter and Saturn as they approach conjunction, however, he was raising a question which *De motu* had not addressed. He was convinced that Kepler's tables erred in describing Saturn's motion. Every time Jupiter approaches conjunction with Saturn, Saturn ought to slow down in its orbit and then to speed up as Jupiter passes conjunction "(by reason of Jupiters action upon him) . . ."[21] *De motu* had postulated only the general presence of inverse-square attractions in the universe. The mutual influence of Saturn and Jupiter implied the

---

[18] Newton to Flamsteed, ca. 12 Jan. 1685; *Corres 2*, 413.   [19] *Corres 2*, 407.
[20] *Corres 2*, 413.   [21] *Corres 2*, 407.

existence of a single universal attraction. Flamsteed was skeptical about such notions. When they are nearest each other, Jupiter and Saturn are separated by a distance four times the radius of the earth's orbit. The aether, which is a yielding matter, would simply absorb any disturbance at that distance. Nevertheless, he did supply exactly the information that Newton wanted to hear about the motion of Saturn and Jupiter in conjunction.[22] Newton accepted the information without bothering to answer Flamsteed's doubts.

> Your information about y^e error of Keplers tables for ♃ & ♄ has eased me of several scruples. I was apt to suspect there might be some cause or other unknown to me, wch might disturb y^e sesquialtera proportion [i.e., the ratio of Kepler's third law: $T \propto R^{3/2}$] . . . It would ad to my satisfaction if you would be pleased to let me know the long diameters of y^e orbits of ♃ & ♄ assigned by your self & Mr Halley in your new tables, that I may see how the sesquiplicate [=sesquialtera] proportion fills y^e heavens together wth another small proportion wch must be allowed for.[23]

Every new question broadened the investigation. When he replied to Aston in February, Newton apologized for his delay in sending the emended treatise. He had intended to complete it sooner, "but the examining severall thinges has taken a greater part of my time then I expected, and a great deale of it to no purpose."[24]

When he wrote to Flamsteed again in September to obtain more data about the comet of 1680–1, he had yet another question. Flamsteed had published some observations about tides in the estuary of the Thames, and Newton wanted additional information about them, indicating to us if not to Flamsteed a significant further extension of the gravitational concept.[25] Against the background of a constantly expanding study, we can appreciate the significance of Newton's comment to Halley in 1686 about a paper of the 1660s in which he tried to calculate how much the centrifugal force due to the earth's rotation decreases gravity. "But yet to do this business right is a thing of far greater difficulty then I was aware of."[26] The greater difficulty measured the difference between *De motu* and the ultimate *Principia*.

The first serious problem that Newton confronted was dynamics itself. If the seventeenth century had taken giant strides in mechanics, it had yet to crown its efforts with a science of dynamics. In order to write the *Principia*, Newton had first to create a dynamics equal to the task. His investigation of impact as an undergraduate offered a point of departure in its new conception of force as an action external to a body which effects a change of motion propor-

---

[22] *Corres 2*, 408–9.   [23] *Corres 2*, 413.   [24] *Corres 2*, 415.   [25] *Corres 2*, 419–20.
[26] Newton to Halley, 14 July 1686; *Corres 2*, 445.

tional to itself. He had employed the concept in his demonstration of the ellipse in 1679–80. "Hyp. 2. The alteration of motion is ever proportional to ye force by wch it is altered."[27] Beyond the statement about force, the demonstration of 1679–80 had proceeded in kinematic terms, using the parallelogram to compound two motions, the tangential and the centripetal, in successive instants of time. That approach did not please Newton in 1684. He was seeking a rigorous demonstration of orbital motion from first principles of dynamics. Part of his heritage, which he could not forget, was his essay "De gravitatione," in which he vehemently rejected the relativism of Cartesian mechanics for an absolutistic dynamics founded on the principle that force is the distinguishing characteristic of true motion. In 1684, Newton had just confirmed his early rebellion against Cartesianism by rejecting Cartesian mathematics as well. It is not surprising that he returned to "De gravitatione" for his statement of first principles. In doing so, he plunged the dynamics of *De motu* into a series of internal contradictions, to the resolution of which he devoted much of the following six months.

[27] *Herivel,* p. 246. On Newton's science of dynamics and its development, see I. Bernard Cohen, *Introduction to Newton's 'Principia'* (Cambridge, 1971); *Isaac Newton, The Creative Scientific Mind at Work* (the Wiles Lectures for 1966, n.p., n.d.); "Galileo and Newton," in *Saggi su Galileo Galilei* (Florence, 1972); "Newton's Second Law and the Concept of Force in the Principia," in Robert Palter, ed., *The 'Annus Mirabilis' Of Sir Isaac Newton* (Cambridge, Mass., 1970); "Newton's Theory versus Kepler's Theory and Galileo's Theory," in Y. Elkana, ed., *The Interaction between Science and Philosophy* (Atlantic Highlands, N.J., 1974); " 'Quantum in se est': Newton's Concept of Inertia in Relation to Descartes and Lucretius," *Notes and Records of the Royal Society, 19* (1964), 131–55; "Isaac Newton, the Calculus of Variations, and the Design of Ships," in R.S. Cohen, J.J. Stachel, and M. W. Wartofsky, eds., *For Dirk Struik. Scientific, Historical and Political Essays in Honor of Dirk J. Struik* (Boston Studies in the Philosophy of Science, *15*), (Dordrecht, 1974); and "The Concept and Definition of Mass and Inertia as a Key to the Science of Motion: Galileo-Newton-Einstein," *Colloquium: Principal Stages in the Evolution of Classical Dynamics (17th–20th Centuries),* XIII International Congress of the History of Science (Moscow, 1974). See also R. G. A. Dolby, "A Note on Dijksterhuis' Criticism of Newton's Axiomatization of Mechanics," *Isis, 57* (1966), 108–15; John Herivel, "Newton's Achievement in Dynamics," in Palter, ed., *The 'Annus Mirabilis',* pp. 120–35, and *The Background to Newton's Principia* (Oxford, 1965); Roderick W. Home, "The Third Law in Newton's Mechanics," *British Journal for the History of Science, 4* (1968), 39–51; W. H. Macauley, "Newton's Theory of Kinetics," *Bulletin of the American Mathematical Society, 3* (1896–7), pp. 363–71; John Nicholas, "Newton's Extremal Second Law," *Centaurus, 22* (1978), 108–30; D.T. Whiteside, "Newtonian Dynamics," *History of Science, 5* (1966), 104–17; Alan Gabbey, "Force and Inertia in Seventeenth-Century Dynamics," *Studies in History and Philosophy of Science, 2* (1971), 1–67; Brian D. Ellis, "The Origin and Nature of Newton's Laws of Motion," in Robert G. Colodny, ed., *Beyond the Edge of Certainty* (Englewood Cliffs, N.J., 1965), pp. 29–68; Margula R. Perl, "Newton's Justification of the Laws of Motion," *Journal of the History of Ideas, 27* (1966), 385–92; and John Price Losee, "Newton's Two Views of Mechanics," *Actes du XIIe congrès internationale d'histoire des sciences, 4,* 103–6.

De motu started with two definitions and two hypotheses. Definition 1 drew upon the lesson about circular motion that Hooke had taught him in 1679 to introduce a new word into the vocabulary of mechanics.

> I call that by which a body is impelled or attracted toward some
> point which is regarded as a center centripetal force.[28]

Newton later explained that he coined the word "centripetal," seeking the center, in conscious parallel to Huygens's word "centrifugal," fleeing the center. No single word better characterized the *Principia,* which more than anything else was an investigation of centripetal forces as they determined orbital motion.

The legacy of *De gravitatione* appeared in Definition 2, about rectilinear motion.

> And [I call] that by which it endeavors to persevere in its motion in a
> right line the force of a body or the force inherent in a body.

Hypothesis 2 extended the definition into a general conception of motion.

> By its inherent force alone, every body proceeds uniformly in a right
> line to infinity unless something extrinsic hinders it.[29]

The combined statement of the definition and the hypothesis is a startling assertion to find in the first draft of the work that established the principle of inertia as the foundation of modern science. Together they indicate the extent of the task Newton faced in autumn 1684.

One cannot stress too much the crudity of the dynamic foundation on which the brilliant structure of *De motu* rested. The tract attempted to derive orbital motion from the interaction of two forces: inherent force, which maintains rectilinear motion, and centripetal force, which continually diverts it. To compound the two, Newton employed the parallelogram of forces, which he inserted later as Hypothesis 3, although he made use of it before he added its formal statement.

> When it is acted upon by [two] forces simultaneously, a body is
> carried in a given time to that place to which it would be carried by
> the forces acting separately in succession during equal times.[30]

The two forces he wanted thus to compound were unable to mate, however, because they differed in species. Inherent force, like the impetus of medieval science, was force internal to a body which

[28] *Herivel,* p. 277; original Latin, p. 257. I have altered the translation.
[29] *Herivel,* p. 277; original Latin, pp. 257–8. Again, I have altered the translation.
[30] *Herivel,* p. 278; original Latin, p. 258. With a slightly altered wording, the parallelogram of forces appeared in this form, asserting that a constant force carries a body with a uniform motion, in the first edition of the *Principia.* Only in the second edition did he revise its statement to the present form, in which the conclusion emerges from a composition of motions rather than forces.

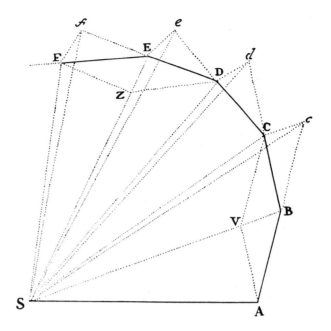

Figure 10.1. The diagram to Proposition I of the *Principia* as it appeared already in *De motu*. The centripetal force acts "at
once with a great impulse" at *B*, *C*, *D*, and *E*, turning
the body aside from its rectilinear path.

operated to maintain the body in uniform motion. Centripetal force
was an external action which operated to alter uniform motion. *De
motu* further compounded the difficulty by employing two different
measures of centripetal force. Theorem 1, substantially identical to
Proposition I of the *Principia,* treated force as a series of discrete
impulses at equal intervals of time. Using the parallelogram of forces
and the elementary geometry of triangles, Newton showed that the
areas swept out by the radius vector in successive intervals are equal,
and so are they also equal when the triangles are infinitely small and
the polygon approaches a curve.[31] Hence Kepler's first law holds for
the path of a body under the action of any centripetal force (Figure
10.1). In contrast to Theorem 1, Theorems 2 and 3 on the measurement of centripetal forces spoke of continuous forces instead of impulses. They employed the geometry of the circle in which the
versed sine ($\frac{1}{2}CD$) of an arc varies as the square of its arc (*BD*) or
tangent (*BC*) when the arcs or tangents are "very small and infinitely

---

[31] *Herivel*, p. 278; original Latin, pp. 258–9. This theorem remained substantially unchanged
   through the revisions of *De Motu;* hence Proposition I, Book I, of the *Principia* contained
   a phrase about a body moving in a straight line "by its innate force" (*vi insita*). For a
   different discussion of this proposition, cf. *Math 6*, 35–7, 539–42.

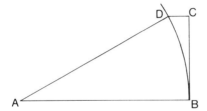

Figure 10.2. Centripetal acceleration generated by a continuous force.

decreased" (Figure 10.2).[32] Where Theorem 1 used the language of infinitesimals, Theorems 2 and 3 used the language of continuous flow on which Newton had chosen to rest his calculus. When a body moves uniformly in a circle, the arc or the tangent represents the uniform flow of time and twice the versed sine represents the distance the body is diverted from its rectilinear path in a vanishingly small moment of time. Since a continuous curve is tangent to its circle of curvature at every point, the same analysis applies to noncircular orbits. The proportion $s \propto t^2$ is the relation that Galileo established for uniformly accelerated motion. In Hypothesis 4, which he did not actually enter in the first version of *De motu,* though he wrote down the number and left a space for it, Newton explicitly asserted the ratio of uniformly accelerated motion.

The space which a body under the action of any centripetal force whatever describes at the very beginning of motion is proportional to the square of the time.[33]

Thus *De motu* attempted to use the parallelogram of motion to compound either of two differing conceptions of force, impulses which produce discrete increments of motion ($f = \Delta mv$) or continuous forces which produce uniform accelerations ($f = ma$), with a third conception of force as an internal propulsion which maintains a uniform velocity ($f = mv$). From such uncertain premises, Newton succeeded in deriving Kepler's three laws and in launching the analysis of motion through resisting media. The derivation could not have survived critical examination, and Newton could not have built the *Principia* on a foundation so uncertain.

Three versions of *De motu* survive. The second version, which is written in Halley's hand and is identical to the copy in the register of the Royal Society, was merely a fair copy of the first with small additions (such as Hypothesis 4) which Newton had indicated his

---

[32] *Herivel,* p. 279; original Latin, p. 259.

[33] *Herivel,* p. 278; original Latin, p. 258. The statement quoted appeared as Hypothesis 4 in the second version, Halley's copy of the first in which he filled in, probably during his second trip to Cambridge, items such as Hypothesis 4 that Newton had indicated but not entered.

intention to insert. The third version, on the other hand, contained the beginning of Newton's reconstruction of his dynamics. The deletion of a reference to the satellites of Saturn and a sentence about the mutual action of planets on each other date it to the period of Newton's correspondence with Flamsteed in December and January. In the third version, Newton moved Hypotheses 3 and 4, the parallelogram of force and the measure of distance described under a centripetal force, to the status of lemmas. To the remaining two hypotheses modestly altered, he added three more and then rejected the name "Hypothesis" in favor of a new one, "Lex." Hence Newton's laws of motion initially numbered five. From its new position at the head of the list, Law 1 continued to assert that a body moves uniformly by its inherent force alone. In Law 2, Newton attempted to define the action of force.

The change of motion is proportional to the impressed force and is made in the direction of the right line in which that force is impressed.

Not wholly satisfied, he altered the initial phrase to read, "The motion generated or the change of motion . . ." Still not pleased, he altered it once more to "The change in the state of moving or of resting is proportional . . ." The concept of impressed force, in contrast to the narrower centripetal force of the first version, marked a step toward a general dynamics concerned with the interaction of inherent force, which maintains uniform motion, and impressed force, which changes it. The addition to the definition of impressed force of the phrase "of resting," which implies the dynamic identity of uniform motion and rest, marked Newton's first step back toward the principle of inertia, which he had abandoned nearly twenty years before. Laws 3 and 4 took further steps in the same direction. Law 3 asserted that the motions among themselves of bodies included in a given space are the same whether the space is at rest or moves uniformly in a straight line without circular motion. Law 4 asserted that the mutual actions of bodies do not change the state of motion or of rest of their common center of gravity. The first draws on the principle of inertia; the second extends it to systems of bodies considered as unities. Newton later demoted the two to the rank of corollaries to the laws of motion, but both made their way into the *Principia*. Law 5, a restatement of Hypothesis 1 of the first version, contained a supposedly empirical statement about the resistance of media.[34] Newton soon realized that such a statement differed in status from the other laws and deleted it. For that matter, he later decided that its assertion of the proportionality of resistance to velocity was mistaken as well. The crux of his dynamics lay in the relation of inherent force and im-

---

[34] *Add MS* 3965.7, f. 40. *Herivel,* p. 299; original Latin, p. 294.

pressed force, what he later called (as he struggled to clarify them), "the inherent, innate and essential force of a body" and "the force brought to bear or impressed on a body."[35] The continuing development of his dynamics hinged on the two concepts.

The third Version of *De motu* contained another feature, a scholium to Problem 5 on "the immense and truly immobile space of the heavens."[36] Like the concept of the inherent force of bodies, which he held to be the distinguishing characteristic of true motion, the scholium looked back to "De gravitatione" and Newton's revulsion from the relativism of Cartesian physics. As he had said at that time, the ultimate absurdity of Cartesian relativism lay in its consequence "that a moving body has no determinate velocity and no definite line in which it moves."[37] In its original form, *De motu* did not contain any reference to absolute space. It had no need to. The inherent force of bodies adequately defined their absolute motions. Law 3 of the third version effectively surrendered that criterion in its admission that the whole solar system may be moving with a uniform motion which is beyond perception. As the demands of an internally consistent dynamics swept Newton inexorably toward the principle of inertia, he began to insist on absolute space. In papers that revised the third version, he considerably expanded its brief statement as he moved still further toward the principle of inertia. The increasing insistence of his assertions looked forward to the well-known scholium on absolute space and time which he included in the *Principia*. He may have capitulated to Descartes on motion. On the issue of relativism, which smacked of atheism in Newton's view, he continued to shout his defiance until his dying day.

In two papers of revisions that followed that third version of *De motu*, Newton completed the transformation of his dynamics into its ultimate form. By now, he had set his mind on logical rigor. Where the third version had defined four terms, the first paper of revisions defined no less than eighteen at one stage or another. Many concerned themselves with absolute space and absolute motion. Accompanying them, alterations in the definitions of motion transformed his concept of inherent force as he moved toward a quantitatively rigorous dynamics.

---

[35] *Herivel*, p. 311; original Latin, p. 306.

[36] *Herivel*, p. 302; original Latin, p. 298. On absolute space and time in Newton, see W. H. Macauley, "Newton's Theory of Kinetics"; Hans Reichenbach, "The Theory of Motion According to Newton, Leibniz, and Huygens," in *Modern Philosophy of Science* (London, 1959), pp. 46–66; Stephen Toulmin, "Criticism in the History of Science: Newton on Absolute Space, Time and Motion," *Philosophical Review*, 68 (1959), 1–29, 203–27; J. E. McGuire, "Existence, Actuality and Necessity: Newton on Space and Time," *Annals of Science*, 35 (1978), 463–508; and Markus Fierz, "Über den Ursprung und die Bedeutung der Lehre Isaac Newtons vom absoluten Raum," *Gesnerus*, 11 (1954), 62–120.

[37] *Halls*, p. 129; original Latin, p. 97.

The implicit idea behind the dynamics of *De motu* was the notion, common in seventeenth-century mechanics, that the force of a body in motion is the sum total of the forces that have acted upon it. Thus one of the new definitions, the force of motion or the force that a body possesses accidentally because of its motion, set it proportional to the motion, which in turn was proportional, of course, to the force that generated the motion.[38] Exactly this issue lay at the heart of the problem posed by Newton's parallelogram of force. As he had learned twenty years before in his study of impact, quantity of motion is not conserved in oblique impacts; the diagonal of his parallelogram, considered as force, was not the numerical sum of the two components that produced it. More generally, in circular motion as Newton now understood it, the continual action of a centripetal force produces no increase whatever in a body's speed, that is, in what he defined as the force arising from its motion. The new insight into circular motion, however, constituted the very heart of the orbital dynamics on which he was at work. Only the principle of inertia, which abandoned the concept of the force of a body's motion, would allow him to treat changes of direction in the same terms as changes of speed. Only the principle of inertia would allow the unique leap of the imagination which made the new dynamics possible, the recognition of the dynamic identity of uniform circular motion and uniformly accelerated motion in a straight line. Heretofore in the history of mechanics, these two motions had been treated as irreducible opposites. What Newton perceived were the possibilities that opened up if he treated the two as dynamically the same.

Though he defined the force arising from a body's motion, he quickly canceled it. He did not cancel the change he introduced in the definition of inherent force.

> The inherent, innate, and essential force of a body is the power by which it perseveres in its state of resting or of moving uniformly in a right line, and is proportional to the quantity of the body. It is actually exerted proportionally to the change of state, and in so far as it is exerted, it can be called the exerted force of a body . . .[39]

The idea of exerted force appealed to him and he experimented with a definition of it. The exerted force of a body is that by which a body strives to preserve "that part of its state of moving or of resting which it loses in single moments and is proportional to the change of that state or to the part lost in single moments . . ."[40] In

---

[38] *Herivel,* p. 311; original Latin, p. 306.

[39] *Herivel,* p. 311; original Latin, p. 306.

[40] *Herivel,* p. 320; original Latin, p. 317. Herivel has misplaced this definition in the second paper of revisions. Newton wrote it on the blank verso side opposite definitions 13 and 14 (which he renumbered 15) of the first paper, with which it belonged. Later he crossed it out.

the second paper of revisions, Newton introduced a further change in the definition of inherent force whereby he assigned it, not to a body, but to a matter, as he did in the *Principia*.[41] Later he suggested another name, *vis inertiae*, the force of inertia. He revised Law 1 in a similar way to state that a body, by its inherent force alone, perseveres in its state of resting or of moving uniformly in a straight line.[42] That is, inherent force ceased to be the cause of uniform motion. An addition to his definition of impressed force ratified the change. "This force consists in the action only, and remains no longer in the body when the action is completed."[43]

A second sentence that Newton initially added to his revised form of Law 1 offers some insight into the magnitude of the intellectual reorientation involved. The law spoke of the state of resting or of moving uniformly in a right line. "Moreover this uniform motion is of two sorts," he continued, "progressive motion in a right line which the body describes with its center which is borne uniformly, and circular motion about any one of its axes which either rests or remains ever parallel to its prior position as it is carried with a uniform motion."[44] By the next time the law appeared, he had recognized that uniform rotation is hardly an example of uniform motion in a straight line. Even in stating it initially, however, he testified to the powerful inclination of everyone to perceive uniform rotation as a state of equilibrium like rest or uniform rectilinear motion. The principle of inertia denied the perceived similarity. With the alterations in the definitions and in Law 1, Newton effectively embraced that principle. In the *Principia* itself, he further eliminated the reference to inherent force from the statement of the first law, thus obliterating the principal record of the path by which he arrived at it.

Once he adopted the principle of inertia, the rest of his dynamics fell quickly into place. He had seized on the essence of his second law twenty years before and had never altered it as he wrestled with the first law. Impressed force alters a body's motion; the change of motion is proportional to it. In this proportionality lay the possibility of a quantitative science of dynamics that would cap and complete Galileo's kinematics. For the final form of the law, he returned to his original statement in the third version, changing it only to the extent of later adding one adjective, "motive," which increased its precision. In the *Principia*, the law did not read as we are accus-

---

[41] *Herivel*, p. 318; original Latin, p. 315.
[42] *Herivel*, p. 312; original Latin, p. 307. Cf. I. Bernard Cohen, " 'Quantum in se est': Newton's Concept of Inertia in Relation to Descartes and Lucretius," *Notes and Records of the Royal Society*, 19 (1964), 131–55.
[43] *Herivel*, p. 318; original Latin, pp. 315–16.
[44] *Herivel*, p. 312; original Latin, p. 307.

tomed to seeing it in modern textbooks of physics. It stated that the change of motion (Δmv) is proportional to the motive force impressed. In a number of propositions of the *Principia*, Newton employed it in this form – for example, Proposition I, which derived the area law exactly as *De motu* did, and propositions on the resistance of media in Book II, in which the physical origin of resistance in Newton's conception, impact with particles of a medium, corresponded to the definition of force. In most of the *Principia*, however, and especially in his treatment of orbital motion, Newton dealt with continuous forces instead of discrete impulses. Though he employed infinitesimal elements of arc, he treated force as uniform within the elements, so that the deviation from the tangent, proportional to the square of time, corresponds to Galileo's formula for uniformly accelerated motion.[45] Much of the story of the failure of dynamics during the seventeenth century before Newton lay in the lack of distinction between dimensionally incompatible conceptions of force. A residue of the problem remained in the *Principia* in the dichotomy between Newton's statement of the second law and his most frequent use of it. What had been a crippling deficiency in his predecessors, however, became no more than a faint haze of logical inconsistency that could be easily removed without dismantling the whole structure. When he used discrete impulses of force, he nearly always specified their separation by equal intervals of time, and when he used impulses otherwise, he made the necessary adjustments. Hence the dimension of time which was absent from the definition of force was present in the problem; and when he stated, as he did in Proposition I, that the centripetal force will act continuously as the breadth of the triangles is diminished *in infinitum,* no one feels inclined to leap up in protest. He smoothed over, if he did not eliminate, the problem by stating in the *Principia* that the same force produces the same motion "whether that force be impressed altogether and at once, or gradually and successively."[46]

The statement of the second law implied a quantity which he

---

[45] Cf. three analyses by Whiteside of propositions employing separate impulses, in which he shows that if, in the limit, we treat an impulse as a constant force exercised over the infinitesimal unit of time dt/n (so that $I = f(dt/n)$) and in successive units of time (n_1, $n_2 \ldots n$) the force becomes effectively continuous, the result is identical to what we obtain by starting with a continuous force (*Math 6*, 35–7, 200–1, 539–42). As Newton used it in his propositions on orbital motion, the differential unit of arc over which the force is treated as constant coincides with the circle of curvature. Hence the proportionality of the versed sine to the arc (or $s \propto t^2$) applies. Because of Corollary II, Proposition VII, and Corollary II, Lemma VII, it is not necessary that the center of attraction coincide with the center of curvature. In Proposition XXX, Book III, when Newton used an impulse of force considered as fully instantaneous, he distinguished it from a continuous force by doubling the space traversed, because of the impulse, in a unit of time.

[46] *Prin,* p. 13.

now defined, quantity of motion. It required in turn a further definition, quantity of matter, which initially he simply included within the first. The quantity of a body, he said, "is calculated from the bulk of the corporeal matter which is usually proportional to its weight."[47] When he revised the definitions, he both separated this one and placed it at the head of the list.

> The quantity of matter is that which arises conjointly from its density and magnitude. A body twice as dense in double the space is quadruple in quantity. This quantity I designate by the name of body or of mass.[48]

Without the concept of mass, here defined satisfactorily for the first time, the second law, the force law, would have remained incomplete. The one required the other. Together they constituted the heart of Newton's contribution to dynamics.

The concept of mass included more than mere quantity of matter, however. As he revised his first law of motion toward the principle of inertia, he transferred his revised notion of inherent force from that law, where it had become an embarrassment, to the concept of mass.

> *The Inherent force of matter* is the power of resisting by which any body, as much as in it lies, perseveres in its state of resting or of moving uniformly in a right line: and it is proportional to its body and does not differ at all from the inactivity of the mass except in our mode of conceiving it. In fact a body exerts this force only in a change of its state effected by another force impressed upon it, and its Exercise is *Resistance* and *Impetus* which are distinct only in relation to each other.[49]

In the classic phrase of the seventeenth century, matter is indifferent to motion. Leibniz was to argue that if matter is wholly indifferent to motion, any force should be able to impart any velocity to a body, and a quantitative science of dynamics would be impossible. Although Newton offered no rationale, it is evident that he responded to similar considerations. As an activity of resistance evoked in changes of state, mass establishes the equation between an impressed force and the change of motion it produces. In the *Principia,* Newton also called it *vis inertiae,* the force of inertia, or perhaps, in a free translation, the activity of inactivity. He arrived at this paradoxical conception of a matter both inert and active early in 1685 and bequeathed it to modern science as an essential aspect of his dynamics.

Thus his final dynamics continued to focus on the interplay of inherent force and impressed force, with the former wholly trans-

---

[47] *Herivel,* p. 311; original Latin, p. 306.
[48] *Herivel,* p. 317; original Latin, p. 315.
[49] *Herivel,* p. 318; original Latin, p. 315. In regard to the discussion of matter, cf. Ernan McMullin, *Matter and Activity in Newton* (Notre Dame, Ind., 1977).

formed from its original form. To replace the parallelogram of forces, he devised a further law of motion which has come down to us in a different wording as the third law.

> As much as any body acts on another so much does it experience in reaction. . . . In fact this law follows from Definitions 12 [the inherent force of a body to persevere in its state of resting or moving uniformly] and 14 [the force brought to bear and impressed on a body to change its state] in so far as the force of the body exerted to conserve its state is the same as the force impressed on the other body to change its state, and the change of state of the first body is proportional to the first force and of the second body to the second force.[50]

We cannot date with precision the sheets that definitively revised the dynamics of *De motu,* but they appear to have belonged to the early months of 1685. Few periods have held greater consequences for the history of Western science than the three to six months in the autumn and winter of 1684–5, when Newton created the modern science of dynamics. With primitive, unexamined concepts ready at hand, he had composed the short tract *De motu,* a treatise revolutionary enough thoroughly to excite Edmond Halley. With the sophisticated dynamics now at his command, much more was possible. The *Principia* was possible. In the preface to the work, Newton distinguished between rational and applied mechanics.

> In this sense rational mechanics will be the science of motions resulting from any forces whatsoever, and of the forces required to produce any motions, accurately proposed and demonstrated. . . . and therefore I offer this work as the mathematical principles of philosophy, for the whole burden of philosophy seems to consist in this–from the phenomena of motions to investigate the forces of nature, and then from these forces to demonstrate the other phenomena . . . [51]

Whereas the invisible mechanisms of orthodox mechanical philosophy, such as Descartes's vortices, had continually diverted attention away from quantitative precision toward picturable images, Newton's new conception of action at a distance invited mathematical treatment. The first stage of Newton's work on the *Principia* was the creation of his dynamics, the instrument that the rest of the task required.

One other development of note took place in the early months. In the mid-1660s, Newton had found the inverse-square relation by substituting Kepler's third law into his own formula for the endeavor of a body to recede from the center. When he had compared the moon's endeavor to recede with the acceleration of gravity at the surface of the earth, the exact inverse-square relation had not

---

[50] *Herivel,* p. 312–13; original Latin, p. 307.   [51] *Prin,* p. xvii.

emerged. Years later, he told at least three men that he had long thought some force arising from the vortex must disturb the motion of the moon.[52] In comparison with the surviving paper, Newton's later account attached a more explicit theory to his early calculation than we can find. Nevertheless, the mere fact that he later referred to it indicates that Newton did not forget his inchoate idea and the calculation it inspired. When he composed *De motu,* he used the word *gravitas* freely, for example in Problem 3, which demonstrated that an elliptical orbit entails that "gravity" vary inversely as the square of the distance from the focus.[53] Since the scholium to the problem spoke of the sun and planets, he was explicitly ascribing gravity to the sun. Newton went back, however, and systematically replaced "gravity" with the neutral phrase "centripetal force." Perhaps the replacement expressed uncertainty, perhaps merely caution. Ideas of gravities specific to systems of related bodies were common in the seventeenth century. Newton could have been extending such an idea to the solar system. His questions to Flamsteed about Jupiter and Saturn may imply such. His inclusion of comets in the orbital dynamics suggests that already he may have transcended that limit. The first version of *De motu* did not contain a correlation of the moon with the measured acceleration of gravity, however. In general terms, the third version did.

> For gravity is one kind of centripetal force: and my calculations reveal that the centripetal force by which our moon is held in her monthly orbit around the earth is to the force of gravity at the surface of the earth very nearly as the reciprocal of the square of the distance from the center of the earth.[54]

According to several accounts, Picard's new measure of the earth, which Newton now used for the first time, yielded the accurate correlation.

Nevertheless, the same scholium that contained the correlation also referred to media whose density (or quantity of solid matter) is "almost proportional to their weight . . ."[55] He made a similar comment in his paper of revisions that followed the third version. The quantity of a body is calculated "from the bulk of the corporeal matter which is usually proportional to its weight." Almost proportional, usually proportional–however broadly Newton's

[52] DeMoivre Memorandum; University of Chicago Library, MS 1075-7. Whiston, *Memoirs,* p. 37. Henry Pemberton, *View of Newton's Philosophy* (London, 1728) Preface. Cf. his notes inside the cover of Vincent Wing, *Astronomia britannica* (London, 1669), Trinity College Library.

[53] *Add MS* 3965.7, f. 57. *Herivel,* p. 282; original Latin, p. 263. Cf. pp. 280, 281, 286; original Latin, pp. 261, 262, 269–70.

[54] *Herivel,* p. 302; original Latin, p. 298.

[55] *Herivel,* p. 302; original Latin, p. 298.

thoughts were ranging, he had not yet arrived at his final conception of universal gravitation if he could use such phrases. In the second case, he even included a practical device to compare the solid matter in two bodies of equal weight: Hang them from equal pendulums; the quantity of matter varies inversely as the number of oscillations made in the same time. Newton crossed the passage out. On a blank verso opposite, he wrote that the weight of heavy bodies is proportional to their quantity of matter as experiments with pendulums can prove: "When experiments were carefully made with gold, silver, lead, glass, sand, common salt, water, wood, and wheat, however, they resulted always in the same number of oscillations."[56]

Now at last, the full implications of an idea only half explored before could open before him. The equal acceleration in free fall of all heavy bodies found its explanation; the pendulums offered a demonstration of the same phenomenon which was far more delicate, since tiny differences would accumulate on successive swings and become manifest. In the heavens, Kepler's third law did the same, unless one accepted the unlikely assumption that the planets are exactly equal in mass. The satellites of Jupiter were so many more celestial pendulums. Not only did their conformance to Kepler's third law reveal the proportionality of their masses to their attractions toward Jupiter, but their concentric orbits about Jupiter demonstrated that the sun attracts both them and Jupiter in proportion to mass.[57] According to the third law, the satellites of Jupiter must attract the sun in return. Apparently every body in the world attracts every other body. We can only imagine the wild surmise that raced through Newton's mind as the principle of universal gravitation silently unfolded. As he pursued orbital mechanics, his new quantitative dynamics had led him to a generalization more inclusive than any that natural philosophy had hitherto proclaimed. The published *Principia* repeatedly insisted that the mathematical treatment of attraction asserted nothing about its physical cause. When he first wrote his discovery down, he was less reserved; the forces proportional to the quantity of matter, he said, "arise from the universal nature of matter . . ."[58] He appears to have arrived at

---

[56] *Herivel*, pp. 311 and 319; original Latin, pp. 306 and 316–17. Herivel misplaces the revision with the second paper; it is found on a verso side of the first (*Add MS 3965.5a*, f. 25ᵛ) opposite the original version, which it corrected.   [57] *Prin*, p. 412.

[58] In 1728, the original concluding book of the *Principia* was published in English translation under the title *A Treatise of the System of the World*. It is also published at the end of *Prin*, from which I quote, p. 571. On the concept of universal gravitation and various associated problems, see Alexandre Koyré, "La gravitation universelle de Kepler à Newton," *Archives internationales d'histoire des sciences*, 4 (1951), 638–53; Florian Cajori, "Newton's Twenty Years' Delay in Announcing the Law of Gravitation," in History of Science Society, *Sir Isaac Newton, 1727–1927* (Baltimore, 1928, pp. 127–88; Curtis A. Wilson,

the concept very nearly at the time when his namesake, Alderman Samuel Newton, proclaimed James Stuart king of England immediately outside his window.

Clearly, a discovery so grandiose required an exposition commensurate to it. The dynamics he had created provided an adequate instrument. Newton began to expand *De motu* into a systematic demonstration of universal gravitation. We know very little about the process except that the nine-page tract of November 1684 had expanded to a treatise in two books of well over ten times that length by roughly the following November. To it he gave the name *De motu corporum* (*On the Motion of Bodies*). After Newton's death, Book II was published under the name *De mundi systemate* (*On the System of the World*), an apt title even though it did not appear on the manuscript, since Book II did expound the system of the world on the basis of the propositions demonstrated in Book I. The manuscript of Book I is frequently called the "Lectiones de motu" (Lectures on Motion), since Newton delivered it to the university as a copy of his lectures in the years 1684 and 1685.[59] He also gave in an incomplete copy of Book II as a record of his lectures in 1687.[60] He may well have begun to lecture on topics relevant to the *Principia* in autumn 1684, after Halley's visit set him to work on it. No one takes the deposited manuscripts as a record of the lectures delivered, however. They were obviously drafts of Books I and III of the *Principia*.

One of the first problems Newton faced as he wrote the expanded work was its mathematical foundation. *De motu* had employed an implicit method based on differential segments of curves. Newton now formalized the approach by drawing on the concept of first and last ratios, which he had developed to eliminate infinitesimals from his fluxional calculus. He embodied it in eleven lemmas which he later designated as Section I of Book I. The mathematics of the *Principia* has always presented something of a puzzle. Why did the inventor of the calculus present his masterwork in the dress of classical geometry? Newton himself compounded the puzzle by several self-serving assertions at the time of the calculus controversy.

By the help of the new *Analysis* Mr. Newton found out most of the Propositions in his *Principia Philosophiae* [he wrote anonymously]:

"From Kepler's Laws, So-called, to Universal Gravitation: Empirical Factors," *Archive for History of Exact Sciences*, 6 (1970), 89–170; Léon Rosenfeld, "Newton's Views on Aether and Gravitation," *ibid.*, 6 (1969), 29–37; and Hélène Metzger, *Attraction universelle et religion naturelle chez quelques commentateurs anglais de Newton* (Paris, 1938).

[59] Cambridge University Library, Dd. 9. 46.

[60] Cambridge University Library, Dd. 4. 18. Cf. Cohen, *Introduction*, pp. 83–92, 109–15, 302–22.

but because the Ancients for making things certain admitted nothing into Geometry before it was demonstrated synthetically, he demonstrated the Propositions synthetically, that the Systeme of the Heavens might be founded upon good Geometry. And this makes it now difficult for unskilful men to see the Analysis by which those Propositions were found out.[61]

No such papers demonstrating propositions of the *Principia* in a form different from that published have ever been found, except for a few in which he later set a couple of propositions over into analytic terms. The problem vanishes, however, when we view the *Principia* against the background of Newton's mathematical development in the years immediately preceding. Around 1680, the study of ancient geometry led Newton to a revulsion from the inelegant demonstrations of modern analysis. As he said in his *Geometria curvilinea,* the method of first and last ratios was the natural method to extend the principles of ancient geometry beyond straight lines and circles to the treatment of curved lines generated by continuous flow. The convictions Newton held, he held strongly. It was inevitable that he chose to present his masterpiece in the form that now alone seemed demonstrative to him.

The problem with the mathematics of the *Principia,* then, is not to look for prior demonstrations in a different form but to see thought patterns of the calculus behind the façade of classical geometry. At one point, Newton thought of making his method more explicit. Among the list of definitions that he drew up after *De motu* he included one of "moments."

The moments of quantities are the principles from which they are generated or altered by a continual flux.[62]

He did not leave the definition among those which introduce the work, but later, when he was composing Book II (in the final numbering), he did insert a longer definition of the same concept as Lemma II. By generated quantities, he explained, he meant quantities that increase and decrease "by a continual motion or flux," and by their moments he understood "their momentary increments or decrements . . . But take care not to look upon finite particles as such. Finite particles are not moments, but the very quantities generated by the moments. We are to conceive them as the just nascent principles of finite magnitudes."[63] The intuitive idea of the instantaneous changes of quantities in continual flux, which furnished the foundation and name of the fluxional calculus, found a different expression in the *Principia*'s method of first and last ratios. He had applied it before to the variables in algebraic equations. For the

[61] "An Account of the Book entitled *Commercium Epistolicum,*" *Philosophical Transactions,* 29 (1714–16), 206.

[62] *Herivel,* pp. 311–12; original Latin, p. 306.   [63] *Prin,* p. 249.

Principia, he set it over into the language of geometrical figures. Whatever the language, the concept of first and last ratios was thoroughly modern; classical geometry had contained nothing similar to it.[64]

Lemma I stated a definition of equality which drew upon his work with infinite series.

> Quantities, and the ratios of quantities, which in any finite time converge continually to equality, and before the end of that time approach nearer to each other than by any given difference, become ultimately equal.[65]

Three further lemmas compared the areas of sets of parallelograms inscribed in and around curves with the areas under the curves themselves. Considering the method as a geometrical version of the calculus, we may call these the integrating lemmas. Lemma V stated a concept of similarity.

> All homologous sides of similar figures, whether curvilinear or rectilinear, are proportional; and the areas are as the squares of the homologous sides.[66]

From this lemma, he demonstrated six more about the ratios of arcs, chords, tangents, versed sines, and the triangles they form as two points on a curve approach each other. We may call these the differentiating lemmas, and in the *Principia* they offered Newton his handiest tool. Thus in Proposition VI, he demonstrated that the centripetal force in the middle of "any arc [*QP*] just then nascent" described in the least time varies as the versed sine ($\frac{1}{2}QR$) and inversely as the time squared. By Kepler's law, demonstrated in Proposition I to hold for all motion under the sole influence of any centripetal force whatever, the time is proportional to the area described by the radius vector ($\Delta SQP = \frac{1}{2}SP \cdot QT$). Hence the centripetal force that holds a body on an incremental segment of any curve is inversely proportional to ($SP^2 \cdot QT^2)/QR$ (Figure 10.3). Newton's strategy in finding the centripetal force required for motion in any curve, such as an ellipse, was to start with the known properties of the curve and to work toward an expression of the distance of any point *P* from the center of force in terms of ($SP^2 \cdot QT^2)/QR$. The strategy required that he implicitly redefine the characteristics of curves in terms of ultimate ratios about the points

---

[64] Cf. D. T. Whiteside, *The Mathematical Principles Underlying Newton's 'Principia Mathematica'* (Glasgow, 1970); published also in *Journal for the History of Astronomy, 1* (1970), 116–38, and the whole of *Math 6.* See also Markus Fierz, "Newtons Auffassung der Mathematik und die mathematische Form der 'Principia'," *Helvetica Physica Acta, 41* (1968), 821–6; and Florian Cajori, *A History of the Conceptions of Limits and Fluxions in Great Britain from Newton to Wodehouse* (Chicago, 1919). Henry Brougham and E. J. Routh, *Analytical View of Sir Isaac Newton's Principia* (London, 1855), set the mathematics of the *Principia* into analytic form as it was current in the mid-nineteenth century.

[65] *Prin,* p. 29.   [66] *Prin,* p. 32.

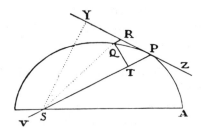

Figure 10.3.   Proposition VI. The general measure of centripetal force in any path.

Q and P as they approach each other. The *Principia* clothed itself in the proportions of classical geometry. The proportions themselves, however, were the ultimate ratios of quantities approaching zero. Euclid would not have recognized his offspring.

Powerful though the instrument was in Newton's hands, it proved insufficient for a number of tasks – perhaps a forecast of the direction that the future development of mathematics would take. In Book II, he had to introduce Lemma II with its concept of a moment to handle problems of resistance. In a number of propositions, he simply employed quadratures without explanation.[67] By the time he revised the original version of the so-called "Lectiones" into their final form, probably early in 1686, he must have become aware that Leibniz had begun to publish his differential calculus, and he wanted to protect his claim to the method he had invented. Since several propositions in Book I assumed the quadrature of curves, he drafted a general exposition of his method of squaring affected equations, though in the end he did not include it.[68] He did insert Lemma II into Book II, however: and lest its thrust be mistaken, he concluded it with a scholium which covertly asserted his priority. The scholium cited his correspondence with Leibniz in 1676 and stated the proposition defining his method, which he had concealed then in an anagram. "The most eminent man [Leibniz] wrote back that he also had fallen into a similar method, and he communicated his method, which scarcely differed from mine except in the words and notations it used.[69] The foundation of both

---

67   E.g., Propositions LXXIX and XC, Book I; *Prin*, pp. 204 and 219. Cf. also his implicit introduction of the second derivative of a function in the Scholium to Section XIII, Book I; *Prin*, p. 225.

68   Three drafts of it have survived; *Math 6*, 450–5.

69   Newton's phrase here, *a mea vix abludentem*, virtually repeated one, *ab his non abludere*, from Leibniz's letter to Oldenburg, commenting on the *Epistola posterior*, of 11 June 1677 (*Corres 2*, 215). Newton must have had his copy of the letter in his hand as he wrote the scholium.

methods is contained in this lemma."[70] It unlikely that Leibniz read the scholium as a conciliatory gesture.

Meanwhile, the expansion of *De motu* into an exposition of universal gravitation proceeded through 1685. Although we do not know at what point he attacked it, we do know that the attraction of a sphere early loomed as a major problem. His experiment with the pendulums, reinforcing his particulate conception of matter, implied that the attraction of a body is simply the sum of the attraction of its particles. The correlation of the moon's orbit with the acceleration of gravity assumed that the inverse-square law holds, not only at the distance of the moon but also at the surface of the earth. All the particles of the vast earth, stretching out in every direction beyond the horizon, combine to attract an apple a few feet above its surface at Woolsthorpe or at Cambridge with a force dependent, not on the apple's distance from the surface, but on its distance from the center of the earth. Was such a notion credible? As Newton told it to Halley in 1686 at least, it was not.

> I never extended ye duplicate proportion lower then to ye superficies of ye earth & before a certain demonstration I found ye last year have suspected it did not reach accurately enough down so low . . . There is so strong an objection against ye accurateness of this proportion, y^t without my Demonstrations . . . it cannot be beleived by a judicious Philosopher to be any where accurate.[71]

The paradox printed itself deeply enough on his mind that he inserted a statement of the dilemma into the *Principia*.

> After I had found that the force of gravity towards a whole planet did arise from and was compounded of the forces of gravity towards all its parts, and towards every one part was in the inverse proportion of the squares of the distances from the part, I was yet in doubt whether that proportion inversely as the square of the distance did accurately hold, or but nearly so, in the total force compounded of so many partial ones; for it might be that the proportion which accurately enough took place in greater distances should be wide of the truth near the surface of the planet, where the distances of the particles are unequal, and their situation dissimilar.[72]

Hence the significance of Proposition XL of the *Lectiones* (corresponding to Proposition LXXI, Book I) in which Newton demonstrated that a homogeneous spherical shell composed of particles that attract inversely as the square of the distance, attracts a particle external to it, no matter at what distance, inversely as the square of its distance from the center of the sphere.

[70] *Var Prin 1,* 369. Since Newton replaced this passage in the third edition with a stronger, though still covert assertion of priority, it is not found in the English translations.
[71] Newton to Halley, 20 June 1686; *Corres 2,* 435–7.   [72] *Prin,* pp. 415–6.

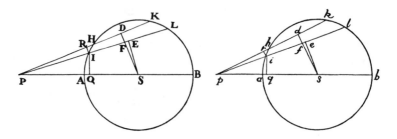

Figure 10.4. The attraction of a homogeneous spherical shell on a particle external to it.

Since he had denied himself his method of quadratures, what would have been (for Newton) a simple exercise in integration became a more difficult enterprise.[73] The symmetry of the circle rescued him, and the comparison of two particles, P and p, placed at different distances from the sphere yielded the sort of proportion in which the *Principia* dealt. From the two particles he imagined two pairs of lines *PK, PL* (and *pk, pl*) to cut differential elements *HI, KL* (and *hi, kl*) from the shell's equator (Figure 10.4). When the lines are rotated about the axis *PS* (and *ps*) each pair cuts two differential rings from the shell, one of radius *IQ* (and *iq*), the other of a radius not drawn on Newton's diagram but similarly placed in relation to *KL* (and *kl*). Newton demonstrated that the attraction on particle P toward the center S by the ring *HI*, is to the corresponding attraction on particle p in the ratio $(HI \cdot IQ)/(hi \cdot iq) \cdot ps^2/PS^2$, that is, directly as the areas of the rings and inversely as the squares of the particles' distances from the center of the shell. So also for the second rings *KL, kl*. By setting arcs *HK* and *IL* equal to arcs *hk* and *il* (or *DS = ds* and *ES = es*) he could maintain the symmetry of the two cases, and by altering the two angles *KPS, kps* (which are not and need not be equal), he could simultaneously exhaust the area of the shell from each point, P and p. Hence a shell of such attracting particles attracts any particle external to it with a force proportional to its area (or quantity of matter when we give it a minimal thickness) and inversely proportional to the square of the particle's distance from the center.[74] The following propositions extended the demonstration to solid spheres composed of homogeneous shells. As Newton went on to remark, the propositions he had demonstrated about motions in conics are therefore equally true when an attracting sphere is located at the focus.[75] As he also realized, the correlation between the moon and the apple ought to be more precise than "very nearly," the words he used in the third version of *De motu*. The calculation he now carried out showed a correlation correct to the last inch (or one part in four hundred) of the acceleration of gravity as Huygens had determined it.[76]

With the demonstration of a sphere's attraction, and with the exact correlation of the moon's motion with the measured acceleration of gravity, the logical foundation of the concept of universal gravitation became secure. Newton was a man of rigor. Though his imagination ranged widely, he could not consider putting so vast an idea as universal gravitation in print before he had satisfied himself of its demonstrability. So far, he had proved the presence of inverse-square attractions in the solar system, indeed the presence of a single inverse-square attraction. Moreover, the earth must attract the moon to hold it in its orbit; and if apples fall to the earth, it must attract them as well. By what right could he extend the ancient word *gravitas* (heaviness), applied to the apple, both to the attraction that holds the moon and to the attraction of the sun? Only the inverse-square correlation of the moon with the apple, plus a liberal dash of the principle of uniformity, permitted the argument. Together with Proposition XI (that elliptical orbits entail inverse-square attractions), the demonstration of a sphere's attraction was one of the two foundation stones on which the law of universal gravitation rested.

The plan of the initial "Lectiones de motu" began, after the definitions, laws of motion, and exposition of the method of ultimate ratios, with a somewhat enlarged set of the orbital propositions that had formed the heart of *De motu*. Newton continued on to a series of propositions, which expanded a single one in *De motu*, on vertical rectilinear motion in an inverse-square force field. Given the mathematical constraints he had imposed on himself, he did not find it a simple task. He solved it by the ingenious device of treating the vertical line as the limiting shape of ellipses and hyperbolas that shrink into their axes. The propositions did not contribute seriously to the argument for universal gravitation. For a brief period, they expanded the range of his dynamics until, in the revision of the "Lectiones" a more powerful proposition superseded them. Newton left them in the published work nevertheless, perhaps as artifacts of the extraordinary exertion that produced the *Principia*.

So far, the exposition had dealt with the abstract problem of bodies in motion about unspecified centers of force, which to New-

---

76 The first version of Book III; *Prin*, p. 560. The more refined correlation, which took account of such factors as the buoyancy of the air and the centrifugal force arising from the earth's rotation, found in the English translation (pp. 425, 483), appeared first in the second edition of the *Principia*.

ton always meant physical bodies. According to the third law, central attracting bodies must themselves be attracted and moved. The "Lectiones" now turned to consider the complications that mutual attractions introduce. First, Newton dealt with the two-body problem. He showed that two mutually attracting bodies describe similar figures around their common center of gravity and around each other, and he derived formulas for their periodic times and major axes.[77] This set of propositions proved that the previous demonstrations on orbital motion remain valid despite the mutual attraction of the sun and a planet. The solar system consists of many bodies, however, a sun and six planets (as far as the seventeenth century knew), three of which (again as far as they knew) have satellites. Could so many mutual attractions fail to upset the orbital dynamics demonstrated for single bodies? The many-body problem presented a more serious challenge than the two-body problem, and one that Newton could not avoid. He had taken as his province the explanation of natural philosophy from the principle of attraction. Abstract demonstrations, however elegant, were one thing. Natural philosophy addressed itself to the real world, and the real world consisted of many bodies in motion, all of them, on Newton's hypothesis, attracting each other. Newton found that a demonstrative solution to the problem exceeded his power. (Indeed, it can now be shown to be impossible.) Nevertheless, he did find an analytic tool by which to attack the simplest form of the problem, three mutually attracting bodies, the conclusions from which he could extend by plausible arguments to a system of many bodies.

The analysis, in Proposition XXXV (Proposition LXVI, Book I), considered the case of a large central body, S (standing for *Sol*, the sun) circled by two planets P and Q. He asked himself what perturbations the attraction of the outer planet Q introduces into the motion of the inner planet P. To answer the question, he expressed the attraction of Q on P by the line LQ (Figure 10.5). In LQ, let KQ be the mean distance of P and Q as P moves in the orbit PAB. KQ provides the metric of the analysis. When P is removed from Q by the distance KQ, KQ also expresses the attraction of Q on P. Since the attraction varies inversely as the square of the distance by assumption, LQ is longer than KQ (in the proportion $LQ/KQ = KQ^2/PQ^2$) when P is closer to Q than the mean distance (as it is in the diagram). The analysis rested on the resolution of LQ into two components: LM parallel to PS, and MQ parallel to SQ. Like the sun's attraction on P, LM is a radial or centripetal force. Hence it does not upset the description of areas in proportion to time. Since

---

[77] *Prin*, pp. 164–7. The corresponding propositions in the "Lectiones" have not survived, but we know from other evidence that they were there.

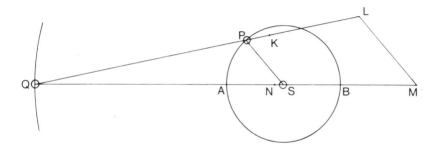

Figure 10.5. No diagram such as this survives. It is reconstructed from the text by analogy with the diagram to Proposition LXVI.

its presence does destroy the inverse-square proportionality of force to distance, it precludes perfectly elliptical orbits. *MQ* is more disturbing since it is neither centripetal nor inversely proportional to the square of the distance. Its disturbing effect is diminished by the fact that *Q* also attracts *S*. Let *NQ* (= *KQ* ≃ *SQ*) express the attraction of *Q* on *S*; not its total attraction, of course, for *NQ* would in that case be as many times larger than *KQ* as *S* is than *P*, but the intensity of its attraction. Newton called *LQ* and *NQ* the attractions on *P* and *S* "in proportion to their weights."

Here Newton began to confront a new issue that may well have taken him back to his definitions. Although the sheets with the definitions in the first version of the "Lectiones" do not survive, we know from the numbers on later sheets that the definitions (including the scholium on absolute space and time) and the laws of motion (including the corollaries and a scholium, if it existed yet) occupied a total of about six pages. In two stages, Newton expanded the six, first to about fifteen and then to nineteen pages (in the version that survives).[78] Probably, part of the expansion was the insertion of three new definitions not found in prior papers: the absolute, accelerative, and motive quantities of centripetal force.

---

[78] Cambridge University Library, Dd.9.46, the "Lectiones," consists of a number of fascicles of eight folios each, with the exception of the first fascicle, which contains twenty folios. Section I (the method of ultimate ratios) begins at the bottom of f.19. In order to join with the following fascicle, f. 20 is written in a small hand and contains roughly twice as much as the other folios; in the original version the same material must have filled about two folios. The folio number of the first sheet of the following fascicle was originally 9, indicating that fascicle 1 contained eight folios also when he first wrote it. Later, someone (probably Halley) read an expanded version of the "Lectiones." His notes (*Add MS* 3965. 9, ff. 94–9) called the second fascicle 2*, as though two fascicles preceded it, and his notes indicate two such fascicles, each of eight folios. His notes included a reference to the verso side of f. 8 of the second, which contained then effectively nine folios in all.

Possibly the issue that Proposition XXXV placed before him first led Newton to realize that he must distinguish among the various quantities under which centripetal force can present itself. The analysis of Proposition XXXV depends on the conscious understanding that one is dealing primarily with the intensity or accelerative quantity of the attraction of Q.

Because Q attracts both P and S, the disturbing effect of the component MQ is only that portion (MN) by which it differs from the accelerative attraction of Q on S (NQ). Nothing can eliminate the disturbance introduced into the orbit of P by the second planet Q. Newton had to acknowledge that the earlier demonstrations are only ideal approximations to reality. However, he could also analyze the conditions that minimize the deviations from the ideal, namely distance, which decreases the quantity of LQ in relation to the attraction of S, and above all the universality of attraction whereby the accelerative force of Q on S becomes a mean among the varying quantities of its force exerted on P. So also, if several planets, P, Q, R, etc., revolve around a great central body S, "the motion of the innermost revolving body P is least disturbed by the attractions of the others, when the great body is attracted and agitated by the others equally in proportion to weights and distances and they by each other."[79]

> And hence if several smaller bodies revolve about the great one, it may easily be inferred that the orbits described will approach nearer to ellipses, and the descriptions of areas will be more nearly uniform, if all the bodies mutually attract and agitate each other equally in proportion to their weights and distances, and if the focus of each orbit is placed in the common center of gravity of all the interior bodies . . . than if the innermost body were at rest, and were made the common focus of all the orbits.[80]

Originally, the analysis of the three-body problem directed itself solely to the problem of multiple planets around the sun and the apparent conflict between universal gravitation and Kepler's laws. Almost at once, Newton saw that it could be turned to quite the opposite use, not to explaining how orbits can so closely approximate Kepler's laws despite mutual attraction, but to explaining the source of observed perturbations. First of all, this meant the moon. By the time Newton drafted Book II, he had composed a new Proposition XXXV (Proposition LXV, Book I) primarily directed to systems, like the earth and moon, that orbit central attracting bodies together. To the analysis of the forces in play (now renum-

---

[79] *Add MS* 3965.3, ff. 7–10. No diagram survives. I have reconstructed one from the proposition by analogy with the diagram for Proposition LXVI of the published work.
[80] *Add MS* 3965.3, f. 11. Cf. *Prin*, p. 190.

bered Proposition XXXVI) he added twenty-one additional corollaries to supplement the original one. The majority addressed themselves to the inequalities of lunar motion. Probably at the same time, he relettered the diagram making the central body T (for *Terra*) and the external disturbing body S. Hence the anomalies that remain in Proposition LXVI, which has "lesser bodies" S and P revolving around the "greatest body" T. The analysis could extend beyond lunar inequalities, however. Let PAB represents a continuous ring of particles instead of the orbit of a single body, and let its radius PT be equal to the radius of the earth. When the ring of particles is a fluid, the analysis deals with the tides. Newton composed Corollaries 18 and 19 about them. When the ring is solid and attached to the earth, the analysis offers a derivation of precession. Corollaries 20 and 21 concerned themselves with that phenomenon.[81]

In the plan of the work, the treatment of the two- and three-body problems led into the examination of the attraction of spheres, in the middle of which the surviving manuscript ends. In Book II, Newton referred to Proposition XLVI (Proposition LXXVI, Book I) on the mutual attraction of spheres, three propositions beyond the last surviving one. Like *De motu,* the "Lectiones" closed with a series of propositions on motion through resisting media which did not contribute directly to the argument for universal gravitation. Book II referred once to Proposition LXXII, which can be identified from the context with the *Principia*'s Proposition XXII, Book II. It is not unreasonable to assume that Newton had by then composed the preceding twenty-one propositions of the ultimate Book II, and the assumption is supported though hardly confirmed by Newton's statement to Halley in the summer of 1686 that he had completed the few propositions he then intended to put in Book II the previous summer.[82] If we do make this assumption, we are still left with four propositions in the original *Lectiones* which we are unable to specify. Nothing of importance hinges on their identification.

In contrast to the starkly mathematical format of Book I, the second book of the expanded work presented a prose essay on the Newtonian system of the world based on the propositions demonstrated in Book I and referring to them for support. We can date the time of its composition with some confidence. In September 1685, Newton wrote to Flamsteed for information about the comet of 1680–1 and about tides, on which Flamsteed had recently published a table of observations in the *Philosophical Transactions.*[83] In the letter, he acknowledged that the two comets of November and

81 Proposition XXXVI with its twenty-two corollaries does not survive. Book II referred to it and to them, however, and the author of the set of suggested corrections (*Add MS* 3965.9, ff. 94–9) did also.
82 Newton to Halley, 20 June 1686; *Corres 2,* 437.   83 *Corres 2,* 419–20.

December 1680 were one and the same. Although it is difficult to imagine how Flamsteed had failed to comprehend *De motu*, nevertheless he entered a triumphant note in the margin beside the explicit concession: "he would not grant it before see his letter of 1681."[84] Newton told Flamsteed that he was undertaking to determine the orbit of the comet. The following summer, in a letter to Halley, he confessed that the theory of comets still remained unfinished. "In Autumn last I spent two months in calculations to no purpose for want of a good method, wch made me afterwards return to ye first Book & enlarge it wth divers Propositions some relating to Comets . . ."[85] Both statements are consistent with the relatively crude treatment of comets in the early Book II; they suggest that Newton composed it during the autumn of 1685. In regard to this request for information on tides, Newton expanded the treatment of them in the very process of composing Book II and, as part of the expansion, inserted a paragraph that referred to Flamsteed's table.[86]

The original system of the world differed from that ultimately published as Book III of the *Principia* in more than format. With the exception of the brief Hypothesis VI (Phenomenon III in later editions), the *Principia* took the Copernican-Keplerian system for granted. The early Book II did not, and it devoted considerable space to justifying a heliocentric universe as the necessary foundation of the system based on universal attraction. Indeed, Newton chose to open the argument with a passage taken directly from his "Origins of Gentile Theology."

> It was the most ancient opinion of those who applied themselves to
> Philosophy, that the fixed stars stood immovable in the highest parts
> of the world, that under them the planets revolved about the sun, that
> the earth, as one of the planets, described an annual course about the
> sun, while by a diurnal motion it turned on its axis, and that the sun
> remained at rest in the center of the universe. This was the philosophy
> taught of old by Philolaus, Aristarchus of Samos, Plato in his riper
> years, the whole sect of Pythagoreans, and that wisest king of the
> Romans, Numa Pompilius. As a symbol of the round orb with the
> solar fire in the center, Numa erected a round temple in honor of
> Vesta, and ordained a perpetual fire to be kept in the middle of it. The
> Egyptians were the earliest observers of the heavens, and from them,
> probably, this philosophy was spread abroad. For from them it was,
> and from the nations about them, that the Greeks, a people more
> addicted to the study of philology than of Nature, derived their first,
> as well as their soundest, notions of philosophy; and in the Vestal

---

[84] *Corres 2*, 421.   [85] Newton to Halley, 20 June 1686; *Corres 2*, 437.
[86] *Add MS* 3990, f.28^v. The paragraph inserted here corresponds to Paragraph 46 in the published version (*Prin*, p. 587).

ceremonies we can recognize the spirit of the Egyptians who concealed mysteries that were above the capacity of the common herd under the veil of religious rites and hieroglyphic symbols.[87]

From the observed phenomena of the heliocentric system, Newton argued the necessity of attractive forces to hold bodies in closed orbits, of inverse-square attractive forces to sustain orbits stable in space and systems that conform to Kepler's third law, and finally of a single inverse-square attractive force that arises, "from the universal nature of matter."[88] Most of Book I had looked directly to that conclusion.

Book II proceeded then briefly to suggest, on the basis of Proposition XXXVI, that the principle of universal gravitation could account for other observed phenomena. One paragraph asserted that it could explain the observed anomalies of the moon's motion and a second that it predicted further anomalies as yet unobserved. He devoted single paragraphs, devoid of quantitative details as the paragraphs on the moon were, to precession and tides, and turned to a long qualitative discussion of comets which led to the conclusion that comets come under the laws of orbital motion developed for planets.

> Therefore, during the whole time of their appearance, comets fall within the sphere of activity of the circumsolar force, and hence are acted upon by its impulse and therefore (by Coroll. 1, Prop. XII [Proposition XIII, Book I]) describe conic sections that have their foci in the center of the sun, and by radii drawn to the sun describe areas proportional to the times. For that force, propagated to an immense distance, will govern the motions of bodies far beyond the orbit of Saturn.[89]

The determination of an orbit was more difficult than the assertion of the general principle. In five lemmas and two problems that concluded Book II, Newton tried to lay down a method by which to determine an orbit from observations. In fact, he was not able to produce a method that he could apply successfully to a specific comet. When he compared his quantitative treatment of comets with his confident promise to Flamsteed in September, he could only have regarded the effort as a failure. His letter to Halley the following summer admitted as much.

In fact, Book II never existed in the form described above. In the very process of composing it, Newton introduced a change pregnant with further possibilities. As soon as Humphrey had copied the paragraph that dealt with tides, before he could proceed to the next paragraph, Newton crossed it out, and in its place he began to

---

[87] *Add MS* 3990, f.1. A different English version is given in *Prin*, p. 549.

[88] *Prin*, p. 571–2.

[89] *Prin*, p. 614; I have altered the translation somewhat. Cf. *Math* 6, 484–5.

compose a detailed, quantitative discussion of tides based on Proposition XXXVI and its twenty-one new corollaries.[90] In the very manuscript, the analysis takes shape before our eyes. Most of it appears in Humphrey's hand, but three times Newton took the pen himself to complete difficult passages.[91] In a similar way, he amended his initial computation of the comparative forces of the sun and the moon to produce the tides, canceling an excessively rough-and-ready calculation before its ink was fully dry and substituting in its place an excessively sophisticated one which later he had to abandon also.[92] He extended the original paragraph to an eighteen-paragraph discussion which has all the elements of the one that finally appeared in the *Principia*.

About this time, Newton submitted his creation to inspection and criticism. Notes with suggested emendations, some of which he adopted, survive among Newton's papers. The notes are confusing since the references in them point to the early version of Book I but to the final versions of Books II and III as they were published. Whoever their author was, the notes indicate that he understood what he was reading. That fact in itself powerfully suggests Edmond Halley as their author, as does the further fact that Halley mentioned Newton's "incomparable *Treatise of Motion* almost ready for the *Press*," in a presentation to the Royal Society before the manuscript of Book I arrived, and briefly summarized its contents.[93] Let us assume that Halley did compose them.[94] In autumn 1685, he received his third shock from the incredible professor in Cambridge. In August of the previous year, he had heard with joy Newton's assertion that he had demonstrated how inverse-square forces entail elliptical orbits. In November, he had received something far more, a sketch of a complete orbital mechanics. As far as we know, he had heard nothing more in the intervening year. Imagine his amazement as he read a manuscript which both presented a developed system of dynamics and argued from it to something Halley would not have dreamed of, a universal principle of gravitation pertaining to all the matter in the universe. The tract *De motu* of November 1684 had excited Halley enough. He may have been fortunate to survive the treatise of 1685. Perhaps we can trace to that reading Halley's extraordinary behavior in 1686 and 1687, when he sacrificed all his own

[90] *Add MS* 3990. The original paragraph on the tides is on ff. 23–4. The expanded treatment occupies ff. 24–33 (*Prin*, pp. 581–95).

[91] *Add MS* 3990, ff.27, 28–9, 29–30.   [92] *Add MS* 3990, f.32.

[93] 21 April 1686; Birch, *History, 4*, 479. Published as "A Discourse Concerning Gravity, and its Properties," *Philosophical Transactions, 16* (1686), 3–8.

[94] *Add MS* 3965.9, ff. 94–9. Cf. Cohen, *Introduction*, pp. 122–4, where these notes were first identified. Cohen is not entirely satisfied that Halley wrote the notes, and neither am I. The hand differs somewhat from that in other papers known to have been written by Halley.

activities to insure the publication of the *Principia*. And perhaps we can trace to that reading the tone of awe with which he henceforth referred to the still growing masterpiece. In 1686, it was "Your Incomparable treatise,"[95] the same adjective he used before the Royal Society; by 1687, it had become "your divine Treatise."[96] Later, in the "Ode to Newton" with which he prefaced the *Principia*, Halley chose a more restrained – or was it more restrained? – description: "Nearer the gods no mortal may approach."[97]

A young Scottish mathematician, John Craig, who was in Cambridge during 1685, apparently also saw at least some of the developing manuscript. In his copy of the *Principia*, he later wrote a note beside Newton's demonstration in Book II of the solid of least resistance that he had suggested the problem.[98] David Gregory's knowledge of what he called in autumn 1686 "a book of Astronomie" probably derived from Craig.[99] Similarly, knowledge of the impending work filtered through the British scientific community during the year before its publication. Flamsteed wrote about it to Towneley in November 1686, basing his information on what Halley told him.[100] For his part, Halley also spread news about it in letters to philosophers on the Continent, letters lavish in praise which were not likely to diminish the recipients' eagerness to see the work.[101]

The two books, *De motu corporum* probably completed in autumn 1685, were not yet Newton's divine treatise, however. The excitement of the pursuit still held him; and as further horizons continued to open before him, he allowed them to lead him on. What he had written so far was an argument, containing a few digressions, for the principle of universal gravitation. He now began to expand it further into an investigation of motions terrestrial and celestial which would of course still conclude in the law of universal gravitation, but would also present a general system of dynamics, propose on its basis a new ideal of science, and demonstrate the impossibility of the rival Cartesian system.

Part of the expansion of Book I was mathematical. Section V, copied directly from his work of the late 1670s on the four-line locus, had very little to do with the argument of the *Principia*.

95 Halley to Newton, 22 May 1686; *Corres 2*, 431.
96 Halley to Newton, 5 April 1687; *Corres 2*, 473.
97 *Prin*, p. xv.   98 *Math 6*, 463.
99 Gregory to Campbell, 2 Oct. 1686; *Corres 2*, 451.
100 Quoted in Cohen, *Introduction*, p. 136.
101 Halley to Johann Sturm, 16 May 1686, and to Dr. Salomon Reisel, 19 July 1686; Eugene Fairfield MacPike, ed., *Correspondence and Papers of Edmond Halley* (Oxford, 1932), pp. 63, 68–9.

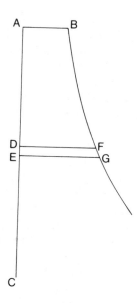

Figure 10.6.   The general treatment of vertical motion under a force
that varies as any function of distance.

Newton appears to have inserted it largely to flex his mathematical
muscles in public. In a work that eschewed the established forms of
modern analysis, he published a solution of the locus that gave the
lie to Descartes's boast by likewise eschewing the help of modern
analysis.

More important were the additions to his dynamics. Whereas he
had written a discourse directed primarily to motion under the
influence of inverse-square forces, he now extended it to the investigation of "the motions of bodies tending to centers by means of
any forces whatsoever . . ."[102] In the generalization of his dynamics, no proposition bore greater significance than Proposition
XXXIX, on the vertical motion of bodies under the action of a
force that varies in any way with respect to distance. Whereas the
early "Lectiones" had solved this problem for inverse-square forces
by an ingenious use of conics, Newton now abandoned all but a
formal obeisance to the mathematical method of the *Principia* and
brought the more powerful artillery of his fluxional calculus to
bear. In his diagram, the line *AC* represents the vertical path of
motion. (Figure 10.6). Mathematically, *AC* is the axis to which the
curve *BFG*, which expresses force as a function of distance, is
drawn. Let *DE* be a minimal increment of distance (Δs). Since $v =$
$\Delta s/\Delta t$ and $f = ma = m \, \Delta v/\Delta t$, "the time in which the body in falling

[102] *Math 6*, 380. *Prin*, p. 147.

describes the very small line *DE*, is directly as that line and inversely as the velocity *v;* and the force will be directly as the increment *I* [Δv] of the velocity and inversely as the time; and therefore if we take the first ratios when those quantities are just nascent, as (*I · V*)/*DE* [$f \propto \Delta v/\Delta t = \Delta v/(\Delta s/v) = v\Delta v/\Delta s$], that is, as the length *DF*." The first ratios in play here returned to Newton's original use of them as replacements for infinitesimal increments of the quantities in analytic equations. Since he decided to omit the theorem he drafted on quadratures, he had to ask the reader to take the critical step on faith; "granting the quadratures of curvilinear figures . . . a force proportional to *DF* or *EG* will cause the body to descend with a velocity that is as the right line whose square is equal to the area *ABGE*."[103] That is, in more familiar terms, the terms in which Newton undoubtedly conceived it himself (the Leibnizian symbolism excepted),

$$F \, ds = \int v \, dv \propto v^2$$

Expressed in this form, the solution assumes, of course, that the body started to fall from rest at *A*. The result is mathematically identical to what we call today the work–energy equation, an equation dynamics frequently employs in problems that express force as a function of distance. The quantity mv^2 implicit in it appeared shortly, under the name *vis viva*, both as the centerpiece and as the foundation stone of a somewhat different science of dynamics developed by Leibniz. For Leibniz, the quantity carried a heavy freight of ontological significance. For Newton, it did not. Force played the major role in his ontology; mv^2 was merely a mathematical expression that appeared in a given problem.

In Proposition XXXIX, Newton proceeded to demonstrate that the time of fall is proportional to the area under another curve ("granting the quadratures of curvilinear figures," once more). The two results provided the foundation for a new Section VIII, which he added now to the original manuscript. Proposition XVII of the early "Lectiones" had demonstrated how the conical orbit of a body set in motion in an inverse-square force field at a given place with a given velocity can always be found. On the basis of Proposition XXXIX, Section VIII universalized the result. For any force defined as a function of distance, Newton demonstrated how to determine the elements of the orbit.[104] Section IX, another addition, investigated the stability of orbits in space when various force laws are in operation.

[103] *Math 6*, 336–8. *Prin*, pp. 125–6.
[104] *Math 6*, 344–8. *Prin*, pp. 130–3. Cf. *Math 6*, 348–51.

He also extended his investigation of attracting spheres to other force laws. The initial "Lectiones" probably included among unidentified propositions the demonstration of the attraction of spheres composed of particles that attract in direct proportion to the distance. Now he added Propositions LXXX and LXXXI, which generalized the result to any force. Section XIII extended the treatment to the attractions of nonspherical bodies. In Section XIV, which he probably added at this time also, Newton applied the dynamics of forces acting at a distance to optics and showed that, under the assumption that light consists of corpuscles attracted within a narrow zone by transparent bodies, the sine law of refraction follows as a necessary consequence.[105]

As Newton generalized his dynamics, he recognized the privileged status of two force laws: forces that decrease in proportion to the distance squared, and forces that increase in proportion to the distance. They had appeared already in *De motu,* which had demonstrated that elliptical orbits are compatible both with inverse-square attractions to a focus and with attractions directed to the center that increase in proportion to the distance. If the first version of the "Lectiones" included the proposition on the attraction of spheres composed of particles that attract in proportion to distance, he found the further analogy between the two laws that in both cases spheres attract by the same law as their component particles.

> I have now explained the two principal cases of attractions [he stated in a scholium]; to wit, when the centripetal forces decrease as the square of the ratio of the distances, or increase in a simple ratio of the distances, causing the bodies in both cases to revolve in conic sections, and composing spherical bodies whose centripetal forces observe the same law of increase or decrease in the recess from the center as the forces of the particles themselves do; which is very remarkable.[106]

He discovered still another remarkable analogy in his investigation of the stability of orbits. Among physically probable laws of attraction, only these two laws yield orbits whose lines of apsides do not move. When he studied the problem of many bodies attracting each other, he did so in terms of the two forces. He found that the presence of any number of bodies, all attracting by forces that increase in proportion to distance, does not disturb the perfect ellipticity of orbits and the proportionality of areas described to time, although it does increase the velocities and decrease the periods. As

[105] See A. I. Sabra, "Explanations of Optical Reflection and Refraction: Ibn-al-Haytham, Descartes, Newton," *Proceedings of the Tenth International Congress of the History of Science* (Paris, 1964), *1,* 551–4, and "A Note on a Suggested Modification of Newton's Corpuscular Theory of Light to Reconcile It with Foucault's Experiment of 1850," *British Journal for the Philosophy of Science, 8* (1954), 149–51.   [106] *Prin,* pp. 202–3.

we have seen, the same does not hold for the inverse-square law. In that case, a third body always destroys perfect ellipticity and the proportionality of areas to time, though certain conditions, identical to those which obtain in the solar system, keep perturbations to a minimum. In the many-body problem, Newton investigated only the two force laws. In the other problems, he investigated various possible force laws and found no others so compatible with a harmonious universe. In the case of attracting spheres, for example, he demonstrated that inverse-cube and inverse-quartic attractions by particles would result in infinite attractions at the surfaces of composite spheres – an inconvenient arrangement for the Creator to have visited upon His creatures. "When I wrote my treatise about our Systeme," Newton wrote to the Reverend Richard Bentley a few years later, "I had an eye upon such Principles as might work wth considering men for the beleife of a Deity . . ."[107] From the composition of the work, it appears more likely that he fastened his eye on the most important principle only in the process of writing it. Only two laws of attraction, the inverse-square law and the direct-distance law, are compatible with a rationally ordered universe. God showed himself to be Newton's equal in mechanics when he built His cosmos upon them.

As far as the cosmos was concerned, Newton did not consider the law of attraction in proportion to distance a likely candidate. In an infinite universe, it entails the impossible consequences of infinite forces, infinite accelerations, and infinite velocities. It is, moreover, empirical fact that attractions decrease with distance rather than increase – further testimony to God's good sense. The law of force in proportion to distance plays a role in natural philosophy, however, primarily in vibratory motions like those of pendulums. Part of the expansion of the "Lectiones" included Section X, in which Newton examined pendular motion and demonstrated that in an isochronous pendulum, that is, a pendulum swinging in a cycloidal arc, the component of force tangent to the path and therefore effective in accelerating the bob is proportional at every point to displacement from the position of equilibrium. This is the dynamic condition of what we now call simple harmonic motion. Although Newton did not use that name for it, he first submitted it to analysis and derived the relations of displacement, force, acceleration, velocity, and time peculiar to it, and he understood its dynamic identity to elliptical orbits in which the force is directed to the center.[108] Before he completed the *Principia,* he put the analysis to startling use in one of the work's unexpected *tours de force.*

[107] Newton to Bentley, 10 Dec. 1692; *Corres 3,* 233.
[108] Propositions XXXVIII, LII, *Math 6,* 334, 394–404. *Prin,* pp.124, 156–8.

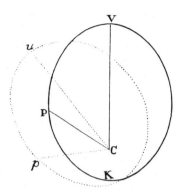

Figure 10.7.   The body *p*, which requires more than 360° to ascend
again to its greatest distance from *C*, can be regarded as
orbiting in the path *up*, an ellipse which rotates uniformly around the focus *C* while the body *p* moves
along it.

Although Section IX, on the stability and motion of the apsides
of orbits, contributed to the generalization of Newton's dynamics,
it concerned itself primarily with a specific problem, the moon's
orbit. Its conception belonged to the same period in which Newton
added the expanded body of corollaries to Proposition XXXVI
(Proposition LXVI, Book I) and began to contemplate an exact
quantitative account, not merely of the gross phenomena of the
cosmos, but of the fine deviations from ideal patterns as well. The
perturbations of the moon offered the prime target for such a program; and the progression of the moon's line of apsides, in contrast
to the virtual stability of planetary orbits, presented the first problem. The original *Lectiones* addressed themselves to the issue only in
a general, largely intuitive discussion placed in a scholium to
Proposition XII (Proposition XIII, Book I).[109] Section IX replaced
the scholium with a sophisticated analysis of the effect that variations from an inverse-square force have on the stability and motion
of apsides. He demonstrated first that an orbit in which the body
returns to its original altitude at an angle other than 360° can be
treated as an elliptical orbit revolving about its focus *C* (Figure
10.7). He demonstrated next that the centripetal force that controls
motion in a revolving orbit must differ from an inverse-square
force. Moreover, he derived a formula that expressed the difference, and using the formula, he proceeded in Proposition XLV to
examine the motion of the apsides in orbits that approximate
circles. If the centripetal force varies as any power of the distance,

[109] *Math 6*, 148–52.

he concluded, "that power may be found from the motion of the apsides; and conversely."[110] Specifically, suppose that an extraneous force 357.45 times less than the primary centripetal force disturbs the inverse-square ratio. Newton introduced the number without explanation. It derived from the squared ratio of the period of the moon around the earth to the period of the earth around the sun, and represented the average reduction during one complete lunar orbit of the centripetal force holding the moon arising from the perturbing presence of the sun. By causing the total centripetal force to depart from the inverse-square ratio, such an additional force would cause the line of apsides to advance 1° 31' 14'' in one revolution.[111] Here Newton left it. He must have viewed the result both as a victory and as a defeat. He had brought a prominent feature of the moon's motion, hitherto recognized as an observed fact which defied explanation, within the scope of his celestial dynamics. The mere fact that he published the analysis indicates that he regarded the result as a considerable achievement. Alas, the quantity that he obtained was only half the observed progression of the lunar apsides. We can measure the extent to which Newton felt the defeat by the fact that he could not bring himself to acknowledge the discrepancy. Only in the third edition, after he had abandoned hope of correcting it, did he insert a brief statement that the progression of the apse is about twice as swift.

According to his letter to Halley in the summer of 1686, Newton completed his expansion of Book I during the winter of 1685–6. The same letter indicated that he also added to the twenty-two propositions that dealt with motions in and phenomena of resisting media as he expanded Book I.[112] When it began to reach excessive size, he decided to divide the book into two, Book I to concern itself entirely with the motions of bodies in spaces free of resistance, and Book II with various problems of resisting media. Deeming Book I complete, he had Humphrey copy it and dispatched the manuscript to the Royal Society via Dr. Vincent, a fellow both of Clare Hall and of the Royal Society, late in April 1686. Although it is impossible to trace his every step with assurance, it appears that he also considered Book II tolerably complete at this time. Even before he expanded Book I, he was satisfied enough with the final book that he set Humphrey to making a fair copy of it, though he stopped him in the middle and later deposited the unfinished manuscript as a copy of his lectures for the year 1687.[113] Regarding his task as nearly finished, he relaxed and resumed alchemical experi-

[110] *Math 6,* 376. *Prin,* p. 145.
[111] *Math 6,* 380. *Prin,* p.147. In the third edition, he altered the numbers slightly.
[112] Newton to Halley, 20 June 1686; *Corres 2,* 437.
[113] Cambridge University Library, *Dd.*4.18.

mentation that spring after he had ignored his laboratory during the whole of 1685.[114] In a famous memorandum written some thirty years later in connection with the calculus controversy, Newton asserted, "The Book of Principles was writ in about 17 or 18 months, whereof about two months were taken up with journeys, & the MS was sent to ye R. S. in spring 1686; & the shortness of the time in which I wrote it, makes me not ashamed of having committing some faults."[115] Perhaps Newton did later regard spring 1686 as the point at which he completed the *Principia*. The last clause indicates that he had an interest in claiming as short a time as possible, however, and too much other evidence points to substantial further work on Books II and III.

Already he had set about revising the manuscript of what would now be Book III. He added a third paragraph to the two that dealt with the moon; in contrast to the original two, it presented a quantitative formula (without the demonstration from which it derived) by which to calculate the distance of the moon from the earth at every point in its orbit. The addition must have been one of the last revisions of the manuscript, since he altered the number of the proposition in Book I on which the discussion of the moon rested to LXVI, the number in the published *Principia*.[116] Another emendation entered a reference to Robert Hooke in the second paragraph. Originally the text had said merely that recent philosophers postulate either vortices or some other principle of impulse or attraction to explain how planets are drawn back from rectilinear paths and constrained to move in closed orbits. Newton altered the statement to read: "More recent philosophers postulate either vortices, as Kepler and Descartes do, or some other principle of impulse or attraction, as in the case of Borelli, Hooke, and others of our nation."[117] In the context of the *Principia,* which also postulated a principle of attraction for that purpose, the addition can only be read as a generous acknowledgment of his debt to Hooke.

On 21 April 1686, in a "Discourse Concerning Gravity" which he was presenting, Halley told the Royal Society that Newton's treatise was nearly ready for the press.[118] A week later, Newton gave Halley's promise concrete support.

> Dr. Vincent presented to the Society a manuscript treatise entitled *Philosophiae Naturalis principia mathematica,* and dedicated to the Society by Mr. Isaac Newton, wherein he gives a mathematical demon-

---

[114] *Add MS* 3973, ff. 19–19^v. *Add MS* 3975, pp. 150–54 (a verbatim copy of 3973). Newton included two dates, 26 April and 16 May 1686, in the notes.

[115] *Corres 2,* 411. Cf. three versions of a slightly different memorandum that all assert the same general chronology: *Add MS* 3968.9, ff. 101–2.   [116] *Add MS* 3990, f.21.

[117] *Add MS* 3990, f.1. *Prin,* p.550.   [118] Birch, *History, 4,* 479.

stration of the Copernican hypothesis as proposed by Kepler, and makes out all the phaenomena of the celestial motions by the only supposition of a gravitation towards the center of the sun decreasing as the squares of the distances therefrom reciprocally.

It was ordered, that a letter of thanks be written to Mr. Newton; and that the printing of his book be referred to the consideration of the council; and that in the meantime the book be put into the hands of Mr. Halley, to make a report thereof to the council.[119]

We can attribute the description of the content of the manuscript to the fact that Halley had recently been appointed clerk to the society and hence wrote the minute himself.

Three weeks passed. Nothing happened. The Royal Society was in disarray at the time, frequently not meeting for lack of an officer present to preside. No council convened to consider Newton's manuscript. Increasingly anxious that Newton should have some response, Halley apparently seized the initiative and raised the question at the society's meeting on 19 May. Although such matters fell within the sole competence of the council, the society voted – nay, ordered –

That Mr. Newton's *Philosophiae naturalis principia mathematica* be printed forthwith in quarto in a fair letter; and that a letter be written to him to signify the Society's resolution, and to desire his opinion as to the print, volume, cuts, &c.[120]

If the resolution was indeed Halley's work, he took a considerable risk in promoting it. Though reared in a wealthy family, Halley had found himself reduced to relative penury by the death of his father in 1684. With a young wife and a family and no means of livelihood, he had accepted the humble post of clerk to the Royal Society at a salary of £50 per annum. Menial staff are not ordinarily expected to initiate projects of major import, neither in the Royal Society of the seventeenth century nor elsewhere. Before the year was out, Halley's audacity nearly cost him his job.

Meanwhile, the vote freed him to write officially to Newton.

S^r

Your Incomparable treatise intituled *Philosophiae Naturalis Principia Mathematica,* was by Dr Vincent presented to the R. Society on the 28th past, and they were so very sensible of the Great Honour you do them by your Dedication, that they immediately ordered you their most hearty thanks, and that a Councell should be summon'd to consider about the printing thereof; but by reason of the Presidents attendance upon the King, and the absence of our Vice-Presi-

---

[119] *Ibid.,* 4, 479–80. The printer's manuscript, the first book of which the Royal Society received at this time, is the first of the successive versions of the *Principia* collated in *Var Prin.*   [120] *Ibid.,* 4, 484.

dents, whom the good weather had drawn out of Town, there has not since been any Authentick Councell to resolve what to do in the matter: so that on Wednesday last the Society in their meeting, judging that so excellent a work ought not to have its publication any longer delayd, resolved to print it at their own charge, in a large Quarto, of a fair letter; and that this their resolution should be signified to you and your opinion therin be desired, that so it might be gone about with all speed.

Halley informed Newton that he was to be in charge of the publication, and he got down to business at once by urging that the diagrams be enlarged.

Halley was the more anxious to write before Newton should hear of something else through other channels.

There is one thing more that I ought to informe you of, viz, that Mr Hook has some pretensions upon the invention of ye rule of the decrease of Gravity, being reciprocally as the squares of the distances from the Center. He sais you had the notion from him, though he owns the Demonstration of the Curves generated therby to be wholly your own; how much of this is so, you know best, as likewise what you have to do in this matter, only Mr Hook seems to expect you should make some mention of him, in the preface, which, it is possible, you may see reason to praefix. I must beg your pardon that it is I, that send you this account, but I thought it my duty to let you know it, that so you may act accordingly; being in myself fully satisfied, that nothing but the greatest Candour imaginable, is to be expected from a person, who of all men has the least need to borrow reputation.[121]

The first paragraph of Halley's letter contained all the praise any man might need to receive. The second paragraph was another matter, and characteristically, Newton's reply focused exclusively on it. "I thank you for w^t you write concerning Mr Hook," he began, "for I desire that a good understanding may be kept between us." Even across the space of three centuries, one can hear Halley's sigh of relief as he read the letter. Annoyed perhaps but well short of anger, Newton recited the events of 1679, softening the account of his own errors a bit but not violating the demands of truth either. He had not mentioned Hooke in Book I because he owed nothing in Book I to Hooke. He did mention him in the part of the work not yet sent. Unwilling to give an inch on the inverse-square law, however, he recalled a conversation with Wren in 1677

[121] *Corres 2*, 431. On the question of Newton and Hooke, see Philip E. B. Jourdain, "Robert Hooke as a Precursor of Newton," *The Monist, 23* (1913), 353–84; Louise Diehl Patterson, "Hooke's Gravitation Theory and Its Influence on Newton," *Isis, 40* (1949), 327–41, *41* (1950), 32–45; and J. A. Lohne, "Hooke *versus* Newton," *Centaurus, 7* (1960), 6–52.

and suggested that Halley consult Wren about it.[122] The expected crisis apparently past, Halley sent a proof of the first sheet for Newton to approve the type. He did not neglect to offer some additional flattery to soothe any feelings still ruffled. He had proof-read the sheet, he told Newton, but he might have missed some errors; "when it has past your eye, I doubt not but it will be clear from errata."[123]

Halley relaxed too soon. Newton's anger required time to ripen fully. For three weeks he fed on Hooke's charge while his fury mounted. Late in June, he was finally ready to write again, a display of pyrotechnics worthy of the ardent young man who in 1672 had cast restraint to the winds rather than swallow Hooke's condescension.

> In order to let you know ye case between Mr Hook & me, [he announced without further ado to a startled Halley who had hoped to hear no more about it] "I gave you an account of w^t past between us in our Letters so far as I could remember . . . . I intended in this Letter to let you understand ye case fully but it being a frivolous business, I shal content my self to give you ye heads of it in short: vizt y^t I never extended ye duplicate proportion lower then to ye superficies of ye earth & before a certain demonstration I found ye last year have suspected it did not reach accurately enough down so low: & therefore in ye doctrine of projectiles never used it nor considered ye motion of ye heavens: & consequently Mr Hook could not from my Letters wch were about Projectiles & ye regions descending hence to ye center conclude me ignorant of ye Theory of ye Heavens. That what he told me of ye duplicate proportion was erroneous, namely that it reacht down from hence to ye center of ye earth. That it is not candid to require me now to confess my self in print then ignorant of ye duplicate proportion in ye heavens for no other reason but because he had told it me in the case of projectiles & so upon mistaken grounds accused me of that ignorance. That in my answer to his first letter I refused his correspondence, told him I had laid Philosophy aside, sent him only ye experim^t of Projectiles (rather shortly hinted then carefully described) . . . to sweeten my Answer, expected to heare no further from him, could scarce perswade my self to answer his second letter, did not answer his third, was upon other things, thought no further of philosophical matters then his letters put me upon it, & therefore may be allowed not to have had my thoughts of that kind about me so well at that time.

As he continued his litany of complaint, Newton made it clear that he had spent some time reviewing his papers. To support his prior knowledge of the inverse-square relation, he cited his early paper

---

[122] *Corres 2*, 433–4.   [123] *Corres 2*, 434–5.

on the endeavor of the planets and the moon to recede, his letter to Oldenburg when he received Huygens's *Horologium,* and the implication of his explanation of gravity in the "Hypothesis of Light." Even if he granted that he got the inverse-square law from Hooke, Hooke only guessed at it, whereas he had demonstrated its truth. So much for the past. He had designed the whole treatise to consist of three books. The second was short and only required copying. The third he had decided to suppress. In a final cry of exasperation, all the pent-up tension of a year and a half of stupendous and unremitting toil burst out. "Philosophy is such an impertinently litigious Lady that a man had as good be engaged in Law suits as have to do with her. I found it so formerly & now I no sooner come near her again but she gives me warning." Before he could mail the letter, he received a report, undoubtedly via Paget, that Hooke was making a stir and demanding that justice be done. Even more enraged now, he added a postscript longer than the letter, in which he recited his complaint once more for Halley's edification. All Hooke had done was to publish Borelli's hypothesis under his own name, and now he claimed to have done everything but the drudgery of calculation.

> Now is not this very fine? Mathematicians that find out, settle & do all the business must content themselves with being nothing but dry calculators & drudges & another that does nothing but pretend & grasp at all things must carry away all the invention as well of those that were to follow him as of those that went before.

Hooke had stated nothing that any mathematician could not have told him once Huygens's work had been published, and in extending the duplicate proportion to the center of the earth, he had fallen into error.

> And why should I record a man for an Invention who founds his claim upon an error therein & on that score gives me trouble? He imagins he obliged me by telling me his Theory, but I thought my self disobliged by being upon his own mistake corrected magisterially & taught a Theory wch every body knew & I had a truer notion of then himself. Should a man who thinks himself knowing, & loves to shew it in correcting & instructing others, come to you when you are busy, & notwithstanding your excuse, press discourses upon you & through his own mistakes correct you & multiply discourses & then make this use of it, to boast that he taught you all he spake & oblige you to acknowledge it & cry out injury & injustice if you do not, I beleive you would think him a man of a strange unsociable temper. Mr Hooks letters in several respects abounded too much wth that humour wch Hevelius & others complain of . . . [124]

---

[124] *Corres 2,* 435–40.

Apparently nothing galled him so much as the demand for acknowledgement. He returned to it in the letter three times. Acknowledge Hooke's priority? Quite the contrary, he went back to the draft of the final book and attacked the reference he had made to Hooke. He slashed out the acknowledgment of Hooke's concept of attraction in the second paragraph. Later on, the discussion of comets had included an observation by "Cl [Clarissimus] Hookius"; one brutal stroke of the pen reduced "the very distinguished Hooke" to mere "Hooke." When he recast the book, he went a step further and eliminated the passage altogether, as he did another one that acknowledged an observation by Hooke.[125]

In trying circumstances, Halley proved himself a diplomat of no small capacity. His letter indicates that he did not fully understand what Newton's threat to suppress Book III meant. Recalling the two-book manuscript he had seen, he thought Newton intended merely to separate the theory of comets from the system of the world. Whatever he understood, he did not intend to stand by silent as Newton castrated his masterwork.

He cajoled. He flattered. He assured Newton that the Royal Society stood on his side. He urged him not to suppress Book III, "there being nothing which you can have compiled therin, which the learned world will not be concerned to have concealed . . ." He took care to follow Newton's suggestion that he question Wren, and he reported on the conversation wholly in Newton's favor. He recounted his meeting with Hooke and Wren in January 1684, again to Newton's advantage. He gave a fuller account of the events at the meeting when Dr. Vincent presented the manuscript, assuring him that Hooke's behavior had been represented to Newton as worse than it was. True, when the members of the Society foregathered as usual at a coffeehouse after the meeting, Hooke claimed he had given the invention to Newton, "but I found that they were all of opinion, that nothing therof appearing in print, nor on the Books of the Society, you ought to be considered as the Inventor . . . what application he has made in private I know not, but I am sure that the Society have a very great satisfaction in the honour you do them, by your dedication of so worthy a Treatise." And finally, Halley did not hesitate to give Newton some straight talk.

---

[125] Cambridge University Library, *Dd*.4.18, f.2 *Add MS* 3990, f. 37. *Prin.* p. 599. He missed another reference to Hooke, who thus remained "very distinguished" in one place for a short time (*Add MS* 3990, f.44). Newton also eliminated this entire reference from the published work, and the later editor of the early manuscript canceled Hooke's "Cl." (*Prin.* p. 609). Hooke's name did appear in the *Principia*, however, in a scholium to Proposition IV, Book I (of which more later), and in a number of observations of comets which Newton required in the revised theory of comets and hence inserted later.

S^r I must now again beg you, not to let your resentments run so high, as to deprive us of your third book, wherin the application of your Mathematicall doctrine to the Theory of Comets, and severall curious Experiments, which, as I guess by what you write, ought to compose it, will undoubtedly render it acceptable to those that will call themselves philosophers without Mathematicks, which are by much the greater number.[126]

Calmed by such ministrations, the storm blew itself out. When he wrote again two weeks later, Newton regretted that he had added the postscript to the earlier letter and even admitted his debt to Hooke in three things. To compose the dispute, he enclosed a revision of the Scholium to Proposition IV. One might search some time to find a less generous acknowledgment. Proposition IV stated the formula for centripetal force in circular motion, and Corollary VI showed that an inverse-square force follows when periods and times correspond to Kepler's third law. The scholium pointed out that the case of Corollary VI obtains in the celestial bodies. To the statement he added a parenthetical comment: "(as our countrymen Wren, Halley & Hooke have also severally concluded)." Not only did he put Hooke's name last, but he added two additional paragraphs, intended both for Hooke's edification and for his own justification. The first indicated how, from the formula for centrifugal force that Huygens had published, centripetal forces can be compared to known forces such as gravity; Newton was confident that Hooke could not handle such a problem. The second asserted his own independence of Huygens by stating another derivation of the formula for centripetal force in circular motion. In his letter, he told Halley he had found it while "turning over some old papers . . ."[127] Newton must have turned over his entire supply of old papers in search of support in the month and a half since he had heard of the charge of plagiary. In the scholium itself, he did not assert the early provenance of the demonstration. To himself–and to Halley–its meaning was unambiguous, however. Without qualification, he denied Hooke's charge. He had found the inverse-square law by himself. And unlike Hooke, he had not guessed at it. He had derived it in a way that mathematicians at any rate would appreciate. Ever the diplomat, Halley quietly altered the order of names so that Hooke's appeared before his own in the published work.

Newton could not get the issue out of his mind. At the end of July, he wrote to Halley again. "Yesterday," he announced, "I unexpectedly struck upon a copy of ye Letter I told you of to Hugenius." Unexpectedly indeed! Apparently he was still poring over his early papers. He noted a passage from the letter to prove

[126] *Corres 2*, 441–3.   [127] *Corres 2*, 444–5.

that he understood the duplicate proportion in 1673, and he cited Hooke's published admission in 1674 that he did not then know by what proportion gravity decreased.

> In short as these things compared together shew that I was before Mr Hook in what he pretends to have been my Master so I learnt nothing by his letters but this that bodies fall not only to ye east but also in our latitude to ye south. . . . And thô his correcting my Spiral occasioned my finding ye Theorem by wch I afterward examined ye Ellipsis; yet am I not beholden to him for any light into y^t business but only for ye diversion he gave me from my other studies to think on these things & for his dogmaticalnes in writing as if he had found ye motion in ye Ellipsis, wch inclined me to try it after I saw by what method it was to be done.[128]

And with that, at last, he left Halley in peace. Since he did not explicitly withdraw his threat to suppress Book III, he also left Halley in doubt as to what else he would receive.

Newton did himself little honor in the incident. He knew how unassailable his position was, and Halley reminded him of it at every turn. Far from injuring him, a modicum of generosity would have further enhanced his reputation. Once aroused, however, he forgot the claims of generosity and equity completely. He interpreted the correspondence with Hooke in 1679–80 in the narrowest sense, fastening upon the increase and decrease of gravity below the surface of the earth and neglecting the question of orbital motion that had been implicit in the exchange. Newton had clashed with Hooke twice before. The charge of plagiary in 1686 was the final straw. He never forgot it, and he never forgave it. Hooke remained his enemy until Hooke's death in 1703. The interest in the incident does not attach to the light it casts on the discovery of universal gravitation. It casts none. The discovery was Newton's, and no informed person seriously questions it. The interest of the incident lies in the glimpse it provides into the tormented soul of a lonely genius. Not all that we see there is attractive.

Nor does the incident cast a better light on Hooke, though for different reasons. He appears never to have understood that the things he claimed, the inverse-square law and the shape of the earth, were consequences entailed by a general mathematical system of the universe. He treated them instead as discrete objects that he believed he had discovered and that Newton must therefore have stolen. He too never forgot or forgave, and the memory of his stolen treasure poisoned the final years of his life. He entered a note in his diary for Friday, 15 February 1689: "At Hallys met Newton – vainly pretended claim yet acknowledged my information. Interest

---

[128] *Corres 2,* 446–7.

has noe conscience; *a posse ad esse non valet Consequentia.*"[129] Later that year, he dragooned John Aubrey into writing a letter, virtually at his dictation, to Anthony à Wood, the Oxford chronicler, stating his claims to the inverse-square proportion and the shape of the earth once more.

> Mr Wood! [Aubrey concluded]
> This is the greatest discovery in nature, that ever was since the world's creation: it never was so much as hinted by any man before. I know you will doe him right. I hope you may read his hand: I wish he had writt plainer, and afforded a little more paper.[130]

It remained instead for the eighteenth-century French scientist Clairaut to do Hooke right. Hooke's examples, he said, "serve to show what a distance there is between a truth that is glimpsed and a truth that is demonstrated."[131]

Meanwhile, the *Principia* proceeded. Though Newton did not utter threats lightly, the work had established a command over him which he could not break. The comparison of 1672 with 1686 is instructive. At the first word of criticism of his paper on colors, Newton began to withdraw into his shell and ended up by severing his connections with the learned world in London. The provocation in 1686 was far worse; the result was quite opposite. If Halley recognized the monumental importance of the manuscript, if the Royal Society greeted it in a similar way, if Hooke called it the most important discovery in nature since the world's creation and tried to claim it as his own, much more did its author understand the significance of his own work. He could make threats in anger. He could not mutilate his masterpiece. The charge of plagiary and the reply to it provided at most an interlude, a momentary release from the tension of composition. The interlude completed, he resumed composition again as though it had never happened.

There was also a new factor in the situation. He had a publisher. The arrival of Newton's manuscript in April 1686 confronted the Royal Society with an unprecedented problem. A year and a half before, it had urged Newton to send his treatise *De motu* to be entered on its register in order to secure the invention to him until he was ready to publish. The society itself was not a publisher, however, and it had not intended to function as such. Unexpectedly, it found itself in receipt of a manuscript of obvious impor-

[129] Diary in R.T. Gunther, *Early Science in Oxford,* 14 vols. (Oxford and London, 1920–45), *10,* 98.
[130] Aubrey and Hooke to Anthony à Wood, 15 Sept. 1689; *Corres 3,* 40–2.
[131] *Exposition abrégée du système du monde,* attached to *Principes mathématiques de la philosophie naturelle,* tr. Madame du Chastellet, 2 vols. (Paris, 1759), *2,* 6.

tance and dedicated to it. Insofar as there was a precedent, it occurred in 1685 and served only to confound the dilemma. In that year, Dr. Tancred Robinson became aware that John Ray had completed the *Historia piscium* (*The History of Fishes,* as the society's journal called it) of his deceased friend and patron, Francis Willughby. Ray was impoverished. Willughby's widow was uninterested. Robinson convinced the Royal Society to publish the work. With many plates, it was an expensive project, and it plunged the society's always tenuous finances into chaos on the eve of the arrival of Newton's manuscript. On 19 May, Halley led the society to order that the *Principia* be printed. Finances belonged to the province of the council, however; and when the council met on 2 June, it pleasantly decided to let Halley stew in the broth he had mixed. "It was ordered, that Mr. Newton's book be printed, and that Mr. Halley undertake the business of looking after it, and printing it at his own charge; which he engaged to do."[132] Later in June, the council asked the president, who was not present, to license the book. As though to underline the connection of the *Principia*'s fate to Willughby's work, it voted on the same day to encourage the measurement of the earth by offering Halley £50 or fifty copies of the *History of Fishes* when he measured the length of one degree to their satisfaction. Before another year had passed, they were paying Halley his salary in copies of the *History of Fishes.* On 5 July the president, Samuel Pepys, signed the Imprimatur, as though the Royal Society thought itself to be in the business of censoring books, and thus the *Principia* appeared as it were under the society's licence. Meanwhile, one can understand the dimension of personal anguish in Halley's successful attempt to calm Newton's rage against Hooke. In effect, he was the publisher.

By Newton's own testimony, he completed Book II in its final form during the autumn of 1686. We know only three things about its earlier form, when he still placed it in Book I: it included the proposition ultimately numbered XXII; it was extensive enough to stand apart as a separate book; it was nevertheless short, according to his letter to Halley in June. All of this is compatible with the assumption that initially it extended through what became Section V. Newton also told Halley that during the winter of 1685–6 he had thought of some additional propositions, though he gave no hint as to what they were. Not to torture a great investigation with excessive erudition, let us simply assume that the initial version extended through Section V. It consisted then of a considerable extension of *De motu*'s two propositions on motion through a resisting medium couched in four sections of formidable mathemat-

---

[132] Birch, *History,* 4, 486.

ics, plus Section V on the physical problem of pressure and density in a heavy medium such as our atmosphere. The fifth section closed with Proposition XXIII, in which Newton demonstrated that Boyle's law follows from the assumption that particles of a fluid repel each other with forces that terminate at the neighboring particles and vary inversely as the distance. Despite the arbitrary nature of the assumption, the proposition was still an important extension of the Newtonian program to reformulate natural philosophy on the premise of forces acting at a distance.

The expansion of Book II into its published form converted it into an attack on Descartes. Sections VI and VII explored the physical cause of resistance. Newton attributed it to the force of inertia of the particles of the medium, which the moving body must set in motion as it collides with them. While lesser components of resistance arise, in his view, from the inner tenacity of fluids (a component which is constant) and friction (proportional to velocity), the major component stems from the medium's force of inertia and hence is proportional to the velocity squared and to the density of the medium. Resistance is also proportional, of course, to the cross-sectional area of the body in motion. Newton demonstrated that the resistance due to the inertia of the medium suffered by a sphere is half that of a cylinder (moving parallel to its axis), and concluded that the resistance to a sphere moving in a continuous, nonelastic medium is to the force that can destroy its whole motion during the time it would move (with its initial velocity) a distance equal to eight-thirds of its diameter as the density of the medium to the density of the body.[133] For Newton's purposes, the role of density in resistance held critical importance. The quantity of matter is proportional to density, and the force of inertia to the quantity of matter. The force of inertia as the source of resistance supplied an apparently irrefutable objection against Cartesian natural philosophy. As Newton said of experiments he performed to test his theory of resistance, "the demonstration of a vacuum depends thereon."[134] He also put it another way: "the resistance that larger and swifter Globes experience in moving through the air can scarcely be diminished much by the infinite division of the air into infinitely subtle particles, and hence all Media in which bodies are resisted much less are rarer than air."[135] If Newton's theory of

---

[133] *Prin*, p. 336. On Book II in general, see Clifford Truesdell, "Reactions of Late Baroque Mechanics to Success, Conjecture, Error, and Failure in Newton's *Principia*," in Palter, ed., '*Annus Mirabilis*', pp. 192–232, and "Rational Fluid Mechanics, 1687–1765," editor's introduction to Leonhard Euler, *Opera Omnia*, published by the Swiss Society of Natural Science (67 vols. in 4 ser.; Leipzig, 1911–), II, 12, ix-cxxv.   [134] *Prin*, p. 321.

[135] *Var Prin 1*, 456. In the second edition, a much stronger statement to the same effect replaced this passage from the first edition (*Prin*, p. 366).

resistance were correct, the Cartesian plenum was flatly incompatible with the observed phenomena of nature.

To test the theory, he performed extensive experiments that he described in a General Scholium at the end of Section VII (moved in the second edition to Section VI, where it is now found). By comparing the motions of pendulums in air, water, and quicksilver, he convinced himself that resistance is indeed proportional to the three factors his analysis had isolated. He also had a vessel nine feet deep constructed for other experiments with bodies falling through water. These experiments did not appear until the second edition. Perhaps an anecdote that Conduitt recorded in rough phrases explains their failure to appear in the first.

> Story of Caton the joiner at Cambridge who made a square hollow piece of wood nine foot deep filled with water, glasses at distance it fall balls of wax [*sic*–a piece of glass at the bottom to which balls of wax fell.] Sr Isaac at bottom looking[,] glass broke–[Caton:] wonder being very thick–Caton you do not know the force of water.[136]

Dry or drenched, always the philosopher.

In case the theory of resistance were not enough, Newton closed the General Scholium on resistance with an account of his experiment to disprove the aether. He had begun the *Principia* with an implicit denial of its existence. Definition I, of the quantity of matter, had included a statement sufficiently pointed. "I have no regard in this place to a medium, if any such there is, that freely pervades the interstices between the parts of bodies."[137] In case the point had nevertheless been missed, he now described the experiment devised several years before to try "the opinion of some that there is a certain aethereal medium extremely rare and subtile, which freely pervades the pores of all bodies," from which "some resistance must needs arise . . ." As we have seen, he concluded that the resistance "is either nil or wholly insensible."[138]

Section VIII proceeded to examine motions propagated through fluids, that is wave motions. In a different way, it opposed Descartes as vigorously as the sections on resistance since it demonstrated, or intended to demonstrate, that the rectilinear propagation of light cannot be reconciled with a theory that light is solely a motion or action. Although Huygens's *Treatise of Light* would shortly undermine the force of that argument, Section VIII contained the first rigorous analysis of wave motion, which Newton based on the treatment of pendulums in Book I. From it he drew one of the minor triumphs of a work whose major thrust aimed at cosmic phenomena, a derivation of the velocity of sound. To check

[136] *Keynes MS* 130.15.   [137] *Prin*, p. 1.
[138] *Prin*, p. 325. The final conclusion appeared in this form only in the first edition (*Var Prin 1*, 463).

his derivation, he measured the time for an echo to return from the far end of a colonnade in Neville's Court (then still under construction in Trinity). Lacking anything resembling a stop watch, he contrived to adjust pendulums to swing in rhythm with successive echoes. On his first attempt, he was able to conclude only that the pendulum had to be shorter than nine inches and longer than four, that is, that sound moves with a velocity more than 866 feet per second and less than 1,272. He was not satisfied with limits that broad, and on a further trial narrowed them to eight and five and a half inches, or a velocity between 920 and 1,085 feet per second. His calculation yielded a velocity of 968, squarely between the two limits.[139] The passage underwent extensive revision in the second edition when new measurements placed the velocity of sound substantially above his upper limit, not to mention his calculated velocity. His basic procedure remained unchanged, however, and it continues today to be the foundation of similar calculations.

With serious wounds already inflicted on Cartesian natural philosophy, Section IX delivered the coup de grâce by attacking that part of Cartesian philosophy most relevant to the *Principia*'s central concern, the theory of vortices. In keeping with the tone of the work as a whole, he delivered the attack by means of a mathematical analysis of the dynamic conditions of vortical motion, something no Cartesian had undertaken. To this end, Newton imagined a sphere set in a fluid medium and turning on its axis. Friction between the sphere and the medium sets a layer of the medium in motion; friction between that layer and the next sets *it* in motion, and so on. Newton's analysis divided the medium into layers of infinitesimal thickness and argued, in accordance with his laws of motion, that a steady state within a vortex requires equal and opposite forces on the two faces of each layer (Figure 10.8). Two damning conclusions followed. First, a vortex cannot sustain itself. Its continuation in a steady state requires the constant transfer of motion from layer to layer until, in Newton's words, it is "swallowed up and lost" in the boundlessness of space.

> Therefore, in order to continue a vortex in the same state of motion, some active principle is required from which the globe may receive continually the same quantity of motion which it is always communicating to the matter of the vortex. Without such a principle it will undoubtedly come to pass that the globe and the inward parts of the vortex, being always propagating their motion to the outward parts, and not receiving any new motion, will gradually move slower and slower, and at last be carried round no longer.[140]

In the second place, the periods of revolution vary in a vortex as the

[139] *Var Prin 1*, 528–30.   [140] *Prin*, p. 390.

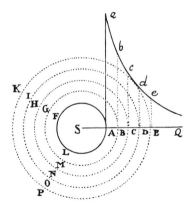

Figure 10.8. The dynamics of a vortex.

square of the radius, whereas Kepler's third law, based on the celestial phenomena, demands the three-halves power. "Let philosophers then see," he concluded, "how that phenomenon of the 3/2th power can be accounted for by vortices."[141]

On the surface, the problem posed to philosophers is not so difficult that even a mere biographer, let alone a philosopher, cannot assay to tackle it. Newton's analysis rested on the assumption that friction within a fluid is proportional to the relative velocity of the parts of the fluid. One has only to assume that friction is proportional to the four-thirds power of the relative velocity and follow the remaining steps in the analysis to arrive at Kepler's third law, $T \propto R^{3/2}$. Setting friction proportional to the four-thirds power of velocity may appear to be an arbitrary and anomalous assumption. It is necessary to insist, however, that Newton's assumption was no less arbitrary. It rested on no empirical foundation whatever. Apparently it derived from a facile mechanical intuition that pictured the frictional force as the sum of the impacts of particles on the two surfaces. Since the entire *Principia* devoted itself to rejecting facile mechanical intuitions, one might be excused for assuming another ratio for friction. The problem that Newton set the vortical theory was more profound, however. A scholium that concluded both Section IX and Book II formulated it in terms of the velocities in a single orbit. The essence of the dilemma lies in the incompatibility between the velocity relations in Kepler's second law and those in his third. No assumption about friction can remove it. One and the same vortex cannot correspond to the different variations in velocity required by Kepler's second and third laws, "so that the hypothesis of vortices is utterly irreconcilable

[141] *Prin,* p. 394.

with astronomical phenomena," Newton concluded, "and rather serves to perplex than explain the heavenly motions. How these motions are performed in free spaces without vortices, may be understood by the first Book; and I shall now more fully treat of it in the following Book."[142]

Meanwhile, as Newton worked on Book II, Halley contracted with a printer, established the book's style including such things as woodcuts of diagrams, which could then appear on the same page with the text, and set publication in motion. Almost at once, Newton's irrepressible doubts and hesitations began to emerge. For no obvious reason, he requested that the work not appear before the end of the Michaelmas term.[143] With the Hooke imbroglio on his hands, Halley had no desire to cross the wishes of his author. Hence the project moved forward slowly and only thirteen sheets had been printed by October. The press then came to a halt for four months; in February, Newton had not received anything to proofread beyond the eleventh sheet.[144] Ignorance about Newton's intentions and Halley's assumption from what he knew that the rest would be short, probably contributed to the halt. After the spate of letters about Hooke, Halley heard almost nothing. He visited Cambridge near the first of September.[145] He must have learned no more than he heard by mail, for in the winter he had no idea what else he might expect to receive from Newton. In December, functioning as the clerk of the Royal Society, he responded to Wallis's hint that he was considering motion through a resisting medium by sending him a copy of Newton's two propositions in *De motu*.[146] A month later, the arrival of Wallis's rather similar solution to the problem, based on the same assumption that resistance is proportional to velocity, put Halley in a ticklish position. The Royal Society resolved to consult Newton as to whether he intended to treat the problem in his treatise.[147] Not caring to precipitate a repetition of the summer's crisis, however, Halley chose not to mention it.

During the same period when he was expanding Book II, Newton was also revising the final book; indeed he was completely recasting it.

---

[142] *Prin*, pp. 395–6.
[143] Perhaps in a note with his letter of 27 May, the original of which we do not have. We know of the request through Halley's response on 7 June (*Corres 2*, 434–5.)
[144] Halley to Newton, 14 Oct. 1686; *Corres 2*, 452. Flamsteed to Towneley, 4 Nov. 1686; quoted in Cohen, *Introduction*, p. 136. Newton to Halley, 13 Feb. 1687; *Corres 2*, 464.
[145] Mentioned by Newton in a letter to Flamsteed, 3 Sept. 1686; *Corres 2*, 448.
[146] MacPike, ed., *Correspondence and Letters of Halley*, pp. 74–5.
[147] Birch, *History*, 4, 521. *Cf. Math 6*, 64–5.

In the preceeding books [he began, in his new introduction] I have laid down the principles of philosophy; principles not philosophical but mathematical: such, namely, as we may build our reasonings upon in philosophical inquiries. These principles are the laws and conditions of certain motions, and powers or forces, which chiefly have respect to philosophy. . . . It remains that, from the same principles, I now demonstrate the frame of the System of the World.
He insisted on the word "mathematical." He had originally composed the final book in a popular form so that many could read it;
but afterwards, considering that such as had not sufficiently entered into the principles could not easily discern the strength of the consequences, nor lay aside the prejudices to which they had been many years accustomed, therefore, to prevent the disputes which might be raised upon such accounts, I chose to reduce the substance of this Book into the form of Propositions (in the mathematical way), which should be read by those only who had first made themselves masters of the principles established in the preceding Books.[148]
Years later, Newton recited a similar story to his friend William Derham. He abhorred contentions, he said. "And for this reason, namely to avoid being baited by little Smatterers in Mathematicks, he told me, he designedly made his Principia abstruse . . ."[149]
It has been almost universally accepted that the recasting of Book III, from the prose essay that constituted the original Book II to the mathematical format that he published, flowed from the clash with Hooke. Hooke was the little smatterer in mathematics; Newton would show him what was what–and for that matter, who was who–by composing Book III, the climax of the work, in a form Hooke could not even follow. Such an account hardly squares with the facts. True, Newton did recast the format in a mathematical style, so that Book III began with a set of nine "Hypotheses" (a mixed bag, consisting of two methodological principles, two philosophical assumptions, and five empirical observations) and proceeded with the paraphernalia of propositions and lemmas. As far as the material found in the original Book II is concerned, however, the change was purely cosmetic. Aside from the decision to take the Keplerian system as an accepted conclusion which required no justification, Book III through Proposition XVIII (together with some of the later propositions) did not differ substantially from the earlier draft. The new Book III began with Proposition XIX, which derived the ratio between the axis of the earth and its diameter at the equator. It included a vastly expanded lunar theory, in which Newton demonstrated quantitatively various inequalities in the moon,

---

[148] *Prin*, p. 397.
[149] Derham to Conduitt, 8 July 1733; *Keynes MS* 133, p. 10.

and a derivation of the precession of the equinoxes which was also quantitative.[150] In a word, it proposed a new ideal of a quantitative science, based on the principle of attraction, which would account, not only for the gross phenomena of nature, but also for the minor deviations of the gross phenomena from their ideal patterns. Against the background of inherited natural philosophy, this was a conception no less revolutionary than the idea of universal gravitation itself. Newton had begun to glimpse its possibility in the winter of 1685–6. It animated the twenty-one new lemmas added to Proposition LXVI, the extended discussion of tides (which went into Book III with no significant addition), the new paragraph about the moon, and Section IX, Book I, on the stability and motion of apsides. Newton made all of these alterations before the episode with Hooke. Since he was at his irritable worst when the excitement of discovery drew him taut, it may well be that the relation between the explosion that disintegrated Hooke and the recasting of Book III was the exact opposite of the accepted account. That is, the tension of the revision may have caused the explosion. I thus find it impossible to believe that Newton's threat to suppress Book III, uttered at the very time when the wider vision of the book was opening itself to his view, was ever more than a passing expression of exasperation.

The revised Book III contained one other thing. Comets finally yielded to his assault. When he wrote to Halley in the summer of 1686, he had not yet succeeded with them. Sometime during the following nine months, he did. He later told Gregory "that this discussion about comets is the most difficult in the whole book."[151] It was more difficult than most of us realize: from observations made on the earth as it moves in an elliptical orbit to determine the conical path of a comet moving in a different plane. What Newton in fact devised was a method to determine, from three evenly spaced observations, the vertex of the projection onto the ecliptic of the segment of a parabola bounded by the first and third observations. From the vertex, he determined the length of the chord, from the length of the chord in the ecliptic the length of the chord in the plane of the orbit, and from the length of the chord the positions of its two ends. Applied to a specific comet, the method depended on placing the middle observation on the orbit by an informed guess.

---

150 In addition to the *Principia*, see Newton's attempt to correct his derivation of the progression of the lunar apogee (*Math 6*, 508–30). Although he mentioned this paper and its result in a scholium that followed Proposition XXXV, Book III, in the first edition, he did not include the computations because they were "too complicated and cluttered with approximations and not accurate enough" (*Var Prin 2*, 658).

151 Gregory's memorandum written about July 1694, based on his visit to Cambridge in May; *Corres 3*, 385.

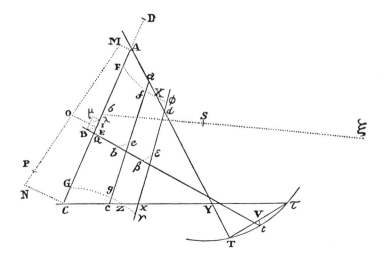

Figure 10.9. Determination of a comet's orbit by successive approximations.

If the calculated length of the chord corresponded to the observed positions, the guess was correct; if it did not, the initial guess was corrected by the error in the length and the entire procedure repeated. Whereas much of the discussion of comets in Book III presented a revision of that in the earlier manuscript, the new method allowed him to cap it with what alone could make it compelling, the reduction of a concrete case to a conical orbit. Newton chose the great comet of 1680–1. Carrying out the correction by trial and error manually with compass and ruler on a large diagram drawn to a scale of 16.33 inches to the radius of the earth's orbit, he pursued the approximation to a result which agrees with modern calculations of the orbit to 0.0017 inch, in the scale of the diagram (Figure 10.9). A modern student of his method has suggested that either he used a supplementary diagram of the central construction drawn to a larger scale, or he guided his construction by calculations he did not mention, or perhaps he possessed as a draughtsman some *vis prope divina*.[152] Those who have followed the precision of Newton's experiments in optics and chemistry will not automatically rule out the last alternative. At any rate, he determined that the comet of 1680–1 moved in a plane inclined at an angle of 61° 20 1/3′ to the ecliptic. By locating its perihelion, he fixed the axis of its orbit, and he calculated its latus rectum. The comet traced a conic (which he treated as a parabola) describing areas in proportion of

[152] A.N. Kriloff, "On Sir Isaac Newton's Method of Determining the Parabolic Orbit of a Comet," *Monthly Notices of the Royal Astronomical Society, 85* (1925), 640–56.

time (Figure 10.10).[153] That is, the laws of planetary mechanics based on the attraction of the sun also governed the motions of the comet of 1680–1 and by inference the motions of all comets.

As the *Principia* became more radical in the new ideal of science that it proposed, Newton began to worry about the reception of the concept of attraction. Initially, he had intended to state his position straightforwardly. In the early versions of Books I and II, he had spoken of attractions without apology and described the gravity of cosmic bodies as a force that arises from the universal nature of matter. He drafted a Conclusio, similar to what he later published as Query 31 of the *Opticks,* in which he suggested the existence of a wide range of other forces between the particles of matter. So far, he began, he had treated the system of the world and the great motions in it that are readily observed.

> There are however innumerable other local motions which on account of the minuteness of the moving particles cannot be detected, such as the motions of the particles in hot bodies, in fermenting bodies, in putrescent bodies, in growing bodies, in the organs of sensation and so forth. If any one shall have the good fortune to discover all these, I might almost say that he will have laid bare the whole nature of bodies so far as the mechanical causes of things are concerned.[154]

By analogy with macroscopic motions and the gravitational forces that control them, he urged that similar forces between particles cause the microscopic motions. As we have seen elsewhere, he proceeded then to cite the phenomena, first of all chemical phenomena that he had observed in his laboratory, then the other crucial phenomena that had figured in his speculations for twenty years which seemed to demand a restructuring of natural philosophy on the concept of forces. He gave up the Conclusio and included the same material in the draft of a preface.[155] In the end, he suppressed both, and in the ultimate preface referred only to his belief in "certain forces by which the particles of bodies, by some causes hitherto unknown, are either mutually impelled towards one another, and cohere in regular figures, or are repelled and recede from one another."[156]

Newton had good reason to be cautious. Weaned on the mechanical philosophy himself, he could not doubt how the concept of

153 *Prin*, p. 512. *Var Prin 2*, 723. In the third edition, following Halley, Newton treated the orbit as an ellipse with a period of 575 years.
154 *Halls*, p. 333; original Latin, p. 321. See J. E. McGuire, "Force, Active Principles, and Newton's Invisible Realm," *Ambix, 15* (1968), 154–208, and "Neoplatonism and Active Principles: Newton and the *Corpus Hermeticum,*" in Robert S. Westman and J. E. McGuire, *Hermeticism and the Scientific Revolution* (Los Angeles, 1977).
155 *Halls*, pp. 305–8; original Latin, pp. 302–5.   156 *Prin*, p. xviii.

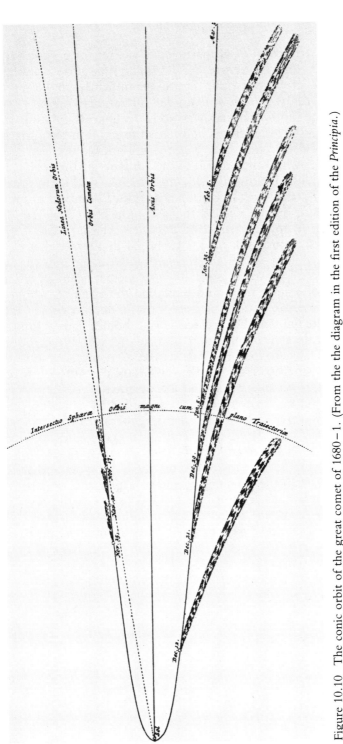

Figure 10.10 The conic orbit of the great comet of 1680–1. (From the the diagram in the first edition of the *Principia*.)

a universal attraction of all particles of matter for each other would be received. Shortly before the *Principia* appeared, a young Swiss mathematician who would figure prominently in Newton's life during the following years, Fatio de Duillier, arrived in London and heard about the work that was soon to appear. In June 1687, he mentioned it to Huygens. The English found him too Cartesian, he reported, and told him that Newton's book would change physics entirely. He went on to describe the contents of the work, some of which he had already studied, and indicated that it proposed a principle of attraction. The book was full of beautiful propositions, he concluded, but he wished that Newton had consulted Huygens about the principle of attraction between celestial bodies. Huygens did not waste much ink on his reply. "I don't care that he's not a Cartesian as long as he doesn't serve us up conjectures such as attractions."[157] Newton must have expected as much. It had not been that long since he would have said the same.

As he worried about the reception of his work, he began to hedge. The line in the preface about bodies being impelled by causes unknown was only part of a more extensive camouflage in which he insisted that his mathematical demonstrations did not entail any assertions about the ontological status of forces. He amended the second paragraph of original Book II with such a statement. As he wrote it first, it concluded by asserting the necessity of some force, which he called in general centripetal force, that draws bodies back from their rectilinear paths and holds them in closed orbits. Now he added a further statement.

> But our purpose is only to trace out the quantity and properties of this force from the phenomena, and to apply what we discover in some simple cases as principles, by which, in a mathematical way, we may estimate the effects thereof in more involved cases . . . We said, *in a mathematical way,* to avoid all questions about the nature or quality of this force, which we would not be understood to determine by any hypothesis . . .[158]

He inserted similar disclaimers at the beginning and end of Section XI.

> I here use the word *attraction* in general for any endeavor whatever, made by bodies to approach to each other, whether that endeavor arise from the action of the bodies themselves, as tending to each other or agitating each other by spirits emitted; or whether it arises

---

[157] Fatio to Huygens, 14 June 1687; Huygens, *Oeuvres Complètes,* pub. Société hollandaise des sciences, 22 vols. (The Hague, 1888–1950) *9,* 168–9. Huygens to Fatio, 11 July 1687; *ibid., 9,* 190–1.

[158] *Add MS* 3990, f. 2. *Prin,* pp. 550–1. The English translation draws on Newton's final introduction to Book III to amplify the Latin text a good deal. It does not violate its spirit, however.

from the action of the aether or of the air, or of any medium whatever, whether corporeal or incorporeal, in any manner impelling bodies placed therein towards each other. In the same general sense I use the word *impulse*, not defining in this treatise the species or physical qualities of forces, but investigating the quantities and mathematical proportions of them . . .[159]

He could have saved his breath. Fifty such disclaimers – or fifty times fifty – would not have stifled the outrage of mechanical philosophers.

Newton apparently finished Book II sometime during the winter of 1686–7. Meanwhile, as we have seen, the process of publication had come to a halt. Whatever Newton's contribution, the primary cause was a crisis in the Royal Society that touched Halley. On 29 November, the council suddenly resolved that Halley's continuation in his office should be put to a vote and that a new election of a clerk in his place should be held. On 5 January following, the council appointed a committee to investigate his performance of his duties.[160] We know nothing about the background of this attack. On the surface at least, it looks political. Halley had revived the *Philosophical Transactions* (which continued to be a private enterprise, to be sure, rather than a publication of the society), and the committee of investigation eventually reported that the society's books and papers were "in a very good condition . . ."[161] The council resolved to elect a new clerk to replace Halley the day before the annual meeting at which elections were held, 30 November, and Hooke was elected to the council that year. In a letter to Newton in February, Halley said that "6 of 38, last generall Election day, did their endeavour to have put me by."[162] On 5 January, when the committee was formed to investigate Halley, the council invited Hooke to bring in a proposal to supply meetings with experiments and discourses – for a salary, of course. Through much of 1687, the two men appeared to compete for the society's limited funds. Thus, on 15 June, the Council voted to pay Halley £50 due him for 1686, but voted further on the same day to execute an earlier order to pay Hooke before it paid anyone else. In July, the Council offered to pay off both men with fifty copies of the *History of Fishes*. Hooke held out for cash, which he eventually got.[163] Though Halley continued in the position of clerk, the society did

---

[159] *Prin*, p. 192. The corresponding scholium in the first version of the "Lectiones" did not contain this passage (*Add MS.* 3965.3, f. 12). The part of the "Lectiones" corresponding to the introduction to Section XI, where the *Principia* had a similar statement, does not survive.   [160] Birch, *History,*, 4, 505, 516.

[161] *Ibid.*, 4, 523.   [162] *Corres 2*, 469.

[163] Birch, *History*, 4, 542, 545, 553, 554, 557.

not pay him until October 1690 and then only for salary earned in 1687 and subsequent years.[164] Thus there are grounds for believing that Hooke's circle of supporters mounted the attack on Halley, perhaps for his leading role in the society's endorsement of the *Principia,* though it is necessary to add that Halley's correspondence gave no hint that Hooke conducted such a vendetta.

Whether such was the case or not, the council's committee reported in Halley's favor on 9 February. It could have been no later than the following day when he wrote to Newton for a copy of a sheet of the manuscript lost by the printer, and indicated his intention to push the publication toward completion. In his reply on 13 February, Newton thanked Halley for "putting forward ye press again . . ."

> The second book I made ready for you in Autumn, having wrote to you in Sommer that it should come out wth ye first & be ready against ye time you might need it, & guessing by ye rate of ye presse in Sommer you might need it about November or December. But not hearing from you & being told (thô not truly) that upon new differences in ye R. Society you had left your secretaries place; I desired my intimate friend Mr C. Montague to enquire of Mr Paget how things were & send me word. He writes that Dr Wallis has sent up some things about projectiles pretty like those of mine in ye papers Mr Paget first shewed you, & y^t 'twas ordered I should be consulted whether I intend to print mine. I have inserted them into ye beginning of ye second book wth divers others of y^t kind: wch therefore if you desire to see you may command ye book when you please . . .[165]

One can imagine Halley swallowing hard as he realized that Newton had heard about Wallis's paper. He sent back a brief description of the paper and volunteered to send a copy, and he assured Newton that Wallis "had the hint from an account I gave him of what you had demonstrated . . ."[166] While the story was not wholly ingenuous, Newton accepted it. The great work was essentially complete now. No longer stretched out, he did not snap, and the problem disappeared. Halley took care that the issue of the *Philosophical Transactions* with Wallis's paper appeared only on the eve of the *Principia*'s publication, and in the same issue he inserted his own review of Newton's work with suitable stress on its treatment of motion through resisting media.[167]

---

[164] The minutes of the council are quoted in MacPike, ed., *Correspondence and Papers of Halley,* p. 239.   [165] *Corres 2,* 464.   [166] *Corres 2,* 469.

[167] John Wallis, "A Discourse concerning the Measure of the Airs resistance to Bodies moved in it," *Philosophical Transactions, 16* (1687), 269–80. Although this was the fascicle for January–March, Halley sent copies of it to Wallis on 25 June (MacPike, ed. *Correspondence and Papers of Halley,* p. 85).

Beyond the question of Wallis's essay, Newton's letter of 13 February, together with the following correspondence, is interesting for the light it sheds on Halley's relation to him. Until he received it early in March, Halley did not know what Book II would contain. Since Newton had not responded to his plea the previous summer, he did not know if there was to be a Book III, let alone its contents. Only with that information can we appreciate the magnitude of the gamble Halley took. And only by appreciating the gamble, can we appreciate the impact of Newton's manuscript on him. Though the first, he was not the last to experience its power.

With the information that Book II was on its way, Halley engaged a second printer to set it while the first completed Book I. Apparently he expected the short manuscript Newton had promised the previous summer. When he received it early in March, he found a treatise nearly as long as the first book and filled, like the first, with new investigations of which he had never dreamed. Holding his breath, he finally chanced an oblique question about the matter Newton had left in doubt the previous June.

> You mention in this second, your third Book *de Systemate mundi,* which from such firm principles, as in the preceding you have laid down, cannot chuse but give universall satisfaction; if this be likewise ready, and not too long to be got printed by the same time, and you think fit to send it; I will endeavour by a third hand to get it all done togather, being resolved to engage upon no other business till such time as all is done: desiring herby to clear my self from all imputations of negligence, in a business, wherin I am much rejoyced to be any wais concerned in handing to the world that that all future ages will admire.[168]

In fact, Book III was virtually ready, and only a month later, Halley wrote to assure Newton that he had received the third part of his "divine Treatise . . ."[169] Since the first printer had nearly completed Book I, he did not have to employ a third, but gave Book III to him.

For Halley, a hellish four months ensued. The eighteenth sheet of Book I, mostly concerned with the difficult Proposition XLV, on the motion of apsides, gave him "extraordinary trouble," and he feared they might need to reset it. In April, he excused himself for not writing to Wallis by his attention to Newton's book; "the correction of the press costs me a great deal of time and paines." In June, he pled the same excuse for the same lapse.[170] He even felt compelled to insert an "Advertisement" at the end of issue number

[168] *Corres 2*, 472.   [169] *Corres 2*, 473.
[170] Halley to Newton, 14 March 1687; *Corres 2*, 473. MacPike, ed., *Correspondence and Papers of Halley,* pp. 80–1, 85.

186 of the *Philosophical Transactions,* immediately after his review of the *Principia,* when the issue finally appeared three months late.

> Whereas the Publication of these Transactions has for some Months last past been interrupted; The Reader is desired to take notice that the care of the Edition of this Book of Mr. Newton having lain wholly upon the Publisher (wherein he conceives he hath been more serviceable to the Commonwealth of Learning) and for some other pressing reasons, they could not be got ready in due time; but now they will again be continued as formerly, and come out regularly . . .[171]

For his devoted labor, which absorbed most of his time for a year, Halley received a handsome acknowledgement in the preface. As far as we know, it never occurred to Newton to thank him as well. Halley told both Newton and Wallis that he hoped to complete the edition by the Trinity term, that is, by 21 June. As it worked out, he missed by two weeks. On 5 July, he announced that the task was finally completed.

> Honoured S^r
>
> I have at length brought your Book to an end, and hope it will please you. the last errata came just in time to be inserted. I will present from you the books you desire to the R. Society, Mr Boyle, Mr Pagit, Mr Flamsteed and if there be any elce in town that you design to gratifie that way; and I have sent you to bestow on your friends in the University 20 Copies, which I entreat you to accept.[172]

Newton sent Humphrey about Cambridge presenting the twenty copies to his acquaintances and to the heads of the colleges, "some of w^{ch} (particularly D^r. Babington of Trinity) said that they might study seven years, before they understood anything of it."[173] A student had the last word. As Newton passed him in the street, he delivered the ultimate benediction of Restoration Cambridge on the genius it harbored: "there goes the man that writt a book that neither he nor any body else understands."[174]

[171] *Philosophical Transactions, 16* (1687), 297.

[172] *Corres 2,* 481. See W. R. Albury, "Halley's Ode on the *Principia* of Newton and the Epicurean Revival in England," *Journal of the History of Ideas, 39* (1978), 24–43.

[173] *Keynes MS* 135.

[174] *Keynes MSS* 130.6, Book 2; 130.5, Sheet 1. Conduitt attributed the story to Martin Folkes, who presumably heard it from Newton.

11

Revolution

NEWTON was hardly an unknown man in philosophic circles before 1687. The very extent to which he had made his capacity in physics and mathematics known had functioned in the early 1680s to destroy his attempt to reconstruct an isolation in which he might pursue his own interests in his own way. Nevertheless, nothing had prepared the world of natural philosophy for the *Principia*. The growing astonishment of Edmond Halley as he read successive versions of the work repeated itself innumerable times in single installments. Almost from the moment of its publication, even those who refused to accept its central concept of action at a distance recognized the *Principia* as an epoch-making book. A turning point for Newton, who, after twenty years of abandoned investigations, had finally followed an undertaking to completion, the *Principia* also became a turning point for natural philosophy. It was impossible that Newton's life could return to its former course.

Rumors of the coming masterpiece had flowed through Britain during the first half of 1687. In Scotland, David Gregory held up the publication of a method of quadratures until he saw if Newton printed his method in the *Principia*.[1] It was a judicious precaution on Gregory's part, since he had only recently plagiarized the method from the notes that John Craig had brought home from Cambridge in 1685. When the young Swiss mathematician, Nicolas Fatio de Duillier, arrived in London in the spring, he found intellectual circles aflutter with expectation of the book which would, he was assured, remodel natural philosophy.[2] For those who had not heard, a long review in the *Philosophical Transactions* announced the *Principia* shortly before publication. Although the review was unsigned, we know that Halley wrote it. With the exception of Newton himself, no one knew the contents of the work better. He insisted on its epochal significance.

> This incomparable Author [the review began] having at length been
> prevailed upon to appear in publick, has in this Treatise given a most
> notable instance of the extent of the powers of the Mind; and has at
> once shewn what are the Principles of Natural Philosophy, and so far
> derived from them their consequences, that he seems to have ex-

---

[1] Gregory to Campbell, 2 Feb. 1687; *Corres 2*, 463.
[2] Fatio to Huygens, 14 June 1687; Huygens, *Oeuvres complètes,* pub. Société hollandaise des sciences, 22 vols. (The Hague, 1888–1950) *9*, 167–9.

hausted his Argument, and left little to be done by those that shall succeed him.

After the body of the review presented a summary of the *Principia*, Halley closed with a further encomium: "it may be justly said, that so many and so Valuable *Philosophical Truths,* as are herein discovered and put past Dispute, were never yet owing to the Capacity and Industry of any one Man."[3]

In mathematical circles, such as the one gathered around David Gregory in Scotland, the fame and influence of the *Principia* spread quickly. Gregory got hold of a copy at once. He had tried previously to establish contact with Newton, on the occasion of the publication of his *Exercitatio geometrica* in 1684. Now he tried again. In a letter filled with soaring praise, he thanked Newton "for having been at the pains to teach the world that which I never expected any man should have knowne."[4] He elicited no more response this time than he had three years before. Gregory was a persevering man, however, and he had recognized Newton's transcendent ability. He did not give up until he had enrolled himself among Newton's disciples – and had made his own fortune by doing so. Within his circle, John Craig also obtained a copy of the *Principia*. It was much in demand. In December, Colin Campbell prevailed upon Craig to loan it before he had mastered it himself. Gregory had warned Campbell that the book was hard work. "I beleeve Newton will take you up the first month you have him."[5] Campbell still had the copy a year later when Craig finally recalled it to loan to another friend who wanted to read it.[6] In 1690, David Gregory's younger brother James published a compendium of the *Principia*.[7]

Across the Channel, a political refugee, John Locke, set himself to mastering the book. Since he was not a mathematician, he found the demonstrations impenetrable. Not to be denied, he asked Christiaan Huygens if he could trust the mathematical propositions. When Huygens assured him he could, he applied himself to the prose and digested the physics without the mathematics.[8] Locke

---

[3] *Philosophical Transactions, 16* (1686–7), 291–7.

[4] Gregory to Newton, 2 Sept. 1687; *Corres 2,* 484.

[5] Craig to Campbell, 29 Dec. 1687; *Corres 2,* 501. Gregory to Campbell, 16 Dec. 1687; quoted in *Corres 2,* 501.

[6] Craig to Campbell, 15 Jan. 1689; *Corres 3,* 7.

[7] There is a copy in the Clare College library; *Edleston,* p. lviii.

[8] J. T. Desaguliers said that he heard this story several times from Newton (*A Course of Experimental Philosophy,* 2 vols. [London, 1734–44], *1,* preface). I assume that Newton heard it from Locke. Conduitt also heard it (*Keynes MSS* 130.6, Book 2; 130.5, Sheet 1). In his tract, *On Education,* Locke stated that nonmathematicians could read the *Principia* with profit and accept Newton's conclusions on the testimony of the mathematicians who have vouched for the demonstrations (*The Works of John Locke,* 10 vols. [London, 1801], *9,* 186–7).

realized that Newton was one of the intellectual giants of the age. He made it his business to meet Newton upon his return to England, and he included an admiring reference to him in the preface to his *Essay on Human Understanding* (1690).[9]

In London, young Abraham DeMoivre met the *Principia* by chance when he happened to be at the home of the Duke of Devonshire at the time (probably in 1688) when Newton came to present a copy. DeMoivre was supporting himself by teaching mathematics. All of twenty-one years old, he considered himself the perfect master of the subject.

> The young mathematician opened the book and deceived by its apparent simplicity persuaded himself that he was going to understand it without difficulty. But he was surprised to find it beyond the range of his knowledge and to see himself obliged to admit that what he had taken for mathematics was merely the beginning of a long and difficult course that he had yet to undertake. He purchased the book, however; and since the lessons he had to give forced him to travel about continually, he tore out the pages in order to carry them in his pocket and to study them during his free time.[10]

Like Gregory, he succeeded ultimately in enrolling himself among the disciples of his new master.

At the other end of the spectrum stood Gilbert Clerke, quondam fellow of Sidney Sussex, now living in retirement in Stamford because of heterodox theological views remarkably similar to Newton's own.[11] Clerke was a philosopher and mathematician as well who had published a number of minor works over the course of a quarter-century. It says something about the rapid circulation of the *Principia* through mathematical channels that already in the autumn of 1687 he was writing to Newton about it. Living as he did in an obscure village, he said, he had never heard of Galileo, and he despaired of understanding the *Principia*. Nevertheless, he wanted to satisfy himself about the tides and some other phenomena. In all, he wrote four letters, which not only put questions but also criticized Newton's mathematical terminology. The draft of one reply survives; Clerke's letters make it evident that Newton replied at least to the first three.[12]

As we have seen, Robert Hooke paid the ultimate compliment by insisting that Newton had stolen the *Principia* from him. Having raised his claim initially in 1686, Hooke became only more aware after the *Principia*'s publication of the richness of the prize that had

[9] *Works of John Locke*, 1, n.p.
[10] "Éloge de Mr. Moivre," *Histoire de l'Academie royale des sciences*, 1754, pp. 261–2.
[11] H. J. McLachlan, *Socinianism in Seventeenth-Century England* (Oxford, 1951), pp. 323–5.
[12] *Corres 2*, 485–98.

been, he was convinced, taken from him. He continued to press his claim, though with no more success than he had at first.[13]

Newton's book was acknowledged and recognized on the Continent as well as in Britain. During the spring and summer of 1688, three of the leading continental journals of opinion carried reviews of it: the *Bibliothèque universelle* in the Netherlands, the *Journal des sçavans* in France, and the *Acta eruditorum* in Germany. The review in the *Bibliothèque universelle*, almost certainly written by John Locke, merely summarized the work and indicated its place in the tradition of mathematical mechanics. The *Journal des sçavans* agreed that it presented "the most perfect mechanics that one can imagine," though it went on to object strenuously to the physical hypothesis it espoused – that is, the concept of attraction. Much the longest review appeared in the *Acta eruditorum*, an eighteen-page summary of the *Principia* expressed in a tone of warm admiration.[14] Halley had also taken care to present copies to the leading philosophers of Europe. The *Principia* was not likely to go unnoticed.

Newton's book took Britain by storm. Almost at once it became the reigning orthodoxy among natural philosophers. On the Continent, its triumph was more protracted. Nevertheless, it refused to be ignored. Its impact can be gauged by the response of two towering figures, Christiaan Huygens and Gottfried Wilhelm Leibniz, both of whom received copies from Newton, and both of whom rejected its central concept. Huygens found the principle of attraction "absurd."[15] For his part, Leibniz was astonished that Newton had not proceeded to find the cause of the law of gravity, by which he meant an aethereal vortex which would reduce attraction to a mechanical cause.[16] Despite their carping, neither could conceal the impression the work had made. According to his own account, Leibniz saw the review in the *Acta* before he received the book itself. Ever mindful of his intellectual capital, he saw Newton forestalling him on several problems on which he had thought in the past. His initial response to the *Principia*, then, was the hasty composition of three papers which he rushed to the *Acta* to defend his own priority: a paper on refraction, a paper on motion through resisting media, and a paper on orbital dynamics set in the framework of a vortical theory. What word of praise could have sur-

---

[13] *Corres 3*, 40–2. In February 1690, Hooke repeated his claim bitterly in a lecture on Huygens's *Discourse on the Cause of Gravity* that he read to the Royal Society (A. R. Hall, "Two Unpublished Lectures of Robert Hooke," *Isis, 42* [1951], 219–30).

[14] I. Bernard Cohen, *Introduction to Newton's 'Principia'* (Cambridge, 1971), pp. 145–57. See James L. Axtell, "Locke's Review of the Principia," *Notes and Records of the Royal Society*, 20 (1965), 152–61.

[15] Huygens to Leibniz, 18 Nov. 1690; Huygens, *Oeuvres, 9*, 538.

[16] Leibniz to Huygens, Oct. 1690; Huygens, *Oeuvres, 9*, 523–4.

passed his action? Huygens told his brother that he admired very much "the beautiful discoveries that I find in the work he sent me," and he made a point of meeting Newton when he visited England in 1689.[17] Both men plied Fatio, their acquaintance, with questions about Newton and his work. Until Huygens's death in 1695, issues raised by the *Principia* – attractions, vortices, the shape of the earth, absolute motion, optics, mathematics – seasoned their correspondence.[18] Titans such as Huygens and Leibniz did not become disciples of Newton. By dominating their correspondence, however, Newton demonstrated that the *Principia* had advanced him in one leap to the front rank of natural philosophers.

Other continental philosophers indicated the same. Late in the 1690s, Dr. John Arbuthnot met the Marquis de l'Hôpital, a prominent French mathematician. L'Hôpital complained that none of the English could demonstrate to him the shape of a body that offered the least resistance to a fluid. When Arbuthnot showed him that Newton had done so in the *Principia*,

> he cried out with admiration Good god what a fund of knowledge there is in that book? he then asked the D^r every particular about Sr I. even to the colour of his hair said does he eat & drink & sleep. is he like other men? & was surprized when the D^r told him he conversed chearfully with his friends assumed nothing & put himself upon a level with all mankind.[19]

Meanwhile, events of a wholly different nature were thrusting on Newton another form of prominence which would affect the rest of his life, if not his enduring role in history, even more than the *Principia*. Early in 1685, just as he was immersing himself in the work, Charles II died and James II succeeded him on the throne. During the following two years, the *Principia* absorbed Newton's consciousness so completely that it excluded everything else. As it happened – providentially, Newton might have said – exactly those two years were required to bring the crisis James precipitated to Cambridge. As one of the first acts obligatory on a publisher, Halley presented a copy of the new book to the king, with a letter which dwelt especially on its treatment of the tides as a topic likely to interest an old naval commander.[20] Probably James did not recognize the author's name. If he bothered to ask his advisers, he learned that the Lucasian professor of mathematics at Cambridge during the previous four months, while Halley shepherded the completed manuscript through the press, had placed himself irrevocably in the ranks of James's enemies.

---

[17] Huygens to Constantyn Huygens, 30 Dec. 1688; Huygens, *Oeuvres*, *9*, 305.
[18] Huygens, *Oeuvres*, *9*, 10, *passim*.
[19] *Keynes MS* 130.5, Sheets 2–3.   [20] *Corres 2*, 483.

The crisis had built up gradually. James understood that control of the universities was essential to his goal of reconverting England to Catholicism. The well-established policy of royal intervention by dispensations and letters mandate gave him a ready instrument. From the beginning of his reign, letters mandate filled almost every vacant fellowship in the colleges of both universities.[21] When Edward Spence of Jesus College dared to satirize the Catholic church in a speech before the university late in November 1686, James had him prosecuted until he recanted publicly in the Senate House. The humiliation upset Spence's mental stability. A month later, the man who prosecuted him, Joshua Bassett of Caius, whom the university considered a secret papist, found himself elevated by letter mandate to the mastership of Sidney Sussex College. This was to strike at the very heart of English Protestantism, for Sidney Sussex had been founded early in the century as one of the two quintessential Puritan colleges. Its statutes required the master to take an oath to "detest and abhor Popery." Needless to say, a royal order dispensed with the oath. Bassett also became a member of the caput, the four heads of colleges who, with the vice-chancellor, controlled the business of the senate. The coming crisis had announced its approach in unmistakable terms. Nevertheless, the crisis was bound to perplex the university, for an institution shaped by royal patronage scarcely knew how to act when royal patronage promoted an end the majority both hated and feared.

The crisis finally arrived on 9 February 1687 in the form of a letter mandate to admit Alban Francis, a Benedictine monk, to the degree of Master of Arts without exercises and without oaths. The university had received many such mandates in the past and had readily conferred degrees on visiting Catholic dignitaries. Only two years before, Cambridge had created the secretary of the ambassador of the Emperor of Morocco Master of Arts without requiring that he take the oath of supremacy, in effect an oath to uphold the established Anglican religion. Everyone understood that the case of Father Francis differed, however. Father Francis had made his presence felt about the university. Two months before, he had carried from London the letter mandate to install Bassett in Sidney Sussex. Unlike the secretary to the ambassador of the Emperor of Morocco, he intended to reside in Cambridge and, as a Master of Arts, to participate in the business of the university. No one could doubt that other fathers stood in the wings ready to descend on the university in Father Francis's wake to catholicize it. If the university were going to make a stand, it must be here.

---

[21] The Conclusion Book, 1646–1811, of Trinity recorded seven mandates in less than a year beginning in September 1686 (pp. 172–5).

John Peachell, the vice-chancellor, held the letter mandate for nearly two weeks while he sought advice. News of its arrival filtered quickly through the university. In February, Newton had not yet completed Book III of the *Principia*. He must have stood very close to its completion, however, and with its burden lifting from him, he was free as he would not have been a year earlier to give his attention to other affairs. On 13 February, he wrote to Halley that, "through other occasions," he would not have time to think further about solar parallax, a problem Halley had put before him.[22] Six days later, he drafted a letter, apparently intended for someone in authority in the university, urging "an honest Courage" which would "save ye University." He presented an argument for refusing to accept the mandate. The law required the oath of supremacy for those who received degrees; "if his Majesty be advisd to require a Matter wch cannot be done by Law, no Man can suffer for neglect of it."[23] Already Newton had resolved the double dilemma with which the affair of Father Francis confronted him. The arch-heretic decided to throw his support to the Anglican university. The beneficiary of a dispensation took his stand on the supremacy of law over royal will. The first dilemma was private, and he kept it so. The second was the dilemma of the whole university, and James did not allow them to pretend it did not exist.

Impartial information on the events in Cambridge in the spring of 1687 is hard to discover. Every discussion that I have found in print is based directly on a pamphlet called *An Account of the Cambridge Case*, which was published immediately after the revolution. A manuscript version of it in Humphrey Newton's hand, differing some from the published pamphlet as well as from two other manuscript versions in the Baker Papers in the Cambridge University Library, survives among Newton's papers.[24] There is every reason to take the pamphlet as a piece of propaganda for the revolution compiled by the very men whose activities it described. Except for a briefer narrative by Gilbert Burnet in his *History*, which may also have used the pamphlet, it is the primary source to which one must turn. A biographer of Newton must be particularly unhappy

---

[22] *Corres 2*, 464.   [23] *Corres 2*, 467–8.

[24] *Keynes MS* 113. Cambridge University Library, Baker MSS, Mm.5.51. The account in Howell's *State Trials* is merely a reprint of this pamphlet, and the account in Cooper's *Annals* is based directly on it. However suspect it may be, it is almost the only source, and I base my account on it also. An amusing mistake in *Keynes MS* 113 may indicate Newton's active role in composing the pamphlet, since it implies that Humphrey copied it, not from another exemplar, but from Newton's dictation. In his testimony, Vice-Chancellor Peachell repeated the words of his oath that he should faithfully administer "munus or officium prochancellarii." Judge Jeffreys interrupted, as he frequently did, to repeat the words ironically: "Ay, munus or officium." Humphrey wrote, "I munus or officium."

to rely almost solely on it since Newton apparently participated in its composition.

The first impulse of the university was to hope that the crisis was a bad dream which would go away with the dawn. The Duke of Albemarle, the chancellor, suggested that the senate petition the king to withdraw the mandate. That proved to be impossible. The senate could only act on matters placed before it by the caput. The caput in turn had to act unanimously, and Joshua Bassett, the new master of Sidney Sussex, belonged to the caput. Hence arrangements were quietly made that when they met on 22 February each house of the senate would send an unsolicited opinion to the vice chancellor that it would be "illegal and unsafe" to admit Father Francis without oaths.

Alas, the crisis would not go away. With Father Francis earnestly plying the roads, exactly two days sufficed to bring a second letter mandate from London, identical to the first except for the further clause that they would refuse "at their peril." Caught between the university and the court, Peachell was beside himself. An all-too-typical product of the Restoration university, distinguished primarily by his red nose (a medal won by heroic attendance at the tavern), he was made a reluctant hero of liberty. Almost frantic with anxiety, he wrote to his friend Samuel Pepys, an adviser to James, to explain his actions. "Worthy Sir, tis extraordinary distresse and affliction to me, after so much indeavour and affection to his Royall person, crown, and succession, I should at last, by the providence of God, in this my station, be thus exposed to his displeasure. . . . "[25] Others shared Peachell's concern. William Lynnet, a senior fellow of Trinity, thought the consequences of resisting the mandate must bring evil upon the university. "Since it may be said that in contesting with the Kings will & pleasure, we call his power & wisedome into question, which will be interpreted as unbeseeming presumption in us, who in our Constitution so immediately depend upon his Maties good Grace."[26] If we can believe Burnet, the anxiety so evident on the part of Peachell and Lynnet animated others; "all the great preferments of the church being in the king's disposal, those who did pretend to favour, were not apt to refuse his recommendation, lest that should be afterwards remembered to their prejudice."[27]

The fears of Peachell and Lynnet help us to understand the events of 11 March, when, after another delay to consult and consider, the senate met again to hear the second mandate. By 11 March, New-

---

[25] Peachell to Pepys, 23 Feb. 1687; *Memoirs of Samuel Pepys, Esq., F.R.S.*, ed. Richard Lord Braybrooke (London, n.d.), pp. 636–7.

[26] Lynnet to Alexander Akehurst, 19 April 1687; Trinity College Library, O. 11a. 1^1.

[27] Gilbert Burnet, *History of His Own Times*, 6 vols. (Oxford, 1823), *3*, 143.

ton was virtually free of the *Principia*. Halley had the manuscript of Book II and Humphrey was copying Book III, which Newton would ship to London in three weeks. If he was free of the *Principia*, he was equally free of the concerns that immobilized others, for he had deliberately withdrawn from the pursuit of preferment nearly two decades before. We know nothing of what went on in the senate meeting on 11 March. We do know the result. The non-regent house (composed of the more senior Masters of Arts) chose Isaac Newton, a fellow of Trinity hitherto known primarily for his aloofness from the university, as one of two representatives to convey to the vice-chancellor their voluntary advice that it would still be illegal and unsafe to admit Father Francis to the degree without the oath. Surely we must assume that Newton spoke out and articulated the common fears when prudential considerations left others mute. And in April, when a king furious at being thwarted summoned Peachell and representatives of the university to appear before the Court of Ecclesiastical Commission, the senate elected Newton (and Humphrey Babington) among the eight it designated to perform that duty.

Newton was not present at the meeting of the senate on 11 April which chose him. The Exit and Redit Book shows that he left the college on 25 March; and though it records no redit, the Buttery Books and the Steward's Book indicate that he returned only about 17 or 18 April. From a letter of Halley, we know that he went to Colsterworth.[28] Among other things, he was engaged in litigation with his tenants, Edward Storer and his sons, as we learn from a letter of the following January.[29] Nevertheless, the trip at this time was peculiar. If we judge by the date on which Halley received Book III, Newton left Cambridge a week before the manuscript was completed. Moreover, the crisis, in which he had assumed a leading role, still hung over the university; he could not have imagined that the matter was settled. One gains the impression that Newton said enough on 11 March that he found it advisable to disappear. He returned to Cambridge only in time to prepare for the trip to London.

Back in Cambridge, Newton threw himself into the preparations for the hearing. His papers contain a number of documents, copied in Humphrey's hand, concerned with the defense.[30] Newton himself told Conduitt that he alone had averted a compromise which would have surrendered the university's position. Before the delegation left for London, the chancellor of Ely drew up a paper in which they agreed to admit Father Francis on condition that he not become a precedent for others,

---

[28] Halley to Newton, 5 April 1687; *Corres 2*, 473–4.   [29] *Corres 2*, 502–4.
[30] *Keynes MS* 114. *Yahuda MS* 32.

to which they all seemed to agree but he disliking it arose from
the table & took 2 or 3 turns & said to the Beadle who . . . was
standing by the fire this is giving up the question, so it is said the
beadle why do not you go & speak to it upon wch he returned to
the table & told them his mind & desired the paper might be
shewn to Council . . . [31]

In the end, he gave the delegation backbone enough to reject it. An
identical account in Burnet, though without Newton's name, adds
credence to the story.[32]

On 21 April, while Halley was seeing the *Principia* through the
press, Newton with eight of his peers stood before the Ecclesiastical
Commission led by the notorious Lord Jeffreys. Newton may have
met him before, for Jeffreys spent the year 1662 as a student in
Trinity. It was not an occasion for renewing acquaintances, however. In all, the delegation appeared before the commission four
times, on 21 and 27 April and 7 and 12 May. The hearing on 7
May, when Jeffreys questioned Peachell about the written answer
they had submitted, explaining why the university had not accepted
the king's mandate, was the critical one. Peachell made a wretched
showing. Lord Jeffreys turned aside every effort of the other delegates to aid his defense with rough humor about their apparent
aspirations to be vice-chancellor. At the end of the hearing, pro-
nouncing Peachell guilty of "an act of great disobedience," the
commissioners deprived him of his office, suspended him from the
mastership of Magdalene, and stripped him of the income of his
position.

Peachell's deposition freed the other delegates to act. In haste,
they drew up a written answer to the questions Peachell had bum-
bled. Five drafts of the answer survive among Newton's papers,
suggesting that he played a leading role in its composition.[33] With
its citation of precedents and command of facts, it was exactly the
task for which Newton had prepared himself in his years of biblical
scholarship. As ably as the case allowed, the paper defended the
tenuous distinction on which the university's case stood, the difference between a multitude of mandates accepted and statutes ignored
and the present mandate for which the university asserted the invi-
olability of law. To one draft, Newton appended a paragraph
which went beyond the dry details of legal precedents to the heart
of the matter.

They [the senate] were influenced also by their religion established &
supported by ye laws they are commanded to infringe. Men of y^e
Roman Faith have been put into Masterships of Colleges. The entrance into Fellowships is as open. And if forreigners be once incor-

---

[31] *Keynes MS* 130.10, ff. 3^v–4.   [32] Burnet, *History*, 3, 142.   [33] *Keynes MS* 116.

porated 'twill be as open to them as others. A mixture of Papist & Protestants in ye same University can neither subsist happily nor long together. And if ye fountains once be dryed up ye streams hitherto diffused thence throughout ye Nation must soon fall of. Tis not their preferments for a time but their religion & Church wch men of conscience are concerned for, & if it must fall they implore this mercy that it may fall by ye hands of others.

The final answer did not contain the paragraph, and the commission refused to receive the paper in any case. In its fervor and eloquence, we can perhaps catch the reason behind Newton's sudden elevation to prominence in the university. Only he recognized the irony behind its appeal to the Anglican religion.

On 12 May, the delegation gathered before the commission one last time to hear if Peachell's fate was theirs as well. Jeffreys informed them that they had shown themselves pernicious and obstinate, but the commission was inclined to attribute their errors to their leader, a shaft of sarcasm that none would have missed.

Gentlemen [he concluded], your best course will be a ready obedience to his majesty's command for the future, and by giving a good example to others, to make amends for the ill example that has been given you. Therefore I shall say to you what the scripture says, and rather because most of you are divines; Go your way, and sin no more, lest a worse thing come unto you.

But for all of that, Father Francis did not receive a degree – not least because Newton had refused to be frightened.

Eighteen months later, as fate began to close in, Jeffreys had more occasion than Newton to heed his own advice. At the last moment, when it was already too late, James tried to undo his disastrous ventures with the two universities. Neither the reversal of policy nor anything else he did sufficed to ward off the inevitable. Early in November 1688, William landed with his army in England. As he approached London in the middle of December, the Protestant mob took over Cambridge. Overt Catholics barely escaped with their lives. Dr. Vincent of Clare Hall, who had carried the manuscript of Book I to the Royal Society two and a half years before, but who had compromised himself by seeking preferment from James, saved his neck by getting out of town. On 14 December a rumor that a troop of Irish soldiers were slitting throats in Bedford and heading for Cambridge, whipped the mob into a frenzy until travelers from Bedford laid the rumor to rest and restored calm. By Christmas, it was all over. James had fled, leaving the Glorious Revolution in power.

We know almost nothing about Newton's activities during the year and a half between the hearing and the revolution. Only two letters written by him survive from the period: one of his replies to Gilbert Clerke and a letter about the suit he had brought against the Storers for their failure to maintain Woolsthorpe. We do know one other fact, in its implications the most important event in Newton's life at this time. In September 1687, he accepted his third and last student, Robert Sacheverell. Sacheverell was the son of William Sacheverell, a leader in Parliament during the reign of Charles II who had led the attempt to exclude James from the throne. Later, William Sacheverell would be a member of the Convention Parliament with Newton. A letter addressed to Newton in Parliament involved a message to him.[34] Robert Sacheverell made no more impression on Newton's life than his two predecessors, though he was more scrupulous in presenting the expected plate to the college. His significance lies in his father's position. In the view of leaders of the opposition to James, Newton had joined their ranks. He had good cause to lie low until the revolution.

Once the revolution had ratified Newton's courage, he found himself one of the prominent figures in Cambridge. When the senate met on 15 January 1689 to elect two representatives of the university to the convention called to settle the revolution, Newton was one of three men put forward, and he was one of the two elected. From this time on, until he resigned his fellowship and chair in 1701, he was invariably one of the commissioners appointed by acts of Parliament to oversee the collection in Cambridge of aids voted to the government. Tax commissions were indexes to the leading citizens in a city or county, and Newton's appearance on them was a measure of his rising importance.

It was not the only measure. As a supporter irrevocably committed to the new government, Newton found himself an object of its patronage. In August 1689, the provost of King's College died. His death must have been expected, for already in July, John Hampden, the grandson of John Hampden of Ship Money fame and a leading Whig in the Parliament, was soliciting the position for Newton.[35] William intended to appoint him by letter mandate. The college protested. Their statutes required that the provost be in orders and that he be chosen from the members of the college. One monarch who had aroused a storm by too free use of letters mandate had

---

[34] William Herbert to Newton, 9 Jan. 1690; *Corres 3*, 66.

[35] Constantyn Huygens, who served William III, recorded that on 10 July his brother Christiaan (who was visiting England), Fatio, Hampden, and Newton went to London from Hampton Court to recommend Newton for the mastership of a college (Huygens, *Oeuvres, 22*, 749). Without dating it, Huygens mentioned the occasion in his journal of his trip in a way that implied the audience to have been at Hampton Court, and on 28 July he recorded further that Hampden talked to the Duke of Somerset about it (*ibid., 22*, 744, 746).

recently departed for France. William chose not to imitate him, and for the time being the matter of a suitable reward to Newton lapsed.[36] It was not forgotten, however.

Newton also began to perceive himself in a new light which was incompatible with the isolation he had striven to maintain for twenty years. Humphrey Newton recalled that at his "seldom Entertainm^ts" his guests were primarily masters of colleges.[37] We must assume that Humphrey referred to the period after the crisis of 1687. In 1695 Newton joined the masters of Christ's and Corpus Christi colleges and of Trinity Hall, the president of Queen's College, the chancellor of Ely, and a large number of persons from beyond the university in loaning money to St Catharine's College to help finance its reconstruction.[38] When David Loggan published his *Cantabrigia illustrata* in 1690, Newton appeared as the patron of the print of Great St. Mary's. To be sure, Loggan had resided in Trinity for more than a decade while he prepared his volume, and as early as 1677 Collins heard from him that he had drawn Newton's picture to illustrate a volume on optics. For all that, Newton placed himself in a circle of some eminence – including such men as the Duke of Lauderdale, the Earl of Westmoreland, Francis North, Baron Guilford, the bishops of Ely and of Lincoln, and Thomas Tenison, the future Archbishop of Canterbury – in accepting the dedication of a plate. His friend Humphrey Babington appeared beside him for the last time in the volume. On the plate, Loggan described Newton as "Lucasian Professor of Mathematics, Fellow of Trinity, Fellow of the Royal Society, Very Accomplished Mathematician, Philosopher, Chemist." Loggan undoubtedly chose the adjective "Very Accomplished" in the time-honored tradition of clients. For the substantives, he surely followed Newton's desires. Even without the adjective, it was not the self-description of a retiring man. Nor was it the act of a retiring man when he arranged in London to have his portrait painted by the leading artist of the day, Sir Godfrey Kneller (Figure 11.1). Since Loggan's drawing has not survived, the Kneller

[36] J. E. Foster, ed., *The Diary of Samuel Newton, Alderman of Cambridge (1662–1717)* (Cambridge, 1890) p. 102. Charles Henry Cooper, *Annals of Cambridge*, 5 vols. (Cambridge, 1842–1908) *4*, 8.   [37] *Keynes MS* 135.

[38] In return for a loan of £25, Newton was to receive an annuity of 50s a year for life. The reconstruction had begun earlier, in 1674, and from the contents of a surviving document it was almost certainly near that time, though possibly at the time of the loan, that Newton also aided the college with a calculation of the extent to which their new building cut off light from the Queen's College chapel (*Corres 7*, 388–9). As far as the annuity was concerned, the college, which was very poor, never paid Newton a penny, and in 1714 he signed a document that legally released it from the obligation (St Catharine's College Muniment Room, and W. H. S. Jones, *A History of St Catharine's College, Cambridge* [Cambridge, 1936], pp. 248–55).

Figure 11.1.   Newton at forty-six. Portrait by Sir Godfrey Kneller,
1689. (Courtesy of Lord Portsmouth and the Trustees of
the Portsmouth Estates.)

portrait is the earliest likeness of Newton we possess – an arresting
presence, instinct with intelligence, caught when his capacities
stood at their height. Without difficulty, we recognize the author
of the *Principia*. It is perhaps also indicative of his new sense of
himself that five years later Newton mentioned to Flamsteed in
passing that he kept a servant whom he was teaching to do astro-

nomical calculations, though in Humphrey he had perhaps already had a similar servant.[39]

Newton set out for London almost as soon as he was elected to the convention. A note by Robert Morrice indicated that, together with Sir Robert Sawyer, the other representative of Cambridge, and Mr. Finch, perhaps the third candidate from Cambridge who was defeated in the election, he dined with no less a personage than William of Orange on 17 January.[40] With the exception of six weeks in September and October, during an adjournment, he spent the whole of the following year in London. Part of the time at least, he rented a chamber in the house of a Mr. More in the Broad Sanctuary near the west end of Westminster Abbey and only a short walk from the House of Commons. During March, some undefined indisposition confined him to his chamber, and again in May "a cold & bastard Pleurisy" did the same.[41] Beyond those brief interruptions, he engaged himself primarily with Parliamentary affairs.

The convention formally assembled on 22 January. Reconstituted as a Parliament after it vested the Crown in William and Mary, it continued to sit without intermission until 20 August, when it adjourned for two months. Having reconvened on 19 October, the Parliament sat until 27 January 1690, when William prorogued and shortly thereafter dissolved it. There is no way to pretend that Newton played a leading role in its deliberations. According to a story that rests solely on anecdotal authority, he spoke only once; feeling a draft, he asked an usher to close a window. It is not merely anecdotal that none of the surviving accounts of the Parliament contain any record of his participation in its debates. Only to a limited extent do we know where he stood on the great issues the Convention Parliament decided. We have no reason to doubt, however, that Newton took a position in harmony with his past actions and with his beliefs. He hated and feared popery. He had dared to defend law against arbitrary will. On 5 February, the House of Commons divided on the assertion in the bill settling the Crown that the throne was vacant as a result of James's "abdication." A

---

[39] Newton to Flamsteed, 17 Nov. 1694; *Corres 4*, 47.

[40] Dr. Williams's Library; Morrice, Entering Books, *2*, f. 429. Cited in Margaret C. Jacob, *The Newtonians and the English Revolution* (Ithaca, N.Y., 1976), p. 33. For whatever it is worth, there does seem to be a conflict between Morrice's note and some of the Trinity records. Although the Exit and Redit Book was silent on this departure, the Buttery Book showed him in Trinity through 18 January (*Edleston*, p. lxxxvii). Perhaps the Steward's Book is compatible with Morrice's note. His total residence of 19 weeks during academic year 1689, as shown in it, required two and a half weeks in January in addition to other documented periods of residence, thirteen and a half weeks after Michaelmas 1688 and three weeks in September 1689.

[41] Newton to Covel, 15 May 1689; *Corres 3*, 23. Newton to Covel, 16 March, 10 May 1689; *Corres 3*, 18, 22.

contemporary pamphlet published a blacklist of the hundred and fifty who opposed.[42] Newton was not among them. Hence we have solid evidence to assert that Newton stood squarely with the majority that declared James had forfeited the crown and that tendered it to William and Mary on 13 February. One day earlier, he wrote to John Covel, the vice-chancellor of the university, with advice on how the university should proclaim the new monarchs, and a week later he urged Covel to be quick in grinding out the obligatory volume of verses.[43] We may assume equally that he supported the convention's conversion into a Parliament once the Crown had been settled and other basic legislation enacted, such as the Bill of Rights, which gave permanent standing in the constitution to the principles the revolution secured. Not the least of these principles, in Newton's eyes, would have been the permanent exclusion of Catholics from the throne. It appears that Newton publicly aligned himself with the more extreme Whig faction in the Convention Parliament. In the summer of 1689, he allowed John Hampden, one of its leaders, to solicit the king on his behalf. In January 1690, in the other recorded division list from that Parliament, he voted for the Sacheverell clause, a proposal by the extreme Whigs to exclude from public office all who had cooperated with James in his remodeling of corporation charters.[44]

Newton saw his principal role in the Parliament as forming a liaison with the university. During the early months of the session, he sent at least fourteen letters to the vice-chancellor, John Covel, whose manuscripts of Revelation he had borrowed ten years before, with information about proceedings relevant to the university and advice on how the university should conduct itself. The new oath of allegiance quickly became an issue of importance. Cambridge had a nucleus of high churchmen, centering in St. John's College, whose consciences did not allow them to renounce allegiance sworn to James, whatever his faults. Even John Billers, the university orator, who with Newton had represented the university before the Ecclesiastical Commission, felt unable to take the new oath. One of Newton's first letters to Covel marshaled arguments to remove "the scruples of as many as have sense enough to be convinced wth reason. . . .

> 1. Fidelity & Allegiance sworn to ye King, is only such a Fidelity & Obedience as is due to him by the law of ye Land. For

---

[42] "A Letter to a Friend upon the Dissolution of the late Parliament," *A Collection of Scarce and Valuable Tracts* (known as the *Somers Tracts*; 2nd ed., ed. Walter Scott, 13 vols.; London, 1809–15), *10*, 254–7.   [43] *Corres 3*, 10, 12–13.

[44] The division list is published in John Oldmixon, *The History of England during the Reigns of King William and Queen Mary, Queen Anne, King George I* (London, 1735), p. 36.

were that Faith and Allegiance more then what yᵉ law requires, we should swear ourselves slaves & yᵉ King absolute:
whereas by yᵉ Law we are Free men notwithstanding those
oaths.

2. When therefore the obligation by the law to Fidelity and
allegiance ceases, that by the oath also ceases. . . . [45]

No one seriously questioned the imposition of the new oath on
those taking it for the first time. The heated issue hinged on forcing
those already in ecclesiastical and academic positions to take a new
oath to William and Mary. Despite the letter to Covel and perhaps
because he himself had known the security of passing under cover
of earlier oaths, Newton took his stand against requiring the new
oath of those already in positions. In March, he assured Covel that
the bill did not impose "the new oaths on all persons in preferments, but only on those who take new preferments. . . . This I
acquaint you wᵗʰ particularly because I would have yᵉ University
satisfied that these new oaths are not designed to be imposed on
them all as I am told they still beleive thô I wrote formerly to
remove this their prejudice."[46] The skeptics in the university understood the forces at work better than Newton. By the end of the
month, he had to inform Covel that the Lords had not accepted the
bill from Commons but had sent down a more severe one; and a
week later, he could not even bring himself to tell Covel what was
happening to it in the House as the more advanced Whigs gained
sway. "They out voted us yesterday by about 50 votes," was all he
would say.[47] In the end, the new oath was imposed on all, and
Newton's final surviving letter to Covel, written jointly with Sir
Robert Sawyer, advised the university on how to administer it.[48]
Twenty fellows of Johns refused to take the oath and by various
strategems succeeded in retaining their fellowships until 1717, when
all but six had gone to their graves. Even then, the college protected them, and Thomas Baker, one of those finally ejected from
his fellowship, died in his chamber in the college in 1740, having
sustained an obstinate loyalty to the departed despot for more than
fifty years.[49]

Another issue of importance for the university arose in the
spring. A bill was drawn to confirm the statutes of the colleges and
universities, and Newton urged Cambridge to bestir itself to suggest changes it wanted made. Oxford sent in its recommendations,

[45] Newton to Covel, 21 Feb. 1689; *Corres 3*, 12.   [46] *Corres 3*, 18.
[47] *Corres 3*, 19, 17. Newton misdated the second of these letters 6 March. It dealt with the
business of early April, however, and he did frank it 6 April.
[48] Newton and Sawyer to Covel, 15 May 1689; *Corres 3*, 24.
[49] James Bass Mullinger, *A History of the University of Cambridge* (London, 1888), p. 165.
James Bass Mullinger, *St. John's College* (London, 1901), pp. 193–216.

but Newton did not succeed in eliciting a peep from his own university. In the end, he acted on his own and drew up proposals to inhibit letters mandate, to grant the library of each university a copy of every book published, and to restore the rights of university preachers. He also proposed to entitle professors explicitly to the incomes attached to their chairs.[50] As it turned out, the sluggards in Cambridge had once more known better. Nothing came of the bill.

On one other matter that also affected the university, the religious settlement, Newton did not say a word to Covel. We may assume that he was equally silent in Parliament; he had had a long lesson on the virtue of holding his tongue on that issue. Three hotly debated bills came before the Parliament: one to tolerate public worship by dissenters, one to repeal the Test Act of 1673, and one to comprehend many dissenters within the Church of England by broadening its definition. In the end, only the first passed into law. By its provisions, nearly all Protestant dissenters received the legal privilege to worship as they chose. Since the Test Act, which required that all public employees take the sacrament according to usage of the Anglican church, remained in effect, they did not receive civil equality. What must have concerned Newton most were the two exclusions from the privileges of toleration, Roman Catholics and "any person that shall deny in his Preaching or Writing the Doctrine of the Blessed Trinity as it is declared in the aforesaid Articles of Religion [The Thirty-nine Articles of the Church of England]."[51] The two exceptions were hardly equal. Protestant Englishmen believed that Catholics threatened the sovereignty of the state. As the memory of James faded, their fears faded with it, and Catholics enjoyed toleration in fact if not in law. No one considered Arians a threat to the state. They were a threat rather to the moral foundations of society. Newton was well aware that the vast majority of his compatriots detested the views he held—more than detested, looked upon them with revulsion as an excretion that fouled the air breathed by decent persons. He had lived silently with that knowledge for fifteen years. The debate in Parliament, or the virtual lack of debate on a provision accepted without serious question, cannot have failed to bring it home to him once more.

Newton's heterodoxy allowed him easy concealment. Catholics aside, the laws had to do primarily with public worship. Newton did not worship in an Arian church. None such existed. As long as he was willing occasionally to take the sacrament of the Church of England, the law required of him nothing at which he need balk.

---

[50] Newton to Covel, 7 May 1689; *Corres 3*, 21.   [51] *Statutes of the Realm, 6*, 74–6.

Only on his deathbed did he venture finally to refuse the sacrament.[52] Nevertheless, Newton had moved a considerable distance since 1674. In that year, he had prepared to vacate his fellowship rather than accept the mark of the beast in ordination. In his defense of the university in 1687 and in his service in Parliament, in both of which he pretended to orthodoxy, he demonstrated that his conscience had grown considerably less tender. Soon he began to court a position in London. It is manifest that he did not intend to let his religious convictions interfere. It is also true, as far as we know, that Newton did not seek reelection to Parliament in 1690. It has been assumed that distaste for the increasing factional strife in the final months of the session led him to withdraw, although his vote on the Sacheverell clause hardly supports that conclusion. It may well be that the threat of discussion on matters he dared not discuss also played a role.

As a member of Parliament, Newton participated in the stirring spectacles by which the revolution bestowed legitimacy upon itself. On 15 February, William and Mary were proclaimed monarchs before Parliament in Westminster, and then a magnificent procession carried them through the City of London, where the proclamation was repeated. Two months later came the coronation. As it happened, a new chancellor of Cambridge University was installed on 30 May. Twice, Newton had traveled down to London to represent Trinity in the installation of chancellors, and though no record survives, we can be sure that he, attired in his professorial gown, was among the seven hundred members and alumni who assembled to lend pomp to the ceremony for the Duke of Somerset. In September, while he was back in Cambridge, he took part in another ceremony. Following the Stuart tradition, William attended the races in Newmarket. On 7 October, after receiving a formal invitation, he visited Cambridge. Alderman Samuel Newton found the visit less splendid than earlier ones. No mace-bearer preceded the king; no kettledrums announced him; not more than twenty-four or twenty-five of his guards accompanied him. Nevertheless, forty-four degrees brightened an occasion mercifully spared the usual formal disputation; and before he returned to Newmarket, William inspected the Trinity library, which at least had a roof on it by then. He did not offer to contribute to it.[53]

The Parliamentary experience left no discernible mark on Newton. The year in London did. Freed from the constrictions of Cambridge society and buoyed by a new sense of confidence, he found new

---

[52] Conduitt recorded this with embarrassment and attempted to explain it away (*Keynes MSS* 130.6, Book 1; 130.7, Sheet 1).   [53] Cooper, *Annals*, 4, 9–10.

acquaintances under whose encouragement his accustomed reserve
began to melt. One of them was Christiaan Huygens. Huygens's
brother Constantyn had accompanied William of Orange on the
expedition. In June 1689, Christiaan came to visit him. He had
received a copy of the *Principia* immediately upon publication, and
he had in any case not forgotten the author of the paper on colors.
It is reasonable to assume that one of his objects in the visit was to
meet Newton. At any rate, he did, within a week of his arrival in
London. On 12 June, Huygens attended the Royal Society and gave
an account of his *Treatise of Light* and his *Discourse on the Cause of
Gravity*, which he was about to publish together. Newton was
present at the meeting. It strains credulity to believe he was present
by accident. The two met at least twice more. On 9 July, Newton
apparently visited Huygens at Hampton Court, where he was staying with his brother. Constantyn recorded in his diary that the next
day Huygens and Newton went to London at seven in the morning
in order for John Hampden to recommend Newton to the king for
the mastership of a college in Cambridge.[54] In August, before he
left for home, Huygens received two papers from Newton on motion through a resisting medium.[55] At some point, they also discussed optics and colors. Huygens told Leibniz that Newton had
communicated to him "some very beautiful experiments" on the
subject – probably his experiments with thin films similar to the
ones Huygens himself had performed less elegantly twenty years
earlier.[56] No continuing correspondence resulted from their meeting, however.

Continuing correspondence did result from another encounter.
Although we do not know exactly when Newton met John Locke,
it was probably during 1689 and probably at the house of the Earl
of Pembroke. As I have indicated, Locke had already encountered
the *Principia* in the Netherlands before the revolution allowed him
to return home early in 1689. He had taken the measure of its
author once and for all, and he never ceased to praise him. Without
knowing the details, we must assume that he made it his business to
meet Newton while both were in London. We do know that they
were already in correspondence before the autumn of 1690, the date
on the first surviving letter. The date on the copy of the early
demonstration of elliptical orbits that Locke's amanuensis took,
"Mar 89/90," provides compelling evidence that they knew each
other well half a year earlier. The Newton of 1689 was a different
man from the Newton of the 1670s. The completion and publication of the *Principia* and his own realization of its significance gave

[54] Huygens, *Oeuvres, 9*, 749. Huygens's own journal appears to place the audience in Hampton Court (*ibid.*, 9, 744).   [55] *Ibid.*, 9, 321–7, 328–9.
[56] Huygens to Leibniz, 24 Aug. 1690; *ibid.*, 9, 471.

him new confidence. Nothing reveals the new Newton more clearly than his relation with Locke. Where he had shunned proferred correspondence with James Gregory, Huygens, and Leibniz in the 1670s, he not only seized the opportunity with Locke, but he did so with alacrity. The two shared many interests, all of Newton's commanding interests indeed with the exception of mathematics. Each recognized in the other an intellectual peer. If Locke testified publicly to his esteem for Newton, Newton reciprocated quietly by preparing a special copy of the *Principia*, in which he had all the corrections he had made so far inserted, as a gift for Locke.[57] Their letters from the early 1690s, exchanging views on the subjects Newton had pursued in isolation for nearly twenty years, marked a new departure in Newton's correspondence. Only his exchange with Boyle on chemical-alchemical matters, little of which survives, offered a precedent.

Religion provided what was easily the dominant theme of the correspondence and apparently of their conversation when they met. Locke later told his cousin, Peter King, that he knew few who were Newton's equal in knowledge of the Bible.[58] A note inside the front cover of Newton's theological notebook, "Fr Massam at Oats Highlaver Parish. near Harlow," also testifies to their religious discussion.[59] Francis Masham was the husband of Locke's friend, Ralph Cudworth's daughter Damaris; Locke spent most of his final years after he returned to England at Oates, their home in Essex. The note may have been connected with Newton's first surviving letter to Locke, even though the letter was directed to him in London, since the letter concerned papers Newton was composing for which he returned to material he had entered in the notebook at the beginning of his theological study. In that letter, Newton told Locke that the papers were taking him longer than he expected.[60] Even new confidence could not wholly submerge profound anxieties that formed part of his very nature.

Two weeks later, he finally dispatched the papers: a treatise, in the form of two letters addressed to Locke on 14 November 1690, with the title *An historical account of two notable corruptions of Scripture, in a Letter to a Friend*.[61] The two corruptions were the prime trinitarian passages in the Bible, 1 John 5:7, and 1 Timothy 3:16.

[57] The copy is now in the Trinity College Library. Cf. G. A. J. Rogers, "John Locke and Isaac Newton," unpublished paper. See also Rogers, "Locke's *Essay* and Newton's *Principia*," *Journal of the History of Ideas, 39* (1978), 217–32; and James L. Axtell, "Locke, Newton, and the Elements of Natural Philosophy," *Paedagogica Europaea, 1* (1965), 235–45.

[58] Locke to King, 30 April 1703; Lord Peter King, *The Life of John Locke*, new ed. (London, 1858), p. 263.   [59] *Keynes MS 2.*

[60] Newton to Locke, 28 Oct. 1690; *Corres 3,* 79.

[61] *Corres 3,* 82 (a covering letter), 83–122.

Newton also composed a third letter about some twenty-six additional passages, all lending support to trinitarianism, that were corruptions too; we do not know if Locke ever received it.[62] Although Newton presented the discourse as the mere disclosure of a pious fraud and not as a theological discourse, it is hard to believe that anyone in the late seventeenth century could have read it as anything but an attack on the trinity.

> By these instances it's manifest [he said near the conclusion of the third letter] that ye scriptures have been very much corrupted in ye first ages & chiefly in the fourth Century in the times of the Arian Controversy. And to ye shame of Christians be it spoken ye Catholicks are here found much more guilty of these corruptions then the hereticks. . . . The Catholicks ever made ye corruptions (so far as I can yet find) & then to justify & propagate them exclaimed against the Hereticks & old Interpreters, as if the ancient genuine readings & translations had been corrupted.[63]

Clearly Locke and Newton had got down to basics quickly and found they shared similar, unmentionable opinions. As far as we know, Newton had never dared to discuss his convictions with anyone before.

They shared as well a rationalistic approach to religion, which Newton had recently embodied in his "Origins of Gentile Theology." In his treatise on corruptions of scripture, Newton argued that 1 John 5:7 made sense without the disputed passage, but no sense with it.

> If it be said that we are not to determin what's scripture & what not by our private judgments, I confesse it in places not controverted: but in disputable places I love to take up wth what I can best understand. Tis the temper of the hot and superstitious part of mankind in matters of religion ever to be fond of mysteries, & for that reason to like best what they understand least. Such men may use the Apostle John as they please: but I have that honour for him as to beleive he wrote good sense, & therefore take that sense to be his wch is the best: especially since I am defended in it by so great authority.[64]

It says much of Newton's confidence in 1690 that he sent such an Arian manifesto to Locke. It says even more that on the very morrow of the debate in Parliament he sent it with the explicit understanding that Locke was to forward it to the Netherlands to be translated into French and published – anonymously to be sure, but still published.[65] Then as now, such matters had a way of not remaining secret. Locke accordingly mailed the treatise to Jean Le Clerc in Amsterdam, though without naming the author. Le Clerc

---

[62] *Corres 3*, 129–42.   [63] *Corres 3*, 138.   [64] *Corres 3*, 108.
[65] Newton to Locke, 14 Nov. 1690; *Corres 3*, 82.

agreed to publish it but suggested that the author consult Richard Simon's recently published critical study of the New Testament. The advice passed through Locke to Newton, who in turn sent back some additions to the text culled from Simon. A year later, Newton began to realize the enormity of the risk he was taking. Though the original understanding had been explicit enough, he now expressed surprise to learn that Locke had sent the manuscript forward, and he entreated him to stop the publication. He would pay any expenses that had been incurred.[66] He was well advised. Le Clerc knew who the author was, and fifty years later, when his manuscript was found in the Remonstrants Library in Amsterdam where he deposited it, it was published under Newton's name. In 1692, such a publication would have led to Newton's ostracism from Cambridge and from society.

Beyond the Trinity, Locke and Newton discussed the prophecies and miracles. Newton remarked that they could carry on the discourse "wth more freedom" when they met. Locke submitted at least part of his *Third Letter on Toleration* to Newton's criticism.[67] If Newton did comment on it, the letter has not survived.

In January 1691, Newton spent about a week with Locke at Oates. He wrote early in February thanking both Locke and Lady Masham for their cordiality.[68] In May 1692, Locke visited Newton in Cambridge.[69]

Early in 1692, another topic entered their correspondence. "I understand," Newton wrote in a postscript to a letter, "Mr Boyle communicated his process about ye red earth & ☿ to you as well as to me & before his death procured some of y^t earth for his friends."[70] A continuing correspondence on alchemy, much of which has been lost, ensued. In July, Newton indicated that Locke had sent him more of the red earth than he expected.

> For I desired only a specimen, having no inclination to prosecute ye process. For in good earnest I have no opinion of it. But since you have a mind to prosecute it I should be glad to assist you all I can, having a liberty of communication allowed me by Mr B. in one case wch reaches to you if it be done under ye same conditions in wch I stand obliged to Mr B. For I presume you are already under ye same obligations to him. But I feare I have lost ye first & third part out of my pocket.[71]

---

[66] Newton to Locke, 16 Feb. 1692; *Corres 3*, 195.

[67] Newton to Locke, 30 June 1691; *Corres 3*, 152 (cf. Newton to Locke, 7 Feb. 1691, 3 May 1692; *Corres 3*, 147, 214). Locke to Newton, 26 July 1692; *Corres 3*, 216.

[68] *Corres 3*, 147. The Exit and Redit Book did not record this absence; the Buttery Book did (*Edleston*, p. lxxxviii).

[69] In fact, all we know is that Newton wrote to Locke on 3 May saying he had just received Locke's letter announcing his intention to come (*Corres 3*, 214).

[70] *Corres 3*, 193.   [71] *Corres 3*, 215.

Locke could take a hint. He explained to Newton that he was one of three persons to whom Boyle left the inspection of his papers. He sent two of them to Newton because he knew that Newton wanted them. Since Newton appeared to want only part of one, that was all he sent, though he assured Newton that he could have it all if he wished.[72]

On 2 August, Newton responded with a letter ostensibly meant to dissuade Locke from wasting time and money in trying the process. He was sure the mercury was the one about which Boyle had written in the *Philosophical Transactions* over fifteen years before, and he knew from Boyle that Boyle himself had never tried it. He had learned of a company in London who were trying the same process; when he enquired about them, he found that two of them had been forced to other means of living and that the chief artist, while still at work, was deeply in debt. Locke was probably not greatly impressed by the arguments, for Newton failed entirely to conceal his own fascination.

But besides if I would try this R_x, I am satisfied that I could not. For M^r B has reserved a part of it from my knowledge. I know more of it then he has told me, & by that & an expression or two wch dropt from him I know that what he has told me is imperfect & useless wthout knowing more then I do. And therefore I intend only to try whether I know enough to make a ☿ wch will grow hot wth ☉, if perhaps I shall try that. For Mr B. to offer his secret upon conditions & after I had consented, not to perform his part looks odly; & that y^e rather because I was averse from medling wth his R_x till he perswaded me to do it, & by not performing his part he has voided y^e obligation to y^e conditions on mine, so y^t I may reccon my self at my own discretion to say or do what I will about this matter tho perhaps I shall be tender by using my liberty. But that I may understand y^e reason of his reservedness, pray will you be so free as to let me know the conditions wch he obliged you to in communicating this R_x & whether he communicated to you any thing more then is written down in y^e 3 parts of y^e R_x. I do not desire to know what he has communicated but rather that you would keep y^e particulars from me (at least in y^e 2^d & 3^d part of y^e R_x) because I have no mind to be concerned wth this R_x any further then just to know y^e entrance. I suspect his reservedness might proceed from mine. For when I communicated a certain experim^t to him he presently by way of requital subjoined two others, but cumbered them wth such circumstances as startled me & made me afraid of any more. For he expected y^t I should presently go to work upon them & desired I would publish them after his death. I have not yet tried either of them nor intend to try them but since you have the inspection of his papers, if you designe to publish any of his remains, You will do me a great favour to let these two be published among ye rest. But then I desire that it may not be known that they come through my hands.[73]

The comments on Boyle recall earlier ones in which he said that he had declined to communicate with Boyle on these matters because he conversed with all sorts of people and was "in my opinion too open & too desirous of fame."[74] Newton closed his letter to Locke by mentioning an argument against multiplication to which he had never found an answer. If Locke were interested, he would send it in his next. Unfortunately, the next letter and all the rest of the correspondence on alchemy has disappeared.

At about the same time he met Locke, Newton made another new acquaintance, Nicolas Fatio de Duillier (Figure 11.2). A brilliant Swiss mathematician, then only twenty-five years old, Fatio had come to England two years earlier after a stop in the Netherlands, where he had met Huygens. He carried an introduction from Henri Justel, a savant in Paris well known to the Royal Society, which promptly elected Fatio to membership.[75] As a friend of Huygens, he attended the meeting on 12 June 1689 at which Huygens discoursed on light and gravity. There at least, if not before, he met Newton. The attraction between the two was instantaneous. Fatio was one of the party of 10 July when Huygens, Hampden, and Newton set out from Hampton Court to petition the king on Newton's behalf. On 10 October, a few days before his return for the second session of Parliament, Newton asked Fatio if there might be a chamber for him where Fatio stayed. "I intend to be in London ye next week & should be very glad to be in ye same lodgings wth you. I will bring my books & your letters wth me."[76] Already the two were very close. By November Fatio, who had arrived in England a Cartesian, had been converted to Newtonianism. Newton was (he wrote to his friend Jean-Robert Chouet) "*le plus honnête homme* I know and the ablest mathematician who has ever lived." He had discovered the true system of the world in a way that left no doubt for those who could comprehend it. The Cartesian system, which had revealed itself to Fatio as only "an

---

[73] *Corres 3*, 218. Cf. Newton to ———, undated; *Corres 7*, 393.

[74] Newton to Fatio, 10 Oct. 1689; *Corres 3*, 45.

[75] The letter from Justel was read to the Society on 8 June 1687; a week later, Fatio was proposed and elected by the council, and still another week later proposed to the society (Thomas Birch, *The History of the Royal Society of London*, 4 vols. [London, 1756–7] 4, 541, 542). Another year passed before he bothered to attend and be formally admitted by signing the book.

[76] *Corres 3*, 45. In three different places, words have been cut out of this letter, although the context does not suggest anything particularly revealing.

Figure 11.2. Nicolas Fatio de Duillier. Artist unknown. (Courtesy of the Bibliothéque Publique et Universitaire de Genève.)

empty imagination," was finished.[77] If Newton had not already sent Humphrey home before he went to Parliament, his acquaintance with Fatio could have decided him.

Despite what he wrote to Chouet, Fatio initially adopted a patronizing attitude toward Newton–at least in his letters to the Continent. Early in 1690, he assured Huygens that Newton would be happy to receive his comments on certain propositions in the *Principia*. "I have found him ready to correct his book on the matters that I have told him about so many times that I cannot admire his facility too much . . ."[78] Later he mentioned the possibility that he himself might do a second edition of the *Principia*. Because of the things he would want to add, it would become a rather large folio. "Nevertheless, it would be possible to read and to understand that folio in much less time than it takes to read or to understand Mr Newton's quarto."[79] In his own copy of the *Principia*, beside Newton's mistaken proposition on water running out of a hole in a tank, Fatio entered a similar comment: "I could scarcely free our friend Newton from this mistake, and that only after making the experiment with the help of a vessel which I took care to have prepared."[80] It did not take him long to learn to sing in a different key, however. By early 1692, in telling Huygens of his own work in mathematics, about which he did not tend to show excessive modesty, he mentioned that he had seen Newton's papers. "I was frozen stiff when I saw what Mr. Newton has accomplished . . ."[81] It was soon one of his proudest claims that Newton "did not scruple to say *That there is but one possible Mechanical cause of Gravity, to wit that which I had found out.*" He did feel constrained to add a qualification: "Thô he would often seem to incline to think that Gravity had its Foundation only in the arbitarary Will of God . . ."[82] In his youthful self-confidence, Fatio would have been shocked to read a note, presumably made a few years later, in a memorandum by David Gregory: "Mr Newton and Mr Hally laugh at Mr Fatios manner of explaining gravity."[83]

[77] Manuscript letter in Geneva quoted in Charles Andrew Domson, "Nicholas Fatio de Duillier and the Prophets of London: An Essay in the Historical Interaction of Natural Philosophy and Millennial Belief in the Age of Newton" (doctoral dissertation, Yale University, 1972), pp. 32–4.

[78] Huygens, *Oeuvres*, *9*, 387.

[79] Fatio to Huygens, 18 Dec. 1691; Huygens, *Oeuvres*, *10*, 213.

[80] Quoted in *Corres 3*, 169. Fatio made something of a career of bringing up that particular correction. He mentioned it to Gregory in 1694 (Gregory's memorandum of 16 May 1694; *Corres 3*, 355).   [81] Huygens, *Oeuvres*, *10*, 271–2.

[82] The comment is found on a sheet among his papers written after 1701 (Bernard Gagnebin, "De la cause de la pesanteur. Mémoire de Nicholas Fatio de Duillier présenté à la Royal Society le 26 février 1690," *Notes and Records of the Royal Society*, 6 [1949], 117).

[83] Gregory's memorandum of 28 Dec. 1691; *Corres 3*, 191. This note is in a very different hand from the rest of the memorandum; presumably Gregory added it later (Cohen, *Introduction*, p. 180).

After Parliament was prorogued on 27 January 1690, Newton stayed on in London for another week. Near the end of February, Fatio wrote that he and John Hampden had planned to come to Cambridge to visit until Newton wrote that he was coming to London. He expected any day to receive the copy of the *Treatise of Light* that Huygens was sending to Newton. He would keep it until Newton told him to send it. "It beeing writ in French you may perhaps choose rather to read it here with me." The Exit and Redit book indicates that Newton left on 10 March and returned again on 12 April. On 13 March, Fatio transcribed a revision of Proposition XXXVII, Book II, from Newton's copy of the *Principia,* and years later he mentioned a list of errata compiled by Newton which he had not had time to copy that March.[84] We have every reason to think that Newton spent the month in London with Fatio, perhaps reading Huygens's *Treatise.*

Fatio's reference to Hampden is intriguing. As I mentioned, Fatio went with the party Hampden organized the previous July to solicit the provostship of King's for Newton. Now Fatio mentioned that Hampden had wholly lost the favor of William, who had secured his defeat in the recent Parliamentary election. Fatio was already at work on other plans for Newton, however, plans that would move him permanently to London. "I did see Mr Lock above a week ago, and I desired him that he should speak earnestly of you to Mylord Monmouth. He promised me he would do it . . ."[85] It is the first known reference to a position for Newton in the metropolis.

About the beginning of June, Fatio went to the Netherlands for an extended stay of fifteen months, much of it with Huygens in The Hague. Newton wrote to Locke in October when he had not heard from him in half a year. When Fatio returned at the beginning of September 1691, he must have let Newton know at once, for writing to Huygens on 8 September, he said he would see Newton soon, "since he is to come here in just a few days." Newton had scarcely returned home from London, where he had spent a month partly in the company of David Gregory and Edward Paget. The Exit and Redit Book indicates nevertheless that he was absent from the college from 12 to 19 September. He did not bother to get in touch with his other London friends. In October, Gregory wrote to tell him that Fatio had returned.[86]

---

[84] Fatio's transcription of the revised proposition is quoted in *Corres 3,* 39–40. His later note, entered in his copy of the *Principia* after Newton's death, is quoted in *Corres 3,* 169.

[85] Fatio to Newton, 24 Feb. 1690; *Corres 3,* 390–1. When Huygens sent seven copies of his *Traité de la lumière* to Fatio that same month, he included Hampden in the list of those to receive them, along with Fatio himself, Newton, Boyle, Halley, Locke, and Flamsteed (*Oeuvres, 9,* 357–8).

[86] *Corres 3,* 79. Huygens, *Oeuvres, 10,* 145–6. *Corres 3,* 170.

Although their correspondence during the following year, until September 1692, does not survive, Fatio's letters to Huygens indicate a steady exchange and at least one visit when Fatio saw some of Newton's mathematical papers. We know that Newton was in London much of January 1692; on 9 January, Pepys entertained him.[87] Newton's letter to Fatio on 14 February 1693 mentioned a recent visit of Fatio in Cambridge.[88] Huygens and Leibniz came to regard Fatio as their intermediary through whom they learned about Newton's opinions on mathematics, gravity, and light. Newton soon taught Fatio to share his other interests as well – heterodox theology, the prophecies, and alchemy – and they may have spent as much time on these matters as they did on mathematics and physics.[89]

Once raised, the question of a governmental appointment did not go away. Fatio may have offered the chief incentive, but as a foreigner, he lacked the connections to make arrangements. Newton looked to other channels. In May 1690, he received a letter from Henry Starkey, whom he later described as his solicitor, which mentioned, among other places, the positions of master, warden, and comptroller of the Mint, "very good places and they [the incumbents] make them as good as they please themselves . . ."[90] The crassness of the information does not appear to have offended Newton; a year later he wrote to Locke to ask him for a letter in regard to "the controulers place of ye M."[91] In fact, Locke with his political connections became Newton's principal agent in the search for an appointment. "I am extremely much obliged to my L^d & Lady Monmoth for their kind remembrance of me," he wrote to Locke in October 1690, "& whether their designe succeed or not must ever think my self obliged to be their humble servant."[92] A year later, after several more references to the same subject, Newton wrote about another place, the mastership of the Charterhouse, for which (he had learned through his friend Laughton) Locke had recommended him to the archbishop.

I have all ye reason imaginable to take very kindly then your remembrances of me in my absence, but whilst you seem still to think on

87 Pepys to Evelyn, 9 Jan. 1692; *Private Correspondence and Miscellaneous Papers of Samuel Pepys,* ed. J. R. Tanner, 2 vols., (London, 1926), *1,* 51–2.   88 *Corres 3,* 245.

89 Fatio to Newton, 30 Jan., 8 March, 11 April, 4 May 1693; *Corres 3,* 242, 261, 392, 265–7. Domson, "Fatio" pp. 47–51. Domson argues that Fatio had shown no interest in religion, the prophecies, or alchemy before he met Newton, and he ascribes his interest in them, especially in the prophecies, to Newton's influence. Heterodox theology did not enter their correspondence, but Fatio was later reputed to be an Arian.

90 Quoted in the Sotheby *Catalogue of the Newton Papers Sold by order of the Viscount Lymington* (London, 1936), p. 49. Newton referred to Starkey as his solicitor in writing to Locke on 30 June 1691 (*Corres 3,* 152).

91 Newton to Locke, 30 June 1691; *Corres 3,* 152.

92 *Corres 3,* 79. Cf. Newton to Locke, 14 Nov. 1690, 7 Feb. 1691; *Corres 3,* 82, 147.

Charterhouse I beleive your notions & mine are very different about
that matter. For by ye information I have had of it, its but 200£ *per*
an besides a Coach (wch I reccon not) & lodgings: the competition is
hazzardous & I am loath to sing a new song to ye tune of King's
College: & the confinement to ye London air & a formal way of life
is what I am not fond of.[93]

The final objection cannot have run very deep; he did continue to
pursue a place in London with all the vigor he could muster. It is
worth recalling that in the early 1690s Trinity lay in the grip of a
financial crisis. In 1688, 1689, and 1690, it paid no dividend at all,
and the two previous years, it paid only half-dividends. A prudent
man looked to his interests.

The growing realization that he had established himself as the
leading intellectual of the land could not have discouraged Newton
from seeking a position in the capital city. From every indication,
he relished a new role of scientific consultant as much as he resented
lesser intrusions on his time in earlier decades. On 30 December
1691, the day before Newton left for London, Robert Boyle died.
Newton apparently attended his funeral. We know that two days
later, at Samuel Pepys's invitation and at his home, Newton joined
with John Evelyn and Thomas Gale, the master of St. Paul's
School, who had preceded Newton in his chambers in Trinity, "in
thinking of a man in England fitt to bee sett up after him [Boyle]
for our Peireskius . . ."[94] Newton may have participated that evening in the selection of Richard Bentley to deliver the inaugural
series of Boyle Lectures.[95]

Two years later, in the autumn of 1693, Samuel Pepys consulted
him again about a problem of chance raised by the lottery newly
proposed by Thomas Neale. Among his multiple preferments,
Neale was master of the Mint, where Newton would shortly meet
him. Newton found some ambiguity in the question that was put
to him. To clarify it, Pepys redefined the question in terms of a
man condemned to death who is offered a chance to roll dice for his
life. If he is given the choice, are his chances better to roll one six
with six dice, two sixes with twelve, or three sixes with eighteen.
In three lengthy and careful letters, Newton explained why the
chance of one six in a roll of six dice is higher.[96] It is quite impossible to imagine him devoting so much time to a similar matter

---

[93] Newton to Locke, Dec. 1691; *Corres 3*, 184.

[94] Pepys to Evelyn, 9 Jan. 1692; Pepys, *Private Correspondence, 1*, 51–2.

[95] Cf. Margaret C. Jacob, *The Newtonians and The English Revolution, 1689–1720* (Ithaca,
     N.Y., 1976), p. 155.

[96] Pepys to Newton, 22 Nov., 9 Dec., 21 Dec. 1693; Newton to Pepys, 26 Nov., 16 Dec.,
     23 Dec. 1693; *Corres 3*, 293–303. See Florence N. David, "Mr Newton, Mr Pepys &
     Dyse: A Historical Note," *Annals of Science, 13* (1957), 137–47.

earlier in his life. To be sure, another issue lay beneath the surface of the correspondence. Earlier that fall, Pepys had received a wild letter from Newton, and he was concerned to find a topic on which to engage Newton in discourse. Newton was concerned to efface the memory of the letter. The additional fact remains, however, that Pepys was a man of importance despite his exclusion from political power by the revolution, and Newton had found he enjoyed being courted by men of such rank.

In the summer of 1694, the governing court of Christ's Hospital consulted him about a proposed revision of the curriculum of their mathematical school. Twelve years earlier, Newton's recommendation had played the major role in the selection of Edward Paget as master of the school. Now, perhaps in an effort to save a position he was about to lose because of neglect and dissolute habits, Paget proposed a revision of the school's curriculum. It was not Paget but the court which consulted Newton, however, and he devoted a great deal of time and trouble, composing several drafts, to his reply. He endorsed the new curriculum for its method and organization, which the old curriculum had notably lacked, but he urged the further inclusion of a course in mechanics as the theoretical foundation of engineering practice.

A Vulgar Mechanick [he argued] can practice what he has been taught or seen done, but if he is in an error he knows not how to find it out and correct it, and if you put him out of his road, he is at a stand; Whereas he that is able to reason nimbly and judiciously about figure, force and motion, is never at rest till he gets over every rub. Experience is necessary, but yet there is the same difference between a mere practical Mechanick and a rational one, as between a mere practical Surveyor or Guager and a good Geometer, or between an Empirick in Physick and a learned and a rational Physitian.[97]

Endorsed by Newton, Wallis, and Gregory, the new mathematical curriculum was adopted, but Paget resigned the following February regardless.

Moved to the forefront of English natural philosophy by the *Principia,* Newton began to be wooed by the younger generation, who sought his patronage. Whenever he was in London, he saw Edward Paget, though Paget was well on the way to forfeiting his position. On a visit in the summer of 1691, Newton finally met David Gregory, or Gregory finally met Newton after two efforts to establish a correspondence had failed. Gregory was the first to realize the potential advantage of Newton's favor, and from the first he courted him shamelessly. In conversation, Newton enquired about

[97] Newton to Hawes, 25 May 1694; *Corres 3,* 357–66. The passage I quote is found on pp. 359–60. Drafts of his reply and other related papers are located in *Add MS* 4005.16, ff. 87–90, 93, 100, and Trinity College Library, R.5.4[22]. Cf. *Math 7,* xxii.

Scottish universities. Almost immediately, while both men were still in London, he received a long essay describing them that Gregory wrote solely for him.[98] No sooner had he left for Cambridge, than Gregory pursued him with another letter, and he took care that the connection, finally established, did not languish for want of exercise. He flattered Newton extravagantly. "Farewell, noble sir," he concluded a letter, intended for publication to be sure, with a garbled line from Virgil, "and proceed as you do to advance philosophy 'beyond the paths of Sun and Sky'." (When the letter was not published, he did not blush to dust off the same line and send it to Huygens two years later.)[99] Gregory's adulation was doubtless genuine. Even in the privacy of his personal memoranda, he referred to Newton only as "Mr. Newton" or, after 1705, "Sr Isaac Newton."[100] Gregory also had a specific end in view, however. As a result of the resignation of Edward Bernard, the Savilian professorship of astronomy at Oxford stood empty. Gregory secured Newton's recommendation for the chair.[101] As it happens, Edmond Halley also applied for the Savilian professorship. Though Newton saw him in London on occasion, we are not aware of any correspondence between the two from 1687 to 1695. Halley's application for the Savilian chair suffered shipwreck on his reputation for irreligion. His opinions had a general resemblance to Newton's. Conduitt later reported, on his wife's authority, that Newton sometimes became angry with Halley for speaking against religion.[102] Though Newton would neither speak against nor think against religion as such, it is not clear how much his anger focused on Halley's audibility as opposed to his opinions. At any rate, despite his debt to Halley, a debt beyond payment, Newton not only refrained from supporting his application but did support Gregory. Gregory got the chair. The lesson was not lost on other aspiring young men. Nor was it lost on Gregory, who continued to court Newton assiduously. As for Halley's irreligion, the story is told of a Scotsman who traveled to London just to meet a man with less religion than Gregory.

Newton's continuing correspondence with Gregory, as also his correspondence with Locke, Fatio, and others, contrasts sharply with his general reticence before 1687. His correspondence during the seven years from the beginning of 1689 until his departure from Cambridge in the spring of 1696 was as extensive as his correspon-

---

[98] Gregory to Newton, 8 Aug. 1691; *Corres 3*, 157–62.

[99] Gregory to Newton, 7 Nov. 1691; *Corres 3*, 179; original Latin, p. 176. Gregory to Huygens, 12 Aug. 1693; *Corres 3*, 278; original Latin, p. 276.   [100] *Hiscock, passim.*

[101] Newton to Charlett, 27 July 1691; *Corres 3*, 154–5.

[102] *Keynes MSS* 130.6 Book 2; 130.7, Sheet 1. Whiston reported that Halley lost the chair simply because he refused to dissemble his disbelief (*Memoirs of the Life and Writings of Mr. William Whiston* [London, 1749], p. 123).

dence during twenty years, during all of which he was a mature man, before he began the *Principia*. His great work together with the revolution marked a psychological watershed in his life. Active communication with the younger generation of scientists, who looked to him as their intellectual leader and patron, formed the major part of his expanded correspondence.

Not long before he left Cambridge, another hopeful young man, William Whiston, took care to make his acquaintance. By his own account, Whiston heard one or two of Newton's lectures on the *Principia* while he was an undergraduate and failed to understand them. He set himself to master Newtonian philosophy in the early 1690s, and in 1694 he submitted the manuscript of his *New Theory of the Earth* to Newton's inspection. According to Whiston, it won approval. For the time being, nothing more that we know of came of their relation, though Whiston may have discussed theology with Newton, who was finding that others were prepared to entertain his doubts about trinitarian orthodoxy. Not long thereafter, at any rate, Whiston became the articulate spokesman for views virtually identical to Newton's. In 1701, when Newton finally resigned the Lucasian chair, he secured Whiston's nomination to succeed him. Probably with that end in view, he had appointed him his deputy a few months earlier.[103]

By 1701, Newton had filled two of the three university chairs devoted to science and mathematics with his disciples. Shortly, he would place Halley (for all his heretical opinions) in the other Savilian chair at Oxford and help to establish another disciple in the new Plumian chair in Cambridge. Despite the debacle of Paget, the governors of Christ's Hospital consulted him about Paget's successor. Though one of the few positions in England by which a man could support himself as a mathematician, the mastership of the Mathematical School at Christ's Hospital fell into a different category from a university chair. Apparently Halley did consider applying for it, though eventually he decided against it. It is unclear how far Newton's opinion mattered in the appointment of Samuel Newton (who was no relation), but he did offer the new master assistance after he was named.[104]

Already Newton was famous enough to attract foreign visitors. In his correspondence with Fatio there was mention of a Swiss theologian and a Dutch visitor who sought letters from Fatio introducing them to Newton; both did in fact go to Cambridge to meet him.[105]

[103] Whiston, *Memoirs*, pp. 43, 293.
[104] Cf. Newton to Flamsteed, 15 March, 23 April 1695; *Corres 4*, 93–4, 106. Samuel Newton visited him in Cambridge that summer (Newton to Hawes, 29 June 1695; *Corres 4*, 133). Cf. Frank E. Manuel, *A Portrait of Isaac Newton* (Cambridge, Mass., 1968) 272–4.
[105] Newton to Fatio, 24 Jan. 1693; Fatio to Newton, 7 Feb. 1693; *Corres 3*, 241, 243.

He was famous even in Cambridge, and a strange event in the spring of 1694, which attracted the attention of the whole town and university, served further to set him apart. Strange noises in a house opposite St. John's had many convinced it was haunted. The excitement extended over several days, while crowds gathered outside and diverse intrepid scholars and fellows ventured into the house.

> On Monday night likewise there being a great number of people at the door, there chanced to come by Mr. Newton, fellow of Trinity College: a very learned man, and perceiving our fellows to have gone in, and seeing several scholars about the door, "Oh! yee fools," says he, "will you never have any witt, know yee not that all such things are meer cheats and impostures? Fy, fy! go home, for shame," and so he left them, scorning to go in.[106]

If Newton was traveling now in more exalted circles, his interest and obligations in Colsterworth remained. His troubles with tenants continued to be endemic. The Storer brothers fell three and a half years behind in their rent. Newton wrote his usual demand for payment, probably with his usual result. In 1694, he told one E. Buswell, possibly another tenant, that he could not extend his debt and gave him until January to pay.[107] There is likewise no record of whether Buswell paid up in January. In the late summer of 1695, his personal affairs took him to Lincolnshire.[108]

There were problems of another sort as well. While he was in London in January 1692, his old patron and friend Humphrey Babington had a stroke and died on the morning of 4 January.[109] Newton did not return to Cambridge for the funeral.

Another death touched him more closely. On 24 August 1693 his half sister wrote to him from Northamptonshire.

> Dear Brother
>   My Dear Husband ever since his return to Brigstock has been very ill, he has Dr Wright with him or Mr Fowller Most days. I find noe hopes of Cure but that hee lossis his flesh and strength very fast. My

---

[106] The *Diary of Abraham de la Pryme*, ed. Charles Jackson (Durham,1870), p. 42.

[107] *Corres 3*, 393, 374.

[108] On 14 September, he wrote to Flamsteed that he had just returned from Lincolnshire (*Corres 4*, 169). Cf. Newton's earlier letter to a friend on 11 Jan. 1687 (*Corres 2*, 502–4). It mentioned at least seven other letters to and fro about his suit against Storer, which was also the topic of this letter. It also mentioned two other "friends" who were keeping him informed: the rector of Colsterworth who was engaged in arbitrating the dispute, and a Mr. Parkins, evidently a lawyer. Probably a little later, he appointed a Mr. Proctor to manage his affairs in Lincolnshire and indicated in the process how extensive they were (*Corres 7*, 365). A letter to a distant cousin, John Newton, also referred to a tenant (*Corres 7*, 364). The property in Lincolnshire absorbed a fair bit of time and attention.

[109] Foster, *Samuel Newton's Diary*, 106.

Daughters and his portions are soe settled that I am to pay them £8
A year Apeec for intrest soe that as I aprehend it I am to Loose all
Taxis, or other hassards, this I wish to advise in but know not with
whom, besides I am overwhelmed in sorrow and wish for you to
Comfort her that is

Your Loving Sister

Han. Barton[110]

As it happened, Newton was under extreme personal stress at that
very moment and in no shape to help or comfort anyone. Two
years later, he sent her £100 and purchased an annuity for the same
amount to be divided, after his own death, among his sister's three
children.[111] And when he moved to London, he brought one of
them, Catherine, to live with him.

In October 1695, he was able to minister more effectively to the
needs of his half brother Benjamin Smith. When Benjamin's wife
fell ill with an infection, Newton prescribed a plaster or fomenta-
tion to apply to her chest. It worked.

> I reced your kind Lre dated the 31st of October last [Benjamin
> wrote], And the fomentation wee applyed as soone as wee could
> possible; The effects have pved very successfull; for the swelling is
> verry much abated, and the blacknes quite gone. Although att cer-
> taine times shee hath still a paine in her brest; after every bathing,
> shee put the same plaster to her Brest againe, you sent, I could not
> perswade her to take the Sowes, for since her being wth child al-
> most every thing goes against her Stomach; But shee is resolv'd to
> try.[112]

Benjamin's wife appears to have recovered.

Medicines, specifically "ye first imperial powder for ye first re-
gion," also entered into Newton's correspondence with Fatio.
Humphrey Newton wrote to ask if he had the recipe for some pills
that Dr. Clark (of Grantham) had left in Cambridge on a visit. For
Flamsteed's perpetual headache, he had a more direct remedy that
John Batteley, a former fellow at Trinity, had employed. Batteley
used "to bind his head strait wth a garter till ye crown of his head
was nummed. For thereby his head was cooled by retarding the
circulation of the blood."[113]

---

[110] *Corres 3*, 278–9.
[111] The draft of a letter from ca. 1695 mentioned the transfer of £100 to his sister (*Add MS*
3966.15, f. 370^v; *Corres 5*, 201). Whiteside places this second letter in the autumn of 1695
(*Math 7*, 673). When he died, Newton held an annuity on his own life for £100 *per annum*
(Huntington Library MS).
[112] *Corres 4*, 187. "Sowes" is probably a mistake for "sowens," a sort of porridge prepared
from the bran of oats.
[113] Newton to Fatio, 14 Feb. 1693; *Corres 3*, 245. Humphrey Newton to Newton, 19 Dec.
1691; *Corres 3*, 190. Newton to Flamsteed, 20 July 1695; *Corres 4*, 152.

As the new dimensions of his existence gradually unfolded, Newton did not forsake his intellectual pursuits. Quite to the contrary, the early years of the 1690's, his final years in Cambridge, were a period of intense, almost manic, intellectual activity. Swept along by the *Principia*'s success, Newton attempted to pick up the scattered loose threads of earlier enquiries and to weave them into a coherent whole worthy of his great completed work. The effort was his last major intellectual endeavor.

Interestingly, theology was not prominent among his pursuits during this period. If we judge by the surviving papers, the *Principia* interrupted the study which, with alchemy, had dominated his attention during the previous fifteen years, and he did not return to it seriously for another two decades. It is true that external factors could elicit theological opinions from him. When David Gregory visited him in May 1694, they discoursed a little on theology, and Newton summarized his "Origins of Gentile Theology" for Gregory.[114] Conversations with John Locke led to the composition of his tract on the corruption of Scripture. There is good reason to think that he also discussed theology with a number of other trusted intimates, such as Fatio, Halley, and Whiston, who were later reputed or known to be Arians. What Newton had to say in such discussions stemmed from his earlier conversion to heresy, not from current study.

In 1693, Newton engaged in a quasitheological correspondence with Richard Bentley, an aspiring young cleric of formidable intellect. Somewhat like Locke, Bentley had earlier set himself the task of mastering the *Principia*. In the summer of 1691, he applied both to John Craig and to Newton himself for a list of reading necessary to comprehend the work. Craig's imposing course of study ought to have frightened anyone away; Newton's list, with readings in mathematics and Copernican astronomy plus Huygens's *Horologium*, was at least possible. As for the *Principia*, Newton suggested that Bentley read the first three sections in Book I and then proceed to Book III, which would show him what other propositions in Book I he needed to master.[115] Not long thereafter, Bentley was named to deliver the first set of lectures in defense of religion established by the will of Robert Boyle. Late in 1692, as he prepared the manuscript of his lectures (which had drawn heavily on Newton) for publication, he applied to Newton for help with several points.

---

[114] Gregory's memorandum of 5, 6, 7 May 1964; *Corres 3*, 336, 338; original Latin, pp. 334, 336.

[115] Craig's list was sent to Bentley in a letter to William Wotton, 24 June 1691; *Corres 3*, 150–1. Newton's recommendations, *Corres 3*, 155–6. For an inquiry by Newton that Bentley provoked, see M. A. Hoskin, "Newton, Providence and the Universe of Stars," *Journal for the History of Astronomy, 8* (1977), 77–101.

In all, Newton wrote four letters on the subject to Bentley. While they dealt with the arguments from natural philosophy for the existence of God, in effect Newtonian versions of the familiar arguments of the day, the letters did not get into the question of the Trinity, where Newton concentrated his theological attention.

When I wrote my treatise about our Systeme [Newton began his first letter] I had an eye upon such Principles as might work wth considering men for the beleife of a Deity & nothing can rejoyce me more then to find it usefull for that purpose. But if I have done ye publick any service this way 'tis due to nothing but industry & a patient thought.[116]

He went on to summarize the reasons that convinced him that the universe as we know it could not have resulted from mechanical necessity alone but required the intelligence of a Creator. "There is yet another argument for a Deity wch I take to be a very strong one," he concluded enigmatically, "but till ye principles on wch tis grounded be better received I think it more advisable to let it sleep."[117] As far as I know, Newton never explained this reference. He probably had the argument from the providential course of history as foretold in the prophecies in mind.

The greatest interest in the Bentley correspondence lies in the light it threw on questions in Newton's natural philosophy. "You sometimes speak of gravity as essential & inherent to matter," he concluded his second letter: "pray do not ascribe that notion to me, for ye cause of gravity is what I do not pretend to know, & therefore would take more time to consider of it."[118] When Bentley incorporated that position into the text of his sermon, Newton expounded more fully on his meaning.

Tis unconceivable that inanimate brute matter should (without ye mediation of something else wch is not material) operate upon & affect other matter wthout mutual contact; as it must if gravitation in the sense of Epicurus be essential & inherent in it. And this is one reason why I desired you would not ascribe innate gravity to me. That gravity should be innate inherent & essential to matter so y^t one body may act upon another at a distance through a vacuum wthout the mediation of any thing else by & through wch their action or force may be conveyed from one to another is to me so great an absurdity that I beleive no man who has in philosophical matters any competent faculty of thinking can ever fall into it. Gravity must be caused by an agent acting constantly according to certain laws, but whether this agent be material or immaterial is a question I have left to ye consideration of my readers.[119]

[116] Newton to Bentley, 10 Dec. 1692; *Corres 3*, 233. [117] *Corres 3*, 236.
[118] Newton to Bentley, 17 Jan. 1693; *Corres 3*, 240.
[119] Newton to Bentley, 25 Feb. 1693; *Corres 3*, 253–4.

He had not left it entirely to his readers, since Book II had become a sustained argument against the existence of a material agent. Newton's letter to Bentley drew on extensive revisions of Book III, which he decided in the end not to publish. He explored the meaning of his declarations to Bentley more fully in that context.

During the early 1690s, Newton was also in touch with John Mill, the Oxford scholar engaged in establishing the correct text of the New Testament by collating more than a hundred manuscripts. Sometime in the fateful year 1693, Mill visited Newton in Cambridge and left his manuscript, which Newton checked against his own extensive collation of versions of Revelation.[120]

A good part of Newton's activity at this time went into the *Principia* itself. He had a year and a half between the crisis in Cambridge in the spring of 1687 and the revolution at the end of 1688 to which we cannot assign any enterprise with assurance. The investigation that produced the *Principia* may well have carried on without interruption into its further perfection. We know that already early in 1690, after a year's interruption while he sat in Parliament, he had compiled a list of errata, which he communicated to Fatio at that time.[121] As we have seen, Fatio intended to undertake a second edition; we can assume that Newton encouraged him. In December 1691, David Gregory spoke of the plan, and Fatio was still toying with it in February 1692.[122] There is no evidence that Fatio ever made a serious start on the project.

By the summer of 1694, it was David Gregory who planned to manage the new edition. He may have been talking about it a year earlier when he visited Huygens in the Netherlands.[123] In May 1694, Gregory spent several days with Newton. He kept notes on their conversations, which ranged over many subjects but focused primarily on the *Principia* and Newton's intended alterations in the new edition. The memoranda offer concrete evidence of the extent to which Newton had continued to work on his masterpiece.[124] Later in the summer, Gregory summarized what he had learned. Newton intended to restructure the demonstrations of centripetal force in Sections II and III. He would eliminate the two extraneous sections (IV and V) on the mathematics of conics and shape them into a separate treatise. In Book II, he planned to add a new prob-

[120] Mill to Newton, 7 Nov. 1693; Newton to Mill, 29 Jan. 1694; *Corres 3*, 289–90, 303–4.

[121] Fatio took a list of errata with him to the Netherlands in 1690 (Cf. Fatio to Huygens, 8 Sept. 1691; Huygens, *Oeuvres*, *10*, 145–6).

[122] Gregory's memorandum of 28 Dec. 1691; *Corres 3*, 191. Fatio to Huygens, 15 Feb. 1692; Huygens, *Oeuvres*, *10*, 259.

[123] Gregory's memorandum of ca. late July 1694; *Corres 3*, 384–6. Cf. Huygens to Leibniz, 29 May 1694; Huygens, *Oeuvres*, *10*, 614.

[124] A number of memoranda and papers stemmed from the visit (*Corres 3*, 311–15, 323, 326, 327–8, 331–2, 334–6). Cf. papers of Newton's that Gregory's visit apparently stimulated (*Math 6*, 435–8, 470–5).

lem on the trajectory of projectiles when resistance is proportional to velocity squared, a problem Huygens had raised with him. Section VII of Book II would undergo extensive changes, especially Proposition 37 (36 in later editions) on water flowing from a tank and the scholium with experimental investigations of resistance at the end of the section. Newton did not trust his experiments with pendulums to measure resistance; he planned to perform new ones with bodies falling from the roof of Trinity chapel into his garden. He also intended to define the velocity of sound within narrower limits. He had many alterations in Book III: extensive recomputations of lunar theory, a new demonstration of precession, changes in the determination of the orbits of comets with the addition of at least two more concrete orbits of observed comets.[125] As far as we can tell from Newton's papers, he had completed only the emendations to Book I at the time Gregory wrote, and he eventually suppressed most of them. Gregory's description of the intended changes in Book II and III corresponded closely to what appeared in the second edition, but Newton put them into their final form only eighteen years later. In a letter of 24 September 1694, Gregory said he was glad to know that plans for the second edition were moving forward.[126] They came to a stop before long, however. The edition did not appear until 1713, and we hear no more of Gregory's role in it.

Newton's papers indicate that during the early 1690s he devoted considerable attention to the laws of motion. Among other things, he concerned himself with attempting to remove the contradiction between the two conceptions of force, a succession of impulses and a continuous force, that the *Principia* employed.[127] In the end, he significantly altered only Corollary I, the parallelogram of forces, removing from it the apparent concept, which appeared in the first edition, of a uniform force maintaining a uniform velocity.

He also worked out a restructuring of Sections II and III along the lines that Gregory indicated in his memorandum. The revised argument shifted the foundation of demonstration from Proposition VI, a general expression of centripetal force in terms of the incremental elements of any trajectory, to Proposition VII, a measure of centripetal force in circular motion. As Gregory stated the rationale, "The determination of curvatures is useful in all physical problems. The force of revolution in any portion of a curve is the same as in its circle of curvature."[128] Ultimately, he decided to let his original plan of demonstration stand, and he confined himself to adding alternate proofs to Propositions X and XI.

[125] Gregory's memorandum of ca. July 1694; *Corres 3*, 384–6.   [126] *Corres 4*, 20.

[127] *Add MSS* 3965.6, ff. 86, 274; 3965.19, f. 731 (cf. fuller discussion in Richard S. Westfall, *Force in Newton's Physics*, (London, 1971) pp. 479–81).

[128] *Math 6*, 568–92. Gregory's memorandum of 4 May 1694; *Corres 3*, 316; original Latin, p. 312 (I have made a small change in the translation).

Newton devoted attention also to the philosophical assertions on which the *Principia*'s demonstrations rested. He knew that many challenged them. Not only had he conversed with Huygens, but he had also received a presentation copy of his *Treatise of Light* with its appended *Discourse on the Cause of Gravity*. Early in 1693, a letter from Leibniz expressed in passing his continued belief in the necessity of an aether as the cause of gravity.[129] It is not surprising then that Newton thought of buttressing his argument against aethereal mechanisms by a stronger statement, insisting on atoms and voids, of his philosophy of nature. To this end, he composed a new set of corollaries to Proposition VI, Book III (on the proportionality of gravity to mass) and a scholium to Proposition VII (the law of universal gravitation). Ultimately, he chose not to include this material in later editions.

The proposed set of corollaries to Proposition VI argued that, in order for weight to be proportional to mass (as the proposition demonstrated), mechanical explanations of gravity had to affirm the identical size and shape of the particles of bodies and thus to admit atoms and voids in order to account as well for differences in specific gravity. They had further to assert two different forms of matter, for if the particles of bodies must always be identical, they cannot break up into smaller particles to form an aether. But the uniformity of matter in the universe was a basic assumption of mechanical philosophies.[130] Not content merely to insist on the existence of voids, Newton began for the first time to spell out the radical consequences of his new philosophy of nature. Populating the universe with forces entailed depopulating it of matter. Matter can fill only the tiniest portion of the whole. Bodies are far rarer than is commonly believed, he asserted. Gold, the densest substance we know, cannot be wholly solid since mercury and acids penetrate it, as do magnetic effluvia. Gold is composed of particles which are themselves composed of particles. If the proportion of each sort of particle is to its corresponding void as sand is to the space between its grains, that is, as seven to six, then the solid matter in gold would scarcely fill one-third its present volume if it were tightly compacted. Since water is nineteen times lighter than gold, only a fifty-seventh part of water can be solid matter. Air near the surface of the earth is nine hundred times lighter than water. Hence its particles can fill only one part in fifty thousand. The rarity of air can increase without limit. Whoever can devise a hypothesis to explain how water can be so rare and yet be incompressible, Newton concluded, will understand how gold can be equally rare and

[129] Leibniz to Newton, 7 March 1693; *Corres 3*, 257–8.
[130] *Add MS* 3965.6, ff. 267, 310–11.

how the rarity of gold, like that of air, can increase without limit. "We must have recourse to some wonderful and very skillfully contrived texture of particles whereby all bodies, on the pattern of nets, lie open and offer unrestricted passage in all directions both to magnetic effluvia and to rays of light."[131] Once again, the alchemical image of the net, transmuted back into its literal form as a web of tenuous threads, offered the model of Newton's mature conception of matter.

In his response to Leibniz's letter, Newton reasserted the same position as it applied to the cosmos. Leibniz had declared the need for some subtle matter, to carry the planets in their orbits. Newton would have none of it.

> For since celestial motions are more regular than if they arose from vortices and observe other laws, so much so that vortices contribute not to the regulation but to the disturbance of the motions of planets and comets; and since all phenomena of the heavens and of the sea follow precisely, so far as I am aware, from nothing but gravity acting in accordance with the laws described by me; and since nature is very simple, I have myself concluded that all other causes are to be rejected and that the heavens are to be stripped as far as may be of all matter, lest the motions of planets and comets be hindered or rendered irregular. But if, meanwhile, someone explains gravity along with all its laws by the action of some subtle matter, and shows that the motion of planets and comets will not be disturbed by this matter, I shall be far from objecting.[132]

Manifestly, he was farther yet from expecting to hear of such an explanation.

Fatio had said that Newton seemed to waver between accepting his mechanical explanation of gravity and attributing it to the will of God. Fatio appears to have exaggerated considerably the appeal of his theory to Newton. Among the new corollaries to Proposition VI that he planned was a direct assertion of God's causative role.

> Corol 9. There exists an infinite and omnipresent spirit in which matter is moved according to mathematical laws.[133]

131 The crucial phrase in Latin is *more retium* (*Add MS* 3965.6, f. 266ᵛ.) Cf. A. R. and Marie Boas Hall, "Newton's Theory of Matter," *Isis*, *51* (1960), 131–44; J. E. McGuire, "Atoms and the 'Analogy of Nature': Newton's Third Rule of Philosophizing," *Studies in History and Philosophy of Science*, *1* (1970), 3–58; "Body and Void in Newton's De Mundi Systemate: Some New Sources," *Archive for History of Exact Sciences*, *3* (1966), 206–48; "Force, Active Principles, and Newton's Invisible Realm," *Ambix*, *15* (1968), 154–208; and "Neoplatonism and Active Principles: Newton and the Corpus Hermeticum," in Robert S. Westman and J. E. McGuire, *Hermeticism and the Scientific Revolution* (Los Angeles, 1977). See further Ernan McMullin, *Matter and Activity in Newton* (Notre Dame, 1977).

132 Newton to Leibniz, 16 Oct. 1693; *Corres 3*, 287; original Latin, p. 286. Cf. Fatio to DeBeyrie for Leibniz, 30 March 1694; *Corres 3*, 308–9.   133 *Add MS* 3965.6, f. 266ᵛ.

Here was the concept expressed more than twenty years earlier in his essay "De gravitatione." Sometime during the early 1690s, Newton repeated the notion to Locke, who inserted an obscure reference to it, as a comprehensible concept of creation *ex nihilo*, in subsequent editions of his *Essay*. Years later, Newton explained the reference in Locke to Pierre Coste, the translator of the *Essay* into French.[134] He must have repeated the idea to others as well. David Gregory noted that "Mr C. Wren . . . smiles at Mr Newton's belief that it [gravity] does not occur by mechanical means, but was introduced originally by the Creator."[135]

Newton may not have known that Wren smiled at this belief, but he understood how radical his philosophy of nature was. Along with the drafted additions to the early propositions of Book III, he included an argument, drawn from his "Origins of Gentile Theology" and from Cudworth, which summoned up the support of an ancient tradition. Both Fatio and Gregory heard about this passage.[136] Newton took care to eliminate any hint of the radical theology that in his view had accompanied the philosophical tradition as one of the twin vehicles of truth.

> That all matter consists of atoms was a very ancient opinion [he asserted]. This was the teaching of the multitude of philosophers who preceded Aristotle, namely Epicurus, Democritus, Ecphantus, Empedocles, Zenocrates, Heraclides, Asclepiades, Diodorus, Metrodorus of Chios, Pythagoras, and previous to these Moschus the Phoenician whom Strabo declares older than the Trojan war. For I think that same opinion obtained in that mystic philosophy which flowed down to the Greeks from Egypt and Phoenicia, since atoms are sometimes found to be designated by the mystics as monads.[137]

The Epicureans made an absurd mistake in thinking that the distinction of nature into atoms and voids denies the existence of God.

> For two planets separated from each other by a long distance that is empty do not attract each other by any force of gravity or act on

---

[134] Locke, *Essai philosophique concernant l'entendement humaine*, tr. Pierre Coste, 3rd ed. (Amsterdam, 1735), p. 521. Cf. Alexandre Koyré, *Newtonian Studies* (Cambridge, Mass., 1965), p. 92.

[135] Gregory's memorandum of 20 Feb. 1698; *Corres 4*, 267; original Latin, p. 266.

[136] Fatio to Huygens, 15 Feb. 1692; Huygens, *Oeuvres*, *10*, 257. Gregory's memorandum of 5, 6, 7 May 1694; *Corres 3*, 338; original Latin, pp. 335–6. A few years later, his niece, Catherine Barton, heard the same theory, which she later repeated to her husband, John Conduitt: "Sʳ Isaac used to say he believed Pythagoras's Musick of the Spheres was gravity, & that as he makes the sounds & notes depend on the size of the strings, so gravity depends on the density of matter. C.C." (*Keynes MSS* 130.6, Book 3; 130.5, Sheet 1: 130.7, Sheet 3).

[137] *Add MS* 3965.6, f. 270. I use the translation of J. E. McGuire and P. M. Rattansi, "Newton and the 'Pipes of Pan'", *Notes and Records of the Royal Society*, 21 (1966), p. 115. See the entire article for the subject discussed here.

each other in any way except by the mediation of some active principle interceding between them by which the force is transmitted from
one to the other. And therefore those Ancients who rightly understood the mystical philosophy taught that a certain infinite spirit
pervades all space & contains and vivifies the universal world; and
this supreme spirit was their numen, according to the Poet cited by
the Apostle: In him we live and move and have our being. Hence the
omnipresent God is acknowledged and by the Jews is called Place.
To the mystical philosophers, however, Pan was that supreme numen. . . . By this Symbol the Philosophers taught that matter is
moved in that infinite spirit and is acted upon by it, not in an irregular way, but harmonically or according to the harmonic ratios as I
have just explained.[138]

Eventually Newton included some of this material in the Queries
attached to the Latin edition (and the second and subsequent English editions) of the *Opticks*.

The charge of plagiary, which Hooke continued to retail to any
willing consumer, furnished a minor aspect of Newton's concern
with the *Principia* at this time. When Gregory visited him in 1694,
he did not neglect to show him a manuscript written before 1669
"where all the foundations of his philosophy are laid: namely the
gravity of the Moon to the Earth, and of the planets to the Sun.
And in fact all these are even then subjected to calculation."[139]
Since the next entry in Gregory's notes took up the correspondence
with Hooke in 1679–80 and Hooke's claim, it is difficult to believe
that Newton came upon the early paper by chance while Gregory
was there. He probably gave Locke the copy of the demonstration
that elliptical orbits entail inverse-square forces for the same purpose. Locke's amanuensis dated the copy March 1690; in February,
Hooke had reasserted his claim with some bitterness in two lectures
delivered to the Royal Society.[140] Sometime later, presumably with
some purpose in mind, Newton passed on a copy of the same paper

---

[138] *Add MS* 3965.6, f. 269. See also the paragraphs on space, time, and God composed about
this time. (*Add MS* 3965.13, ff. 541–2, 545–6.) Cf. McGuire and Rattansi, "Newton and
the 'Pipes of Pan.'" On various philosophical issues in Newton's science and his enquiry
into them at this time, see also McGuire, "Atoms and the 'Analogy of Nature'," "Body
and Void in Newton's De Mundi Systemate," "Existence, Actuality and Necessity:
Newton on Space and Time," *Annals of Science*, *35* (1978), 463–508, "The Origin of
Newton's Doctrine of Essential Qualities," *Centaurus*, *12* (1968), 233–60, and "Transmutation and Immutability: Newton's Doctrine of Physical Qualities," *Ambix*, *14* (1967),
69–95; David Kubrin, "Newton and the Cyclical Cosmos: Providence and the Mechanical Philosophy," *Journal of the History of Ideas*, *28* (1967), 325–46; I. Bernard Cohen,
"Hypotheses in Newton's Philosophy," *Physis*, *8* (1966), 163–84; and Markus Fierz,
"Ueber den Ursprung und die Bedeutung der Lehre Isaac Newtons vom absoluten
Raum," *Gesnerus*, *11* (1954), 62–120.
[139] Gregory's memorandum of May 1694; *Corres 3*, 332; original Latin, p. 331.
[140] Hall, "Two Unpublished Lectures."

to William Whiston.[141] As far as Hooke himself was concerned, Newton expressed only contempt. William Derham felt obliged on one occasion to tell Newton that Hooke still repeated the charge of plagiary. "To wch he (wth greater warmth & peevishness than was usual in him) gave me this Answer, That *he believed Dr Hook could not perform y^t wch he pretended to: let him give Demonstrations of it: I know he hath not Geometry enough to do it.*"[142]

During the early 1690s, Newton also gave thought to putting his mathematical achievement into publishable form. As far as any evidence informs us, professorial duties ceased to play a role in his mathematical activity. By 1687, virtually every chair in Cambridge had become a sinecure. Newton deposited the first draft of the *Principia*'s final book as the text of lectures given in 1687. If in fact he gave them, they were among the few still delivered in the university, and Newton apparently accepted the prevailing mode thereafter. He held the chair and drew its income without further exercise until 1701. He did work at mathematics, however. Significantly, he devoted an important part of his effort to a restitution of Greek geometrical analysis, as if to demonstrate that his fluxional method, like his philosophy and his theology, rested on ancient exemplars instead of corrupted modern practices. "The resolved locus of the ancients," he told Gregory, "is a Treasury of Analysis."[143] In a series of papers composed at this time, Newton attempted to do in a different way what the whole of seventeenth-century analysis had tried to do, that is, "to restore the technique of the resolved locus in regard to the solution of problems."[144] In keeping with his revulsion from modern analysis, Newton directed his effort toward the restoration of the lost books of ancient analysis, such as Euclid's *Porismata*. Although he actively planned to publish his work in the early 1690s, nothing came of it in the end. He never ceased, however, to praise the ancient geometers. In 1705, when Halley published a Latin translation of a newly discovered Arabic version of a treatise by Apollonius, Newton declared that "this treatise is yet the more valuable in that it is a geometrical analysis."[145] Likewise he found Hugo de Omerique's *Analysis geo-*

---

[141] William Whiston, *Praelectiones astronomicae Cantabrigiae in scholis publicis habitae* (Cambridge, 1707), pp. 137–45.   [142] *Keynes MS 133*, p. 6.

[143] Gregory's memorandum of May 1694; *Corres 3*, 332; original Latin, p. 331. The line is given incorrectly in the *Correspondence*: cf. *Math 7*, 189.

[144] *Math 7*, 259. See *Analysis geometrica* (pp. 200–12), *In analysi veterum observandae sunt hae regulae* (pp. 212–16), *Problematum solutiones juxta sequentes regulas* (212–20). *Inventio porismatum* (pp. 230–46), *Geometriae libri tres. Proemium* (pp. 248–76), *Geometriae liber primus* (pp. 286–338), *Geometriae liber 1* (pp. 352–82), and *Geom. lib 2* (pp. 382–400).

[145] Reported in DeMoivre to Johann Bernoulli, March 1705; quoted in *Math 7*, 198.

metrica valuable. "For therin is laid a foundation for restoring the Analysis of the Ancients wch is more simple more ingenious & more fit for a Geometer then the Algebra of the Moderns."[146]

Nevertheless, he could not escape seventeenth-century mathematics. He had his own treasury of analysis to protect from poachers, as a letter from David Gregory reminded him in the autumn of 1691. Gregory was attempting to extricate himself from a ticklish position. In the summer of 1691, he had ingratiated himself enough to win Newton's support in his application for the Savilian chair of astronomy. He needed to expunge from his record a most incautious act performed in 1688. When John Craig had returned to Scotland with two quadratures copied from Newton's papers, Gregory had succeeded in deducing the binomial expansion that stood behind them. In 1688, he had allowed his friend Archibald Pitcairne to publish it as Gregory's discovery with never a mention of Newton. Further to enhance his credentials for the Oxford chair, Gregory now wanted to publish a general paper on quadratures. He also wanted to retain Newton's friendship. We may be sure that he gave careful thought to his line of approach. He asked that Newton allow him to publish the paper as a letter addressed to Newton, which he would submit for revision and approval. "and since, by what you have told me, I know that ye have such a series long agoe I entreat ye'l tell me so much of the historie of it as ye think fitt I should know and publish in this paper. Sir since I am resolved to doe in this according to your opinion, I hope ye will freely allow me it on the whole mater." When he heard nothing, he persisted a month later with a second request, in which he included the paper itself addressed to "the best of philosophers Isaac Newton . . ."[147]

Gregory's paper was a minor masterpiece in dissimulation, which tried to conceal his failure to mention the communication from Craig and tried to disguise his pretense of independent discovery by a flattering address to Newton. He got in return less than he wanted but more than he deserved. Newton's letter, drafted to accompany Gregory's, spelled out his own early discovery of the series and Craig's communication to Gregory. Although he acquiesced in Gregory's claim of independent discovery, Newton's account left Gregory with few scraps of credit to cover his dignity.[148] With such an accompaniment, he chose not to publish his paper in

---

[146] Bodleian Library, New College MS 361.2, f. 14^v. Like the previous one, I owe this reference to Whiteside (*Math 7*, 198). Cf. Newton to Grandi, 26 May 1704; *Corres 7*, 434.

[147] Gregory to Newton, 10 Oct. 1691; *Corres 3*, 170. Gregory to Newton, 7 Nov. 1691; *Corres 3*, 176–9; original Latin, pp. 172–6.

[148] Newton to Gregory, ca. Nov. 1691; *Corres 3*, 181–2. The letter, of which only an incomplete draft exists, may never have been sent. Gregory's subsequent use of his paper seems to indicate that it was.

the *Philosophical Transactions*, though he did not hesitate to send it to Huygens. Eventually, he published the paper in John Wallis's *Opera*, where again he did not mention the communication from Craig but took care to assert that Newton had known the series before he did.[149] It is evident from Newton's papers that Gregory succeeded in avoiding his wrath primarily because Leibniz acted as a lightning rod to draw the charge off in another direction.

Leibniz had begun to publish his differential calculus in the autumn of 1684 just as Newton began serious work on the *Principia*. In his first paper, he referred to methods of drawing tangents hitherto published which required the rationalizing of surds; every mathematician would have recognized the reference to the publication of René de Sluse in 1673. Neither in that nor in subsequent papers did Leibniz mention Newton. Leibniz can be excused. Newton had nothing on mathematics in print, and a reference to him would not have meant anything to most European mathematicians. Nevertheless, in view of the correspondence of 1676, one would not care to cite Leibniz's silence as a lesson in generosity. Nor would one call it judicious – as Leibniz must later have reminded himself ruefully many times. Although no previous allusions to Leibniz's publications appeared in Newton's papers, he seems to have nursed a growing sense of grievance. For as soon as he began to write the letter that Gregory requested in the fall of 1691, he forgot Gregory, returned forthwith to the correspondence of 1676, and began to compose a defense of his priority vis-à-vis Leibniz. Fatio had come back to England from the Netherlands, and Newton had spent a week with him in London shortly before Gregory's first letter. We do not know what Newton may have heard from Fatio, who did not love Leibniz. Before the year was out, Fatio sent Huygens a strong defense of Newton's priority with a scarcely veiled accusation of plagiary by Leibniz. We cannot be sure that Fatio did not bring stories from the Netherlands which aroused Newton, who had hitherto said nothing on the issue. Newton's papers do not suggest that Fatio played the provocateur, however. In 1684, before Leibniz's publication, Gregory's *Exercitatio* had reminded Newton of Leibniz. Now Gregory reminded him again, and now he had a grievance as well. Fatio was probably expressing what he heard from Newton instead of instigating it.

Suffice it to say that the letter drafted for Gregory rapidly converted itself into a full-scale exposition of Newton's fluxional method, "De quadratura curvarum" (On the Quadrature of Curves), which began with a recitation of his exchange with Leibniz in 1676 and included the translation of the fluxional anagram in

---

[149] Gregory to Huygens, 12 Aug. 1693; *Corres 3*, 275–6. Wallis, *Opera*, 2,337–80.

the *Epistola posterior*.[150] In passing, the treatise defended Newton against Gregory, partly by mentioning the communication to Craig, but far more by a breathtaking extension of quadratures that made Gregory's paper look like an elementary exercise. Where Gregory treated the expansion of binomials into series, Newton gave a general method by which to transform trinomials into binomials that could be squared. Using the proposition that the areas under two curves, whose ordinates vary reciprocally as the fluxions of their abscissas, are equal to each other, he developed rules by which, from any trinomial, to find the simplest binomial to which its area is equal. Not content with trinomials, he proceeded to the quadrature of polynomials by series expansion.[151] When Gregory finally saw the treatise (or perhaps only the later, truncated version of it) in 1694, he was stunned; "he [Newton] develops that matter astonishingly," he recorded in a memorandum, "and beyond what can readily be believed . . . "[152] So much for Gregory's pretensions. Leibniz was another matter. To him the treatise primarily addressed itself. It began by quoting a passage from the *Epistola posterior* in which Newton asserted that his method of drawing tangents did not stick at equations involving surds – an unmistakable reference to Sluse, the one predecessor to whom Leibniz had alluded in his published paper. It included two propositions on the extraction of fluents from fluxional equations, what Leibniz called the inverse problem of tangents and we call the solution of differential equations. It went into an extended set of problems that the fluxional method could solve, problems like those to which Leibniz was addressing his calculus.[153] As though in conscious competition, it developed for the first time on Newton's part a systematic notation as an alternative to Leibniz's. It was "De quadratura" that adopted the familiar dot notation for fluxions, and it experimented with Q (for *quadratura*) as a substitute for Leibniz's ∫ (for *summa*) as a symbol for the operation of squaring. Let Newton rail against modern analysis as he might, he was still the titan of modern analysts. In the proposition on the extraction of fluents by infinite series, almost in passing, he set down the first Taylor expansion of a function.[154] Like much else in the treatise, this conclusion was rediscovered twenty years later, in this case by Brook Taylor, whose name it bears. Whiteside urges, with primary reference to "De quadratura," that Newton was able "to duplicate the combined expertise and output of his contemporaries in the field of calculus."[155]

[150] Two successive versions of the treatise survive; *Math 7*, 24–48 and 48–128.
[151] *Math 7*, 56–62, 68–70.
[152] Gregory's memorandum of 5, 6, 7 May 1694; *Corres 3*, 338; original Latin, p. 336.
[153] Propositions XI and XII; *Math 7*, 70–98. *Math 7*, 100–22.
[154] *Math 7*, 98.   [155] *Math 7*, 20.

The treatise contained one personal entry. After his solution of a fluxional equation, Newton commented, "This rule, or one like it, was communicated to me some while ago by Mr. Fatio."[156] One will look in vain for a similar passage elsewhere in Newton's mathematical papers.

By the end of 1691, Newton's circle of young friends in London had become aware of his treatise. On 28 December, Gregory wrote down various bits of information that he had picked up from Fatio, Halley, Paget, and others. "Mr Newton to publish two parts of Geometry," he noted, "the first the Geometry of the Ancients, the second of Quadratures." Sometime early in January, Fatio repeated a similar message to Constantyn Huygens, who immediately forwarded it to his brother. Much later, Joseph Raphson claimed that he and Halley had both seen the treatise in 1691 and had found it already worn by being handled. He may well have confused the year, however, for Halley wrote of having it to copy in 1695.[157] At any rate, Newton's interest waned as rapidly as it had waxed. By March, Fatio reported to Huygens that Newton's enthusiasm had passed and that he had begun to think he might better avoid the embarrassments that its publication was bound to cause. "We shall assuredly lose a great deal if this Treatise does not appear," he added. "It is certain that until now nothing in abstract Geometry as beautiful as this writing has ever appeared . . . " As late as 1694, Gregory found him toying still with the notion of publishing "De quadratura," this time in conjunction with Sections IV and V of the *Principia*.[158] Not only did he not publish it; he did not complete it, although in the mid-1690s he did produce a severely truncated version that confined itself to quadratures and eliminated both the references to his correspondence with Leibniz and such matters as the solutions of fluxional equations. Eventually he published this version as an appendix to his *Opticks*.

His friends in London had also picked up the priority issue. On 18 December 1691, Fatio stated the issue to Huygens bluntly.

> It seems to me from everything that I have been able to see so far, among which I include papers written many years ago, that Mr. Newton is beyond question the first Author of the differential calculus and that he knew it as well or better than Mr. Leibniz yet knows it before the latter had even the idea of it, which idea itself came to him, it seems, only on the occasion of what Mr. Newton wrote to

[156] *Math* 7, 79.

[157] Gregory's memorandum of 28 Dec. 1691; *Corres 3*, 191. Constantyn Huygens to Huygens, 26 Jan. 1692; Huygens, *Oeuvres, 10*, 236. Raphson, *The History of Fluxions* (London, 1715), pp. 2–3. Halley to Newton, 7 Sept. 1695; *Corres 4*, 165.

[158] Huygens, *Oeuvres, 10*, 271. Gregory's memoranda of 4 May and 5, 6, 7 May 1694; *Corres 3*, 318, 338; original Latin, pp. 313, 336.

him on the subject. (Please Sir look at page 235 [Lemma II, Book II]
of Mr. Newton's book). Furthermore, I cannot be sufficiently surprised that Mr. Leibniz indicates nothing about this in the Leipsig
Acta [his papers that published the differential calculus].[159]
The note in Gregory's memorandum of 28 December—"the Calculus differentialis first hinted at by Mr Newton, without it geometry
not improveable."—probably derived from a conversation with Fatio.[160] By February, Fatio was more explicit.

> The letters that Mr. Newton wrote to Mr. Leibniz 15 or 16 years ago
> speak much more positively than the place that I cited to you from
> the Principles which nevertheless is clear enough especially when the
> letters explicate it. I have no doubt that they would do some injury
> to Mr. Leibniz if they were printed, since it was only a considerable
> time after them that he gave the Rules of his Differential Calculus to
> the Public, and that without rendering to Mr. Newton the justice he
> owed him. And the way in which he presented it is so far removed
> from what Mr. Newton has on the subject that in comparing these
> things I cannot prevent myself from feeling very strongly that their
> difference is like that of a perfected original and a botched and very
> imperfect copy. It is true Sir as you have guessed that Mr. Newton
> has everything that Mr. Leibniz seemed to have and everything that I
> myself had and that Leibniz did not have. But he has also gone
> infinitely farther than we have, both in regard to quadratures, and in
> regard to the property of the curve when one must find it from the
> property of the tangent.[161]

Although Newton decided to avoid the unpleasantries that he
saw his treatise would entail, the priority issue did not wholly
disappear. In the summer, John Wallis opened the pages of his
forthcoming *Opera* to whatever Newton might wish to insert. Wallis's letter is lost; we know what he said only from Newton's replies, one of which is also lost. Wallis probably brought up Gregory, whose method of quadrature was to appear finally in the first
of Wallis's volumes. The précis of "De quadratura" that Newton
sent him defended his priority in that direction as the treatise itself
had done. Before long, however, Wallis began to pester Newton
without interruption about Leibniz. He reported then that he had
heard from Holland that "your Notions (of *Fluxions*) pass there
with great applause, by the name of *Leibnitz's Calculus
Differentialis.*"[162] Hence it is unlikely that he failed to mention Leibniz in 1692. Even if Wallis did, Newton did not forget him. Like
"De quadratura," the précis that he sent to Wallis addressed itself

[159] Huygens, *Oeuvres*, *10*, 214.   [160] *Corres 3*, 191.
[161] Fatio to Huygens, 15 Feb. 1692; Huygens, *Oeuvres*, *10*, 257–8.
[162] Wallis to Newton, 10 April 1695; *Corres 4*, 100.

primarily to Leibniz. It began with the same reference to Sluse in the *Epistola posterior*. It translated both of the anagrams in that letter. And in presenting briefly Newton's solutions to fluxional equations in connection with the second anagram, it pretended that already in 1676 Newton had developed methods that did not in fact appear in his papers before the 1690s. Newton also asked Wallis to insert his series for the circle, as given in the letter of 1676, beside Leibniz's series for the circle, which was also to appear in the volume.[163] Manifestly Newton still had the matter in mind.

So did Leibniz. Huygens had passed on some of Fatio's assessments of Newton's achievements, though not his comments on Newton's priority, and mathematicians all over Europe heard of his plans to publish. When the plans boiled down to an exposition of his method in Wallis's *Opera*, they waited with anticipation for that, Leibniz among them and for his own reasons. In March 1693, he wrote to Newton – a gracious letter seeking, as he had sought nearly twenty years before, to initiate a philosophical correspondence, but also a nervous letter.

> How great I think the debt owed to you, by our knowledge of mathematics and of all nature, I have acknowledged even in public when occasion offered [he began with something less than complete candor]. You had given an astonishing development to geometry by your series; but when you published your work, the *Principia*, you showed that even what is not subject to the received analysis is an open book to you.

He went on to refer to his own work in mathematics. "But to put the last touches I am still looking for something big from you . . ." From mathematics, he proceeded to the *Principia* and to what he had heard from Huygens about Newton's optics. In conclusion, he acknowledged his long silence, though he might well have complained instead of Newton's, and excused himself by saying he had not wanted to burden Newton with letters.[164]

Personal problems postponed Newton's answer until October, and he opened with an apology for the delay. "For although I do my best to avoid philosophical and mathematical correspondences, I was nevertheless afraid that our friendship might be diminished by silence . . . " He feared it all the more, he continued, since Wallis had just inserted some new points from their earlier correspondence in his forthcoming work. At Wallis's request, he was revealing the method concealed before in the anagram, the translation of which he included in the letter, a limited favor since he had already pub-

---

[163] Wallis, *Opera*, 2, 390–6 (vol. 2 (1693) appeared before vol. 1 (1695) of the *Opera*). *Math* 7, 170–80. English translation in *Corres 3*, 222–8. Cf. Newton to Wallis, 27 August 1692; *Corres 3*, 219.

[164] *Corres 3*, 258–9; original Latin, pp. 257–8. I have changed one word in the translation.

lished it in the *Principia*. "I hope indeed that I have written nothing to displease you, and if there is anything that you think deserves censure, please let me know of it by letter, since I value friends more highly than mathematical discoveries." Although he made a brief response to Leibniz's other comments, the letter was the thinnest answer possible, and Leibniz made no attempt to continue correspondence.[165]

The veiled threat in the coming publication could not be missed. In June 1694, Leibniz had still not seen Wallis's volume, and he wrote impatiently to Huygens to send it as soon as possible.[166] When he finally received it in September, he expressed his disappointment that it contained so little on the inverse problem of tangents, but the disappointment sounded more like relief. The two methods were similar, he remarked, but his own was clearer. All Newton presented on the inverse problem of tangents was a means of expressing a given ordinate by an infinite series, which he had understood at the time – that is, in 1676.[167] In a word, Leibniz's attention also focused on the priority issue, and the publication did not seem to undercut him as he had feared. Johann Bernoulli read it in the same light. Writing to Leibniz, he questioned whether Newton had not in fact plundered Leibniz's publications to fashion the method which he only now presented.[168] If the potential dispute which flickered and nearly burst into flame seemed to die down, it was far from dead.

Nor was it in England. Among the memoranda that Gregory made of his conversations with Newton in the summer of 1694, he recorded an ominous question: "Whence the differential calculus of Leibniz."[169] And in the fall, he sketched out a treatise on the calculus that he intended to write in which he would show that Leibniz's calculus reduced to Newton's, which alone had been fully demonstrated.[170] Wallis also was unwilling to let the issue die. Disappointed with what he had been able to publish in the first two volumes of his *Opera*, he began to press Newton in 1695 for permission to publish the two *Epistolae* in full. In May, he sent Newton the copies of them that he had had since the 1670s and asked Newton to correct them for publication; when Newton did not reply, he wrote again in July.[171] This time Newton complied. He also thanked Wallis "for your kind concern of right being done me

---

[165] *Corres 3*, 286–7; original Latin, pp. 285–6. I have made a small change in the translation. Leibniz did try to get information indirectly (cf. Leibniz to DeBeyrie, n.d.; Huygens, *Oeuvres*, *10*, 605–6, and Fatio to DeBeyrie for Leibniz, 30 March 1694; *ibid.*, *10*, 606–8).

[166] Huygens, *Oeuvres*, *10*, 646.   [167] *Ibid.*, p. 675.

[168] Bernoulli to Leibniz, 15 Aug. 1696; quoted in *Math 7*, 181.

[169] A note on Gregory's memorandum of ca. July 1694, which he crossed out; *Corres 3*, 387.

[170] Printed in *Corres 4*, 15–16.   [171] *Corres 4*, 129–30 and 139–40.

by publishing them." To indicate that he was also concerned, he quoted a passage from Collins's letter of 18 June 1673 to insert as a note to the reference to Sluse. If the *Epistola posterior* were not enough, the note should do the job. The letter it quoted, dated before the publication of Sluse's paper, asserted Newton's possession of the method at that time.[172] "I would willingly see Leibnitz's answer," Wallis remarked to Halley.[173]

During the autumn of 1695, while Wallis set the stage for later conflicts, Newton undertook what proved to be his last creative effort in mathematics. Taking up his enumeration of cubics, originally composed over twenty-five years earlier and dabbled with intermittently since, he put it into the final form which he later published in the *Opticks*.[174] About this time he also revised and reduced "De quadratura" to the form in which he published it in the same volume.

Along with the *Principia* and mathematics, Newton returned to his optics during the late 1680s and early 1690s after an interval of nearly two decades. A reference in passing in the published work, that he made two reflecting telescopes about sixteen years before, suggests that he turned to the *Opticks* in 1687 as soon as he completed the *Principia*.[175] Huygens made optics the primary topic of his discourse to the Royal Society in June 1689 which Newton attended, and the following year, in commenting to Leibniz on their conversation, Huygens said that Newton had promised something on the subject.[176] Newton wrote one of the manuscripts related to the promised volume over a letter addressed to him at the House of Commons, and in August 1691, he asked Flamsteed whether the satellites of Jupiter appeared colored immediately before and after eclipses.[177] The question to Flamsteed concerned the possibility of differing velocities attached to separate colors. Apparently he discussed the same notion with Gregory in London at the time he wrote to Flamsteed, for in November Gregory sent along information on the satellites of Jupiter that he had received from Cassini.[178]

Newton's papers do not reveal important optical experimentation at this time, though he did extend his investigation of diffraction.[179]

---

[172] *Corres 4*, 140–1. Wallis did insert it at the point Newton indicated; *Opera 3*, 636.
[173] Wallis to Halley, 26 Nov. 1695; *Corres 4*, 188.
[174] *Math 7*, 588–644. See also his preliminary sketch of a classification, pp. 580–6.
[175] *Opticks*, p. 103.
[176] Huygens to Leibniz, 24 Aug. 1690; Huygens, *Oeuvres*, *9*, 471.
[177] *Add MS* 3970.3 ff. 348–9. *Corres 3*, 164.
[178] *Corres 3*, 171. (cf. Gregory to Newton, 27 Aug. 1691; *Corres 3*, 165).
[179] *Add MS* 3970.10, ff. 645^v–6^v.

Most of his effort went into ordering earlier essays into a complete whole. He completely rewrote the relevant parts of the "Lectiones" for Book I. Most of Book II simply embodied verbatim the paper on thin films, sent to the Royal Society in 1675, and added to it the new section on thick films. Apparently he intended initially to put the work into Latin; though he must have given that up rather quickly since not very much of the text survives.[180]

The *Opticks* that Newton first laid out differed from the one he ultimately published. He planned four books instead of three; and in Book IV, he intended to marshal the evidence of optical phenomena in support of the concept of forces that act at a distance. Thus it would mesh harmoniously with the *Principia*, and together the two would propound the new philosophy of nature. He planned to start with a set of propositions similar but not identical to the sixteen queries that he ultimately attached to his *Opticks*. A "Conclusion" was to contain a number of "Hypotheses" based on the propositions.

> Hypoth. 1. The particles of bodies have certain spheres of activity wthin wch they attract or shun one another. . . .
>
> Hypoth 2 As all the great motions in the world depend upon a certain kind of force (wch in this earth we call gravity) whereby great bodies attract one another at great distances: so all the little motions in ye world depend upon certain kinds of forces whereby minute bodies attract or dispell one another at little distances.

How the great bodies of ye Earth Sun Moon & Planets gravitate towards one another what are ye laws & quantities of their gravitating forces at all distance from them & how all ye motions of those bodies are regulated by their gravities I shewed in my Mathematical Principles of Philosophy to the satisfaction of my readers. And if Nature be most simple & fully consonant to her self she observes the same method in regulating the motions of smaller bodies wch she doth in regulating those of the greater. This principle of nature being very remote from the conceptions of Philosophers I forbore to describe it in that Book leas[t it] should be accounted an extravagant freak & so prejudice my Readers against all those things wch were ye main designe of the Book: & yet I hinted [at them] both in the Preface & in ye book it self where I speak of the [refraction] of light & of ye elastick power of ye Air: but [now] the design of y^t book being secured by the approbation of Mathematicians, [I have] not

---

[180] *Add MSS* 3970.3, ff. 302, 394–426; 3970.4, ff. 583–4; 3970.10, ff. 647–8. For some of his optical work at this time, see Zev Bechler, "Newton's Search for a Mechanistic Model of Colour Dispersion: A Suggested Interpretation," *Archive for History of Exact Sciences*, 11 (1973), 1–37.

scrupled to propose this Principle in plane words. The truth of this Hypothesis I assert not because I cannot prove it, but I think it very probable because a great part of the phaenomena of nature do easily flow from it wch seem otherways inexplicable: such as are chymical solutions precipitations philtrations, detonizations, volatizations, fixations, rarefactions, condensations, unions, separations, fermentations: the cohesion texture firmness fluidity & porosity of bodies, the rarity & elasticity of air, the reflexion & refraction of light, the rarity of air in glass pipes & ascention of water therein, the permiscibility of some bodies and impermiscibility of others, the conception & lastingnesse of heat, the emission & extinction of light, the generation & destruction of air, the nature of fire & flame, ye springinesse or elasticity of hard bodies.[181]

In the 1670s, Newton had called upon the aether and its varying densities to explain such optical phenomena as refraction. Now, as the *Principia* had already shown, with the aether's sad demise, he referred them to forces of attraction. Thus Proposition 1 of Book IV asserted that the refractive power of bodies is proportional to their specific gravities. To this he added a condition reminiscent of his alchemical studies – as indeed the list of natural phenomena in Hypothesis 2 was.

Note that sulphureous bodies caeteris paribus are most strongly refractive & therefore tis probable y^t ye refracting power lies in ye sulphur & is proportional not to ye specific weight or density of ye whole body but to that of ye sulphur alone. For ♁^{s} do most easily conceive ye motions of heat & flame from light & ye action between light & bodies is mutual.[182]

Not everything went as smoothly as refraction, however. Part of Newton's enduring legacy to optics was the demonstration of periodicity in the phenomena of thin films. With an aether in hand, he had related periodicity to vibrations in it. When the aether departed, he found himself left with the phenomena and bereft of any vehicle capable of periodic vibrations.

He tried two expedients, both of which found their way into the *Opticks*. One assigned the vibrations to the transparent medium itself, however unsuitable for such vibrations it may have appeared.[183] The other device was a good deal more subtle. Every ray in its passage through a refracting surface, he said, "is put into a

---

[181] *Add MS* 3970.3, ff. 336, 338–8^v. The brackets fill in words lost by damage to the paper; cf. similar materials on ff. 339–40, 342, 348, 377. Cf. also a folded sheet in a hand I have not been able to identify, headed "De Lumine et Calore," and containing what purport to be the properties of light and heat as Newton recounted them (*Add MS* 3970.7, ff. 594–5).
[182] *Add MS* 3970.3, f. 337. Cf. f. 335 and *Opticks*, pp. 275–6.
[183] *Add MS* 3970.3, ff. 335^v, 342, 348. *Opticks*, pp. 280–1.

certain transient constitution or state wch in the progress of the ray returns at equal intervals & disposes the ray at every return to be easily transmitted through ye next refracting surface & between the returns to be easily reflected by it." The alternate reflection and transmission depends on both surfaces of a film, Newton continued, because it depends on their distance. "It is influenced by some action or disposition prop[ag]ated from the first to ye second because otherwise at the second it would not depend upon the first. . . . But what kind of action or disposition this is, whether it consist in a. circulating or a vibrating motion of the ray or of the medium or something else, I do not here enquire."[184] In the *Opticks* he used the phrases, fits of easy transmission and fits of easy reflection, perplexing phrases born of his own perplexity which have puzzled historians ever since. Newton's problem was simplicity itself, however. He had demonstrated the periodicity of certain phenomena too convincingly to forget it, but he had also demonstrated to his own satisfaction that the only medium to which he could assign periodic vibrations with equal conviction does not exist.

By 1694, Newton had abandoned the initial plan for the *Opticks* and recast it in the form we know. When Gregory visited him in May, he saw " 'Three Books of Opticks': if it were printed it would rival the *Principia Mathematica*. . . . He sets forth unheard-of wonders about colours." Newton told Gregory that he intended to publish the work within five years of retiring from the university; if he published it while still there, he would translate it into Latin. Probably as a result of Gregory's information, the Royal Society resolved to write Newton urging him to communicate the treatise to them for publication.[185] Apparently he did not even bother to reply.

A year later, John Caswell told Wallis about the same treatise, which he had also seen recently in Cambridge. Wallis added optics to his list of grievances against Newton.

> I understand (from Mr Caswell) you have finished a Treatise about Light, Refraction, & Colours; which I should be glad to see abroad. 'Tis pitty it was not out long since. If it be in English (as I hear it is) let it, however, come out as it is; & let those who desire to read it, learn English."[186]

When Newton replied that he had not published the *Opticks* lest it cause him trouble, Wallis returned to the attack with still more vigor.

> You say, you dare not *yet* publish it. And why *not yet*? Or, if not now, when then? You adde, least it create you *some trouble*. What

---

[184] *Add MS* 3970.3, ff. 339–40.

[185] Gregory's memorandum of 5, 6, 7 May 1694; *Corres 3*, 338–9; original Latin, p. 336. *Corres 3*, 340.

[186] Wallis to Newton, 10 April 1695; *Corres 4*, 100.

trouble *now*, more then at another time? . . . And perhaps some other may get some scraps of ye notion, & publish it as his own; & then 'twil be His, not yours; though he may perhaps never attain to ye tenth part of what you be allready master of. Consider, that 'tis now about Thirty years since you were master of those notions about *Fluxions* and *Infinite Series* . . . But if I had published the same or like notions, without naming you; & the world possessed of anothers *Calculus differentialis*, instead of your *fluxions*: How should this, or the next Age, know of your share therein? . . . I own that Modesty is a Vertue; but too much Diffidence (especially as the world now goes) is a Fault.[187]

He even wrote to the Royal Society on the same day urging that perhaps Halley could convince Newton to publish.[188]

Newton proved to be adamant. In the early 1690s, he appears to have had every intention to put the work beside the *Principia* as part of his philosophic legacy. Perhaps Hooke's continuing clamor about the *Principia* changed his mind. In 1693, he told Leibniz that he refrained from publishing books "for fear that disputes and controversies may be raised against me by ignoramuses."[189] In fact, he did hold the manuscript until the particular ignoramus in question had died.

As he drew his philosophic legacy together, Newton did not omit alchemy. On the contrary, a good half of his extensive alchemical papers, which were interrupted for two or three years by the *Principia*, came from the period immediately following it. While he was in London for the Ecclesiastical Commission in the spring of 1687, he purchased chemicals, including antimony, sublimate, sal ammoniac, and double aqua fortis, from Mr. Stonestreet at the sign of the Queen's Head by Bow Church near St. Paul's. Six years later, he purchased quicksilver, sal ammoniac, and single aqua fortis from Mr. Langley, who had succeeded Mr. Stonestreet. He also bought antimony from a Mr. Box.[190] If we can judge by the quantity of manuscripts, Newton devoted more of his time to alchemy in the early 1690s than he did to everything else put together.

He continued to read at length in the literature, both modern and traditional. Thus, on the one hand, he went back systematically through all the writings of Michael Maier taking extensive notes.[191] On the other, he returned also to the treatises of the mythical

---

[187] Wallis to Newton, 30 April 1695; *Corres 4*, 116–7.
[188] Wallis to Waller, 30 April 1695; *Corres 4*, 115.
[189] Newton to Leibniz, 16 Oct. 1693; *Corres 3*, 287; original Latin, p. 286.
[190] *Add MS* 3975, p. 270 and inside of back cover.
[191] *Keynes MS* 32; it is 88 pages long and written in a fairly small hand. Cf. *Babson MS* 417 for similar systematic notes (not so extensive) on the works of Pierre Jean Fabre.

Hermes and Maria the Jewess.[192] Continuing an enterprise begun earlier, he concerned himself extensively with synthesizing the tradition in a coherent statement of the Art. To this end he gathered two collections of "Notable Opinions."[193] A number of papers attempted to formulate the Work as the various authorities testified to it.[194] By now, Newton had immersed himself so deeply in alchemical literature that he could bring the information of one to the criticism of another. In expounding Ripley, he listed three different proportions of materials to use in calcination. "NB," he added, "Philaletha in S.R. [*Secrets Reveal'd*] makes these three proportions belong to 3 works, but since Ripl. bids take wch you will of ye two first, those must belong to ye same."[195] Much the most extensive of these synthetic works was a treatise, apparently intended to contain nine "Works" drawn from the writings of the masters, for part of which he left as many as seven drafts.[196] Habits of mathematical exposition crept into alchemical writing; having expounded the second work once, he set out to give an alternative rendition and headed the new one "Idem aliter."[197]

Closely connected with these synthetic expositions was the extraordinary Index chemicus, which Newton put into the final form in which it survives in the early 1690s. What had begun as his own index to the literature became a general guide to the Art with introductory essays (which he drafted and corrected in his characteristic style) to the more important entries. If we can judge by the manuscript, Newton never completed the Index chemicus. He appears to have continued to add to it as long as his concern with alchemy remained active. In the final form it attained, the Index filled more than a hundred pages with 879 separate entries. It cited more than 150 separate alchemical works stretching from the mythical alchemists of the ancient world to contemporaries such as Dickinson and Mundanus, and gave about 5,000 separate page references to them. It

---

192 "The seven Chapters"; *Keynes MS* 27. English and Latin versions of the *Tabula smaragdina* together with Newton's *Commentarium* on it; *Keynes MS* 28. *Practica Mariae Prophetissae in Artem Alchemicum*; *Keynes MS* 45.

193 *Notanda chemica* and *Sententiae notabiles*: *Keynes MS* 38. The MS is published in F. Sherwood Taylor, "An Alchemical Work of Sir Isaac Newton," *Ambix*, 5 (1953–6), 59–84. *Sententiae luciferae et Conclusiones notabiles*; *Keynes MS* 56. *Keynes MS* 57, which is untitled, is similar.

194 "The three fires"; *Keynes MS* 46. *Keynes MS* 48, especially the section with the title "Decoctio," ff. 26–52, with a preliminary draft, ff. 53–60. Two manuscripts entitled "The Regimen" and one, "Of ye Regimen"; *Burndy MS* 15. "Of ye Regimen" is written over the remains of a receipt dated 11 Sept. 1689. The writing of the other "Regimen" MSS is very similar.   195 *Keynes MS* 54, f. 2.

196 Drafts of the various chapters, one of which exists in seven drafts, one in five, one in three, and four others in two, are found in *Keynes MSS* 40 and 41, *Babson MS* 417, and *Burndy MS* 17. Cf. a fuller discussion in Westfall, "Isaac Newton's Index Chemicus," *Ambix*, 22 (1975), 174–85.   197 *Keynes MS* 41, f. 3.

ransacked the great collections, the huge six-volume *Theatrum chemicum,* the *Ars aurifera,* and the *Musaeum hermeticum.*[198] No similar index to the vast literature of alchemy has ever been compiled. Its composition, beginning in the early 1680s, absorbed a large fraction of Newton's time for about a decade. From the point of view of his alchemical studies, the *Principia* was an intrusion. As I have indicated, Newton interrupted its composition in the spring of 1686 in order to carry out alchemical experiments.

Newton maintained his contacts with the clandestine alchemical network during this period. Among his papers are an anonymous treatise, "Sendivogius Explained," and three transcripts of an otherwise unknown work by an English spagyrical writer of Dutch extraction, William Y-Worth.[199] Early in March 1696, scarcely two weeks before Montague's letter appointing him warden of the Mint, Newton received a visit from "A Londoner acquainted wth M[r] Boyle & D[r] Dickinson," who discoursed with him for two days on the Work according to Jodocus a Rhe, an early seventeenth-century alchemist whom Newton had studied. Every indication in Newton's account of the visit suggests that the man knew exactly for whom he was looking when he came to Cambridge. As for Newton, he composed two drafts of a memorandum recording their conversation.[200]

Newton also continued to experiment. Considering the crisis in the university and the Ecclesiastical Commission, it is unlikely that he found time to tend his furnaces in the spring of 1687, but it is equally unlikely that some of the undated experimental notes that continued the record of 1686 with a barely perceptible change of ink did not contribute to Humphrey's recollections during the year and a half, of which we know so little, following the spring of 1687. The year 1689 would have presented little chance for experimentation, but he dated notes on 5 March 1690, a bare month after his return from London (and less than a week before his sudden decision to return to the metropolis to see Fatio again). Other dated records belonged to December 1692, January 1693, June 1693, April 1695, and February 1696. The final series, performed less than two

---

[198] *Keynes MS* 30. See the breakdown of citations in Westfall, "Newton's Index Chemicus."
[199] "Sendivogius Explained"; *Keynes MS* 55. There are two copies of this treatise in unknown hands in the British Library: Sloane MSS 3630 and 3778. William Y-Worth, "Processus mysterii magni philosophicus or An open Entrance to y[e] great Mysteries of y[e] Ancient Philosophies," in an unknown hand; *Keynes MS* 65. Newton had two other differing transcripts of this work, neither one in his own hand; Lots 116 and 117 of the Sotheby *Catalogue.* The catalogue dates them to 1701 and 1702, hence this work may have stemmed from a later time.
[200] *Keynes MS* 26 (published in *Corres 4,* 196–8). Joseph Halle Schaffner Collection, University of Chicago Library, MS 1075–3. See D. Geoghegan, "Some Indications of Newton's Attitude toward Alchemy," *Ambix, 6* (1957), 102–6.

months before he set out for London and the Mint, might well have given the Lords Commissioners of the Treasury pause had they been informed.[201]

We know no more of the rationale behind these experiments than we do for those of the 1680s. Newton continued to use the same materials; the net, the oak, the sophic sal ammoniac were common ingredients. The notes retained their quantitative character. With his usual capacity to extract results of lasting value from unexpected sources, Newton devised a thermometric scale, which he later published in the *Philosophical Transactions,* from his laboratory work.[202] On his visit in 1694, Gregory described Newton's method of measuring temperatures. He used a thick glass bulb about an inch and a half or two inches in diameter filled with linseed or olive oil, which he plunged into dry earth heated by the sun, boiling water, and even directly into the fire. The tube was not stopped except for a wad of cotton to contain the oil. As Gregory noted, he described degrees of heat in relation to what was required to melt various substances including metals.[203] In the paper itself, Newton used the melting point of a number of alloys of tin, lead, and bismuth, which appeared also in his experimental notes, to establish points on the scale. With its primary attention focused thus on higher temperatures, Newton's scale did not offer a practical device to thermometry. It did help to establish the concept of a thermometric scale, from which more practical alternatives emerged in the early years of the eighteenth century.

From his alchemy, Newton drew a paper, "De natura acidorum" (On the Nature of Acids), which he allowed Archibald Pitcairne, Gregory's friend, to have when he visited Cambridge early in March 1692. Pitcairne also took extensive notes on their conversations, which supplement the short essay. The particular value of "De natura acidorum" lies in the glimpses it affords of Newton translating the activities of alchemical principles into the vocabulary of forces. "The particles of acids," he asserted, " . . . are endowed with a great attractive force and in this force their activity consists by which they dissolve bodies and affect and stimulate the organs of the senses." "Water has no great power to dissolve because it rejoices in little acid," Pitcairne added in a note: "For what attracts and is attracted strongly, we call acid." Newton went on to explain that sal alkali is composed of acid and earthy particles so combined that the acid prevails.

[201] *Add MS* 3973, ff. 21–44. *Add MS* 3975, ff. 154 (where a change of ink seems to mark the end of the experiments in 1686)–158, 267–83.

[202] "*Scala graduum caloris,*" *Philosophical Transactions, 22* (1700–1), 824–9.

[203] Gregory's memoranda from the visit in May 1694; *Corres 3,* 318, 346; original Latin, pp. 313, 344. Gregory's *Notae principiorum,* quoted in *Corres 3,* 322.

If the acid particles are joined with the earthy ones in a lesser proportion they are so closely held by the latter that they are, as it were, suppressed and hidden by them. . . . But the acid, suppressed in sulphureous bodies, by attracting the particles of other bodies more strongly than its own, causes a gentle and natural fermentation and promotes it even to the stage of putrefaction in the compound. This putrefaction arises from this, that the acid particles which have for some time kept up the fermentation do at length insinuate themselves into the minutest interstices, even those which lie between the parts of the first composition, and so, uniting closely with those particles, give rise to a new mixture which cannot be done away with or changed back into its earlier form.

The comment on sulfur recalls Newton's conviction that sulfur is the seat of the attractive power in bodies that inflects rays of light. Indeed, he appears to have endowed sulfur and its attractive force with many of the functions ascribed to the universal active agent by spagyrical writers. Pitcairne noted that Newton believed the lungs take sulfur from the air to support life and sulfur analogously sustains fire. To Newton's allusion to his retiform conception of matter, Pitcairne added a further note.

Gold has particles which are mutually in contact: their sums are to be called sums of the first composition and their sums of sums, of the second composition, and so on. Mercury can pass, and so can Aqua Regia, through the pores that lie between particles of last order, but not others. If a menstruum could pass through those others or if parts of gold of the first and second composition could be separated, it would be liquid gold. If gold could ferment, it could be transformed into any other substance.[204]

In one of his letters to Richard Bentley, Newton mentioned his belief in the universal presence of fermentations and their role in nature. Bentley asked Newton if the heat of the sun might cause the planets to differ in density by concocting and condensing their matter in proportion to their distance. Newton did not think so. The earth is more heated in its bowels, he said, "by subterraneous fermentations of mineral bodies" than by the sun on its surface, and he assumed that the same held true for the other planets.[205]

Alchemy was another interest that Newton shared with Fatio, or taught Fatio to share with him. It came up in Newton's first letter to him; by 1693 it filled Fatio's letters to Newton. Among New-

---

[204] The original of Newton's essay survives, but only Gregory's copy of Pitcairne's notes; *Corres 3*, 209–12; original Latin, pp. 205–9. There is another copy of the essay in *Add MS 3974*, ff.4–4ᵛ. Cf. Proposition 17 of the proposed Book IV of the *Opticks; Add MS 3970.3*, ff.337ᵛ–8.

[205] Newton to Bentley, 10 Dec. 1692; *Corres 3*, 235.

ton's papers is a Latin translation, written out in Newton's hand, of Didier's *Triomphe hermétique,* a work published in 1689 and known to exist at that time only in French.[206] In the light of Fatio's offer to help him read Huygens's *Traité de la lumière,* one cannot avoid wondering if they also read Didier together. "But no man can find ye work in his whole life without a Master," Newton copied about that time from the *Golden Calf* into his collection of "Notable Opinions."[207] As far as we know, Newton himself never had an alchemical father. He may have played out the other half of the relation with Fatio, however, and he may have intended such compilations as the Index chemicus primarily for Fatio's eyes.

During the early 1690s, Newton continued to compose alchemical essays. His own compositions, filled with references to alchemical literature, form a continuous spectrum with papers that appear to be essentially compilations of notes, so that any distinction between them is arbitrary. "Ripley expounded" may fall in the category of notes.[208] Newton's commentary on the *Tabula smaragdina* has more the character of an original composition.[209] In all of them, one recognizes alchemical analogues to the Newtonian philosophy of nature based on the passivity of matter and the activity of force. Thus, the commentary on the *Tabula* asserted that the tincture, the "Fountain of all perfection in the whole world," combines "the penetrating force of spirit and the fixed force of body."[210]

The most important alchemical essay Newton ever wrote, "Praxis," started as a set of notes on Didier's *Triomphe hermétique,* which showed that the process Didier expounded was identical to that in Sendivogius, Basil Valentine, Philalethes, Mundanus, Pierce the Black Monk, Hermes, and others. In its second version, it was a commentary on Ripley, which collated his process with that in Didier, Maier, Snyder, Artephius, and the others.[211] Under the title of "Praxis," it went through two drafts and included among its collated authorities Flamel, Grasseus, and the *Turba.* It also cited Fatio's letter of 4 May 1693. I have not met any other similar reference to an identifiable contemporary in the vast corpus of Newton's alchemical papers.

In its final form, "Praxis" consisted of five chapters in which Newton employed all the imagery of the alchemical tradition. Using Philalethes' striking phrase, which united the themes of purification and sexual generation, he announced that "o^r crude sperm flows from a trinity of immature substances . . . of wch two . . . become a pure milky virgin-like Nature drawn from ye menstruum

206 *Keynes MS 23.*   207 *Keynes MS 38, f. 8.*   208 *Keynes MS 54.*
209 *Keynes MS 28.* "The Regimen" (*Burndy MS 15*) and "The three fires" (*Keynes MS 46*), cited before, were compositions similar to *Keynes MS 28.*   210 *Keynes MS 28, f.6ᵛ.*
211 "The method of ye work"; *Keynes MS 21.* "Of ye first Gate"; *Keynes MS 53.*

of o^r sordid whore."[212] Newton brought in most of the substances of his earlier experimentation – the net, the oak, sophic sal ammoniac, the doves of Diana, the star regulus of Mars – and he summoned up the images of his climactic experiments of 1681, Saturn eating a stone and vomiting it up, and Jupiter grasping his scepter and commanding peace. True to his constant effort, he indicated the prosaic laboratory equivalents of his arcane language. When the doves come flying to be enfolded in the arms of Venus and to assuage the green lion, he added to the last a discreet parenthesis of explanation "(as alcalies do acids)."[213] At its culmination, "Praxis" described a process in the wet way that ended in multiplication.

> Thus you may multiply each stone 4 times & no more for they will then become oyles shining in ye dark & fit for magicall uses. You may ferment it wth ☉ by keeping them in fusion for a day, & then project upon metalls. This is ye multiplication in quality. You may multiply it in quantity by the mercuries of wch you made it at first, amalgaming ye stone wth y^e ☿ of 3 or more eagles & adding their weight of ye water, & if you designe it for mettalls you may melt every time 3 parts of ☉ wth one of ye stone. Every multiplication will encreas it's vertue ten times &, if you use y^e ☿ of ye 2^d or 3^d rotation wthout ye spirit, perhaps a thousand times. Thus you may multiply to infinity.[214]

From the reference to Fatio's letter of May 1693, which he added to the first draft above the line, Newton appears to have composed "Praxis" in the spring and summer of 1693, a time of great emotional stress. We should probably read its extravagant claim in the light of that stress. We should recall also that disillusionment apparently as full as his earlier expectations followed not long after and may have heightened the stress in 1693. In 1681, Newton had crossed out his two climactic exclamations of success. He did not literally cancel "Praxis," but he did so implicitly by abandoning it. It is tempting to connect the disillusionment with the fateful year 1693, though evidence such as the dated experiments of 1695 and 1696 tells us that it could not have been complete in that year, if indeed it began then at all. That there was a disillusionment, however, and that it came not long after 1693 cannot well be ignored. I have found only four alchemical notes, all fragmentary, that can be dated with assurance to the period

---

[212] *Babson MS* 420, p. 8. Newton liked this line from *Ripley Reviv'd* (London, 1678), p. 28, as much as I do, and he quoted it in at least seven other places: "Index Chemicus" (*Keynes MS* 30); *Keynes MS* 34, f.1^v; *Keynes MS* 35, Item 4, Sheets 4, 7; *Keynes MS* 48, ff.16–16^v; *Keynes MS* 51, f.1^v; *Burndy MS* 9, p.40.

[213] *Babson MS* 420, p. 12. Cf. a similar passage on pp. 11^a–13^a.

[214] *Babson MS* 420, p. 18^a. In the final draft of this passage (p. 17), Newton toned it down a little.

after Newton's move to London.[215] Whereas Newton extended all of his other major interests into the London residence, he devoted no further significant time to alchemy. He did continue to acquire some alchemical books; he did not, as far as surviving documents inform us, take notes from them. Conduitt later recorded one nostalgic comment, that "he would if he was younger have another touch at metals."[216] It suggested as little sustained interest as the unread books. Nearly thirty years of intense devotion to the Art left their permanent mark on Newton's intellect. Nevertheless, they did come to an end.

Meanwhile, in 1693, more than Newton's alchemy reached a climax. His relations with Fatio did also. Fatio visited Newton during the autumn of 1692. Presumably not long after he left, Newton received a letter from him written on 17 November.

> I have Sir allmost no hopes of seeing you again. With coming from Cambridge I got a grievous cold, which is fallen upon my lungs. Yesterday I had such a sudden sense as might probably have been caused upon my midriff/diaphragm by a breaking of an ulcer, or vomica, in the undermost part of the left lobe of my lungs. . . . My pulse was good this morning; It is now (at 6. afternoon) feaverish and hath been so most part of the day. I thank God my soul is extreamly quiet, in which you have had the chief hand. . . . Were I in a lesser feaver I should tell You Sir many things. If I am to depart this life I could wish my eldest brother, a man of an extraordinary integrity, could succeed me in Your friendship.[217]

A reply frantic with concern arrived at once.

---

[215] In all cases, notes relevant to the Mint establish the date. *Keynes MS* 13, f. 1 bis^v and *Keynes MS* 56, f. 1, both have short notes connected with the Mint which could have been added after the time of original composition. Two Mint papers (*Mint* 19.5, ff. 42, 54^v) have alchemical scraps jotted on them. The second of these, from around 1702, is the latest alchemical piece of which I am aware. His note in *Sanguis naturae* (1696) about Stacey Sowles, a Quaker widow who was a well-known dealer in alchemical books, must have been after the move (John Harrison, *The Library of Isaac Newton* [Cambridge, 1978], p. 232). Between 1701 and 1705 he purchased sixteen books for which the bill survives; twelve of them were alchemical (*ibid.*, p.9). Sometime after the accession of Queen Anne, he also received a letter from William Y-Worth, the alchemist whose treatise Newton had in several manuscript copies. In its mention of a "wanted [sic] Allowance," the letter seems to imply that Newton contributed to his support. It seems more certainly to imply conversation about alchemy (*Corres 7*, 441). Nevertheless, only three of the alchemical books in his library had an imprint after 1700, and two of those, bound together, were the gift of their author, William Y-Worth, in 1702 with the letter cited above (Harrison, *Library of Newton*, items 1138, 1302, 1644).

[216] *Keynes MSS* 130.6, Book 3; 130.5, Sheet 3.

[217] *Corres 3*, 230. There is a draft of a letter to Fatio in which Newton proposed that he spend the winter in Cambridge (Bodleian Library, New College MS 361.3, f. 34). It was probably written in the autumn of 1692, before Fatio's visit.

S^r

I . . . last night received your letter wth wch how much I was affected I cannot express. Pray procure ye advice & assistance of Physitians before it be too late & if you want any money I will supply you. I rely upon ye character you give of your elder brother & if I find y^t my acquaintance may be to his advantage I intend he shall have it . . . S^r wth my prayers for your recovery I rest

Your most affectionate
and faithfull friend to serve you
Is. Newton[218]

Fatio had dramatized a cold excessively. He was recovering by the time Newton's letter arrived, and in fact he lived for another sixty-one years.

Nevertheless, the cold did linger. In January, a Swiss theologian then in England, Jean Alphonse Turretin, brought Newton word of Fatio's continuing illness, and by the same hand Newton sent a radical proposal back to Fatio.

I feare ye London air conduces to your indisposition & therefore wish you would remove hither so soon as ye weather will give you leave to take a journey. For I beleive this air will agree wth you better. Mr Turretine tells me you are considering whether you should return this year into your own country. Whatever your resolutions may prove yet I see not how you can stirr wthout health & therefore to promote your recovery & save charges till you can recover I [am] very desirous you should return hither. When you are well you will then know better what measures to take about returning home or staying here.[219]

Fatio confirmed his intention to return to Switzerland. The recent death of his mother made the trip more necessary. With the estate she had left him, he added, he could live for some years in England, chiefly in Cambridge, "and if You wish I should go there and have for that some other reasons than what barely relateth to my health and to the saving of charges I am ready to do so; But I could wish in that case You would be plain in your next letter."[220]

The correspondence continued on through the winter and into the spring, focusing on Fatio's health and finances and warily circling the question of his possible move to Cambridge. As to the saving of charges, Newton mentioned his plan "to make you such an allowance as might make your subsistence here easy to you."[221] At the time he offered the allowance, Newton had not received a full dividend from Trinity for seven years. Convinced that Fatio

---

[218] *Corres 3*, 231.   [219] *Corres 3*, 241.
[220] Fatio to Newton, 30 Jan. 1693; *Corres 3*, 242–3.
[221] Newton to Fatio, 14 March 1693; *Corres 3*, 263.

was destitute, he forced an extravagant sum upon him for a couple of books and some medicine that Fatio had left in Cambridge, and in May he offered him more money.[222] Beyond the issue of money he could not be plain, however. Perhaps Fatio tried. "I could wish Sir," he wrote in April, "to live all my life, or the greatest part of it, with you, if it was possible, and shall allways be glad of any such methods to bring that to pass as shall not be chargeable to You and a burthen to Your estate or family." He went on to add that Locke had just been there pressing a proposal that the two of them settle with him at the Masham estate in Essex. "Yet I think he means well & would have me to go there only that You may be the sooner inclinable to come."[223]

In May, Fatio's correspondence turned to alchemy. He had met a man who knew a process by which gold amalgamated with mercury vegetates and grows. His description of the vegetating material sounded much like that which Newton had written in the "Clavis" fifteen years earlier. With his usual sense of drama, he told Newton to burn the letter as soon as he had read it.[224] Less impressed, Newton kept the letter, though he did insert a reference to the process in his "Praxis." Two weeks later, Fatio's new friendship had flourished as well as the matter in the glass. His friend also made a medicine from his mercury; it had finally cured him. The friend now proposed that they become partners in its production. Fatio would need to spend a couple of years picking up a medical degree. With the degree in hand together with the medicine, which was very cheap, he could cure thousands of people free in order to make the medicine known. It was good for consumption and smallpox, and it freed the body of atrabile (black bile), which was known to cause nine diseases out of ten. He could make a fortune. There was a hitch, however. He would need between £100 and £150 per year for at least four years; and in a hesitant backhanded way, he let Newton know that now was the time to help him financially. He urgently requested also that Newton come to London to advise him.[225]

On 30 May, Newton signed out of Trinity for a week. Undoubtedly he went to London, as concerned with Fatio's new friend as he was with his finances. Apparently he left Trinity for another week late in June, no doubt to go to London again.[226] Two other pieces

[222] Newton to Fatio, 7 March, 2 May 1693; *Corres 3*, 260–1, 267. The second reference is to Fatio's reply of 4 May to Newton's lost letter.
[223] *Corres 3*, 391. [224] *Corres 3*, 265–7.
[225] Fatio to Newton, 18 May 1693; *Corres 3*, 267–70.
[226] *Edleston*, pp. lxxxii, lxxxv, lxxxix–xc. It is impossible to reconcile the conflicting evidence of the Steward's Book, the Buttery Book, and the Exit and Redit Book. The Buttery Book showed Newton gone for four weeks in all, including a week in early

of information complete our meager budget of knowledge about this critical summer. On 30 May, he started a letter to Otto Menke, the editor of the *Acta eruditorum,* which he set aside and did not complete for six months. He experimented in his laboratory in June.[227] Beyond that, silence covered nearly four months.

He broke silence with a letter to Samuel Pepys on 13 September.

Sir,

Some time after Mr Millington had delivered your message, he pressed me to see you the next time I went to London. I was averse; but upon his pressing consented, before I considered what I did, for I am extremely troubled at the embroilment I am in, and have neither ate nor slept well this twelve month, nor have my former consistency of mind. I never designed to get anything by your interest, nor by King James's favour, but am now sensible that I must withdraw from your acquaintance, and see neither you nor the rest of my friends any more, if I may but leave them quietly. I beg your pardon for saying I would see you again, and rest your most humble and most obedient servant,

Is. Newton[228]

Three days later, now at an inn in London, he wrote to John Locke.

S^r

Being of opinion that you endeavoured to embroil me wth woemen & by other means I was so much affected with it as that when one told me you were sickly & would not live I answered twere better if you were dead. I desire you to forgive me this uncharitableness. For I am now satisfied that what you have done is just & I beg your pardon for my having hard thoughts of you for it & for representing that you struck at ye root of morality in a principle you laid down in your book of Ideas & designed to pursue in another book & that I took you for a Hobbist. I beg your pardon also for saying or thinking that there was a designe to sell me an office, or to embroile me. I am

your most humble & most
unfortunate Servant
Is. Newton[229]

Eighteen months earlier, the search for a position in London had given rise to paranoid passages in Newton's correspondence with

June, another in late June and early July, and two weeks in September. The Exit and Redit Book showed one absence of about a week in early June. It showed nothing for late June and early July and nothing for September. The Steward's book indicated a total absence for the year of two and a half weeks. We know he was absent for about two weeks in September and was in London at least part of the time.
[227] *Corres* 3, 270–1. Add MS 3973, f.28.   [228] *Corres* 3, 279.   [229] *Corres* 3, 280.

Locke. "Being fully convinced that M^r [Charles] Mountague upon an old grudge wch I thought had been worn out, is false to me," he had written in January 1692, "I have done wth him & intend to sit still unless my Ld Monmouth be still me friend." Three weeks later, he had expressed his pleasure at Lord Monmouth's continued friendship and excused himself with exaggerated concern for an imagined infelicity the last time he saw Monmouth—a letter expressing groveling subservience which is embarrassing to read as a product of Newton's pen.[230] In withdrawing his essay on the corruption of Scriptures, which he had confidently given to Locke for publication a year earlier, the same two letters had marked the end of the manic euphoria that had borne Newton forward since the twin triumphs of the *Principia* and the revolution. Now, in the autumn of 1693, what Frank Manuel has aptly named Newton's black year, he plumbed the depths of the ensuing depression.[231]

The following May, a Scotsman named Colm told Huygens that Newton had had an attack of frenzy that had lasted eighteen months. It was thought that, in addition to excessive study, a fire that had destroyed his laboratory and some of his papers had contributed to troubling his mind. Friends had confined him until he had recovered enough to recognize his *Principia* again, but Huygens assumed he was lost to science.[232] Huygens's story of the fire may appear to confirm the passage quoted earlier from the diary of Abraham de la Pryme, although the dates impose a major problem of reconciliation since Pryme recorded his account in February 1692. In any case, Huygens repeated the story to Leibniz, and it spread rapidly through the whole European community of natural philosophers. By the summer of 1695, John Wallis received from Johann Sturm in Germany an account, which he bluntly denied, that Newton's house and books had all burned and that Newton himself was "so disturbed in mind thereupon, as to be reduced to very ill circumstances."[233]

The story of the fire carries doubtful authority. The letters to Pepys and Locke were undoubtedly authentic, however, and it is impossible to pretend that Newton did not go through a period of mental derangement—though not necessarily one like that described to Huygens and not one that lasted, in its most acute stage, for eighteen months. One cannot sufficiently admire Pepys and Locke. Confronted without warning by such letters, neither gave thought to taking offense. Rather both assumed at once that Newton was ill

---

[230] Newton to Locke, 26 Jan. and 16 Feb. 1692; *Corres 3*, 192, 195.

[231] Manuel, *Portrait*, "The Black Year 1693," pp. 213–25.

[232] Huygens to Constantyn Huygens, 6 June 1694; Huygens, *Oeuvres*, 10, 616. Huygens recorded the conversation in his journal for 29 May; *ibid.*, 10, 616.

[233] Wallis to Waller, 31 May 1695; *Corres 4*, 131.

and acted accordingly. Through his nephew, who was a student in the university, Pepys enquired discreetly of John Millington, the fellow of Magdalene College mentioned in Newton's letter, and eventually wrote directly to Millington. Pepys questioned Newton's sanity. After the visit from Pepys's nephew, Millington, who assured Pepys that he had not delivered any message at all to Newton, much less the one Newton alleged, had tried to call on Newton but found him out. He met him finally on 28 September in Huntingdon,

> where, upon his own accord, and before I had time to ask him any question, he told me that he had writ to you a very odd letter, at which he was much concerned; added, that it was in a distemper that much seized his head, and that kept him awake for above five nights together, which upon occasion he desired I would represent to you, and beg your pardon, he being very much ashamed he should be so rude to a person from whom he hath so great an honour. He is now very well, and, though I fear he is under some small degree of melancholy, yet I think there is no reason to suspect it hath at all touched his understanding . . .[234]

Pepys waited two months until Neale's lottery offered an occasion to write. After a brief and dignified reference to Newton's letter, he assured him of his readiness to serve him and proceeded forthwith to test his understanding for himself by putting the question of chance.[235] The response, discussed above, would not have left him in doubt that Millington's assessment was correct.

Locke, who had no nephew to press into service, waited for two weeks before he replied early in October.

> I have ben ever since I first knew you so intirely & sincerly your friend & thought you so much mine y^t I could not have beleived what you tell me of your self had I had it from anybody else. And though I cannot but be mightily troubled that you should have had so many wrong & unjust thoughts of me yet next to the returne of good offices such as from a sincere good will I have ever done you I receive your acknowledgm^t of the contrary as ye kindest thing you could have done me since it gives me hopes I have not lost a freind I soe much valued.[236]

Newton answered with much the same explanation that he gave to Millington.

> The last winter by sleeping too often by my fire I got an ill habit of sleeping & a distemper wch this summer has been epidemical put me further out of order, so that when I wrote to you I had not slept an hour a night for a fortnight together & for 5 nights together not a

---

[234] Millington to Pepys, 30 Sept. 1693; *Corres 3*, 281–2.
[235] Pepys to Newton, 22 Nov. 1693; *Corres 3*, 293.   [236] *Corres 3*, 283–4.

wink. I remember I wrote to you but what I said of your book I remember not.[237]

Later in the fall, when he answered letters from Leibniz and Menke that he had received six months earlier, Newton told them both that he had mislaid their letters and had not been able to find them.[238]

Over the years a variety of explanations have been offered for Newton's breakdown, such as the strain of composing the *Principia* and anxiety over the university's crisis with James II. The chronology involved renders such explanations most unlikely. To them a new one has recently been added. The symptoms found in Newton's letters—sleeplessness, digestive upset, loss of memory, paranoid delusions—are some of the classic symptoms of chronic mercury poisoning.[239] I have already remarked how Newton exposed himself wantonly to a wide range of dangerous chemicals. Suffice it here to quote an experiment from 1693.

> June 1693. The two serpents ferment well wth salt of ♄ ♃ & ♀ better wth salt of ♄ & ♀ best wth salt of ♄ alone. ☿ added ferments much more in all three cases & volatizes ye mass: but better in ye 2^d case y^n in ye 1st & best in ye last. To ye 2 serpents 24^gr I added ☿ of ♄ 24^gr by degrees & when ye fermentation was over I added ☿ 16^gr & ye matter swelled much wth a vehement fermentation then before & in two or three hours sublimed all to ye top except 3 grains wch remained below spongy in form of a dark cinder, & there was 9¼^gr of running ☿ besides a little that stuck in ye neck of ye glass wch might amount to a grain more so y^t ye 2 matters dissolved about 1/8 of their weight of ☿[240]

Repeatedly exposed for nearly thirty years to mercury both in vapors and in the compounds he tasted, Newton would have been fortunate to avoid mercury poisoning. One needs to add, however, that other symptoms of mercury poisoning, such as tremors and loss of teeth, find no mention. Seitz and Lettvin profess to find his writing tremulous in the mid-1690s. Though I have been very attentive to Newton's hand, I am unable to discern such. It is, moreover, especially significant that Newton recovered rather quickly, though the effects of mercury poisoning are known to persist for a long time.

---

[237] *Corres 3*, 284.

[238] Newton to Leibniz, 16 Oct. 1683; Newton to Menke, 22 Nov. 1693; *Corres 3*, 285–6, 291–2.

[239] P. E. Spargo and C. A. Pounds, "Newton's 'Derangement of the Intellect.' New Light on an Old Problem" (unpublished). Cf. R. Seitz and J. Y. Lettvin, "Mercury and Melancholy: The Decline of Isaac Newton," *Bulletin of the American Physical Society,* ser. II, *16* (1971), 1400, and a letter from Laurence Johnson in *Physics Today,* 30 (May 1977), 100–1; Louis S. Goodman and Alfred Gilman, *The Pharmacological Basis of Therapeutics,* 5th ed. (New York, 1975), pp. 485, 937.   [240] *Add MS 3973,* f. 28.

Moreover, a purely physiological explanation of his aberrant behavior leaves out too many factors. A variant of the old theory of
exhaustion does appear plausible to me. As I have argued, the early
1690s were a period of intense intellectual activity for Newton as he
strove to weave the various strands of his disparate endeavors into a
coherent fabric. Intellectual excitement always stretched him to the
very limit and, on occasion, beyond it. His breakdown in 1693 was
not altogether different from his behavior in 1677–8. To be sure, he
was experimenting in the earlier period as well, but neither time
was incommensurable with what we know of an isolated student,
alienated from his peers, who had not yet discovered alchemy. If
1693 was a climax, it was a climax long in preparation. In the early
1690s, there was the added stress of a manifest sense of humiliation
in soliciting an appointment in London, though Spargo and Pounds
remark that we could treat this plausibly as symptom rather than
cause. Add also the growing doubts that had begun to assail him,
which are also impossible to distinguish into symptom or cause. By
1693, he had withdrawn his theological publication, scuttled plans
to publish works in mathematics and optics, and begun to hesitate
on a second edition of the *Principia*. The apparent climax and disillusionment in alchemy in the summer of 1693 only confirmed his
doubts. There may also have been a fire–another fire, as it appears
to me–which could well have distracted him when he was already
in a state of acute tension. Charred papers survive from the early
1690s, though it is difficult to fit them satisfactorily with the other
dated events.[241]

It is also impossible to leave Fatio out of the account. Newton
was not the only one who was in turmoil. Fatio was also going
through a period of acute personal and religious tension.[242] The
sense of approaching crisis became almost palpable in their correspondence in early 1693. It is unlikely that we will ever learn what
passed between them in London. Their relation ended abruptly,
however, never to be resumed. The primary focus of Newton's
attention for four years, from the time they met in 1689, Fatio
simply dropped out of Newton's life. The rupture had a shattering
effect on both men. Newton rebounded from his breakdown, but
Fatio effectively disappeared from the philosophic scene forever.

---

[241] One of Newton's copies of the *Principia* has badly charred edges. A manuscript probably
written in 1692 and kept inside the folded sheet of a letter of August 1692 has charred
edges (*Corres 3*, 370, 375–7). Many of the manuscripts from the early 1690s for revisions
of the *Principia* are damaged (now gathered in *Add MS* 3965.6). While these fit easily into
the hypothesis of a fire in 1693, the badly burned draft of a letter to Gregory in July 1694
does not (*Corres 3*, 380–2). Neither do papers on lunar theory from 1695 (*Add MS*
3966.1, .2, .3, .4), but papers on lunar theory connected with the first edition are also
charred (*Add MS* 3966.12, ff. 102–11).

[242] Domson, "Fatio," pp. 44–63.

Huygens's brother Constantyn had great difficulty in delivering a letter to him early in 1694, never did see him personally, and heard he had been reduced to the status of a tutor to the children of some lord whose name he forgot. Apparently it was the Duke of Bedford; Fatio finally wrote from Woburn Abbey and was later connected with the Duke's heir. After wondering why Fatio did not write, Huygens finally received a reply nearly a year after his own letter; or perhaps he received one; the surviving copy is found among Fatio's papers, not among Huygens's.[243] In the letter, Fatio mentioned that he had not had any news from Newton for more than seven months. More than a year was probably closer to the truth. At the end of March 1694, Fatio did respond to a letter from Leibniz via DeBeyrie, the Resident for Brunswick in London. He feigned continued intimacy with Newton, but Leibniz sensed something strange in the letter, as he wrote to Huygens, and did not attempt to continue the correspondence.[244] In 1699, Fatio reappeared on the stage briefly with a mathematical tract which, in a reference to Leibniz possibly intended to regain Newton's favor, fanned the languishing priority dispute over the calculus back into flame. David Gregory mentioned him in personal memoranda several times in the early eighteenth century, suggesting that he hovered still on the fringes of intellectual circles, and Gregory noted that Newton was trying watches with jewel bearings that Fatio had made. By Gregory's note again, Fatio was in the group whom Newton promised in 1702 to publish the *Opticks*. Among Newton's papers is a list of expenditures from sometime after 1710 which seems to indicate he gave Fatio £30 at that time.[245] Nothing suggests a resumption of close relations or even an approach to it. By then, Fatio had plunged into the fanatical Camisard prophets from France and disappeared for good from the community of natural philosophers among whom his star had seemed destined to blaze. Beyond his role in rekindling the calculus controversy, he played no further part in Newton's life.

One of the puzzling aspects of Newton's breakdown is the silence of contemporary Englishmen about it. Beyond the responses of Pepys and Locke, who apparently took care never to mention the letters to others, one looks in vain for any mention of it. The Scot, Colm, told Huygens that a conversation Newton had with the Archbishop of Canterbury showed he was out of his mind. Colm

[243] Huygens to Fatio, 30 Nov. 1693; Constantyn Huygens to Huygens, 5 March, 13 April 1694; Huygens to Constantyn Huygens, 19 March, 2 April 1694; Huygens, *Oeuvres*, *10*, 567–9, 581, 583, 598, 599. Fatio to Huygens, 29 Sept. 1694; *ibid.*, *22*, 162–3.

[244] *Corres 3*, 308–9. Cf. Leibniz to Huygens, 22 June 1694; *ibid.*, *10*, 643.

[245] *Hiscock, passim.* Fatio appeared about ten times in Gregory's memoranda between 1702 and 1706. The list of expenditures is published in *Corres 7*, 368.

had better sources than the gossips of London, who never mentioned such an incident. In itself, the silence demonstrates, as Newton's papers and correspondence also do, that the reports that spread on the Continent were greatly exaggerated. Several indirect references do suggest strongly, however, that philosophic circles knew of a breakdown. On 31 October 1694, the Royal Society urged Newton to publish his improvements on the *Principia* and his other discoveries lest they be lost by his death.[246] Shortly, as we have seen, Wallis was using the same argument with him. During the early months of 1695, Flamsteed heard two separate reports of Newton's death that spread among his scientific friends in London.[247] It is hard to believe such reports circulated at this particular time entirely by chance.

When Newton's breakdown first became known to historians early in the nineteenth century, Jean Baptiste Biot interpreted it as the turning point in his life that brought his scientific activity to an end and inaugurated his theological studies. In the form that Biot presented it, the interpretation cannot stand. When David Gregory visited Newton in May 1694, he could hardly write fast enough to take down notes of projects on which Newton was at work or at least pretended plausibly to be. Newton's mathematical papers do not indicate a break in 1693; significant ones certainly dated from the following years. In 1694, he took up anew one of the most difficult problems in the *Principia*, the lunar theory.[248] Manifestly, Newton was still capable of clear scientific thought on the most perplexed questions. As for theology, we have seen that his greatest sustained attention to it belonged to the fifteen years that preceded the *Principia*. Nevertheless, a revised version of Biot's thesis does appear correct. The year 1693 witnessed the climax of the intense intellectual effort that followed the *Principia*; and if Newton had by no means lost his mental coherence after 1693, it remains true that he did not again inaugurate any new investigation of importance. He was no longer a young man. The crisis of 1693 terminated his creative activity. In theology as well as in natural philosophy and mathematics, he devoted the remaining thirty-four years of his life to reworking the results of earlier endeavors – insofar as he did not take refuge in administrative activity to absorb his time.

The possible exception to the assertion above was Newton's work on lunar theory, which dominated his attention for a year beginning in the summer of 1694. His effort deserves careful attention

---

[246] *JB*; quoted in *Corres 4*, 24.

[247] With his customary delicacy, he repeated both reports to Newton (Flamsteed to Newton, 7 Feb., 27 April 1695; *Corres 4*, 84, 110).

[248] Gregory's memorandum of 4 May 1694; *Corres 3*, 317–18; original Latin, p. 313.

against the background of 1693. In order to perfect the lunar theory, Newton needed better observations. To that end, he visited the Greenwich Observatory in the company of Gregory on 1 September 1694. Newton believed, Gregory recorded, "that the theory of the Moon is within his grasp." Flamsteed showed him fifty observed positions of the moon reduced to a synopsis and promised to supply him a hundred more.[249] Thus began a renewed exchange between the two men which constituted virtually the whole of Newton's correspondence during the following eight months, as he applied himself, in his usual way, wholly to the task. It was not an easy correspondence to carry on from either end as two prickly men confronted each other. Each needed the other in his own way. Newton needed Flamsteed's observations; without them his enterprise could not move. Flamsteed's needs were less concrete but no less real: acceptance as a philosophic peer. "All the return I can allow or ever expected from such persons wth whom I corresponded," he responded to Newton's offer to pay him for the trouble of copying observations, "is onely to have ye result of their Studies imparted as freely as I afford them the effect of mine or my paines."[250] He assured Newton that his "approbation is more to me then the cry of all the Ignorant in ye World."[251] Years later, long after their inevitable split, Conduitt heard a story about them: "Machin told me that Flamsted said 'Sr I. worked with the oar he had dug–to wch Sr I. said if Flamstead dug the oar he had made the gold ring."[252] Beside the unmistakable contempt in Newton's retort, Flamsteed's thirst for acceptance acquires a certain pathos. Nevertheless, Newton's need was more immediate, and he found himself as a consequence, at first, in the unfamiliar role of diplomat repeatedly soothing Flamsteed's wounded feelings, as for a time he stifled his own impatience with unremitting complaints, tirades against Halley, and self-righteous posturing.

Even at the meeting in Greenwich, Flamsteed imposed conditions. Newton could have the observations only if he promised not to show them to anyone else, and promised further to show Flamsteed the reformed theory first. In practice, anyone else meant Halley, and accordingly Newton refused to let Halley see the observations when they met shortly thereafter in London. Flamsteed promised more observations whenever he needed them, and already in October Newton started asking. He had spent several days comparing the observations with his theory, "& now I have satisfied

---

[249] Gregory's memorandum, *Corres 4*, 7. Flamsteed also recorded the visit; *Baily*, p. 61.
[250] Flamsteed to Newton, 10 Dec. 1694; *Corres 4*, 58.
[251] Flamsteed's memorandum for a lost letter of ca. 2 Feb. 1695; *Corres 4*, 80.
[252] *Keynes MSS* 130.6, Book 3; 130.5, Sheet 3; Cf. Newton's comment on Flamsteed in his letter to Hawes, 14 June 1695; *Corres 4*, 133.

my self that by both together the Moons Theory may be reduced to a good degree of exactness perhaps to ye exactness of two or three minutes." He went on to specify exactly what data he needed. "By such a set of Observations I belive I could set right the Moons Theory this winter . . . "[253] Unfortunately, what Newton wanted included observations yet to be made, and through the winter he continued to exhort Flamsteed to use his time to that end. "A little diligence in making frequent Observations this month & another month or two hereafter," he wrote in December, "will signify more towards setting right ye Moons Theory then ye scattered observations of many years."[254] Flamsteed had his own plans for those three months, however. For years he had prepared for what he called his "great worke," a new star catalogue.[255] Astronomer Royal to a Crown with no money to spend on astronomy, he had had to supplement his meager salary with a church living twenty miles away and to equip his observatory for his great work from his own resources. Only five years earlier, at his own expense, he had completed the large mural arc that would be his basic instrument. Eager to get on with his catalogue, he hardly welcomed Newton's requests, often barely distinguishable from imperious demands, for fresh observations of the moon. As he wrote plaintively to John Caswell a few years later, "I know no obligations I lay under to spend all my time to serve Mr. N., who would needs question the observations when they agreed not with his theories . . . "[256] It tells us much about Newton that he lacked the sympathy to understand Flamsteed's position, and intent on his own goal, he was not able to believe Flamsteed's insistence that reliable lunar positions would have to wait on the star catalogue, since observations of the moon measured its location in respect to the fixed stars.

Not that Flamsteed readily evoked sympathy. With his constant moralizing, a more annoying correspondent is hard to imagine. "when you have determined what corrections or additions are to be made to that theory which it was my Good fortune to Meet with & usher into ye light," he said, referring to Horrox's lunar theory, "I doubt not but you will impart them to me as freely as I did the observations Whereby you limit or confirm them to you."[257] When Newton's impatience to get more observations burst through, he retreated behind an impenetrable wall of righteousness. He had to visit his church for Christmas and could not send the lunar positions until he returned. They would be his first task.

[253] *Corres 4*, 24–5.   [254] *Corres 4*, 53.
[255] Flamsteed to Newton, 18 Jan. 1695; *Corres 4*, 72.
[256] Flamsteed to Caswell, 25 March 1703, *Baily*, p. 214.
[257] Flamsteed to Newton, 3 Nov. 1694; *Corres 4*, 45–6.

But I am displeased wth you not a little for offering to gratifie me for my paines either you know me not so well as I hoped you did or you have suffered your selfe to be possest with that Character which y[e] malice & envy of a person from whom I have deserved much better things has endeavord to fix on me & which I have disguised because I knew he used me no other ways then he has done the best men of y[e] Ancients nay our Savior his Apostles & the Scriptures permit me to give you a truer Charecter of my selfe & which you shall allways find me answer. I dare boldly say I was never tempted with Covetousness . . . My freinds are allwayes heartily welcome to me whenever they please to oblige me with their good Company but I am I profess it a freind to frugality not for the avoyding of expence so much as the preserveing my often endangered health, to avoyd ostentation & pride & for spareing my time which considering my small assistance you must needs be sensible is precious wth me.[258]

A little of that could suffice for a long time. Newton had to swallow it with almost every letter.

The person to whose malice and envy he alluded was Edmond Halley, and few letters passed without a denunciation of him.

I find you understand him not so well as I doe [Flamsteed assured Newton]. I have had some years experience of him & very fresh instances of his ingenuity with which I shall not trouble you tis enough that I suffer by him I would not that my freinds should & therefore shall say no more but that there needs nothing but that he shew himselfe an honest man to make him & me perfect freinds & that if he were candid there is no body liveing in whose acquaintance I could take more pleasure: but his conversation is such y[t] no modest man can beare it no good man but will shun it.[259]

Though in 1694, Newton had not outdone himself in friendship to Halley, he did stand deeply in his debt, and he received the flood of abuse with growing displeasure.

As the correspondence proceeded, Newton wrestled with the most complex problem in his *Principia*. The first edition had defined an attack on the three-body problem and had made a start at bringing the manifold perturbations of the lunar orbit, hitherto known only empirically insofar as they were known at all, under quantitative treatment within the theory of gravitation. Neither he nor Halley had been satisfied. Now he tried to make his treatment more precise, so that, as he told Flamsteed and Gregory, the discrepancy between theory and observation would not exceed two or three minutes of arc. It was excruciating work. The three-body problem does not admit a general analytic solution, and he had to deal with a

[258] Flamsteed to Newton, 10 Dec. 1694; *Corres 4*, 58.
[259] Flamsteed to Newton, 29 Jan. 1695; *Corres 4*, 77.

set of corrections which could be distinguished from each other and
defined quantitatively only with difficulty. He later told Machin
that "his head never ached but with his studies on the moon."[260]
Whatever else he might be, Flamsteed was not a remedy for a
headache. As the problem baffled Newton, he projected his frustra-
tions and headaches onto Flamsteed and his failure to supply the
needed observations.

A subsidiary problem arose at once. To give accurate positions,
the observations had to be corrected for atmospheric refraction, and
no decent theory by which to order the corrections existed. Though
Newton, who was not above a little self-righteousness himself,
pretended that he only worked on refraction "that I might have
something to present you with for the pains that you have taken for
me about your Observations," the proper correction was essential
to his own interests as well.[261] It turned out to be more of a job
than he had expected, "a very intricate & laborious piece of work,"
as he complained to Flamsteed.[262] In November, he sent a table of
refractions. Flamsteed questioned its accuracy at once, and Newton
had to agree that it was based on a false premise. By early January,
he thought he had the correct approach, but it was the middle of
March, after more than two months of labor, before he had the
table completed. Based on an observed refraction of 13' 20'' at 3°
altitude, the table was accurate, Newton assured Flamsteed, to 1''
for altitudes above 10°.[263] Apparently he also gave a copy to Flamsteed's malicious friend Halley, who later published it (as Newton's
work) in the *Philosophical Transactions* in 1721.[264]

The table of refractions was not enough, however. "Your con-
cealeing the foundations on which you computed it from me,"
Flamsteed wrote upon receiving the first table, "has caused me to
bestow some time & paines on y^t subject." Flamsteed wanted to be
accepted as a peer, not as a drudge. Hastily, Newton sent him the
reasoning on which he had based the table, though even then he
said that the demonstration of the theorem was too intricate to
include in a letter. Flamsteed would not be put off. As Newton
worked on the new Table, he let him know that he would be glad
to have "not onely ye table it selfe but the ground of it . . ." Unfortunately, Flamsteed was incapable of carrying on the philosophical
discourse he sought. While they discussed refraction, he came upon
Proposition XXII, Book II, of the *Principia*, on the density of the

---

[260] *Keynes MSS* 130.6, Book 3; 130.5, Sheet 3.
[261] Newton to Flamsteed 26 Jan. 1695; *Corres 4*, 74.
[262] Newton to Flamsteed, 16 Feb. 1695; *Corres 4*, 86.
[263] Cf. the papers on which it was founded; *Math 6*, 431–4.
[264] "*Tabula refractionum siderum ad altitudines apparentes,*" *Philosophical Transactions*, 31 (1720–
1), 172.

atmosphere at differing heights, as a fresh discovery, and confessed that he had never read Book II. Later, when Newton tried to explain his equation of lunar parallax, Flamsteed wholly misunderstood the vectorial analysis on which Newton's approach to lunar theory rested.[265] Given the gulf between his aspiration and his capacity, and given the finite limits of Newton's patience, we can see that a rupture between them was inevitable.

They had not proceeded far before Newton understood that Flamsteed misapprehended what he hoped to do.

> I beleive you have a wrong notion of my method in determining the Moons motions. For I have not been about making such corrections as you seem to suppose, but about getting a generall notion of all the equations on wch her motions depend & considering how afterwards I shall go to work wth least labour & most exactness to determin them. For the vulgar way of approaching by degrees is bungling & tedious. The method wch I propose to my self is first to get a general notion of the equations to be determined & then by accurate observations to determin them. If I can compass the first part of my designe I do not doubt but to compass the second . . . And to go about ye 2^d work till I am master of ye first would be injudicious . . . [266]

It was a method foreign to Flamsteed's empirical background, and he never comprehended why Newton could not send him a string of corrected parameters in the existing theory based on his observations.

When he did not get the corrections, he began to think that Newton was violating his pledge that Flamsteed should see them first, and he began to hold back. In January, as Newton's impatience began to overflow, he asked Flamsteed to send the raw observations without calculating positions from them. Flamsteed became more suspicious. He continued to furnish observations, but Newton's impatience with the whole enterprise grew faster than his supply of data.

> When I set my self wholly to calculations (as I did for a time last Autumn & again since Christmas in making the Table of Refractions) I can endure them & go through them well enough [he told Flamsteed in April]. But when I am about other things, (as at present) I can neither fix to them wth patience nor do them wthout errors. Which makes me let the Moons Theory alone at present wth a designe to set to it again & go through it at once when I have your materials. I reccon it will prove a work of about three or four months & when I have done it once I would have done with it forever.[267]

[265] Flamsteed to Newton, 10 Dec. 1694, 29 Jan., 2 March, 6 May 1695; *Corres 4*, 57, 78, 89, 122.

[266] Newton to Flamsteed, 17 Nov. 1694; *Corres 4*, 47.

[267] Newton to Flamsteed, 23 April 1695; *Corres 4*, 106.

Flamsteed made the mistake of not heeding Newton's tone; he questioned the accuracy of the table of lunar parallaxes that Newton had sent with his letter. He got for his pains a curt response that called both his observations of parallax and his calculations into question. "But yet if my Table satisfy you not," Newton added, "you may use your printed one . . . "[268]

Flamsteed did not mistake the tone this time, and now he tried to soothe Newton's feelings. Nevertheless, nearly two months passed before he heard again; and when the letter came, the tone was still more angry. Would Flamsteed propose some practicable way of supplying him with observations?

> For as your health & other business will not permit you to calculate the Moons places from your Observations, so it never was my inclination to put you upon such a task, knowing that the tediousness of such a designe will make me as weary wth expectation as you with drudgery. I want not your calculations but your Observations only.[269]

When Flamsteed gathered his cloak of righteousness about him and mentioned among other things a story circulating about London that the second edition of the *Principia* would lack the lunar theory because he refused to supply observations, Newton finally exploded.

> After I had helped you where you stuck in ye three great works, that of the Theory of Jupiters Satellites, that of your Catalogue of the fixt stars & that of calculating the Moons places from Observations, & in all these things freely communicated to you what was perfect in it's kinds (so far as I could make it) & of more value then many Observations & what (in one of them) cost me above two months hard labour wch I should never have undertaken but upon your account, & wch I told you I undertook that I might have something to return you for the Observations you then gave me hopes of, & yet when I had done saw no prospect of obteining them or of getting your Synopses rectified, I despaired of compassing ye Moons Theory, & had thoughts of giving it over as a thing impracticable & occasionally told a friend so who then made me a visit.[270]

For all intents and purposes, Newton's furious letter, which tried to shift his own failure onto Flamsteed, marked the end of his effort with the moon. True, there was a brief reconciliation, and Flam-

[268] Newton to Flamsteed, 25 April 1695; *Corres 4*, 120–1.

[269] Newton to Flamsteed, 29 June 1695; *Corres 4*, 133–4.

[270] Newton to Flamsteed, 9 July 1695; *Corres 4*, 143. Nevertheless, Newton complained enough that the story found its way to the Continent as well as the coffeehouses of London. "Flamsteed withheld his observations of the moon from Newton," Leibniz wrote to Roemer in 1706. "On that account they say he has as yet been unable to complete his work on the lunar motion" (Leibniz, *Opera omnia*, 6 vols. [Geneva, 1768], 4, Pt. II, 126).

steed, who in fact had sent him several hundred positions already, capitulated and started sending him the raw observations. Five years later, he asserted that he spent the whole of 1695, when he was not sick, in furnishing lunar positions to Newton.[271] Newton quickly tired of the exchange, however, and stopped answering Flamsteed's letters. When Flamsteed wrote in January, he had not heard from Newton in four months. He had it from others that Newton had finished his theory and had established six new inequalities. He was glad to know but wondered why Newton had not told him himself "as you promised both when I imparted the 3 Synopses of lunar Calculations & observed places to you . . . "[272] Newton did not even bother to reply. For that matter he did not bother to mention Flamsteed's observations in his additions to the lunar theory, and in the second edition he deleted two references to Flamsteed that had been in the first edition. Knowing full well Flamsteed's paternal pride in Horrox's lunar theory, he did include a favorable reference to Halley's improvement on it.[273]

Perhaps it is wrong to speak of Newton's effort to perfect the lunar theory as a failure. He did define a number of new inequalities, which he drew up in two different forms. One, a paper called "A Theory of the Moon," listed rules for computing seven corrections without discussing their theoretical foundation. Despite the number seven, it was probably what Flamsteed heard about. Several years later, Newton allowed David Gregory to take a copy of it and to publish it in his *Astronomiae physicae & geometricae elementa* (1702).[274] He also composed a scholium, which he inserted in the second edition at the end of the lunar theory, which discussed the theoretical foundation of the new inequalities, although he altered them somewhat from the earlier paper.

Nevertheless, Newton himself regarded the effort as a failure. The set of new corrections cut a poor figure beside the complete revamping of the *Principia*'s lunar theory in nineteen propositions that he had projected.[275] Indeed, much the most important of the corrections was a kinematic theory of the motion of the center of the moon's orbit which had no foundation in gravitational dynamics.[276] He did

[271] Flamsteed to Dr. Smith, 26 Oct. 1700; *Baily*, p. 744.   [272] *Corres 4*, 191–2.

[273] *Prin*, pp. 473–7. Cf. an early text of the scholium, probably from 1695, that did mention Flamsteed; *Corres 4*, 1–3.

[274] *Corres 4*, 322–6. I. Bernard Cohen, *Isaac Newton's 'Theory of the Moon's Motion'* (London, 1975), pp. 6, 12. In addition to Cohen's discussion of the text, this handy volume contains facsimiles of the paper from Latin and English publications.

[275] A list of the propositions is in *Add MS* 3966.8, f. 54, and another in 3966.4, f. 34ᵛ. Cf *Add MSS* 3966.1, ff. 1–7; 3966.2, f. 22; 3966.3, ff. 24–7; 3966.4, ff. 28–34; 3966.7, ff. 41–51; 3966.8, ff. 56–8.

[276] D. T. Whiteside, "Newton's Lunar Theory: From High Hope to Disenchantment," *Vistas in Astronomy, 19* (1975–6), 317–28.

not even touch the most glaring defect in the first edition's treatment of the moon, inability to account for the observed value of the progression of the line of apsides. Flamsteed reported with relish that the revised Newtonian theory erred from observed positions by as much as 10′, despite claims by Halley and Gregory which repeated Newton's initial aspiration, of accuracy within 2′ or 3′.[277] In a different context Newton virtually agreed with Flamsteed's assessment. Commenting in 1725 on a method proposed to determine longitude at sea, he stated that Halley's observations showed that the theory of the moon was accurate only to about 6′.[278] A man such as Clairaut, who later in the eighteenth century contributed a large share of the reformation of lunar theory on the basis of gravitational theory, professed himself to be at once baffled by and unimpressed with the new corrections.[279] His revision of lunar theory had to start anew from first principles; Newton's work in 1694–5 contributed nothing at all. Newton felt the defeat keenly enough that he recalled it more than once in his old age – and still blamed it on Flamsteed. When Halley had made observations for six years, he told Conduitt, "he would have t'other stroke at the moon . . . ," though in a moment of greater realism he also told Stukeley that he would leave the lunar theory to others.[280] He might have felt the defeat still more keenly had he known that Flamsteed also found the table of refractions unsatisfactory.[281]

The failure seemed to confirm the breakdown of 1693. Late in 1695, Newton entered into a vigorous correspondence with Halley on comets. For eight years there had been little commerce between the two men. Reading the lesson of Gregory's triumph at his expense, Halley appears to have decided that he must cultivate Newton more explicitly. From Newton's point of view, the correspondence did not break significant new ground and did not compensate for the lunar debacle.

If he had not realized the high hopes of the early 1690s, Newton's achievement during these years had been considerable. He had put the *Opticks*, the second of the twin pillars of his reputation in science, into the form that he later published. He had completed the

277 *Baily*, pp. 213–17, 304–5, 309–10.
278 Newton to the Admiralty, 26 Aug. 1725; *Corres* 7, 331.
279 Clairaut, *Exposition abrégée du système du monde*, attached to *Principes mathématiques de la philosophie naturelle*, tr. Madame du Chastellet, 2 vols. (Paris, 1759), 2, pp. 104–10.
280 *Keynes MS* 130.5, Sheet 3. *Stukeley*, pp. 14–15. Using slightly different language, he said the same thing to DeMoivre, that if he were younger he would "have another pull at the moon" (Quoted from Matthew Maty, *Mémoire sur la vie et sur les écrits de Mr. Abraham De Moivre* [The Hague, n.d.], by Augustus DeMorgan, "On a Point Connected with the Dispute between Keill and Leibnitz about the Invention of Fluxions," *Philosophical Transactions*, *136* [1846], 109).
281 Flamsteed to Colson, 10 Oct. 1698; *Baily*, p. 163.

two mathematical papers that he himself published, which gave substance to his reputation as a mathematician. Nevertheless, he had not achieved the great synthesis at which he had grasped. The turning point of the *Principia* had come too late. Newton was now well over fifty. He knew that his powers had begun to fade. If it was too late to crown the triumph of the *Principia*, his academic sanctuary ceased to have meaning. He had a second string to his bow, however. He had not yet received his reward for the triumph of the Glorious Revolution.

Against this background, we can evaluate his decision of 1696, so incomprehensible to the 20th-century academic mind, to leave Cambridge for a relatively minor bureaucratic post in London. The notion was not new to him. He had pursued an appointment fruitlessly in the early 1690s. Now, with his friend Charles Montague advancing in power, it became a possibility once more. What attraction would hold him in Cambridge? Certainly not the intellectual community. He had never found such there and had consciously held himself aloof from his peers. London had supplied his first real experience of an intellectual community; and if a desire for such played any role in his decision, it must have tilted the scale decisively for London. Cambridge's advantage for him had always been the uninterrupted leisure it provided to pursue his studies. As he felt his creative energy ebbing, that advantage evaporated. Indeed, his failures in the 1690s may well have driven him to escape from unproductive leisure into concrete activity.

What is equally relevant, Cambridge supplied a rationale. Newton had not separated himself so far from the university that he remained untouched by its ethos. To the Restoration don, the institution existed, not to be served, but to be exploited for personal advantage. In the early 1690s, Newton's peers in Trinity were at last approaching the ultimate status of senior fellows. George Modd, Patrick Cock, Nicholas Spencer, and William Mayor had never among them tutored a student, taken an advanced degree, or produced a line of scholarship. Nevertheless the system of seniority had carried them always abreast of Newton, and now they were beginning to reap their rewards. All now enjoyed lucrative college livings in the environs of Cambridge to supplement their fellowships. The college nominated them in their turns for university offices such as taxor and scrutator, which provided additional income. They had used the university to build their fortunes and had given nothing in return. Newton had stood resolutely aside from the scramble for ecclesiastical preferment, but he indicated that he understood the reigning mores when he silently converted his chair to a sinecure following 1687. He would now demonstrate to those who in amusement repeated anecdotes about the strange fellow who lived by the gate that ecclesiastical advancement was not the only – or the richest – reward one could extract.

Stories about an appointment circulated in London. In November 1695, Wallis heard Newton was to be master of the Mint.[282] Newton denied another report roundly to Halley on 14 March 1696.

> And if the rumour of preferment for me in the Mint should hereafter upon the death of Mr Hoar or any other occasion be revived, I pray that you would endeavour to obviate it by acquainting your friends that I neither put in for any place in the Mint nor would meddle wth Mr Hoar's place were it offered me.[283]

Fortunately, Halley did not have to waste much effort denying the stories. At the very time Newton wrote, Montague was completing the arrangements to appoint him, not to Hoare's post of comptroller, to be sure, but to the better one of warden. Montague dated his letter confirming the appointment 19 March. Newton accepted without pausing even to reflect. After a residence in Trinity of thirty-five years, he contrived to depart, bag and baggage, in less than a month, part of which he spent in London. While he continued to hold both fellowship and chair, and to enjoy their incomes, for another five years, he returned only once for half a week to visit. As far as we know, he wrote not a single letter back to any acquaintance made during his stay.

[282] Wallis to Halley, 26 Nov. 1695; *Corres 4*, 188.   [283] *Corres 4*, 193–4.

12

The Mint

CRISES racked the institution to which Newton moved in the spring of 1696. Indeed, the Mint was an institution within an institution within an institution, all three of which faced crises. The recoinage engaged every pinch of energy at the Mint. The Treasury, of which the Mint was a relatively minor department, devoted equal energy to devising temporary expedients and new machinery to cope with overwhelming financial needs. The English state and the revolutionary settlement it embodied balanced precariously on the outcome of the Treasury's efforts. In proclaiming William of Orange as its king in 1689, England had perforce embraced his foreign policy of resistance to French expansion. Although the ensuing war was known as King William's war, England would have found it impossible to stand aloof in any case, for the France of Louis XIV threatened its security only less than the security of William's native Netherlands. On a scale that exceeded previous wars, with a large English army in permanent operation on the Continent in addition to the naval operations England preferred, the war placed financial demands far beyond any precedent on the state. In 1696, it was not clear that the demands would be met. If they were not, if national bankruptcy ensued, the revolutionary settlement would undoubtedly collapse before a second Stuart restoration. In the larger crises of the government and its finances Newton was not involved beyond his concern as an Englishman committed to the revolution.

The narrower monetary crisis, which bedeviled the financial crisis by reaching a climax when it could least be tolerated, occupied him almost completely for more than two years. Its roots stretched back to a decision taken early in the reign of Charles II to coin by machine. Hitherto coins had been struck by hand. The resultant product, uncertain in outline, had presented a likely object to criminal activity. If one clipped a bit of metal from the edge of a coin, the coin looked no different than it had before, and one had the clipped metal for his pains. Machine coinage made the practice of clipping impossible (Figure 12.1) Rolling mills produced bars of precise thickness from which other machines cut round blanks. Huge presses sank an image no hammer could match. Most important of all was the edging, graining on small coins, a printed legend on the larger ones – *Decus et Tutamen* (A Decoration and a Safeguard). The edging put the clipper out of business since one could not clip an edged coin without mutilating it. Peter Blondeau, the

Figure 12.1   Machinery at the Mint. Above, the coining press. Workmen (at *A* and *B*) pull on the lines causing the die (*N*) to
crash down on the blank. The two heavy weights, or
flies (*F*), set in motion by the workmen, provide the

Frenchman who designed the Mint's new machinery, kept the edger as his own secret and personally edged every coin the Mint produced until his death. Of necessity, others had to learn its secret at that time, but still in 1696, when Newton came to the Mint, and indeed for more than a century after, employees including the warden had to take a special oath that they would not reveal the secret of the edging machine, though anyone able to read French could find it described.

Since the government had not called in the old hammered money when it began to coin the new, the reform had been fatally flawed. Between 1663 and 1696, England offered a classic example of Gresham's law in operation. No one with his wits about him would surrender a new coin of full weight for an old coin that contained less silver. By the 1690s William Lowndes, the Secretary of the Treasury, estimated that not more than one coin in two hundred circulating came from the new coinage.[1] A further reform in 1666, introducing the free coinage financed by a duty on certain imported liquors that Newton would administer for thirty years, served at first only to compound the problem. Since one could have bullion coined at no charge, there was no incentive not to melt down the new coin, least of all since one could, for the modest price of swearing it was not melted coin, export bullion to the Continent or the Orient, where silver fetched a higher price than the Mint paid. If that did not work out, it could be recoined without loss. In 1691, Dudley North (the brother of the onetime master of Trinity) in exasperation called free coinage "a perpetual Motion found out, whereby to Melt and Coyn without ceasing, and so to feed Goldsmiths and Coyners at the Publick Charge."[2]

The crisis had begun to loom by the time North wrote. By then, the hammered silver coins in circulation, thirty years old at the least

[1] Hopton Haynes, *Brief Memoires Relating to the Silver & Gold Coins of England: With an Account of the Corruption of the Hammerd Monys, And of the Reform by the Late Grand Coynage, At the Tower, & the Five Country Mints,. In the Years 1696, 1697, 1698, & 1699* (British Library, Lansdowne MS DCCCI), f. 44.

[2] Dudley North, *Discourses upon Trade* (London, 1691). p. 14.

force behind the blow. The moneyer sitting in the pit (C) flicks the coin out and inserts a new blank. According to Newton, the presses struck at a rate of fifty to fifty-five times a minute. Below, the edging machine, which protected against clipping by marking the edge of each coin. This is a French engraving from the second half of the eighteenth century. Within England the edging machine was treated as a state secret. (From the *Recueil de planches*, which accompanied Diderot's *Encyclopédie*.)

and badly worn, began more urgently to invite the allied skills of the clippers and coiners. Only silver, the coinage of everyday commerce, mattered. Gold coins, which came in higher denominations, passed by weight. Indistinct in any case, the hammered silver coins offered no evident limit to clipping, and counterfeiting an eroded image with an alloy of clippings and copper did not present a challenge beyond the skill of enterprising hoodlums. The Exchequer began to note a serious decline in weight about 1688; once the word got abroad, willing hands set to, and the decline from that time was precipitous. The records of the recoinage showed that nearly £5 million received and melted down at the Exchequer weighed less than 54 percent of the legal weight. Newton estimated that nearly 20 percent of the coins taken in were counterfeits.[3] Finally people quit accepting coins at face value. In 1695, the value of the guinea (a gold coin) inflated to 30s, and other pieces climbed similarly. That is, the exchange value of silver coins declined by more than one-fourth. Foreign markets ceased to accept English coins at full value. In his diary for 12 January 1696, John Evelyn recorded, "Great confusion and distraction by reason of the clipp'd money, and the difficulty found in reforming it."[4] After an escalation of punitive measures against clipping had produced no effect, the government under the leadership of Newton's friend, Charles Montague, then Chancellor of the Exchequer, began to consider a recoining of silver as the only effective remedy.

In 1695, the government sought what advice it could find. Although historians of economic thought point to the 1690s, when the financial crisis stimulated a spate of pamphleteering, as the birthplace of the modern discipline of economics, no acknowledged body of experts stood ready at hand to advise a sorely tried government. William Lowndes, the Secretary of the Treasury, produced a plan to accept part of the current devaluation of silver as permanent by reducing the silver content of coins by 20 percent. Troubled by the implications of a devaluation for the government's credit, the Regency Council resolved to consult a number of leading intellectuals and London financiers, "Mr. Locke, Mr. D'Avenant, Sir Christopher Wren, Dr. Wallis, Dr. Newton, Mr. Heathcote, Sir Josiah Child, and Mr. Asgill, a lawyer."[5] Along with most of the others, Newton replied in the autumn of 1695 with a short essay "concerning the Amendm^t of English Coyns." A general consensus among the respondents accepted the need to recoin, with Gilbert Heathcote, a governor of the recently established Bank, alone in question-

---

[3] *Mint* 1.6, 19.2, f. 618. An earlier draft of the paper with this estimate set the proportion of counterfeits at 10 percent.

[4] *Diary of John Evelyn*, ed. Henry B. Wheatley, 4 vols. (London, 1906), *3*, 126.

[5] 27 September 1695; *CSPD* 1695; p. 71.

ing it. Newton was one of the few who agreed with Lowndes's plan to devalue. Prices would tend to rise an equivalent amount, but he thought that strict governmental controls exercised through the livery companies in London could prevent the inflation. He was willing, however, to let rents rise so that landlords would not suffer permanent loss. Holders of government annuities would suffer such loss, but he assumed that Parliament would assuage their lot to maintain the government's credit. Locke was the most articulate of those who insisted that only recoinage at the old standard could salvage the currency.[6]

By December, the decision was made, though already in November the Mint had begun to receive orders to prepare. On 19 December, a royal proclamation announced the recoinage and set dates early in 1696 beyond which hammered coins would cease to pass at face value. A later act of Parliament had to extend the deadline to June. A government of the wealthy instituted a thoroughly iniquitous procedure which worked to insure their own class against loss and to place the major burden of the recoinage on the poor. The recoinage did not establish any machinery for the general populace to exchange old coins directly for new. The new money came into circulation via payments by the government, which also took the old out of circulation by taxes and loans. Only those who paid direct taxes or made loans to the government could turn in clipped money at face value. The wealthy alone did either. The poor were left to fend for themselves, and most had eventually to sell their coin as bullion and absorb a loss approaching 50 percent. The records of the recoinage indicate that less than half the money in circulation came into the Exchequer before the deadline; the Mint received the rest as bullion. In a last flurry of clipping in the period of virtual immunity between the announcement and the deadline, the populace recouped what it could. Meanwhile, on 21 January 1696, the definitive recoinage act passed through Parliament. The first melting of old coins at the Exchequer commenced the following day. Hence the recoinage was both decided and inaugurated well before Newton's appointment as warden of the Mint. His opinion on the recoinage did not determine the policy, for the government rejected Lowndes's proposal, which Newton had supported, and recoined at the old standard. By the time Newton had moved to London and taken up his new duties, the grace period during which hammered money passed at face value

---

[6] University of London, Goldsmiths' Library, MS 62, ff. 34–6. This manuscript volume in the Goldsmiths' Library contains the replies to the government's query, all copied by the same amanuensis. The original of Newton's essay does not survive, as far as we know. For other discussions of Newton's essay, see John Craig, *Newton at the Mint* (Cambridge, 1946), pp. 8–9, and J. Kieth Horsefield, *British Monetary Experiments, 1650–1710* (London, 1960), p. 52.

had nearly expired. In no sense did Newton bear responsibility for the recoinage. He did accept responsibility for carrying it through to completion.

Montague dated his letter offering the appointment 19 March. Newton went to London immediately. The Exit and Redit Book shows that he left on 23 March, and we know he was in London on 25 March.[7] He did not spend long agonizing over his decision; the warrant for his appointment was drawn up that same day.[8] He had packed his belongings by 20 April, when he left Trinity for good. One aspect of the Mint establishment was the provision of housing for all its salaried employees within the Tower of London, where it was located. Probably Newton initially made the warden's house his home. In a letter of 16 June about his salary addressed to the lords commissioners of the Treasury, he mentioned that it included a house worth £40 a year.[9] It took him nearly as long to settle in as he had taken to leave Cambridge, but on 2 May he was ready to swear the special oath. Four days later, in conjunction with Thomas Neale and Thomas Hall, the master and his assistant in charge of the recoinage, he signed his first memorial to the Treasury.[10]

Newton did not find the Mint a satisfactory place to live. The narrow street with buildings on either side was crammed into a space scarcely one hundred feet wide at its broadest. The warden's house backed up against the outer wall of the Tower and faced the blank wall of the inner bailey, which loomed forty feet above it. Years later, in defending the Mint's domain against the encroachment of the Board of Ordnance housed in the Tower, he explained why the Mint houses stood empty. The noise, the smoke of the forges, and the neighborhood of soldiers made them undesirable.[11] During the recoinage, when work began at 4:00 A.M. and left off at midnight, they must have been trebly so. The horses that worked the mills turned it into a virtual stable. The Mint spent nearly £700 during the recoinage in hauling manure away.[12] By August, Newton had located a house on Jermyn Street, near St. James's Church, Westminster, and there he lived for more than a decade.

Newton's decision to engage himself closely in the recoinage was his own free choice of a duty not necessarily incumbent on the warden. Under the old constitution of the Mint, the warden had exercised the ultimate authority as the agent of the Crown, which

[7] The minutes of the court of Christ's Hospital for 25 March 1696 mention that Newton was then in London (*Edleston*, p. 299).   [8] *CTB 10*, 1358.
[9] *Corres 4*, 205–6. "An accompt of what belongs to the Warden of his Mats Mint within the Tower of London," a paper in Newton's possession though not in his hand, listed the buildings at the Mint that were considered to belong to the warden: a large house with a garden, a house of four rooms, and formerly a coachhouse and stable (*Mint 19.1*, f. 25).
[10] *Corres 4*, 201, 202–3.   [11] *Mint 19.3*, f. 426ᵛ.   [12] *Mint 19.1*, ff. 19–20.

had monopolized coinage as a profit-making venture. The Crown had contracted with the master worker, who had undertaken the coinage under the terms of a contract, which was called the Indenture. The reorganization of 1666, which abandoned the principle of coinage as a royal monopoly and treated it as a public service financed by a special duty, retained all of the ancient forms of the Mint. The reality behind them changed. From this time, the wardenship became a sinecure. The master and worker became the true authority. He received a salary (£500) higher than the warden's (£400), in addition to the profit on coinage allowed him under the Indenture. Indeed, the new constitution reduced the warden nearly to the status of the master's hired servant, since the support of the Mint, including the warden's salary, passed through the master's hands. Not long after Newton took up his position at the Mint, he received a memorandum from Thomas Fowle, who had been clerk there for twenty-four years. Sir Anthony St. Leger, who had been the warden in 1672 when Fowle began to work in the Mint, had seldom come there at all, Fowle said, and when he had done so, it had only been to ask how things were going. Sir Thomas Wharton and his son Philip, who jointly succeeded St. Leger, had treated the position similarly. While other wardens had been more active, none had considered the office a serious job.[13] Fowle wrote to excite Newton to more vigorous action in defense of the special legal immunities that employees of the Mint enjoyed; what he described was a quintessential sinecure.

Years later, the Earl of Halifax (as Montague became) used to remark that he could not have carried on the recoinage without Newton.[14] At the time, however, Montague offered the position as a sinecure. It was worth, he wrote, five to six hundred pounds (a deliberate exaggeration) "and has not too much bus'nesse to require more attendance then you may spare."[15] We can only speculate on Montague's motives. Since his letter also mentioned a possible place for John Laughton, one could treat the appointment as the simple payment of a personal obligation from his days in Trinity College. Montague had entered Trinity as a fellow commoner in November 1679, proceeded M.A. by royal mandate less than two years later, and become a fellow, also by royal mandate, shortly thereafter. The presence of his third cousin, John Montague, in the master's lodge would not have hindered his progress. We do not know what brought him and Newton together, but early in 1685, Newton mentioned to Aston that the two had cooperated in attempting to organize a philosophical society in Cambridge, and two years later

---

[13] *Mint* 19.1, ff. 21–3; summarized in *Corres* 7, 411.
[14] *Keynes MSS* 130.6, Book 2; 130.7, Sheet 2.   [15] *Corres* 4, 195.

he called Montague his "intimate friend."[16] The desire to satisfy a friend would have been consistent with the justification Montague later offered: "He would not suffer the lamp that gave so much light to want oil."[17] If such were indeed his motive, he could hardly have misunderstood the trimming of lamps more grossly, but that is another matter. The story does fit with another that Conduitt recorded, that Montague employed John Machin, though in vain, to teach him enough mathematics to enable him to understand the *Principia*.[18]

The politics of revolutionary England did not ordinarily operate on that friendly basis, however. Patronage was the very marrow of power. Montague had only recently arrived at a position of power, and it is unlikely that he would have expended such a ripe plum as the wardenship of the Mint wantonly. The Whig Junto was known for quite the opposite. It is true that Montague apparently did find the promised place for Laughton, whose income a prebend in Lichfield Cathedral began to sweeten in 1696. Small pickings in the church went to different uses than significant sinecures in London, however, and Laughton was apparently related to Montague.[19] In appointing Newton, Montague did not gratify anyone else. What advantage could he have expected to extract from the appointment? Most wardens sat in Parliament and with other holders of similar places supported the government. In Newton's case at least, we need not pitch the expectations on a servile plane. He had made his opinions known. He had also stood for Parliament successfully on one occasion. The Steward's Book at Trinity showed that he spent half a week there in 1698, and his ballot in that year's election survives.[20] We may speculate that he canvassed the possibility of standing himself in future elections. At least he did so in the next election, in 1701, when he was returned. He considered it strongly in 1702, and stood once more in 1705, when his defeat wrote finis to his Parliamentary career.

When Montague offered a position that would not absorb much of his time, he did not take Newton's need to escape from intellectual activity into account, or his inability to do anything half way. From the first, Newton flung himself fully into the recoinage. Years later, presumably recording what Newton told him, Stukeley noted that he gave up intense study when he moved to London.[21] The Mint

---

[16] Newton to Aston, 23 Feb. 1685, and Newton to Halley, 13 February 1687; *Corres 2*, 415, 464.   [17] *Keynes MS* 130.7, Sheet 4.

[18] *Keynes MSS* 130.6, Book 3; 130.5, Sheet 2.

[19] In his will, Halifax left several bequests to nieces and nephews named Lawton. De Morgan thought it probable that they were related to John Laughton (Augustus De Morgan, *Newton: His Friend: and His Niece* [London, 1885], p. 54). Certainly the clan flocked to Trinity.   [20] *Edleston*, p. lxviii.   [21] *Stukeley*, p. 58.

records support this observation. Newton's name began to appear regularly on communications from the Mint as soon as he had taken the oath on 2 May. On 22 May, in the company of the other officers, he attended the lords commissioners of the Treasury. He returned twice on 25 May and again on 27 May.[22] Thomas Neale, the master, a political adventurer who dabbled in any enterprise likely to return a profit, from the postal service in the American colonies to the lotteries designed to make war taxes more palatable, was too distracted to give the recoinage the attention it required. In February, the government had appointed Thomas Hall, a commissioner of the Excise with previous experience in the Mint, as his special assistant for the duration of the recoinage. Neale had also appointed John Francis Fauquier, a recently naturalized Huguenot refugee, as his deputy. Newton found both men to his liking; he continued to associate with them and to employ Fauquier as deputy until Fauquier's death not long before his own. Hall and Fauquier had arrived too late wholly to correct Neale's mismanagement, however, and when Newton arrived the recoinage was floundering, compounding the crises of 1696 as the strain of the war stretched the revolutionary government almost to the breaking point. The Mint needed all the energetic and intelligent leadership it could find. Newton gave it his all. It is instructive that Halley, for whom Newton obtained a post in the Chester mint, complained that the work was drudgery at best.[23] Not only did Newton never make a similar complaint, but he contrived to make his position, which for others was temporary, into a permanent one.

In the memoir that he wrote on the recoinage, Hopton Haynes remarked that Newton's skill in numbers enabled him to comprehend the Mint's accounting system immediately.[24] Undoubtedly Haynes was correct, but the gifts that Newton brought to the position were not confined to understanding accounts. He possessed an innate tendency to order and to categorize. As we have seen, his first step in every new intellectual undertaking was the composition of some sort of index to help him organize his knowledge. That same tendency served him well at the Mint. In the perspective of history, the recoinage appears a mean thing in comparison to the *Principia*. Be that as it may, Newton had made his choice. He was a born administrator, and the Mint felt the benefit of his presence.

If the recoinage itself fell within the province of the master, the warden oversaw the maintenance of the premises. During the recoinage, this duty involved the construction of five temporary mints in different corners of the country as well as the provision of

22 *CTB 11*, 16, 17, 18.
23 Halley to Sloane, 25 Oct. 1697; Eugene Fairfield MacPike, ed., *Correspondence and Papers of Edmond Halley* (Oxford, 1932) p. 103.   24 Haynes, *Brief Memoires*, f. 68.

additional facilities at the Tower. Newton supervised the expenditure of £13,054 for such purposes.[25] He undertook such additional duties as the provision of weights that would serve as standards of money in Ireland.[26] Strictly speaking, this was an obligation of the master; and when Newton became master, he did not expect or allow the warden to perform such functions. It is informative about his role in the recoinage to see him doing such things.

One aspect of the recoinage that had not progressed well was the erection of temporary country mints to speed the diffusion of the new coin throughout the realm. The plan had called for five of them: in Norwich, York, Chester, Bristol, and Exeter. Although the grace period during which the old coin continued to pass had nearly expired when Newton assumed his position, the country mints remained far from operational. They were one of the tasks to which he turned his hand. The extent of Newton's responsibility for their successful operation cannot be demonstrated precisely; evidence that he played a role consists primarily of chronological coincidence. When he arrived, the country mints were badly behind schedule, and the Treasury was pressing hard for them. A number of items concerned with the country mints, which would not ordinarily have fallen within his province, appeared among his papers, and in fact they did begin to function less than three months after he arrived. There were others at the Mint also involved, however, especially Hopton Haynes, a clerk at the Mint for about ten years at that point, to whom special responsibility for the country mints was delegated. Newton took a special interest in Haynes, whose permanent patron he became. Suffice it to say in regard to the country mints, that when they finished their job in 1698, it was Newton who instructed them how to draw up their final accounts. He noted specifically that this would not ordinarily have been his responsibility, but he did it at the request both of Neale and of the lords commissioners.[27]

As a result of his energy, he found himself in a position finally to repay some of his debt to Halley. Each of the country mints needed its full delegation of officers, who were appointed as deputies of the officers of the Tower Mint. Newton arranged for Halley's appointment as the deputy comptroller in the Chester mint, at a salary of £90 per annum.[28] As it turned out for Halley, headaches worth twice that sum accompanied the post.

[25] Neale's Accounts; Public Record Office, E351, Roll 2103. Cf. the list of vouchers on Newton's account in *Mint* 19.1, ff. 19–20.

[26] Newton to the Treasury, 19 Oct 1697; *Corres 4*, 250.

[27] Newton to the country mints, 16 April 1698; *Corres 4*, 271–2.

[28] In a letter to Newton on 30 Dec 1697, Halley acknowledged that he owed his position "to your particular favour . . . " (*Corres 4*, 254).

The recoinage was a trauma for the body politic of England which reached its climax in the period of late spring and early summer 1696, just as Newton arrived at the Mint. Not enough of the new coin had issued forth to allow the necessary transactions of daily life. On 13 May, John Evelyn noted, "Money still continuing exceeding scarce, so that none was paid or receiv'd, but all was on trust, the Mint not supplying for common necessities." He had not observed much improvement a month later: "Want of current money to carry on the smallest concerns, even for daily provisions in the markets. . . . and nothing considerable coin'd of the new and now onely current stamp, cause such a scarcity that tumults are every day fear'd, nobody paying or receiving money . . ."[29] In a hasty effort to provide an alternative, the government tried to issue Exchequer bills, but this early experiment with a paper currency was not initially received with confidence. Bills from the new Bank of England found more acceptance, and did something to alleviate the problem. What coinage did issue tended to be hoarded in the expectation of a devaluation, until Parliament publicly rejected such on the same day it reconvened in October. With the Mint now functioning better, the extreme shortage began to ease, and the worst of the crisis had passed.

Within the Mint, all was frantic activity. Lord Lucas, the governor of the Tower, thought that five o'clock in the morning was early enough to open the gates; the Treasury ordered him to open them at four o'clock. Work continued until midnight. According to Haynes, nearly three hundred workmen crowded into the narrow confines of the Mint, and fifty horses turned the ten mills that operated. A man was killed on one of them. Nine great presses worked, each striking according to Newton's calculation between fifty and fifty-five times a minute with what must have been an incredible din. By heroic efforts, the Mint managed to push its production up to £100,000 per week during the summer of 1696, and by the end of the year it had coined £2,500,000. Haynes attributed considerable credit for the performance to Newton. He pointed especially to careful studies Newton made of each operation so that "he could judge of the workmens diligence . . ." Since Newton appears to have commissioned Haynes's *Brief Memoires*, we should treat his testimony with caution. Nevertheless, the studies of individual operations do exist among his papers.[30] The warden of the Mint did not report at four in the morning, of course. He did fling himself into the task with equivalent vigor.

[29] Evelyn, *Diary*, 3, 130, 131.
[30] *CTP 11*, 19, 31. *CTB 13*, 300. Haynes, *Brief Memoires*, ff. 70^v–72, 78^v, 83. Newton's "Observations concerning the Mint"; *Corres 4*, 255–8.

Early in 1697, the House of Commons appointed a special committee to enquire into alleged abuses by the officers of the Mint. It became the occasion for Newton to enhance his position. Apparently impressed by his testimony, the committee drew the substance of its report directly from him.[31] They underlined the impression by calling him "Dr. Newton."

The committee's report ended with alarming information about the engraver, John Roettiers (whose name the report bowdlerized as Old Rotter) and his sons. Charles II had brought three brothers Roettiers to engrave in the Mint in 1661. One of them, John, still occupied the engraver's house in the Mint, though he no longer engraved. The family were, the committee stated, "violent Papists & refuse to take the Oaths or to subscribe the Association . . ." One of John Roettiers's sons was charged with high treason for participation in a plot to assassinate William. The governor of the Tower thought Old Rotter too dangerous a person to have on the premises. In the summer of 1697, the continuing suspicion of Roettiers gave rise to the absurd story that he was hiding James II in his house, and this in turn precipitated a trial of strength between Newton as warden and Lord Lucas as lieutenant or governor of the Tower, both of whom claimed jurisdiction over the Mint. What with opening the gate at four in the morning, Lucas had not found the Mint during the recoinage an easy tenant to have within his command. When he heard the report of James's presence, he ordered a search of Roettiers's house, not once but twice, violating Newton's authority over the Mint. Both immediately started gathering stories. Newton heard of drunken officers assaulting employees of the Mint and of sentries ordered to shoot on sight, which Newton referred to as "bloody discipline." Lucas, in turn, waxed indignant at the "scandalous and untrue paper" that Newton submitted to the Treasury against him. Not only were Newton's stories false, but he had his own atrocity to recite of a drunken horsekeeper who

> came on a Gallop wth 2 horses in the sentinells post, at the Deputy Governours doore, & ridd ag^t and thrust him to a post, and call'd him many foul names, without any manner of provocation or reason, and the sentinell endeavouring to keepe his post, the horsekeeper was encouraged & directed, (by one Fowles a Clerk in the Mint) to dragg the Centinell by the Eares from his post, & shoot him

[31] "The Report of the Committee of the House of Commons upon the State of the Mint, 8 April 1697"; *Mint* 2.11 (not part of Newton's Mint papers). Also in *Journals of the House of Commons*, 11, 774–7. Much of the report comes verbatim from Newton's "State of the Mint" (*Corres 4*, 207–8; drafts in *Mint* 19.1, ff. 2–4, 6–7). His "Account of the Mint in the Tower of London" (*Corres 4*, 233–5) was an earlier draft of this paper.

through the head, and others of the horsekeepers in like manner threatened the Centinell . . .[32]

Fowle was the clerk who urged Newton to greater energy in the defense of the legal immunities of Mint employees. Eventually the tempest in a teapot calmed down, and the recoinage continued. Tension with the military in the Tower, who coveted parts of the Mint, remained endemic until Newton's death. Though in ordinary circumstances the Mint had no use for much of its space, which stood empty and unused, Newton defended it as his domain with all the force at his command. His failure over the years marked the victory of common sense over historic precedent.

Lieutenants of the Tower were not the only new species Newton learned to know. He became familiar also with the financiers of the City. The recoinage might be straining the government, from which the financiers derived great benefit, to the limit. Its extremity did not stop them from attempting shamelessly to fleece it in every way possible. Peter Floyer and Charles Shales, respected goldsmiths who also contracted with the Mint as melters and refiners, offered to melt the clipped money at the Exchequer for 12½d and half a farthing per pound troy. Newton and Hall informed the Treasury that the job could be done for 7½d and half a farthing, a sum that included a profit of 1d per pound for their effort and risk.[33] Shales and Floyer had proposed for themselves the considerable gratuity of £7,000, give or take a few shillings, for the entire recoinage. Jonathan Ambrose, another goldsmith-melter, tried to engineer a similar coup.[34] The trenchant criticism of Ambrose in the report of the committee of the House in April 1697 probably derived from Newton's observations of his mode of operation. No one would venture to compare coping with such strategems or defending the prerogatives of the Mint with his intellectual achievements. It is revealing of Newton, however, that with no prior experience at all beyond his unremitting struggle with tenants, he was able to shift to the one from the other and to carry out new duties with manifest success.

Despite his broader experience, Halley was having less success in Chester. By late 1696, the deputy master in Chester had begun to absent himself, leaving the deputy warden and Halley to shoulder a heavier load. By early summer, they were suspecting fraud on the part of Edward Lewis, Halley's clerk, who appears to have been the deputy master's creature, since Lewis constantly took his side and became his subdeputy when dismissed as Halley's clerk. With the suspicion of fraud and the general concern not to be caught holding

[32] Newton's copy of Lucas's report; *Mint* 19.3, ff. 388ᵛ–9. Newton's complaint is published in *Corres 4*, 242–5. It was presented to the Treasury on 13 July 1697; Lucas presented his reply on 3 Aug. (*CTB 12*, 58, 67).   [33] *Corres 4*, 236.   [34] *Corres 4*, 216–17.

the bag in the final accounting, what began as petty annoyance degenerated quickly into bitter hostility. The officers of the Mint wrote endless letters in an attempt to restore peace. They counseled restraint. Let there be no further quarreling, "for the Mint will not allow of the drawing of Swords, & assaulting any, nor ought such Language, Wee hear has been, be used any more amongst You." It was all to no avail. The conflict continued to the very end, when the deputy master attempted to make off with all the salaries for the final quarter. Newton found it impossible to stand aloof. In addition to the burden of correspondence that the quarrel imposed on all officers alike, he functioned as Halley's confidant and protector, and in the end came into direct conflict with Neale, who encouraged his deputy to defy Newton's orders and defended him before the lords commissioners. Although few men of the time displayed more skill than Neale at administrative infighting, the recently transformed natural philosopher carried the day.[35]

It did not take Newton long to size up realities in the Mint. He had accepted the position under the impression that the warden held the highest authority. By June he understood otherwise and petitioned the Treasury for an increase in salary. The warden's salary, he complained, "is so small in respect of the Salaries & Perquisites of the other Officers of the Mint as suffices not to support the authority of his Office."[36] He sized up Neale as well – "a Gentleman who was in debt & of a prodigal temper & by irregular practices insinuated himself into ye Office [of master] . . . ," as he remarked later.[37] Tension between them had developed by the time of the Parliamentary investigation in 1697; the open dispute over the Chester mint was only a further installment in their growing discord. In the papers that he composed for the committee of the Commons, Newton tried again to reassert the authority of the warden: "The Warden is . . . by his Office a Magistrate & the only Magistrate set over the Mints to do Justice amongst the members thereof in all things . . . The Workers (one of wch is Master of ye rest) are they who melt refine allay & run the standarded gold & silver into Ingots to be coyned." Originally, no workers were standing officers, but the reorganization of 1666

---

[35] Mint to Chester mint, early Aug. 1697; *Corres 7*, 400–1. Cf. Weddell's memorandum of 25 Sept. 1697 on Newton to the Chester mint, ca. Sept. 1697; *Corres 7*, 402; also *Mint* 10.2, the record of the country mints. On 2 Feb. 1698, Neale, Mason, and Molyneaux, the two comptrollers, and Newton appeared before the lords commissioners of the Treasury, with Newton and Neale taking opposed positions. The lords commissioners sided with Newton and ordered Neale to dismiss Lewis. (*CTB 13*, 59–60).

[36] Newton to the Treasury, 16 June 1696; *Corres 4*, 205–6. Also on 16 June, there was a meeting at the Treasury, with all of the officers of the Mint present, at which the contract of Neale's indebtedness to James Hoare, the comptroller, was aired (*CTB 11*, 28–9).

[37] *Mint* 19.1, f. 407. Cf. draft, ff. 405^v–6. The letter he sent on 21 June 1700 toned down the reference considerably (*Corres 4*, 348).

gave the master a salary higher than the warden's and appointed him to receive and distribute the income of the Mint from the coinage duty. The power of the purse brought the Mint under the master's control. The moneyers also resisted the authority of the warden by pretending to be a corporation. The comptroller, the other officer of the Mint, had got the office of master into his hands – a reference to the loan of the comptroller, James Hoare, to Neale, in return for which Hoare (and his associates) received half of Neale's income from the Mint – and thus raised himself to equality with the master and warden.

> And thus the Wardens Authority which was designed to keep the three sorts of Ministers in their Duty to ye King & his people, being baffled & rejected & thereby the Government of ye Mint being in a manner dissolved those Ministers act as they please for turning the Mint to their several advantages. Nor do I see any remedy more proper & more easy then by restoring the ancient constitution.[38]

The attempt to remodel the constitution of the Mint never got into motion. Adopting an alternative strategy, Newton set about making himself master in fact if not in name. By the methods he had applied in enterprises of a wholly different nature, he undertook a systematic study both of the history of the Mint and of its present operation, so that he might stand on a footing of incontestable knowledge. He collected copies of proclamations and warrants relevant to the Mint stretching back to the reign of Edward IV.[39] He took care to inform himself of Neale's indebtedness to Hoare and of the contractual relation to each other in which the two men stood.[40] He pored over the old accounts to become familiar with the level of payments for various services.[41] He studied each of the operations of the Mint in detail, recording the various expenses it involved, such as the cost of a melting pot and the number of times it could be used. "By experim^t I found that a pound Troy of ½ crown blancks lost 3½ gr. [in blanching]," he noted.[42] With his multiple ventures, Neale had never acquired a similar knowledge of the inner working of the Mint.

---

[38] *Corres 4*, 207–8. In his complaint against Lord Lucas, the governor of the Tower, in the summer of 1697, Newton stated that the workers in the Mint are incorporated into one body "under ye Government of a Warden . . ." (*Corres 4*, 242).

[39] *Mint* 19.1, ff. 28–61; 19.3, ff. 344–85; 19.5, ff. 13–27.

[40] He obtained a copy of a warrant of 1680 whereby John Buckworth, Charles Duncomb, and James Hoare were appointed commissioners to exercise the office of master and worker (*Mint* 19.1, f. 192). The warrant resulted from their loan to Neale. Cf. his copy of a report of 1684 on the office of the warden (*Mint* 19.1, f. 13).

[41] *Mint* 19.3, ff. 413–14. Cf. his copy of a warrant of 1662 about certain expenses in the Mint (*Mint* 19.1, f. 15).

[42] *Corres 4*, 257. Cf. the whole paper, "Observations concerning the Mint," *Corres 4*, 255–8.

One fascinating aspect of Newton's habits emerges from these papers. It remained characteristic of all his papers at the Mint and helps to illuminate his other papers, among which multiple drafts are very common. Newton was an obsessive copier. The Halls have suggested that he could not read attentively without a pen in his hand.[43] By confirming this habit with material that required no creative thought, the Mint papers serve to underscore it. With a stable of amanuenses at his command, Newton copied a report of 1675 on the state of the coinage, and then copied it again a second time.[44] He copied the record of the amount coined, both by weight and by tale, both in gold and in silver, year by year from 1659 to 1691 – and again copied it all a second time.[45] In part, the copying stemmed from the conviction that he could rely with confidence on himself alone. As he advised the officers of the country mints in the instructions for their accounts, "trust not the computation of a single Clerk nor any other eyes then your own."[46] The matter went beyond trust, however. Even a minor letter could extract two drafts and two fair copies from him.[47]

By the completion of the recoinage in the summer of 1698, Newton had mastered the operation of the Mint to the extent that he had virtually assumed the title of master. The lords commissioners asked him to take charge of drawing up the final accounts of the country mints, which Neale would ordinarily have done.[48] He also performed Neale's task of composing the final report of the recoinage.[49] Indeed, Neale's records stood in such impossible shape that he did not succeed in clearing his accounts before his death at the end of 1699, and they remained as an additional task for Newton – a symbol of the true situation within the Mint well before his death. The tangled affairs of Anthony Redhead, Neale's deputy in the Norwich mint, dragged on into the summer of 1704, when he apparently secured his release after five years in prison for failure to make good his accounts.[50]

[43] *Halls*, pp. 397–8.   [44] *Mint* 19.2, ff. 261, 271–2.

[45] *Mint* 19.2, ff. 257, 260. At some point, Newton extended the account on f. 257 to 1714. What makes the two copies more interesting is the fact that Newton also had three others in the hands of amanuenses (ff. 256, 258, 295).   [46] *Corres 4*, 271.

[47] In 1724, Newton was asked to supervise the trial of the pyx for Wood's copper coinage for Ireland. He wrote out two drafts and two fair copies of a letter which merely suggested that the pyx be brought to London (*Corres 7*, 273–4; drafts and fair copies in *Mint* 19.2, ff. 466, 468, 471). I use this letter only as an example ready at hand. I am convinced that I could easily find at least a hundred similar ones from every part of his tenure at the Mint (cf. the letter on Anthony Redhead cited in n. 50).

[48] Newton to all country mints, 16 April 1698; *Corres 4*, 271–2.   [49] *Mint* 19.2, f. 273.

[50] The last record of Redhead that I found was a petition in July 1704 to be released from prison (*CTP 1702–7*, p. 279). I could not find a record of any action on his petition (cf. four drafts of a letter Newton wrote in January 1703 recommending Redhead's release). Although Newton believed Redhead's imprisonment was his own fault, he did not take a vindictive attitude toward him (*Corres 4*, 396–7; drafts in *Mint* 19.2, ff. 481, 485, 491, 493–4).

According to Newton's records, the Mint (including the country mints) succeeded in recoining £6.8 million from the beginning of 1696 to the summer of 1698 – nearly twice the total coinage, measured by the number of coins, of the previous thirty years.[51] Through no fault of his, it all came to naught. Despite the government's willingness to impose a sudden deflation on an economy strained to the breaking point by the war and to risk social upheaval by its inequitable provisions, it did nothing to correct the basic undervaluation of silver. Almost as fast as it issued from the Mint, the new coinage went into the melting pots of the same goldsmiths who tried to fleece the government for its legal melting. During the rest of Newton's life, the Mint coined silver only at those times when special acts of the government brought plate and bullion into the Mint, and even then it coined very little. Within two decades, the Mint was experimenting with quarter-guineas, that is, gold coins, to relieve the shortage in small denominations. John Conduitt, the husband of Newton's niece and his successor at the Mint, noted in his "Observations" on the coin in 1730 that very little silver remained in circulation.[52]

The office of warden involved one dimension for which Newton may not have bargained initially. The warden was charged with the apprehension and prosecution of coiners and clippers. With the escalation of coining and clipping in the late 1680s, this aspect of the warden's duties had also grown; by 1689, the warden had received a second clerk because his first clerk had become wholly engaged in the pursuit of criminals.[53] The Treasury Papers and Treasury Books are filled to overflowing with orders and petitions relevant to coiners and clippers and their prosecution in the early 1690s, and such continued after the recoinage in lesser quantity on into the eighteenth century. As the destruction of the currency accelerated during the 1690s, new legislation against coiners and clippers poured out of Parliament. Beginning with the "Act to prevent counterfeiting and clipping the coin of this Kingdom" in 1694, five new statutes against "the wicked and pernicious crime of Clipping" and the associated and perhaps more pernicious crime of coining appeared during the following four years. When the recoinage put clippers out of business, many of them turned to coining as an equivalent trade. Counterfeiting had always been a form of petty treason. With an act of 1697, coining and the making and repairing

---

[51] *Mint* 19.2, f. 264. Slightly different figures are found on two other sheets (ff. 275, 294). In his journal for 19 July 1698, Locke noted: "Mr. Newton told me that there had been coined in all the mints since the calling in of the clipped money 6,500,000 or thereabouts in [silver]" (quoted in Maurice Cranston, *John Locke, a Biography* [London, 1957], 439n.).
[52] William A. Shaw, ed. *Select Tracts and Documents Illustrative of English Monetary History, 1626–1730* (London, 1896), p. 191.   [53] *CTP 1557–1696*, p. 65.

of tools for it became high treason punishable by death in the pleasant manner reserved for traitors.[54]

Newton's first reaction was to shrink from the job. Apparently in the summer of 1696, he wrote to the Treasury in complaint. He found himself exposed to censure in prosecuting for men's estates (which were forfeited to the warden to finance the pursuit of coiners). Judges and juries did not believe witnesses who would receive rewards if the prosecutions ended in conviction.

> And this vilifying of my Agents & Witnesses is a reflexion upon me which has gravelled me & must in time impair & perhaps wear out & ruin my credit. Besides that I am exposed to the calumnies of as many Coyners & Newgate Sollicitors as I examin or admit to talk with me if they can but find friends to beleive & encourage them in their false reports & oaths & combinations against me.

He ended by requesting that a duty "so vexatious & dangerous" not be required of him any longer. The job belonged more properly to the Solicitor General.[55] Initially, he concluded with the request that the task not be required, "at least not without enabling me to go through it with safety credit & success."[56] In fact, the Treasury decided to proceed in this direction by authorizing additional funds for the employment of another clerk.

It was not only the Treasury, but also the lords justices, the council that governed England while William campaigned on the Continent, who insisted on Newton's obligation. Just as he arrived at the Mint, they were pursuing testimony from several coiners in custody, which led on 29 July to the charge that dies had been stolen from the Mint. Summarily, they ordered that "Doctor Newton" attend them the next day "and that he have directions to enquire into it." Newton appeared as ordered on 30 July. He appeared before the lords justices nine more times in the next two months to inform them of the continuing investigation. They also recommended that he have an additional clerk.[57]

Unable to slough the obligation off, Newton plunged into it with customary thoroughness. Professor Manuel has argued that Newton's prosecution of coiners gave expression to his suppressed aggressions and that in coiners he found socially acceptable objects on which vicariously to wreak vengeance on his stepfather.[58] In the context both of the times and of Newton's career at the Mint, his prosecution of coiners looks a great deal less peculiar. He served as warden at the time of maximum concern about the debasement of the currency. The new legislation of the 1690s and the immense

[54] *Statutes of the Realm, 6,* 598–600; *7,* 1–4, 94–7, 269–71, 381–2.   [55] *Corres 4,* 209–10.
[56] *Mint* 19.1, f. 439. *Corres 4,* 210, mistakenly includes this canceled conclusion in the letter.
[57] *CSPD* 1696, pp. 302, 306, 320, 343, 353, 354, 362, 372, 398, 403, 406.
[58] Frank E. Manuel, *A Portrait of Isaac Newton* (Cambridge, Mass., 1968) pp. 234–5.

attention to the problem before his appointment make it clear that Newton did not invent the pursuit of coiners by a quirk of his own tortured psyche. Moreover, the prosecution of coiners continued under other wardens after Newton's elevation to the mastership, though less and less connected with the warden himself because it was increasingly professionalized in the hands of his special assistants, men such as Calverly Pinckney, who filled the position with energy equal to Newton's during the 1720s. Nor is it evident that Newton invested more concern in this aspect of his job than in others. Typically, he began with a study of its background; he collected papers about earlier prosecutions and composed a short essay on the history of such during the past twenty-five years.[59] He had done the same for the Mint as a whole, and he proceeded to devote to counterfeiters the same undivided intensity he gave to everything he did.

It is certainly true, as Professor Manuel contends, that he invested great energy in the task. Conduitt, who as an officer of the Mint would have appreciated its significance, commented on the trouble Newton took with coiners and clippers—"took all informations of wch wee burnt boxfuls–& attended all the trials." He also gave Newton credit for the act of 1697 against those who made or repaired tools for coining.[60] Suggestions for legislation on counterfeiting do exist among Newton's papers, though not for that particular act.[61] Newton's accounts as warden also confirm Conduitt's statement in an entry of £120 for his expenses "in Coachire and at Taverns, Prisons and other places in the Prosecution of Clipp^rs and Coyners . . ."[62] His correspondence and surviving depositions fill in the details of the expenditure. The conviction of a coiner required witnesses. Newton was industry itself in finding them. The surviving book of depositions contains as many as fifty-eight depositions before him in the space of two months, depositions taken in the taverns mentioned in his account and in the less salubrious atmosphere of Newgate and other prisons in London.[63] We can only speculate how many more those boxfuls that Conduitt helped him burn contained. Among his correspondence, some of the end-

---

59 *Mint* 19.1, ff. 430–4, 425.

60 *Keynes MSS* 130.6, Book 2; 130.7, Sheet 2.

61 *Mint* 19.2, f. 92. In December 1701, the officers of the Mint sent a letter with proposed changes in legislation against coiners (*Mint* 1.7, p. 37; summary in *Corres 7*, 422–3). Newton mentioned his suggestions for the act of 1697 in a memorandum of 1698 on Chaloner (*Corres 4*, 261).

62 Public Record Office, E351, Roll 2073. Newton had requested that allowance of £120 on 1 Oct. 1699 (*Corres 4*, 317); it was authorized on 9 November.

63 *Mint* 15.17. Cf. John Craig, "Isaac Newton and the Counterfeiters," *Notes and Records of the Royal Society*, 18 (1963), 136–45, and "Isaac Newton–Crime Investigator," *Nature*, 182 (1958), 149–52.

less letters required to transfer prisoners from jail to jail in order to have them present in court survive, only an indication of the much larger number he must have written.[64] He had himself commissioned as a justice of the peace in all of the home counties.[65] His accounts showed the operation of agents in eleven counties. He bought special clothes for Humphrey Hall "to qualify him for conversing with a Gang of coyners of Note in order to discover them." The accounts named twenty-eight coiners whom he prosecuted successfully. In addition they included several payments for the discovery of coiners (in the plural) plus one payment to a man who prosecuted twenty-six persons in addition to seventeen whom he convicted.[66] With the additional names that appear in the Treasury Books, Treasury Papers, and State Papers, there is evidence for the pursuit of far more than a hundred coiners, not all by Newton personally, of course.

The world of coiners and clippers had never known a prosecutor so pertinacious; and if the testimony Newton heard were correct, they soon came to regard him with special venom. On 16 September 1698, Samuel Bond deposed that a month earlier, when he was in Newgate, he heard one of the objects of Newton's attention, Francis Ball,

> complain of ye Warden of the mint for severity ag^t Coyners and say Damne my blood I had been out before now but for him (meaning out of Newgate) and Whitfield who was also there in prison made answer y^t the Warden of the mint was a Rogue and if ever King James came againe he would shoot him and the s^d Ball made answer God dam my blood so will I and tho I dont know him yet Ile find him out.[67]

Though Newton showed unexpected zeal in their apprehension, he was not the only one interested in the coiners, as we need constantly to remember in assessing his activity. One of his biggest catches was Captain William Wintour of Gloucestershire. In February 1697, Wintour was reported in the House of Commons to have worked in collaboration with several receivers of revenue. The House instructed its committee investigating the Mint to interview him in prison.[68] The State Papers reveal that through 1696 and 1697 the highest councils of the government concerned themselves al-

---

[64] *Corres 4*, 211–12. Cf. pp. 218, 219–20, 247, 248, 249.

[65] As far as I know, there is solid evidence, in letters about prisoners, of his being a justice of the peace only in Surrey and Middlesex (*Corres 4*, 211, 212). He possessed an account of the disbursements of his predecessor, Benjamin Overton, to be made J.P. in all seven home counties; it is highly probable that he did as Overton had done (*Corres 4*, 217).

[66] *Mint* 19.1, ff. 475–8.

[67] *Mint* 15.17, no. 27. Cf. the similar threat of Gibbons in 1697; *CSPD 1697*, p. 439.

[68] *Journals of the House of Commons, 11*, 701.

most on a daily basis with coiners. As I have indicated, the lords justices virtually compelled Newton to take that part of his responsibility seriously.

If the financiers of London were a breed foreign to Newton's experience, the assortment of riffraff, desperadoes, and unfortunates whom he interviewed in Newgate were stranger yet. By and large, the women fell into the final category. Terrorized and brutalized by the men, who did not hesitate to deliver them to the scaffold for the reward when occasion arose, they eked out a sorry existence on the far side of the law. Newton heard of a Mrs. Ilbury who confessed that the profit she gained by clipping and filing "was so small that she must find some other Invention or way of living for that would not do." Edward Jones, alias Ivy, belonged to the riffraff. Terrified to find himself face to face with the ultimate penalty, he was, in words that the Mayor of Bristol applied to another coiner, "mightily desirous of meriting his life . . ." That is, he attempted to save his own neck by putting the noose around the necks of as many others as he could name.[69] Prominent among them was the sinister porter at Whitehall, John Gibbons, who lived on the blackmail he took from coiners and clippers, apparently mostly female, who lived in mortal fear of him.[70] The world of coiners and clippers that Newton now learned to know had little of the celestial majesty that had recently been his sole concern.

Of all the coiners, none was more colorful or more resourceful than William Chaloner, one of those whom the lords justices sent to Newton for questioning in the summer of 1696. Before he had done with Chaloner, Newton had collected most of his life history, which he set out in a memorandum for Parliament. "A japanner in clothes threadbare, ragged, and daubed with colours," Chaloner "turned coiner and in a short time put on the habit of a gentleman." He took up his new trade about 1690, working at first mostly in foreign coins, which passed continually in the daily life of London. An artist among counterfeiters, he was the author of a new method of coining, which Newton found the most dangerous he had met. The precise nature of the new method nowhere appears in the surviving records, although it apparently involved casting. Edward Jones, who could not thrust vanity away even in his hour of peril, recommended his own skill to Newton by comparing himself favorably with Chaloner. He also had flair; and alone among the ominous and distasteful denizens of the coining world, Chaloner saw the possibilities in playing both sides of the street. He started, not with coinage, but with Jacobite propaganda. In the early 1690s, he

---

[69] *Mint* 15.17, nos. 5, 106, 31.

[70] See especially *Mint* 15.17, no. 53. Many other depositions concerned Gibbons: nos. 31, 42–6, 52, 88–91, 96–7, 99, 104.

cajoled a couple of printers into producing some declarations in favor of James, and then turned them in for a reward of £1,000. In his own private terminology, which even his associate had to interpret, he "funned" (this is, deceived) the king of £1,000. Impressed by the ease of his gain, he discovered a plot to cheat the Bank, and funned it of £200.[71] Unfortunately for Chaloner, he lacked the sense to realize that he could not play the same act to the same audience forever. In February 1696, just before Newton arrived, he tried it a third time by submitting two papers to the Privy Council on abuses in the Mint and methods of preventing counterfeiting.[72] Shortly thereafter, he made the acquaintance of Newton by suggesting one of his associates, Thomas Holloway, as a man suitable to be his special clerk to pursue coiners. All was not merry deception of the establishment, however. The unfortunate printers went to the gallows, and Chaloner contrived to have two coining associates "hanged out of the way" when they informed on him under duress.

Newton's serious concern with Chaloner began in 1697 when he decided to fun the Parliament. Chaloner testified to the committee investigating abuses in the Mint that he could improve the coinage at no extra cost in a way that would prevent counterfeiting. His imagination now running riot, he proposed that he should be installed as supervisor of the Mint to oversee his improvements. Though Neale, the master, was a member of the committee (perhaps as regular in attendance there as he was at the Mint), Chaloner was plausible enough that Newton received an order from the chairman to provide him with the tools he needed to demonstrate his method.[73] Since one of the changes proposed involved edging, the machine for which Newton had sworn he would not reveal, he refused the order, though not without raising further problems with Commons. The committee's report stated that Chaloner had given "undeniable Demonstrations" of a much better way of coining that would prevent counterfeiting.[74] He even published a flyer on the subject, filled with reflections on the officers of the Mint, and he claimed to be writing a book about it.

He also thought he would fun the government once more with the Jacobite gambit. It was once too often. Skeptical lords justices heard his charges in June; they sought further information. In August, they heard Newton's testimony of Chaloner's ongoing enter-

---

[71] Newton's composition, "Chaloner's Case"; *Corres 4*, 261–2.

[72] *Mint* 1.1, no. 197 (not part of Newton's papers). Three years earlier, on 5 Jan. 1693, Chaloner had petitioned for a grant to set a seal on shears to prevent clipping. In October 1693, he had petitioned to coin farthings and halfpence from copper and pennies and twopence from an alloy of silver and copper (*CTB 10*, 4, 415).   [73] *Corres 4*, 231–2.

[74] *Journals of the House of Commons*, 11, 774–7.

prise in coining; and early in September, they charged Newton not to let Chaloner, then in prison, be bailed.[75] Had Chaloner known that the highest council of the government was hearing his case on a regular basis during the summer of 1697 and expressing its desire to execute him for treason if it could collect enough evidence to convict him, he might have chosen to lie low. Not knowing, he rushed on to his fate. Ever audacious, he petitioned Parliament, charging that the Mint was trying to destroy him in revenge for this testimony against them the previous session.[76] Chaloner's petition exercised Newton greatly, and he devoted considerable attention to a reply that would enlighten the House on Chaloner's real character.[77] He might have spared himself the worry. The committee appointed to investigate the petition included Secretary Vernon, Montague, and Lowndes, who were all aware of Chaloner's activities. Chaloner was not aware of theirs.

He had one last act to play, for he succeeded in gaining his release from prison sometime early in 1698. As Newton later learned by diligent enquiry, he bought off the chief witness against him, Thomas Holloway, and got him to flee to Scotland. One Henry Saunders told Newton that when he visited Chaloner in Newgate to tell him Holloway had gone, he "seemed to be very joyfull and said a fart for ye world." As soon as he was released, he organized or joined a flatulent enterprise to forge malt tickets, one of the new devices of paper currency, issued in connection with a duty on malt enacted the previous year. As he told his associate Carter, "it was as good a time as ever if people were but true to one another." True to one another was what they could not be, neither Chaloner nor his confederates. As soon as anyone was arrested (and jail was a second home to them all), he sought to save himself by implicating the others. Already in May, Montague and Secretary Vernon began to receive evidence about the plan.[78] It was not Newton but Vernon, who had become convinced the previous autumn that Chaloner was too dangerous to be at large, who engineered his doom and issued the final warrant for his arrest. It was Newton, however, who sealed his fate by weaving a net of evidence about him from which he could not escape. When Holloway returned to London from Scotland, Vernon sent news of him to Newton, who took him into custody. Already in 1698, in one draft of his memorandum to Parliament on Chaloner's petition, Newton cited fourteen witnesses against him. In late 1698 and early

---

75 *CSPD 1697*, pp. 202, 340, 351. The lords justices continued active concern with him on into October.

76 *Corres 4*, 259–60. The House of Commons heard the petition on 18 February (*Journals of the House of Commons, 12*, 119).

77 *Corres 4*, 261–2. Drafts in *Mint* 19.1, ff. 496, 499, 501–2, 503.

78 *Mint* 5.17, nos. 98, 116. *CTB 13*, 87–91.

1699, he took more than thirty further depositions as he pieced together the story both of the malt-ticket caper and of earlier activities. When Chaloner returned to Newgate, Newton constructed a circle of spies to inform him of Chaloner's every stratagem. With his customary aplomb, which had not yet deserted him, Chaloner wrote to Newton from prison.

> I have been close Prisoner 11 weeks and no friend sufferd to come neer me but my little child I am not guilty of any crime, and why am I so strictly confined I do not know   I doubt Sr You are greatly displeased with me abo^t the late business in Parliam^t but if you knew the truth you would not be angery with me for it was brought in by some persons ag^t my desire
>
> S^r I presume you are satisfied what ill men Peers and ye Holloway's are who wrongfully brought me into a great deal of trouble to excuse their Villanys   Wherefore I begg you will not continue your displeasure ag^t me for I have sufferd very much So I wholy throw my self upon your great Goodness[79]

At much the same time, he was telling his confederate Carter that "he would pursue that old Dogg the Warden to the end so long as he lived . . ." One of the "old Dogg's" spies, John Whitfield, who needed to atone for his own earlier threats, duly passed it on.[80]

Despite a bevy of lawyers in daily attendence and a pretense of madness, Chaloner was convicted of high treason on 3 March 1699. He had cut a sufficient figure and could afford lawyers good enough that the king himself heard his petition for pardon on 17 March.[81] As he faced the terrible punishment that bloody imaginations had devised for high treason, he finally collapsed.

> Most mercifull S^r
>
> I am going to be murther^d allthough perhaps you may think not but tis true I shall be murdered the worst of all murders that is in the face of Justice unless I am rescued by yor mercifull hands.
>
> S^r
>
> pray considr my unprecedented Tryall 1 That no person swore they ever saw me actually coyn y^t I should own it . . . 6^{ly} It was hard for me to be taken out of my 5 weeks Sick Bedd the last 3 weeks light headed So y^t I was not provided for a tryall nor in my Senses w^n tryd 7^{ly} W^t Mrs Carter swore ag^t may appear direct mallice I have 3 yeares before Convicted her husband of Forgery and discoverd where he and she were coyning for wch he is now in Newgate But I desire God Allmighty may Damne my Soul to eternity if every word was not false that Mrs Carter and her Maid swore ag^t me abo^t

[79] Chaloner to Newton, late Jan. 1699; *Corres 4*, 305.
[80] Whitfield's deposition of 22 Feb. 1699; *Mint* 15.17, no. 121.
[81] *CSPD* 1699–1700, p. 103.

coyning and Mault Tickets for I never had any thing to do with her in coyning nor ever intended to be concerned in Mault Tick^{ts} nor ever spoke to her abo^t any such things Mrs Holloway swore false ag^t me or I desire never to see the Great God and I desire the same if Abbot did not swear false ag^t me so y^t I am murder^d O God Allmighty knows I am murder^d Therefore I humbly begg yor Wor^p will consid^r these Reasons and y^t I am convicted without Preced^{ts} and be pleased to speak to Mr Chancell^r to save me from being murtherd O Dear S^r do this mercifull deed O my offending you has brought this upon me O for Gods sake if not mine Keep me from being murder^d O dear S^r no body can save me but you O God my God I shall be murder^d unless you save me O I hope God will move yor heart with mercy and pitty to do this thing for me I am

Yo^r near murder^d humble Servt
W. Chaloner[82]

It wasn't possible. The government that existed only for Chaloner to fun wasn't really going to execute its unthinkable sentence on him. In fact it was, and Newton could not have stayed the remorseless machine of justice had he wished.

> Thursday, 23 March [Luttrell recorded]. Yesterday 7 of the criminals, condemned last sessions at the Old Baily, were executed at Tyburn; Chaloner, for coining, drawn on a sledge; Mr. John Arthur, for robbing the mail, was carried in a coach; and 5 other men, for robbery and burglary.[83]

Further entertainment awaited Chaloner at Tyburn before death put an end to his suffering and to his career in coining.

Before we judge Newton too facilely for his evident unwillingness to hear such cries for mercy, it is worth recalling that within six months he witnessed a deposition that John Lawson had returned to coining. Lawson was one of his informants in Newgate who purchased his freedom by spying on Chaloner and by protesting eternal obedience. Chaloner's confederate Carter, who tried as eagerly to save himself at Chaloner's expense as Chaloner tried at his, evaded the law somewhat longer but was eventually apprehended again in 1704. Newton's experience indicated that recidivism was universal and pleas for mercy unfailingly mendacious. Indeed Chaloner's effusion may have been his final effort to fun the Mint. On the day after his execution, Carter, in jail and hoping to get out, wrote, "I hear Chaloner dyed no otherwise then he lived but persisted to the last how insolent he was for w^t he dyed."[84]

---

[82] *Corres 4*, 307–8.

[83] Narcissus Luttrell, *A Brief Historical Relation of State Affairs*, 6 vols. (London, 1857), 4, 496–7.

[84] *Mint* 15.17, nos. 225, 130. *CTB 19*, 386.

Although the Mint dominated Newton's early life in London, it did not constitute its whole. For a decade before the move, his closest friends had been in London and Oxford, and he continued to see them. From time to time David Gregory wrote memoranda to himself about his conversations with Newton.[85] He sought Newton's help on a proposition of Euclid, whose works he was publishing.[86] Since Gregory's personal concerns were never far from his own consciousness, it is not surprising that he sought Newton's support for an "establishment in London that is consistent with what I have here" – to wit, the position of mathematical tutor to Anne's son, the Duke of Gloucester.[87] Having found one place for Halley, Newton also continued to look out for his interests. Early in 1697, he informed him of two posts with the military engineers for which he had recommended him.[88] Although Fatio was in London at least part of the time and attended some meetings of the Royal Society, no suggestion indicates that Newton saw him.

He maintained his connection with the Mathematical School at Christ's Hospital. At the beginning of July 1697, he examined five boys, and at a meeting of the court on 13 July reported that he found them fit to be placed at sea as apprentices. Twice more during 1697, the minutes of the court of the Hospital indicated his presence when matters concerned with the school came up.[89]

The Royal Society did not figure prominently in Newton's life. Charles Montague was its president when he arranged Newton's appointment to the Mint, and he was reelected in November in both 1696 and 1697. In the latter year, the society also elected Newton to the council. He did not attend a single meeting either of the council or of the society during the year, nor did he attend one between his move and November 1697. Montague as president attended an equal number. During 1699, Newton did finally make an appearance. When the society received a copy of a work on *Analysis geometrica* by a Spaniard (undoubtedly Hugo de Omerique), they sent it to Newton for an opinion. At the following meeting, on 19 April, Newton reported in person that the author was "of the same opinion with y^t of the ancients." In August, Newton appeared a second time to display a new sextant which he had invented. At the following meeting, Hooke claimed the invention as his own. Encouraged by his presence, the society elected Newton to the council again in 1699, but he repeated his former

---

[85] He wrote one such on 20 Feb. 1698; *Corres 4*, 265–6. Another on 21 May 1701, with questions he wanted to ask Newton, appeared to look forward to an expected meeting (*Corres 4*, 354–5).

[86] Gregory to Newton, 30 Sept. 1701; *Corres 4*, 391–2.

[87] Gregory to Newton, 23 Dec. 1697; *Corres 4*, 253.

[88] *Corres 4*, 229.   [89] *Edleston*, p. 299.

record of attendance at its meetings. During the year, he did show up at another meeting of the society, however, and he did respond to a couple of enquiries the society addressed to him.[90] No one would wish to argue that the Royal Society was a significant factor in Newton's decision to move to London.

If he maintained his old friendships with the younger generation of scientists, Newton found himself plunged for the most part into a society new to him, a society made up in part of the upper tier of the political nation, in part of the upper stratum of the London financial world. Four of the five men who followed him as warden, during the period when he served as master of the Mint, were members of Parliament. They tended to hold other places in the government's gift as well as the wardenship. Newton's immediate successor, Sir John Stanley, the only one not in Parliament, made up for the deficiency by even more appointments, which came his way via family connections. He moved on from the Mint to the Customs Commission, and became among other things lieutenant of the Tower. All of them, with the exception of Stanley again, came from Oxbridge, more recent copies of the pensioners Newton had known as an undergraduate. He had not liked them then; there is no evidence that he found them more palatable now. With one, Craven Peyton, he lived at swords' points. Beyond the appearance of their names with his in official business of the Mint, they did not enter his life. Thomas Hall, whom Newton found at the Mint on temporary assignment to supervise the recoinage, resembled the wardens except in one thing. Where they were essentially place-men, he had proved himself as a man of ability. Cashier of the Excise, comptroller of the Salt Office, commissioner of Customs, he was one of the new breed of professional civil servants, of which Newton became another. Some sort of enduring relation existed between them. In 1700 and 1702, Hall stood security for Newton at the signing of his Indentures, and his son did so in 1718. Newton in turn was an executor of Hall's will.[91]

In 1702 and 1718, another man whom Newton found at the Mint when he arrived, John Francis Fauquier, Neale's deputy, also stood security for him. Newton retained him as his own deputy for more than twenty-five years until Fauquier's death. He also employed

90 *JB 10,* 53, 114–15, 118, 145, 167, 175, 178, 181, 193; *CM 2,* 159. His plan of an improved sextant was published years later: "A Description of an Instrument for observing the Moon's Distance from the Fixt Stars at Sea," *Philosophical Transactions, 42* (1742-3), 155–6. Hooke was not without some justification since he had described a similar instrument (*The Posthumous Works of Robert Hooke* [London, 1705], p. 503). Apparently he had not published it by 1699, however.
91 Sir Charles Godolphin to Newton, 17 Dec. 1718, and Francis Hall to Newton, 2 Sept. 1719; *Corres 7,* 24–7, 59–60. Trinity College, R.16.38, f. 433.

Figure 12.2. Newton at fifty-nine. Sir Godfrey Kneller, 1702. (Courtesy of the Trustees of the National Portrait Gallery.)

him as his financial agent. The title of deputy is apt to mislead. Fauquier cut an important figure in the London financial world, in which a circle of Huguenot refugees was prominent. A governor of the Bank, a man of substance, he was able to subscribe £26,400 to the doubling of the Bank's capital in 1709. Through the Mint, Newton came in touch with other men prominent in the financial world. James Hoare, long-time comptroller when Newton arrived (though he would soon die), had founded Hoare's Bank. Leading goldsmiths, such as Peter Floyer, Philip Shales, Jonathan Ambrose, and John Cartlich, served the Mint as melters and refiners; that is, Newton as master contracted with them for these services. Newton was to have his conflicts with the goldsmiths, who wanted to dominate the Mint to their own advantage. Apparently he distrusted Ambrose, whom he never employed. On the other hand, Floyer stood security for Newton in 1700. Grasping and hard the financiers may have been, but they were also men of accomplishment. The slender evidence that exists, especially the case of Fauquier, suggests that Newton found them more to his taste than the sinecurists who played at being warden.

Most of what we know about Newton's daily life in London comes from a later time—a few miscellaneous bills, the inventory of his goods and chattels after his death, the bills paid by his estate, and the comments of John Conduitt, who lived with him for a number of years. While we should exercise some caution in using them, there is no reason to think that Newton later changed his habits in any marked degree. Conduitt's summary fits the evidence admirably. "He always lived in a very handsome generous manner tho^u without ostentation or vanity, always hospitable & upon proper occasions gave splendid entertainments."[92] That is, without attempting magnificence, he lived in a style consistent with his new dignity. In summarizing the inventory of his goods, Villamil contrived to paint a picture of spartan utilitarianism in his household furnishings.[93] The inventory showed a well-furnished house, however, and I do not see how one can judge the quality and artistry of furniture, as Villamil did, from bare descriptions as tables and chairs and the like. One surviving bill recorded the purchase of four landscapes to decorate the walls and twelve Delft plates.[94] The inventory, with three dishes, three salvers, a coffeepot, and two candlesticks (all of silver), forty plates, a full set of silver flatware, about ten dozen glasses and six and a half dozen napkins, demonstrated that he had the equipment for the occasional splendid entertainments that Conduitt mentioned. For other needs, he possessed no fewer than two silver chamberpots, which no one would characterize as spartan utilitarianism. He owned clothes valued at only £8 3s 0d; but by the time of the inventory, Newton had been a semi-invalid for five years afflicted with inconti-

---

[92] In the account he sent to Fontenelle; *Keynes MS* 129A, p. 18.   [93] *Villamil*, pp. 15, 35.

[94] *Yahuda MS* 7.3p. There is an uninformative receipt for miscellaneous small objects from a brazier in *Mint* 19.5, f. 41, and a bill from William Grindall, a turner, for various items and services (which appear to be for Newton's house rather than the Mint) from June 1712 to February 1713, in *Babson MS* 734.

nence of urine, which would take its toll of any wardrobe. It is true that the total assessed value of the household furnishings came to scarcely more than £400, whereas Newton left securities valued at more than £30,000. Since we do not know the accepted practice in assessing such used property, it is impossible to reach any conclusions about Newton's style of life from the two sums.[95] As Villamil remarked, he had a penchant for crimson – crimson draperies, a crimson mohair bed with crimson curtains, crimson hangings, a crimson settee. Crimson was the only color mentioned in the inventory, and Villamil suggested with justice that he lived in an "atmosphere of crimson."[96] Before his late illness, he apparently kept a coach; and he maintained a stable of servants, six at the time of his death.[97] Years before, Newton had resented his servile status as a sizar. There is every reason to believe that he now seized the opportunity to adopt the style of the better circles of London society, and that he relished doing so.

As for his table, Conduitt reported that he was always very temperate in his diet. One note spoke of him living on vegetables, though another denied that he abstained from meat.[98] Perhaps Conduitt's information was consistent with the judgment of the Abbé Alari, the instructor of Louis XV, who dined with him in 1725 and found the repast detestable. He complained that Newton was stingy and served poor wines which had been given to him as presents.[99] Since French visitors invariably commented on English cuisine in a similar vein, we cannot conclude too much from Alari's dyspepsia. A bill showing the delivery of one goose, two turkeys, two rabbits, and one chicken to the household in the space of a single week reminds us that Conduitt was applying eighteenth-century standards when he described Newton's diet as temperate. After his death, his estate settled a debt of £10 16s 4d with a butcher and two others, which totaled £2 8s 9d, with a poulterer and a fishmonger. In contrast, he owed the "fruiter" only 19s and the grocer £2 8s 5d.

---

[95] The inventory of Newton's "Goods, Chattels and Credits" taken 21–27 April 1727 is published in *Villamil*, pp. 50–61.   [96] *Villamil*, pp. 14–15.

[97] *Mint* 19.3, f. 170, from about 1708, indicates that he kept two maids and a manservant in the house on Jermyn Street. The inventory of the St. Martin's Street house shows rooms for the same number. His executors paid back wages due for about four and a half months to six servants and gave them all a year's wages. Mrs. Rogers, the housekeeper, received £16 per annum, the cook £10, a housemaid £6, and another maid, Ann Wallis, who tended the house on St. Martin's Street after Newton moved to Kensington, £6. Mr. James Woston, whose duties were not specified, received £12 per annum, and Adam the footman, £6 (Huntington Library MS of the executors' accounts). Citing no source, Brewster asserted that Newton kept three male and three female servants (David Brewster, *Memoirs of the Life, Writings, and Discoveries of Sir Isaac Newton*, 2 vols. (Edinburgh, 1855) 2, 410–11).

[98] *Keynes MSS* 129A, p. 23; 130.6, Book 2; 130.7, Sheet 4.

[99] *Edleston*, pp. lxxviii–lxxix.

A bill of £7 10s 0d for fifteen barrels of beer again suggests less than heroic temperance.[100]

The move to London did not alter his habits. He hardly sampled the recreations afforded by a great metropolitan center. According to Conduitt, he never diverted himself with music or art. "Never" was too strong, for Newton did tell Stukeley that he went to the opera once. He found it too much of a good thing, like a surfeit at dinner; "the first act, said he, I heard with pleasure, the 2^d stretched my patience, at the 3^d I ran away." As for art, all he could think to say about the Earl of Pembroke's famous collection of statues was that Pembroke "was a lover of stone Dolls." His penchant for study did not leave him even if he did little but reshuffle old ideas. After he came to London, Conduitt reported, "all the time he had to spare from his business & the civilities of life in wch he was scrupulously exact & complaisant was employed the same way [in study] & he was hardly ever alone without a pen in his hand & a book before him."[101] His papers indicate that he continued most of his earlier pursuits, though not necessarily in the same proportion. Villamil remarked the absence from his library of any of the English classics – Chaucer, Spenser, Shakespeare, and Milton – and of poetry in general. Years later, on doubtful authority, he was reported to have described poetry as "a kind of ingenious nonsense."[102]

In keeping with his new station in life, he began to assume the role of patron. Along with Montague, Somers, and the Earl of Dorset, and also with Martin Lister, Tancred Robinson, Hans Sloane, and Francis Aston of the Royal Society, he financed the publication of Edward Lhuyd's *Lithophylacii britannici ichnographia* in 1699.[103]

If Newton continued former studies, they in turn refused to leave him alone. On 29 January 1697, he received two challenge problems issued by Johann Bernoulli. Bernoulli had published one of them originally in the *Acta eruditorum* the previous June – to find the path by which a heavy body will descend most quickly from one point to another that is not directly beneath it – and he had set a limit of six months to the challenge. When December came, he had not yet received a satisfactory answer, though he had received a letter from Leibniz with both the assertion that he had solved the problem and the request that the time be extended to Easter and the

---

[100] *Yahuda MS* 7.3p. Huntington Library MS.

[101] *Keynes MSS* 136, p. 12 (cf. *Stukeley*, p. 14); 130.7, Sheet 4; 129A, pp. 15–16.

[102] Joseph Spence, *Anecdotes, Observations, and Characters, of Books and Men*, ed. Samuel S. Singer (London, 1820), p. 368.

[103] Lhuyd, *Ichnographia*, new ed. (Oxford, 1760), n.p.

problem republished throughout Europe. In accepting Leibniz's request, Bernoulli added a second problem – to find a curve such that the sum of two segments, *PK* and *PL*, on a line drawn at random from a point *P* to cut the curve in two points *K* and *L*, though the two segments be raised to any power, is a constant. Bernoulli had copies of the problems sent to the *Philosophical Transactions* and the *Journal des sçavans*. He also sent copies to Wallis and Newton. Recall that earlier in 1696 Bernoulli had expressed the opinion that Newton had filched the method that he first published in Wallis's *Opera* from Leibniz's papers. Manifestly, both Bernoulli and Leibniz interpreted the silence from June to December to mean that the problem had baffled Newton. They intended now to demonstrate their superiority publicly. In case the direct mailing to him were not pointed enough, Bernoulli inserted a scarcely veiled reference in the announcement itself. He and Leibniz would publish their solutions at Easter, he stated.

> If geometers carefully examine these solutions, drawn as they are from what may be called a deeper well, we are in no doubt but that they will recognize the narrow limits of the common geometry, and will value our discoveries so much the more as there are fewer who are likely to solve our excellent problems, aye, fewer even among the very mathematicians who boast that by the remarkable methods they so greatly commend, they have not only penetrated deeply the secret places of esoteric geometry but have also wonderfully extended its bounds by means of the golden theorems which (they thought) were known to no one, but which in fact had long previously been published by others.[104]

Whatever Bernoulli – and Leibniz – had in mind, Newton saw the problems as a challenge issued personally to him. It etched its recollection deeply enough on his consciousness that two years later, when Flamsteed exasperated him, he used the incident of 1697 to browbeat the Astronomer Royal: "I do not love . . . to be dunned & teezed by forreigners about Mathematical things . . ."[105] In 1697, he accepted the challenge by recording on the paper the time at which it arrived. "I received the sheet from France, Jan. 29. 1696/7." He dated a letter to Charles Montague, president of the Royal Society, which set down the answers to both problems, 30 January.[106] His sense of triumph was great enough that the story made its way, via his niece Catherine, into Conduitt's collection of anecdotes. "When the problem in 1697 was sent by Bernoulli – Sʳ I. N. was in the midst of the hurry of the great recoinage did not come home till four from the Tower very much tired, but did not

---

[104] *Corres 4*, 225–6; original Latin, p. 222. I have altered the translation slightly.
[105] Newton to Flamsteed, 6 Jan. 1699; *Corres 4*, 296.   [106] *Corres 4*, 228, 220–4.

sleep till he had solved it wch was by 4 in the morning."[107] In addition to Leibniz's solution, Bernoulli received two others, one from the Marquis de l'Hôpital in France and an anonymous one from England. Disabused on Newton's skill in mathematics, Bernoulli recognized the author in the authority the paper displayed – "as the lion is recognized from his print," in his classic phrase.[108] Leibniz found himself sufficiently embarrassed by the debacle that he wrote to the Royal Society denying that he had been the author of the problem.[109] Newton underlined his understanding of the incident by planning to issue his own challenge problem to Leibniz and Bernoulli, though in the end he chose not to.[110]

Even without the stimulus of a challenge problem, Newton also allowed the moon to occupy him somewhat. Already by September 1697, a letter from Flamsteed indicated that the two had discussed the moon in London; and in December of the following year, Newton visited Greenwich.[111] By now, the basic misunderstanding between the two men had become too deep to be healed. Unfortunately, Flamsteed discovered repeated errors in the computed places of the moon that he continued to furnish to Newton, and he sent the corrections with obvious embarrassment. The errors had been mere inadvertence, he protested. "I must entreat you to lay by all your apprehensions of any Intended practise for I assure I had none whatever you may suppose or hath been suggested to you to ye Contrary."[112] As repeated corrections continued to arrive, Newton must have wondered about the reliability of the data he received even if he did not suspect "Intended practise." For his part, Flamsteed also continued to receive information, reports of what Newton was telling others about the moon, and he could not suppress his fear of betrayal. "In your letter you say these corrections will Answer all my observations within 10 minutes," he wrote in December 1698. "Mr Halley boasts that those you have given him will

107 *Keynes MS* 130.5, Sheet 1; 130.6, Book 1. In the sketch he sent to Fontenelle, Conduitt left out the "4 in the morning" to give the impression that Newton had solved it with greater speed (*Keynes MS* 129A, pp. 10–11). D. T. Whiteside suggests that his need for twelve hours to solve the problems was an indication that his decline had set in ("Newton the Mathematician," unpublished paper).

108 In Latin, of course: "tanquam ex ungue leonem." It is indicative of Newton's pride in his solution that he believed that l'Hôpital did not succeed without help.

109 *JB 10*, p. 52.   110 Gregory's memorandum of 20 Feb. 1698; *Corres 4*, 266.

111 *Baily*, pp. 63, 65. On the margin of his own copy of his letter to Newton of 29 Dec. 1698, Flamsteed mentioned the visit he received on 4 December. (*Baily*, p. 165). In addition to the correspondence on the moon cited below, cf. tables of lunar positions that can be dated with certainty to the late 1690s; *Add MSS* 3966.13, ff. 117, 120; 3966.14, f. 173; 3966.15, ff. 198, 299, 302.

112 Flamsteed to Newton, 4 Sept. 1697; *Corres 4*, 249–50. Alas, he had to send further corrections on 10 December, 10 Oct. 1698, and 29 Dec. 1698, and according to his own note he supplied others when he met Newton (*Corres 4*, 251–2, 286, 290–1. *Baily*, p. 166).

represent them within 2 or 3 or Nearer."[113] Had Flamsteed been privy to all that passed between Newton, Halley, and Gregory, his paranoia would have flourished still more luxuriantly. Already in July 1698, before he visited Greenwich to get more observations from Flamsteed, Newton told Gregory that the theory of the moon would not be completed because of the Astronomer Royal, "nor will there be any mention of Flamsteed . . ."[114] He went so far as to accuse Flamsteed of plagiarizing his lunar tables from Halley.[115]

Small wonder then that their relations came to a head once more at the end of 1698. Under constant pressure to justify himself by publishing, Flamsteed yielded that year to Wallis's entreaty that he print an account of his supposed observation of stellar parallax in the final volume of Wallis's *Opera*, which was due to appear in 1699.[116] As he put the matter later to Newton, he used his account "to silence some busy people y^t are allwayes askeing, *why I did not print?*"[117] Hence he recited his accomplishments as Astronomer Royal, a new catalogue of the fixed stars, for example, and rectified solar tables. Assailed as he was by repeated rumors, which he could not reasonably doubt, of Newton's complaints in regard to lunar observations, including one story that Newton claimed to have rectified the lunar theory with Halley's observations, Flamsteed also added a paragraph on what he had done in that respect.

> I had also become closely associated with the very learned Newton
> (at that time the very learned Professor of Mathematics in the Uni-
> versity of Cambridge) to whom I had given 150 places of the Moon,
> deduced from my earlier observations and her places at the times of
> the observations as computed from my tables, and I had promised
> him similar ones for the future as I obtained them, together with my
> calculations, for the purpose of improving the Horroxian theory of
> the Moon, in which matter I hope he will have success comparable
> to his expectations.[118]

---

[113] *Corres 4*, 291. Writing to Colson on 10 Oct. 1698, Flamsteed denied a story Colson had heard that Newton was saying he had used Halley's observations to perfect the theory of the moon! (*Baily*, pp. 162–3.) He later took it more seriously.

[114] *Corres 4*, 277; original Latin, p. 276.

[115] A memorandum by Gregory in the margin of his copy of the *Principia*, referring especially to a conversation in December 1698; printed in Brewster, *Memoirs, 2*, 165–6.

[116] Stellar parallax, a displacement in the angular position of a fixed star in observations made from opposite sides of the earth's orbit, would have offered direct evidence of the annual motion of the earth around the sun, something no competent natural philosopher questioned any longer. Thirty years later, James Bradley showed that Flamsteed mistook aberration for parallax, which is small enough that it had to wait for more refined instruments in the nineteenth century before it could be observed. Aberration, an optical phenomenon, also derives from the motion of the earth.

[117] Flamsteed to Newton, 2 Jan. 1699; *Corres 4*, 292.

[118] *Corres 4*, 293. I have altered one phrase in the translation. On a margin of Newton's reply (6 January) to his letter next to a comment that Newton did not want to have work that

Since Wallis's colleague David Gregory was in London, Flamsteed sent the paper to Wallis via him. As Flamsteed later pieced the story together, Gregory read the paper, saw the paragraph that mentioned Newton, and sent Newton an account of it as he forwarded the paper to Wallis. Flamsteed assumed that Gregory was deliberately attempting to foment trouble to aid his pursuit of the tutorship to the Duke of Gloucester, a position Gregory knew was promised to Flamsteed. In the light of Gregory's letter to Newton about that appointment, the suspicion is credible. Be that as it may, Flamsteed returned home to Greenwich on the night of 31 December, after a visit to London, to find a letter from Wallis awaiting him. Wallis had heard from an unnamed correspondent in London, a friend both of Newton and of Flamsteed, who asked Wallis without stating his reasons not to print the paragraph. Flamsteed recognized Gregory's work, and noted on the letter that he was clearly no friend of his.[119] He wrote to Newton at once.

> S^r My observations lie ye King & Nation in at least 5000 *lib*, I have spent above 1000 *lib.* out of my own pocket in building Instrum^{ts} & hireing a servant to assist me now neare 24 yeares; tis time for me (& I am now ready for it) to let the World see I have done something that may answer this expence. & therefore I hope you will not envy me the honor of haveing said I have been usefull to you in your attempts to restore the Theory of the Moon I might have added the Observations of the Comets places given you formerly of the superior planets & refractions at y^e same time wth y^e $\mathcal{D}$s. but this I thought would look like boasting & therefore forbore it.[120]

He received no response. He wrote a second time and again heard nothing. On 7 January, he wrote to Wallis that Newton apparently was unconcerned and that the paragraph might stand.[121]

He wrote too soon. Newton's letter, dated 6 January, arrived as soon as he mailed his own to Wallis. In it, Newton cast aside the mask of friendship and assaulted Flamsteed with all the brutality of which he was capable.

> S^r
>
> Upon hearing occasionally that you had sent a letter to Dr Wallis about ye Parallax of ye fixt starrs to be printed & that you had mentioned me therein with respect to ye Theory of ye Moon I was concerned to be publickly brought upon ye stage about what perhaps

might never be perfected mentioned publicly, Flamsteed justified himself with a bitter reference to the rumors he heard: "When Mr Halley boast 'tis done & given him as a secret tells ye Society so & forreigners see Mr Colsons letter to me" (*Baily*, p. 166). Colson's letter, mentioned above, repeated a story that Newton said he had perfected the lunar theory with Halley's observations.

[119] *Baily*, pp. 164–5.   [120] *Corres* 4, 294.   [121] *Baily*, pp. 166–8.

will never be fitted for ye publick & thereby the world put into an
expectation of what perhaps they are never like to have. I do not love
to be printed upon every occasion much less to be dunned & teezed
by forreigners about Mathematical things or to be thought by our
own people to be trifling away my time about them when I should
be about ye Kings business. And therefore I desired Dr Gregory to
write to Dr Wallis against printing that clause wch related to that
Theory & mentioned me about it. You may let the world know if
you please how well you are stored wth observations of all sorts &
what calculations you have made towards rectifying the Theories of
ye heavenly motions: But there may be cases wherein your friends
should not be published without their leave. And therefore I hope
you will so order the matter that I may not on this occasion be
brought upon the stage. I am

> Your humble servant
> Is. Newton[122]

In his reply, Flamsteed pointedly mentioned Newton's own
willingness to advertise his lunar theory orally, and took exception
to the implication that his own work, and Newton's as well, were
trifling. But he bowed to the inevitable and told Wallis to remove
"ye Offensive Innocent Paragraph . . ."[123] For a time, Newton's
relations with the Astronomer Royal, which oscillated between
Newton's unremitting contempt for Flamsteed and Flamsteed's un-
quenchable need to win Newton's respect, passed through a trough
on their cycle. In 1700, Flamsteed tried pathetically once more to
restore communication by starting an extended demonstration of
the nutation of the earth's axis based on his observations. Apparently he neither completed nor mailed it.[124]

By this time, Newton was a famous man, and visitors to London
from abroad who were informed about natural philosophy made it
their business to meet him. Early in 1698, Peter I, Tsar of Russia,
was in England on his great embassy to the West. Montague
showed him through the Mint in February. From the letter Newton
received instructing him to be present, it appears that the tsar,
whose turn of mind was wholly technical and scientific, had let it
be known that he was as interested in the warden of the Mint as in
its operations.[125] In 1702, Frans Burman, a Dutch minister, served
as chaplain to the embassy sent by the United Provinces to con-
gratulate Anne on her succession to the throne. A pupil of de
Volder, he sought Newton out on 13 June and discussed the Newtonian system, vortices, and comets. Newton assured Burman that

---

[122] *Corres 4*, 296.   [123] *Corres 4*, 302–3. Baily, p. 168.
[124] *Baily*, pp. 176–87.   [125] *Corres 4*, 265.

he would like to introduce him to the Royal Society were it not for the pressure of business at the Mint; he did send a letter to Halley to perform that office in his stead. The embassy spent an unconscionable time congratulating Anne, with the result that Burman was around London, bumping into Newton, for the rest of the year. In September, he contrived that Newton show him about the Mint and demonstrate its processes (edging presumably excepted); after the tour Newton entertained the party sumptuously.[126] He could please Dutchmen more readily than Frenchmen.

No visit held more significance than that of Jacques Cassini in the spring of 1698. According to Conduitt, who must have repeated what he heard from Newton much later, Cassini offered Newton a large pension from Louis XIV.[127] This could have referred only to an appointment in the Academy of Science, which was then being reorganized. Newton declined. The reorganization also established eight foreign associates, of which the King appointed three and the academy elected five. If Conduitt's story was correct, we can understand why Louis chose not to include Newton among his nominees of foreign associates (Leibniz, Tschirnhaus, and the relatively obscure Italian physicist, Domenico Guglielmini). Newton did accept election by the academy, however, along with Hartsoeker, Roemer, and the two Bernoullis. James Gregory reported that he would receive a pension of 1,500 livres as a foreign associate, a sum of about £125. Since 1,500 livres was the average pension of the twenty full members, whereas no money was allotted to foreign associates, Gregory's report was probably confused. Nothing in Newton's papers referred to a pension, then or later. At the time, Newton did not mention Cassini's offer, so that David Gregory reported with indignation that he had not been named by the king and was chosen last of all by the members.[128]

It was not only France that recognized Newton. The English government and ruling class manifested in their own ways that they too recognized his eminence. References to him as "Dr. Newton," which continued for well over a year after the move to London, suggest the respect his learning commanded.[129] The Earl of Pembroke, who had perhaps not heard the slighting reference to his collection of statues, pressed him to take the mastership of St.

---

[126] Burman's journal of his trip is printed in J. E. B. Mayor, *Cambridge under Queen Anne* (Cambridge, 1870). The references to Newton are on pp. 314–315, 319, 322, 323.

[127] *Keynes MS* 130.7, Sheet 1. On 6 April 1698, on his trip home, Cassini wrote to Newton from Dover, sending some observations of the satellites of Saturn that he had just received from his father in Paris (*Corres 4*, 268–9).

[128] James Gregory to Campbell, 29 May 1699; *Corres 4*, 311. Hiscock, p. 9. Cf. I. B. Cohen, "Isaac Newton, Hans Sloane and the Académie Royale des Sciences," *Mélanges Alexandre Koyré, 2*, 61–116.

[129] As late as 12 Oct. 1697, the Treasury was still referring to him as Dr. Newton (*CTB 13, 3*).

Figure 12.3. Newton in his sixty-fourth year. Portrait by William
Gandy, 1706. The Gandy portrait has apparently been
lost. It survives in the lithograph made in the middle of
the nineteenth century. (Permission of the David Eugene Smith Collection, Rare Book and Manuscript Library, Columbia University.)

Catherine's Hospital to supplement his income at the Mint.[130] When the mastership of Trinity fell open in 1700, it was offered to him if he would take orders. This was to broach dangerous issues, though Newton had learned a thing or two about evading them. "Tenison [the archbishop of Canterbury] importuned him to take any preferm^t that fell said why will not you? – you know more than all of us put together – Why then said S^r I, I shall be able to do you the more service by not being in orders."[131]

Although he served at the Mint, the government utilized him as its resident expert in science on those few occasions when it needed such, thus wasting his time (or filling it) with trivia. In the summer of 1699, he responded to a request from the navy, similar to others he would receive later, to judge a proposal for determining longitude.[132] Although we do not know what the proposal was, it apparently did prick Newton's imagination, and a few days later he appeared before the Royal Society to exhibit his improved sextant. The Royal Society attempted to utilize his gifts in a similar way, and on several occasions referred questions in mathematics and astronomy to him.[133]

London also offered the opportunity of freer theological discussion than Cambridge permitted – a dangerous opportunity, to be used only with care. If the importunities of Archbishop Tenison that he be ordained did not adequately remind Newton of the dangers, the Act of 1698 for the Suppression of Blasphemy and Profaneness, which disabled from pubic office anyone who denied a Person of the Trinity to be God, would have done so. When Newton arrived in London, the ambitious and rising young divine, Richard Bentley, was installed in an apartment in St. James's Palace as keeper of the king's libraries. Bentley had established his reputation in 1691 with an essay on the Byzantine chronicler, John Malalas, a topic sure to attract Newton's eye. As we have seen, Newton may have contributed to his selection as the first Boyle lecturer, and he certainly contributed to the published version of the lectures. The lectures had confirmed Bentley's reputation and insured his future. The sinecure with the king's libraries was only part of the patronage he already enjoyed by 1696. Montague was said to be his

---

[130] *Keynes MSS* 130.6, Book 1; 130.7, Sheet 1.
[131] *Keynes MSS* 130.6, Book 2; 130.7, Sheet 1. Conduitt referred this story to the authority of his wife. In 130.7, the later copy of his collection of anecdotes, Conduitt placed the offer of the mastership in 1683, at the elevation of John Montague. It is inconceivable at that time. Newton was then an unknown figure, a wholly unlikely candidate for rich patronage, and he had been most reticent about communicating his theological learning. The appearance of Tenison in the story, whom Conduitt did not, it is true, link explicitly to the offer of Trinity, makes 1700 more probable yet. In 1700, Tenison was the Archbishop of Canterbury; in 1683, he was merely a rising young cleric.
[132] *Corres 4*, 314. [133] *JB 10*, 114–15, 118–19, 167, 178, 181, 193.

patron, a status which could have been either the effect or the cause of Newton's friendship. Ever ambitious, Bentley thrust himself into the center of London intellectual life by organizing a club which met or at least was intended to meet, as frequently as twice a week in his apartment. We hear of it through a letter of 21 October 1697. Bentley meant to include Sir Christopher Wren, John Evelyn, John Locke, and Newton – a considerable galaxy of talent for a young man to gather about him.[134] Nothing among Newton's papers and nothing in Evelyn's diary referred to the club, about which we know only from Bentley's letter. Whiston set down a story about Newton and Bentley, whom he claimed to have introduced about 1696, which could account for Newton's silence. Bentley engaged at that time in a passionate dispute about the meaning of a "day" in the prophecies. He was convinced that a day did not signify a year; and when Newton maintained the opposite, he bluntly asked if he could "demonstrate" it. Newton took this to be a slighting reference to his being a mathematician and was so offended that he refused to speak to Bentley for another year.[135] With Evelyn present, discussions in the club could not have turned to forbidden topics. It seems likely that at some time Newton and Bentley opened up somewhat more, for Bentley, despite his ecclesiastical and academic positions, eventually made himself known as a freethinker. On one occasion, he preached a sermon that questioned the received reading of 1 John 5:7, one of Newton's two notable corruptions. Indeed, Bentley belittled religion enough, Catherine Conduitt told her husband, to lessen Newton's affection for him.[136]

Although after 1693, Newton never resumed a correspondence with Locke as intimate as their earlier one, the two did remain in contact. In the autumn of 1702, Newton visited Oates and saw there Locke's newly completed commentary on 1 and 2 Corinthians. Since he did not have time to study it carefully, he asked Locke to send him a copy. When Locke heard nothing in reply, he wrote in March. Still awaiting a reply, he wrote a second time late in April and sent the letter to his cousin Peter King with the request that he deliver the note in person.

> The reason why I desire you to deliver it to him yourself is, that I would fain discover the reason of his so long silence. I have several

---

[134] James H. Monk, *The Life of Richard Bentley, D.D.*, 2nd ed., 2 vols. (London, 1833), *1*, 96.

[135] William Whiston, *Memoirs of the Life and Writings of Mr. William Whiston* (London, 1749) p. 107.

[136] *Keynes MS* 130.7, Sheet 1. In his original notation of this anecdote, Conduitt put it rather differently: "Angry even with Halley for talking against religion. C. C. [Catherine Conduitt] with Bentley for not. Dº [ditto?]" In the light of 130.7, the apparent epigram was probably only Conduitt's sloppy expression of the story.

reasons to think him truly my friend, but he is a nice man to deal with, and a little too apt to raise in himself suspicions where there is no ground; therefore, when you talk to him of my papers, and of his opinion of them, pray do it with all the tenderness in the world, and discover, if you can, why he kept them so long, and was so silent. But this you must do without asking him why he did so, or discovering in the least that you are desirous to know. . . . . Mr. Newton is really a very valuable man, not only for his wonderful skill in mathematics, but in divinity too, and his great knowledge in the Scriptures, wherein I know few his equals. And therefore pray manage the whole matter so as not only to preserve me in his good opinion, but to increase me in it; and be sure to press him to nothing, but what he is forward in himself to do.[137]

Stimulated by King's visit, Newton finally wrote on 15 May, apologizing for his long silence and commenting at length on 1 Corinthians 7:14, on the meaning of which he differed from Locke.[138]

Apparently the two never met again. Confined mostly to Oates, Locke seldom came to the city, and Newton indicated that he would probably not be able to visit him in Oates that summer. Early in 1704, Newton presented him a copy of his newly published *Opticks*. Later that year, in the autumn, Locke died. With his valedictory letter to Peter King on 25 October, he sent a sealed packet and a brass ruler, which he asked King to deliver to Newton.[139] The brass ruler raises questions to which no answers exist. A decade earlier, brass rulers also figured in Newton's relation with Fatio. "you put your self to ye expence of 5£ for three Rulers wch will now be of no use to you," Newton had written in March 1693, when he was pressing money on Fatio by every device he could think of. "Possibly they may be sometime or other of use to me & therefore I have sent you 5£ for them if you please to send them to me. For I reccon them bought upon my account."[140] No word of explanation of the rulers, of why Fatio would not need them any more, or of why Locke felt impelled on his deathbed to return his, emerged. Perhaps they were only a coincidence.

In the spring of 1698, Newton engaged in a brief correspondence with an Oxford undergraduate John Harington, of the family of the political philosopher of the same name who was active at the time of the Commonwealth. Writing, he said, at Newton's request, Harington outlined his theory that the same harmonic ratios that melodious sounds embody stand behind sensations pleasing to all the

[137] Lord Peter King, *The Life of John Locke*, new ed. (London, 1858), p. 263.
[138] *Corres 4*, 405–6. [139] Cranston, *Locke*, p. 479.
[140] Newton to Fatio, 7 March 1693; *Corres 3*, 260–1. Fatio sent them the next day (*Corres 3*, 261–1).

senses. He mentioned specifically architectural beauty and the proportions in sacred architecture and in the ark. This was to raise themes that Newton had explored in his "Origins of Gentile Theology," which undoubtedly explained Newton's interest. Although Harington covered himself with apologies for his presumption in writing to the great man, Newton took the time to compose a full answer. He assured Harington that he had examined similar ideas at an earlier time and that he wished he had time to do so again, for they deserved investigation. "In fine, I am inclined to believe some general laws of the Creator prevailed with respect to the agreeable or unpleasing affections of all our *senses*; at least the supposition does not derogate from the wisdom or power of God, and seems highly consonant to the macrocosm in general."[141] Although he urged Harington to continue to send his researches to him, no further letters survive.

Somewhat unexpectedly, Newton also found theological companionship at the Mint in the person of a young clerk, Hopton Haynes. Haynes must have been extremely able. He had come to the Mint in 1687 at the age of about fifteen. He educated himself enough to undertake the translation of Newton's *Two Notable Corruptions of Scripture* into Latin early in the eighteenth century, and his theological discourses revealed a knowledge of Greek. Though he was still not twenty-five when the recoinage began, officers at the Mint placed him in charge of training the officials and clerks of the country mints.[142] It was possibly in this connection that Newton met him, was impressed, and assumed forthwith the role of patron. In order to undertake the task of the country mints, Haynes had had to resign his previous appointment, and officers of the Mint had encouraged him to do so. Probably that decision antedated Newton's arrival, but he gladly picked up the obligation to find him a new position once the recoinage was completed. Haynes was not destitute; in the accepted practice of the day, he held positions both at the Excise office and at the Exchequer. Nevertheless, Newton intended to ease his lot with a post at the Mint as well. Until one should open up, he set him to work composing a history of the recoinage. Haynes's *Brief Memoires of the Recoinage*, a panegyric to Newton, drew heavily on material found among Newton's

---

[141] Harington to Newton, 22 May 1698, and Newton to Harington, 30 May 1698; *Corres 4*, 272–3, 274–5.

[142] I take these details of Haynes's life from Newton's account (Newton to Treasury, September 1701; *Corres 4*, 375–6). Whiston stated that he translated *Two Notable Corruptions* (*Collection of Authentick Records, Belonging to the Old and New Testament*, 2 vols. [London, 1727–8], *2*, 1077). The translation in Newton's possession of the first half of it into Latin was in Haynes's hand (*Yahuda MS 20*).

papers, which he must have furnished.[143] Newton did not hesitate to ask for favors in return. In 1701, when he was working on a report on foreign coins and needed a copy of an earlier one prepared in 1692, which was available in the Excise office, he got Haynes to copy it. To the copy, Haynes appended a personal note which helps to clarify their relation.

I am sorry I had not ye good fortune to see You ysday at the Excise, tho' I hope You are so kind, as stil to continue y^r good inclination to favour my pretension, if occasion be.

But I have rec^d such demonstration of y^r friendship, allready, for which I never pretend to return, & You, I dare say, never expect any other requital than my gratitude that I cannot but assure myself of Y^r good Offices when a fair occasion presents, by which You'll extremely add to the many obligations You have already layd upon

S^r Y^r most Obed^{nt}
& most H. Servt
H. Haynes [144]

A fair occasion did in fact present itself the following year when the post of weigher and teller fell empty. Newton favored Haynes's pretensions to the extent of composing six successive drafts of his recommendations.[145] Needless to say, Haynes became the weigher and teller at the Mint; he held the position until 1723, when Newton secured his appointment as assaymaster. In 1714, Newton consulted Haynes about the design of the coronation medal for George I.[146] Haynes married about 1698, and like Humphrey Newton, he named his fourth son, born some time near 1705, Newton Haynes.

What we know of their theological relationship is confined to Haynes's assertions and to reports of Haynes's assertions. Richard Baron, an abrasive unitarian who described Haynes years later as "the most zealous unitarian" he had ever known, reported that Haynes had told him Newton held the same views, and Haynes himself criticized Newton for not leading a new Reformation.[147] Known in the eighteenth and nineteenth centuries, these reports clashed with the accepted image of Newton and were not believed. In an age less concerned with religious orthodoxy and more informed about Newton's manuscripts, we recognize that the stories agreed with Newton's own papers, and we can only conclude that

---

[143] For example, Newton's computations in his papers addressed to Pollexfen, of the amounts of coin in circulation, his careful calculations of the rates of various operations at the Mint, and his account of the Chaloner affair. [144] *Mint* 19.2, f. 227.

[145] Newton to Treasury, Sept. 1701; *Corres 4*, 375–6. Drafts in *Mint* 19.1, 111–12, 117–25. Haynes was appointed on 30 September 1701 (*CTB 16*, 368).

[146] *CTP 1720–1728*, p. 241. *Mint* 19.3, f. 309.

[147] In his preface to Thomas Gordon, *A Cordial for Low Spirits*, 3rd ed., 3 vols. (London, 1763). A. B. (Haynes), *Causa Dei contra novatores* (London, 1747), p. 58.

Haynes was correctly informed. Beyond that conclusion, speculation alone exists. From the time of the Convention Parliament, Newton had discovered the possibility of discreet discourse in London on matters left untouched in Cambridge. With Locke, with Fatio, with Halley, and with Bentley, at various times and in various ways, he exchanged honest theological opinions. Near the time of his move, he apparently did so as well with a young man in Cambridge, William Whiston. Without venturing to imagine the circumstances, we must assume that Newton early recognized in Haynes one who held similar views, or was capable of holding them. It seems more than mere speculation that Newton's patronage of Haynes, like his patronage of Whiston, rested in good part on theological agreement. It also seems more than mere speculation that both young men learned the greater portion of their heresy from Newton.

As for Haynes's criticism of Newton for not leading a new Reformation, one is surprised, in the light of Whiston's debacle, at his naiveté. He was not similarly naive in his own behavior. Haynes conformed outwardly to the established church by attending its services, though he did venture to express disagreement by sitting down at certain times. Only at the very end of his life did he dare to publish a statement of his beliefs, though without his name attached, and his full unitarian credo, still anonymous, appeared only after his death.[148] Newton chose to disguise his heterodoxy more fully. He allowed himself to be appointed as a trustee of the Golden Square Tabernacle, a chapel endowed by Archbishop Tenison to relieve the crowding at St. James's, in the parish of which Newton's house on Jermyn Street stood.[149] He also became a member of the commission to complete St. Paul's until he disputed one day with Archbishop Wake about hanging pictures in the cathedral, "told a story of a Bishop who said on that subject that when this snow (pointing to his grey hairs) falls, there will be a great deal of dirt in churches . . ."[150] According to Catherine Conduitt, he never attended another meeting of the commission.

Catherine Conduitt, who supplied the story of the bishop and much else to her husband John Conduitt, was Newton's niece, née Catherine Barton. According to Conduitt, she lived with Newton

---

[148] *The Scripture Account of the Attributes and Worship of God: and of the Character and Offices of Jesus Christ.* By a Candid Enquirer after Truth (London, 1750).

[149] See his draft of the Tabernacle accounts in 1701; *Corres 4*, 377–80. Three letters summoning him to meetings of the trustees, in 1702, 1713, and 1721 survive (*Corres 4*, 424; *Corres 6*, 381; *Corres 7*, 182). In 1708, he contributed £20 for the building (*Sotheby Catalogue*, item 202). Newton was one of nine original trustees appointed in 1700 and continued to serve for twenty-two years.

[150] *Keynes MS* 130.7, Sheet 1. Cf. 130.6, Book 2, where it is clear that the bishop told the story.

twenty years before and after her marriage. The daughter of Newton's half sister, Hannah Smith, who married Robert Barton, a cleric in Northamptonshire, Catherine was born in 1679. Her father died in 1693, leaving her mother nearly destitute, if we can judge by Hannah Barton's letter to Newton at the time, and as we have seen, Newton purchased an annuity for the three children around 1695. Once he settled into the house on Jermyn Street, he made arrangements for Catherine to live with him. It is usually stated, on the basis of no authority whatever, that Newton supervised her education. Since she had reached at least seventeen years by the time she came to London, that is highly improbable. No evidence establishes the time at which she joined Newton. The first sign of her, a letter that Newton addressed to her in August 1700 at Mr. Gyre's near Woodstock in Oxfordshire, where she had retired to recover from smallpox, breathed an air of familiarity which suggests she had lived with him already some time.

I had your two letters & am glad ye air agrees wth you & th[ough y^e] fever is loath to leave you yet I hope it abates, & y^t y^e [re]mains of ye small pox are dropping off apace. Sr Joseph [Tily] is leaving Mr Tolls house & its probable I may succeed him [I] intend to send you some wine by the next Carrier w^{ch} [I] beg the favour of Mr Gyre & his Lady to accept of. My La[dy] Norris thinks you forget your promis of writing to her, & wants [a] letter from you. Pray let me know by your next how your f[ace is] & if your fevour be going. Perhaps warm milk from ye Cow may [help] to abate it.

I am
Your very loving Unkle,
Is. Newton[151]

Her face was very well; the smallpox did not disfigure it.

By every account, Catherine Barton possessed unlimited charm, a woman of beauty and wit. She was the one member of Newton's family who seemed to share his talents, though being a woman, she had to exercise hers in rather different channels. Quickly, she became the toast of London. The Kit-Kat Club commemorated her charms.

At Barton's feet the God of Love
   His Arrows and his Quiver lays,
Forgets he has a Throne above,
   And with this lovely Creature stays.
Not Venus' Beauties are more bright,
   But each appear so like the other,
That Cupid has mistook the Right,
   And takes the Nymph to be his Mother.

[151] *Corres* 4, 349.

For a time, Swift was her intimate friend who dropped in frequently for gossip and risqué stories, which he relayed home in his *Journal to Stella*. "I love her better than any one here . . . ," he allowed.[152] Rémond de Monmort, a member of the French Regency Council who met her at Newton's table in 1716, gushed all over the page when her compliments were sent to him. "I have retained the most magnificent idea in the world of her wit and her beauty."[153] In eighteenth-century terminology, Newton's niece was the famous witty Mrs. Barton.

When Voltaire visited England in the 1720s, he too heard of Catherine Barton, and what Voltaire heard all of Europe heard.

> I thought in my youth that Newton made his fortune by his merit. I supposed that the Court and the city of London named him Master of the Mint by acclamation. No such thing. Isaac Newton had a very charming niece, Madame Conduitt, who made a conquest of the minister Halifax. Fluxions and gravitation would have been of no use without a pretty niece.[154]

When the story to which Voltaire's sally referred surfaced in the nineteenth century, any number of brows furrowed deeply in its refutation. Charles Montague (that is, Halifax before his elevation to the peerage in 1700) had won respect as a statesman who had helped to fashion the new financial machinery set in place during the 1690s and had guided the recoinage. Newton might properly associate with Montague at the Mint. Halifax, on the other hand, had gained notoriety as a high-flying roué. The younger son of a younger son of the first Earl of Manchester, destined originally for the church, Montague commanded a ready pen and a ready tongue but not, initially, ready cash. The last he acquired in 1688 by marrying the Duchess Dowager of Manchester, the widow of his cousin, a woman old enough to have a son older than her new husband. She lived until 1698, easily long enough to launch his career. Apparently he was already establishing his reputation at the Kit-Kat Club, where the Whig leaders assembled, before her death, which did nothing to hamper his exploits. Baron Halifax of Halifax was known as a lover – or, to his enemies, as a would-be lover. With Halifax the libertine, Victorian eulogizers could not

[152] John Dryden, *Miscellany Poems*, 5th ed., 6 vols. (London, 1727), 5, 61. Editions of Dryden's *Miscellany Poems* after the third contained poems, such as the Kit-Kat poems, ascribed to 1703, composed after his death. Jonathan Swift, *Journal to Stella*; *The Works of Jonathan Swift*, 19 vols. (Edinburgh, 1814), 2, 215.

[153] Monmort to Taylor, 12 April 1716; Brook Taylor, *Contemplatio philosophica* (London, 1793), pp. 93–4. Cf. Monmort to Newton, 14 Feb. 1717; *Corres* 6, 380. If these two letters were typical, Monmort had a weakness for extravagant flattery which approached the ridiculous; too much should not be made of his apostrophes to Mrs. Barton.

[154] Voltaire, *Dictionnaire philosophique; Oeuvres complètes de Voltaire*, 70 vols. (Paris, 1785–9), 42, 165. This famous jibe first appeared in 1757 in a collected edition of Voltaire.

bear to associate Newton. Nor could they bear the thought, the point of Voltaire's gibe, that Newton used the degradation of his niece to advance his own career. For better or for worse, the obsessions of the late twentieth century differ from those of the mid-nineteenth. As Professor Manuel sums it up, "Victorians could not suffer the intrusion of illegitimate sex in the family of the hero; and we cannot tolerate the idea of mere aimless dalliance."[155] The *Principia* remains the *Principia* for us whatever the relation of Catherine Barton to Halifax and whatever Newton's role in the affair. Nevertheless, the skein is jumbled enough not to admit of ready untangling.

The second issue, Newton's personal profit from his niece's charms, seemed absurd on the face of it and was refuted several times by the mere recital of relevant dates. When Halifax had Newton appointed as warden, Catherine Barton had not reached her seventeenth birthday, and there is no reason to think that Halifax had yet laid eyes on her. Voltaire's specific words, however, were master of the Mint, not warden; and while we may legitimately doubt that Voltaire understood the distinction between the two offices, the charge can only be refuted if we insist on the later appointment, at the end of 1699, when Catherine had reached both the blooming age of twenty and, almost certainly, the city of London. Even in this case, the circumstances overwhelmingly imply that the charge was mere salacious gossip. Charles Montague had fallen from power during 1699, resigning as chancellor of the Exchequer in May and as first lord of the Treasury in November, a month before the death of Thomas Neale, when alone the nomination of his successor could be made final. Neale had been in failing health for several years. Maneuvering to appoint his successor may well have begun before his death. Nevertheless, we know nothing about it, and the suggestion that Newton owed his elevation to the mastership to Catherine's influence appears so wildly improbable as not to merit serious consideration.

The wider ramifications of the story, that is, her supposed affair with Halifax and Newton's involvement in it, do not evaporate with equal ease, however. By 1703, if not before, Halifax (as he now styled himself) had made her acquaintance. It was a custom at the Kit-Kat Club to inscribe a lady's name on a toasting glass with a diamond and to mark the occasion with verses to her. A group of them are assigned to the year 1703. Along with the one cited above were two others that are usually ascribed to Halifax. One underlined her relationship to Newton by its repetition of imagery drawn from the Mint.

---

[155] Manuel, *Portrait*, p. 262.

> Stampt with her reigning Charms, this Standard Glass
> Shall current through the Realms of Bacchus pass;
> Full fraught with beauty shall new Flames impart,
> And mint her shining Image on the Heart.[156]

Though opinions vary, most commentators hold that Catherine should have taken umbrage at the quality of the verse whatever her response to the proposition behind it. As Dr. Johnson later put it, "It would now be esteemed no honour, by a contributor to the monthly bundle of verses, to be told, that, in strains either familiar or solemn, he sings like Montague."[157] Suffice it to say she left no judgment on the verses, perhaps because Halifax toasted the charms of at least six other ladies at the Kit-Kat Club that year. By 1706, their acquaintance had progressed well beyond toasts. In that year, Halifax drew up his will. Two days later, he added a codicil which bequeathed £3,000 and all his jewels to Catherine Barton "as a small Token of the great Love and Affection I have long had for her." In October 1706, he added to the bequest by purchasing for her, in the name of Isaac Newton, an annuity of £200 per annum for the remainder of her life.[158]

The official life of Halifax commissioned by his heir and published soon after his death felt obliged to mention their relation.

> I am likewise to account for another Omission in the Course of this History, which is that of the Death of the Lord Halifax's Lady; upon whose Decease, his Lordship took a Resolution of living single thence forward, and cast his Eye upon the Widow [rather, sister] of one Colonel Barton, and Neice of the famous Sir Isaac Newton, to be Super-intendant of his domestick Affairs. But as this Lady was young, beautiful, and gay, so those that were given to censure, pass'd a Judgment upon her which she no Ways merited, since she was a Woman of strict Honour and Virtue; and tho' she might be agreeable to his Lordship in every Particular, that noble Peer's Complaisance to her, proceeded wholly from the great Esteem he had for her Wit and most exquisite Understanding . . . [159]

Whatever the Victorians made of it, most people in the twentieth century take £3,000 and all his jewels plus the annuity as a stiffer price than wit and understanding normally command.

There were those in the Augustan age who did so also. In 1710,

---

[156] The other verse ran as follows: "Beauty and Wit strove each, in vain, / To vanquish Bacchus and his Train; / But Barton with successful Charms / From both their Quivers drew her Arms; / The roving God his Sway resigns, / And awfully submits his Vines." / (Dryden, *Miscellany Poems*, 5, 61)

[157] Samuel Johnson, *The Lives of the Most Eminent English Poets*, 4 vols. (London, 1781), 2, 300.

[158] *The Works and Life of the Right Honourable Charles, Late Earl of Halifax* (London, 1715), p. iv. The entire will of Halifax including the codicils, is printed in these pages as an appendix to the *Works and Life*.   [159] *Ibid.*, pp. 195–6.

Mrs. Mary de la Rivière Manley, a Tory scandalmonger, published a satire on the Whigs entitled *Memoirs of Europe, Towards the Close of the Eighth Century. Written by Eginardus*. Under the guise of Eginard's description of society in Constantinople, Mrs. Manley gave the Whig establishment what for. Halifax appeared as Julius Sergius, complaining of his mistress Bartica, on whom he had "lavish'd Myriads . . . besides getting her worthy ancient Parent a good Post for Connivance." Of all things, Bartica now wanted to marry him! " 'twas ever a proud Slut . . . "[160] Voltaire probably got his story from Mrs. Manley.

In the light of Halifax's bequest, Mrs. Manley's presentation of Catherine Barton as his mistress is believable, but Dean Swift's *Journal to Stella* complicates the issue. Between 28 September 1710 and 16 December 1711, roughly the year following Mrs. Manley's book, Swift frequently saw both Mrs. Barton and Halifax, but never both together. On 28 November 1710, he dined with Halifax, apparently at his house. Two days later, he dined with Mrs. Barton, apparently at her house, alone. Living near Leicester Fields, he described Mrs. Barton as his "near neighbor." By that time, Newton lived on St. Martin's Street, just off Leicester Fields. One can only conclude that in 1711 Catherine was living with Newton, not with Halifax.

Wherever she was in 1711, it is difficult to believe that Mrs. Barton was not in Halifax's house by 1713. On 1 February of that year, Halifax drew up a second codicil to his will which revoked the first and replaced it with one nothing short of magnificent. To Isaac Newton he left £100 "as a Mark of the great Honour and Esteem I have for so Great a Man." To his niece, Mrs. Catherine Barton, he bequeathed £5,000 with the grant during her life of the rangership and lodge of Bushey Park (a royal park immediately north of Hampton Court) and all its furnishings and, to enable her to maintain the house and garden, the manor of Apscourt in Surrey. "These Gifts and Legacies, I leave to her as a Token of the sincere Love, Affection, and Esteem I have long had for her Person, and as a small Recompence for the Pleasure and Happiness I have had in her Conversation."[161] When Flamsteed heard about the bequest after Halifax's death, he spitefully wrote to Abraham Sharp that it was given Mrs. Barton "for her *excellent conversation*."[162] Flamsteed placed a value of £20,000 or more on the house and lands – that is, he valued the whole bequest at £25,000 or more plus

---

[160] Mary de la Rivière Manley, *Memoirs of Europe, Towards the Close of the Eighth Century. Written by Eginardus* (London, 1710), p. 294.

[161] *Works and Life*, pp. v–vi. As innumerable commentators have pointed out, "conversation" held a broader meaning in the eighteenth century than it does today. We might say instead "company."   [162] *Baily*, 314.

the annuity, a fortune by the standards of the early eighteenth century. Flamsteed also remarked that Halifax was reported to have left a total estate of £150,000, handsome testimony to what an enterprising lad could accomplish with a mere five years in office.

The problem with the codicil of 1713 is sheer size. If the bequest of 1706 makes it impossible to believe in a platonic relation between Halifax and Mrs. Barton, the later one makes it difficult to believe she was merely a mistress. If not a mistress, what? About two years later, four days after Halifax's sudden death from an inflammation of the lungs, Newton wrote a note to his distant cousin, Sir John Newton, excusing himself from a minor social obligation. "The concern I am in for the loss of my Lord Halifax & the circumstances in wch I stand related to his family will not suffer me to go abroad till his funeral is over."[163] On the basis of that one sentence, Augustus De Morgan wrote a book arguing for a secret marriage between Halifax and Mrs. Barton. No one else ready to load that much meaning on a single sentence has yet come forward. The size of the bequest argues a marriage better. So does the deliberate reference to Newton in the codicil in introduction to the bequest to Mrs. Barton. Arguments which no one has been able to refute argue against it, however. If there had been a marriage, why was it kept secret? The motive proposed – Halifax's fear of ridicule for marrying beneath his class – carries no conviction. Even if we allow it, why conceal the marriage after his death, when the bequest left his supposed widow exposed to slander? When Catherine Barton married John Conduitt in 1717, she recorded herself in the presence of her groom, who could not have been ignorant of her liaison with Halifax or of its material legacy, as a spinster.[164] Since there are reasons to doubt that she was either wife or mistress, perhaps she occupied some intermediate status. If such exists, I have not heard of it.

The issue here is Newton's role in and attitude toward the affair, which began in the early years of his residence in London. It has been felt that his acquiescence in his niece's relation to Halifax, which was clearly not a legal marriage, must somehow diminish his stature. Merely to pose the issue appears to me to assume that Newton stood on a moral plane separate from and superior to the society in which he lived. And yet, for all his genius, he was a man like all of us, facing similar moral choices in terms not altered by his intellectual achievement. His success at the Mint, in administering the recoinage, in dealing with the goldsmiths, above all in maneuvering himself into position to become master when he real-

---

[163] *Corres* 6, 225.

[164] Joseph Lemuel Chester, ed., *The Marriage, Baptismal, and Burial Registers of the Collegiate Church or Abbey of St. Peter, Westminster* (London, 1876). p. 354.

ized the wardenship was a sham, does not suggest an otherworldly saintliness out of touch with hard reality. He knew what compromise was. His pretense of religious conformity for social acceptance and material benefit was not utterly incommensurable with acquiescence in a most advantageous liaison. For that matter, he knew what sexual attraction was – and from every indication, its gratification, of necessity outside the bonds of holy wedlock. Bishop Burnet said that he honored Newton "for something still more valuable than all his Philosophy for being the *whitest* soul he ever knew."[165] The close examination of his treatment of Hooke and Flamsteed to name no more, has revealed some dark stains on the erstwhile whitest soul. There is no reason to separate sexual behavior from other moral conduct. Newton's role in history was intellectual not moral leadership. From the point of view of the late twentieth century, after the barbarities we have witnessed, the charge against him does not seem oppressively heavy, but even if it could be proved beyond doubt that Newton was the leading whoremonger of London, the immensity of his impact on the modern intellect would remain unaltered. For me at least, the recognition of his complexity as a man helps in understanding the price his genius exacted. I find it hard to reconcile the *Principia* with a plaster saint.

This much is certain: Newton did not propose that his niece should lose the fortune Halifax bequeathed her. Among his papers are notes on and suggested changes in the legal documents between Catherine Barton and Halifax's heir that carried the bequest into effect.[166] Catherine Barton returned to Newton's house off Leicester Fields. Fragments of a letter to him from about 1717 survive.

> I desire to know whether you would have me wait here . . . or come home . . .
>
> Your Obedient Neece
> and Humble serv^t
> C. Barton[167]

Far from feeling shame, Newton kept a portrait of Halifax in his room, as the Abbé Alari noted when he visited him ten years later.[168]

Although Catherine Barton was much closer to Newton than any other member of the family, he maintained cordial relations with his other half sister and with his half brother. According to Humphrey Newton, he supported the family of his sister Mary when her hus-

---

165 *Keynes MS* 130.7, Sheet 1. Cf. 130.6, Book 2, where Conduitt ascribed this story to W. Montague.

166 *Keynes MS* 127. Lehigh University Library, MS 731. *Add MS* 3968, f. 139ᵛ. It appears nevertheless that she did not enjoy much of the bequest (*Corres 5*, xlv). For Newton's participation in her financial affairs, see also George Watson to Newton, 4 March–; *Corres*, 7, 383.

167 *Sotheby Catalogue*, lot 176, p. 41.   168 *Edleston*, pp. lxxviii–lxxix.

Figure 12.4. Newton sometime during his London years. Two ivory
medallions probably sculpted by David Le Marchand.
They are not dated. Dated likenesses of Newton done
by Le Marchand in 1714 and 1718 exist. Le Marchand
established himself in London about the same time
Newton did, and there is no reason to assume these
medallions could not have been executed earlier. (Courtesy of the Royal Society and the Provost and Fellows of
King's College, Cambridge.)

band, Thomas Pilkington, died.[169] His brother Benjamin Smith, for whose wife he prescribed plasters, named one of his sons Newton Smith.[170] He also maintained cordial relations with his native village

[169] *Keynes MS* 135.

[170] See the genealogy, *Corres* 7, 485–9. Stukeley mentioned him in 1727; at that time he had one of the Kneller portraits of Newton and some of his furniture. (*Keynes MS* 136).

of Colsterworth. As an official of the government, he found himself consulted in 1700 by the parish authorities about their right to taxes from the Lordship of Twyford. He composed a memorandum that outlined the legal recourse open to Colsterworth.[171]

On 23 December 1699, Thomas Neale, the master of the Mint, died. It had not taken Newton long to understand realities at the Mint, and it had not taken him much longer to realize that his effort to reverse the shift of true authority from the warden to the master was not going to float. As he observed the recoinage, the disparity between the seats of formal and real authority, not to mention their disparity in remuneration, must have struck him more sharply. Although he shouldered great burdens in the recoinage, he received the same salary, £400 per annum, he would have received had he acted like former wardens and done nothing. Neale did very little, leaving all to his assistant, Thomas Hall, and to Newton. Not only did he receive his salary of £500 per annum, but he got, according to the terms of his Indenture, a set profit on every pound weight troy that was coined. In addition to his salary, Neale earned more than £22,000 during the recoinage. Newton perceived and digested this fact. He schooled himself in the operations of the Mint that a master should know. And he waited, for Neale was a sick and failing man.

At first glance, it appears that Neale lasted too long, for Montague had fallen from power by the time he died. We can only speculate on why it did not matter. As Neale's profits on the recoinage indicated, the mastership could be a considerable item of patronage. Nevertheless, even with Montague out of office, Newton was allowed to take it, and quickly. Only three days after Neale's death, on 26 December, Luttrell heard the news. "Dr. Newton, the mathematical professor, is advanced from being warden to master of the mint in the room of Mr. Neal, deceased; and sir John Stanley succeeds the Dr. as warden, a place worth 500*l.* per annum"[172] Luttrell's use of the title "Dr." may provide the best clue to Newton's promotion. Recognized as England's leading intellectual, he was a figure in his own right who was able to command the post he wanted. Though Luttrell dated his news 26 December, the Mint accounts showed that he took office on 25 December, in which case the position was a birthday present. The letter patent confirming his appointment was finally sealed on 3 February 1700.[173]

In assessing Newton's appointment as master, we need to keep two facts in mind. First, the progression from warden to master had no precedent at the Mint and was not repeated. Second, Newton still held both his fellowship and his chair in Cambridge. In fact, three and a half years had sufficed to convert him into a civil servant. Far from wanting to return to Cambridge, he pursued the better position in order to insure his continuation in London. He finally resigned both places in Cambridge in 1701, a year of heavy coinage when he took in nearly £3,500 as master, a sum that must have made the Cambridge income look too paltry to matter. Three and a half years in London had been long enough also to teach him the facts of political life. Even with his patron out of office, he was able to secure the mastership he desired. And people worry about Catherine Barton's affair with Halifax!

Heretofore the mastership had always been granted for life. In 1697, the committee of the House of Commons to examine the Mint had suggested that no appointment should be for life but that all should depend on satisfactory performance.[174] Since much of the report drew directly on Newton's testimony, it is probable that this recommendation did also and that it reflected Newton's disgust with Neale. As a result, he found himself the first master appointed, not for life, but during the king's pleasure. It did not matter. His appointment was automatically renewed with each new monarch and extended to his death.

Upon appointment, Newton faced the need to post a bond. Historically, masters had posted bonds of £2,000, but the perplexity of Neale's finances had led the government to demand of him a bond of £15,000. Newton petitioned to have his bond set at the former level. In typical fashion, he consulted at least twenty earlier Indentures to be sure of his facts. His petition was granted without extended discussion. Montague and Thomas Hall stood surety for him.[175]

The Indenture was more complicated, and a full year passed before its sealing on 23 December 1700. The amount of time involved suggests serious negotiation, but in fact the Indenture, and also the new one that had to be drawn up on Anne's succession, repeated in identical terms the Indenture under which Neale had minted. A formal contract between the master and the monarch, the Indenture embodied the supremacy of the master over the Mint by specifying the duties and remuneration of the other officers in a document to which they were not parties. It also specified the terms under which Newton as master would mint—for example, the coins he would issue, their weight, the standard of their metal, and the remunera-

---

[174] *Journals of the House of Commons, 11*, 778.
[175] *Corres 4*, 348. Mint 19.1, ff. 395, 397, 399–401, 405–7, contain his researches on earlier Indentures. On 1 July 1700, his petition was accepted (*CTB 15*, 396).

tion that most of those involved in the process of minting would receive.[176]

From the Indenture and from Newton's annual accounts, it is possible to determine his income as master. The Indenture set a payment to the master of 17½d per pound weight troy of silver minted. From this sum, Newton had to pay the moneyers 10½d per pound weight. His papers indicated two other payments set by separate contracts which were not part of the Indenture: 3½d to the melter who produced the bars the moneyers received to mill, cut, size, and coin, and ¼d to the smith who made the dies for the presses. Hence Newton made a profit of 3¼d per pound weight of silver coined. In fact, silver coinage netted him very little profit. During the whole of his tenure as master, there were only four years of significant silver coinage: 1701, when there was heavy coinage which Newton attributed to the peace (an explanation I do not understand), 1709 and 1711, when bounties were offered for plate brought in to be coined, and 1723, when the South Sea Company had to import a large volume of silver into the Mint as a condition of the government's salvage operation. Newton earned more than three-quarters of his total profit on silver in those four years. During the other twenty-three, he averaged £28 per year on silver. Gold coinage took up the slack. The Indenture set the payment on gold at 6s 6d per pound weight troy. From that he paid out 4s 8d in the three payments above, leaving a profit of 22d per pound weight. Over the period of twenty-seven years when he was master, his total profit from gold and silver coinage averaged £994 per annum. Beginning in 1703 and continuing at least through 1717 and probably longer, he received £150 per annum for managing the storage and sale of tin. For seven years, 1718–24, he earned an additional £100 in round numbers, from the coinage of copper halfpennies and farthings. He received in addition an annual salary of £500. According to statements he made in 1713, he had certain necessary expenses which he could not avoid: £52 per year for "diet," that is, meals for the officers at the Mint (an expense the Indenture specified), an annual gift to the Treasurer, and an annual salary to his deputy. He set the sum of the three at £180. Neale had paid his deputy by appointing him as the master's assayer, an office no longer exercised; it is not clear whether Newton did the same. The gift to the Treasurer reminds us that Newton undoubtedly received gifts in his turn. We know that the stationer to the Mint gave him almanacs at the beginning of the year, and twice he thanked someone for "a collar of very good

---

[176] I have not been able to locate his Indenture of 1700. His Indenture under Anne, sealed on 14 Jan. 1703, was identical to Neale's Indenture and hence in all probability to Newton's of 1700 (*Mint* 19.1, ff. 382–94). Some minor changes appeared in the Indenture under George I, in 1718 (*Mint* 4.57, which also has a copy of the 1703 one).

brawn."[177] The Abbé Alari consumed wine that Newton had received as a gift. Such gifts should not be construed as corruption. They did augment his salary on a regular though unmeasurable basis. It seems probable that at least they equaled the annual expenses he mentioned. They may have far exceeded it. His average income as master probably came to about £1,650. Individual years varied wildly, from £663 in 1703 (when his profit amounted to £13) to £4,250 in 1715 (when his profit swelled to £3,606). The average is misleading, since Newton's twenty-seven years as master included eleven years of the War of the Spanish Succession, which depressed the coinage. For the other sixteen years, his income as master averaged nearly £2,150, £2,250 during the years of copper coinage.[178] To put his income in perspective, recall that the Treasurer's salary under Charles II was £8,000 per annum, though it was later reduced somewhat. No other official received half that. Most officials had various devices of dubious ethical status, such as Montague must have used in building his fortune, to increase their real incomes. The master of the Mint had less opportunity in similar enterprises. Newton would have disdained to engage in such in any case.

As I have mentioned, 1701 was a big year for the Mint. With a profit of £2,959, Newton's income almost reached £3,500. Already at the beginning of the year, nearly five years after his departure from Cambridge, he appointed William Whiston as his deputy in the Lucasian chair with the enjoyment of its full income. On 10 December, Newton formally resigned, enabling Whiston to become his successor, and about the same time he also resigned his fellowship at Trinity. He stood then eleventh in the order of seniority.[179] He must soon have regretted his decision to surrender the income at Cambridge, for his profits at the Mint fell off quickly almost to nil. For five years, from 1703 through 1707, they did not get close to £100. Only with the peace did they recover to the level he must have expected when he pursued the position. At a time when £1,200 per annum for a single man was described as not merely easy but splendid, Newton's income as master was always considerable, and once the war ended, it ensured a life of material plenty, even under the more expensive conditions of London.

As master, Newton faced the repeated ordeal called the trial of the pyx, when both his honesty and the accuracy of his coinage underwent solemn and public scrutiny. Under the terms of the Indenture, he was to produce coins of a certain fineness and of a certain face value from each pound weight troy of precious metal.

---

[177] Charles Rawson to Newton, 31 Dec. 1713 and 1725, *Corres* 7, 252, 341. *Corres* 7, 365. *Yahuda MS* 7.3h, f. 2ᵛ. As the editor of *Corres* points out, these last two may be drafts for the same letter.   [178] His accounts are in *Mint* 19.4.

[179] The Senior Bursar's Accounts show that he received the first quarter's stipend for academic year 1702, that is, for the term that ended on Christmas 1701.

The Indenture permitted a total variation in weight and fineness of two pennyweight (48 grains) per pound of silver and of one-sixth carat (40 grains) per pound of gold. This variation was called the master's remedy. To exercise control, the Indenture demanded that a certain number of coins from every batch made be put into a special box (the pyx) with three locks, the keys to which the master, warden, and comptroller held severally. Periodically, in ordinary times every three or four years, the two pyxes (the one for gold, the other for silver coins) were opened in the presence of the king or his representatives, and the coins were examined "by fire by water by touch or by weight or by all or by some of them, in the most just & exact manner that can be thought of, it being a business of a very publick concern."[180] Though the king himself did not come, the lords of his Council assuredly did, together with members of the episcopal bench and the governors of the Gold-smiths Company, who functioned as the jury. The first trials New-ton attended took place in awesome Star Chamber, though later they moved to the Exchequer. The master had to deliver in his bond; and if his coinage did not fall within the remedy, he stood exposed to fine or ransom. Newton had gone through a trial of the pyx on 16 July 1696, almost as soon as he reached London, though seated in the relative security of the warden's chair, and at short intervals thereafter two more checked the recoinage.

The first trial of Newton's coinage came on 6 August 1701. At ten in the morning the officers of the Mint, the lords of the Coun-cil, and the governors of the Goldsmiths gathered in the house of the usher of the Exchequer. They opened the two pyxes and poured their contents out on the table. The Lord Chancellor administered the oath to the jury. The lords of the Council then retired, and the goldsmiths set to work, watched by the officers of the Mint to protect their interests. They told and weighed the money. They found the gold coins to be deficient by a twentieth of a carat per pound, the silver by 23 grains, both well within the remedy. The jury then selected a few coins at random, melted them into ingots, and checked their fineness in comparison to the trial pieces kept in the Exchequer. The coinage passed again. In the 1707 trial, some excitement intervened at this stage when the melting set fire to the Exchequer.[181] The goldsmiths drew up their formal verdict and presented it at the appointed time to the lords of the Council. Though Newton could not seriously have doubted the outcome, his reputation in London hung on it, and his papers make it evident that he invested considerable concern in preparing for it.[182]

[180] From a sheet in Newton's hand, about 1700; *Mint* 19.1, f. 232.   [181] *CTB 22(2)*, 72.

[182] For his own instruction, he drew up "Directions about the Triall of the monies of Gold & Silver in the Pix," *Corres 4*, 371–3. Drafts in *Mint* 19.1, ff. 232, 266–7, 271–2.

The 1701 trial witnessed some tension as the goldsmiths tried to exert themselves and Newton resisted. The goldsmiths insisted that they should try the coins in private. Newton insisted just as strongly on the right of officers of the Mint to be present. He later devoted much attention to a memorandum to the Treasury supporting his position. The trial of the pyx did not assume any wrongdoing, he argued. It existed to determine a matter of fact; and the more open it was, the more confidence its verdict would engender. If it became private, the Goldsmiths Company would have placed themselves in position to control the Mint to their own advantage.[183] Ensuing silence on the subject implied that Newton won his point. 1701 was not his last trial of strength with the Goldsmiths Company.

With the verdict returned, it was time to relax. The warden had arranged a magnificent feast at the Dog Tavern, at the glorious expense of £146, roughly Newton's annual income in Cambridge. As always, the auditor graciously allowed the warden to put the expense on his account and thus pass the bill on to the public. Among Newton's papers is a scrap in preparation for a later trial of the pyx with two lists of guests, fourteen in all, including the Duke of Buckingham, the Earl of Oxford, and the Bishop of London. It also contained a list of the dishes to be served: "Fish, Pastry. fricasy of chickens & a dish of puddens. Q^{tr} Lamb. Wild foul. Peas & Lobsters."[184] The steady filling out of Newton's face in successive portraits suggests that the menu for the feast may have reflected his diet better than Conduitt's note that he lived on vegetables.

According to Newton's story, he lived in constant friction with the goldsmiths from the very beginning. In the past, they had been allowed the full advantage of the remedy in the coin they received in return for bullion. Hence they took home £4 0s 2d in gold for every pound weight of bullion, whereas they should have received only £3 19s 8d plus 3 farthings. Moreover, inexactness in sizing gave them a profitable business in culling out heavy gold coins to melt down and return to the Mint to be coined again. The heavy coins were known as "Come-again Guineas," and as Newton told it, the game of repeated coining and melting had served to enrich the masters, moneyers, and goldsmiths at the public expense. In a paper he wrote soon after he became master, he estimated that an eighth of the guineas coined in the decade 1689–99 were culled, melted, and returned to be recoined.[185] As soon as he became master, the Mint stopped giving the goldsmiths the advantage of

[183] *Mint* 19.1, ff. 307–8. Perhaps four previous drafts suggest the intensity of feeling behind this paper (*Mint* 19.1, ff. 259–64). I have not found any evidence of its receipt by the Treasury, however.

[184] *CTB 16*, 425. *Mint* 19.1, f. 344. [185] *Mint* 19.2, f. 609^{v}.

the remedy, and he bragged that he improved the sizing of gold coins to the point that culling ceased to be profitable. The records of trials of the pyx showed a high correlation between the weights of the whole and the weights of arbitrary samples in relation to tale; since earlier trials had not bothered to weigh and count samples, but had merely weighed and counted the entire contents of the pyx, such tests were probably Newton's device to demonstrate his accuracy. The records also showed a remarkable consistency from trial to trial. Both testimonies bore out Newton's claim.[186] Possibly the argument in 1701 reflected the desire of the goldsmiths to bring the new master to heel. When that failed, they bided their time. In 1707, at the time of the union with Scotland, a new trial piece was made. A goldsmith, Cartlich, made it, and Newton later became convinced that it was finer than the legal standard. When they employed the new trial piece at the 1710 trial of the pyx, the jury of goldsmiths found Newton's coin deficient in fineness. To be sure, the deficiency was small; his coinage passed the trial. Nevertheless, Newton treated the matter as a challenge to his control of the Mint, and he counterattacked with all the vigor he commanded.

He carried out his own assays to check the 1707 trial pieces.[187] With the information he gathered, he composed a memorandum to the Treasury stating his case. We can measure Newton's intensity on a given subject by the number of drafts he composed. More than ten drafts of this memorandum, plus other papers as well, indicate that his intensity stood at a very high level indeed.[188] The goldsmiths refused to test their own plate by the new trial pieces, he argued. In applying it to the coinage, they were trying once more to bring the Mint under their control and reinstitute their guaranteed profit, which he had earlier abolished, in dealings with it. When his memorandum of 1711 failed to achieve its effect, he began to plan for the next trial. He composed a "Petition of the Merchants & Goldsmiths underwritten in behalf of themselves & other Importers of Bullion into Her Ma^tys Mint" against the use of

---

[186] Cf. the reports on the trials of 1707, 1710, 1713, and 1715; *Mint* 7.130, ff. 64, 66ᵛ, 68ᵛ, 69ᵛ–70.

[187] He was not as accurate in assaying as he thought. Cf. the comparison of his results on various trial pieces with modern tests in E. G. V. Newman, "The Gold Metallurgy of Isaac Newton," *Gold Bulletin, 8* (1975), 90–5.

[188] *Corres 5*, 84–8. Drafts in *Mint* 19.1, 242–52, 273–6, 288, 291–2, 295–8. He sent his memorandum to the Treasury on 8 Jan. 1711 (*CTB 25(2)*, 3). He also composed five drafts of a paper on how new trial pieces should be made (*Mint* 19.1, ff. 284–7, 290).

Figure 12.5.   A map of the Mint prepared for Newton early in 1701, almost immediately after his appointment as master. (From the Twenty-first Annual Report of the Deputy Master of the Mint, 1890.)

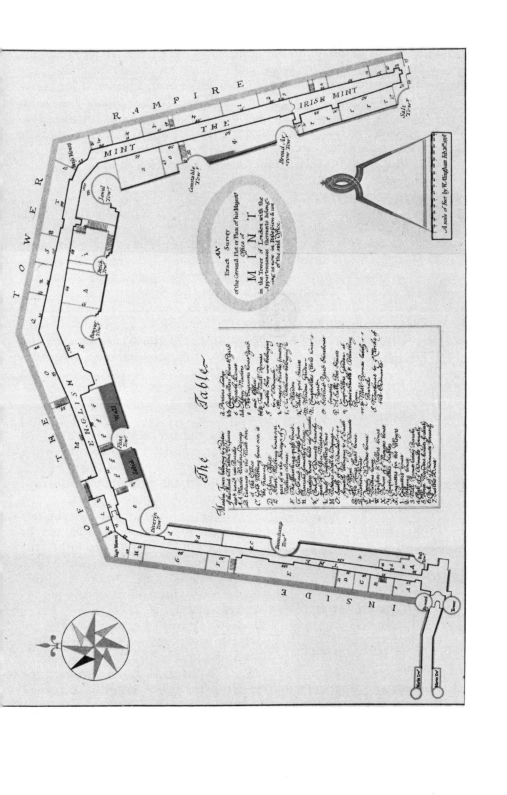

the 1707 trial piece.[189] Newton must have been at work on the petition about the time when Halifax composed the new codicil, the worldliness of which is supposed to have been a problem for Newton's ethics. Apparently he never found the merchants and gold-smiths on whom he had hoped to father the petition, for the Treasury never received it. Though the trial of the pyx in 1713 used the objectionable trial piece again, Newton eventually prevailed. His new Indenture of 1718 under George I specified the sole use of the earlier ones.[190]

From Thomas Neale, Newton inherited an administrative mess. Although the country mints had closed down and the recoinage concluded a year and a half before his death, Neale did not manage to straighten out their affairs, and Newton had to take charge of clearing their accounts. He was still writing memoranda about Anthony Redhead, the deputy master of the Norwich mint, in 1704. He had also to clear Neale's accounts for the final four years of his mastership, including of course the recoinage. So perplexed were the accounts that another year and a half passed before Newton could straighten them out minimally, and even then their irregularities were such that only a letter patent under the great seal could carry them through the audit.[191] Disorder was antithetical to Newton's soul. He resolved that under him the Mint would run in a different fashion. Whereas accounts under Neale were slipshod and erratic, during Newton's mastership the officers of the Mint filed their accounts promptly year by year, so that a clear picture of the administrative machinery of the Mint emerges from them. Alexander Pope, an acquaintance of Conduitt and presumably informed by him, later reported (or was reported to have reported) that Newton, "though so deep in Algebra and Fluxions, could not readily make up a common account; and, when he was Master of the Mint, used to get somebody to make up his accounts for him."[192] Whatever the truth of Pope's statement, it could have referred at most to the mechanics of adding sums. Orderly accounts came naturally to Newton, and the surviving record, in sharp contrast to previous practice, testifies as much.

Orderly accounts were indicative of broader issues. Newton attempted to bring similar order into the whole administrative machine of the Mint. Shortly after he assumed office, he inaugurated a minute book of meetings of the Mint Board (the master, the warden, and the comptroller). They met every Saturday to review the daybooks, examine the records of the pyx, supervise mainte-

---

[189] There are two drafts of the "petition" in Newton's hand (*Mint* 19.1, f. 278).
[190] *Mint* 5.47, p. 33.   [191] *Mint* 1.6, ff. 71ᵛ–2. Public Record Office, E351, Roll 2103.
[192] Spence, *Anecdotes*, p. 175.

nance, and take up whatever business arose.[193] The absence of any similar record from earlier times leads one to conclude that Newton inaugurated the meetings. Truth to tell, they did not survive. After about a year, the minute book became the file for orders of reference received from the Treasury. During the first ten months, Newton did not miss a meeting. Other officers attended less regularly, and it is likely that the practice foundered on the unwillingness of sinecurists to endure such rigors.

The meetings may also have foundered on the crisis that confronted Newton almost as soon as he took office. The two men who held the comptrollership jointly, Thomas Molyneux and Charles Mason, were probably already in conflict. The conflict became open in the summer of 1700 with charges of corruption against Molyneux, who was said to take gifts to pass accounts, and countercharges against Mason of treating the post as a sinecure and fomenting trouble in the Chester mint in order to replace Halley with his own man.[194] It is impossible to untangle the threads of the quarrel at this distance. None of the dirt rubbed off on Newton. A new comptroller, James Ellis, of a family already well established in the Mint, replaced the two in May 1701. By then, the recorded weekly meetings of the board had ceased.

Newton intended to be master within the Mint as well as titular master. Control of the other two officers, who were usually gentlemen with connections and members of Parliament, was difficult. It is possible that he influenced the choice of Ellis and obtained thereby an officer likely to follow his lead. In the 1690s, a Christopher Ellis had functioned as his special clerk in charge of the prosecution of coiners. John Ellis, who was also undersecretary to the Secretary of State, cannot be regarded as Newton's creature, however. There is no evidence that Newton ever controlled the warden, but the appointment of Hopton Haynes in 1701 as weigher and teller may have given him leverage on the warden. The administrative arrangements of the Mint divided its personnel into two groups, one under the master, the other under the warden. Although the warden received the money he controlled from the master, he paid the salaries of a specified number from the sum, and he filed his own accounts. The weigher and teller stood on the warden's account. It is highly probable that Newton's zeal in the appointment of Haynes to that position bore some relation to his desire to have a representative in the warden's camp. From the lesser figures in the Mint, he expected obedience. When Sylvester, the smith, died in 1712, Newton vigorously petitioned the Lord Treasurer for a change in the smith's status. By serving as well as

---

[193] *Mint* 1.7.   [194] *CTP* 1696–1701/2, p. 415. *CTB* 16, 3–9.

smith to the Ordnance, Sylvester had gained a degree of independence and had behaved with insolence. Newton wanted to have the
salary of the next smith reduced. The smith was to be his servant;
he wanted him employed on terms that insured as much.[195]

Another threat to his authority came from the Tower. Once the
recoinage was completed, much of the area of the Mint stood
empty. The officers of the Ordnance, housed in the Tower and
pressed for space, naturally turned an eye toward it. Solely, it appears, because the area between the walls belonged to the Mint's
historic domain, Newton resisted. By and large he did not succeed,
as good sense prevailed over prescriptive right. The struggle continued throughout Newton's tenure.[196]

With constant vigilance to maintain his authority went equal concern to maintain and improve professional standards. In 1707, he
drew upon his experience in the recoinage to suggest procedures in
the Scottish recoinage then underway that would avoid earlier
abuses.[197] Nearly twenty years later, a report to the royal Council
on the copper coinage in Ireland credited Newton with the controls
inserted in its patent which had insured its quality, controls unknown to similar projects in the past.[198] Newton himself placed
great stock in his level of professional performance. About 1718,
when two disappointed applicants for the contract to furnish copper
for that coinage tried to blame their failure on him, he cited his
record in his defense;

> as I have brought the sizing of the gold & silver moneys to a much
> greater degree of exactness than ever was known before, & thereby
> saved some thousands of pounds to the government, & am not
> blameable for any accidental errors within the remedy of those
> moneys, & have found out a new assay for stopping the importation
> of copper not sufficiently refined or dammaged by the Sea-coal: so I
> hope that I shall not be blamed for want of better assays, or for such
> errors in the assays of copper as are only accidental & in value do not
> exceed the Remedies allowed in weight, especially while I am en-
> deavouring to set up this coinage in the best manner.[199]

One cannot miss the pride behind his recital of achievements.

Along with his central concern for coining went a number of other
administrative responsibilities which he shouldered gladly. Strictly
speaking, maintenance of the buildings fell to the warden. When a
great storm in November 1703 did immense damage to the Mint,
which consisted largely of timber structures in a ruinous condition,

[195] *Corres 5*, 186–7, 352.

[196] Cf. letters in 1699, 1702, 1715, and 1720, *Corres 4*, 393–4; *Corres 6*, 235; *Corres 7*, 86–7, 94–5, 410–11, 412. *CTP* 1697–1701/2, p. 510. *CTB 15*, 74; *16*, 95; *27(2)*, 184.

[197] Newton to Gregory, autumn 1707; *Corres 4*, 502–3.   [198] *Mint* 1.1, pp. 218–19.

[199] *Corres 7*, 58.

Newton could not stand aloof. According to the terms of the Coinage Act, the Mint received a fixed sum each year for salaries and maintenance, and Newton had to scrape his accounts for years to repair the damage. The Mint made medals as well as coins. Upon the death of William, Newton in his typical style took the initiative in reminding the Treasury to order the coronation medals. When the rate of coinage fell off almost to zero during the War of the Spanish Succession, he also took the initiative in gaining the right for the moneyers to make medals as a private enterprise to sustain their fortunes. In all, they produced forty medals during his tenure, only three of which the government commissioned.[200] He had to defend the traditional liberties of workers at the Mint from arrest, taxation, and militia duty. As in the case of the Mint's domain, he did not wholly succeed in preserving anachronistic rights. In 1705, the courts struck down immunity from service in the militia.[201]

He even got into the tin business. Since Cornwall returned more members to the House of Commons than London, Godolphin, the Lord Treasurer, perceived the possible advantages in a special contract whereby the government agreed to buy a set amount of tin each year well above the market rate. The contract went into effect late in 1703. Newton agreed to receive the tin at the Mint, to store it, and to sell it. In his usual professional manner, he proposed careful arrangements to insure the quality of the tin and the integrity of the operation.[202] Since the Cornish mines were the principal source of tin in Europe, and since the contract purchased more than the annual demand, Newton found himself both storing a pile of tin which mounted remarkably and retailing it through agents over much of Europe. Tin became a major item of concern. When William Tyndale proposed that the metal be used as security for paper money, Newton had to compose a memorandum on the subject. When John Williams proposed to manipulate its price, Newton had to advise the Treasury.[203] The Treasury consulted him on proposals to ship tin to Leghorn and Amsterdam, invariably to the large profit of the proposer and equal loss to the Crown.[204] As the stock of tin steadily increased, he composed memoranda on means to minimize the government's

---

[200] *Corres 4*, 384–9. *Corres 6*, 233. Craig, *Newton at the Mint*, pp. 53–4.

[201] In 1700, Newton invoked the franchise of the Mint in response to a summons to furnish two men from the Mint to the militia (*Mint* 1.7, p. 9). In 1705, the Attorney General turned down another petition and ruled that the Mint was not exempt from militia duty (*Mint* 1.7, pp. 50–1). Cf. *Corres 7*, 370.

[202] Newton to Godolphin, 20 Oct. 1703; *Corres 4*, 409–10. Draft in *Mint* 19.3, f. 473. Cf. Newton to Godolphin, April 1708, on assaying tin; *Corres 4*, 516–18.

[203] *Mint* 19.3, f. 534.

[204] *Corres 4*, 466–7. Newton made at least four drafts of this letter (*Mint* 19.3, ff. 563–6). *Corres 5*, 93–4.

loss.[205] The business dragged on into the 1720s, virtually to the end of his life.

In his study of *Newton at the Mint*, John Craig has given a negative assessment of Newton's efforts there. A stickler for precedents, he accepted anything with the sanctity of tradition, but he did not challenge arrangements, such as the office of warden, that had ceased to fulfill useful purposes. Thus he failed to centralize the administration of the Mint, which reverted, after his death, to an internal balance of power between the officers. He failed likewise to change coinage policy, so that England continued to undervalue silver and hence to lose it.[206] I find Craig's judgment undeserved. As it appears to me, he has saddled Newton with responsibility for the shortcomings of an age and charged him with failures which were beyond his power to correct. Monetary policy was not his province. If the government sought his advice, it never proposed to let him determine policy. As we know, he favored lowering the value of gold to retain silver as a currency. Higher authority decided otherwise. In weighing Newton's inability to institutionalize the efficiency that he himself brought, one has constantly to recall the context and the times. In 1701, Newton faced a problem about the Engineer's salary. Charles II had awarded the office to one Thomas D'Oyley for successfully prosecuting a coiner. A linen draper, D'Oyley had no knowledge of the job, but for nearly twenty years he collected the salary.[207] Newton did succeed in having the payment to him canceled. Consider equally an entry in his accounts for the same year. To pass his Indenture, Newton had to pay fees at several offices totaling £222 16s 6d.[208] Although he had only three Indentures during his tenure as master, every year saw analogous fees. He was allowed to enter the expense on his account; the fees were *douceurs* paid by the government to its own servants, and rather generous ones at that. The encompassing web of sinecures, represented by D'Oyley, and subventions, represented by the fees, which bound the operations of the state together, certainly interfered with administrative rationalization. It is not clear that the government of the early eighteenth century could have dispensed with such devices. It is wholly clear that the master of the Mint, a minor department of the Treasury, could not have abolished them on his own authority. Newton faced the issue of sinecures in the Scottish Mint after the union. For a little over a year,

---

[205] "Considerations upon the present state of the Tin affair," *Mint* 19.3, f. 478. Cf. ff. 517–18, 542.

[206] Craig, *Newton at the Mint*, pp. 119–22.

[207] I cite from Newton and Stanley to the Treasury, 30 April 1701; *Mint* 1.7, pp. 33–4; summarized in *Corres* 7, 421.

[208] *Mint* 19.4, f.6. Cf. a list of fees paid at various offices (*Corres* 7, 267–8).

the Edinburgh mint recoined the Scottish currency to the English standard. It never coined again, but its officers continued to draw their annual salaries. Newton made it clear that he regarded the payments as total waste, but he showed his good sense in not wasting much time on a hopeless cause. He showed it again in not trying to abolish the wardenship. He showed less good sense in proposing and pushing a reorganization of the Mint (though not along the lines of Craig's proposal) in 1715. He should have saved his energy and his ink. It never came close to realization.

Studies of administrative history have pointed to the period around Newton's service in the Mint as the seedtime of a professional civil service in England. They indicate the Treasury especially as the locus of such a development.[209] The accounts from the Mint were not the only ones that began to arrive annually. In my opinion, Newton deserves to be recognized as a distinguished civil servant in the first age when such appeared. No doubt he did not transform the Mint. He did make it operate with more efficiency by far than it had shown in the past or would show again for another century. The record may be as much as one can ask.

No one, I think, would care to measure Newton's work at the Mint with the same scale applied to the *Principia*. If, however, we accept the proposition that Newton's creative energy had exhausted itself, as every creative energy must in the end, and if we look at his post at the Mint in terms, not of its contribution to civilization, but of the light it casts on Newton, attention to it yields some fruit. It tells us much about the man that without prior experience, after a cloistered life devoted solely to study, he was able to pick up the threads of a practical pursuit and to perform it with distinction. Craig berates his defense of precedents. Precedents were his only protection against the rapacity of others in the bloody arena of Augustan political life. In that arena, he held his own and more. One may not admire the activity as such, but one cannot ignore the force of character necessary to sustain it.

Newton's work as an administrator also casts light back on his earlier life as a scholar. The concern for order that he brought to the Mint illuminates the indices he compiled in every new study he undertook. A concern for order is not equivalent to genius. In Newton's case at least, it did enhance genius. What he learned, he learned for good by entering it in its proper place. The same instinct served him well at the Mint.

209 Stephen B. Baxter, *The Development of the Treasury, 1660–1702* (London, 1957), pp. 167–264. Henry Roseveare, *The Treasury. The Evolution of a British Institution* (London, 1969), pp. 57–81, and *The Treasury 1660–1870. The Foundations of Control* (London, 1973), pp. 78–82.

Administration as such did not exhaust Newton's activities at the Mint. They also involved an intellectual dimension, for the government sought and expected advice on monetary questions from the master. His first serious exercise in this direction, which may have occurred soon after his arrival, probably resulted from a request of John Locke and probably circulated only privately. A colleague of Locke's on the Board of Trade, John Pollexfen, composed a *Discourse of Trade, Coyn, and Paper Credit*, which he dated 15 July 1696. It was published the following year without Pollexfen's knowledge, according to the publisher, who indicated that it had been shown in the meantime to "some Persons of great Judgment . . ." Newton's comments on what he called Mr. Pollexfen's discourse on paper credit seem to fit this description, though other evidence suggests 1700, when Pollexfen reissued his pamphlet, for Newton's reply.[210]

Dominated by mercantilist preoccupations, Pollexfen was convinced that the current pattern of trade was stripping the nation of its bullion and hence of its wealth. Paper credit encouraged the loss by creating the illusion of prosperity and encouraging the importation of foreign luxuries for which bullion had to be paid. Newton subscribed to most of the mercantilist prejudices that Pollexfen held. He too felt that the importation of luxuries, "of shapeless China pictures [pitchers] & other superfluities . . ." was pernicious.[211] He disagreed, however, with Pollexfen's objection to paper credit as the cause of such importation. "But if this be a good objection, we must reject wine because it occasions drunkenness, & all the best things because by corruption they become the worst. Rather let us suppress drunkenness & keep o^r wine."[212] The crux of Newton's reply, which contended that the loss of currency was a temporary phenomenon due to the war, was to urge the benefit of a properly controlled use of credit to maintain the supply of money, which would in turn promote prosperity by keeping the rate of interest low.

> Credit is a present remedy against poverty & like the best remedies in Physick works strongly & has a poisonous quality. For it inclines the nation to an expensive luxury in forreign commodities. But good physitians reject not strong remedies because they may kill but study how to apply them with safety & success. Let that Expense [of

---

[210] John Pollexfen, *A Discourse of Trade, Coyn, and Paper Credit* (London, 1697). Newton's response is in *Mint* 19.2, ff. 608–11, with drafts in ff. 624–32. Cf. Craig, *Newton at the Mint*, pp. 40–3. Craig places Newton's response in 1700. A draft on the back of a memorandum of advice to the parish of Colsterworth in 1700 probably establishes the date beyond question (*Corres 4*, 319–20). However, Newton's reference to the war as though it were in progress does suggest that he was one of the persons to whom it was shown in 1696 or 1697.

[211] He crossed out this comment, which appears in a draft; *Mint* 19.2, f. 624^v.

[212] *Mint* 19.2, f. 613.

luxuries] be corrected by good Orders & laws well executed & then credit becomes a very safe & soveraign remedy.[213]

Although Newton's attention in his paper focused on the issue of paper credit, he did mention in passing that the loss of silver currency arose from the much higher valuation of silver in relation to gold in the Orient. The question of the value of silver was to remain as the basic theme of his monetary advice to the government. An exchange which continued briefly followed, with Pollexfen replying to Newton's observations and Newton commenting on the reply before it died away.[214]

Newton's stress on management of the money supply may have been ahead of the time, but one must not try to elevate him, either for this paper or for others, to the rank of major economic thinker. Under the pressure of the fiscal crisis induced by the war, the 1690s witnessed a considerable publication of pamphlets on economic policy, which historians have come to view as the source of the discipline of economics.[215] Newton must have read and digested this literature, especially the contributions of John Locke to it. For the most part, his views simply repeated accepted opinions. Very little that is original or important appeared in his memoranda on monetary policy. Shirras and Craig have concluded that as an economist he was an outstanding mathematician.[216] For our understanding of Newton, it is important that he did school himself in the best current opinion. Part of his distinction as a civil servant lay in the fact that the government could get from him an intelligent reflection of the best current opinion on monetary policy.

In January 1701, Newton took an initiative by sending a memorandum to the government on the overvaluation of French and Spanish pistoles in England, which was leading to their excessive importation. The government accepted his advice and lowered the value of these gold coins, which had been exchanging for 17s 6d, to 17s. In a draft of his memorandum, Newton expressed his growing distrust of the city financiers. The price of the coins in question, he asserted, "seems raised above ye standard value by the Exchangers & Bankers imposing upon the nation for their private advantage."[217] The exchangers and bankers might well have retorted in kind. As a result of the government's action, the French and Spanish coins flooded into the Mint to be coined into English money,

213 *Mint* 19.2, f. 611.
214 Drafts of Newton's "Observations on M^r P's Reply"; *Mint* 19.2, ff. 612–22, 633–41.
215 William Letwin, *The Origin of Scientific Economics*, (London, 1963). J. Kieth Horsefield, *British Monetary Experiments, 1650–1710* (London, 1960).
216 G. Findlay Shirras and J. H. Craig, "Sir Isaac Newton and the Currency," *Economic Journal*, 55 (1945), 217–41.
217 *Corres 4*, 352–3. *Mint* 19.2, f. 148^v.

and Newton profited to the extent of £2,500 from his own advice. It was later in 1701 that he had his first clash with the goldsmiths at the trial of the pyx.

Newton's memorandum of January also dealt with a second matter. Silver bullion, he noted, carried a higher value than silver coin which was, as a result, melted down and exported. He returned to that theme in September in a further memorandum which advised the government that the French had altered the relative values placed on gold and silver, putting their monetary policy into line with that of the Netherlands. With the exception of Spain, where he noted that practice differed from policy, England was left alone in its undervaluation of silver. Implicitly he proposed a change in policy to preserve English silver coins.[218] The government asked instead for a thorough study of the relative valuations of gold and silver throughout Europe, a topic to which Newton devoted much of his time during the following nine months. He assayed, or caused to be assayed, coins from every important country of Europe. One table he drew up contained the assays of 119 different coins.[219] He drew up the results of the study into a memorandum submitted to the Treasury on 7 July 1702. Newton urged several changes. The value of the guinea, which passed then at 21s 6d, should be reduced by 6, 9, or even 12d in order to bring England's valuation of silver into agreement with the Continent's and to stop the melting down of silver coin and its exportation as bullion. He argued for a change in the law that forbad the exportation of coin but not of bullion. Allowing merchants to reexport foreign bullion even in the form of English coin would encourage its coinage, and some of the coin might remain in England. Finally, realizing the ultimate responsibility of the eastern trade for the exportation of silver, he urged a law that would compel the East India Company to conduct a certain proportion of its trade in English goods. He specifically opposed alteration of the weight or standard of silver coins; he did think that a coarse alloy in small coins to make them wear better would do no harm.[220] In connection with the memorandum, Newton also drew up a proposal for an act to preserve and increase the coin of the kingdom which embodied his recommendations on the exportation of coin and bullion. It suggested that exportation be confined to London and placed under the control of the Mint, which alone would be authorized to melt coin of what-

---

[218] *Corres 4*, 373–4.

[219] *Mint* 19.2, ff. 51–2. Cf. many other papers on the value of various coins; *Mint* 19.2, ff. 2–64, 155, 157, 165, 168, 174–7, 183–7, 197–8, 208. His report to Godolphin, including the table of the value of European coins, is printed in Shaw, *Select Tracts*, pp. 136–49.

[220] *Corres 4*, 388–90. There were, of course, several drafts; *Mint* 19.2, ff. 74–6, 78–9, 99, 101, 104–5.

ever source into bullion. One version of the proposed act forbade the use of gold and silver in clothes, coaches, gilding and the like. He wanted even to ban large vessels of gold and silver, since coin was melted down to make them and their export gave it a means of escape. Warming to his theme, he went on to give full vent to familiar mercantilist views.

The like limitations for China earthen ware would save ye Nation much money & so would be a prohibition of importing Cabinets & other laquered wooden ware from Japan & other parts of ye Indies. For these things serve for nothing but a useless & expensive sort of luxury maintained by the exportation of o^r gold & silver to the Indies.[221]

It is unlikely that Newton's proposal had any chance of acceptance. The advent of the War of the Spanish Succession destroyed whatever chance there was, and the loss of silver currency continued, though it would no doubt have happened regardless.

During the war, the government frequently turned to Newton for advice on the rate at which to exchange money used to pay troops on the Continent. In the fall of 1711, England prepared to send commissioners to Spain to receive 800,000 Mexican dollars to be coined into two-real pieces to pay the English troops there. Upon the government's request, Newton prepared instructions for the commissioners to protect them from being cheated by the Spanish mint. He prepared a sort of catechism in eight questions for the commissioners to follow. What was the total weight of the 800,000 dollars? What was the weight of the ingots into which they were melted? When the various losses of the melting were taken into account, he explained, about 1/300 of the weight of the coins should be lost. What was the total weight of the two-real pieces? He gave them the weight they ought to receive. What was the number of two-real pieces after the charges for melting and coining were subtracted?[222]

On a different plane, Newton offered the design for Anne's coronation medal. He summoned up his knowledge of ancient history and suggested the portrayal of Anne as Pallas destroying a giant with thunder.

It alludes to an ancient warr between the ancestors of the Egyptians represented by Gods (Jupiter, Palls &c) & their enemies represented by Giants [he explained]. The Giants to denote that they were not single persons but great bodies of men were painted with many heads & hands, & to express their hostile force & terrour they had

---

221 A memorandum on the subject; *Mint* 19.2, ff. 591–2. Drafts in ff. 595–6. The proposed act is in ff. 576–83, with drafts in ff. 574–5, 589, 593–4, 597, 599, 600, 602–3.

222 *Mint* 19.2, ff. 193–4. Draft in ff. 199–200. An earlier form of his advice is published in *Corres 5*, 197–8.

Figure 12.6. A medal of Newton engraved by John Croker, the Mint's engraver during his tenure as master.

scaly & snaky leggs as Pallas had a snaky shield. When Jupiter was weary Pallas came in & carried on the warr.

The whole signifies that her Mas^ty continues ye scene of ye last reign [since William's coronation medal depicted him as Jupiter].[223] He even included annotations to justify the various images. In fact, the coronation medal did embody his design, though it is doubtful that many penetrated its meaning. A puzzled catalogue of medals struck by the Mint describes it as Pallas (Anne) hurling thunder against a double-headed monster with four arms and snaky ex-

[223] *Mint* 19.3, f. 291. A second draft (ff. 292–3) gave two full pages of annotations on the images. Newton loved this assignment; he composed seven more drafts on the medal (ff. 294–301).

tremities. With his credentials established, Newton contributed suggestions for the other official medals struck while he was master.

The office involved a further obligation, or potential obligation – membership in the House of Commons, where he could lend his support to the government, or perhaps to Halifax. As I have pointed out, Newton visited Cambridge in the election year 1698 though he did not stand. In 1701, he did. He was elected and served in the Parliament that began to sit on 20 December. As before in the Convention Parliament, he was not prominent in any respect. The one division of the House that a contemporary recorded in print saw him vote in support of Halifax and the other Junto lords on a motion related to their impeachment which was then before the Lords.[224] That is, on a partisan measure he voted with the slender Whig majority. The death of William led to a prorogation in May 1702 followed soon after by dissolution. Newton did not formally stand for the following election later that year. There had been unpleasantness at the former one. The defeated candidate, Anthony Hammond, had composed a pamphlet, *Considerations upon Corrupt Elections of Members to serve in Parliament*. While it contained no explicit reference to Cambridge, it did argue that the New East India Company was engaged in an extensive program of electoral corruption to insure a governmental policy favorable to its interests. Halifax was associated with the New East India Company, having introduced the bill for its incorporation in 1698. Both Newton and the public at large could easily have read Hammond's pamphlet as a charge that he was a paid lackey. Furthermore, it hinted darkly that radical religious groups might subvert the Anglican church by the same device.[225] In a letter to a friend (probably Bentley) in the summer of 1702, Newton indicated that he refused to come to Cambridge to stand openly in the new poll. If elected, he would serve, though he did not expect to be. "To solicit and miss for want of doing it sufficiently, would be a reflection upon me, and it's better to sit still."[226] Why would he not stand openly? Perhaps the role of religious conformity which Anne herself injected into the election with her closing speech to the previous Parliament played some part in his decision. A pamphlet on the election of 1702 by the Jacobite James Drake, which referred specifically to Cambridge and to Halifax as a patron powerful in the

---

[224] James Drake, "Some necessary Considerations relating to all Future Elections of Members to serve in Parliament" (1702), *Somers Tracts, 12,* 198–218.
[225] The pamphlet was reprinted in Anthony Hammond and Walter Moyle, *The Honest Elector; or Unerring Reasons for the Prevention of Chusing Corrupt Members to serve in Parliament: With Instructions for the Choice of a Speaker* (London, n.d.). [226] *Corres 4,* 382–3.

election there, placed the issue of hypocrites, who were destroying the church by pretending to be true Protestants, at the center of attention.[227] Drake's message was more unsettling than Hammond's. Newton instinctively ran for cover whenever similar arguments appeared.

Halifax, whose fortunes depended on a block of support in Commons, had a more aggressive attitude in mind, and he took care to prepare Newton's mind for the next contest. It came in 1705. Parliament was prorogued on 14 March. An election had to be held that year, and Newton had already been to Cambridge once to test the water.[228] On 17 March, Halifax described his plans for the election, as far as Cambridge was concerned, to Newton. "Mr Godolphin will go down to Cambridge next weeke and if the Queen goes to Newmarket, and from thence to Cambridge she will give you great Assistance the Torys say she makes that tour on purpose to turn Mr Ansley out."[229] "Mr. Godolphin" referred to Francis Godolphin, son of the Lord Treasurer, who stood with Newton. Arthur Annesley, one of the incumbents, who had replaced Newton in 1702, was a staunch Tory. Newton was less convinced than Montague. Writing to his friend in Cambridge, he indicated his hesitancy whether or not to stand. Nevertheless, "moved by some friends of very good note," he asked his correspondent to reserve his vote until Newton wrote again or came to visit.[230]

In the end, Halifax prevailed, and Newton fairly wore out the road to Cambridge with three more visits during the following month. About the first, we know only that he returned to London on or by 5 April.[231] He must have given the word that he would stand. He was back in Cambridge on 16 April when the queen made the visit foretold by Halifax. At the final royal visit to the university that he would attend, Newton occupied a place on the stage. "The whole University lined both sides of the way from Emanuel College, where the Queen enter'd the Town, to the public Schools," Stukeley, who was an undergraduate then, recalled. "Her majesty dined at Trinity college where she knighted Sir Isaac, and afterward, went to Evening Service at King's college chapel . . . The Provost made a speech to her Majesty, and presented her with a bible richly ornamented. Then she returned amid the repeated

[227] Drake, "Some necessary Considerations."

[228] A note by Flamsteed on the draft of a letter to Newton indicated his trip to Cambridge between 5 January and 26 February; *Corres 4*, 435. He presided at the Royal Society on 17 January and then was absent until 21 February.   [229] *Corres 4*, 439.

[230] *Corres 4*, 441. Cf. a similar letter, Newton to ———, n.d.; *Corres 7*,

[231] Flamsteed wrote him on 5 April, then added a note that he had not sent the letter since Newton returned too soon (*Baily*, pp. 238–9).

acclamations of the scholars and townsmen."[232] The queen's "great Assistance" to Newton's election was his knighting, an honor bestowed, not for his contributions to science, not for his service at the Mint, but for the greater glory of party politics in the election of 1705. Halifax, who had organized the visit, orchestrated it as a political rally. Besides Newton, the queen also knighted Halifax's brother and mandated the university to confer an honorary doctorate on Halifax himself. In a nonpartisan gesture, he allowed her also to knight Newton's old friend, John Ellis, a mere academic then vice-chancellor of the university. Having gone back to London once more, Newton returned a third time about 24 April and stayed on, soliciting votes, until the election on 17 May. To Flamsteed he had expressed his pessimism over the outcome. "'Tis something doubtful whether he will succeed or no," Flamsteed repeated to Sharp, "by reason he put in too late."[233] In fact, he could hardly have put in, or been thrust in, earlier. It is even more doubtful that he would have stood at all had he been as pessimistic as he said; Newton always liked to prepare an excuse just in case.

In the event, he needed one, for things did not go well. Early in May, Halifax tried to encourage him.

I have writ to My Ld Manchester to engage Mr Gale for Mr Godolphin, but I am affraid his Letter will not come time enough. There can be no doubt of Ld Manchesters sentiments in this affair, Mr Gale may be sure He will oblige Him, and all his friends by appearing for Mr Godolphin, and He can do you no good any other ways. I am sorry you mention nothing of the Election, it does not look well but I hope you still keep your Resolution of not being disturbed at the event, since there has been no fault of your's in the Managemt, and then there is no great matter in it: I could tell you some storys where the Conduct of the Court has been the same, but complaining is to no purpose and now the Die is cast, and upon the whole Wee shall have a good Parliamt.[234]

Newton in turn described how bad the situation was to Halifax. Annesley appeared to have twenty-six or twenty-eight votes more than he had. If Windsor would withdraw, Newton would pick up votes, and Godolphin would have enough to spare him ten or fifteen. If Windsor had not entered the election, he would not have needed assistance. "For the opposition of W & the vogue against me of late have discouraged my friends & checkt & diminished my interest & inclined indifferent persons against me." He had a tense spat with his running mate Godolphin, and told him, "I have done more for you then you have done for either me or your self." Supporters struck a hard bargain and insisted on a return for their

[232] *Stukeley*, p. 9.   [233] *Baily*, p. 239.   [234] *Corres 4*, 445.

votes. Newton had to contribute £60 to Trinity and agree to let Whiston publish his lectures on algebra.[235]

As it turned out, nothing helped, and all of Newton's resolution could not prevent his being disturbed. Not only did he run dead last and not even close in a field of four, but the unpleasantness of 1701, which he had feared to face in 1702, repeated itself and took on a form which could not have upset him more. Simon Patrick, Bishop of Ely, described the scene to the House of Lords the following December as he urged an investigation into the corruption of youth – by Anglican zealots of all things! – at the university: "at the election at Cambridge it was shameful to see a hundred or more young students, encouraged in hollowing like schoolboys and porters, and crying, No Fanatic, No Occasional Conformity, against two worthy gentlemen that stood candidates."[236] Occasional conformity was the accepted practice by which dissenters qualified for full civil rights by taking the sacrament in the established church once a year. The move to revoke it, which was pushed by extreme Tories, struck at the heart of Newton's security. No scene could have shaken him more. No amount of encouragement from Halifax could tempt him to chance its repetition. The year 1705 marked the end of his Parliamentary career.

Administrative duties continued at the Mint, however. What with a constant flow of references from the Treasury and memoranda in reply, the business of the Mint included much more than coining. It formed the pervasive background of Newton's life in London. With the advent of the war in 1702, its demands did slacken, however. Beginning with 27 May 1703, the Mint did not coin a single day for nine months. It coined a total of seventy-five days during the ensuing six years.[237] Under such circumstances, Newton was finally free, after seven years dominated by the administrative demands of his new institution, to consider other activities.

[235] *Corres* 7, 437, 438. *Edleston*, p. lxxiii. *Hiscock*, p. 36.
[236] William Cobbett, *The Parliamentary History of England*, 36 vols. (London, 1806–20), *6*, 496.
[237] From the Comptroller's Book; *Mint* 9.48.

13

President of the Royal Society

T HE Royal Society, to which Newton had dedicated his *Principia* in 1687 only to ignore it steadfastly when he moved to London, stood at a low ebb during the early years of his residence in the capital city. Membership, which had reached more than two hundred in the early years of the 1670s, now scarcely numbered more than half that figure, and few of the members attended the weekly meetings. In light of what the minutes indicate about the meetings, one cannot blame them for not attending. Unplanned and unstructured, given over mostly to miscellaneous chit-chat devoid of serious scientific interest, the meetings suggested little of the interests that had brought the society together forty years earlier. In 1699, a concerned council gave thought to stimulate members to perform experiments at meetings. On 8 March, they drew up a list of thirty-eight members (including "M^r. Newtoun") who might be charged with that function, and a week later seventeen of them ("M^r. Newtoun" not included) showed up at the council and promised to undertake the obligation. The following week, some of the seventeen suggested that the council pay them £5 per year for their pains.[1]

The flurry of interest died out as quickly as it had arisen, however, and later that spring Mr. Van de Bemde's remark "y^t cows piss drank to about a pint, will either purge or vomit with great Ease," led to an all-too-typical discussion of the medicinal value of bovine urine.[2] On another occasion, an account of what I take to be lemmings stimulated a penetrating exchange.

Dr Woodward said that it is not water that Nourishes but Earth.

The V. Presid^t said the best time to smell at flowers is in ye Morning.

The D^r. was thanked & ye V. P. commended.

Dr Sloane desired it might be printed, and said it was a very proper discourse for this place.

The V.P. said if every body would make their discourses methodically as this, all would be well.

The Society thanked the D^r.[3]

Again in the summer of 1701, when the society adjourned for the vacation after a series of disastrously barren meetings, they instructed their servant Hunt, at their reconvening in October, "to

---

[1] *CM 2*, 148–50. [2] 31 May 1699; *JB 10*, 128. [3] 25 April 1699; *JB 10*, 120.

summon the Members to attend the Meetings, as usually has been accustomed." As before, it was to no avail, and in the spring of 1702 Sir John Hoskins, the presiding officer, edified the hardy few who were present with an account of a woman in Gloucester who had finally succeeded in poisoning her husband with arsenic after "she had first tried Sow-bread, Nightshade, Mad-nips, Spiders, & Toads without Effect."[4]

Behind the emptiness of the meetings lay a failure of leadership. In January 1699, when the council was meeting once a week to consider the crisis both financial and intellectual of the society, Mr. Bridgman reported that he had given the Lord Chancellor an account of their recent proceedings, "who approved thereof and was pleased to Say That he Should be always ready by his presence or otherwise to give all the encouragement he could to the Society."[5] The society had elected the author of these noble sentiments, Lord Somers, a leader of the Whig Junto and thus a colleague of Montague, both member and president the previous November. Somers demonstrated his devotion by attending a meeting, at which he presided, the following August; and during the five years he served as president, he attended two other meetings, at one of which he presided, and summoned the society to attend him on one occasion. Somers had succeeded Montague, who bore the title of president of the Royal Society for three years (1695–8), during which he performed equal prodigies on its behalf. In Dr. Hans Sloane, the society had a vigorous secretary who, by all accounts, dominated its proceedings. The experiments in awarding the presidency to prominent politicians uninterested in the society's aims proved a disaster, however, and it did not prosper again until a natural philosopher committed to its goals occupied the presidential chair.

A further crisis confronted the society in March 1703 after the death of Robert Hooke. Long the commanding presence in the society intellectually, the man who as curator of experiments had provided what substance meetings had had, Hooke had ceased to figure prominently in its affairs as sickness ravaged his final years. Nevertheless, he had remained essential to the society since they met in his chambers in Gresham College. On 24 March 1703, the committee in charge of Gresham College asked the Royal Society to hand over the keys to the chambers and to clear out their library and repository forthwith.[6] The society begged for time. In the end, the committee did not insist – because of the intervention of Dr. Woodward, another Gresham professor, according to one account – but the threat of eviction now hung over them and influenced their proceedings during the following years.

---

[4] 6 Aug. 1701; *JB 10*, 227. 22 April 1702; *JB 10*, 246.   [5] *CM 2*, 143.   [6] *JB 11*, 16.

The same Robert Hooke, not Newton's favorite natural philosopher, may well have determined his absence from the weekly meetings. Hooke was usually there. He was a member of the council nearly every year. When Newton put in one of his rare appearances to show a "new instrument contrived by him," a sextant, which would be useful in navigation, Hooke reminded him of past antipathies by claiming he had invented it more than thirty years before.[7] Hooke's death in March 1703 removed an obstacle and prepared the way for Newton's election as president at the next annual meeting on St. Andrew's Day, 30 November.

Obscurity covers the background to Newton's election. Spontaneous expressions of popular will did not govern the selection of officers of the Royal Society. The two previous presidents, for example, had had to be elected to membership on the same days when they were elected president. With prominent personalities involved, someone had clearly made prior arrangements. In all probability, that someone was Dr. Hans Sloane, the secretary, and we may assume that Dr. Sloane, recognizing the disastrous result of *fainéant* figureheads, likewise arranged Newton's election. If Conduitt's story was correct, Newton was not Sloane's first choice. Rather, he approached Sir Christopher Wren, who desired the society to elect Newton instead.[8] At the meeting on 30 November something nearly went awry. Unlike Montague and Somers, Newton was not a political leader who had only to be proposed to be elected. Neither had he extended himself, in the seven and a half years of his residence in London, in serving the society. At least thirty members attended the meeting. Only twenty-two of them voted to place Newton on the council, a necessary preliminary to election as president. At least half of the other members chosen for the council received more votes than he did.[9] Once elected to the council, he still received only twenty-four votes for president. Clearly, a group within the Royal Society did not rush to welcome England's preeminent natural philosopher to the presidential chair. Truth to tell, they did not rush to reelect him the next year, and the absence of vote totals in the Journal Book the following two years strongly implies a continuing want of universal enthusiasm.[10]

[7] 16 August and 25 October (the first meeting after that on 16 Aug.) 1699; *JB 10*, 145.

[8] *Keynes MSS* 130.6, Book 2; 130.5, Sheet 2.

[9] Hill received 30 votes, the highest number recorded. The Journal Book failed to get the votes for four members elected; of the seventeen counts given, ten were higher than Newton's (*JB 11, 36*).

[10] *JBC 10*, 89, 116, 148. For the election of 1706, the last of these, the Journal Book gave the votes for everyone except Newton. For 1707 and 1708, votes were again included; in both years others received more votes than Newton did (*JBC 10*, 168, 199).

Less than two years after Newton's election, Queen Anne knighted him in Cambridge. Master of the Mint and president of the Royal Society, Sir Isaac Newton had become a personage of consequence. The attention he devoted to his coat of arms testifies that he recognized as much.[11] A year before his election, he had sat for another portrait by Kneller and upon his election one by Charles Jervas (Figure 13.1). Now, as Sir Isaac Newton, he was painted by William Gandy and by Sir James Thornhill.[12] The rebel of yore had allowed the establishment wholly to coopt him.

Newton did not attend the meeting of the society on 8 December, the first one after his election. On 15 December, he did appear and immediately took command. To the Royal Society he brought the same qualities he exercised at the Mint, administrative talent and a constitutional inability to slough off an obligation he had agreed to shoulder. In a history of the administration of the Royal Society, Sir Henry Lyons has stressed the paramount importance at this time in the society's affairs of vigorous and continuous leadership.[13] After the interlude of absentee presidents chosen for their political prominence, the society watched with surprise as a man who had devoted his life to its announced goals seized its helm and bent his energy to steering it on a determined course. Newton made a point of running the council. It almost never met without him. Whereas Montague had attended one meeting of the council during the three years of his presidency and Somers none during his five, Newton failed to preside at a total of three meetings during the next twenty years before age began to restrict him.[14] A society which had seen its president in the chair three times during the preceding eight years now saw him there more than three meetings out of four. When they received letters in 1704 about burning glasses and burning mirrors, the president set to at once to construct similar ones for the society. The glass, a composite of seven glasses, each 11 1/2 inches in diameter, arranged to form a segment of a sphere, would vitrify brick or tile in a moment and melt gold

---

[11] *Keynes MS* 112. *Babson MSS* 439–44. Manuscript in the Humanities Research Center, University of Texas. Cf. LeNeve to Newton, 24 Nov. 1705; *Corres 4,* 460.

[12] Gandy painted Newton in 1706. Conduitt placed the Thornhill portrait, of which two versions exist, in 1705 (*Keynes MS* 130.6, Book 3). One of the copies has the year 1710 on it, however, and Bentley, who appears to have commissioned it, mentioned a sitting in a letter of Oct. 1709 (*Corres 5,* 8). Either year was well before Conduitt met Newton, so that his testimony does not need to be taken seriously in this case.

[13] Henry Lyons, *The Royal Society 1660–1940: A History of Its Administration under Its Charters* (Cambridge, 1944), pp. 117–58.

[14] Lyons states that Newton attended 161 of 175 council meetings during his tenure (*ibid.,* p. 121). I count differently: 167 of 177. As I mentioned, 7 of the 10 absences fell during the final three and a half years of his life.

Figure 13.1. Newton at sixty. Portrait by Charles Jervas, 1703, which Newton presented to the Royal Society in 1717. (Courtesy of the Royal Society.)

in half a minute.[15] Newton frequently participated in the discussion at the meetings. His contribution to the Royal Society as its president was administrative rather than intellectual, however. It was

[15] The two were mentioned numerous times between 12 Jan. and 12 July 1704 (*JBC 10*, 58–82). Cf. a description of the glass (*Edleston*, p. lxxi).

not coincidence that the society's fortunes began to revive at much the time when he took charge of its affairs.

Administration did involve intellectual matters, of course. Newton was aware that meetings lacked serious content, and he came to the presidency armed with a "Scheme for establishing the Royal Society," intended to cure the disease. "Natural Philosophy," the "Scheme" declared, "consists in discovering the frame and operations of Nature, and reducing them, as far as may be, to general Rules or Laws, – establishing these rules by observations and experiments, and thence deducing the causes and effects of things . . ." To this end, it might be convenient that one or two and ultimately perhaps three or four men skilled in the major branches of philosophy be established with pensions and bound to attend the weekly meetings. He went on to set down five major branches of natural philosophy, for each of which, presumably, he looked forward to appointing a pensioned demonstrator: mathematics and mechanics; astronomy and optics; zoology (to use our word), anatomy, and physiology; botany; and chemistry. He stated explicitly his intention that the society appoint only men who had established reputations in the sciences.[16] It would be a mistake to read Newton's "Scheme" as a proposal to remodel the Royal Society along the lines of the French Académie des sciences. He did not suggest any limitation of membership to professionals. Rather, he proposed the extension of an institution native to the society, the curator of experiments, to provide at the weekly meetings solid substance for the body of amateur members. Newton's old nemesis, Robert Hooke, had filled that post with distinction for many years and by his efforts had kept the society afloat when the formless garrulity of the members had threatened to ground the meetings on utter banality. Newton did not mention Hooke, but his "Scheme for establishing the Royal Society" testifies adequately that he recognized what Hooke had done for the society. With Hooke dead, he set his first priority in replacing him – preferably in the plural.

He did more than hope. He found Hooke's replacement, Francis Hauksbee. We know nothing of Hauksbee's origin and background or of how Newton made his acquaintance. We only know that on 15 December 1703, at the first meeting over which Newton presided, Hauksbee appeared at the Royal Society for the first time, and though he was not then a member, performed an experiment with his newly improved air pump. He continued to attend the meetings; nearly every week, he showed an experiment using the air pump. In February, the council voted to pay him two guineas for his trouble,

---

16 There are seven drafts of the "Scheme" (*Add MS* 4005.2). One is printed in David Brewster, *Memoirs of the Life, Writings, and Discoveries of Sir Isaac Newton*, 2 vols. (Edinburgh, 1855) *1*, 102–4.

and in July, as they broke up for the summer, five guineas more. Hauksbee presented himself at the latter meeting of the council and proposed to regularize the arrangement. He would show an experiment every week – either one the society directed him to make or one that he himself contrived with the air pump. That is, Hauksbee proposed that he would function as one of the salaried demonstrators that Newton's "Scheme" called for, though he had not at that time established a reputation as an investigator. Surely he came before the council with Newton's encouragement. Newton must have regarded the council's response as a partial rebuff; for though they accepted Hauksbee's offer for the coming year, they refused officially to appoint him or to set a salary. They would, they assured Hauksbee, "gratify him accordingly to the proportion of his Services."[17] Hauksbee continued to serve the society, providing much of the scientific content of its meetings, for ten years until his death in 1713. The society never did give him an official position. Each year the council voted him a gratuity, £15 for 1704–5, as much as £40 in some years, though the council occasionally reduced the sum when he had performed less energetically.

The precise nature of Hauksbee's relation to Newton cannot be defined with assurance. Left to himself, Hauksbee did not appear to have immense intellectual initiative. He devoted his first year and a half at the society to experiments with his air pump which showed little imagination and largely repeated experiments done earlier by Boyle and others. Later, he did get on to new themes, especially electricity and capillary action, in which his experiments exercised considerable influence on Newton. Of course, neither topic was previously unknown to Newton. No evidence tells us how far he may have guided Hauksbee's new ventures, though we must beware of attributing to Newton what was not his. There is no reason to think that he suggested mounting a globe of glass on an axle and thus nearly inventing the static-electric machine, though the effects that Hauksbee generated thereby powerfully stimulated Newton's imagination. Equally there is no reason to think that Newton proposed the experiments on capillarity in vacuums, since these experiments first taught him that he had been mistaken over forty years in asserting that capillary phenomena do not take place *in vacuo.* The rapid extension of the capillary experiments into a quantitative investigation of forces of attraction, which Newton inserted into the second English edition of the *Opticks,* bore the mark of the lion's claw, however, and he probably directly guided this work. As he began to prepare the second edition of the *Principia,* Newton probably also suggested experiments that Hauksbee performed in 1710 to

---

[17] *CM 2,* 169, 174.

test the theory of air resistance by dropping balls of blown glass from the top of St. Paul's. The *Philosophical Transactions* published a steady stream of Hauksbee's experimental papers, and in 1709 he collected them together in his *Physico-Mechanical Experiments*. As a result, he became a famous scientist in his own right. In 1710, the German traveler Zaccharius von Uffenbach made certain to visit "the natural-philosopher Hauksbee."[18] However, Newton appears to have treated him not as an intellectual peer, but as a servant. In September 1705, he wrote Sloane to have Hauksbee bring his air pump to Newton's house on a certain evening to show his experiments to "some philosophical persons . . . & I will give him two guineas for his pains." The audience for whom he was summoned to perform included such well-known philosophers as Halifax, the Earl of Pembroke, and the Bishop of Dublin.[19]

In 1707, Newton appeared for a time to have found a second demonstrator to supplement Hauksbee at meetings of the society. Dr. James Douglas frequently performed dissections at meetings, and in July of that year the council voted him a gratuity of £10. For reasons not recorded, the arrangement with Douglas failed to solidify. Although Douglas continued as an active participant in meetings, he never received another gratuity for his efforts.

There is no way to pretend that on the morrow of Newton's election the meetings of the Royal Society suddenly transformed themselves into profound discussions bubbling with philosophical ferment. The society's appetite for monstrosities brooked no satisfaction. Between dissections, Dr. Douglas showed them "a puppy Sufficiently Nourished whelpt alive about 10 Days ago which had no Mouth," and a week later he brought in its skull.[20] In 1709, "There was shewed four Piggs all Growing to One Another taken out of a Sow after she was killed. M^r. Hunt was Ordered to give the Bearer three half Crowns and to preserve them in Spirits of Wine."[21] On another occasion, Mr. Cowper "Shewed the penis of the possum which belonged to the Society and died Lately." He was desired to draw up an account of it, which accordingly he did.[22] Newton contributed his own share to the miscellaneous reflections which threatened always to swamp serious scientific discussion, telling the members on one occasion of a man who died

[18] J.E.B. Mayor, *Cambridge under Queen Anne* (Cambridge, 1870) p. 402.
[19] Newton to Sloane, 14 and 17 Sept. 1705; *Corres 4*, 446–8. Cf. Henry Guerlac, "Francis Hauksbee: Expérimentateur au profit de Newton," *Archives internationales d'histoire des sciences*, 16 (1963), 113–28; "Sir Isaac and the Ingenious Mr. Hauksbee," *Mélanges Alexandre Koyré*, 2 vols. (Paris, 1964), *1, L'aventure de la science*, ed. I. Bernard Cohen and René Taton, 228–53; *Newton et Epicure*. Conférence donnée au Palais de la Découverte le 2 Mars 1963 (Paris, 1963).
[20] 22 and 29 Oct. 1707; *JBC 10*, 163–4.   [21] 21 Dec. 1709; *JBC 10*, 228.
[22] 12 Jan. and 2 Feb. 1704; *JBC 10*, 58, 60.

from drinking brandy and on another of a dog at Trinity that was killed by Macasar poison. At one meeting, he informed the society "that Bran Wetted and Heated would bread Worms, which are Supposed to Proceed from Eggs Laid therein."[23] When Uffenbach visited the society in 1710, his disappointment fell little short of contempt.

> Miserable state of the Royal Society's apparatus. The guide, if asked for anything, generally said: 'a rogue has stolen it away;' or he shewed fragments of it, saying: 'it is broken.' The 'transactions' of the first six years of the society are worth all the rest together. The entire series can be had bound for £12. The society never meets in summer, and very little in autumn. The present secretary, Dr. Sloane, is indeed a very learned man, but engrossed with his practice and his own large cabinet. The president, Newton, is an old man, and too much occupied as master of the mint, with his own affairs, to trouble himself much about the society. For the rest, excepting Dr. Woodward and a couple more Englishmen, and the foreigners, there remain only apothecaries and the like, who scarce understand latin.[24]

Uffenbach's comments appear too severe, both on the society and on Newton. The weekly meetings showed a steady improvement during his presidency. From the low point in the 1690s, membership steadily grew and more than doubled during his years in office. No doubt many things contributed to the renewed vitality, but the enhanced level of the meetings, which Newton actively fostered, was not least among them.

Beyond the basic question of intellectual substance, a number of other problems confronted Newton in the presidency. Financial crises had been endemic in the society since its beginning, and Newton found himself forced to deal with finances. In the 1670s, the society had used a bequest from John Wilkins to purchase fee farm rents – survivals from the manorial system – worth £24 per annum. By 1703, the rents were far in arrears; early in Newton's presidency the council initiated legal action to recover the arrears, though ten years elapsed before the society received the money due. More significant was the perennial problem of unpaid dues, which the council began to discuss in 1706. In December, Newton proposed that they restore the old practice of requiring new members to sign a bond for their weekly payments. In a series of meetings in January, the council not only restored the practice but voted to require that current members also post bonds. No one who had failed to sign a bond or who stood in arrears with his payments

---

[23] 7 Nov. 1705, 3 Dec. 1708, 15 Nov. 1704; *JBC 10*, 113, 200, 85.
[24] Mayor, *Cambridge under Queen Anne*, p. 365.

could serve on the council. They even empowered the president to order Henry Hunt, the society's clerk, to wait on members to collect arrears.[25] He may have done so, since at least two old members tendered their resignations. The Journal Book reveals that the society did begin to require the bond of new members. Only under Newton's successor, Sir Hans Sloane, did it sue in court to enforce the bond, however.

There is some question, which the surviving evidence does not resolve, about the respective roles of Newton and Sloane (who became secretary in 1693 and continued in office until 1713) in the effort to restore the society's finances. Sloane had worked at the task to some effect before Newton appeared on the scene. Moreover, Sloane was an aggressive man, not prone to follow the direction of others. In the view of some, who did not love him to be sure, he dominated the society in a dictatorial manner. It is impossible to reconcile that account with what we know of Newton, who was no more prone to suffer Sloane's domination than Sloane was to suffer his. In the matter of finances, it is likely that the two worked together, since neither had anything to gain from the society's poverty. As far as the general government of the society is concerned, there is at least one piece of evidence that the two strong-willed men found it difficult to tolerate each other. In 1710, on the occasion of a conflict between Sloane and Dr. John Woodward, Newton received a letter from one of Woodward's supporters. The author of the letter, probably William Derham, reminded the president that he had complained a great deal about Sloane's dictatorial methods, had declared to more than one friend how "little qualified he was for the Post of Secretary," and had called him "a Tricking Fellow" and even "a Villain, & Rascal" for his deceitful ways.[26] Over against this letter is Flamsteed's comment in his diary a year later that Sloane was one of Newton's "assertors . . . in all cases, right or wrong."[27] In fact, Sloane did eventually resign his secretaryship, though he was eager to assume the presidency upon Newton's death, a circumstance which suggests the first testimony was more perceptive than the second.

As president, Newton had also to deal with the threat that the Royal Society would find itself thrust from its home in Gresham College out onto the street. The danger was real. Like all buildings, Sir Thomas Gresham's once-magnificent Elizabethan house had aged. About the turn of the century, the committee responsible for Gresham College began to consider the desirability of pulling it down and rebuilding. For that, Parliamentary authorization was

[25] *CM 2*, 185–91. Having approved the various measures twice, on 22 and 29 January, the council then approved them all a third time on 5 March.
[26] *Corres 5*, 19.   [27] *Baily*, p. 228.

required because of the college's status. In 1701, such a bill was nearly passed by Parliament before a petition by Robert Hooke, who was a Gresham professor, killed it. Hooke's opposition probably stimulated the notice to vacate that the committee for Gresham College sent to the Royal Society immediately after his death. The council had already appointed a committee to locate other accommodations when Newton became president. Soon thereafter, the committee for Gresham College decided to try once more to get a bill to authorize rebuilding through Parliament.

Nevertheless, the situation remained well short of desperate. Two recent presidents of the society were powerful politicians, and the current president was a close friend of one of them. In Dr. Woodward, another Gresham professor, they still had an advocate in the college. Thus it was not surprising that Newton could report to the council in February 1704 that Halifax had informed him indirectly that the Parliament committee in charge of the bill was willing to accommodate the society with a meeting room and space for the repository and library. The council immediately delegated Sir Christopher Wren to consider what accommodations the society needed.[28] Wren drew up plans for an extravagant structure, forty by sixty feet, which the society would graciously allow the City and Mercers Company to build and lease to them gratis for 999 years. Newton, Wren, and Sloane met with the committee for Gresham College twice during the year. When it became apparent that the committee did not consider charity to the Royal Society its top priority, the society returned to the second meeting with reduced plans. Nevertheless, they still expected the committee to provide a building for them free; on 13 December, the committee voted unanimously to decline that privilege. Convinced that the society could block their bill in Parliament, they also voted not to proceed with it.[29] The action of the committee was reported to the council on the same day in the garbled form that they would not grant the society any room.[30] Though it is impossible to believe that men such as Newton, Wren, and Sloane failed to inform themselves correctly, the society proceeded in January to petition the queen for a plot of land in the Royal Mews in Westminister on which to build.[31] When that fell through, the council, still fearing ejection from Gresham College, tried to arrange the purchase of a house in Whitehall; and in the years ahead they considered first

---

[28] *CM 2*, 167–8.

[29] The details of the negotiations were reported in the minutes of the committee for Gresham College (*Corres 5*, 62–4). Cf. Lyons, *Royal Society*, p. 131, and Dorothy Stimson, *Scientists and Amateurs, a History of the Royal Society* (New York, 1948), p. 122.

[30] *CM 2*, 177.

[31] *JBC 10*, 92–3. Newton's draft of the petition is in *Add MS 4006*, f. 36.

removal to the Cottonian Library and then the purchase of various parcels of land.[32]

As he bent his energies to the countless miscellaneous minutiae of administration, Newton also reminded the Royal Society of its basic purpose in the most effective way possible. On 16 February, 1704, from the chair, he presented it with his second great work, the *Opticks*. The society fittingly appointed Halley to peruse it and give them a summary.[33] Unlike the printing of the *Principia*, we know nothing about the details of the *Opticks'* publication. Newton's election to the presidency may have figured in his decision finally to publish it. As we have seen, John Wallis had been hectoring him about the book for nearly a decade. More recently, David Gregory had taken up the cry, and he recorded on 15 November 1702 that Newton "promised Mr. Robarts, Mr. Fatio, Capt. Hally & me to publish his Quadratures, his treatise of Light, & his treatise of the Curves of the 2^d Genre."[34] He did not say when he would do so, however, and his elevation to the presidency may have provided the crucial stimulus. Though he could not have seen the book through the press between 30 November and 16 February, he undoubtedly knew of his coming election far enough ahead to allow time. Newton did not dedicate the *Opticks* to the society, as he had the *Principia*. Nevertheless he did identify it with the society by letting the title page state that Samuel Smith and Benjamin Walford, printers to the Royal Society, were the publisher, and by mentioning the society in the "Advertisement" that functioned as preface.

Newton's election as president was not the only, or for that matter the primary cause of the *Opticks'* publication in 1704, however. In the "Advertisement" he described briefly how he had composed most of it many years before. "To avoid being engaged in Disputes about these Matters, I have hitherto delayed the Printing, and should still have delayed it, had not the importunity of Friends prevailed upon me."[35] Not many members of the Royal Society would have missed the veiled reference to Hooke, whose death in 1703 removed an obstacle both to Newton's presidency of the society and to his publication of the *Opticks*.

The "Advertisement" contained two more paragraphs, each of which alluded to further incentives to publish. One mentioned the "Crowns of Colours" that sometimes appear around the sun and the moon. The publication of Huygens's *Dioptrica* with his posthumous works in 1703 had included an account of such crowns. New-

[32] 13 June 1705 to 23 June 1708; *CM 2*, 179–200.
[33] *JBC 10*, 63.   [34] *Hiscock*, p. 14.   [35] *Opticks*, p. cxxi.

ton apparently wanted to assert the independence of his own account of them.[36] More significant was the third paragraph, which introduced the two mathematical papers, "Tractatus de quadratura curvarum" (A Treatise on the Quadrature of Curves) and "Enumeratio linearum tertii ordinis" (Enumeration of Lines of the Third Order), which Newton appended to the *Opticks*. Some years ago, he stated, he had loaned out a manuscript with some general theorems about squaring curves, "and having since met with some Things copied out of it, I have on this Occasion made it publick . . ."[37] What he had met was a book published in 1703 by George Cheyne, *Fluxionum methodus inversa (The Inverse Method of Fluxions)*. According to David Gregory, in a memorandum of 1 March 1704, "Mr. Newton was provoked by Dr. Cheyns book to publish his Quadratures, and with it, his Light & Colours, &c."[38] The issue was rather more complicated than Gregory's note implied. In his collection of anecdotes, Conduitt included a story he heard from Peter Henlyn that when Cheyne came to London from Scotland, Dr. Arbuthnot introduced him to Newton and told him about the book Cheyne had written but could not afford to publish. Cheyne also reported later that he had shown the manuscript to Newton who "thought it not intolerable."[39] As Conduitt heard it, Newton offered Cheyne a bag of money; Cheyne refused to take it. Both were thrown into confusion by the impasse, and Newton refused to see him any more.[40] Cheyne must have been startled to read the extent of Newton's resentment of his assertion of independence in the "Advertisement." It possibly explains why Cheyne quickly dropped out of the Royal Society and chose to pursue his career in medicine rather than mathematics and natural philosophy.[41]

As far as the mathematical papers were concerned, Newton merely published expositions, composed a decade earlier, which summed up work that stretched back nearly forty years. Moreover, since Leibniz and his followers had been publishing a method identical to the basic concepts in "De quadratura" for a number of years, the appearance of the short treatise in 1704 did not constitute a major event in the history of mathematics. The papers marked an epoch for Newton nevertheless. At long last, after more than thirty years of delay and evasion, he published some mathematical work. If it was too late now to head off the battle with Leibniz, at least he

---

[36] Cf. *Math 3*, 441.   [37] *Opticks*, p. cxxii.   [38] *Hiscock*, p. 15.

[39] Quoted in H. W. Turnbull, *The Mathematical Discoveries of Newton* (London, 1945), p. 21.

[40] *Keynes MSS* 130.6, Book 2; 130.7, Sheet 1.

[41] Cheyne appeared only twice in the Journal Book of the Royal Society. On 23 Feb. 1704, on the eve of the publication of the *Opticks*, he gave a solution to a geometric problem, and he presented a copy of his account of natural religion to the Society in June 1705 (*JBC* 10, 65, 107). I did not see his name again during Newton's presidency.

could show the world something of the substance behind his reputation as a mathematician. Despite the wide dissemination that
Leibniz's method already enjoyed, "De quadratura" was appreciated. In 1709, Rémond de Monmort, who would later admire
Catherine Barton so extravagantly, wrote Newton that he had had
a hundred copies of the treatise printed in Paris to distribute to
those unable to obtain a copy of the *Opticks*.[42]

The situation in regard to the *Opticks* partially repeated that in
regard to the mathematical papers. As far as our understanding of
Newton's scientific thought is concerned, the *Opticks* contained
nothing new. With only the smallest exceptions, it presented work
he had completed more than thirty years earlier, and the exceptions
belonged to the early 1680s. Unlike the fluxional method, however,
the *Opticks* had not been duplicated by another investigator. In
1704, only a few men had digested the import of Newton's published paper of 1672. Hence the impact of the *Opticks* virtually
equaled that of the *Principia*. Indeed, it may have exceeded it, for
the *Opticks,* written in prose rather than geometry, was accessible to
a wide audience as the *Principia* was not. Through the eighteenth
century, it dominated the science of optics with almost tyrannical
authority, and exercised a broader influence over natural science
than the *Principia* did. More than one of Newton's younger contemporaries, including his disciple John Machin, told Conduitt
there was more philosophy in the *Opticks* than in the *Principia*.[43]
The work remains permanently one of the two pillars of Newton's
imperishable reputation in science.

The *Opticks* that Newton published in 1704 was not the *Opticks*
he had planned in the early 1690s. That work had reached its climax
in a Book IV dedicated to the demonstration that forces that act at a
distance exist. As he had done on other occasions, he finally now
shrank from laying so much bare in public. In the *Opticks* he did
publish, he eliminated Book IV and focused the work sharply on
optical problems – the theory of colors and the allied concept of the
heterogeneity of light. There is no need to repeat his demonstration
of the two again. Suffice it to say, they constitute an enduring
legacy to the science of optics. Somewhat unobtrusively, Newton
did slip in assertions that optics requires forces acting at a distance
similar to the *Principia*'s force of gravity. In Book II, he pointed out
that reflection cannot be caused by light impinging on the solid
parts of bodies. The reflection of light from the back side of a glass
in a vacuum argues against the impact theory of reflection; the

[42] *Corres 4*, 533–4.
[43] *Keynes MS*. 130.5, Sheets 1 and 2. Much has been written about the influence of the
Opticks on eighteenth-century science; see especially I. Bernard Cohen, *Franklin and Newton* (Philadelphia, 1956).

uniformity of reflection from a surface, which would require the perfect alignment of all its particles, argues the case more strongly.

And this Problem is scarce otherwise to be solved, than by saying, that the Reflexion of a Ray is effected, not by a single point of the reflecting Body, but by some power of the Body which is evenly diffused all over its Surface, and by which it acts upon the Ray without immediate Contact. For that the parts of Bodies do act upon Light at a distance shall be shewn hereafter.[44]

He proceeded then to argue that bodies reflect and refract light by one and the same power, and from a table that compared refractive power to density, both for bodies in general and for the special class of "fat sulphureous unctuous Bodies," he concluded that the reflecting and refracting power arises from the sulfurous parts bodies contain.[45] The argument assumed the corpuscular conception of light, of course.

Book III, Newton's still brief investigation of diffraction, contained the promised demonstration that a body "acts upon the Rays of Light at a good distance in their passing by it."[46] In a set of sixteen Queries, the first incarnation of the famous Queries that conclude the *Opticks* and Newton's substitute in the published work for the suppressed Book IV, he went on to consider forces further in an explicitly speculative context. The Queries of 1704–6 were Newton's last major publication of hitherto unknown scientific work, the culminating assertion of the Newtonian program in natural philosophy before the timidity of age and his progressive domestication as he settled ever more comfortably into the seat of authority and power led the quondam rebel to compromise some of his bolder positions. We read the Queries today as they were published in the third English edition, really as they were published a few years earlier in the second English edition, since he altered little after it. Among the final thirty-one, Queries 17–24 assert the existence of a universal aether and offer an explanation of forces in terms of such. To understand the original set of Queries it is necessary to remind oneself that they ended with number 16 and contained no suggestion of an aether to modify the assertions they made under the guise of rhetorical questions. Not all of the Queries concerned forces; the first ones did.

> *Query 1.* Do not Bodies act upon Light at a distance, and by their action bend its Rays; and is not this action (*caeteris paribus*) strongest at the least distance?
>
> . . .

---

[44] *Opticks*, p. 266. The citation of evidence for the conclusion was identical to the 1675 paper on thin films; the conclusion quoted had not appeared in the 1675 paper, however.

[45] *Opticks*, pp. 269–75. See J. A. Lohne, "Newton's Table of Refractive Powers. Origins, Accuracy and Influence," *Sudhoffs Archiv*, 61 (1977), 229–47.   [46] *Opticks*, p. 319.

Qu. 4. Do not the Rays of Light which fall upon Bodies, and are reflected or refracted, begin to bend before they arrive at the Bodies; and are they not reflected, refracted, and inflected [diffracted], by one and the same Principle, acting variously in various Circumstances?

Qu. 5. Do not Bodies and Light act mutually upon one another; that is to say, Bodies upon Light in emitting, reflecting, refracting and inflecting it, and Light upon Bodies for heating them, and putting their parts into a vibrating motion wherein heat consists?

. . .

Qu. 7. Is not the strength and vigor of the action between Light and sulphureous Bodies observed above, one reason why sulphureous Bodies take fire more readily, and burn more vehemently than other Bodies do?[47]

Though he couched them as questions, no one is apt to mistake the positive answers Newton intended. It was a less explicit form than he once had planned; nevertheless the *Opticks* did set forth the Newtonian program in natural philosophy.

"My Design in this Book," Newton began the *Opticks,* "is not to explain the Properties of Light by Hypotheses, but to propose and prove them by Reason and Experiments."[48] The statement was all that remained of a projected introduction in which he flung down a methodological gauntlet to mechanical philosophers to match the metaphysical one. There is a twofold method (he argued in the suppressed introduction), of resolution and composition, which applies to natural philosophy as well as to mathematics, "& he that expects success must resolve before he compounds. For the explication of Phaenomena are Problems much harder then those in Mathematicks." He described the method in terms nearly identical to those he later used in Query 31, with echoes also of a passage later embodied in his third rule of reasoning in philosophy.

Could all the phaenomena of nature be deduced from only thre or four general suppositions there might be great reason to allow those suppositions to be true: but if for explaining every new Phaenomenon you make a new Hypothesis if you suppose y^t ye particles of Air are of such a figure size & frame, those of water of such another, those of Vinegre of such another, those of sea salt of such another, those of nitre of such another, those of Vitriol of such another, those of Quicksilver of such another, those of flame of such another, those

---

[47] *Opticks,* pp. 339–40. See Alexandre Koyré, "Les Queries de l'Optiques," *Archives internationales d'histoire des sciences, 13* (1960), 15–29; and A. R. and Marie Boas Hall, "Newton's 'Mechanical Principles'," *Journal of the History of Ideas, 20* (1959), 167–78.
[48] *Opticks,* p. 1.

of Magnetick effluvia of such another, If you suppose that light consists in such a motion pression or force & that its various colours are made by such & such variations of the motion & so of other things: your Philosophy will be nothing else then a systeme of Hypotheses. And what certainty can there be in a Philosophy wch consists in as many Hypotheses as there are Phaenomena to be explained. To explain all nature is too difficult a task for any one man or even for any one age. Tis much better to do a little with certainty & leave the rest for others that come after, than to explain all things by conjecture without making sure of any thing.[49]

Even as president of the Royal Society, Newton did not find it easy to express fundamental convictions in public. He feared criticism. He preferred silence to the risk of controversy in which he might find himself made an object of ridicule. It tells us much about his unassuageable sense of insecurity that even from the pinnacle of renown he occupied in 1704 he suppressed the polemical introduction and did not dare to publish the once projected Book IV, though the suggestions he did publish argue that it expressed his beliefs. The Queries that he did allow to pass were brief, filling only two and a half manuscript pages at first. Nevertheless, they represented a considerable step for Newton, who until this time had not allowed any more than hints about his convictions on the ultimate nature of things to appear in print. And having taken one step he found it possible to take others. Perhaps I should say rather that he found it impossible not to, for the Queries acquired an independent existence and took command over Newton as other things had done in their turn. Apparently his original intention had been a set of short, staccato questions of one or two sentences, such as Queries 1–7 permanently remained. By the time he reached Query 12, however, he felt impelled to write a bit more, about a third of a sheet in all, and so also for Queries 13 and 15. Queries 10 and 11 no longer seemed adequate, and he returned to expand them.[50] Here he did manage to stop for the first edition, but the preparation of that edition nearly merged with the preparation of the Latin edition, which followed two years later. In the Latin edition, he further expanded Query 10, and more importantly, he added seven new Queries all but one of which were longer than the original sixteen

[49] *Add MS* 3970.3, f. 479. There are two drafts of the introduction (ff. 479–80). Cf. a passage in what I take to be the first draft: "But if wthout deriving the properties of things from Phaenomena you feign Hypotheses & think by them to explain all nature you may make a plausible systeme of Philosophy for getting your self a name, but your systeme will be little better then a Romance" (f. 480^v). See J. E. McGuire, "Newton's 'Principles of Philosophy': An Intended Preface for the 1704 *Opticks* and a Related Draft Fragment," *British Journal for the History of Science, 5* (1970), 178–86.

[50] *Add MS* 3970.3, ff. 231–3. F. 359, which has been separated from the others, contains the final lines of the original set of Queries.

taken together. In the new Queries, Newton expressed fundamental views on the nature of light, on the nature of bodies, on the relation of God to the physical universe, and on the presence in nature of a whole range of forces which furnish the activity necessary for the operation of the world and for its permanence. At the last moment, he dared even a bit more, and inserted three further speculative passages in the Addenda to the volume.

The new Queries were the most informative of the speculations that Newton ever published. In the second English edition, he added eight more Queries, which he inserted as numbers 17–24 between the first and second sets. Hence the seven Queries added to the Latin edition appear in all later editions, including those in general circulation today, as Queries 25–31. To avoid confusion, I shall refer to them by their final numbers, though they carried different ones when they first appeared. A point important for understanding the Queries new to the Latin edition, especially Queries 29 and 31, lies behind the difference in number. The third and final set of Queries, 17–24, are the ones that assert the existence of an aether which pervades all space. When he added them to the second English edition, Newton also put in passages about a second subtle fluid found in the pores of bodies that causes electrical phenomena (static electrical, of course) when it is agitated. In his final years, a growing philosophic caution led Newton to retreat somewhat toward more conventional mechanistic views, although his subtle aethers composed of particles repelling each other always remained more sophisticated than the clumsy fluids of standard mechanical philosophies. When he originally published Query 31, no suggestion of an aether or fluid modified its rhetorical questions about the prevalence of forces between bodies at every level of phenomena.

Query 31 was an extended version of the speculations on forces that Newton had once planned to insert in the *Principia*. Heavily, indeed overwhelmingly, chemical in content, it was arguably the most advanced product of seventeenth-century chemistry. What had Newton drawn from chemistry? The conviction that its phenomena require the presence of forces between particles for their explanation.

> Have not the small Particles of Bodies certain Powers, Virtues, or Forces, by which they act at a distance, not only upon the Rays of Light for reflecting, refracting, and inflecting them, but also upon one another for producing a great Part of the Phaenomena of Nature? For it's well known, that Bodies act one upon another by the Attractions of Gravity, Magnetism, and Electricity; and these Instances shew the Tenor and Course of Nature, and make it not improbable but that there may be more attractive Powers than these. For Nature is very constant and comfortable to her self.[51]

[51] *Opticks*, pp. 375–6.

The body of the Query detailed the evidence both from chemistry and elsewhere on which his reasoning rested. The remarkable continuity in Newton's lifelong investigation of the nature of things revealed itself in the appearance of all the crucial phenomena that had captured his attention more than forty years earlier when, as an undergraduate, he composed an earlier set of "Quaestiones." Together with the chemical phenomena, they now seemed to require the admission of forces of attraction and repulsion between particles. Thus nature will be very comfortable to herself, he concluded, performing all her great motions by the attraction of gravity and her small ones by the forces between the particles. Moreover, nature requires the presence of active principles. Inertia, the basic concept of conventional mechanical philosophies, is a passive principle by which bodies persevere in their motions. Phenomena reveal, however, that nature contains sources of activity, active principles which can generate new motions. Still worried by the hostile reception of his dynamic conception of nature, Newton felt obliged to add a disclaimer.

These Principles I consider, not as occult Qualities, supposed to result from the specifick Forms of Things, but as general Laws of Nature, by which the Things themselves are form'd; their Truth appearing to us by Phaenomena, though their Causes be not yet discover'd. For these are manifest Qualities, and their Causes only are occult. . . . To tell us that every Species of Things is endow'd with an occult specifick Quality by which it acts and produces manifest Effects, is to tell us nothing: But to derive two or three general Principles of Motion from Phaenomena, and afterwards to tell us how the Properties and Actions of all corporeal Things follow from those manifest Principles, would be a very great step in Philosophy, though the Causes of those Principles were not yet discover'd: And therefore I scruple not to propose the Principles of Motion above-mention'd, they being of very general Extent, and leave their Causes to be found out.[52]

Query 31 hinted in passing at Newton's retiform conception of matter, which nearly purged the universe of solid matter in order to populate it with forces binding punctiform particles into delicate nets. An addendum to Proposition VIII (Book II, Part III), which he once thought of making Query 17, expressed this notion more fully.[53] In another draft, he argued that God must have created

[52] *Opticks*, pp. 401–2.
[53] *Add MS* 3970.3, f. 296. Cf. the same conception of matter as Gregory recorded it in a memorandum of 21 Dec. 1705 of a conversation with Newton (*Hiscock*, pp. 30–1). See A. R. and Marie Boas Hall, "Newton's Theory of Matter," *Isis, 51* (1960), 131–44; Arnold Thackray, "'Matter in a Nutshell': Newton's *Opticks* and Eighteenth-Century Chemistry," *Ambix, 15* (1968), 29–53; and S. I. Vavilov, "Newton and the Atomic Theory," in The Royal Society, *Newton Tercentenary Celebrations*, pp. 43–55.

matter in textures nicely contrived for uses just as He created animals with the limbs and organs they need. Once again his conception of nature in which forces replaced particles as the basic reality led him to depart from conventional mechanical philosophies. We cannot, he insisted, consider matter as "irregular particles casually laid together like stones in a heap, but as formed wisely for all those uses."[54] In Query 30, which helped to explain his view on the role of sulfurous particles in the refracting power of media, he argued that gross bodies and light convert into each other and that bodies receive much of their activity from the particles of light in their composition. In a remarkable paragraph which did not survive into subsequent English editions he compared the force of attraction in proportion to size in particles of light and gross bodies by comparing velocities and radii of curvature of rays of light and projectiles. He concluded that the force of attraction in particles of light is more powerful by a factor of 10^{15}, that is, the short-range forces are immensely more powerful than gravity.[55]

In the first paragraph on Query 31, Newton inserted a caution about his assertion of the existence of forces. "How these Attractions may be perform'd, I do not here consider. What I call Attraction may be perform'd by impulse, or by some other means unknown to me. I use that Word here to signify only in general any Force by which Bodies tend towards one another, whatsoever be the Cause."[56] We read the passage today after the Queries on the aether and after a paragraph, which he inserted in the second English edition to conclude Query 29, which refers the meaning of the word "attraction" to those Queries. In the Latin edition of 1706, the reservation recalled rather the conclusion of Query 28. In Query 28, the refutation of wave theories of light led Newton into an argument against the possibility of a dense, Cartesian aether filling the heavens, and thence into an explication of his ultimate objection against conventional mechanical philosophies, their tendency to make nature self-sufficient and thus to dispense with God. Some ancient philosophers, he argued, took atoms, the void, and the gravity of atoms as the first principles of their philosophy and attributed gravity to some other cause than matter.

Latter Philosophers banish the Consideration of such a Cause out of natural Philosophy, feigning Hypotheses for explaining all things mechanically, and referring other Causes to Metaphysicks: Whereas

---

[54] *Add MS* 3970.3, f. 234ʳ.

[55] *Optice*, pp. 320–1. Cf. Gregory's memorandum of 21 Dec. 1705, which mentioned Newton's conviction that the rays of light enter into the composition of bodies (*Hiscock*, p. 31). See Zev Bechler, "Newton's Law of Forces Which Are Inversely As the Mass: A Suggested Interpretation of His Later Efforts to Normalise a Mechanistic Model of Optical Dispersion," *Centaurus*, 18 (1974), 184–222.   [56] *Opticks*, p. 376.

the main Business of natural Philosophy is to argue from Phaenomena without feigning Hypotheses, and to deduce Causes from Effects, till we come to the very first Cause, which certainly is not
mechanical; and not only to unfold the Mechanism of the World, but
chiefly to resolve these and such like Questions. What is there in
places empty of Matter, and whence is it that the Sun and Planets
gravitate towards one another, without dense Matter between them?
Whence is it that Nature doth nothing in vain; and whence arises all
that Order and Beauty which we see in the World? . . . How do the
Motions of the Body follow from the Will, and whence is the Instinct in Animals? Is not infinite Space the Sensorium of a Being
[Annon Spatium Universum, Sensorium est Entis] incorporeal, living, and intelligent, who sees the things themselves intimately, and
thoroughly perceives them, and comprehends them wholly by their
immediate presence to himself . . . [57]

David Gregory, who held an extensive discussion of the new Queries with Newton on 21 December 1705, recorded the interpretation of this passage in a memorandum.

His Doubt was whether he should put the last Quaere thus. *What the
space that is empty of body is filled with.* The plain truth is, that he
believes God to be omnipresent in the literal sense; And that as we
are sensible of Objects when their Images are brought home within
the brain, so God must be sensible of every thing, being intimately
present with every thing: for he supposes that as God is present in
space where there is no body, he is present in space where a body is
also present. But if this way of proposing this his notion be too bold,
he thinks of doing it thus. *What Cause did the Ancients assign of Gravity.* He believes that they reckoned God the Cause of it, nothing els,
that is no body being the cause; since every body is heavy.[58]

At the last moment, after the last moment really, Newton decided
that he had indeed been too bold. He tried to recall the whole
edition; and from all the copies he could lay his hands on, he cut
out the relevant page, and pasted in a new one which asserted, not
that infinite space is the sensorium of God, but that "there is a
Being incorporeal, living, intelligent, omnipresent, who in infinite
Space, as it were in his Sensory, [tanquam Sensorio suo] sees the
things themselves intimately . . ."[59] Alas, he failed to alter every

[57] *Optice,* pp. 314–15. I have used the English in the *Opticks* (pp. 369–70), altering it only
where the original Latin demands it.

[58] *Hiscock,* p. 30. Whiston gave an identical account of Newton's view (William Whiston, *A
Collection of Authentick Records Belonging to the Old and New Testament,* 2 vols. [London,
1727–8], *2,* 1072–3).

[59] Cf. Alexander Koyré and I. Bernard Cohen, "The Case of the Missing *Tanquam,*" *Isis, 52*
(1961), 555–66. In drafts written in English, of Query 23 (later 31) from about this time,
Newton expressed the same concept. "Life & will are active Principles by wch we move
our bodies, & thence arise other laws of motion unknown to us. And since all matter duly

copy, and one of the originals made its way to Leibniz, who did not fail to hold up to ridicule the concept of space as the sensorium of God. In its initial form the passage recalled "De gravitatione," the beginning of Newton's rebellion against Cartesian philosophy because of its atheistical tendencies.[60] Following the implications of the rebellion, he had traveled far. In the Latin edition of the *Opticks,* he gave the fullest exposition of his own conception of nature he would ever put in print before, in his old age, he tried to placate critics by seeming retreats to more conventional positions.

In addition to its importance for Newton's philosophy, the Latin edition of the *Opticks* also provided the occasion for a graceful personal relation. Abraham DeMoivre saw it through the press. Every evening, according to the story, Newton would wait for him in a coffeehouse where DeMoivre would go as soon as he finished the mathematical lessons with which he supported himself. Newton would take him home, and the two would spend the evening in philosophical discussion.[61] DeMoivre was one of the young men in London, disciples really, with whom Newton found companionship possible in a way it had never been in Cambridge. Another young disciple, Samuel Clarke, translated the *Opticks* into Latin and received £500 for his pains: £100 for each of his five children.[62]

In 1707, the year following the Latin edition, another work by Newton appeared through the efforts of a third young disciple. William Whiston, his successor in the Lucasian chair in Cambridge, published the deposited manuscript of his lectures on algebra as *Arithmetica universalis.* Newton did not like much of anything about the project. He had not wanted the publication in the first place. In the summer of 1706, when it was in the press, he told Gregory about the origin of the edition. "He was forced seemingly to allow of it, about 14 months agoe, when he stood for Parliament-man at that University." This could only have referred to pressure by Richard Bentley, the master of Trinity, who was always anxious to

formed is attended with signes of life & all things are framed wth perfect art & wisdom & Nature does nothing in vain; if there be an universal life & all space be the sensorium of a thinking being who by immediate presence perceives all things in it as that wch thinks in us perceives their pictures in the brain, the laws of motion arising from life or will may be of universal extent" (*Add Ms* 3970.9, f. 619). Cf. also other drafts on f. 620^v and on 3970.3, f. 252^v. Cf. J. E. McGuire, "Force, Active Principles, and Newton's Invisible Realm," *Ambix, 15* (1968), 154–208, and Ernan McMullin, *Matter and Activity in Newton* (Notre Dame, Ind., 1977).

[60] Newton also brought up atheism in drafts of Query 23 (31); e.g., *Add MS* 3970.9, f. 619.

[61] DeMoivre's *Éloge; L'Histoire de l'Académie royale des sciences* 1754, p. 268. Apparently DeMoivre was quite close to Newton at this time; nearly every one of his letters to Bernoulli mentioned that he had seen Newton (Karl Wollenschläger, "Der mathematische Briefwechsel zwischen Johann I. Bernoulli und Abraham DeMoivre," *Verhandlungen der Naturforschenden Gesellschaft in Basel, 43* [1931–2], 151–317).

[62] William Whiston, *Historical Memoirs of the Life of Dr. Samuel Clarke* (London, 1730), p. 13.

capitalize his connection with Newton and would soon propose a second edition of the *Principia*. Newton claimed that he could not even remember what was in his manuscript and indicated his intention to go to Cambridge, where it was printed, to buy up the whole edition if it did not please him.[63] He did not buy it up, but it did not please him either. He objected to the title Whiston put on the book, although it was identical to the title Newton himself had placed on his aborted revision of the manuscript in 1684 – a coincidence so astonishing as to raise some doubt about the ingenuousness of his pretense of noninvolvement. Committed now to finding faults for the sake of finding faults, he objected even to the running heads ("Algebrae elementa") at the top of each page.[64] He did not allow his name to appear in the edition, although his authorship was common knowledge from the day of publication. He later told Conduitt that Whiston printed the work with so many errors that he had to republish it to correct them.[65] However, the changes that he introduced in the edition of 1722 were primarily reorderings of his own manuscript, not corrections of Whiston's edition.[66] If his relation with DeMoivre in the Latin *Opticks* was graceful, he could not have made his relation with Whiston in the *Arithmetica universalis* more petty.

As far as we know, Newton's discussions with DeMoivre confined themselves to mathematics and natural philosophy. With Clarke and Whiston, he shared – or perhaps came to share – extensive theological interests as well. Both men delivered Boyle lectures; Newton may have participated in their selection for that function.[67] Both identified themselves publicly as Newton's disciples. Before long, both became notorious as the two leading Arians in England, or at least the two leading known Arians.

What information we have about their theological interchange we learn mostly from Whiston. Whiston suffered greatly for his beliefs and as a result published extensive *apologiae pro vita sua*. He kept Newton's name out of them while Newton was alive, though references to "a very great Man" who held similar views may have been, coming from Whiston, less opaque than the great man desired. On the morrow of Newton's death, Whiston wrote an explicit account which he later embellished in his *Memoirs*.

> . . . Sir I. N. was one who had throughly examined the state of the Church in its most critical Juncture, the fourth Century. He had

[63] *Hiscock*, p. 36.   [64] *Hiscock*, p. 37.
[65] *Keynes MS* 130.5, Sheet 1.   [66] *Math 5*, 13–14.
[67] Margaret C. Jacob, *The Newtonians and the English Revolution, 1689–1720* (Ithaca, N.Y., 1976), chap. 4, "The Church, Newton, and the Founding of the Boyle Lectureship," pp. 143–61.

early and throughly discovered that the Old Christian Faith, concerning the Trinity in particular, was then changed; that what has been long called Arianism is no other than Old uncorrupt Christianity; and that Athanasius was the grand and very wicked Instrument of that Change. This was occasionally known to those few who were intimate with him all along; from whom, notwithstanding his prodigiously fearful, cautious, and suspicious Temper, he could not always conceal so important a Discovery.[68]

By his own account, Whiston was among the favored few. In his *Memoirs,* he mentioned a discussion on Revelation which lasted four hours, and in various places, including the passage above, he indicated a knowledge of Newton's theological papers which accords with the papers themselves.[69] Whiston never claimed that Newton instructed him in his heresy. Whiston nourished a mammoth ego, however, and it is possible that he could not bring himself to admit, even to himself, that positions that cost him so much he held second hand. He did concede that he learned some things that were unorthodox from Newton, and from his various accounts it seems plausible that Newton, of whom he always stood in awe, at least gave him the courage at a critical moment to embrace opinions he might otherwise have shunned.[70] His Arianism appears virtually identical to Newton's.

Samuel Clarke, who decided at the last moment not to follow Whiston into the wilderness, consequently said a good deal less. Whiston certainly hinted that Clarke learned Arian views at Newton's knee, and he mentioned frequent exchanges between Clarke and himself.[71] Brief references also suggested a somewhat wider circle of Arians, all connected with Newton. Whiston called Hopton Haynes his "intimate Friend."[72] At times, he seemed to place Richard Bentley in the circle. In 1709, when Clarke performed his Act and took his D.D. at Cambridge, Dr. James, the Regius professor of divinity, was suspicious and pressed him hard to reject Arianism straightforwardly. That evening, Whiston and Clarke dined with Bentley, who made jokes about the episode and composed a Latin jingle on it.[73] A few years later, upon Bentley's election to the same Regius chair, he fluttered orthodox circles by delivering his probationary lecture on the corruption of 1 John 5:7, one of New-

---

[68] Whiston, *Authentick Records, 2,* 1077. Appendix IX (pp. 962–1082) of *Authentick Records* was "A Confutation of Sir Isaac Newton's Chronology," the last twelve pages of which Whiston devoted to an account of his personal relations with Newton and of Newton's religious views.

[69] William Whiston, *Memoirs of the Life and Writings of Mr. William Whiston* (London, 1749), pp. 40, 365; *Authentick Records, 2,* 1078.   [70] *Authentick Records, 2,* 1071–2.

[71] Whiston, *Clarke,* pp. 12–13, 155–6; *Authentick Records, 2,* 1067, 1075–6.

[72] Whiston, *Memoirs,* p. 206.   [73] Whiston, *Clarke,* p. 21.

ton's *Two Notable Corruptions*. Bentley, who had no intention whatever of giving up his comforts in the defense of truth, succeeded in
passing the lecture off as a mere exercise in textual criticism.[74]
Whiston appears to have felt himself obligated to the group, and he
maintained his silence on each one until he was either dead or
beyond injury. In his recollections, one catches a glimpse–is it a
true image or is it a mirage? – of one of the most advanced circles of
free thought in England grouped about Newton and taking its inspiration from him.

Whiston did not share Newton's "fearful, cautious, and suspicious Temper." Having found the truth, he mounted the barricades
to defend it. At the end of 1707, he made public confession in his
Boyle lectures on the prophecies. Once he had taken the step, he
forgot every semblance of prudence. He preached his views openly
in Cambridge. He published a spate of theological works designed
to shock orthodoxy and personally arranged to have them dissemi-
nated around the university.[75] One of them, a collection of *Sermons
and Essays upon Several Subjects,* reprinted a sermon of 1705 which
ended with a prayer to Christ, "To whom with the Father and the
Holy Ghost, Three Persons, and One God, be all Honour,
Glory . . . henceforth and for evermore." At the end of the volume, a single erratum directed the reader to the prayer: "*r* [read] in
the Holy Ghost, *and dele* Three Persons and One God."[76] He could
not have picked a more inauspicious time. The notorious Dr. Sacheverell, who was then whipping up the paranoia of the orthodox
for the greater glory of conservative politics, seized on the erratum
as an example of blasphemy and irreligion. In the autumn of 1710,
Whiston appeared before the vice-chancellor and heads to answer a
charge of violating the statutes of the university by espousing Arianism. He chose to defend himself by attacking the proceedings.
For his pains he found himself summarily stripped of his chair and
banished from the university. Whiston appears to have seen himself
as a new Luther who would lead a final Reformation. He spent the
rest of a long life protesting the injustice done him, somewhat
puzzled at the failure of the church to follow his lead. Newton had
never deluded himself on that score. Now he sat by quietly, saying
never a word as Cambridge drove out the successor he had selected.
Bentley, master of Trinity, chose to absent himself from Whiston's
trial.

Clarke's turn came not long after Whiston's. His friendship with

---

[74] James H. Monk, *The Life of Richard Bentley, D.D.* 2nd ed., 2 vols. (London, 1833), *2,*
15–18.

[75] William Whiston, *An Account for Mr. Whiston's Prosecution at, and Banishment from, the
University of Cambridge* (London, 1718), pp. 2–3.

[76] William Whiston, *Sermons and Essays upon Several Subjects* (London, 1709), pp. 123, 412.

Whiston was known; suspicions that he too was an Arian abounded. In 1714, Convocation set out to censure him for his *Scripture Doctrine of the Trinity*. As Whiston saw it, Clarke wriggled out of the charge ignominiously and managed to hold on to his preferments. Whiston maintained silence until Clarke's death, but his sense of grievance festered as he watched Clarke enjoy the advantages he had foresworn in the interest of truth. Soon after Clarke died, he published a bitter biography which laid out the whole story of their relationship.[77] Though he commented pungently on Newton at times, he never expressed a similar bitterness in his regard.

Meanwhile, Newton watched with growing alarm the public identification of men closely associated with him as heretics. He made no move to protect Whiston. He did not immediately break with him, however.[78] It was probably in the years soon after 1710 that he befriended Whiston by employing his nephew in his household. One day Newton missed £3,000 in bank bills, and he harbored strong suspicions that the nephew, who purchased an estate of that value soon after the loss, had stolen the money. He also lost one hundred guineas that were taken from his desk and believed that the nephew had slipped the key to the desk out of his breeches. When he was asked how much he had lost, he replied too much, but he refused to prosecute Whiston's nephew.[79] According to Whiston, Newton did finally reject him. For twenty years after their meeting in 1694, he claimed, he enjoyed Newton's favor.

> But he then perceiving that I could not do as his other darling Friends did, that is, learn of him, without contradicting him, when I differed in Opinion from him, he could not, in his old Age, bear such Contradiction; and so he was afraid of me the last thirteen Years of his Life.[80]

Whiston went on to assess Newton's character.

> He was the most fearful, cautious, and suspicious Temper, that I ever knew: And had he been alive when I wrote against his Chronology, and so throughly confuted it . . . I should not have thought proper to publish it during his Life Time; because I knew his Temper so well, that I should have expected it would have killed him.[81]

One may reasonably question Whiston's analysis of the cause of the break. Clarke's trial by Convocation took place in 1714. Newton

---

[77] Whiston, *Clarke*, pp. 4, 70–86.
[78] Cf. an incident they shared in 1712 (Whiston, *Memoirs*, p. 206). Also, Whiston to Cotes, 7 April 1715; Trinity College, MS R.4.42, f. 10.
[79] *Keynes MSS* 130.6, Book 2; 130.7, Sheet 1. Although Conduitt recorded these two incidents together, they may have involved different people. He called the nephew "W." Whiston and the second suspect "J.W."   [80] Whiston, *Memoirs*, pp. 293–4.
[81] *Ibid.*, p. 294. He also spoke of Newton's "cautious Temper & Conduct" a third time (*Authentick Records*, 2, 1070–1).

appears to have wanted to separate himself from a notorious apostate such as Whiston. Two years later, in May 1716, Martin Folkes proposed Whiston for membership in the Royal Society.[82] The council never considered the nomination, which just disappeared. In his *Memoirs,* Whiston said that Newton had blocked it. Earlier, at a coffeehouse discussion, Halley had asked Whiston why he was not a member. Whiston replied that the society would not elect anyone who held his religious views, whereupon Halley said that if Sir Hans Sloane, who was also present, would nominate him, he would second the nomination. Either Whiston or the Journal Book was in error about the agent, but nominated he was. When Newton heard, he talked with members privately and threatened to resign unless the nomination were dropped.[83] The following year, perhaps in spite, Whiston dedicated his *Astronomical Principles of Religion, Natural and Reveal'd* to the president, Newton, the council, and the members of the Royal Society.

A few bits of evidence suggest that for all his caution Newton did not wholly escape suspicion. Whiston spoke of "those Surmises or Reports which have been sometimes spread abroad" that he was an infidel, and he also mentioned a pamphlet published anonymously in 1719, *The History of the Great Athanasius,* so like Newton's views that he was suspected as the author. He added that a jesting paragraph in the pamphlet proved it could not have been Newton's, "He being ever grave and serious, and never dealing in ludicrous Matters at all . . ."[84] Likewise, Flamsteed entered an observation in his diary for 26 October 1711, the day of a painful scene with Newton at the Royal Society. Whipped to fury, Newton charged that Flamsteed had called him an atheist. "I never did," Flamsteed told his diary; "but I know what other people have said of a paragraph in his *Optics;* which probably occasioned this suggestion. . . . I hope he is none."[85] Undoubtedly the paragraph in question was the conclusion of Query 28, in which Newton spoke of infinite space as the sensorium of God. Whereas he saw this concept as the perfect antidote to atheism since it entailed the active participation of God in every phenomenon of nature, others saw it as the assertion that God requires organs in order to perceive.[86] Aside from a few such passing references which did not touch on Arianism, however, Newton carefully kept his opinions out of print. He had no inclination whatever to call his position in London into question by challenging orthodoxy openly. The rumors may have pained him. He was too firmly established to be endangered by them.

[82] *JBC* 11, 124.   [83] Whiston, *Memoirs,* pp. 292–3.
[84] Whiston, *Authentick Records,* 2, 1078, 1080.   [85] Baily, p. 229.
[86] So Leibniz interpreted it. Cf. *The Leibniz-Clarke Correspondence,* ed. H. G. Alexander (New York, 1956), *passim.*

The Royal Society brought Newton into occasional contact with another young friend, or former friend, of heterodox religious views, Fatio de Duillier. Periodically, Fatio would appear at the meetings of the society. He was frequently present in the spring of 1706, arranged for the election of his brother, and later read a couple of letters from him to the society.[87] Gregory's memorandum from the early years of the century also contain evidence that Newton and Fatio saw each other now and then, but there was no suggestion of a return to former intimacy. In 1707, Fatio committed himself to the wild Camisard prophets driven to England by the persecution of Protestants in France and to their prophecies of the imminent fulfillment of Revelation. When the authorities, in an effort to control the sensation they were creating, sentenced them to stand twice in the pillory, at Charing Cross and at the Royal Exchange, Fatio was one of those who had to endure the humiliation. Rumors held him to be the head of the fanatical sect. Later in the century, Joseph Spence recorded two stories that linked Newton to the prophets. Dr. Lockier told him it was probable that Newton "might have had a hankering after the French prophets." Ramsay, who had known Fatio well, had a fuller account. "Sir Isaac himself had a strong inclination to go and hear these prophets, and was restrained from it, with difficulty, by some of his friends, who feared he might be infected by them as Fatio had been."[88] Since we know from a memorandum by Gregory that Newton once discussed the prophets with Fatio, there may have been some foundation for the rumors.[89] Nevertheless the stories are not believ-

---

[87] John Christopher Fatio de Duillier was proposed and elected between 27 Feb. and 3 April 1706 (*JBC 10*, 124–7). On 12 and 19 June, "Mr. Fatio," that is, Nicolas Fatio, read accounts by his brother of astronomical observations in Geneva (*JBC 10*, 136–7). Cf. Charles Andrew Domson, "Nicholas Fatio de Duillier and the Prophets of London: An Essay in the Historical Interaction of Natural Philosophy and Millennial Belief in the Age of Newton" (doctoral dissertation, Yale University, 1972) p. 84. Fatio was mentioned in connection with a watch and was apparently present on 10 May 1704 (*JBC 10*, 74). Letters from him were read on 27 May and 2 Dec. 1714 (*JBC 10*, 574; *JBC 11*, 32). With Derham he gave an account of their observations of an aurora borealis on 8 March 1716 (*JBC 11*, 108). He was present and gave various accounts three times between 21 March and 11 April 1717 (*JBC 11*, 168, 172, 174–5). Newton presided on all the occasions when Fatio was present.

[88] Joseph Spence, *Anecdotes, Observations, and Characters, of Books and Men*, ed. Samuel S. Singer (London, 1820), pp. 56–7, 72. Cf. the excellent discussion of Fatio and the prophets in Frank E. Manuel, *A Portrait of Isaac Newton* (Cambridge, Mass., 1968) pp. 206–12.

[89] University of Edinburgh Library, Gregory MSS Folio c DC, 1.61, f. 707. The memorandum is dated London, 30 Jan. 1707. In it Gregory mentioned what Newton had told him he had heard from Fatio about the prophets. In the same memorandum, Gregory recorded his own conversation with Fatio the day before. I wish to thank Margaret Jacob, who called this memorandum to my attention and furnished me with a photo-copy of it (see Margaret Jacob, "Newton and the French Prophets," *History of Science, 16* [1978], 134–42).

able in the form in which they come down, with their evident meaning that Newton himself had some affinity with the prophets. Enthusiasts who spoke in tongues and believed in their own direct inspiration, the Camisards embodied an approach to the prophecies antithetical to that of Newton, who held that the written prophecies were the objective revelations of God, which carried the key to their own interpretation, and that their interpretation must accordingly be purged of the vagaries of private fancy. The stories testify to an enduring tendency of some people to connect the two men, but they do not give a likely account of Newton's activities in 1707.

Meanwhile, his presidency of the Royal Society had led Newton to an unhappy resumption of relations with another former acquaintance, John Flamsteed. Only a few months after his election, on 12 April 1704, Newton went down to Greenwich, enquired about the state of Flamsteed's observations, and when he was shown them asked to recommend them to Prince George, Queen Anne's consort, for his financial support of their publication. According to Flamsteed's later account, a friend of his had earlier acquainted the prince with the progress of his work and his need of assistance. Newton heard about the prince's interest through Halifax and thrust himself into the business. In the light of Newton's later actions, there is only one reasonable interpretation of the visit. Still tormented by his failure with lunar theory and still convinced that Flamsteed had caused it, he determined to exercise his authority as President of the Royal Society to get Flamsteed's observations in order, as he put it, to have another stroke at the moon. A perfected lunar theory would crown a second edition of the *Principia*. He was sure that Flamsteed had a trove of lunar observations that he had not communicated. He would need the star catalogue as well, for he had finally absorbed Flamsteed's protests that observations of the moon in relation to stars were only as accurate as the known stellar positions. Moreover, it was not necessary that he have the information privately. If the world had it in published form, he too would have it. Thus Newton was all benevolent philanthropy on the visit in April despite Flamsteed's suspicions. "Do all the good in your power," he told Flamsteed as he left, and Flamsteed, typically, noted in his recollections that such had always been the rule of his life, "though I do not know that it ever has been of his."[90] Newton's philanthropy had given way to despotic ire long before Flamsteed made his sour comment.

Through 1704, Flamsteed carried on hesitant negotiations with the prince, negotiations in which Newton had no part. At some

[90] *Baily,* p. 74.

point, perhaps even before Newton's visit, the prince came to Greenwich and saw Flamsteed's papers and plans first hand.[91] In October, Flamsteed informed his friend Sharp that the prince had made offers to print his works, but he dared not promise himself too much from the overtures yet.[92] On 15 November he told Pound that the prince had seen six constellations he had drawn and apparently would pay for the publication.[93] Already, however, Flamsteed had taken another action that passed the initiative from his hands to Newton's. Flamsteed was all too aware that he had been at the Observatory then for thirty years with precious little in print to show for his efforts. According to him, "false and malicious suggestions of some few arrogant and self-designing people" continually implied that he had not used his opportunity well.[94] The "people" referred to were Halley and Gregory. Their malicious suggestions survive only in Flamsteed's account of them, but whatever the reality, his letters from the period amply confirm that he was extremely sensitive to the criticism. Already, five years earlier, his sensitivity had led him to write the public letter for Wallis's works which infuriated Newton. In an effort to justify himself, he implied then that the catalogue of fixed stars, his great work, which was always in his own eyes the ultimate criterion of his accomplishment, was virtually completed, and a number of letters from the following years said the same.[95] We do not know exactly what Flamsteed said to Newton on 12 April 1704, but Newton's request to recommend the work to Prince George implied that Flamsteed again gave the impression that the catalogue was complete or at least nearly so. Early in November, responding once more to the same rumors, Flamsteed drew up an "Estimate" of what the projected *Historia britannica coelestis* (*British History of the Heavens*) would contain, in effect an objective account of his achievement at Greenwich. It stated that the charts of the constellations could be drawn at once and that he could complete the third part of the *Historia,* of which his catalogue of the fixed stars was to be the centerpiece, while the first two parts, which were already finished, went through the press.[96] Flamsteed gave the "Estimate" to James Hodgson, a former assistant who had married his niece, to show around the Royal Society as evidence of his accomplishment. Unfortunately, as it passed around on 15 November, the Secretary,

---

[91] Flamsteed to ———. 24 Oct. 1715; *Baily,* p. 316.
[92] *Baily,* p. 218.   [93] *Corres 4,* 427.   [94] *Baily,* p. 75.
[95] Flamsteed to John Lowthorp, 10 May 1700; *Baily,* p. 174. In this letter, Flamsteed mentioned that he had told Newton of his progress with the catalogue a week earlier. Flamsteed to Dr. T. Smith, 26 Oct. 1700; *Baily,* pp. 744–6. Flamsteed to Wallis, 24 June 1701; *Baily,* pp. 196–7. Flamsteed to Sharp, 9 February 1703; *Baily,* p. 212. Flamsteed to the Reverend S. Thornton, 18 Feb. 1703; *Baily,* p. 748.   [96] *Corres 4,* 420–2.

Dr. Sloane, saw it and read it to the society. They thanked Flamsteed and promised to encourage the work as much as they could.[97] Newton, who was in the chair that day, could not resist the opportunity thus presented to seize control and insure his access to the precious observations. By the annual meeting two weeks later, the society had approached the prince with Flamsteed's estimate of the number of pages involved, and the prince had expressed his interest in very positive terms. To facilitate matters, the society proceeded at once to elect Prince George to membership, and before December was out they received a letter from his secretary in which "the President was desired to take what Care in this Matter he shall think Necessary Towards the most Speedy publication of so usefull a Work . . ."[98] More than ten years passed, years filled with inexpressible bitterness, before Flamsteed succeeded on the eve of his death in pushing Newton back out of his affairs.

Flamsteed cannot escape his own share of responsibility for the debacle that followed. In view of what he had said, Newton concluded that the star catalogue was complete, and he bent every effort to get it into print at once, as an item of maximum value to himself. The catalogue was still five years away from the "great work" Flamsteed dreamed of. Consequently, it immediately became a focus of controversy. Moreover, Flamsteed had lost none of his unique capacity to be insufferable. When Newton visited Greenwich in April, Flamsteed thanked him for the copy of his *Opticks* that he had received.

He said then he hoped I approved it. I told him loudly "no", for it gave all the fixed stars bodies of 5 or 6 seconds diameter, whereas four parts in five of them were not 1 second broad. This point would not bear discussion; he dropped it, and told me he came now to see what forwardness I was in. . . . I showed him also my new lunar numbers fitted to his corrections, and how much they erred, at which he seemed surprised, and said "it could not be", but when he found that the errors of the tables were in observations made in 1675, 1676, and 1677, he laid hold on the time, and confessed he had not looked so far back, whereas if his deductions from the laws of gravitation were just, they would agree equally in all times.[99]

One can only sympathize with Newton's unwillingness to discuss his *Opticks* with a man who could not understand it. Flamsteed based his objection that all stars ought to appear in telescopes with a diameter of five or six seconds if Newton were correct, an objection he thought undermined the whole *Opticks,* on a distinction

[97] *JBC* 10, 86. Cf. Flamsteed's account in a letter to Sharp, 24 April 1705; *Baily,* p. 239.

[98] *JBC* 10, 91.

[99] Flamsteed to Sharp, 4 May 1704; William Cudworth, *Life and Correspondence of Abraham Sharp* (London, 1889), p. 80.

between "native light" and its modification into colors. The *Opticks* devoted itself to disproving just such a distinction. Flamsteed contended that "only native light contributes to the forming of the distinct picture of the object; the coloured rays are not perceived whilst mixed with it, but only fringe its limbs with an edging of colours."[100] How could Newton pursue such a discussion? Why not drop it? With lunar theory Flamsteed was on solid ground, but he never understood how touchy a subject it was with Newton or the special responsibility for the failure that he bore in Newton's eyes. He would have done better not to mention the moon. In fact, he never missed an opportunity to bring it up.

Whatever Flamsteed's faults, however, Newton was the primary cause of the unhappy episode that followed. At the beginning, Flamsteed was conscious of the service being done for him in gaining Prince George's support. "Mr. Newton is become exceeding kind of late," he wrote to Sharp as he described how Newton had become the solicitor to the prince on behalf of his work. "I may allow him to do himself the honor, and regard his own interest in it too, since he saves me the labor of attendance and solicitation."[101] His letters to Newton attempted to revive a spirit of friendly cooperation as he wished him good health and good fortune in the Parliamentary election. When he received back only notes of the most curt variety, and when disagreement about the catalogue began to develop, he drafted a pathetic memorandum to Wren. "Mr. Halley has set the Master of the Mint at distance from me by false suggestions. . . . Entreat him for the future to do by me as I have done by him, and reject all their imputations till proved."[102] Meanwhile, he remained suspicious, as well he might. Though Flamsteed was a member of the Royal Society, it never occurred to Newton to include him in the delegation that waited on Prince George about his work. Worse, the referees appointed to survey Flamsteed's papers and make recommendations, referees of whom Newton was the leader of course, systematically ignored Flamsteed's carefully thought-out plan of publication. The plan was neither arbitrary nor silly. Eventually, after Flamsteed's death two of his devoted assistants did complete the *Historia coelestis* according to his plan, and it is recognized by qualified experts as one of the great landmarks of the science of astronomy. Flamsteed wanted to set his catalogue squarely in its historical tradition by publishing in the same volume all the significant prior catalogues from Ptolemy to Hevelius. Knowing that the catalogue he wanted to give to the world, the monument of a lifetime's devoted labor, was not yet done, he asked

---

[100] Flamsteed to Sharp, 2 Sept. 1704; *Baily*, p. 217. Cf. Flamsteed to Sharp, sometime after 30 March 1704; Cudworth, *Sharp*, p. 79.

[101] Flamsteed to Sharp, 30 Dec. 1704; *Baily*, p. 232.   [102] *Baily*, pp. 245–6.

for money to hire calculators to complete the reduction of his observations. The referees said not a word about the earlier catalogues; Newton never even replied to Flamsteed's letters about the need for a new, correct translation of Ptolemy's. They recommended £180 for calculators to compute "the places of ye moon and planets & comets" – that is, the information Newton wanted. Implicitly they treated the catalogue of stars as complete in its present state by allowing nothing for further computations on it, and Newton's demand to proceed with it at once became a bone of contention. Flamsteed envisaged a set of sixty large charts of the constellations, twenty-four by twenty inches, to accompany the catalogue. They would be "the glory of the work, and next the catalogue, the usefullest part of it . . ."[103] Eventually, Flamsteed's wife and former assistant did publish an abbreviated *Atlas coelestis* which was much admired and frequently republished both in England and on the Continent. Newton never showed the slightest interest in the atlas, and the referees left it out of their recommendation. Flamsteed heard (it is impossible to know how rightly) that Prince George was prepared to devote £1,200 to the work. The referees turned in a budget of £863. And yet they told the prince, "This set of observations we Repute the fullest & Complatest that has ever yet Been made: and as it tends to the perfiction of Astronomy & Navigation: so if it should be Lost, the Los would Be irreparable . . ."[104] One would think Newton might have considered that the man who had given his life to making the observations could be trusted to present them properly. Quite the contrary; convinced that he alone fully understood, Newton pressed forward and succeeded in needlessly depriving the world of the observations and the catalogue for another twenty years.

It is sometimes represented that Newton acted in the affair in the public interest to secure general access to observations made by a public official, while Flamsteed tried to treat the observations as private property. Such was Newton's self-justification. Unfortunately, it will not bear scrutiny. Flamsteed was eagerness itself to see his work completed and published. The *Historia coelestis* would embody irrefutable proof to his God and to his critics that he had been a faithful steward of the talent entrusted to him. As for Newton, his eye always on the data he hoped to obtain, he consistently attempted to mold Flamsteed's work into a form that would best serve his own ends. The angry temper which, in the young man, could not cope with contradiction, now manifested itself in the president of the Royal Society as imperious autocracy. He and he alone knew what was best. He found it impossible to grant to

---

[103] Flamsteed to Sharp, 4 May 1704; *Baily*, pp. 216–17.   [104] *Corres 4*, 436.

Flamsteed what he had required – and demanded – for himself: the right to define his own work in his own way. "But he knows well," he had written to Oldenburg in fury about Hooke's comments of 1672 on perfecting refracting telescopes, "y^t it is not for one man to prescribe Rules to ye studies of another . . ."[105] What he once knew in his own regard he could not grant in another's. No one questions Newton's intellectual superiority over Flamsteed or the higher level of scientific achievement in the *Principia* compared to the *Historia coelestis*. Genius does not excuse the abuse of power, however. Ultimate judgments aside, intelligent understanding of his own self-interest in obtaining the observations and catalogue ought to have dictated another approach to Flamsteed. What Newton succeeded primarily in doing was delaying the publication.

After they had reviewed Flamsteed's papers, the referees appointed by the Royal Society pursuant to the instructions in the letter from Prince George's secretary – Newton, Wren, Gregory, Francis Robartes, and Dr. John Arbuthnot – proceeded with arrangements. Flamsteed was almost incandescent with excitement, traveling to London at least once a week during the winter and spring of 1705 to inquire and discuss. Newton had to attend the election in Cambridge. "My work stands, by reason of Sir I. Newton's absence," he wrote in his diary on 2 May: "pray God it may go on at his return." And a month later: "God send a good issue to this business, and me peace."[106] Rather quickly he began to realize that Newton did not much care for his opinion. On 5 March, he met the referees at the Castle Tavern and looked over specimen sheets that Awnsham Churchill, a bookseller being considered as publisher, had had printed. Two weeks later, "Mr. Ashton told me he dined with Mr. Newton, Mr. Roberts, Drs. Arbuthnot and Gregory, at Mr. Churchill's, when we met at the Castle Tavern, March 5th last, after I was gone thence: and that all things, as he thought, were agreed but paper." Flamsteed went home and ordered his servant to start keeping a file of all the letters he received from Newton.[107] Aston's revelation was almost the end of his last effort at friendliness with Newton. He soon became convinced that Newton had taken up the business only for his own purposes. Toward the end of the year, he noted down two meetings at which the referees consulted him. "I do not remember that I was present at any more of their meetings but one, where nothing material was determined whilst I was present; though I considered that being the person chiefly concerned, I ought either to have been present at all, or at least to have had their resolves immediately signified to me, of which I had nothing but at second hand, and sometimes knew less

---

[105] *Corres 1*, 172.   [106] *Baily*, p. 220.   [107] *Baily*, p. 219.

than those that were altogether unconcerned."[108] By that time, his long effort to win Newton's approbation had given way finally to unremitting shrill hatred.

He resented Churchill's role in the project. There was no need to have a publisher (or "undertaker" in the language used), who would only suck up the prince's money into his own profit. On 5 March, he found the specimens poorly done, and he rushed about London to locate another printer who prepared sheets at Flamsteed's expense – much better specimens in his opinion. The referees paid no attention and proceeded with Churchill. In June, Flamsteed refused to sign an agreement that guaranteed Churchill 34s per sheet.

What particularly galled him about Churchill's profit was the unwillingness of the referees even to consider what Flamsteed called "an honorable recompense for my paines, and 2000*lib.* in expense."[109] His salary of £100 as Astronomer Royal was not munificent. From it he had to hire his own assistants. While the king had provided the Observatory building, he had had to outfit it with instruments at his own expense, except for some given to him by Sir Jonas Moore. Now he expected some return for thirty years of labor.

> 'Tis very hard, 'tis extremely unjust [as he expressed it in one memorandum to himself], that all imaginable care should be taken to secure a certain profit to a bookseller, and his partners, out of my pains, and none taken to secure me the re-imbursement of my large expenses in carrying on my work above 30 years.[110]

Another memorandum made Newton the villain in Churchill's place.

> I have plowed sowed reaped brought in my Corne, with my own hired Servants & purchased Utensills. S^r I N. haveing been furnished from My Stores would have me thrash it all out my selfe & charitably bestow it on ye publick that he may have ye prayse of haveing procured it. I am very desirous to supply the Publick wth my Stores if it will but afford me what I have layd out in tillage and harvesting in Utensills & help. & afford me hands to work it up since the labor is both too hard & much for me. for an adequate recompense I doe not expect but I must stand upon a reasonable one since God has blest my labors with large fruites & not to doe it were not to acknowledge his goodness; & my Countries Ingratitude would be attributed, by S^r I N himselfe, to my *Stupidity*.[111]

One can understand Flamsteed's point, but it is impossible wholly to sympathize with him and impossible not to have some sympathy with Newton in having to deal with him. True, his salary had not

[108] *Baily*, pp. 254–5.   [109] *Baily*, p. 219.   [110] *Baily*, p. 246.   [111] *Corres 4*, 490–1.

been large, but he also had the income of a rich benefice and a substantial inheritance from his father. If he paid his own expenses, he did not live in poverty, and he had the enormous privilege of spending his life at the work he loved. How could he weigh a lifetime against £2,000 expenses and haggle over the paper and the number of copies to be printed because they would influence his return in selling the books? He had an unhappy tendency, moreover, to drown his claims in bathos as he brought up long cold nights in the observatory which undermined his health. Newton conspicuously lacked the tolerance to deal with such an exasperating man. Quickly he seized the importance of the recompense to Flamsteed, and for no apparent reason other than spite refused to hear of it – behavior no less painful than Flamsteed's eternal harping on money. The issue acquired a further dimension. As soon as the prince accepted the budget with an item for calculators, Flamsteed hired two and set them to work – on the fixed stars, to be sure, not on planets, comets, and the moon as the referees intended. In a short time he ran up a bill of £173; Newton kept him waiting three years before he released £125 to him.

Negotiations on the articles of agreement used up nearly all of 1705. In June, Flamsteed refused to sign a proposed set of articles. He submitted his own set in the autumn. Newton would have none of them. In fact, the articles that were finally signed on 17 November did embody many of Flamsteed's demands, such as a strict limitation to four hundred copies, which would be his property, and a right for him to inspect the press at any time to be sure that Churchill did not print more. His suspicion that Churchill would secretly print extra copies and sell them, undercutting Flamsteed's sales, almost reached the level of paranoia. On other matters, such as Churchill's role, he did not prevail, however. Among the latter was the decision to print the catalogue in volume 1. That is, the articles of agreement effectively specified that the catalogue of fixed stars to be published would be the currently existing catalogue, not a future catalogue yet to be completed.[112] In a letter to Sharp written three days later, Flamsteed said, "Sir Isaac Newton has at last forced me to enter into articles for printing my works with a bookseller, very disadvantageously to myself."[113] In his autobiography, he claimed that he heard the articles read once and had then to sign, or Newton would have called off the whole project and placed the onus on him.[114] Flamsteed signed, but he treated the parts he did not like, especially the agreement about the catalogue, as null.

[112] *Corres 4*, 454–8.
[113] Cudworth, *Sharp*, p. 59. Cf. Flamsteed to Sharp, 28 Nov. 1705; *Baily*, p. 256.
[114] *Baily*, pp. 80–1.

Three months later, Newton had not even supplied him with a copy of the agreement.[115]

As the articles demanded, Flamsteed immediately gave Newton the manuscript copy for volume 1, the catalogue excepted. Seven weeks later, all impatience, he heard that Churchill had not even agreed with a printer yet. On 2 February 1706, he told Sharp that he expected to have the first sheet in a fortnight, and hoped then to move forward vigorously.[116] He did not have it in a fortnight. Instead, Newton asked to see his original observation notes, and Flamsteed heard on 23 February that Newton had four or five sheets of corrections to the printer's manuscript.

> I visited him last Monday [25 February] & desired to see them he told me Dr Gregory had collected them. the D^r soon came, when we sat down to examine them S^r Is. told me he did not beleive them to be errors but desired that himselfe & the D^r might be informed of my ways of observeing. they were proper judges in the meantime I ordered James [Hodgson] to come to me for I have resolved not to talk with them without good Witness he came in good time we got to work & found a great many differences but all of the Drs Makeing he had formed a table for turning ye revolves & parts [of the screw] into degrees minutes & seconds. & supposing the thrids of the screw every where equall wonderd that his aequipollent degrees minutes & seconds agreed not with mine. I told them I wonderd he should adventure to make this table. wipt out his emendations from my Margin. engaged to give them an account of the other differences dined wth them & returned home next day I caused my own larg table to be copyed & the day following sent it them . . . I told them I had been at great expense in this work & expected a recompense but I fear S^r Is: had rather stop it then give himselfe any further trouble for he finds I doe not court him, & his temper wants to be cried up & flattered.[117]

The list of errors—the "pretended faults" as he called them— exercised Flamsteed as much as the errors he found in the lunar theory exercised Newton, and he spent much of the following month proving that there was nothing of importance in them. "Apr 4 at London. hear y^t all ye mistaken errors are quitted & y^t ye first Sheets will goe to ye presse this Week."[118] In fact it was yet another month before he had the first sheet.

Meanwhile, the issue of the catalogue had also helped to delay the press. The articles called for the catalogue of fixed stars to appear in volume 1. Newton would not allow work to begin until he had a

115 Flamsteed to Newton, 26 Feb. 1706; *Corres 4*, 473.
116 Flamsteed to Sharp, 17 Jan. and 2 Feb. 1706; *Baily*, p. 257.
117 Flamsteed to Sharp, 2 March 1706; *Baily*, p. 259. Cf. his entry for 25 February in his diary (*Baily*, p. 222). 118 *Baily*, p. 224.

copy of it in his hands. After some backing and filling, Flamsteed agreed early in March to give Newton a copy of the catalogue as it then stood. He insisted that the copy be sealed. Later, Flamsteed loudly protested Newton's perfidy in breaking his word and opening the catalogue. There is no evidence whatever that Newton ever accepted Flamsteed's condition, however. From the time he received it, Newton knew what the catalogue lacked. Although he could have received that information from Flamsteed by word of mouth, he probably treated the catalogue as an open manuscript from the beginning. There is no point in pursuing the issue of the sealed catalogue at great length. Too much attention has focused on it. Flamsteed had grievances aplenty. It appears to me that he grasped at the alleged breaking of the seal to give specific content to his wholly justified sense of outrage at the way he was treated.

On 23 March 1706, Flamsteed met with the referees and Churchill at Newton's home. The various obstacles had been cleared from the path. The referees agreed to draw the prince's money and begin. "Sr Issack askt me if things went not now to my Content I returned y^t it was strang that I should be so little taken notice of who was the person mainly concernd at which he seemed chagrin."[119] Nevertheless, Newton unbent sufficiently to ask him how much he had spent in copying the manuscript for the first volume, an obvious offer to issue money to him. Flamsteed was so far committed to hostility that he could only reply he did not know since he used his amanuensis for other jobs as well. Newton was not one to persist in unrequited gestures of friendship, and the moment passed unseized.

Finally, on 16 May, the first sheet was printed off. Flamsteed could not contain his excitement. On 19 May, he went to Newton's house to collect his observation notes for use in correcting the proofs. Newton told him they must go slowly at first. By 24 May, when he had not yet received the second sheet, he wrote Churchill in admonishment.[120] He wrote Churchill again on 6 June, strongly dissatisfied both with the rate of progress and with the accuracy. On 7 June, a Friday, he went to Churchill's office; though the printer did not come when summoned, he did send the fourth sheet, signature D. Flamsteed returned it corrected Monday morning, 10 June. "11 June recev'd the 4th Sheet D to correct ye 2d time returnd it by ye coach, was at ye gate for horsback. found it sterk lockt, it began to rane continued violent I returnd home sent up ye correct Sheets by ye coachman at 8½ ☿ [Wednesday] *mane* . . ."[121] And so it continued, with Flamsteed in hot haste to forward the press, while Churchill,

[119] *Baily*, p. 223.   [120] *Baily*, p. 224.   [121] *Baily*, p. 225.

despite the agreement to print five sheets per week, barely produced one.

Still Flamsteed did not receive the money for his calculators. Newton's records show that he drew £375 from the prince's treasurer and immediately advanced £250 to Churchill. By the articles of agreement, the other £125 was meant for Flamsteed, and yet when Flamsteed went to him on 19 April for the money, Newton entertained him with a lugubrious account of Prince George's financial overextension.[122] In the fall, Flamsteed approached him again, telling Newton that he had had to dismiss his calculators. There is a draft in Newton's hand of an order dated 28 November to pay Flamsteed £50 to defray his expenses. Apparently he thought better of it, for the money was not handed over.[123]

The press continued at its glacial pace until December 1707, when all of Flamsteed's observations taken before the completion of his great mural arc in 1689, that is, all the material intended for volume 1 except for the catalogue, had been printed – 97 sheets in 89 weeks by Flamsteed's count. All the evidence indicates that relations with Newton were not smooth. Irritable and irritating as always, Flamsteed was not mellowed by the gout which had now bent him over.[124] Nor did the episode of the Plumian professorship of astronomy at Cambridge improve his temper. Flamsteed had provided the immediate cause of Dr. Plume's bequest by loaning him Huygens's *Cosmotheoros*. Plume's will directed the trustees, of whom Bentley was the principal one, to consult Newton and Flamsteed about the statutes of the new chair. When he first heard about it in 1706, Flamsteed found that the trustees had already framed the statutes and chosen the first professor without approaching him.[125] He did not believe they had similarly ignored Newton.

For his part, Newton did not find Flamsteed any easier. Flamsteed knew how to nettle him as no one else did. Newton might control the purse and refuse to issue money rightfully due. He could not bend a man filled with the righteousness of God to his will. On 15 April 1707, as the printer neared the end of the manuscript available, Newton and Gregory visited Greenwich, viewed Flamsteed's papers, and delivered an ultimatum. Flamsteed must promise to hand over his observations with the mural arc and "give up" a copy of the catalogue of fixed stars. When he had done so, he would receive the money for calculators. The referees, they informed him, "are positively resolved not to print any further after

[122] *Baily*, p. 224.
[123] Flamsteed to Newton, 14 Sept. 1706, *Corres 4*, 477. The draft of the order; *Corres 4*, 480.
[124] Uffenbach reported him so (Mayor, *Cambridge under Queen Anne*, p. 351).
[125] Flamsteed to Newton, 26 Feb. 1706; *Corres 4*, 473. Cf. the entry in his diary on 15 March (*Baily*, p. 223).

this day, nor to give Mr Flamsteed any money until these conditions are fullfilled."[126] Flamsteed did not relinquish the papers,
however, and the press continued nevertheless. He also apparently
got in one or two of his special jibes. Something recently happened,
he wrote to Sharp in May, "that has discovered his [Newton's]
proud and insolent temper, and exposes him sufficiently. He has
been told calmly of his faults, and could not contain himself when
he heard of them. My affair was not forgot."[127] In his autobiography, he described the same scene, or another similar one. "I . . .
always found him insidious, ambitious, and excessively covetous of
praise, and impatient of contradiction. I had taken notice of some
faults in the 4th book [*sic*] of his *Principia:* which, instead of thanking me for, he resented ill. Yet was [so] presumptuous that he
sometimes dared to ask 'why I did not hold my tongue.' "[128] Two
more incompatible colleagues have never been yoked together for
their sins in a single enterprise: Newton, imperious, unable to participate without domineering; Flamsteed, righteousness personified,
unable to imagine that his corrections could give offense.

With the completion of the manuscript Flamsteed had delivered,
the question of the catalogue could no longer be evaded. Would it
go in volume 1 as Newton wanted or in volume 3 as Flamsteed
wished? The press came to a halt. Finally, on 20 March 1708, all
parties concerned met at the Castle Tavern and agreed that Flamsteed would hand over his observations with the mural arc and
another catalogue of the fixed stars that he brought to the meeting,
that he would further correct the deficiencies of the catalogue he
had given two years before, that Newton would issue £125 to him,
and that upon delivery of the catalogue of the fixed stars "as far as it
can be Compleated at this Time" he would receive the rest of the
money due him.[129] He did in fact finally receive £125 in April. A
clause in the agreement further stipulated that Flamsteed would
maintain a corrector in town in order that the press not be delayed.
Since it is abundantly clear that Flamsteed was punctuality itself
with the corrections, the clause is peculiar. He thought that Churchill inserted it to blame delays on him. In the event, it appears
more likely that the clause was Newton's idea, for in July, when
not a single sheet had been printed for more than half a year,
Flamsteed received a resolution from the referees that if he did not
take care of corrections with dispatch they would appoint other
correctors.[130] From every indication, the resolution was the prelude
to a power play in which Newton would order the publication to

---

[126] A manuscript in Gregory's hand entitled "A Paper offered to Mr Flamsteed to be signed
♂ [Tuesday] 15 Aprile 1707, at Greenwich by Sir Isaac Newton & Dr Gregory sent by
the Referees"; *Corres 4*, 487–8.   [127] *Baily,* p. 264.   [128] *Baily,* p. 73.
[129] *Corres 4*, 513.   [130] *Corres 4*, 524.

proceed along his lines because Flamsteed had violated the agreement. In alarm, Flamsteed wrote an eloquent letter to Christopher Wren which recited the history of the affair, recognized that the question of the catalogue was central, and gave his argument for not printing it then.

> I am not only willing [he concluded] but desirous that the Press should proceed to finish the first Volume of Observations . . . Sr Isaac Newton has 175 [manuscript] sheets of the 2^d Volume in his hands, that the press may proceed with whilest I am Compleating the Catalogue, So there need be no stop on my Account as there never was, nor hereafter shall be, God spareing me Life & Health & prospering, as I firmly believe he will, my Sincere Endeavours.[131]

Perhaps Wren slowed Newton down. At any rate, nothing more had been done by October, when Prince George died and the project of necessity came to a halt. Newton vented his frustration by having Flamsteed's name stricken from the list of the Royal Society in 1709 for nonpayment of dues–though he ignored plenty of others no less in arrears.[132] Flamsteed used the respite to do what he wanted to do anyhow; he finished his catalogue.[133]

Newton's responsibilities at the Mint continued, of course, while he was president of the Royal Society. The early years of his tenure, for nearly a decade, were a period of very low coinage; and while various administrative and advisory duties remained, the demands of the Mint were not heavy. The Royal Society met on Wednesdays, the day of the week specified in the Indenture for the Mint to receive bullion and issue coin. Even with the conflict, he was able to preside at more than three-quarters of the meetings.

In 1707, the union with Scotland brought a flurry of activity with it. The Act of Union specified that the Edinburgh Mint would continue to exist, though remodeled to the plan of the London Mint, and union entailed a Scottish recoinage to provide a single currency for the United Kingdom. Inevitably, Newton found himself deeply involved in the Scottish recoinage even though he did not leave London. A flood of memoranda began in the spring as the government sought his advice on the many details that required decisions: trial pieces, weights, dies, tools.[134] The problem of procedures was more difficult, for the Edinburgh Mint had to learn the ways of the London Mint, which differed from theirs. Newton suggested in March that officers of the Scottish Mint should come to London and that an officer from the English Mint be sent there.[135] Less thorough mea-

---

[131] Flamsteed to Wren, 19 July 1708; *Corres 4*, 525–7.   [132] *CM 2*, 210.
[133] He stated this in his autobiography (*Baily*, p. 89).
[134] *Corres 4*, 485–94. *Corres 7*, 447–8.
[135] Newton to Godolphin, 23 March 1707; *Mint* 19.3, f. 72.

sures prevailed, and as a result, because Scottish officials did not understand English practices, Newton later found himself drafting letters for his Edinburgh counterpart, George Allardes, to send to the Lord Treasurer.[136]

Although the government did not pick up his suggestions that officers be exchanged, it did in the end accept a later proposal that David Gregory and a clerk from the Mint go to Edinburgh to instruct them in the English procedures, especially the methods of keeping books.[137] This was a nice plum of patronage which Newton was able to throw Gregory's way. For four month's service, one of which he spent in transit, he received £250. A clerk from the Mint, Richard Morgan, and three moneyers accompanied him. The warrants that made the appointments official were dated 12 July. They left London on 21 July, arrived on 1 August, and immediately set to work putting the recoinage in order. During the autumn, Gregory sent Newton a series of harassed letters which recall the atmosphere of the Tower Mint a decade earlier. Supplies did not arrive on time, and he wrote in haste to secure them. Melting presented difficulties. The Edinburgh Mint melted with pit coal, which heated the metal to a higher temperature and caused the alloy to fume away. It proved impossible in this instance to reduce Scottish practice exactly to the English, but by trial and error Gregory worked out a passable alternative which maintained the requisite degree of fineness.[138] Once the melting was adjusted and an adequate rate of coinage attained, Gregory got leave to return to London. He left Edinburgh on 21 November and dated his report to Godolphin 13 December.[139] The moneyers stayed on until the completion of the recoinage early in 1709. The Edinburgh assignment was the last benefaction Gregory received from Newton. He died in October 1708, less than a year after he returned.

Newton's involvement in the recoinage did not confine itself to Gregory and did not end with Gregory's return. In a very real sense, he managed the Scottish recoinage from London. Nearly every question that arose got referred to him for advice – problems of engraving, refining, coining, and bookkeeping.[140] When the collector of bullion died in the middle of the recoinage, leaving the bullion on hand tied up in his estate, the warden wrote, not to his counterpart in London, but to Newton for help in freeing bullion to

[136] *Mint* 19.3, f. 119.
[137] Newton to Godolphin, 24 June 1707; *Corres 4*, 494–5.
[138] Gregory to Newton, 12 August, 16 September, 9 October; Newton to Gregory, n.d. and ca. 15 Nov. 1707; *Corres 4*, 497–9, 502–3; *Corres 7*, 452–4.
[139] *Corres 4*, 503–4.
[140] James Clark to ?Newton, 9 Sept. 1707; *Corres 7*, 451. Patrick Scott to Newton, 31 Jan. 1708; *Corres 4*, 510–12.

allow the recoinage to continue.[141] Above all, when the recoinage was at last complete and accounts had to be cleared through the Treasury in London, the sinecurist master in Edinburgh, Allardes, who had learned nothing, turned to Newton for guidance. He ran through categories of expenses and, referring to himself in the third person, put a question that must have caused shudders in the model administrator who read it. "If on these accounts he ought not to be allowed So much upon the pound Weight of the Coinage. Without being obliged to give in particulars which may be impracticable."[142] Allardes died before he could clear his accounts, and Newton had to do it for him, as he had done it for Neale, in order that the moneyers be paid.[143]

Newton also received communications of another sort from the Edinburgh Mint. Its officers seemed to regard him as a fellow-conspirator who could be counted upon to use his influence to grease their palms. Complaining that his salary was too low, the comptroller, Boswell, tried to enlist Newton's aid in having it increased. While he was about it, he suggested raises for several clerks as well.[144] In August 1709, Allardes wrote from his sickbed. He feared he was dying, as indeed he was, and sought Newton's assistance in securing his position, worth £200 per annum, for his son, a lad of seventeen.[145] Suffice it to say that Allardes's successor was John Montgomery. Newton had a copy of Montgomery's Indenture, and his papers contain several drafts of proposed emendations to it.[146] He would have been outraged had Montgomery been invited to comment on his Indenture. It took the Edinburgh Mint several years to realize that the master of the London Mint was an enemy who regarded them with fear and suspicion as a potential drain on the limited finances of his own institution. He never attempted to have the Edinburgh Mint dissolved; such an effort would have been, as he realized, a hopeless undertaking. He did use all his influence to have its expenditure closely defined and fixed. The officers who blithely thought to use him to increase their salaries and to secure sinecures for their sons could not have mistaken him more. The Edinburgh Mint continued to exist for another century with an annual payroll of about £1,000. It never spent another penny coining another shilling, however. Not least among the causes of its nonoperation was Newton.

---

[141] Drummond to Newton, 12 July 1708; *Corres 4*, 522–3.

[142] Allardes to Newton, 12 Feb. 1709; *Corres 4*, 531–2. Cf. Lauderdale to Newton, 10 Aug. 1709; *Corres 4*, 542.

[143] *Mint* 19.3, ff. 31^v–2, 185. Only on 24 March 1712 was a warrant drawn to pay the moneyers £2,693 for their services in Scotland (*Mint* 1.8, pp. 98–9).

[144] Boswell to Newton, 10 July 1708; *Corres 7*, 463.   [145] *Corres 4*, 541.

[146] *Mint* 19.3, ff. 80, 82–93, 202^v–3.

Figure 13.2. The house on St. Martin's Street, immediately south of Leicester Fields, as it appeared in the middle of the nineteenth century. Newton lived here after September 1710. (Courtesy of the Warden and Fellows of New College, Oxford.)

In the autumn of 1709, after more than a decade on Jermyn Street, Newton moved to Chelsea. He must not have found it satisfactory, for in only nine months he moved once more, to a house in St. Martin's street south of Leicester Fields (what is now Leicester Square), which he rented for £100 per annum (Figure 13.2).[147] Among manuscripts for the second edition of the *Principia* is a bill for household furnishings, probably furnishings for the new house.

The Brockadilla hangings wth ye matt fringe & picture	2.3.0
The Cistern & sink in ye Passage	1.0.0
The Tapestry in the Hall Closet 40 ells	3.5.0

There were also a water tub with iron hoops, a bell, two bottle racks and three beer stands, glasses, a range, a marble mortar and

[147] His estate paid £25 rent for one quarter (Huntington Library MS).

pestle, equipment for the fireplace, a napkin press, a bedstead, and so forth – in all twenty-one items with a total price of £27 8s 0d plus four other items without prices.[148] The new house should have been well furnished. There he resided for fifteen years until ill health led him to seek purer air in Kensington two years before his death.

The same year that he moved to St. Martin's Street, 1710, saw a pair of critical problems in the Royal Society. The council of the society met on 22 March. The minutes of the meeting recorded nothing unusual except that the president called another meeting for the following week. On 28 March, the day before that meeting, Newton received from a member of the council, probably William Derham, a long letter which makes it evident that the previous meeting had been anything but ordinary.[149] Summoned precipitously on 21 March, the council had considered disciplinary action against one of its members, Dr. John Woodward, for an incident at the society's public meeting on 8 March.

Dr. Woodward was the stormy petrel of the society, a man not well loved by Dr. Hans Sloane, the secretary. When a satire, called the *Transactioneer,* on Sloane and the *Philosophical Transactions* that he had edited, had appeared anonymously in 1700, Sloane had suspected Woodward and John Harris of being its authors, though the council had accepted their protestations of innocence.[150] Again in 1706, upon complaint of "unjust reflections" by Woodward on some members at meetings, the council had voted that the society be admonished when he was present "that if any Members of the Society Should hereafter cast reflections on the Society or any of the Fellows thereof the Statute concerning Ejection Shall be taken into Consideration . . ."[151] Chastened neither by the warning nor by the events of 1710, Woodward went on in 1717 to clash with Newton's friend, Dr. Richard Mead, with whom he fought a famous

---

[148] *Add MS* 3965.12, f.209ᵛ.

[149] *Corres 5,* 17–22. The letter survives in an apparent copy (*Keynes MS* 151) without a signature. I assume from its content that it was a real letter intended to influence Newton and therefore written by a member of the council and signed. By comparison of the letter with those present on 22 March and elimination of those it mentioned, one arrives at two possible authors, Derham or John Van de Bemde. Van de Bemde, elected to the society in 1678 and to the council in 1698, was a shadowy figure of whom little more than the name survives. He never published in the *Philosophical Transactions*. The letter appears to come from an active natural philosopher, such as Derham was. In a list of the council, Newton included Derham among those to be purged at the next election, but not Van de Bemde (*Yahuda MS* 15.7, f. 180ᵛ). There were references to Derham in at least two of Woodward's letters from this time, one of them calling him an "ingenious gentleman." However, the references were not nearly explicit enough to allow any conclusions to be drawn from them (Woodward to Thoresby, 8 Nov. 1709 and 10 Jan. 1710; *Letters of Eminent Men, Addressed to Ralph Thoresby, F.R.S.,* 2 vols. [London, 1832], *2,* 201, 216). Hence there are grounds (well short of conclusive to be sure) for thinking Derham the more likely author. [150] *CM 2,* 154–5.

[151] *CM 2,* 181. The order was carried out at the meeting on 15 May 1706 (*JBC 10,* 133).

duel. Losing his footing, he fell and lay defenseless beneath Mead's sword. "Take your life," said Mead, disdaining to kill him, whereupon Woodward seized immortality instead with the deathless reply, "Anything but your physick."[152] In the event, Mead was sufficiently overcome by the retort that Woodward escaped with his mortality as well.

His antagonist in 1710 was Sloane rather than Mead. The letter to Newton on 28 March and later minutes of the council provide all the information we have about their clash. Newton was not in the chair on 8 March. It was a thin meeting. Sloane was providing what substance he could by reading a translation of a memoir to the French Academy on gallstones, to the small pleasure of at least some present. In passing, Sloane remarked that gallstones cause colic. Woodward disagreed – civilly according to Derham. Unable to sustain his assertion, Sloane had recourse to "Grimaces very strange and surprizing," not once, but two or three times. Walter Clavell later confirmed this before the council. Woodward's temper did not begin to stretch that far, and he told Sloane in anger to "Speak Sense or English and we Shall understand you. If you understood anatomy you would know better."[153] In Derham's account, Woodward's language was somewhat more diplomatic and Sloane's discomfiture in the medical argument absolute. The incident was unsettling enough that the society did not meet on 15 March. On 22 March if Derham's account is true, Sloane tried to take his revenge.

Derham wrote as a frank partisan of Dr. Woodward, almost as full of antagonism for Sloane as the doctor himself was. He spoke of the low level to which the society had sunk and of "how loud Men of Sense all over the Town are in their Complaints," a state of affairs which he attributed directly to Sloane, whom he pictured as a tyrant over the society's affairs. According to Derham, Sloane controlled the society by packing the council, and he had tried to destroy Woodward by summoning a meeting without adequate notice and without letting Woodward know his intentions. There is evidence in a letter that Woodward wrote in January 1710 (evidence which Derham's letter further supports) that Woodward had engineered the election of John Harris, at the annual meeting in November 1709, to replace Richard Waller as one of the two secretaries.[154] Woodward understood Harris's election as the first step toward supplanting Sloane. Undoubtedly Sloane understood it in the same way and seized the present occasion to counterattack. At the council meeting on 22 March, according to Derham, Sloane threatened to resign if the

[152] Joseph M. Levine, *Dr. Woodward's Shield* (Berkeley, 1977), p. 16.   [153] *CM 2*, 213–14.
[154] Woodward to Thoresby, 10 Jan. 1710; *Letters to Thoresby, 2*, 217. *Cf.* Levine, *Woodward's Shield*, pp. 84–91. It should be obvious that I cannot agree with the implication of Levine's account, that Newton was a cipher in the society's affairs.

council did not support him. It is impossible to reconcile Derham's portrait of Sloane's dominance either with Newton's character or with the records of society affairs since his election to the presidency. As Flamsteed knew, Newton domineered; he was not accustomed to suffer the domination of others. True, Derham quoted Newton's own complaints about Sloane, and since the letter attempted to sway him, we must assume he quoted correctly. It is not surprising if two strong-willed men clashed on occasion. On the whole they appear to have worked together, however, to govern the society, a dictatorial duumvirate, instead of wasting their energies in internecine struggles. Derham's letter assumed Newton's support of Sloane at the meeting on 22 March. He failed utterly to budge him from that position for the future. Woodward suffered the consequences.

Sloane's stratagem on 22 March failed solely because of the chance presence of Walter Clavell and John Lowthorp, according to Derham. Not much was left to chance on 29 March. The chambers in Gresham College held the largest council they had witnessed in many years. Newton and Sloane summoned up the establishment, such men as the treasurer, Alexander Pitfield, and the former secretary, Richard Waller, neither of whom was accustomed to burden the council with frequent attendance. Harris was not present, being away at his benefice for Holy Week. Woodward opened by requesting that the inquiry be deferred until Harris's return. The council ignored the request and proceeded, and this time they censured Woodward by voting that "the Said Reflecting Words intended to the detriment of the Royal Society."[155] At a subsequent meeting, they reinforced the decision by ordaining that the presiding officer could order an offending member of the society to withdraw from a meeting on pain of expulsion.[156] A final showdown came on 24 May. It must have been prepared; an even fuller council, all but three members, was present. Sloane declared that he had meant no offense to Woodward by any gestures he made. The council decided that his declaration was sufficient satisfaction. Woodward refused to declare that he was sorry he misunderstood, however, and the Council solemnly voted

> Whether D^r. Woodward for creating disturbances by the Said reflecting words after a former admonition upon the Statute of Ejection and for restoring the peace of the Society be removed from the Council
>
> Carried in the Affirmative that he be removed:
> Whether the Thanks of the Council Shall be returned to D^r. Sloane for his pains and fidelity in Serving the Society as Secretary.
> It was Carried in the Affirmative & D^r. Sloane was Thanked.[157]

---

[155] *CM 2*, 213–14.   [156] *CM 2*, 215.   [157] *CM 2*, 218.

Newton told Woodward, "We allow you to have *natural* philosophy, but turn you out for want of *moral*."[158] Woodward did not
humbly swallow such preachments. He brought a suit against the
society for his restoration, but without success.

For Newton, as president, the incident had meaning beyond its
revelation that grown men in the eighteenth century were as prone
to adolescent behavior as they are in the twentieth. Both Derham's
letter and the records of the council indicate the presence in the
society of a coherent nucleus of opposition to the Newtonian establishment. Derham's letter defined it to some degree: Woodward,
Harris, Clavell, Lowthorp, and himself, a significant cadre in a
council of twenty-one. In fact, three others may have adhered to
them: Van de Bemde, Dr. William Cockburn, and Sir John Percival. All had been active in the society and the council for a number
of years; all now dropped out. Some members of this group later
resumed activity in the society after Sloane stepped down as secretary at the end of 1713; others left for good. Moreover, Sloane was
not their sole target, for the first vote at the meeting on 29 March
was on a motion that Newton should not occupy the chair while
the issue was under discussion.[159] Furthermore, Newton understood himself to be involved. Before the next election, he went
through a list of the council putting an **X** by each man he wanted to
see eliminated, the same group that lined up against Sloane.[160] Harris apparently knuckled under; he allowed himself to be named to
the committee appointed to handle the defense against Woodward's

---

[158] *Stukeley*, p. 67.   [159] *CM 2*, 213–14.

[160] *Yahuda MS* 15.7. f. 180ᵛ. Newton's "hit list" started with an incomplete roster of the
1709–10 council. He wrote down nineteen of them, forgetting two, Foley and Moreland.
He then tried to dredge up the other two names from his memory, putting down "Dʳ Mʳ
———" and Pettiver (who was not on the council). Initially he went down the list putting
checks by those of whom he approved and **X**'s by those to be eliminated. The list was
exactly the group I have defined with the exception of Van de Bemde ("Bembdy" as
Newton put him). Seven, himself included, got checks the first time. He then went down
the list a second time underlining those of whom he approved – all those without **X**'s. To
the side he compiled a list of six new members of the council. Foley appeared on this one.
Thus twelve of the other twenty members of the council had Newton's approval, and
seven his disapproval. Moreland was simply forgotten. Pettiver received an **X,** but incon-
sistently Newton also included him in the list of possible new members.

Below the original list, Newton compiled a second one from the eleven names he had
underlined, the members of the council to be returned to the next one. He listed Wren
(Sir Christopher's son) as an alternative, thus effectively approving of twelve, himself
and eleven others. Once again he put Foley down as a new member, to raise the number
approved really to twelve others. On the bottom list, Pettiver did not appear as a new
member.

Cf. *Add MS* 3965.11, f.148ᵛ on which Newton in his systematic manner set down four
points about the quarrel: "What occasion Dʳ Sloane gave / What words past between
them / Whether the Society be injured thereby & by whom / Whether was it a meeting of
the Society."

suit.[161] He was dropped from the secretaryship in November, nevertheless, and did not again serve on the council. However ludicrous the incident appears, it shook the society enough that it adjourned for the summer in the middle of June, more than a month before the usual time.

Flamsteed watched the events from a distance and reported them to Sharp with relish.

> Sir I. Newton has put our Royal Society into great disorder by his partiality for E. Halley and Dr. Sloane, upon a small and inconsiderable occasion: so that they have broke up some few weeks before their time. Dr. Harris has lost all his reputation by actions not fit for me to tell you.[162]

Flamsteed was not an impartial observer, of course, Already the previous autumn, with what justice we do not know, he had told Sharp that Newton was "much talkt of but not much to his advantage. Our Society is ruined by his close politick & cunning forecast, I fear past retreiveing . . ."[163]

The second critical issue the society faced in 1710 may have stemmed from the first. Derham's letter claimed that Dr. Woodward had intervened in 1703 to prevent Gresham College from evicting the society. If that were true, the officers may have felt that they had no option but to find other quarters after they ejected him. The society had been looking constantly since 1703, however, and it could have been sheer accident that they finally found a satisfactory place at the end of the summer. Be that as it may, Newton summoned a meeting of the council on 8 September, in the middle of the recess, to inform them that the house of the late Dr. Edward Browne (son of the celebrated Sir Thomas Browne, author of *Religio Medici*) in Crane Court off Fleet Street had been put up for sale. The council appointed a committee, which included Newton, to look into the matter.[164] The society's first move, when it considered a house, was invariably an examination of it by Sir Christopher Wren. Wren and his son (who was also a member of the society) both approved in this case. Newton moved swiftly. He called another council for 16 September to authorize the purchase. When only five members showed up, he set another meeting for 20 September. This time thirteen members attended and found themselves with an alternative. Hearing the society was about to buy, a Mr. Heggs came forward to offer a house in Westminster for £3,000. They could purchase Browne's for £1,450, however, and voted to do so. One member opposed the purchase. (According to a contemporary pamphlet, this was Lowthorp. Dr. Cockburn dis-

---

[161] *CM 2*, 219.   [162] Flamsteed to Sharp, 14 July 1710; *Baily,* p. 276.
[163] Flamsteed to Sharp, 25 Oct. 1709; *Baily,* p. 272.   [164] *CM 2*, 220.

sented *viva voce* but refused to vote.) Harris and Clavell attended the following meeting on 26 October, when Newton informed the council that the purchase had been completed; they thereupon withdrew, presumably to express their opposition.[165]

On 22 November, someone published an anonymous pamphlet, *An Account of the Late Proceedings in the Council of the Royal Society, In order to Remove from Gresham-College into Crane-Court, in Fleet-Street.* With its reference to Hans Sloane, "who had engross'd the whole Management of the Society's Affairs into his own Hands, and despotically Directs the President, as well as every other Member . . . ," it appears to have been the product of Woodward or one of his supporters. Needless to say, it presented the purchase as the work of Sloane's cabal abetted by Newton, who did as Sloane told him. The pamphlet described an extraordinary meeting of the society with the council on 16 September, a meeting not otherwise recorded. The members were surprised and stunned, it claimed, by the news of an impending move. After seven years of search, it is difficult to imagine how they could have been surprised. Members remonstrated with the president that there was no reason to move from Gresham College, the pamphlet asserted, and that an affair of such importance should not be settled at a meeting called on short notice.

> This the President would not hear of. They therefore offer'd to give him their Opinion either by Ballotting or Voting *viva voce* . . . But in vain; his Scruples were unmoveable. So that some of the Gentlemen with warmth enough ask'd him, To what purpose then had he call'd them thither? Upon which the Meeting broke up somewhat abruptly, and not only the Members of the Society, but most of the Council also, left the President with Dr. Sloan, Mr. Waller, and one or two more, to take such Measures at the Councill as they best lik'd.[166]

There may have been such a meeting; the council minutes for 16 September are consistent with the pamphlet's account. Clearly there was opposition to the move. The later votes in the council and the financial support of the purchase by the membership belie the claim that opposition was general.

To pay for the house, the society put £550 down and took a mortgage for £900 at 6 percent. Before they could move in, exten-

---

[165] *An Account of the Late Proceedings in the Council of the Royal Society, In Order to Remove from Gresham-College into Crane-Court, in Fleet-Street. In a Letter to a Friend* (London, 1710), p. 18. CM 2, 221–3.

[166] *Ibid.*, pp. 6–8. Extracts from the pamphlet are published in *Corres 5*, 76–8. Cf. Charles R. Weld, *A History of the Royal Society*, 2 vols. (London, 1848), *1*, 391–4. The pamphlet was inconsistent in its treatment of Newton, sometimes (as in one passage quoted) picturing him as Sloane's puppet, sometimes (pp. 8, 17, 30) implying that he himself was the society's tyrant.

sive repairs had to be made, at a further expense of £310. Having taken the plunge, they decided to go all the way and authorized £200 more to build a repository, which Wren designed. Alas, it cost exactly twice the sum. Nevertheless, the financial situation stood well short of disaster. Gifts poured in, including £100 from Newton. By the middle of January, before the bills for either the repairs or the repository arrived, the society retired the mortgage, though partly by borrowing £464 from their clerk, Henry Hunt. By June, further gifts reduced the debt to £200. For the repairs and the repository, the society borrowed an additional £450 from Hunt and stood £650 in his debt when he died in June 1713.[167] Apparently Newton, Sloane, and Waller assumed Hunt's loans at that time. In 1715, when at long last the society won its suit to collect arrears on the fee farm rents purchased in 1675, the council paid £200 each to Newton and Sloane to retire bonds they held, and it settled a debt of £250 with Waller's heirs a year later.[168] In less than six years, the society had fully paid for its new home and stood free of debt. When Francis Aston left them a handsome bequest in 1716, the society was able to invest it in Bank annuities. Not everyone responded to the appeal for gifts, however. John Van de Bemde, who after twelve years of conscientious service on the council had not attended a single meeting following the Woodward affair, answered the appeal by canceling his membership.[169]

Repairs went on at Crane Court through the spring of 1711. The society informed Mr. Tooke of Gresham College that they would not require his room beyond June. Actually it was 2 August before the council ordered Hunt to move to Crane Court and take possession of the house.[170] For the first time in the fifty years since its establishment, the Royal Society had its own home (Figure 13.3).

What with both the Woodward incident and the purchase of Crane Court during the year, the annual meeting on 30 November promised to give trouble. Indeed, the pamphlet that described the purchase of Crane Court was an election manifesto which ended with an appeal to the members to turn out the present council and officers. Worried about it, the council devoted some attention to

---

[167] *CM 2,* 232–56. Gifts totaling £1,101 10s had been received by May 1713 (some of these are recorded in the Journal Book; *JBC 10,* 297, 298, 300, 305, 325, 394, 486). Earlier, in January 1709, Newton had given £20 to the society, and in 1718 he gave them £70 more.

[168] *CM 2,* 291–2. *JBC 11,* 123.   [169] 28 June 1711; *JBC 10,* 306.

[170] *CM 2,* 245. The pamphlet against the purchase of Browne's house stated that the president gave notice that meetings would resume at Crane Court on 8 Nov. 1710 after the annual recess (*Account of Proceedings,* p. 20). Whatever Newton may have said, the society cannot have moved in at that time. Extensive repairs were first carried out. Not only did the order to Hunt on 2 Aug. 1711 imply a later move, but earlier, on 20 January, the council appointed a committee to supervise the transfer and on 26 June they ordered Hunt to move the repository (*CM 2,* 234, 241).

Figure 13.3. Crane Court, the home of the Royal Society, purchased in the autumn of 1710. (Courtesy of the Royal Society.)

preparing, and among other things repealed the "Statute or Supposed Statute" passed in 1707 to limit membership on the council to those who had posted bonds for their weekly payments or were not in arrears.[171] The meeting saw the largest attendance in years, at least eighty members, the number who voted for Thomas Foley. Newton received sixty-nine votes, Sloane seventy. Compared with Foley's vote, the figures provide a rough measure of the size of the opposition group. Another measure was the failure of any of the Woodward faction to win reelection to the council.[172]

In the years following 1710, Newton steadily consolidated his position within the society. At the end of 1713, Sloane decided to step down after twenty years as secretary. Apparently he made the decision under some pressure. More than a month before the annual meeting, John Chamberlayne wrote to Sloane about the move to retire him.[173] Flamsteed, who did his best never to miss a piece of malicious gossip about Newton, also heard that Sloane's retirement was not wholly voluntary, and that some tension accompanied it.

> On St. Andrew's day, Dr. Sloane laid down his Secretaryship of the Royal Society: but either he, or another, had so managed the business, that Sir Isaac Newton had like to have been left out of the Presidency. There were high and furious debates. Dr. Halley is Secretary in Dr. Sloane's room; and Dr. Keill is brought into the Council. Sir I. Newton sees now that he is understood.[174]

A letter from John Chamberlayne to Newton just before the election tends to confirm Flamsteed's account of a struggle.

> Honored S^r
> I beg the Favor of you to mark the inclos'd List for me between this & Munday next, just as you intend to do your own both for New Councelors & New Officers all but one, whom I desire to choose Freely, & whom I would make Perpetual Dictator of the Society, if that depended only on the vote of
> Your most faithful Humble Serv
John Chamberlayne[175]

If Newton had his opponents, he had his partisans as well. The partisans prevailed. The election of 1713 was one of the few in which Newton received the highest number of votes for the coun-

---

[171] 13 Nov. 1710; *CM 2*, 227. A prior meeting (*CM 2*, 226) had already begun preparations for the election. It is not apparent what use the council made of the repeal of the earlier statute. [172] *JBC 10*, 249.

[173] British Library, Sloane MS 4043, f. 195; quoted in Levine, *Woodward's Shield*, p. 88.

[174] Flamsteed to Sharp, 8 Dec. 1713; *Baily*, p. 306.

[175] *Corres 6*, 43. I do not know how to reconcile this letter with Chamberlayne's earlier one to Sloane, which seemed to offer Sloane his support.

cil, which casts some doubt on Flamsteed's account. He was correct, however, about the position of secretary. Halley, a man clearly identified as a Newtonian, who would not be an independent factor in the society, succeeded Sloane. Sloane himself continued on the council with only two votes less than Newton. Whatever the grumbling about Sloane's tyranny, the society reelected him to the council every year during the rest of Newton's tenure, always among those with the most votes, and without hesitation they chose him as president upon Newton's death.

Scarcely a year after Sloane's resignation, the other secretary, Richard Waller, died in January 1715. Another identifiable Newtonian, Brook Taylor, succeeded him. Taylor resigned at the end of 1718 to be succeeded by John Machin, whom Newton's support had earlier helped to place in the Gresham chair of astronomy. Three years later, James Jurin, a protégé of Bentley from Trinity, replaced Halley.[176] Alexander Pitfield remained treasurer during the whole of Newton's presidency, a fact one can only interpret to mean that he cooperated with Newton. With average attendance at council meetings low, a cadre of four, who were of course better informed about business than their colleagues, found themselves able to dominate affairs. Beyond the officers, the society in these years acquired an increasing number of young, active members, natural philosophers and mathematicians, who can only be described as Newtonians – John Craig, William Jones, John and James Keill, John Freind, Roger Cotes, Robert Smith, Colin Maclaurin, J.T. Desaguliers, Henry Pemberton – and if Newtonian is not a meaningful description of physicians, there were nevertheless two who identified themselves with him, Richard Mead and William Cheselden. Several passages in Stukeley's *Memoirs* suggest the extent of Newton's domination of the society during his final years. The council was composed, Stukeley said, of the older members who had been of some service and who were chosen by rotation to acquaint them with the society's administration.

> He regarded the choice of useful members, more than the number, so that it was a real honor. Nor did any presume to ask it, without a genuine recommendation, and having given some proofs of their abilities. They were then previously to be approved of by the Council, where their qualifications were freely canvased; therefore less lyable to be balloted for with partiality or prejudice.[177]

Elsewhere, Stukeley made it clear just who canvassed qualifications.

> In November 1725 I was again auditor of the accounts of the Royal Society: we din'd with Sir Isaac, and after dinner we desired him to recommend the Council to be elected on S. Andrews day approach-

---

[176] *CM 2*, 287. *JBC 11*, 40, 260, 269; *JBC 12*, 171–2.   [177] *Stukeley*, p. 80.

ing: which he did. I have now the paper of his own writing, . . . the names of the Council for the ensuing year, among which he put down mine.[178]

Stukeley also mentioned that in 1721 when Halley retired, a number of members, including Hans Sloane and Lord (formerly Sir John) Percival, induced him to stand for secretary against Newton's opposition. Stukeley lost, though by a small majority. "Sir Isaac show'd a coolness toward me for 2 or 3 years, but as I did not alter in my carriage and respect toward him, after that, he began to be friendly to me again."[179] Flamsteed said more than he realized when he told Sharp that Newton saw he was understood. Newton did and he was. He was understood to be in control, and he saw it.

It was symbolic that the society should change its meeting day to accommodate Newton. 1711 was a busy year at the Mint as a result of a special bounty to attract silver plate. For the first time, it seriously conflicted with the society. On 1 March, after Newton had missed all but one meeting so far that year, the council voted to shift the weekly meetings to Thursday. The society resolved likewise on 7 March and began to meet on Thursday the following week.[180]

An almost imperial tone crept into the society following 1710. At the Council meeting on 20 January 1711, four proposals "were thought fit to be made Orders of the Council" and were read at the next meeting of the society.

1. That no Body Sit at the Table but the President at the head and the two Secretaries towards the lower end one on the one Side and the other Except Some very Honorable Stranger, at the discretion of the President.
2. If any Paper be read before the Society, it Shall be minuted, and all the principal parts entred in the Journal, to be read to the Society the next Meeting which will give an Opportunity to the Society to debate of the particulars.
3. That no person or persons talk to one-another at the Meetings, or So loud as to interrupt the business of the Society but address themselves to the President
4. That all papers to be read be translated into English.[181]

At some point, Newton also established a practice that the mace be placed on the table only when the president was in the chair. Sloane's first act after his election was to decree that the mace appear on the table at every meeting regardless of the presiding officer.[182] Stukeley's description of how Newton conducted meetings used terms consistent with the decrees of 1711.

---

[178] *Stukeley*, p. 17.　　[179] *Stukeley*, p. 17.　　[180] *CM 2*, 236. *JBC 10*, 262.
[181] *CM 2*, 233–4. *JBC 10*, 256.　　[182] 29 March 1727; *CM 2*, 385.

Whilst he presided in the Royal Society, he executed that office with singular prudence, with a grace and dignity – conscious of what was due to so noble an Institution – what was expected from his character. . . . There was no whispering, talking, nor loud laughters. If discussions arose in any sort, he said they tended to find out truth, but ought not to arise to any personality. . . . Every thing was transacted with great attention and solemnity and decency; nor were any papers which seemed to border on religion treated without proper respect. Indeed his presence created a natural awe in the assembly; they appear'd truly as a venerable *consessus Naturae Consiliariorum*, without any levity or indecorum. [183]

The style had been captured perfectly in the portrait of himself by Jervas that Newton presented to the society in 1717. Flat, expressionless, formal, the portrait presents us, not with a man, but with the reigning monarch of the philosophic world.

The style was imperious as well as imperial. In 1712, Newton reported that one of the auditors had hesitated to sign at the last audit unless he saw the original orders of the council for expenditures and determined that the councils in question were regularly summoned and had a quorum present. "It was Ordered that hereafter the accounts Shall be passed as usual and that the President's Signing an Order in Council Shall be a Sufficient Order to the Treasurer as it hath been heretofore." [184]

In keeping with the imperial style, the society under Newton continued, as it had always done, to court membership by the prominent figures of the day, men such as Lord Chief Justice Trevor, Lord Chief Justice Parker, and Lord Harley. As soon as the Hanoverian dynasty succeeded to the throne, the society desired Newton, no doubt at his own suggestion, to wait on the king and prince to ask the honor that the king sign as patron and the prince as fellow. [185] Mr. Williams, "a Black Native of Jamaica," was, on the other hand, the first candidate since the society's inception who was denied election. [186] In 1713, when the Duke D'Aumont attended the society, the whole meeting was devoted to his entertainment; he sat at the table with the president. A year later, the council summoned a special meeting after the summer adjournment to elect and admit Prince Menshikov, who had signified his desire of such. [187]

No doubt the special attention Newton received from visitors encouraged him to assume airs. When Claude Joseph Geoffroy, the chemist, and Pierre Rémond de Monmort visited the Royal Society, it was primarily Newton they wanted to meet. When Signor

---

[183] *Stukeley*, pp. 78–81.   [184] 20 March 1712; *CM 2*, 252–3.
[185] *CM 2*, 177, 252. *JBC 11*, 122.   [186] *CM 2*, 299.
[187] *JBC 10*, 483. *CM 2*, 284. *JBC 11*, 17. Cf. Menshikov to Newton 23 Aug. 1714 and Newton to Menshikov, 25 Oct. 1714; *Corres 6*, 171–2, 183–4.

Figure 13.4. Newton at seventy-one. A miniature by Christian Richter, 1714. The original painting appears to have been lost; it survives in this engraving of 1785. (By permission of the David Eugene Smith Collection, Rare Book and Manuscript Library, Columbia University.)

Bianchi, the Resident of Venice, had to return home before his election to the society could be completed, he let it be known that it would be some satisfaction to learn of the council's approval of him from Newton, and much greater satisfaction "to see the renown'd President" before he left. John Chamberlayne wrote to arrange, not a meeting, but an "Audience."[188] The Dutch physicist Willem 'sGravesande and the French astronomer Jacques-Eugène d'Allonville both arranged to meet Newton and to be elected to the Royal Society when they visited England.[189] Newton stood alone among English natural philosophers, without a peer and without a rival. It was natural that he dominated the Royal Society. Only the fact that there was opposition to his dominance was strange.

Financially, the Newtonian Royal Society was secure and intended to stay so. Shortly before Crane Court was paid for, bequests began to pour in. Francis Aston left them his library, lands in Lincolnshire, and £445. Dr. Thomas Paget bequeathed two London houses worth nearly £100 per annum. Robert Keck left £500. The society was able to invest £1,500 in Bank annuities and South Sea bonds and to settle a salary on each of its secretaries.[190] When the council considered suing for arrears in dues, the society's manifest solvency may have convinced members not to take the threat seriously.[191]

In the years following 1710, the level of the meetings continued to rise. Hauksbee usually had an experiment, on electricity or capillary action or the refraction of light. Denis Papin, who had served as curator of experiments for a time in the seventeenth century, frequently contributed. Papin's primary goal was to gain the financial support of the society for one of his projects, such as a new furnace or a joint stock company to encourage inventions, primarily those of Papin. Although the society paid him £10 in both 1711 and 1712 for his "Services," they refused to become financially involved in his projects, and Papin eventually disappeared.[192] Immediately after the climactic annual meeting of 1710, whether by coincidence or by design, Dr. Douglas reappeared and began to present anatomical demonstrations on a regular basis – always a popular topic in a society heavily loaded with doctors. Whatever the topic, the papers began to sound more prepared, as far as one can judge them through the minutes, and the meetings consequently had less rambling miscellaneous commentary.

[188] 21 April–5 May 1715; *JBC 11*, 60–2. Chamberlayne to Newton, 21 Aug. 1710; *Corres 5*, 59.

[189] 'sGravesande to Newton, 28 May 1714; Fontenelle to Newton, 9 June 1714; *Corres 6*, 144–6.

[190] *JBC 11*, 79, 194. *CM 2*, 297, 307, 318, 323, 353. Cf. Lyons, *Royal Society*, pp. 126–7.

[191] *CM 2*, 281–3.

[192] *JBC 10*, 214, 278–9. *CM 2*, 240, 252.

In pursuit of point 2 in the four principles of January 1711, the minutes suddenly became much fuller than they had been. This policy may have derived from Derham's letter which complained of Sloane's management of the society through the minutes, which were never read before that time.[193] Whatever the source of the policy, meetings began to open with the reading of minutes, and the practice did foster more consideration of papers and more thoughtful discussion.

Some of the topics deemed suitable for discussion are surprising to the more reticent twentieth century. In 1712, an account was read of what was observed in opening Lady Packington's body. Lady Packington had died in 1679; perhaps a suitable interval had passed. The dukes of Queensbury and Leeds, however, had their intestines described in detail before they had time to settle comfortably into their graves.[194] In 1716, Dr. Sloane triumphantly displayed the stone found in the bladder of the late Dr. Hicks, sometime Dean of Worcester. "It was more than Ordinary Solid and on both sides was Rough and has Several protuberant Asperities which had made themselves Cells or Receptacles in the Substance of the Blader by which means they were Connected."[195]

With the sickness and death of Hauksbee in 1713, the meetings declined anew. Newton proposed the appointment of four committees "for the more Effectual promoting of the Ends of the institution of this Society . . .": one to survey new books for extraction and presentation at the meetings, one to plan experiments, one for anatomical subjects, and one to inquire into extraordinary phenomena such as earthquakes and meteors.[196] The proposal recalled his earlier "Scheme for establishing the Royal Society." The committees were duly appointed only to sink immediately from sight, but early in 1714 Newton discovered Hauksbee's replacement, J. T. Desaguliers. Really he found two replacements, for in William Cheselden he appeared to have the anatomical demonstrator he had sought in vain. In the summer of 1714, the council voted to excuse both from the weekly payments in view of their expected usefulness to the society. Cheselden never worked out, probably because his surgical practice thrived too well. Now and then, rather infrequently, he performed demonstrations at meetings, but the society under Newton never did acquire the anatomical demonstrator he wanted. Desaguliers, however, became a fixture at the meetings, where he carried out sets of experiments intimately related to various aspects of Newtonian natural philosophy. Some of his experiments, such as the transmission of heat through a vacuum, influ-

[193] *Corres 5*, 21.   [194] *JBC 10*, 314, 355–6, 418–19.
[195] *JBC 11*, 104.   [196] 30 March 1713; *CM 2*, 261–2.

enced Newton's views, and others found their way into the third edition of the *Principia*. With Desaguliers as the mainstay and other young Newtonians such as Jurin, Taylor, and Keill frequently contributing papers, the meetings picked up again for a few years. In an amateur society their vitality was always tenuous, however, and in Newton's old age, when a firm hand no longer grasped the helm, they slackened once more.

Newton himself fairly expanded during these years as he entered vigorously into the weekly discussions. Many topics engaged him – watches, clepsydras, barometers, thermometers, magnets and the law of magnetic attraction, the saltiness of the sea, the mechanism of adjustment of the eye, the parallax of Mars. When Hauksbee constructed a device to measure the refractive power of fluid media, Newton urged him to try it on butter of antimony and told him of a cement for the hollow glass prism which butter of antimony would not dissolve. Once started, he could not stop and suggested further that Hauksbee make a tiny prism from the pellucid parts of arsenic to measure its refractive power, something he wanted to know because of arsenic's "great Specifick Weight."[197] Chemical matters tended to seize his imagination and stir his tongue. In 1714, he proposed an experiment to make a digestive heat by the fermentation of bran and water. Such a heat was suitable for chemical experiments and both cheaper than fire and easier to use. By the next week, heat was already observable in a sample of fermenting bran, and Newton told the society it would last ten to twelve days, and by the substitution of successive vessels it could be continued indefinitely. As it turned out, there was one hitch, which Newton himself had mentioned earlier, an overabundance of maggots.[198] The subject of gold reminded him of a recent struggle with the Goldsmiths, and he told the society that he had some gold more than 24 carats fine.[199] On occasion, he even invaded the doctors' learned discussions. "The President said that he had Seen a Large Worm in a Dogs Kidney, Rouled up Spirall and takeing up the Greatest part of the Kidney." A week later he recalled for their edification "Severall Worms with Severall Legs that had been Lodged near the Nose in a Dog."[200] Perhaps from his days as a sizar in the kitchen at Trinity, he also remembered when he had seen the heart taken from a live eel and cut into pieces. The pieces continued to beat in unison; but when a drop of vinegar was put on one, it stopped immediately.[201]

Amid Newton's various successes in the Royal Society, one failure continued to nag him, the edition of Flamsteed's observations. With

[197] 12 April 1711; *JBC 10*, 271.
[198] 21 and 28 Jan. 1714; *JBC 10*, 540–1.   [199] 26 June 1712; *JBC 10*, 410.
[200] 22 and 29 March 1711; *JBC 10*, 264.   [201] 13 Nov. 1712; *JBC 10*, 428.

Figure 13.5. Newton at seventy-one. An ivory bust sculpted by David Le Marchand in 1714. Le Marchand carved at least two busts, which are nearly identical to each other, in 1714. (By kind permission of Sotheby Parke Bernet & Co.)

the death of Prince George in 1708, the authority of the referees had lapsed and the project had come to a halt. Nothing happened for two years except that Flamsteed seized the respite finally to complete his catalogue of fixed stars to his own satisfaction.[202] As far as we know, the two men did not communicate. From Flamsteed's point of view, Newton had used his power to obstruct the publication of a work which would detract from his renown. Newton saw the episode with the colors reversed, and in the end he could not tolerate the thought that Flamsteed had denied him the observations he needed. Indeed he now needed them even more, for he had finally engaged himself to publish a second edition of the *Principia*. On 14 December 1710, before a special meeting of the council, Dr. Arbuthnot, one of the former referees who was also physician to Queen Anne, suddenly produced a warrant by which the queen appointed the president of the society and such others as the council thought fit to be "constant Visitors" of the Royal Observatory.[203] The word "visitor" in this usage derived from ecclesiastical sources and referred to one authorized to visit an institution officially for the purpose of inspection and supervision in order to prevent or remove abuses or irregularities. Visitors could be appointed either for specific occasions or for continuing supervision. Hence the adjective "constant" placed the Observatory permanently under the control of the Royal Society. The Journal Book captured the essence of the new relation when it summarized the warrant to state that the president and those joined with him were made "Visitors and Directors" of the Observatory.[204] The warrant required the visitors to demand yearly copies of the observations made by the Astronomer Royal and authorized them to direct him from time to time to make what observations they wished. In the years ahead, Newton did not neglect to exercise both powers over Flamsteed. We know nothing of the background of the warrant. Flamsteed never doubted that Newton engineered it to put him and the Observatory at the president's mercy.[205] Newton consistently used it to that end, and it stretches credulity to question Flamsteed's account.

On the virtual morrow of the warrant, on 14 March 1711, Dr. Arbuthnot wrote to Flamsteed that the queen had "commanded" him to complete the publication of the *Historia coelestis*. He asked Flamsteed to deliver the material still lacking, primarily the catalogue of the fixed stars.[206] A question exists about the order Ar-

---

[202] Flamsteed to Sharp, 24 March 1709; *Baily*, p. 270.

[203] *CM* 2, 229–31. The warrant is published in *Corres 5*, 79–80.

[204] 20 Dec. 1710; *JBC 10*, 252–3.

[205] Flamsteed to Sharp, 23 Jan. 1711, and Sharp to Flamsteed, 8 March 1711; Cudworth, *Sharp*, pp. 99–100. Cf. Flamsteed's diary for 29 March 1711; *Baily*, p. 227.

[206] *Baily*, p. 280.

buthnot claimed to act under. No order, either to him or to anyone else, to resume the publication has ever been found. There is no reason to doubt that an order in some form existed, however. Flamsteed did not question it, and the Treasury later accepted the bill for the costs of publication. To whom was the order directed? On 21 February, the society instructed Sloane to write to Flamsteed for the missing part of the catalogue "now Printing by Order of the Queen . . ." Sloane wrote the letter with Newton's help, but the society later substituted Arbuthnot's letter for it.[207] These events suggest that the order was directed to the Royal Society, probably pursuant to its new role as visitor to the Observatory, and that Arbuthnot was drawn in as a cover since he was known to be on friendlier relations with Flamsteed than others were. Certainly it was Newton, not Arbuthnot, who ran the ensuing show, and it was Newton whom the Treasury eventually consulted about the bill.

Flamsteed did not respond to Arbuthnot immediately. He must have inquired discreetly what was happening. According to his letter to Sharp in the middle of May, he learned on 25 March that the printing of the catalogue had already commenced.[208] That same day, he finally sent to Arbuthnot a reply he had composed earlier, welcoming the news that publication had resumed and informing him that he had completed the catalogue as far as he thought necessary. Meanwhile, "ye good providence of God (that has hitherto conducted all my Labours; and I doubt not will do so to an happy conclusion)" had led him to further fresh discoveries. Finding how far the planets' observed places differed from existing tables, he had begun to construct new ones. He needed help to complete them and make the work worthy of the queen's patronage and the memory of her consort. He asked Arbuthnot to meet him to discuss the matter.[209] The mask now dropped, and Flamsteed received a furious letter, not from Arbuthnot but from Newton, whose patience did not survive even the first stage of the renewed trial.

S^r

By discoursing wth Dr Arbothnot about your book of observations wch is in the Press, I understand that he has wrote to you by her Ma^{ts} order for such observations as are requisite to complete the catalogue of the fixed stars & you have given an indirect & delatory answer. You know that the Prince had appointed five gentlemen to examin what was fit to be printed at his Highness expence, & to take care that the same should be printed. Their order was only to print what

---

[207] 21 and 28 Feb. and 7 and 15 March 1711; *JBC 10,* 260–2.   [208] *Baily,* pp. 291–2.
[209] *Baily,* 280–1. The amanuensis dated the letter 23 March, but at the bottom Flamsteed entered the date 25 March, indicating that he held it until then. (*Corres 5,* 100–1).

they judged proper for the Princes honour & you undertook under your hand & seal to supply them therewith, & thereupon your observations were put into the press. The observatory was founded to the intent that a complete catalogue of the fixt stars should be composed by observations to be made at Greenwich & the duty of your place is to furnish the observations. But you have delivered an imperfect catalogue wthout so much as sending the observations of the stars that are wanting, & I heare that the Press now stops for want of them. You are therefore desired either to send the rest of your cataloge to Dr Arbothnot or at least to send him the observations wch are wanting to complete it, that the press may proceed. And if instead thereof you propose any thing else or make any excuses or unnecessary delays it will be taken for an indirect refusal to comply wth her Maj^{ts} order. Your speedy & direct answer & compliance is expected.[210]

Although Flamsteed was certainly stalling for time, his letters to Sharp during 1710 demonstrate that the new planetary tables were not a pretence.[211]

Arbuthnot also replied, he all suave diplomacy, eager to aid Flamsteed's efforts but meanwhile anxious to have the catalogue that the queen had ordered him to print. Flamsteed again suggested a meeting, and finally they did meet, at Garraway's coffeehouse on 29 March.[212] According to Flamsteed's diary, he agreed to deliver the rest of the catalogue. He asked if the catalogue was already in the press. Arbuthnot assured him that it was not. As they left, Flamsteed said he wished to proceed with the edition *"provided that I might have just, honourable, equitable and civil usage:* which he assured me I should . . ."[213] Flamsteed underlined the condition, which stated the essence of the problem in its most succinct form.

Four days later, he confirmed the rumor that the catalogue was already printing by getting a completed sheet into his hands, and shortly thereafter a second. More than that, he found that there were errors and alterations in them, and he soon learned that his enemy Halley was the author of the changes. Flamsteed claimed there were errors as great as fifteen seconds in the positions of the stars – in a work that sought to be accurate to five seconds. No one has undertaken the drudgery of comparing the two catalogues in question with a modern one; without such it is impossible to judge what foundation his claim had. The alterations consisted of a few changes in the Ptolemaic names of stars. For Flamsteed, the long

---

[210] *Corres 5,* 102. Perhaps this letter was never sent. We only know of Newton's draft, and Flamsteed did not mention it anywhere.
[211] Flamsteed to Sharp, 11 Feb., 14 July, 20 Sept. 1710; *Baily,* pp. 273, 276, 277.
[212] *Baily,* pp. 281–2.   [213] *Baily,* p. 226.

tradition in stellar catalogues held great importance. Astronomers had used Ptolemy's names for nearly two thousand years, translating them into every European tongue. If one changed the names, one would render the generally received forms of the constellations and the observations of the ancients incomprehensible.[214] Halley and Flamsteed did exchange letters in which they disputed the alterations in two constellations. It is difficult to avoid the conclusion that it was a tempest in a teapot. Moreover, Halley had corrected quite a few copying errors in Flamsteed's manuscript, an improvement by any criterion.[215] Flamsteed had stirred himself beyond the possibility of rational discussion by that time, however. He had heard – whether rightly or wrongly it is again impossible to say – that Halley was displaying sheets of the catalogue in coffeehouses and boasting of how many faults he had corrected.[216] For Flamsteed, any chance for accommodation had passed.

He confronted Arbuthnot with his lie about the catalogue and with Halley's errors and alterations. Arbuthnot evaded the lie, told Flamsteed the alterations were made to please him, argued that they did not matter, and concluded by bluntly informing him that they would calculate the rest of the catalogue from his observations if he did not send it.[217] More than the catalogue was involved. Though it had not come out in earlier correspondence, Flamsteed knew that Newton did not intend to publish all of his observations with the mural arc, observations which Flamsteed regarded as the empirical foundation of the catalogue and the voucher to its accuracy.[218] After thirty-five years of labor, he faced the brutal fact that his enemies held the power to publish his life's work in a form which, in his eyes, mutilated and spoiled it. Confronted with *force majeure,* Flamsteed gathered his righteousness about himself and for once transformed it into dignity and indeed heroism.

> I have now spent 35 years in composeing & Work of my Catalogue which may in time be published for ye use of her Majesty's subjects and Ingenious men all ye world over [he wrote to Arbuthnot]: I have endured long and painfull distempers by my night watches & Day Labours, I have spent a large sum of Money above my appointment, out of my own Estate to compleate my Catalogue and finish my Astronomical works under my hands: do not tease me with banter by telling me y^t these alterations are made to please me when you are sensible nothing can be more displeasing nor injurious, then to be told so.

---

[214] John Flamsteed, *Historia coelestis britannica*, 3 vols. (London, 1725), *1*, 156.
[215] *Baily*, pp. 287–90.   [216] *Baily*, p. 95.
[217] Arbuthnot to Flamsteed, 16 April 1711; *Baily*, pp. 282–3.
[218] Flamsteed to Sharp, 13 June 1709; *Baily*, p. 271. Cf. the judgment of Baily, an expert on the subject of catalogues, on the importance of publishing the observations (*Baily*, p. xli).

Make my case your own, & tell me Ingeniously, & sincerely were you in my circumstances, and had been at all my labour charge & trouble, would you like to have your Labours surreptitiously forced out of your hands, convey'd into the hands of your de[c]lared profligate Enemys, printed without your consent, and spoyled as mine are in ye impression? would you suffer your Enemyes to make themselves Judges, of what they really understand not? would you not withdraw your Copy out of their hands, trust no more in theirs and Publish your own Works rather at your own expence, then see them spoyled and your self Laught at for suffering it. . . . I shall print it alone, at my own Charge, on better paper & with fairer types, then those your present printer uses; for I cannot bear to see my own Labours thus spoyled . . .[219]

Withdraw his copy he could not. Newton had it. Withdraw cooperation he could, though only at the risk that his life's work might never see the light of day. The biographer of Newton unhappily searches in vain through the whole dismal chronicle of closet tyranny for an action one tenth as creditable.

The die was cast. Newton, with Halley following his direction, proceeded directly to complete the publicaton of the *Historia coelestis* early in 1712. A large folio volume, it began as Newton had always intended it should begin, with the catalogue of fixed stars, the copy Flamsteed handed to Newton in 1708 with the missing constellations filled in by 500 stars that Halley calculated from Flamsteed's observations. Thus the catalogue, like the one Flamsteed planned, increased the number of stars charted by astronomy from approximately 1,000 to approximately 3,000. The volume went on to the observations made before 1689, which had been printed in 1706–7, and then to observations of the planets, the sun and moon, and eclipses of the satellites of Jupiter made with the mural arc. Nowhere did Newton's determination to bend Flamsteed's work to his own purposes appear more clearly. He eliminated the observations of the fixed stars, the results of which appeared of course in the catalogue, and printed those of use to him. A shameful preface asserted that Flamsteed had not wanted to give his observations out and that only the order of Prince George and the industry of the referees had secured their publication. Flamsteed had planned a long Prolegomena to describe his methods of observation to justify the degree of accuracy the catalogue claimed: units of five seconds, an increase in accuracy of one order over previous catalogues. Not only did the preface say nothing to this effect, but it cast doubt on the whole enterprise by suggesting that Halley had had to correct various deficiencies. It even blamed Flamsteed for the lack of maps of the constellations.[220] Newton arranged for the government to pay Halley £150 for his work, £25 more than Flamsteed would have received for more work.[221] Newton was then revising the *Principia* for the second edition. As he had formerly done with Hooke, he went through the first edition systematically removing every reference to Flamsteed that he could. Since his successful attack on the comet of 1680–1 rested primarily on Flamsteed's observations, for which he had no substitute, he was not quite able to reduce him to nonexistence. He did cancel his name in fifteen places.[222]

Nor was Newton done yet. He intended to bring Flamsteed to heel. As Visitor of the Observatory, he summoned him to a meeting at Crane Court on 26 October 1711, with Sloane, Mead, and himself, to inform them if the instruments were in order and fit to carry on observations.[223] The meeting was a mistake. Of all the men alive, Flamsteed knew best how to drive Newton wild. He was never in better form. Crippled with gout so that he could manage the stairs only with help, he was well enough to bring his implacable righteousness to bear with devastating effect. "Y^e P^r ran himself into a great heat & very indecent passion," he reported to Sharp, not without satisfaction. He demonstrated that all the instruments were his own private property and thus not subject to the Visitor's authority. This nettled Newton, who said, "as good have no Observatory as no Instruments." Flamsteed complained of the publication of his catalogue, that they robbed him of the fruits of his labors – not an excessive description of the edition.

at this he fired & cald me all the ill names Puppy &c. that he could think of. All I returnd was I put him in mind of his passion desired him to govern it & keep his temper. this made him rage worse, & he told me how much I had receaved from ye Goverm^t in 36 yeares I had served. I asked what he had done for ye 500 ^{lb} per Annum y^t he had receaved ever since he setled in London. this made him calmer but finding him goeing to burst out againe I onely told him: my Catalogue half finished was delivered into his hands on his own request sealed up. he could not deny it but said Dr Arbuthnot had procured ye Queens order for opening it. this I am persuaded was false, or it was got after it had been opened. I sayd nothing to him in return but with a little more spirit then I had hitherto shewd told them, *that God (who was seldom spoke of with due Reverence in that Meeting) had hitherto prospered all my labours & I doubted not would do so to an happy conclusion,* took my leave & left them.[224]

220 John Flamsteed, *Historia coelestis libri duo* (London, 1712), "Praefatio ad lectorem," pp. i–vi.
221 Newton to Oxford, 14 Feb. 1712; *Corres 5*, 224–5. Cf. Arbuthnot to Newton, early 1712; *Corres 7*, 479–80.   222 *Var Prin 2*, 557–717.   223 *CM 2*, 246.
224 Flamsteed to Sharp, 22 Dec. 1711; *Baily*, pp. 294–5.

"God forgive him," he added ineffably to his account of the same meeting in his autobiography. "I do."[225]

Newton had one more try at Flamsteed. When he presented a copy of the second edition of the *Principia* to the queen in July 1713, she expressed her desire that the Royal Society "take care of M^r. Flamsteads Observatory at Greenwich."[226] The minutes of the Royal Society, to whom Newton immediately repeated the queen's instructions, state that she said it "of her own accord . . ." No doubt the queen always spoke of her own accord. Nevertheless it is hard to believe that Flamsteed's Observatory was the first thing on her mind and that she spoke with complete spontaneity. At any rate, the society immediately sent a committee to Greenwich to survey the instruments, and upon their report sent a letter to the Ordnance, the office responsible for the maintenance of the Observatory, recommending the repair and replacement of certain instruments.[227] If instruments were repaired or replaced, Flamsteed would lose his claim to immunity. In this situation, however, Flamsteed found himself with allies. The officers of the Ordnance, housed in the Tower, had a long history of conflict with Newton. Flamsteed took care to keep his "very good friends at the Tower" informed.[228] To the Royal Society's letter the Ordnance replied that it had no money for repairing instruments, and responses to the Ordnance were still being drafted in vain when the queen's death removed the mandate.[229]

Meanwhile Flamsteed was not idle. He had told Arbuthnot that he would print his catalogue as he wanted it at his own expense, and this he set out to do. In hot haste, he put the catalogue into final form, commenced printing in the summer of 1712, and had it done by the end of the year.[230] Not yet satisfied, he began printing his observations with the mural arc. At this point, fate stepped in. Queen Anne died in 1714. The Tory government fell, and the Whigs returned to power. The death of Halifax in the summer of 1715 removed Newton's chief contact with the new regime, while Flamsteed knew the Lord Chamberlain, the Duke of Bolton. Someone near the Lord Chamberlain indicated to Flamsteed that he could get the copies of the *Historia coelestis* that had not been distributed if he wished. He did wish. On 30 November 1715, Bolton signed a warrant to Newton, the other referees, and Churchill to deliver the three hundred remaining copies of the work to the author.[231] The referees replied that their

[225] *Baily*, p. 97.

[226] Newton's report to the Royal Society on 30 July 1713; *JBC 10*, 511.

[227] *Corres 6*, 24.

[228] The phrase comes from an earlier time and issue also related to Newton. Flamsteed to Sharp, 13 June 1709; *Baily*, p. 271.

[229] *Corres 6*, 69–70.   [230] *Baily*, pp. 98–100.   [231] *Corres 6*, 255–6.

Figure 13.6. Newton at seventy-five. Portrait by Thomas Murray, 1718. (Courtesy of the Master and Fellows of Trinity College, Cambridge.)

authority over the work had ended with the queen's death. Although Flamsteed had received a gratuity of £125, they added disingenuously, and had caused a great deal of trouble by handing in imperfect copy, "yet we designed to have begged of her Majestie the Remainder of the Copies for him."[232] No large degree of cynicism is

[232] *Corres 6,* 267–8.

required to doubt that statement. Finally, on 28 March 1716, Flamsteed got the copies in his hands. He separated out the catalogue and
the hundred and twenty pages of excerpts from his observations with
the mural arc and "made a *Sacrifice of them to Heavenly Truth.*"[233]
That is to say, he burned them. He devoted the brief time he had left
to printing his observations with the mural arc, and had virtually
completed them when he died in 1719. His two former assistants,
Joseph Crosthwait and Abraham Sharp, oversaw their completion
and the printing of the Prolegomena and other material that went
with the catalogue in volume 3 of the new work. In 1725, the *Historia
coelestis britannica* finally appeared in three volumes: volume 1, the
early observations salvaged from the 1712 edition; volume 2, the
complete observations with the mural arc; volume 3, the catalogue
and attendant materials. This was substantially the publication Flamsteed had always planned, and it is the form in which the *Historia* is
known and honored today. Eventually, Flamsteed's widow and
another assistant, James Hodgson, even published a brief version of
the charts of the constellations, the *Atlas coelestis*. Newton's effort to
subject Flamsteed to his will ended in complete failure, and the only
solace he could find was the presentation of a copy of the 1712
edition, bound in red leather and gilt, to the Royal Society in 1717, a
futile act of defiance which could not reverse his defeat.[234] Flamsteed
also found it impossible permanently to maintain his posture of lofty
dignity. In June 1716 he told Princess Caroline, who was visiting his
Observatory, that Newton was a great rascal who had stolen two
stars from him. Unfortunately, the princess could not restrain herself
from giggling.[235]

The publication of the *Historia coelestis* was the most unpleasant
episode in Newton's life. It was important both for the light it cast
on his character and for what it revealed of his relations with the
British scientific community. As for Newton, impatience with contradiction, which manifested itself in the young man in a readiness
to throw caution to the winds in challenging established authorities
such as Hooke, had become in his old age tyrannical will to domi-
neer, an unlovely trait which one cannot ignore. Perhaps the most
interesting facet of the Flamsteed episode, however, was its revelation of the extent to which Newton had failed to make himself the
dictatorial voice of British science. In 1709, two positions needed to
be filled, the minor post in the Mathematical School at Christ's
Hospital vacated by Samuel Newton's resignation, and the Savilian
chair of astronomy at Oxford left empty by Gregory's death. Newton recommended William Jones for the first. Although we do not

---

[233] *Baily*, p. 101.   [234] 21 Nov. 1717; *JBC 11*, 205.
[235] Princess Caroline to Leibniz, 15 June 1716; Gottfried Wilhelm Leibniz, *Die Werke von
Leibniz*, ed. Onno Klopp, 11 vols. (Hannover, 1864–88), *11*, 115.

know that he meddled with the Oxford appointment, Halley apparently pushed John Keill, who had already established his credentials as a Newtonian. Neither one obtained the appointment in question. Flamsteed's former assistant, James Hodgson, became the mathematical master at Christ's Hospital. John Caswell was appointed Savilian professor of astronomy, to Flamsteed's delight. When Caswell died three years later, Keill did succeed him.

One must not imagine that Flamsteed challenged Newton's position in the British scientific world. No one challenged it. He dwelt on a level apart, and by his intellectual attainment alone he should have dominated it more than he did. The annual elections at the Royal Society revealed as much as the two positions he failed to fill. Year after year, others obtained more votes for membership on the council than Newton did. The minutes recorded fifty-odd members present in 1714; Newton received 45 votes. In 1715, forty-nine members were present; Newton received 35 votes. William Derham returned to the council with 46. Newton did not attend the 1716 election. In 1723, when his senescence may have become a factor, only one of the eleven members of the council elected to continue received fewer votes.[236] Year after year, approximately a fifth or a quarter of the members voted against him. The membership was made up mostly of nonscientists, of course, and elections had nothing to do with scientific achievement. We can take the elections as a rough measure of the extent to which his despotic domination of the society alienated those who would not have hesitated to acknowledge his intellectual primacy.

[236] *JBC 11*, 30–1, 90–1, 145. *JBC 12*, 406.

14

The priority dispute

WELL before the contested edition of Flamsteed's *Historia coeles-tis* in 1712 brought that episode to a temporary conclusion, two new concerns, which would dominate Newton's life for more than five years, had imposed themselves upon him. In 1709, work began in earnest on a second edition of the *Principia*. In the spring of 1711, a letter from Leibniz to Hans Sloane, secretary of the Royal Society, inaugurated a heated controversy over claims of priority in the invention of the calculus. Moreover, a fourth problem of great import for Newton was also taking form. Already an ugly scene with Craven Peyton, the warden of the Mint, had signaled a deterioration of their relations which culminated in a major crisis in the Mint in 1714, when the battle with Leibniz was reaching its highest pitch. The Mint was the bedrock on which Newton's existence in London stood. Trouble there had to affect his whole life. In its intensity, the period from 1711 to 1716, succeeding more than a decade of relative calm, matched the great periods of stress at Cambridge, when his relentless pursuit of truth stretched him to the limit. The coincidence of these events, the demands they placed on Newton, may help to explain the furious scene with Flamsteed at Crane Court on 26 October 1711 and much else from these years not yet mentioned.

The second edition of the *Principia* began first. Newton had been talking about it almost from the day the first edition appeared, first with Fatio, then with Gregory. At that time he had begun to compile lists of corrections and alterations he intended to make. He had had a copy of the book specially bound with blank sheets between the printed pages on which he could enter corrections, and he had written emendations into at least one other copy.[1] Nevertheless, neither the hopes of Fatio to supervise the new edition nor Gregory's later ones had materialized, and with the move to London active plans languished for a time. They never died, however. In March 1704, about the time his *Opticks* appeared, Newton told Gregory that he intended to republish the *Principia,* although he did not say when, and in the summer Gregory reported that he was correcting the work at his leisure and had already advanced a good way. By July 1706, as Gregory also learned, Newton had reached Section VII in Book II, which he took to be the hardest part in the

---

[1] Cambridge University Library, Adv.b.39.1. Trinity College Library, NQ.16.200.

work. At other times, he had accorded that honor both to the theory of comets and to the theory of the moon. It is true, however, that Section VII, on the resistance of fluids to moving bodies, gave him the most trouble in the preparation of the second edition. In 1706 Gregory was hopeful that the new edition would go to press the following winter. He was disappointed. In April 1707, Newton had still not satisfied himself on the resistance of fluids, for which he would need new experiments, and he had still to revise his treatment of precession, though he told Gregory he felt himself to be in command of the problem. The following March, Gregory recorded that Newton was in fact engaged in experiments on resistance. In 1707 Newton told him that he did not intend to proceed with the edition for two years, but on 25 March 1708, in some excitement, Gregory reported that the new edition was finally in the press in Cambridge.[2]

There were certainly sound reasons not to put off the new edition any longer. Although Newton, with his inability ever to feel secure and to stop revising, could have continued to tinker with the text forever, others had need for a new edition. Copies of the first edition were hard to come by and consequently expensive. About 1710, William Browne, a student, paid two guineas for a copy, and another student who could not afford two guineas copied the entire book by hand. Cotes's preface to the second edition justified it for the same reason.[3] It is doubtful, however, that such considerations swayed Newton, if indeed he was aware of them. What did sway him were the manipulations of Richard Bentley, the academic entrepreneur now installed as master of Trinity College. Bentley had long made it his business to cultivate Newton and now aggressively maneuvered him into acquiescence in an edition which Bentley himself published through the university press. Someone asked Newton later why he had let Bentley, who did not understand the *Principia*, publish it: "– why said he [he] was covetous & loved mony & therefore I lett him that he might get mony."[4] Bentley's career at Trinity left no doubt as to his greed. Nevertheless, he was capable of acting on other motives, for he also tried desperately to reform the college and shake it out of its academic lethargy even while he was attempting to suck it dry. A great classical scholar

[2] *Hiscock*, pp. 16, 19, 36–7, 40, 41. In July 1708, DeMoivre also reported to Bernoulli that the edition was under way in Cambridge (Karl Wollenschläger, "Der mathematische Briefwechsel zwischen Johann I. Bernoulli and Abraham De Moivre," *Verhandlungen der Naturforschenden Gesellschaft in Basel, 43* [1931–2], 257).

[3] Browne mentioned his purchase in a speech before the Royal Society in 1772 (John Nichols, *Literary Anecdotes of the Eighteenth Century,* 9 vols. [London, 1812–15], *3,* 332). On the other student see I. Bernard Cohen, *Introduction to Newton's 'Principia'* (Cambridge, 1971), p. 200. *Prin,* p. xxxiii.

[4] *Keynes MSS* 130.6, Book 2; 130.7, Sheet 1.

himself, he valued learning, and his interest in the edition undoubtedly sprang as much from scholarly as from financial concern.

What is equally germane, Bentley wanted Newton's support in
the upheavals that his measures, both of reform and of finance,
were bound to cause in the college. It is relevant in this respect that
he always threw his weight behind Newton's candidacy for Parliament and that he commissioned a portrait of Newton by Sir James
Thornhill–"my Picture," as he called it–at about the time when
the edition commenced printing (Figure 14.1).[5] At the minimum,
Thornhill must have consumed a fair portion of Bentley's profit on
the edition. Meanwhile, in 1709 Bentley proposed to revise the
Trinity dividend, a typical project, which sought at once to enrich
him by increasing the master's proportion and to encourage learning by making advanced degrees a factor in the division which pure
seniority had hitherto governed. When the plan raised a storm in
the college, Newton himself, the doyen of English scholars, drew
up a memorandum justifying the proposal.[6]

Two years later, after the university ejected Whiston, Bentley
succeeded in enlisting Newton to support, however lukewarmly,
his candidate for the Lucasian chair, one of his protégés in Trinity, a
Mr. Hussey. Newton hardly knew Hussey. He had talked to him
once for about an hour to see if his mathematical learning was
enough for a teaching position in London and had found him "very
well qualified." Surely this was the Christ's Hospital post, and it
was hardly an enthusiastic endorsement when Newton proceeded
on those grounds to pronounce him fit to fill the Lucasian chair at
Cambridge. He was unable, he said, to comment on his other
qualifications.

> S^r [he added, in a postscript to Bentley]
> I have made a resolution not to meddle with this election of a
> Mathematick Professor any further then in answering Letters & have
> given this answer to some who have desired a certificate from me. In
> answer to your Letter & another I received from Mr Cotes I send you
> the two inclosed & give you both & your friends leave to shew them
> to any of the Electors.[7]

The letters were not warm enough to help anyone. Hussey failed to
win the nomination, which went instead to Nicholas Saunderson.
Another matter, one of more concern to Bentley, arose in 1714.

---

[5] Bentley to Newton, 20 Oct. 1709; *Corres 5*, 8. Thornhill produced two versions of this magnificent portrait.

[6] "The Case of Trin. Coll."; Trinity College Library, R.4.48^a. The manuscript is in New-ton's hand.

[7] Newton to Bentley, ca. November 1711; *Corres 7*, 479. Cf. Cotes to Newton, 25 Oct. 1711; *Corres 5*, 202.

Figure 14.1. Newton at sixty-seven. Portrait by Sir James Thornhill, 1710. (Courtesy of the Master and Fellows of Trinity College, Cambridge.)

When the reaction to his attempted revision of the Trinity dividend culminated in a hearing by the Bishop of Ely to expel him from office, he enjoyed the advantage of having Newton sit as one of the assessors.[8] Although we know nothing about Newton's appraisal of the charges, it does appear that only the death of Bishop More saved Bentley from ejection. In the meantime, the second edition of the *Principia* fit neatly into Bentley's sustained effort to win Newton's favor.

Whatever Gregory heard in March 1708, it is far from clear that Newton had fully committed himself to the Cambridge edition at that time. From Bentley he received in June a sample of the first sheet (eight pages) printed as Bentley intended the whole edition to be, with the added information that he had purchased a hundred reams of the paper, which came from Genoa.

I hope you will like it & ye Letter too. wch upon trials we found here to be more suitable to ye volume than a greater, & more pleasant to ye Eye. I have sent you like wise ye proof sheet, y^t you may see, what changes of pointing, putting letters Capital, &c I have made, as I hope, much to ye better. This Proof sheet was printed from your former Edition, adjusted by your own corrections and additions. The alterations afterwards are mine: which will shew & justify themselves, if you compare nicely the proof sheet with ye finishd one. The old one was without a running Title upon each page, wch is deformd. Y^e Sections only made wth Def. I. Def. II. which are now made full & in Capitals DEFINITIO. I. &c. Pray look upon *Hugenius de Oscillatione*, wch is a book very masterly printed, & you'l see that is done like this. Compare any period of ye Old & New; & you'l discern in ye later by ye chang of Points and Capitals a clearness and emphasis, y^t the other has not: as all y^t have seen this specimen acknowledg. Our English compositors are ignorant & print Latin Books as they are used to do English ones; if they are not set right by one used to observe the beauties of ye best printing abroad. In a few places I have taken ye liberty to chang some words, either for ye sake of ye Latin, or ye thought it self as y^t in p. 4. *motrices, acceleratices* et *absolutas*. I placed so because you explain them afterwards in y^t order. But all these alterations are submitted to your better judgment; nothing being to be wrought off finally without your approbation.[9]

Those who frequent used-car lots will recognize the tone of the letter, the attempt to determine the decision by acting as though it were already made. It may be that the tone, and what it implied for

---

[8] Sir James Montague, Halifax's brother, and Samuel Clarke were also assessors. *Edleston*, pp. lxxv–vi. James H. Monk, *The Life of Richard Bentley, D.D.*, 2nd ed., 2 vols. (London, 1833), *1*, 357–65.

[9] *Corres 4*, 518–19. An account of the press for the year ending 3 November 1708 showed the cost of setting the sheet and printing twenty-five copies of it. (*Hiscock*, p. 41).

the edition if he acquiesced, put Newton off, because nothing happened for another year. When the edition did in fact begin to move in 1709, a new figure, Roger Cotes, not a stylist like Bentley but a mathematician like Newton, was found interposed between them and actively in charge while Bentley stayed in the background.

Roger Cotes, twenty-seven years old in 1709, was one of Bentley's young followers in his struggle to inject new life into seniority-ridden Trinity. Through the tutoring of his uncle, the Reverend John Smith (father of Robert Smith, who would succeed Cotes at Cambridge and establish a modest reputation as a scientist), Cotes had made great progress in mathematics before his admission to Trinity in 1699. Bentley spotted him early and tapped him in 1705 for the newly created Plumian professorship of astronomy before Cotes had yet incepted M.A. Needless to say, Cotes gave his full support to Bentley's attempted reforms of college life. Through Bentley he met Newton. To accompany the Plumian chair and to make Trinity a center of scientific life, Bentley set out to construct an observatory above the great gate of the college. Early in 1708, on a visit to London, Cotes received Newton's promise of a pendulum clock, to be constructed at a cost of at least £50, for the observatory.[10] He had to wait seven years before he finally received it. Meanwhile, in 1709, Bentley, who could command rather than ask Cotes, put him in charge of the edition.

In March, Bentley sent Cotes an amended copy of the *Principia* to study, and in May he wrote that Newton would be glad to receive him in London, where he would hand over a large part of the book corrected as he wanted it to be printed.[11] Accordingly, Cotes traveled to London in July, but Newton, repeating a well-established pattern, found that he was not after all ready yet to part with the copy. He promised to send it in a fortnight. Cotes restrained himself a month before he finally wrote on 18 August.

S[r].

The earnest desire I have to see a new Edition of Y[r] Princip. makes me somewhat impatient 'till we receive Y[r] Copy of it which You was pleased to promise me, about the middle of the last Month, You would send down in about a Fourtnights time. I hope You will pardon me for this uneasiness from which I cannot free my self & for giveing You this Trouble to let You know it.

He went on to tell Newton that he had been checking the second corollary of Proposition XCI and found by the table in "De quadratura" that it was correct, though he did find two errors in the

---

[10] Cotes to John Smith, 10 Feb. 1708; *Edleston*, pp. 197–8.
[11] Bentley to Henry Sike, 31 March 1709; *Edleston*, p. xvi. Bentley to Cotes, May 1709; *Edleston*, p. 1. See A. R. Hall, "Correcting the Principia," *Osiris*, 13 (1958), 291–326.

Figure 14.2.   Newton at sixty-seven. Portrait by Sir James Thornhill, 1710. (Courtesy of Lord Portsmouth and the Trustees of the Portsmouth Estates.)

table of quadratures.[12] Cotes had to wait nearly two months more before he received for his pains the manuscript for roughly the first half of the book, and soon thereafter a reply.

> I thank you for your Letter & the corrections of yᵉ two Theorems in yᵉ treatise de Quadratura. I would not have you be at the trouble of

[12] *Corres* 5, 3–4.

examining all the Demonstrations in the Principia. Its impossible to print the book w^{th}out some faults & if you print by the copy sent you, correcting only such faults as occurr in reading over the sheets to correct them as they are printed off, you will have labour more then it's fit to give you.[13]

It was Bentley who replied to this.

You need not be so shy of giving Mr Cotes too much trouble: he has more esteem for you, & obligations to you, than to think y^t trouble too grievous: but however he does it at my Orders, to whom he owes more than y^t. And so pray you be easy as to y^t; we will take care y^t no little slip in a Calculation shall pass this fine Edition.[14]

Bentley's orders or no, Cotes had his own ideas of what his task entailed, and before he was done, he forced on Newton detailed reconsideration of the work far beyond what Newton had had in mind.

All of this came later, however. Book I involved little serious alteration. Cotes found nothing on which to comment, and it moved forward swiftly. Newton's letter to Cotes in October seemed to assume that printing had not yet started. Bentley, who did not intend to give him a chance to hesitate further, wrote back to let him know that in fact they had already set page 50 and expected to continue at that pace. He admonished Newton to take care to have material ready for them. The woodcuts had been prepared in London under Newton's supervision. Newton suggested that Mr. Livebody, who made them, come to Cambridge to supervise their setting in the pages during composition. Bentley would have none of it.

I proposed to our Master Printer to have Lightbody [*sic*] come down & compose, which at first he agreed to; but the next day he had a character of his being a mere sot, & having plaid such pranks y^t no body will take him into any Printhouse in London or Oxford; & so he fears he'll debauch all his Men.[15]

No letters between Newton and Cotes during the following six months survive, and from the content of Cotes's letter on 15 April 1710 it is unlikely that any were sent. He had been making small alterations for the sake of clarity and elegance, he informed Newton, and also for the sake of truth, but he had not wanted to trouble him with every particular. At that point they had completed page 224 – well into Book II and nearly half of the edition's ultimate 484 pages – and they were approaching the end of the copy Newton had supplied. If Book I had presented few problems, Book II, on motions in and of fluid media, offered more. Cotes discussed several points in the Scholium to Proposition X.[16] The letter revealed his careful attention to the technical details of a difficult text,

13 Newton to Cotes, 11 Oct. 1709; *Corres 5*, 5.
14 Bentley to Newton, 20 Oct. 1709; *Corres 5*, 7.   15 *Corres 5*, 7–8.   16 *Corres 5*, 24–6.

though as it happened – a matter of some significance – he failed to find a major error in the proposition itself. Subsequent letters, which now became quite frequent, bombarded Newton with problems in the first half of Book II to which he did not immediately respond. It was a subject he confessed, "from wch I have of late years disused myself . . ."[17] Though he accepted some of Cotes's corrections, he rejected others, only to have Cotes return again to the attack. This was not a treatment to which Newton had accustomed himself over the years; and even though he had finally to acknowledge Cotes's reasoning, he clearly did not enjoy being made to run the gauntlet.

> Mr Professor
> I have reconsidered the 15th Proposition with its Corollaries & they may stand as you have put them in your Letters.[18]

By the middle of May, when a problem in Section IV hung them up, the press caught up with the correction of the text and came briefly to a halt. After they set Section IV to rights, Section VI, the beginning of Newton's theory of the resistance of fluids to the motion of projectiles, stalled both them and the press again. "You need not give your self the trouble of examining all the calculations of the Scholium [at the end of the section]," Newton advised Cotes. "Such errors as do not depend upon wrong reasoning can be of no great consequence & may be corrected by the Reader."[19] Cotes gave himself the trouble nevertheless. In a short time they solved the problems of Section VI, and on 30 June Cotes announced that the press had set all the copy Newton had furnished and he was going home to Leicestershire for six weeks for a vacation.[20] Section VI of Book II had been completed. In nine months, they had set 296 pages, not far short of two-thirds of the total.

The correspondence of the previous three months had been a new experience for Newton. Unwillingly at first, he had allowed Cotes to draw him into a genuine scientific exchange which finds no parallel, except for later discussion with Cotes, in the whole of his correspondence. His letters at first were brusque and cold, even to the point of being curt. By June, however, he was beginning to enjoy the discussion as Cotes exposed him again to the original excitement of his great work. He did not repeat the derisory greeting to "Mr Professor." Indeed, he thanked Cotes more than once for his corrections, and his reply to the letter of 30 June, besides its promise to have the rest of the corrected text ready, concluded with a warm salutation.

[17] Newton to Cotes, 1 May 1710; *Corres 5*, 31–2.
[18] Newton to Cotes, 13 May 1710; *Corres 5*, 35–6.
[19] Newton to Cotes, 15 June 1710; *Corres 5*, 51.　　　[20] *Corres 5*, 54–5.

I am wth my humble service to your Master [Bentley] & many thanks to your self for your trouble in correcting this edition

S^r

Your most humble servant

Is. Newton.[21]

In fact Cotes extended his vacation to more than two months, finally returning to Cambridge early in September. True to his promise, Newton immediately dispatched a further batch of the text which included the rest of Book II and Book III as far as Proposition XXIV. It began with Section VII, the heart of Newton's theory of the resistance of fluids, the topic, he had told Gregory, which was giving him the most trouble. No part of the *Principia* had been more imperfect. It was an important topic, moreover, for it provided the heart of his argument against Cartesian natural philosophy. In the first edition, a dubious analysis had arrived at the conclusion that the resistance to a spherical projectile in a medium such as water is to the force that would generate or destroy its motion during the time it moves a distance equal to two-thirds its diameter as the density of the medium to the density of the projectile.[22] Alas, his experiments with pendulums had seemed to reveal that the resistance was only about one-third his calculation, and he had concluded that two-thirds of the motion impressed on the medium by the front of the projectile must be restored to it by the pressure of the medium on its rear – a curious return to the outdated doctrine of antiperistasis.[23] Fundamental to the analysis was Proposition XXXVII, on the velocity with which water runs from a hole in the bottom of a tank. Newton's strategy was to imagine a body suspended in the middle of the stream and to calculate the pressure on it. Invoking the relativity of motion, he then concluded that an equal body moving with equal velocity through water at rest experiences a resistance equal to the pressure. The analysis of water running from a tank, the foundation of the whole theory, had arrived at the conclusion that its velocity equals that of a body falling from half the height of the water's surface within the tank. A simple experiment in which he had measured the quantity of water that flowed in a given time had seemed to confirm the conclusion.

Not much of anything in Section VII was satisfactory. Newton did not trust the experiments with pendulums. An anecdote by Conduitt (cited above) indicates that he devised a different experiment to test resistance before he left Cambridge. He had a square

---

[21] Newton to Cotes, 1 July 1710; *Corres 5*, 56.

[22] In the first edition he formulated his conclusion in a different manner from this second-edition form (*Var Prin 2*, 781–2). [23] *Var Prin 1*, 460.

vessel 9 inches across and 9 feet 6 inches deep constructed, and in it he timed the fall through water of wax balls weighted with lead shot. Gregory reported that Newton was making further experiments on resistance in 1708, perhaps those with a vessel 15 feet 4 inches deep, which the second edition included. In May 1710, he indicated to Cotes that he was waiting on the outcome of still more experiments. Undoubtedly he referred to those Hauksbee reported to the Royal Society on 7 and 14 June.[24] Hauksbee set up a small platform in the dome of St. Paul's, 220 feet above the floor. A mechanism allowed someone on the floor to pull a latch, which permitted the platform to turn on hinges dropping balls that rested on it. The same pull started a pendulum that timed the fall. The minutes of the Royal Society mentioned three substances of different densities, glass balls, cork, and quicksilver. In the *Principia*, Newton cited the results only for small glass balls filled with mercury and glass bubbles 5 inches in diameter.

As for the theory, Fatio, who could never sufficiently admire his own brilliance in this regard, had convinced Newton early in the 1690s that water must flow out of a tank with a velocity equal to that, not of a body falling from half the height, but of one falling from the full height of the water in the tank. Fatio supported his conclusion by an experiment in which a small jet directed upwards from the bottom of a tank of water rose to the level of the surface. In the copy he sent to Cotes in September 1710, Newton completely reconstructed Section VII to yield a theory which accounted with considerable precision for the results of the various new experiments, both in water and in air. Central to it was a new Proposition XXXVI (as the first edition's Proposition XXXVII was now numbered).

We do not know exactly what the manuscript of Proposition XXXVI that Cotes received contained, but we can make an informed guess from Cotes's reply and from the final version. To attain the higher velocity of efflux, Newton simply imagined the water to fall in a contracting cataract from the surface to the hole. To explain his meaning, Newton later used a thought experiment which was probably in the text Cotes received. Imagine that a cylinder of ice, *APQB*, of the same dimensions as the vessel, descends with a slow uniform motion, melting instantly at the surface *AB*. The downward velocity of the water at *AB* is equal to that of a body falling from the plane *KIL* (Figure 14.3). When the diameter of the tank is large in comparison to the diameter of the hole, the height *IH* is small. In any case, *KIL* represents the surface of water in the tank as far as the calculation of the efflux is concerned.

[24] *JBC 10*, 243. *Prin*, pp. 361–2.

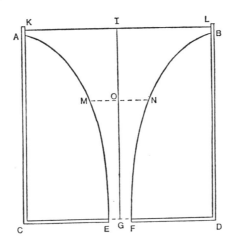

Figure 14.3. The cataract as a device to calculate the velocity of water flowing from a hole. (From Westfall, *Force in Newton's Physics.*)

Imagine further that water *AMEC* and *BNFD* is frozen to form a funnel of ice through which the remaining water flows without friction. Every drop of water falls freely with a uniformly accelerated motion. The shape of the funnel of ice represents the smaller circles through which equal volumes of accelerating water can pass in a unit of time. Obviously the water flows from the hole *EF* with the velocity of a body falling through the height *IG*, since it is in fact such a body, and Newton had the result he desired. Though he arrived at the higher velocity, the analysis made use of the same concepts drawn from forces applied to discrete bodies that he had tried on the problem in 1686. The concepts were no more adequate to the problem in 1710 than they had been before.

For that matter, his ensuing treatment of resistance continued to use the similarly inadequate model of impact. "The pressure arising from the motion of the body," he asserted, "is spent in generating a motion in the parts of the fluid, and this creates the resistance."[25] Neither the analysis of water flowing from a tank nor the theory of resistance long survived. We may take it as a measure of his continuing facility at the age of seventy that from two faulty premises he succeeded in extracting an exact agreement between theory and experimental results.

Cotes did not have our advantage of hindsight, but he had other reasons for disliking what he received.

I confess I was not a little Surprized upon yᵉ first reading of Prop: 36 . . . One of my greatest difficultys was an Experiment of Monsr.

25 *Prin*, p. 348.

Marriotte which he says (pag. 245 Traite du Mouvm: des Eaux) he often repeated wth great care. By his Experiment I concluded y^t the Velocity of the effluent water was equall to y^t gotten by a heavy Body falling but from half the height of the Vessel, as You had determined in ye former Edition of Y^r Book.[26]

Newton replied that he had not seen Mariotte's book but that something must have been wrong with his experiment.

For I have seen this experiment tried & it has been tried also before the Royal Society, that a vessel a foot & an half or two foot high & six or eight inches wide with a hollow place in the side next the bottom & a small hole in the upper side of the hollow, being filled with water; the water wch spouted out of the small hole, rose right up in a small streame as high as the top of the water wch stagnated in the vessel, abating only about half an inch by reason of the resistance of the air.[27]

Cotes refused to retire. When he answered on 5 October, he had just performed Mariotte's experiment himself. It was in fact the same experiment on which Newton had relied in 1686. From the measured volume of water that flowed from a given hole in a given time he calculated the velocity; it agreed roughly with the velocity of a body falling from half the height of the surface. "I wish You would be pleased Y^r self to make the Experiment before You resolve to print the 36 Proposition as it now stands," Cotes concluded.[28] When he received no reply in three weeks, he wrote again.

S^r [Newton responded]

I received both your Letters & am sensible that I must try three or four Experiments before I can answer your former.[29]

Six months passed before Cotes heard again.

This was a busy time for Newton. Not only was the Royal Society completing its negotiations for Crane Court, but Newton himself was relocating. After only nine months in Chelsea, where he had moved from Jermyn Street late in 1709, he had rented the house on St. Martin's Street, immediately south of Leicester Fields, which would be his home nearly until his death. In December, the appointment as Visitor to the Observatory made possible the renewed effort to print Flamsteed's observations. The Mint also imposed demands, for the government was preparing to offer the

[26] Cotes to Newton, 21 Sept. 1710; *Corres 5,* 65–7.
[27] Newton to Cotes, 30 Sept. 1710; *Corres 5,* 70.
[28] *Corres 5,* 73–4. On 24 March 1720, Desaguliers performed this experiment before the Royal Society (*JBC 11,* 468–9). A week later he mentioned that on the twenty-fourth he had also performed it privately for Newton in his home (*JBC 11,* 472).
[29] Newton to Cotes, 27 Oct. 1710; *Corres 5,* 75.

bounty on silver plate which would make 1711 a year of heavy coinage. When Newton finally wrote in March offering a solution, Cotes found objections to it.[30] By now still another matter had arisen which would distract Newton even more from the new edition. On 22 March, the Royal Society received the letter from Leibniz which effectively opened the priority dispute. Only in June 1711 did the next letter from Newton finally settle Section VII, but even then the edition hardly moved during the rest of that year.[31]

As far as Proposition XXXVI was concerned, Newton resolved the discrepancy between the two experiments by the principle of the *vena contracta*, which he sent to Cotes initially in March. He observed that a stream flowing out of a hole in a tank does not have the full diameter of the hole. It contracts into a narrower vein. He had a hole bored exactly 5/8 of an inch in diameter. The diameter of the stream "measured with great accuracy at the distance of about half an inch from the hole, was 21/40 of an inch."[32] That is, the diameter of the stream was 21/25 of the diameter of the hole. As it happened, 21/25 was a very convenient fraction. The areas of two circles with diameters in that proportion are as $(21/25)^2$, and $(21/25)^2$ is nearly equal to $1/\sqrt{2}$. The velocity of a body that has fallen $\frac{1}{2}h$ is to the velocity of a body that has fallen h in the ratio of $1/\sqrt{2}$. Cotes immediately understood the nature of the resolution thus offered.

I find that 25 & 21 express the proportion of $\sqrt{\sqrt{2}}$ to 1 as nearly as it is possible for so small numbers to do it, whence it is probable y^t the exact proportion of the diameter of the Hole to y^e diameter of y^e Stream is that of $\sqrt{\sqrt{2}}$ to 1, & then y^e proportion of 44 to 37 will be much nearer the truth than y^t of 25 to 21.[33]

Possibly Newton feared he could not with a straight face claim to have "measured" a proportion of 44 to 37 "with great accuracy." At any rate he remained content with 25 to 21. It brought the two experiments into agreement but reduced the demonstration to a sham. The cataract in the funnel of ice remained though now, for no announced cause, the water reached the hole with a velocity equal to that gained in falling half the distance from the surface. Newton then reversed the principle of the cataract whereby a stream contracts because it accelerates, to explain that the stream is further "accelerated by converging" to increase its velocity, in a space roughly equal to the diameter of the hole, by a factor equal to the root of 2. The proposition did not demonstrate the velocity as

30 Cotes to Newton, 31 March 1711; *Corres 5*, 107–10.
31 The considerable volume of manuscripts related to Section VII testifies that Newton did indeed devote extended attention to it. They include *Add MSS* 3965.8, ff. 72–8; virtually all of 3965.10, and more than fifty folios (204–18, 221–62) in 3965.12. 3965.13 has many more connected with the third edition.   32 *Prin*, p. 339.
33 Cotes to Newton, 31 March 1711; *Corres 5*, 107.

the corresponding proposition in the first edition had tried to do. It merely announced it as an experimental result, and devoted eight pages of opaque prose to concealing what it had done.[34] The principle of the cataract did still function in the calculation of the weight sustained by a body suspended in the effluent stream. Newton showed that the weight must be more than one-third of the cylinder of water above it and less than two-thirds. Hence he concluded that it is "very nearly" equal to the weight of half the cylinder. He tended to forget the figure was only an approximation as he proceeded triumphantly to demonstrate that the resistance to a sphere in a compressed fluid medium such as water or air – or by implication a Cartesian aether – is to the force that can destroy or generate its motion in the time it moves a distance equal to eight-thirds of its diameter as the density of the fluid to the density of the sphere. He found the consequent theory to agree, to an average of about one part in fifty, with the results (verbally adjusted in several cases) of thirteen different experiments. No one would care to deny that it was a virtuoso performance. Few would care to dignify it with the title of scientific investigation.

Meanwhile, the letter from Leibniz had arrived. As Newton contemplated its implications, it began to dominate his consciousness to the point of virtually sweeping every other topic out of his mind. Documents of every sort from the following years tend to be interrupted by furious paragraphs against Leibniz, as Newton, in his typical way, honed his prose with infinite care to razor sharpness. The very completion of the second edition seems in the circumstances something of a miracle, perhaps possible only to the extent that the battle spread out beyond the field of mathematics to cover natural philosophy as well, so that the *Principia* sustained one part of the front.

As we have seen, the storm that burst in 1711 had been gathering for many years, indeed from the moment in 1684 when Leibniz chose to publish his calculus without mentioning what he knew of Newton's progress along similar lines. The publication in 1699 of volume 3 of Wallis's *Opera*, with the full texts of the two *Epistolae* of 1676 plus other letters that testified to Newton's progress by the year 1673 and Leibniz's status in 1674 and 1675, all set in print with Newton's aid and cooperation, was the decisive event which made the public dispute inevitable. Before 1699, Newton had alluded to the correspondence of 1676 in the *Principia* and in the truncated "De quadratura curvarum" published in Wallis's volume 2 in 1693.

[34] *Prin*, pp. 337–45. Cf. *Add MS* 3965.12, ff. 230, 232ᵛ. Also a draft of the following Proposition XXXVII; *Corres 7*, 475–6.

Only Newton and those in communication with him – and Leibniz – understood the allusions. No one on the Continent knew the extent of Newton's communication in 1676. Pierre Varignon interpreted the fluxion scholium in the *Principia* as Newton's acknowledgment of Leibniz's invention of the calculus.[35] There is every reason to think that his reading of it was typical. When Johann Bernoulli saw the piece in volume 2 of Wallis's *Opera,* he suggested that Newton had concocted it out of Leibniz's papers. To be sure, Bernoulli repeated that charge later, but it is equally true that when he saw *Commercium epistolicum* in 1712, with the same letters Wallis printed plus some more, he advised Leibniz that his best defense was to prove that the letters had been altered.[36] It does not appear that Wallis's *Opera* circulated widely on the Continent. The *Acta* (probably Leibniz writing anonymously) reviewed the volume as though the letters devoted themselves mostly to the further celebration of Leibniz's early genius.[37] Nevertheless, in 1699 the full text of a correspondence that Leibniz had concealed became public knowledge, and from that time on Leibniz's conduct changed.

In the same year 1699, Fatio de Duillier made his last significant impression on Newton's life with the publication of a mathematical tract, *Lineae brevissimi descensus investigatio geometrica duplex* [*A Twofold Geometrical Investigation of the Line of Briefest Descent*]. Six years had passed since the rupture with Newton. We have no reason to think that Newton had any hand in the treatise. Leibniz, who knew of their former intimacy, undoubtedly thought Newton did, all the more so since Fatio used the tract, as its title hinted, to attack Leibniz and Bernoulli for seeking to embarrass Newton with the challenge problem of 1696. Since Fatio advertised himself as a fellow of the Royal Society on the title page, Leibniz wrote to the society in complaint.[38] In any event, Fatio had gone out of his way to insult Leibniz and had passed beyond insult to something much more. Fatio had invented his own calculus beginning in 1687, he stated, and owed nothing whatever to Leibniz, who would have to pride himself on his other disciples but not on Fatio.

However, driven by the factual evidence, I recognize that Newton was the first and by many years the most senior inventor of this calculus: whether Leibniz, the second inventor, borrowed anything

---

[35] Varignon to Leibniz, 9 Aug. 1713; *Corres 6,* 27.

[36] Bernoulli to Leibniz, 23 May 1714; G. W. Leibniz, *Mathematische Schriften,* ed. C. I. Gerhardt, reprint ed., 7 vols. (Hildesheim, 1962), *3/2,* 931–3.

[37] "Johannis Wallis, S.T.D. Geometriae Professoris Saviliani, operum mathematicum volumen III," *Acta eruditorum* (May 1700), pp. 193–8. The review immediately preceded Leibniz's reply to Fatio's charge of plagiary.

[38] He sent the letter to Wallis, who laid it before the society on 31 January 1700 (*JB 10,* 164).

from him, I prefer that the judgment be not mine, but theirs who have seen Newton's letters and his other manuscripts. Nor will the silence of the more modest Newton, or the active exertions of Leibniz in everywhere ascribing the invention of the calculus to himself, impose upon any person who examines these papers as I have done.[39]

Leibniz both reviewed the publication anonymously and published a signed reply to the charge in the *Acta.* In denying it, he carried his silence in regard to the correspondence in 1676 a step closer to outright falsehood by asserting that in 1684, when he first published his method, he knew only that Newton had a method of tangents.[40]

Nor was Leibniz solely the passive object of aggression from England. In 1699, he also published an attack on David Gregory's attempted demonstration of the catenary. Leibniz had an unfortunate preference for mounting his assaults anonymously, a tactic that did much to compromise his position after 1711, since his authorship was generally recognized. His animadversions on Gregory likewise appeared anonymously. Since Gregory's demonstration was erroneous, it was fair game. However, Leibniz went beyond criticism of the demonstration, and treating Gregory as a representative of Newtonian mathematics, he slyly implied that the fault of the demonstration followed from the deficiency of the fluxional method.[41] No one would wish to attach the adjective "fair" to this stratagem, especially when he shielded himself in anonymity. A similar underhanded attack on Newtonian mathematics appeared in 1701 in an appendix to Johann Groening's *Historia cycloeidis.* The appendix contained a list of errors in the *Principia* presented as the work of Huygens. In fact, the list was Newton's own, given by him to Fatio, carried by Fatio to the Netherlands in 1691, and now published surreptitiously from Huygens's papers. Among Newton's friends there was no doubt that Leibniz had arranged the publication.[42]

In this atmosphere of rising suspicion and hostility, Newton finally decided that the time had come to publish a fully fleshed specimen of his mathematics. In a memorandum of 1702, Gregory recorded Newton's promise given to Halley, Fatio, Robartes, and himself to publish the *Opticks* and with it two mathematical treat-

[39] Fatio, *Lineae brevissimi descensus investigatio geometrica duplex* (London, 1699), p. 18.

[40] "Nicolai Fatii Duillierii R.S.S. Lineae brevissimi descensus investigatio geometrica duplex . . . ," *Acta eruditorum* (November 1699), pp. 510–13. "G.G.L. Responsio ad Dn. Nic. Fatii Duillierii imputationes," *ibid.* (May 1700), pp. 198–208.

[41] "Animadversio ad Davidis Gregorii schediasma de catenaria," *ibid.* (February 1699), pp. 87–91.

[42] Gregory dated notes to this effect 4 July 1705 and 7 February 1706; *Hiscock,* pp. 26–7, 32.

ises.[43] Accordingly the work appeared two years later with "De quadratura curvarum" and the "Enumeratio linearum tertii ordinis" appended. Newton's introduction made it clear that he did not publish them merely as a whim. "In a Letter written to Mr. Leibnitz in the year 1676, and published by Dr. Wallis [he stated in the preface], I mention'd a Method by which I had found some general Theorems about squaring Curvilinear Figures . . ."[44] Thus he contrived to imply that "De quadratura," actually composed in the 1690s, stemmed from the early 1670s, a claim on a par with Leibniz's reply to Fatio about what he knew in 1684. The introduction to "De quadratura" addressed itself no less to Leibniz and fortunately stuck closer to the truth. "I gradually fell upon the Method of Fluxions which I have used here in the Quadrature of Curves," he asserted, "in the Years 1665 & 1666."[45] An anonymous review of "De quadratura" appeared in the *Acta eruditorum* for January 1705. Newton later claimed that he first saw the review in 1711, a claim there is some reason to question. Whenever he saw it first, he never doubted that Leibniz wrote it. As we know from Leibniz's papers, he was correct. In describing the content of the work, the review transposed it into the language of the differential calculus, invented by Mr. Leibniz.

> Instead of the Leibnizian differences, therefore, Mr. Newton employs, and has always employed, *fluxions, which are almost the same as the increments of the fluents generated in the least equal portions of time.* He has made elegant use of these both in his *Mathematical Principles of Nature* and in other things published later, just as Honoré Fabri in his *Synopsis geometrica* substituted the progress of motions for the method of Cavalieri.[46]

Although he never acknowledged his authorship, Leibniz did argue that the passage did not imply plagiary by Newton. To be sure, it was as artfully worded as the review of Gregory had been, but only a dull ear can miss the intended ring of the words.

Among those with sharper ears was a young Oxford don, John Keill, who published a paper on centrifugal forces in issue no. 317 of the *Philosophical Transactions* for September and October 1708. Toward the end of his paper, Keill inserted a blunt rejoinder to Leibniz's insinuation.

> All of these [propositions] follow from the now highly celebrated Arithmetic of Fluxions which Mr. Newton, beyond all doubt, First Invented, as anyone who reads his Letters published by Wallis can easily determine; the same Arithmetic under a different name and

---

[43] *Hiscock*, p. 14.   [44] *Opticks*, p. cxiv.   [45] *Opticks*, ed. 1, p. 166 bis.
[46] "Isaaci Newtoni tractatus duo, de speciebus & magnitudine figurarum curvilinearum," *Acta eruditorum* (January 1705), pp. 30–6. I quote from p. 35.

using a different notation was later published in the Acta eruditorum, however, by Mr. Leibniz.[47]

Journals did not circulate with lightning speed in the early eighteenth century. Although the issue was published in 1709, presumably early, Leibniz did not see Keill's article for some time. When he finally did see it, he complained to the Royal Society, in the letter that arrived in March 1711.

Newton later protested that he was unaware of Keill's paragraph and indeed was upset when he did learn of it until Keill showed him Leibniz's review of "De quadratura." It is necessary to note, however, that Keill was the protégé of Gregory and Halley in Oxford and that his article took the form of a letter addressed to Halley. Furthermore, he presented it to the Royal Society, with Newton presiding, on 3 November 1708, when the society judged it fit for publication in the *Philosophical Transactions.*[48] All things considered, one cannot seriously believe that Newton was not informed of the riposte – or perhaps did not participate in its formulation. Leibniz was on his mind. Earlier in 1708 he had told Gregory that he intended to reprint "De quadratura" with the new edition of the Principia.[49]

What Keill said was no more than the common currency that passed through the whole learned world of Great Britain. Undoubtedly he merely repeated what he had heard at Oxford from Wallis, Gregory, and Halley, not to mention Fatio, who was there during the early years of the century. In the 1690s Gregory had seen the correspondence that would come out in Wallis's volume and had formulated for himself the argument the letters made for Leibniz's plagiary. "These letters are to be printed," he concluded, "in the folio that Dr. Wallis is now aprinting, in the order of their dates, without any notes or commentaries or reflections: but let the letters themselves speake."[50] George Cheyne's *Inverse Method of Fluxions* (1703) contended that nothing in mathematics had been published in the last twenty or thirty years, a period carefully chosen, that was not a repetition or trivial corollary of an earlier discovery by Newton.[51] Writing to William Jones early in 1711, before the arrival of Leibniz's letter, Cotes spoke of Joseph Raphson's neglect of the Collins papers in Jones's possession, "which I

[47] "Epistola ad clarissimum virum Edmundum Halleium Geometriae Professorem Savilianum, de legibus virium centripetarum" *Philosophical Transactions,* 26 (1708-9), 174-88. I quote from p. 186. See J. O. Fleckenstein, *Der Prioritätstreit zwischen Leibniz und Newton"* (Basel and Stuttgart, 1956).   [48] *JBC 10,* 195.

[49] *Hiscock,* p. 41.   [50] *Hiscock,* p. 7.

[51] So Bernoulli summed up the book for Leibniz (Bernoulli to Leibniz, 29 July 1713; *Mathematische Schriften, 3/2,* 916). Cheyne himself set the period at twenty-four years, long enough to take in Leibniz's earliest publication (George Cheyne, *Fluxionum methodus inversa* [London, 1703], pp. 59-60).

do now very much wonder at if his Intention was to do justice to Sr Isaac."[52] Cotes need not have worried about Raphson's book. If the one published in 1715 after his death bore any resemblance to what Raphson was writing in 1711, justice to Mr. Newton, in the English understanding of the phrase, was its primary purpose, as he stated explicitly in the preface.[53] The book itself was as belligerently partisan as anything Keill ever wrote, to the extent that Newton blocked its publication for four years, or so he claimed at least.[54] Keill chanced to become the vehicle of Newton's revenge. It could easily have been someone else.

Collins's papers, to which Cotes referred, may have been instrumental in moving Newton to assert his case more strongly. William Jones, a young mathematician who taught in London, had obtained them in 1708, the year of Keill's paper. They offered Newton documentary evidence, independent of his own papers, of his early mathematical development. In the end, he made the papers central exhibits in the whole controversy. Among them Jones found a copy of Newton's early treatise, "De analysi," which he had transmitted to Collins in 1669, not to mention numerous letters referring to it which amply established its early date. Jones had little trouble, in the climate that had built up by then, in winning Newton's cooperation in the publication of "De analysi" together with three other papers ("De quadratura," the "Enumeratio," and the "Methodus differentialis," Newton's treatise on interpolation) and excerpts from a number of letters. Dr. Mead presented a copy of the small volume to the Royal Society on 31 January 1711, nearly two months before they received Leibniz's letter, which cannot then have provoked the publication.[55] Jones's preface bore the heavy mark of Newton's influence in its marshaling of historical evidence, and its explicit reference to Newton's own papers placed his cooperation beyond doubt. Jones recounted his acquisition in 1708 of Collins's papers, among which he found "De analysi," the date of which he set by citing letters of Collins and Barrow from 1669.

> From certain papers of Newton that I have seen [he continued] I understood that he first deduced the Quadrature of the Circle, Hyperbola, and certain other Curves by means of Infinite Series from the *Arithmetica Infinitorum* of our countryman Wallis, and that he did so in 1665; then he devised a method of finding the same Series by Division & Extraction of Roots, which he made general the following Year.

Although the preface contained no mention of Leibniz, its insistence on dates in the 60s was sufficiently explicit, as was also its praise of

---

[52] *Corres 5*, 95.   [53] Joseph Raphson, *The History of Fluxions* (London, 1715), Preface.
[54] A draft of Newton to Bernoulli, early 1720, and a draft of another letter of about that time; *Corres 7*, 80, 82.   [55] *JBC 10*, 257.

the Newtonian reliance on the limits of nascent and ultimate ratios instead of "that harsh Hypothesis of Infinitely small quantities or Indivisibles . . ." In case the preface and "De analysi" were not enough, Jones also included a section of "Excerpts from the Letters of the most learned Newton relating to the Method of Fluxions and Infinite Series," excerpts drawn from the two *Epistolae* of 1676, from Newton's letter to Collins on 8 November 1676 in which he laid out the full extent of the problems his method could solve, and finally from a letter to Wallis in 1692, an exposition of his method in explication of the letters from 1676.[56] Keill was brutally direct, but Jones's more suave message carried an identical meaning.

With Leibniz's letter complaining of Keill, which arrived shortly after the Jones volume appeared, the covert preliminaries came to the end, and the open battle long in preparation began at last. My account of the controversy has nothing to do with the issue of priority. As far as I am concerned, that issue has been settled by examination of the papers left by the two principals. Newton invented his fluxional method in 1665 and 1666. About ten years later, as a result of his own independent studies, Leibniz invented his differential calculus. Newton contended – with his endless redrafting, contended repeatedly – that second inventors have no rights. A more patently absurd claim could not have been advanced. The first inventor clutched his discovery to his breast and communicated almost nothing. The second inventor published his calculus and thereby raised Western mathematics to a new level of endeavor. Newton realized as much in the end, and a good half of his fury was flung at Leibniz as a surrogate for his former self who had buried such a jewel in the earth.

My account of the quarrel, I repeat, does not concern itself with the question of priority. Varignon once remarked that the glory of the invention was sufficient for both men.[57] My account of the quarrel is the story of their inability to share it amicably. If the glory of the invention was enough for both, so was the blame for the contest. This they did succeed in dividing in rough equality. In Newton's case, the cluster of inhibitions and neuroses that had prevented him from publishing his method in the first place likewise held him back from a forthright assertion of his claims. For nearly thirty years, from the time of Leibniz's first paper, he confined himself in print to opaque references to a correspondence with Leibniz in 1676. Beyond a small circle of Newton's intimates, no one except Leibniz understood what he was hinting at. Meanwhile, he did grumble privately to his intimates, and through them he

---

[56] *Analysis per quantitatum series, fluxiones, ac differentias: cum enumeratione linearum tertii ordinis* (London, 1711).

[57] Varignon to Leibniz, 9 Aug. 1713; *Mathematische Schriften*, 4, 195.

poisoned the minds of a whole generation of English mathematicians. According to Newton's own testimony, he filled John Craig's ear with his complaints already in 1685.[58] Later Fatio, Gregory, and Wallis, who did not to be sure require any convincing, all heard the same, and he supplied the letters for Wallis's early *commercium epistolicum*. It was no accident that John Keill, who finally precipitated the open battle, came from Oxford, where a group of Newton's epigoni gathered. Newton liked to say that he hated controversy and tried to avoid it. In the matter of the calculus such does appear to have been the case. Though he protested in private, he drew back time and again from a direct challenge to the perceived injustice, as though he understood too well where his passion would carry him if it were once unchained. Well he might hold back, for once Leibniz's letter aroused him he was fury incarnate.

As for Leibniz, his eagerness for acclaim, his need to gather and guard the intellectual capital which assured his livelihood, led him in 1684 to the fatal error of trying to seize the undivided credit for his stupendous invention by neglecting to mention the correspondence of 1676. To Newton, this was his original sin, which not even divine grace could justify. Leibniz's paper had said that his method reached to the most sublime problems, which could not be resolved without it or another like it [*aut simili*].

> What he meant by the words AUT SIMILI was impossible for the Germans to understand without an Interpreter. He ought to have done Mr. Newton justice in plain intelligible Language, and told the Germans whose was the *Methodus* SIMILIS, and of what Extent and Antiquity it was, according to the Notices he had received from England; and to have acknowledged that his own Method was not so ancient. This would have prevented Disputes, and nothing less than this could fully deserve the Name of Candor and Justice.[59]

Perhaps such a statement by Leibniz in 1684 would have prevented the dispute. Perhaps nothing would have. At any rate he had not mentioned what he knew of Newton's work, and by 1699, when Wallis's volume appeared, it was already too late. By then he had posed as the sole inventor of the calculus for fifteen years. Even to his friends he had not mentioned that exchange. Fontenelle's *éloge* seventeen years later reveals the extent of his initial success in passing as the sole inventor.[60] What Leibniz feared was the conclusion

---

[58] "Mr Craige is a witness that in those days I looked upon the method as mine" (a draft of his "observations" on Leibniz's second letter, 1716; *Add MS* 3968.29, f.422^v. Cf. a slightly different version of the same; f.117^v bis).

[59] "Account of Commercium epistolicum," p. 220. Cf. his draft of a letter to Varignon that was never sent, late 1718; *Corres 7*, 18.

[60] "Éloge," *Histoire de l'Académie royale des sciences*, 1716, pp. 115–56. Cf. Fontenelle to Chamberlayne, 24 June 1717; *Corres 6*, 394–5.

the English drew. Freely to acknowledge the exchange in its full scope at that late date might lead to suspicions about his own independent discovery. When one recalls Bernoulli's response to *Commercium epistolicum* – that Leibniz's best defense was to prove that the letters were frauds – we can appreciate his dilemma.

Nothing is more revealing of Leibniz's problem than his silence about "De analysi." As he and he alone knew, the *Epistolae* of 1676 were not the half of the matter. In London he had also read "De analysi," the treatise so prominent in the dispute because independent evidence established its date. Leibniz never breathed a word about "De analysi," and Newton, who for polemic effect hinted vaguely at Leibniz's stop in London, did not know he had seen it and initially never mentioned what he would have loved to shout across Europe. Only when Leibniz, shortly before his death, inadvertently revealed the extent of Collins's liberality in the fall of 1676, did Newton begin to realize that he might have seen the tract as well.

Leibniz's strategy in the dispute flowed from the mortal danger to which the venial sin of 1684 exposed him. Hence the response to Fatio – that in 1676 he learned only of a method of tangents – to which he later added that he also learned about infinite series. Hence the review of Wallis, which stressed the evidence of Leibniz's own accomplishment by 1676, leading on to the argument that Newton was the pupil in the exchange. Hence the curiously restrained review, also anonymous, of "De analysi," which never said a word about its date but presented an argument, or perhaps only a suggestion of an argument for the reader to flesh out, that "De analysi" merely employed Archimedes' method of exhaustion and Fermat's method of infinitesimals, which differed from the new concepts found in the calculus invented by the "illustrious Leibniz."[61] His friend Christian Wolf, who also did not know that Leibniz had read "De analysi" in 1676, urged that he needed to carry that argument further and prove that the tract did not contain the algorithm of the calculus – another unconscious illumination of Leibniz's dilemma.[62] What would Wolf have said if he had read "De analysi" and known that Leibniz had also read it in 1676? Leibniz chose rather to mention "De analysi" as little as possible.

Until the final act of the ensuing drama at least, Leibniz was all urbane sophistication. He never missed a chance to praise Newton in public even while he attacked him anonymously and by indirect insinuation. His signed "Response" to Fatio's attack could not

---

[61] "Analysis per quantitatum series, fluxiones ac differentias cum enumeratione linearum tertii ordinis," *Acta eruditorum* (February 1712), pp. 74–7.

[62] Wolf to Leibniz, 4 May 1715; C. I. Gerhardt, ed., reprint ed., *Briefwechsel zwischen Leibniz und Christian Wolff* (Hildesheim, 1963), pp. 164–5.

sufficiently express his pained surprise that anyone should think the challenge problem was a trap set for Newton, for whom he had only the highest regard. "Indeed I have freely acknowledged his immense merit whenever occasions have presented themselves . . ."[63] In 1701, Sir A. Fontaine found himself at dinner with Leibniz at the royal palace in Berlin. When the Queen of Prussia asked him his opinion of Newton,

> Leibnitz said that taking Mathematicks from the beginning of the world to the time of Sʳ I. What he had done was much the better half – & added that he had consulted all the learned in Europe upon some difficult point without having any satisfaction & that when he wrote to Sʳ I. he sent him answer by the first post to do so & so & then he would find it out.[64]

Two years before he had implied, anonymously, that Gregory's mistake with the catenary flowed from the defects of Newton's method, and in the same year 1701, if Gregory and Fatio were correct, he arranged for the publication of the list of errors in the *Principia*.

Newton also hid behind a shield of anonymity and spoke through the mouth of another. Subtle insinuation was not the idiom of John Keill, a crude and abrasive man who did Newton's cause much harm before the learned world, which quickly learned to despise him. Bernoulli, who spoke of him as "Newton's toady" and a hired pen, expressed his contempt in print by never referring to him by name, but only as "a certain individual of Scottish race."[65] It has become fashionable of late to push the blame for the controversy onto Keill. No doubt he did nothing to calm the storm, but Keill's belligerence was also Newton's style. If Newton were going to speak through the mouth of another, he could not have found an instrument more fit. Leibniz always understood that Keill was Newton's mouthpiece. A man of the world, who would have understood a response of overt praise and covert thrust, he gasped in astonishment when he found that the author of the *Principia* and *Opticks* was a wild bull whose only impulse was to lower his head and charge – and that he had locked himself into the same pen with this embodiment of rage.

The goad that finally incited Newton was Leibniz's letter to the Royal Society dated 4 March 1711 in complaint of Keill's "imperti-

---

[63] "Responsio," p. 203. [64] *Keynes MS* 130.7, Sheet 2.
[65] Bernoulli to Leibniz, 6 Feb. 1715; *Mathematische Schriften, 3/2,* 936; translation in *Corres 6,* 203. Bernoulli to Burnet, 15 May 1714; *Corres 6,* 125. "Joh. Bernoulli responsio ad nonneminis provocationem, ejusque solutio quaestionis . . .," *Acta eruditorum* (May 1719), pp. 216–26; I quote from p. 217. Cf. Varignon to Leibniz, 9 Aug. 1713; *Mathematische Schriften, 4,* 195.

nent accusation." Protesting his innocence, Leibniz expressed his
fear that dishonest and impudent people would repeat the charge
unless it were withdrawn. The accusation had appeared in the *Philosophical Transactions*. Both he and Keill were members of the Royal
Society. He applied to them for a remedy, to wit, that Keill publicly testify that he did not mean to charge Leibniz with what his
words seemed to imply.[66] Newton presided at the meeting of the
Royal Society on 22 March when the letter was read. The minutes
record only that the society ordered Sloane, the secretary, to write
an answer.[67] If Sloane ever drafted such a reply, no evidence of it
exists. Instead, after a pause of two weeks, the Royal Society took
up the matter again. Newton was in the chair. Keill had come
down from Oxford.

> M[r]. Keill Observed that in the Lipsick Acta Eruditorum for the year
> 1705 there is an unfair Account Given of Sir Isaac Newtons Dis-
> course of Quadratures Asserting the Method of Demonstration by
> him there made use of to M[r]. Leibniz &c, upon which the President
> Gave a Short Account of that Matter with The perticular time of his
> first mentioning or discovering his Invention, referring to some Let-
> ters published by Doctor Wallis, upon which M[r]. Keill was Desired
> to draw up an Account of the Matter in Dispute and Sett it in a Just
> Light and also to Vindicate himself from a perticular reflection in a
> Letter from M[r]. Leibnitz to Doctor Sloane.[68]

When the minutes were read the following week, Newton ampli-
fied his remarks, mentioning "his Letters many years ago to M[r].
Collins about his Method of Treating Curves &c . . ." The society
asked Keill, who was present again, to draw up a paper that asserted Newton's rights.[69]

The manuscript evidence indicates a flurry of activity during the
two weeks while Sloane was not writing the letter that the society
had ordered initially. Newton had a copy of Leibniz's letter sent to
him.[70] He spoke with Keill, only recently returned from a trip to
New England, and Keill showed him the review of "De quadratura." Upon reading it, he wrote to Sloane, "I found that I have
more reason to complain of the collectors of y[e] mathematical papers
in those Acta then M[r] Leibnitz hath to complain of M[r] Keil."[71] On
3 April, in a letter which appears to assume an earlier discussion,
Keill sent him a copy of the *Acta* plus a couple of other items from
the 1710 volume not calculated to soothe his feelings.[72] His notes
on the offending review survive.[73] Newton must have spent a good
portion of the period looking through his own papers. As he never
published without extensive revision, so he did not speak extempo-

[66] *Corres 5*, 96.   [67] *JBC 10*, 264.   [68] *JBC 10*, 266–7.   [69] *JBC 10*, 270.
[70] *Add MS* 3968.18, f.262.   [71] *Corres 5*, 117.   [72] *Corres 5*, 115.
[73] *Add MS* 3968.4, f. 16[v].

raneously on important issues. When he recounted the history of his invention of the fluxional method and of his correspondence with Collins for the edification of the Royal Society on 5 and 12 April, he undoubtedly spoke with his facts well in hand. On the same sheet with the draft of his letter to Sloane about his discourse with Keill he began to compose a rebuttal to Leibniz, leaving blanks for dates and page references he had not yet checked.

> That by your Letter to Mr Collins dated   & published by Dr Wallis   he was convinced that you did not then use the methodus differentialis. That the next year in my letter dated   . . . I represented that I had a method of solving direct & inverse Problemes of tangents . . . That in my first Letter wch was sent to you by Mr Oldenburg & dated   1676, I described at once the foundation of the differentiall method . . .[74]

The continued elaboration of his case, as he piled detail upon detail from the manuscript record, absorbed Newton's life during the coming years. No more obscure allusions to a correspondence unknown to any reader; the priority dispute was war declared.

A month and a half went into the preparation of Keill's reply, the period during which Cotes waited for Newton's final resolution of Proposition XXXVI. We do not have manuscript evidence of Newton's extensive participation in the letter's composition, but everything about it, its intimate knowledge of Newton's early papers and correspondence, details of its argument about Leibniz's progress as revealed by his correspondence in 1675 and 1676, above all its style, which treated the issue as a historical question to be settled by empirical evidence from the manuscript record, cries aloud of the hand that shaped it. Newton did have an amanuensis's copy of the completed letter, which he went through making a few final corrections – his unavoidable procedure with his own writings, but something he never did with others'.[75] The Royal Society, with Newton presiding, heard the finished letter on 24 May and ordered that it be sent to Leibniz but not published in the *Philosophical Transactions* until Leibniz's answer showed that he had received it. Newton even drafted Sloane's covering letter.[76] In fact, Keill's reply to Leibniz's complaint never appeared in the *Philosophical Transactions;* when Leibniz's response arrived, it gave rise to a more extended publication.

---

[74] *Add MS* 3968.30, f. 439^v. Partly published in *Corres 5,* 117n. The blanks are in the original.

[75] *Add MS* 3968.22, ff. 333–8. The letter that was sent did incorporate Newton's alterations (Keill to Sloane for Leibniz, May 1711; *Corres 5,* 133–41). Keill's original draft of it is in Cambridge University Library, Res. 1894(a), Packet 11.

[76] *JBC 10,* 290–1. Sloane to Leibniz, May 1711; *Corres 5,* 132. The draft of this letter, in Newton's hand, is in *Add MS* 3968.30, f.422^v.

Leibniz devoted some thought to his answer, which he did not send until 29 December and which the Royal Society did not receive until 31 January 1712. Though he could not have failed to understand that Newton, the president of the Royal Society, stood behind a letter transmitted to him by the society, he chose carefully to distinguish Newton from Keill, whom he treated as an upstart before whom he had no need to justify himself. Keill had now attacked his sincerity more openly. Leibniz rejected his appeal to the *Acta* for justification,

> for in them I find nothing that detracts anything from anyone; rather I find that in passages here and there everyone receives his due. I, too, and my friends have on several occasions made obvious our belief that the illustrious discoverer of fluxions arrived by his own efforts at basic principles similar to our own. Nor do I lay less claim to the rights of a discoverer . . . Thus I throw myself on your sense of justice, [to determine] whether or not such empty and unjust braying should not be suppressed, of which I think Newton himself, a distinguished person who is fully acquainted with past events, disapproves; and I am confident that he will freely give evidence of his opinion on this issue.[77]

Leibniz was justified in his confidence that Newton would express his opinion, though rather more covertly and with far more violence than Leibniz imagined. Again he obtained a copy of the letter, on which he made extensive annotations. He started to draft a reply addressed to Sloane, protesting that he had no intention of entering the dispute, but he abandoned it when the subject dragged him into the argument despite his intentions. He drew up notes for a further address to the Royal Society on the history both of his invention and of the quarrel.[78] We do not know when he obtained a copy of Leibniz's review of "De analysi," which was published in February. It probably arrived considerably later than Leibniz's letter. Whenever he saw it, the review further inflamed him. He took his usual notes on it and composed at least three drafts of a reply, which he never published.[79] Finally he lit upon an unexpected riposte. Leibniz had thrown himself on the justice of the society. Very well, let the society sit in judgment on the question. On 6 March 1712, the Royal Society appointed a committee to inspect the letters and papers relating to the matter: Arbuthnot, Hill, Halley, Jones, Machin, and Burnet, to which

---

[77] *Corres 5,* 208; original Latin, p. 207.
[78] *Add MS* 3968.18, f. 261^v. *Corres 5,* 212–14. These are two drafts of his intended comments; *Add MSS* 3965.8, f. 79^v, and 3965.13, f. 465.
[79] His notes are in *Add MS* 3968.4, ff. 17, 19. The three drafts of a reply are published in *Math 2,* 263–73.

they later added Robartes, DeMoivre, Aston, Taylor, and Frederick Bonet, the minister in London of the King of Prussia.[80] Newton liked to refer to this committee as "numerous and skilful and composed of Gentlemen of several Nations . . ."[81] Rather it was a covey of his own partisans to which Herr Bonet, to his undying shame, allowed himself to be coopted to provide some minimal veneer of impartiality. The veneer was so minimal, in fact, that the published report did not name the committee, whose membership remained concealed in the society's records until the nineteenth century.

Neither Herr Bonet nor the others found the task difficult. Newton had already done the spadework in a year of intensive investigation. The full advantage of Jones's lucky recovery of Collins's papers now emerged. Together with Oldenburg's correspondence in the records of the society, Collins's papers provided the factual foundation of a report which did not need to call upon Newton's own papers at all. Leibniz later complained – with full justice – that the committee conducted a judicial proceeding without informing him of the fact or allowing him to present evidence. The committee did not call upon Newton to present evidence either. It did not need to. Newton carried out its investigation, arranged its evidence, and wrote its report, which explains why the committee was able to submit the report, which presumed to survey the whole history of the calculus, on 24 April, a full month and a half after its appointment. The final three members had been appointed on 17 April, one week earlier. The procedure may explain why the committee did not sign the report, but Newton's creatures were too tame to balk at submitting it. Of his own central participation there can be no doubt. Beyond the extensive manuscript remains of his researches, which went into the volume later published, his draft of the report still exists. In his later "Account of *Commercium epistolicum*," Newton waxed indignant over Leibniz's claim to have invented his calculus before he received the letters of 1676. "But no Man is a Witness in his own Cause," he thundered.[82] With his own words he passed judgment on himself.

Not surprisingly, the committee, or court, found in Newton's favor, in a condemnation of Leibniz beside which Keill's paragraph looks like praise. "Wee Have Consulted the Letters and Letter Books in the Custody of the Royall Society and those found Among the Papers of Mr. John Collins," the report began, and they had confirmed the hands so that they believed the letters to be

[80] *JBC 10*, 369, 375, 377, 386.
[81] "Account of Commercium epistolicum," p. 221.
[82] Ibid., p. 194. Biot and Lefort used the Latin version of Newton's proclamation of principle on the title page of their edition of *Commercium epistolicum* (Paris, 1856).

authentic. By the letters and papers they found, first, that Leibniz was in London in 1673, that he maintained a correspondence with Collins after he returned to Paris, that he visited London again in 1676 on his way to Hanover, and that Collins was very free in communicating the things he had. Second, that when Leibniz was first in London he claimed the invention of another differential method though it was proved to be that of Mouton, and that no evidence showed him to have any differential method other than Mouton's before his letter of 21 June 1677, more than a year after he received a copy of Newton's letter of 10 December 1672 and more than four years after Collins began to communicate its contents to correspondents, "In which Letter the Method of Fluxions was Sufficiently Described to any Intelligent Person." Third, that Newton's letter of 13 June 1676 (the *Epistola prior*) showed that he had the method of fluxions more than five years before, and that "De analysi," which he communicated in 1669, proved that he had it still earlier.

4^{thly} That the Differential Method is One and the same with the Method of Fluxions Excepting the name and Mode of Notation . . . and therefore wee take the Proper Question to be not who Invented this or that Method but who was the first Inventor of the Method, and wee beleive that those who have reputed M^r. Leibnitz the first Inventor knew little or Nothing of his Correspondence with M^r. Collins, and M^r. Oldenburg Long before, nor of M^r. Newtons haveing that Method above Fifteen Years before M^r. Leibnitz began to Publish it, in the Acta Eruditorum of Leipsick.

For which Reasons we Reckon M^r. Newton the first Inventor and are of Opinion that M^r. Keill in Asserting the same has been noways Injurious to M^r. Leibnitz and wee Submitt to the Judgment of the Society whether the Extract of the Letters and Papers now Presented Together with what is Extant to the same Purpose in Doctor Wallis's third Volume may not deserve to be made Publick.[83]

Leibniz used to say, and others have repeated, that the report only expressed the judgment of the committee, which the society never endorsed. That is not true. The minutes record that the society "agreed" with the report unanimously *(nemine contradicente)* and ordered it to be printed together with the supporting documents.

[83] *JBC 10,* 389–91. The report, with the parts of it that are found in Newton's draft italicized, is published in *Corres 5,* xxvi. That draft, which is in private possession, is not generally accessible. Another draft of part of the report, or perhaps a draft for the later republication of it, is in *Add MS* 3968.4, ff. 15–16. As the report makes clear, the supporting documents, which make up the bulk of *Commercium epistolicum*, were submitted with it. There are many other papers in *Add MS* 3968 with such things as extracts from Collins's papers that are related to the total report. I may not have located all of them, but among them are 3968.10, ff. 121–8, 140–4; 3968.14, ff. 246–7; 3968.19, ff. 263–95.

The society further paid the full expenses of the publication.[84] The report could scarcely have been more official.

The resulting volume, *Commercium epistolicum D. Johannis Collins, et aliorum de analysi promota* (*The Correspondence of the Learned John Collins and Others Relating to the Progress of Analysis*), appeared early the following year. Copies of it were shown at the society's meeting on 8 January 1713, though Newton ordered members of the committee to examine them before the work was issued.[85] Only a small number were printed. They were not generally offered for sale, although the society sent twenty-five copies to a Scots bookseller in The Hague, T. Johnson, who was making himself useful to Newton at that time. Johnson had earlier announced that he had a copy at his shop.[86] Rather, the Royal Society sent copies to key figures in the learned world of Europe, to mathematicians "who were able to judge these matters," as the preface to the second edition stated.[87] As the report was Newton's, so was the book. Numerous drafts of the preface, many of them going beyond the one finally printed to offer an extended commentary on the dispute, exist among his papers.[88]

Not much needs to be said about *Commercium epistolicum,* since it merely repeated the earlier report of the committee, undoubtedly further polished during the intervening months according to Newton's habit. The overwhelming bulk of the volume consisted of letters and papers and extracts chronologically arranged, leading up to the judgment of the committee, which it printed in full in its original English. For those who might have missed Jones's volume, it reprinted "De analysi," with which it effectively began, together with the correspondence that established the date of that treatise. Footnotes of the most partisan kind, pointing out the fluxional method in Newton's papers and letters, and denigrating the letters of Leibniz, provided a running commentary. The footnotes painted Leibniz as a compulsive plagiarist. The content of his first letter in the volume turned out (if one accepts the note) to have been stolen from Pascal.[89] His third contained another theft, a series for the quadrature of the circle. Collins had communicated similar series discovered by Newton and Gregory three and four years earlier, an indignant footnote explained. Leibniz had been in England before

---

[84] The council voted to pay the expense of publication on 29 Jan. 1713 (*CM 2*, 259). On 11 June they paid Halley, who had overseen the publication, £22 2s 6d for it (*CM 2*, 264).

[85] *JBC 10*, 438.

[86] 17 June 1714; *JBC 11*, 4. *Journal literaire de la Haye, 1* (May and June, 1713), 2nd ed., 209.

[87] "Ad lectorem" to the second edition; *Comm epist* (Biot-Lefort ed.), p. 8.

[88] *Add MS* 3968.37, ff. 539–57. Cf. "Annotationes in Commercium Epistolicum," *Add MS* 3968.14, ff. 236–43. With page references to the published work, these annotations had to come after it and were possibly intended for a later edition, which Newton did in fact finally issue.   [89] *Comm epist,* p. 37.

he wrote the letter in question, "and did not communicate series of this sort, nor did he begin to communicate them with his friends before he had received them from Oldenburg, as will soon appear; nor did he communicate anything except what he received."[90] With Leibniz's letter of 21 June 1677, his response to the *Epistola posterior* in which he expounded his differential method for the first time in the correspondence, a similar note inevitably appeared. The climax of *Commercium epistolicum,* this note pointed out that Leibniz did not send his method before he had read "what Newton had written of this method in two letters and also other papers of Newton when he returned home through London toward the end of 1676."[91] The passion Newton invested in the argument as he took his revenge burst out in mathematical notes where one scarcely expects to find it. Leibniz's letter of 12 July 1677 remarked that he had found some of Newton's series among old papers of his own that he had forgotten. Newton's indignation erupted. Leibniz had received the series two years earlier, asked for the method behind them, received it from Oldenburg, had trouble understanding it, and now found he had discovered it earlier himself! So it had been with other series. "Thus the method which earlier he wanted, asked for, received, and understood with difficulty, he discovered forsooth either first or at least by his own effort."[92] Borne forward by his own passion, Newton did not hesitate even to assert that he had sent the principle propositions of the *Principia* to the Royal Society in 1683.[93] Since it was part of his argument that he used the fluxional method to work out the *Principia*'s propositions, which he later set over into synthetic demonstrations for publication, 1683 placed a major exercise of his method before Leibniz's publication of the calculus. The argument had only one defect, namely that it was false. Such did not seem significant in the climate of rage.

Commercium epistolicum was a brilliant exercise in partisan polemics which testified to Newton's continuing mental vigor as he approached the age of seventy. The total impact of the notes, the total impact of the whole volume in the absence of anything in Leibniz's defense, is devastating. Perhaps it is too devastating. Swept along by his own fury, Newton failed to recognize the utility of moderation. No doubt the volume informed a particular public of events that Leibniz had not been forward to advertise. It is not clear that it was convincing, however. Leibniz had made too deep an impression on the learned world to be hustled off the stage as a fraud at this point in his career. "Since it does not appear that M. Leibniz is satisfied with this decision," the *Journal des sçavans* commented

90 *Comm epist*, p. 38.   91 *Comm epist*, p. 88.
92 *Comm epist*, p. 96.   93 *Comm epist*, p. 97.

dryly, "the public will no doubt receive further information on the subject from him."[94]

The second edition of the *Principia* hardly moved during 1711 as the dispute with Leibniz, not to mention Flamsteed's *Historia coelestis* and a busy year at the Mint, distracted Newton. In June, the problems with Proposition XXXVI and Section VII, Book II, as a whole, which had stalled the press for nearly a year, were finally resolved. Bentley brought all the rest of the copy from London, and the press seemed to move once more. Alas, Cotes found new problems with Proposition XLVII in Section VIII, on the propagation of pulses such as sound through elastic media. After he had waited nearly a month for a reply, he wrote again.

> If You have received those sheets, You will perceive by them that y[e] Press is now at a stand. But having no Letter from You I fear the sheets have miscarried. The Compositor dunn's me every day, & I am forc'd to write to You again to beg Y[r] Resolution.[95]

Ten days later, Newton finally replied. He offered the pressure of other affairs by way of apology. Now he was home sick for a few days and could find time to write. As to Cotes's problem with Proposition XLVII, he rejected his argument. Cotes was sufficiently unwise to press the objection. A month of silence greeted him. Early in September he wrote again, repeating needlessly that the press was at a stand.[96] Newton replied on 2 February 1712.

> S[r]
>
> I have at length got some leasure to remove the difficulties w[ch] have stopt the press for some time, & I hope it will stop no more. For I think I shall now have time to remove the rest of your doubts concerning the third book if you please to send them.[97]

As for Proposition XLVII, he stood his ground again and refused to budge. At this point, the edition, which had progressed swiftly through 296 pages in the first nine months, had managed to add a scant 40 more in the following nineteen. Cotes's enthusiasm for correcting the perceived defect in the proposition on pulses had waned. "I have received Your Letter," he wrote, "& as to ye buisness of Sounds I do intirely agree with You upon considering that matter over again."[98]

As Cotes quickly learned, Newton meant to be as good as his word about proceeding with Book III, even though Leibniz's sec-

[94] *Journal des sçavans*, 1713, p. 286.
[95] Cotes to Newton, 23 June and 19 July 1711, *Corres 5*, 167, 174.
[96] Newton to Cotes, 28 July 1711; Cotes to Newton, 30 July and 4 Sept. 1711; *Corres 5*, 179, 183, 193.   [97] *Corres 5*, 215.
[98] Cotes to Newton, 7 Feb. 1712; *Corres 5*, 220.

ond letter had arrived only three days before he wrote and he was already contemplating the committee and its report, which would fill part of the coming three months. Perhaps a year's labor over the documentary record had left him confident in his command of the facts, or perhaps he now saw more fully the relevance of the new edition to the dispute. Whatever the cause, he set his mind seriously on his masterwork for the last time, and the edition proceeded to its conclusion, though not without two more minor interruptions.

The philosophic differences of Newton and Leibniz were of long standing. Leibniz could not accept the concept of attractions, which he considered a return to the occult qualities of scholastic philosophy. As the calculus controversy heated up, the philosophic issues inevitably became involved. In his *Theodicy* (1710), Leibniz expressed his disagreement with the notion of action at a distance. Many other references to their differences, some of the references more acceptable than others, soon followed. One of the pieces that Keill showed Newton in 1711 was the review in the *Acta* – again anonymous – of John Freind's application of Newtonian concepts of forces to chemistry as a return to "a certain fantastic scholastic philosophy, or even the enthusiastic philosophy of Fludd." Freind responded in the *Philosophical Transactions* with a vigorous defense of Newtonian philosophy and the concept of attractions, a defense that frequently referred to Leibniz, and the question held enough interest to elicit a reply in the *Acta*. In 1710, Keill also exchanged letters in the *Acta* with Christian Wolf on the same topic. About the time when the Royal Society appointed its committee in March 1712, the *Mémoires de Trévoux* published a philosophic exchange between Leibniz and Hartsoeker in which Leibniz went out of his way to insert a slighting reference to people who, following Roberval, believed that bodies attract each other and thus abandoned natural causes for miracles.[99] Newton's reply to the philosophic criticisms, to which he began at this time to devote more attention, contrasted his experimental philosophy with the hypothetical phi-

---

[99] "Praelectiones chymicae . . . a Johanne Freind, M.D. . . . ," *Acta eruditorum* (September 1710), pp. 412–16; I quote p. 412. "Johannis Freind, M.D. Oxon. Praelectionum chymicarum vindiciae, in quibus objectiones . . . contra vim materiae attractricem allatae, diluuntur," *Philosophical Transactions, 27* (1710–11), 330–42. "Responsio ad imputationes Johannis Freindii . . ." *Acta eruditorum* (July 1713), pp. 307–14. "Johannis Keill . . . , epistola ad clarissimum virum Christianum Wolfium . . . ," *ibid.* (January 1710), pp. 11–15. "C. W. Responsio ad epistolam viri clarissimi Johannis Keill . . . ," *ibid.* (February 1710), pp. 78–80. "Lettres de Monsieur le Baron de Leibnits à Mr. Hartsoeker, avec les responses de Mr. Hartsoeker," *Mémoires pour l'histoire des sciences et des beaux arts* (known as the *Mémoires de Trévoux*) (March 1712), pp. 494–523. The thrust at Newton is on p. 502. The Leibniz-Hartsoeker correspondence was reprinted in *Memoirs of Literature, 2* (1712), 5 May 1712, pp. 137–43, and in the Amsterdam edition of the *Journal des sçavans* (December 1712), pp. 603–25.

losophy of his antagonists. He did not, he insisted, attempt to teach the causes of phenomena except insofar as experiments revealed them. He did not wish to fill philosophy with opinions that experiments could not prove. His critics found it a fault that he did not offer some hypothesis about the cause of gravity, as though it were an error not to dilute demonstrations with speculations.[100] Thus the major thrust of the new edition was to further emphasize that feature of the *Principia* which had opened itself to him in his final expansion of the work. The second edition changed little in Book I, where the consideration of the gross features of the universe led to the recognition of inverse-square attractions. The classic demonstrations of the dynamic foundation of Kepler's laws required no revision. He concentrated rather on further enhancing the work's derivation of the quantitative details of physical phenomena, that revolutionary feature which endless pejorative incantations against occult qualities could not conjure away. Such was the issue behind Section VII of Book II, on which the edition lay marooned for a year, whatever the success of the outcome. Such were the major revisions of Book III, the final polishing of which still lay ahead. The dispute about the calculus did not cause these changes. Some of them were already present in the copy Newton sent to Cotes in September 1710, before Leibniz's letter intervened. The changes only repeated Newton's deepest convictions about the nature of the scientific enterprise, that if its business began with observation, only the derivation of exact quantitative relations concealed in observations deserved the name of science. James Jurin once reminded Martin Folkes of a saying frequently in Newton's mouth: "That Natural History might indeed furnish materials for Natural Philosophy; but, however, Natural History was not Natural Philosophy."[101] The conflict with Leibniz, and its philosophic overtones, loaned these issues new urgency.

Book III opened with a declaration of philosophic principle in a new rule of reasoning. In the first edition, a set of nine "Hypotheses" had introduced the book. Newton now broke them up into three groups, Rules of Reasoning in Philosophy, one Hypothesis (which he no longer placed at the beginning of Book III), and Phaenomena. To the two former hypotheses (relabeled "Rules") he now added a third, perhaps his most important statement of epistemology.

> The qualities of bodies, which admit neither intensification nor remission of degrees, and which are found to belong to all bodies within the reach of our experiments, are to be esteemed the universal qualities of all bodies whatsoever.

100 "Account of Commercium epistolicum," pp. 222–3.
101 Jurin to Folkes, 1728; Nichols, *Literary Anecdotes, 3,* 320.

The extended discussion of Rule III, addressed to Cartesians, to mechanists in general, and to Leibniz in particular, contrasted his empirical experimental philosophy with hypothetical philosophy. "For since the qualities of bodies are only known to us by experiments, we are to hold for universal all such as universally agree with experiments . . . We are certainly not to relinquish the evidence of experiments for the sake of dreams and vain fictions of our own devising . . ." Experience, not reason, teaches us that bodies are hard, extended, impenetrable, mobile, and endowed with inertia, and from our experience with gross bodies we assign the same qualities to their ultimate particles.

> Lastly, if it universally appears, by experiments and astronomical observations, that all bodies about the earth gravitate towards the earth, and that in proportion to the quantity of matter which they severally contain; that the moon likewise, according to the quantity of its matter, gravitates towards the earth; that, on the other hand, our sea gravitates towards the moon; and all the planets one towards another; and the comets in like manner towards the sun; we must, in consequence of this rule, universally allow that all bodies whatsoever are endowed with a principle of mutual gravitation.[102]

In the third edition he added a cautionary note that he did not affirm gravity to be essential to bodies.

Cotes had no comment to make on Rule III. In the intense correspondence about the details of Book III that now commenced between the two, the first issue that detained them was a new scholium to Proposition IV, the correlation of the moon's orbit with the measured acceleration of gravity at the surface of the earth, the linchpin in the argument for universal gravitation. In the first edition, Newton presented a modest correlation which found that the centripetal acceleration of the moon corresponds to a measured value of ½ g of 15 feet 1 inch (Parisian feet, as used in Huygens's measurements, one of which equals 1.068 English feet), a degree of accuracy of one part in one hundred and eighty. In the new correlation he allowed not only for the disturbing force of the sun, which had been concealed in the original calculation, but also for the loca-

---

[102] *Prin*, pp. 398–400. On Rule III and the general issue of Newton and method, see A. C. Crombie, "Newton's Conception of Scientific Method," *Bulletin of the Institute of Physics, 8* (1957), 350–62; Ralph M. Blake, "Sir Isaac Newton's Theory of Scientific Method," *Philosophical Review, 42* (1933), 453–86; H. R. Burke, "Sir Isaac Newton's Formal Conception of Scientific Method," *New Scholasticism, 10* (1936), 93–115; J. E. McGuire, "Transmutation and Immutability: Newton's Doctrine of Physical Qualities," *Ambix, 14* (1967), 69–95; Robert Kargon, "Newton, Barrow and the Hypothetical Physics," *Centaurus, 11* (1965), 46–56; A. J. Snow, "The Role of Mathematics and Hypothesis in Newton's Physics," *Scientia, 42* (1927), 1–10; and E. W. Strong, "Newton's 'Mathematical Way'," *Journal of the History of Ideas, 12* (1951), 90–110, and "Newtonian Explications of Natural Philosophy," *ibid., 18* (1957), 49–83.

tion of the common center of gravity of the earth and the moon and for the centrifugal effect arising from the rotation of the earth, arriving at a correlation correct to 15 feet 1 inch 5½ lines (1 line = 1/12 inch), a degree of accuracy now of one part in about four thousand, if we take the fraction seriously, as Newton did.[103] Newton did not present any new data to justify the new calculation, however. He started from exactly the same body of data used in the first edition, Huygens's accurate measurement of g by means of a pendulum, and several observations of the distance of the moon. Indeed, the first edition had cited one more observation, Kircher's value of 62½ earth radii, which Newton eliminated now because it would have set the distance of the moon too high if averaged in. The accuracy of the correlation depended on the distance of the moon that he used. In order to be valid, the correlation had to start from two independently determined measurements, g and the distance of the moon. Since the correlation used the common center of gravity of the earth and the moon, Newton eventually shifted it to Corollary 7 of Proposition XXXVII, where consideration of the tides led to the determination of the common center of gravity. There it emerged as what it truly was, a calculation of the distance of the moon from the earth that started with a very precise value of g.[104] When he thought originally of putting it with Proposition IV, Newton was engaging in a form of quantitative polemics, bolstering his philosophy by pretending to a degree of accuracy beyond any legitimate claim.

Cotes, who was thoroughly committed to the Newtonian school and understood the argument being advanced, had a suggestion to make. "In the Scholium to the IVth Proposition I think the length of the Pendulum should not be put 3 feet & 8²⁄₅ lines; for the descent will then be 15 feet 1 inch 1⅓ line. I have considered how to make that Scholium appear to the best advantage as to the numbers & I propose to alter it thus."[105] He proceeded to suggest specific values for the size of the earth and the latitude of Paris that led to a correlation accurate not merely to ½ line but to 1/6 line. In the end Newton settled for slightly different numbers and an accuracy of one part in a mere two thousand. Nevertheless he too was concerned to make the edition appear to the best advantage as to the

103 The draft of the intended scholium is printed in *Corres 5*, 216–18. As I have argued, it is interesting that he intended originally to make the correlation a scholium to Proposition IV. He later broke it up between Propositions XIX and XXXVII. In Proposition XXXVII it is really a calculation of the distance of the moon. In Proposition IV the distance of the moon was a given, and the calculation pretended to give an exact correlation.

104 *Prin*, pp. 482–3. The numbers in the third edition differed a little from those in the second, which differed in turn from those in the sheet Newton sent to Cotes in February 1712. Cf. Proposition XIX, pp. 424–5, where he inserted part of the correlation.

105 Cotes to Newton, 16 Feb. 1712; *Corres 5*, 226.

numbers. Quantitative precision provided the unanswerable argument for his philosophy. He was not above placing it in the most favorable light possible. Among his papers on the dispute with Leibniz is one that contains a calculation of the correlation on the same sheet.[106]

Already in Book II, he had carried out another computational sleight-of-hand to give a similar pretense of precision. The derivation of the velocity of sound from first principles of dynamics had been one of the *Principia's* triumphs. Newton had arrived at a calculated velocity of 968 (English) feet per second, a figure which settled easily between the limits of 600 and 1,474 feet per second in the received measurements, and between the narrower limits of 920 and 1,085 feet per second established by his own experiments. Newton had not been satisfied with this aspect of his work. In 1694, Gregory found him planning to improve it, and sometime he entered the results of new experiments performed in the arcade at Trinity, which raised the limiting velocities to 984 and 1,109.[107] The lower limit was now higher than his own calculated velocity of sound. A sheet of errata in the *Principia*, which probably stood in close relation to the new experiments, included computational experiments in which he tried the effects of differing densities of air on the result. The ratio of the density of water to the density of air entered into the derivation of the velocity. In the first edition he had used the figure 850. In the light of available measuring techniques the figure was arbitrary, and Newton now tried 900 and 950, raising the derived velocity to 996 and 1,023 feet per second.[108] Before the second edition came out, new measurements by William Derham of the Royal Society and Joseph Sauveur of the Académie des sciences raised the measured velocity higher yet.[109] They were much the most careful measurements so far; Newton could not ignore them. He chose Derham's figure, 1,142 feet per second, and cited Sau-

---

[106] *Add MS 3968.30, f. 441ᵛ.*

[107] *Corres 3*, 384. *Var Prin 1*, 531–3. Nearly all of the figures pertaining to the new experiments were written over and altered, and it is difficult to know what numbers Newton meant at any given stage (Cambridge University Library, Adv.b.39.1, f. 370A).

[108] *Var. Prin 1*, 528.

[109] On 28 Feb. 1705, the Royal Society thanked Lord Granville of the Ordnance for firing guns at Black Heath for Derham's benefit (*JBC 10*, 97–8). Derham's measurements for different distances, some as much as twelve miles, timed the arrival of the sound of a shot after the flash of the cannon was seen. He published his results in the *Philosophical Transactions* in 1708, arriving at an average velocity of 1,142 feet per second ("Experimenta & observationes de soni motu, aliisque ad id attinentibus," *Philosophical Transactions, 26* [1708–9], 2–35). Sauveur correlated the lengths of organ pipes with frequencies and found that a frequency of 100 vibrations per second corresponded to a pipe 5.12 French feet long, yielding a calculated velocity of 1,024 French feet (= 1,094 English feet) per second (Joseph Sauveur, "Sur la détermination d'un son fixe," *Histoire de l'Académie royale des sciences*, 1700, pp. 166–78.)

veur's slightly lower measurement as confirmation of the general range. No ratio of the densities of water and air within reason could adjust that discrepancy.

The new derivation of the velocity of sound in the second edition differs from the new correlation of the moon with g because new empirical data did enter in. Since Newton did not alter his basic analysis, however, the new experimental figures served, not to increase his precision, but only to perplex him. His calculated velocity of sound was roughly 20 percent too low. More than a century later, Laplace demonstrated that the necessary correction follows from the heat generated in the compressions of sound waves and is equal to the square root of the ratio of the specific heat of air at constant pressure to its specific heat at constant volume. With this correction, Newton's analysis remained valid. That scarcely helped Newton as he prepared the second edition in an age still struggling to separate the concepts of temperature and heat.

Meanwhile, he had no intention of exposing a 20 percent deficiency to the barbs of continental critics. He began to look for other adjustments. The same sheet that recorded his own new measurements of velocity contained the first mention of one that he eventually included, the "crassitude" of the particles of air.[110] From the ratio of densities, he argued that the dimensions of particles of air occupy between one-ninth and one-tenth of any linear distance in air (i.e., $9^3 < 850 < 10^3$). Hence the diameter of a particle is between one-eighth and one-ninth the distance between particles, a figure he then silently set at one-ninth. Since sound travels instantaneously through solid particles, the calculated velocity must be increased by 11 percent. Later he found another adjustment. Vapor in the air does not vibrate with the air, he reasoned, and the calculated velocity must be increased further in proportion to the square root of the amount of air the vapor displaces. If the atmosphere holds one part of vapor to ten parts of air, the velocity will be increased by $(11/10)^{1/2}$ or about 21/20. Together the two adjustments raised the calculated velocity to 1,128 feet per second. Here he could bring the ratio of densities into play. Setting it at 870, he arrived, *mirabile dictu*, at a velocity of 1,142 feet per second, perhaps a supernumerary effort, since Derham's figure was merely the average of a number of measurements.[111] The passage is one of the most embarrassing in the whole *Principia*, since the adjustments rested on no empirical grounds whatever, and in their manifest hollowness served only to cast undeserved doubt on the basic analysis. In its very flagrancy,

---

110 Cambridge University Library, Adv.b.39.1, f. 370A. *Var Prin 1*, 532–3.
111 *Prin*, pp. 382–4.

however, the adjusted derivation gives us insight into the polemic goal behind the pretense to a higher degree of precision.

As he completed Book III, Newton doctored still another demonstration in his effort to create an illusion of great accuracy. Proposition XXXIX, on the precession of the equinoxes, was one of the demonstrations drawn from the new corollaries to Proposition LXVI, Book I. By considering the third body, not as a satellite, but as a ring of solid matter contiguous to the earth, Newton calculated from the regression of the lunar nodes to the motion of the ring of matter about the earth's equator produced by the centrifugal effect of rotation. In the first edition, he had arrived at a derived precession of 49″ 58‴ per year, in good agreement with the observed value of 50″. Aside from observed astronomical quantities, three factors entered into the calculation. The attractions of both the sun and the moon contribute to precession, with the moon several times more effective because of its proximity. Newton established the ratio of their attractions from measurements of spring and neap tides made in 1668 by Samuel Sturmy at Bristol and Samuel Colepresse at Plymouth.[112] Sturmy's measurements were more accurate, and in the original version of Book III Newton had used them. The combined attractions of the moon and sun [*L(una)* + *S(ol)*] generate the spring tide; the excess of the moon's attraction over the sun's produces the neap tide, which occurs when the moon is in quadratures. Hence $(L + S)/(L - S) = 9/5$, Sturmy's measurement of the ratio of spring tide to neap. However, this equation required modification since the moon in quadratures declines 23½° from the plane of the equator. Hence he had set $(L + S)/(.841\ L - S) = 9/5$, $L = 5\frac{5}{11}\ S$, a figure he had adjusted to $5\frac{1}{3}$.[113] "The System of the World" did not attempt to calculate precession. The final version, Book III, did, and the ratio $5\frac{1}{3}$ would not work. Hence Newton had averaged Sturmy's observations with those of Colepresse to compute a ratio of $6\frac{1}{3}$. This had led to a precession of about 41″ 43‴. Here he had introduced the second factor, the varying density of the earth, which causes the diameter at the equator to exceed that at the pole even more than the calculation from centrifugal force alone reveals. He had decided that this adjustment amounted to an increase of exactly one-fifth and had arrived consequently at his final figure for precession.[114] The third factor had entered earlier. The initial comparison of the ring of the matter

---

[112] "An Account of some Observations, made this present year by Capt. Samuel Sturmy in Hong-road within four miles of Bristol, in Answer to some of the Quaeries concerning the Tydes," *Philosophical Transactions, 3* (1668), 813–17. "Of some Observations made by Mr. Samuel Colepresse at and nigh Plymouth, An. 1667. by way of Answer to some of the Quaeries concerning Tides," *Philosophical Transactions, 3* (1668), 632–3.

[113] "The System of the World"; *Prin*, p. 593.   [114] *Var Prin 2*, 666–7, 683–5.

about the earth to the moon assumed that the ring is concentrated entirely at the equator whereas it is in fact spread unevenly over an oblate spheroid. In the first edition, a faulty lemma had concluded that the effect of the matter spread over an oblate spheroid is one-quarter the effect of the same matter in a ring. By 1694, he had realized that the lemma contained an error.[115] It was apparently one of the two things in the *Principia* that still bothered him in 1708. In the second edition, two new lemmas replaced it. The essence of Newton's problem with the precession in the second edition lay in the fact that the new lemmas increased the effect of the matter in an oblate spheroid to turn the earth from one-quarter to two-fifths. Without further change, it would have left him with a precession more than 50 percent too large.

To adjust the difference, Newton turned back to the tides. In February 1712, Cotes began to comment on Proposition XXXVII, which derived the ratio L/S. From his comments it appears that Newton's first step had been to drop Colepresse's observations and thus to reduce the ratio again to about $5\frac{1}{3}$. Cotes remarked that this figure would alter those in the corollaries to the proposition, especially the comparative masses of the earth and moon: "This alteration will very much disturb Your Scholium of y^e IVth Proposition as it now stands; neither will it well agree with Proposition XXXIX . . ."[116] Newton replied that he could not make the corrections because he had lost his copy of the amended Proposition XXXIX. "If you can mend the numbers so as to make ye precession of the Equinox about 50″ or 51″, it is sufficient."[117] After what had gone before, the laxity is surprising. Cotes had no intention now of settling for so little. He brought up the factor of the earth's density used in the first edition. Newton had simply omitted it in the new demonstration as one means of gaining a reduction, even though the edition explicitly stated that the diameter at the equator exceeds that at the pole, not by seventeen miles, but by thirty-two. Cotes thought the issue needed explicit comment. Newton proposed a brief statement that the increase in precession due to the added height at the equator was compensated by the decrease due to the greater rarity. "You have very easily dispatch'd the 32 Miles in Prop XXXIXth . . . ," Cotes noted wryly.[118]

Newton had scarcely begun to exercise his skills, however. He turned his attention now to the ratio L/S. In its original form, it had been rather crude. He set out now to refine it. The highest tides do not occur exactly at syzygies but when the luminaries are about 15° beyond conjunction or opposition. Hence he must reduce the force

[115] Newton to Gregory, 14 July 1694; *Corres* 3, 380.
[116] *Corres* 5, 220–1.   [117] *Corres* 5, 222–3.
[118] Cotes to Newton, 23 Feb. 1712; *Corres* 5, 233.

of the sun at both its positions in the equation. The new equation yielded a ratio of 4⁵/₇, from which he calculated a precession of a little under 52″.

> If the force of the Moon in moving the Sea [he continued in a draft of the proposition] were to the force of the Sun as 4⁴/₇ to 1 (for the proportion of these forces cannot yet be determined very accurately from the phaenomena), an annual precession of the Equinoxes of 50″ 40‴ 43^{iv} would result.[119]

Later he carried his exploration a step further. If the ratio were 4½, the precession would be 50″ 14‴ 45^{iv}.[120]

At this point, Cotes introduced a new difficulty. According to Newton's lunar theory, the moon approaches the earth more closely in syzygies than in quadratures. Were he to allow for this, it would increase the ratio so much as to upset Proposition XXXIX unless, he suggested, "You think fit to ballance it some other way for there is a latitude in that XXXVIIth Proposition."[121] When Newton expressed some skepticism, he did the arithmetic, showing that it would increase the ratio to 5²/₇.[122]

Over a month passed while Newton coped with the new problem, not to mention the Royal Society's committee to consider Leibniz's complaint. The revision that he finally produced both incorporated Cotes's new adjustment and attained the ratio of 4½ by exploiting further the argument, which became critical to the whole demonstration, that the highest tides occur sometime after syzygies. The empirical information was sufficiently vague that he could set the angle at any value he chose within the limits with which he was working. In February, Newton was using an angle of 15¼°. By increasing it to 17½°, he got the further reduction he needed. Beginning now to see the possibility of another impressive demonstration, he shifted from fractions to decimals of seven places.[123] Cotes was not yet satisfied, however; he still found fault with the correction for the moon's distance. "I could wish when the whole is settled that the proportion of 4½ to 1 may be retained for the sake of Proposition XXXIX," he added. "I think there is no Proposition in Your Book which does more deserve Your care than ye XXXIXth."[124] Newton was willing. He had found the key. He not only accepted Cotes's new correction; but in compensating for it by a further adjustment of the angle, he pushed the ratio below 4½, with of course a further improvement in the calculated preces-

---

[119] *Corres 5*, 228.   [120] Newton to Cotes, 19 Feb. 1712; *Corres 5*, 230–1.

[121] Cotes to Newton, 23 Feb. 1712; *Corres 5*, 235–6.

[122] Cotes to Newton, 28 Feb. 1712; *Corres 5*, 242–4.

[123] Newton to Cotes, 8 April 1712; *Corres 5*, 263–4. The draft of Proposition XXXVII; *Corres 5*, 264–7.

[124] Cotes to Newton, 15 April 1712; *Corres 5*, 271–2.

sion. Cotes was delighted. "I am very glad to see the whole so perfectly well settled & fairly stated, for without regard to y^e Conclusion, [*sic*!] I think the distance of 18½ degrees ought to be taken being much better than 17½ or 15¼. And y^e same may be said of y^e other changes in the principles from which the conclusion is inferr'd."[125] The equation which had started out as $(L + S)/(.841 L - S) = {}^9/_5$ now read[126]

$$\frac{1.017342\ L + .7986355\ S}{.9828616 \times .8570328\ L - .7986355\ S} = \frac{9}{5} \qquad L = 4.4815\ S$$

Using this ratio, he carried through the correlation of the moon with g, now moved to its position in Corollary 7 of the proposition, and calculated a precession of $50''\ 00'''\ 12^{iv}$, a degree of precision better than one part in three thousand. The ratio ${}^9/_5$ on which it depended came from the measurements over a period of one year of a retired sea captain, who stressed how difficult it was to make such measurements and who summed up two unequal columns of figures to "45 feet circiter."

Only when they had settled the precession did Cotes and Newton come to grips with lunar theory, although it occupied Propositions XXV–XXXV of Book III. It was probably inevitable that nothing important would come of it. Newton had exhausted himself on lunar theory in the 1690s and was incapable now of seriously amending it. The changes in lunar theory in the second edition, especially the new scholium that concluded it, had been worked out at the earlier time. Another issue that entered their correspondence at the same time also worked against a fruitful exchange about the moon. Cotes introduced it in the same letter that greeted the final resolutions of Propositions XXXVII and XXXIX.

> I am glad to understand by Dr Bentley [Cotes ventured, at the end of the letter] that You have some thoughts of adding to this Book a small Treatise of Infinite Series & ye Method of Fluxions. I like the design very well, but I beg leave to make another Proposal to You. When this Book shall be finished I intended to have importun'd You, to review Your Algebra for a better Edition of it, & to have added to it those things which are published by Mr Jones & what others You have by You of the like nature. . . . Your Treatise of ye Cubick Curves should be reprinted, for I think ye Enumeration is imperfect, there being as I reckon five cases of Aequations. . . . I should have acquainted You with this before Mr Jones's Book was published if I

---

[125] Cotes to Newton, 26 April 1712; *Corres 5*, 278–80.

[126] *Corres 5*, 274–5. In his usual way, Cotes went through it all again and made some small changes in the numbers that did not alter the final ratio. On the pretended level of precision in the second edition, see Richard S. Westfall, "Newton and the Fudge Factor," *Science, 179* (1973), 751–8.

had known any thing of yᵉ Printing of it, for I had observed it two
or three Yeares ago. I think there are some other things of lesser
moment amiss in yᵉ Treatise.[127]

Cotes had presumed too far. He was mistaken as well as presumptuous for his additional case of cubics reduced to two of Newton's
four cases. For nearly three months, the two men had carried on an
intense intellectual exchange. Cotes's letter of 26 April terminated
it. Newton never even answered to the proposal here made, and he
let three letters accumulate before he replied at all. Though Newton
did write twice during May, both letters were brief, even curt. He
did not write again for three months while Cotes waited impatiently for an answer to questions on lunar theory that had been left
dangling. In May, when he had apparently not yet sensed the
change in tone, Cotes sent Newton a treatise he had composed on
logarithms, virtually begging Newton to comment on it and tell
him what might need correction before he ventured to publish it.
Eventually, after two months, Newton sent it back with no more
than a laconic note that he was not displeased with it, a comment
which he repeated later in as few words.[128] Cotes did manage
briefly to revive the dialogue, now about the scholium to the lunar
theory, in September. Newton was clearly impatient with it. By
and large, he refused to give ground to Cotes's objections, though
he did accept corrections to a number of hasty errors and thanked
Cotes for insisting on one.[129] He tired of the new exchange
quickly, and with another curt letter on 23 September he simply
ended it.

> The alterations made in the last Paragraph of the Scholium were
> advisedly. The description of the Variatio secunda is derived only
> from phaenomena & wants to be made more accurate by them that
> have leasure & plenty of exact observations. The public must take it
> as it is.[130]

Among Newton's papers there is a draft of a preface for the edition,
with a handsome tribute to "the very learned Mʳ Roger Cotes," his
collaborator, who corrected errors and advised him to reconsider
many points.[131] At some point, probably about this time, he suppressed it, and the new edition appeared without any mention of
Cotes other than his name at the end of the preface he wrote.

Three weeks after they finished with the moon, on 14 October,
another letter from Newton abruptly announced. "There is an error

---

[127] Cotes to Newton, 26 April 1712; *Corres 5*, 279–80. Cf. note 19, p. 281: as Whiteside
shows, Cotes's additional case reduces to Newton's first and third cases.
[128] Cotes to Newton, 25 May and 10 August; Newton to Cotes, 12 Aug. 1712; *Corres 5*,
305–6, 318–19, 320–2.
[129] Newton to Cotes, 13 Sept. 1712; *Corres 5*, 337–8.
[130] *Corres 5*, 341.   [131] *Corres 5*, 112–13.

in the tenth Proposition of the second Book, Prob. III, wch will require the reprinting of about a sheet & an half."[132] The origin of this blunt message was a visit of Nikolaus Bernoulli, the nephew of Johann Bernoulli, to London that fall. The best account of what happened is found in a letter of DeMoivre to Johann Bernoulli on 18 October, four days after Newton's letter to Cotes. DeMoivre told Bernoulli that he had introduced his nephew to Newton and Halley.

> We have together seen Mr. Newton three times, and he was so amiable as to invite us twice to dine with him; I must not omit a remarkable circumstance; your nephew having told me that he had an objection against a result in Mr. Newton's book concerning the motion of a body describing a circle in a resisting medium [Proposition X, Book II], and having imparted this objection to me, I thereupon on his behalf, showed it to Mr. Newton: Mr. Newton said he would examine it and two or three days later, when I had gone to his house, he told me that the objection was valid, and that he had corrected the result; indeed, he showed me his correction, and it proved agreeable to the computation made by your nephew; thereupon he added that he intended to see your nephew in order to thank him, and begged me to bring him to his house, which I did. Moreover, Mr. Newton affirms that this error is the simple consequence of his having considered a tangent at the wrong end, but that the foundation of his calculation and the conclusions he has drawn from it may still stand.[133]

In Proposition X, on the motion of a projectile under uniform gravity through a medium that resists in proportion to velocity squared, Newton derived a general expression for the ratio of resistance to the force of gravity at any point on a trajectory. Using o to represent an infinitesimal increment of the abscissa, he had employed his binomial expansion to develop a series in powers of o to express the ordinates at $x - o$ and $x + o$, terms of which he used in deriving the ratio of resistance to gravity. As DeMoivre said, in the first edition he used a tangent to the curve at x,y and mistakenly calculated the deviation of the tangent at $x - o$, y $_{(x - o)}$, when he should have calculated the deviation at x,y from the tangent to $x - o$, y $_{(x - o)}$ (Figure 14.4). As a result, his expression of the ratio was only two-thirds the value that Bernoulli found by a different method. Apparently Bernoulli could not locate the error in Newton's demonstration, however. It was rather his alternative demonstration which Nikolaus Bernoulli showed to Newton. One would like to know more about the circumstances. Ordinarily Newton was not receptive to suggestions that he had erred. However, the

---

[132] *Corres 5*, 347.   [133] Wollenschläger, "Briefwechsel," pp. 270–1.

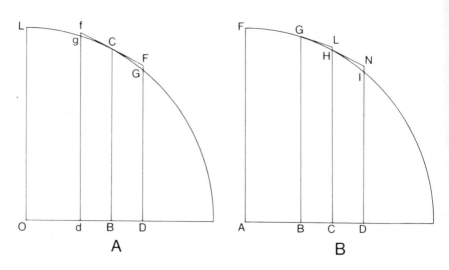

Figure 14.4. *A.* Proposition X, first edition. *B.* Proposition X, second edition.

available evidence seems to indicate that on this occasion he immediately accepted Bernoulli's demonstration as correct.[134]

Bernoulli had discovered the mistake by 1710.[135] By accident, the precise size of Newton's error was that which would have resulted had he used successive terms of the expansion in place of higher-order differentials. For Bernoulli, the mistake became evidence that Newton had not understood second derivatives when he wrote the *Principia,* a matter of obvious relevance to the priority dispute. Bernoulli wrote up his discovery for both the *Acta* and the *Mémoires* of the Académie. At much the same time, he decided that Newton had not demonstrated the inverse problem of central forces; that is, specifically, while he had demonstrated that conic orbits entail inverse-square forces, he had not demonstrated that inverse-square forces entail only conic orbits.[136] It seems clear from subsequent correspondence that Bernoulli, knowing a second edition was in progress, tried to time his two papers to come out after the new edition, in the expectation that the same defects, uncorrected, would not only display his own brilliance but would also become

---

[134] See the discussion in D. T. Whiteside, "The Mathematical Principles Underlying Newton's *Principia Mathematica,*" *Journal of the History of Astronomy, 1* (1970), 126–9.

[135] Bernoulli to Leibniz, 12 Aug. 1710; *Mathematische Schriften, 3/2,* 854–5.

[136] "De motu corporum gravium . . . autore Joh. Bernoulli" *Acta eruditorum* (February and March 1713), pp. 77–95, 115–32. "Extrait d'une lettre de M. Bernoulli . . .," "Addition de M. [Nikolaus] Bernoulli," *Mémoires de l'Académie royale des sciences* (1711; published 1714), pp. 59–72. "Extrait de la réponse de M. Bernoulli à M. Herman . . .," *ibid.* (1710; published 1713), pp. 685–703.

effective arguments that the Leibnizian calculus was a different and more powerful instrument than Newton's fluxional method.[137] He had not reckoned on the delays in publication or on his nephew. His own papers appeared in print a few months before the completion of the second edition. In the fall of 1712, before he had seen Bernoulli's two papers, Newton understood the correction as a friendly gesture. He immediately secured Bernoulli's election to the Royal Society and heightened the intended distinction by delaying the election of Bernoulli's nephew until a later time.[138] Newton's attitude changed as soon as he read the two articles and perceived their intention. As for Bernoulli, who had tried to set the trap, he could not contain his anger when he found that Newton did not attribute the correction to him.

Newton saw immediately that much more than a particular result was at stake. He had just told Cotes that the public would have to take the lunar theory as it was. He had no intention that the public receive, much less take, Proposition X as it was. Cognizant of the relevance of the *Principia* to the controversy with Leibniz, he had planned to include a republication of "De quadratura" in the new edition, together with a preface to it that would stress the role of fluxions in the composition of the work and insist on his early invention of the method.[139] He also worked on a new scholium to Lemma II, the fluxional lemma, with a detailed history of the events of the 1670s.[140] Possibly the progress of *Commercium epistolicum* led him to drop these plans. Proposition X, an apparent demonstration of mathematical weakness, was another matter. Four years later, Newton told the Abbé Conti that he had "left off Mathematicks 20 years ago."[141] This does appear to have been true. In October 1712, moreover, he was less than three months removed from his seventieth birthday, not anyone's prime age for mathematics. Nevertheless, he not only fell to the problem with a will, quickly consuming some twenty sheets of paper in its consideration, but he found the source of his error as Bernoulli had not been able to do. He dashed off a letter to Nikolaus Bernoulli, sending

[137] Cf. Bernoulli to Leibniz, 6 Feb. 1715; *Mathematische Schriften 3/2*, 936–7. Also, Bernoulli to Wolf, 8 April 1716; *Corres 6*, 302.

[138] DeMoivre to Bernoulli, 17 Dec. 1712; Wollenschläger, "Briefwechsel," p. 277. In February 1714, Newton proposed Nikolaus Bernoulli for election to the Royal Society (*JBC 10*, 545).

[139] Seven different drafts of a preface that explained the inclusion of "De quadratura" in the second edition are published in Cohen, *Introduction*, pp. 347–9.

[140] *Add MSS* 3968.5, ff. 30–36; 3968.6, ff. 37–48; 3968.7, ff. 49–66, 113–16. I am convinced that Newton composed these drafts in 1712, even though that part of the second edition had already been printed.

[141] Newton to Conti (for Leibniz), 26 Feb. 1716; *Corres 6*, 285. Cf. Newton to————[John French], 22 March 1715; *Corres 6*, 212.

him the corrected proposition, and asking him to send it to his uncle.[142] Undoubtedly DeMoivre's letter to Johann Bernoulli at this time was dictated by Newton with the same end in mind, as the final sentence in the portion quoted suggests. Subsequently he worked out no fewer than four other alternative demonstrations of it.[143] Work on the proposition appears on a draft of a preface for *Commercium epistolicum*.[144]

On 14 October, Newton told Cotes that there was an error and the proposition would need to be corrected. By 18 October, according to DeMoivre, he had amended it. Nevertheless, he let Cotes wait another three months before he sent the revised proposition.[145] It was necessary to reprint one whole sheet, signature Hh, plus the final two pages of the previous sheet, signature Gg, which were inserted as a cancel pasted to the stub of the original page, telltale evidence of a late correction that Bernoulli did not fail to notice.[146]

To his brief letter that accompanied the revised Proposition X, Newton added another surprise for Cotes: "I shall send you in a few days a Scholium of about a quarter of a sheet to be added to the end of the book: & some are perswading me to add an Appendix concerning the attraction of the small particles of bodies."[147] About the same time, Bentley told Cotes orally that he should compose a preface to the volume. Thus were conceived the two most visible additions to the second edition. Newton and Newtonians were highly aware of the mounting tide of criticism of his natural philosophy and its concepts of attractions and repulsions, a tide rising with the priority dispute and found in such things as the review of Freind's *Chemistry* in the *Acta* and the slighting references in the Hartsoeker-Leibniz exchange. Newton himself had been agitated enough by the latter to compose an answer addressed to the editor of the *Memoirs of Literature,* one of the journals in which it appeared.[148] Both Cotes's preface and Newton's General Scholium replied to the criticism. Placed at the front and rear, like symbolic covers, they gave to the new edition a polemic tone in harmony with the modifications designed to enhance the work's appeal to natural philosophers.

---

[142] *Corres 5,* 348.

[143] *Add MSS* 3965.10, ff. 103, 109, 135–6; 3965.12, ff. 190–201, 219–20; 3965.13, ff 373, 377. Cf. Whiteside, "Mathematical Principles Underlying *Principia*," p. 129.

[144] *Add MS* 3968.37, f. 540ᵛ. Among the papers devoted to the priority dispute, there is also at least one other sheet concerned with Prop. X (*Add MS* 3968.18, f. 261).

[145] DeMoivre to Bernoulli, 18 Oct. 1712; Wollenschläger, "Briefwechsel", pp. 270–1. Newton to Cotes, 6 Jan. 1713; *Corres 5,* 361.

[146] The review of the edition in the *Acta eruditorum* (March 1714), pp. 131–42, took pains to point it out.   [147] *Corres 5,* 361.   [148] *Corres 5,* 298–300.

The appendix on the attractions of particles fell by the way, but various drafts of similar essays do exist among Newton's papers.[149] Although the papers bore some resemblance to earlier discussions, which had found their classic exposition in Query 31 of the *Opticks,* important modifications distinguished the drafts of 1712–13. Newton insisted more on the differences between gravity on the one hand and electric and magnetic attractions on the other. Gravity is always proportional to the quantity of matter; electric and magnetic forces are not. Gravity acts at long distances, they at short. Gravity is not interrupted by the interposition of other bodies; they are. Taking all the differences into account, Newton began, in the case of electric attractions, his principal interest, to talk of an electric spirit which lies hid in the pores of bodies, a spirit by which light and bodies act mutually upon each other, a material spirit which is agitated by friction. It is, he said, "exceedingly subtle & easily permeates solid bodies." It is also "exceedingly active and emits light."[150]

Newton's concept of an electric spirit bore the imprint of the experiments of Francis Hauksbee, the demonstrator whom he had brought in to add substance to meetings of the Royal Society. In 1705–6 Hauksbee's attention turned to electricity, a topic Newton had touched on briefly thirty years earlier in his "Hypothesis of Light." Hauksbee carried the subject much further. Mounting a glass globe on an axle, he turned it rapidly while he held his hand against it – the immediate forebear of the static electric machine. When he evacuated the air from the globe, striking optical displays occurred within it. He could make sparks strike against objects outside the globe. In 1708, a Mr. Wall showed some other electrical experiments at a special night meeting where the optical phenomena could be more readily perceived.[151] In such experiments, effluvia excited by the friction seemed to become visible and palpable. Newton allowed himself to be convinced. On 6 November 1706, Hauksbee performed a demonstration before the society. "The president said he thought these Experiments Evinced that Light proceeded from the Subtle Effluvia of the Glass," the minutes related, "and not from the Grosse body."[152]

Nor was electricity all. Hauksbee also turned his hand to capillary experiments, another old interest of Newton. Especially in 1712,

---

[149] *Add MS* 3965.12, ff. 350–1, 356–7, 361–2. *Add MS* 3968.18, f.260^v. Some of the drafts of the General Scholium are published in *Halls,* pp. 348–64. "De vi electrica," *Corres 5,* 362–5.   [150] *Add MS* 3965.12, f. 361^v.   [151] 2 June 1708; *JBC 10,* 189.

[152] *JBC 10,* 140. Cf. Henry Guerlac, "Francis Hauksbee: Expérimentateur au profit de Newton," *Archive internationales d'histoire des sciences,* 16 (1963), 113–28; "Sir Isaac and the Ingenious Mr. Hauksbee," *Mélanges Alexandre Koyré,* 2 vols. (Paris, 1964), 1, *L'Aventure de la science,* ed. I. Bernard Cohen and René Taton, 228–53.

Figure 14.5. The experiment to measure forces of attraction in capillary action. A drop of orange juice is placed between
two sheets of glass 24 inches long. The entire apparatus
can be tilted at an angle to bring the weight of the juice
into equilibrium with the attraction between the glass
and the juice.

these experiments took more than one novel turn. Hauksbee exhibited capillary phenomena inside the exhausted receiver of his air
pump.[153] For fifty years, Newton's explanation of capillary action,
the reduction of air pressure in narrow tubes because of the unsociability of air with glass (or whatever material), had depended on the
assertion that capillary phenomena do not occur in a vacuum. As a
result of Hauksbee's experiments, he scrapped the old explanation
for a new one based on attractions between the fluid and the material of the tube. Other experiments by Hauksbee reminded him that
capillarity offered a possibility of exact mathematical treatment of
the forces between particles. Hauksbee stood two sheets of glass,
inclined to each other to form a narrow wedge, in a pan of water.
Water rose between the sheets, taking the shape of a hyperbola with
the surface in the pan as one asymptote and the point of the wedge
as the other. With narrow tubes, he found a similar inverse relation
between the height of the water and the diameter of the tube.
Finally, he devised an apparatus with two sheets of glass set together so as to form a narrow wedge of air between them. A drop
of orange juice between the glasses moved toward the point of the
wedge. The whole apparatus could be tilted to bring the capillary
attraction into equilibrium with the weight of the drop (Figure
14.5).[154]

The last experiment especially grasped Newton's interest, and he
appears to have entered into its elaboration. On 22 May 1712,
Newton directed Hauksbee to perform it in vacuo "so as to Ascer-
tain the Proportion of the Power of Gravity and Congruity or
Agreement of the Parts by Observing at what Angle the Drop is
Observed to be Stationary and not to move Towards the Edge of
the Wedge formed by the two planes."[155] Hauksbee had used sheets
of glass 6 inches long. Newton made his 20–25 inches long, the
better to measure the force exactly. "On the Electric Spirit," which

[153] 28 Feb. 1712; *JBC 10*, 368.
[154] 24 Jan.–6 March 1712; *JBC 10*, 356, 358, 368, 369.   [155] *JBC 10*, 400.

was possibly the essay he once intended to append to the *Principia*, used the experiment to calculate the force at a point of equilibrium where the sheets were 1/80 inch apart and found it equal to the weight of a cylinder of the fluid that had the same diameter and was a bit over 3/8 inches high. Since experiments showed that the force varied inversely as the distance, he calculated that at a distance of 10^{-7} inches it would be sufficient to support a column of the fluid 52,500 inches long. Newton did not choose the distance of 10^{-7} inches at random. His optical experiments had shown it to be about the distance that separates two sheets of glass pressed together – that is, the distance at which the cohesive forces between particles effectively operate. "But so great a force," he argued, "is abundantly sufficient to provide the cohesion of bodies." "Moreover," the essay concluded, "these very great forces by which the particles of bodies attract each other mutually and cohere will effect remarkable results in fermentation, putrefaction and chemical reaction."[156]

Newton's electric spirit diverged so radically from the fluids and effluvia of earlier mechanical philosophies that the differences outweigh the similarities. As he stressed, the electric spirit was exceedingly rare and exceedingly elastic. When he reintroduced an aether of similar qualities into his philosophy a few years later, he argued from its properties that it was composed of particles that repelled each other powerfully. Thus it embodied the very concept of action at a distance that conventional mechanical aethers were summoned up to explain away. Nevertheless, the electric spirit and later the new aether involved profound problems. They were two distinct fluids in his view, the one found in the pores of bodies and concerned with attractions of particles, the other pervading the empty spaces of the cosmos. Whatever their properties, such subtle fluids had produced over the years an abundance of verbal explanations of phenomena but never an exact quantitative account. Indeed, they had invariably conflicted with exact quantitative accounts. No one had succeeded, for example, in extracting Kepler's laws of planetary motion from a vortical theory. Even in introducing the electric spirit, however, Newton indicated his primary concern with quantitative descriptions. There is no satisfactory explanation of his return to such fluids, even with their differences from conventional ones. It is hard to imagine Newton acquiescing in such a retreat at an earlier, more vigorous age. In any event, a full essay on such an electric spirit would have raised severe problems in an edition designed to defend his philosophy against conventional mechanical ones, which probably explains his decision to omit it. He did allow

---

[156] *Corres 5*, 367–8; original Latin, pp. 364–5. He inserted the same experiment in Query 31 in the second English edition of the *Opticks* (pp. 392–4).

himself to insert a final paragraph in the General Scholium which referred obscurely to "a certain most subtle spirit which pervades and lies hid in all gross bodies . . ."[157]

In January 1713, Newton warned Cotes that a General Scholium would conclude the work. He finally sent it in March. Despite the reference to an electric spirit in the final paragraph, the General Scholium opened with a ringing challenge to mechanical explanations of the heavens. "The hypothesis of vortices is pressed with many difficulties," he began. As he proceeded to detail them— comets, resistance to motion, and the contradiction, for vortices, between Kepler's second law and his third–the verb "pressed" hardly seemed adequate to express the desperate straits of vortical theory. In more general terms, the order of the cosmos is incompatible with mere mechanical necessity.

> This most beautiful system of the sun, planets, and comets, could only proceed from the counsel and dominion of an intelligent and powerful Being. . . . This Being governs all things, not as the soul of the world, but as Lord over all; and on account of his dominion he is wont to be called *Lord God*, παντοκρατωρ, or *Universal Ruler;* for *God* is a relative word, and has a respect to servants; and *Deity* is the dominion of God not over his own body, as those imagine who fancy God to be the soul of the world, but over servants.

Newton continued, expounding his conception of God and of absolute space and time as the consequences of his infinite extension and duration.

> He is omnipresent not *virtually* only, but also *substantially;* for virtue cannot subsist without substance. In him are all things contained and moved; yet neither affects the other: God suffers nothing from the motion of bodies; bodies find no resistance from the omnipresence of God.

God is devoid of all body and ought not to be worshiped through any material image. We have ideas of his attributes; we cannot know his substance. We know him by his works; we admire him for his perfections; "but we reverence and adore him on account of his dominion: for we adore him as his servants; and a god without dominion, providence, and final causes, is nothing else but Fate and Nature."

So far, he concluded, he had explained the phenomena of the heavens by the force of gravity, but had not shown the cause of the force–a dubious statement after what he had just said about the dominion of God. He explained what the cause must account for: the action of gravity, not in proportion to the surfaces of bodies

---

[157] *Prin*, p. 547. For a different treatment of the electric spirit, see R. W. Home, "Newton on Electricity and the Aether," unpublished manuscript.

"(as mechanical causes used to do)" but in proportion to quantity of matter, its penetration to the very center of all bodies without diminution, its propagation to immense distances decreasing in exact proportion to the square of the distance. "But hitherto," he proceeded, in one of his most frequently quoted passages, "I have not been able to discover the cause of those properties of gravity from phenomena, and I feign no hypotheses . . . And to us it is enough that gravity does really exist, and act according to the laws which we have explained, and abundantly serves to account for all the motions of the celestial bodies, and of our sea."[158] Composed virtually at the end of his active life, the General Scholium contained a vigorous reassertion of those principles which Newton had adopted in his rebellion against the perceived dangers of Cartesian mechanical philosophy. The same principles had continued to govern his scientific career as he followed the consequences of his rebellion into a new natural philosophy and a new conception of science. He regarded it as an important statement, and true to habits he could not change now, he sent further revisions of it after it had already been printed, so that the final page, which had to be reset, was a cancel pasted in.[159]

Meanwhile Cotes worried over his assignment. He tried to get Newton and Bentley to advise him on it. He saw the preface in purely polemic terms directed against Leibniz. He would be glad, he said, "to speak out the full truth" on the dispute over the calculus. He thought Leibniz's *Tentamen*, his essay on celestial dynamics published immediately after the *Principia*, deserved censure. He realized the advantage of having such a piece in his name, but he thought Bentley and Newton should write it. "You may depend upon it," he told Bentley, "that I will own it, & defend it as well as I can, if hereafter there be occasion." All Bentley would say in reply was that Leibniz's name should not be mentioned. Newton declined even to see Cotes's preface because he feared being taxed with writing it. Cotes did eventually get Samuel Clarke to read over what he wrote and to comment on it.[160] Left to his own devices, he produced a

---

[158] *Prin*, pp. 543–7. On various issues concerned with the General Scholium, see E. W. Strong, "Newton and God," *Journal of the History of Ideas, 13* (1952), 147–67, and "Hypotheses non fingo," in Herbert M. Evans, ed., *Men and Moments in the History of Science* (Seattle, 1959), pp. 162–76; N. R. Hanson, "Hypotheses fingo" in Robert E. Butts and John W. Davis, eds., *The Methodological Heritage of Newton* (Toronto, 1970), pp. 14–33; and I. Bernard Cohen, "Isaac Newton's Principia, the Scriptures, and the Divine Providence," in Sidney Morgenbesser, Patrick Suppes, and Morton White, eds., *Philosophy, Science, and Method: Essays in Honor of Ernest Nagel* (New York: 1969), pp. 523–48.

[159] Cohen, *Introduction*, pp. 249–50.

[160] Cotes to Bentley, 10 March 1713; *Corres 5*, 389. *Corres 5*, 390–1. Newton to Cotes, 31 March 1713; *Corres 5*, 400. Cotes to Clarke, 25 June 1713; *Corres 5*, 412–13.

lucid exposition of the argument for universal gravitation pitched at a popular level and an acid reply to the critics of Newtonian philosophy worthy of his mentor's polemical style. As Bentley ordered, Leibniz's name did not appear. The references were sufficiently pointed that no one mistook the target, however, and Leibniz, who had not exerted himself unduly to avoid cutting references to Newton's philosophy, took the preface as fresh provocation.

As spring passed into summer in 1713, the edition finally approached completion. On 3 May, Cotes thought it would be done in two or three weeks. On 25 June, he informed Clarke that it had been finished a week before.[161] It was only on 30 June that Bentley announced it to Newton. "At last Your book is happily brought forth; and I thank you anew y^t you did me the honour to be its conveyor to ye world." Bentley gave Newton only six presentation copies, although he and Cotes did take up some of the slack with their own presentations, which Bentley listed for him to avoid duplication.[162] Newton used the small number as an excuse not to send one to Johann Bernoulli. Desirous of wooing the Académie, he sent copies to the Abbé Bignon, Fontenelle, and Varignon. To Bignon he said that the work would not be very popular because it was difficult, ran counter to commonly received notions, and lacked hypothetical explanations.[163] Voltaire had at least the first in mind when he penned one of his epigrams on the philosophic scene. "In London, very few people read Descartes, whose works have become quite useless," he observed; "neither do many read Newton, because one must be very learned to understand him."[164]

Bentley's accounts for the edition show that 700 copies were printed at a cost of £117. By the end of 1715, he had only 71 copies left and had already realized nearly £200 of his expected profit. He could have realized even more. The work was in such demand on the Continent that a reprint appeared in Amsterdam already in 1714 and a second in 1723.[165]

As for Cotes, he neither received any payment for his labor from Bentley's profits nor even any word of thanks from Newton. As I have mentioned, Newton suppressed the preface that contained a tribute to him and expunged a reference to him from the text. Six months after the book had been published, he suddenly sent Cotes a list of Errata, Corrigenda, and Addenda which he apparently expected to have printed up and bound with the volume. Somewhat chagrined, Cotes got his back up far enough to say that he had

---

[161] *Corres 5*, 412. Cf. Cotes to Jones, 3 May 1713; *Corres 5*, 403.   [162] *Corres 5*, 413–14.

[163] Newton to Abbé Bignon, late 1713; Varignon to Newton, 5 Dec. 1713; Fontenelle to Newton, 24 Jan. 1714; *Corres 6*, 40–1, 41–2, 57–8.

[164] *Lettres philosophiques; Oeuvres*, ed. Beuchot, *37*, 191.

[165] *Corres 5*, 417. Cohen, *Introduction*, pp. 252, 256–7.

found several of the errata himself, "but I confess to You I was asham'd to put them in the Table, lest I should appear to be too diligent in trifles." He knew of at least as many more, he added, that Newton had missed.[166] He did not get his back up far enough to substitute plain "you" for the divine mode of address, however.

Distant relations did continue between the two. In 1715, Cotes sent an artful letter to Newton which began with an account of an eclipse of the sun he had observed, proceeded to describe a new instrument he had devised to correct the times, arrived at the fact that the clock in his observatory was most imperfect, and finding himself on that subject, begged Newton to have the clock he had ordered for the Cambridge observatory back in 1708 sent down. It worked. Two weeks later, he wrote in thanks.[167] In the fall, Cotes consulted Newton about a preferment possibly offered in London and got back blunt though friendly advice that if he wanted the place, the mastership of the Charterhouse, which Newton himself had declined to pursue nearly twenty-five years earlier, he would need to make his support of the Hanoverian succession explicit.[168] Nothing more was heard of the affair. Cotes remained nervous in his relations with Newton. He thought he had a correction to "De quadratura," but remembering the result of his earlier venture in that direction, he was afraid to present it.[169] Suddenly, on 5 June 1716, Cotes died at the age of thirty-three. His cousin, Robert Smith, reported a splendid tribute by Newton. "If He had lived we might have known something."[170] Unfortunately, Newton's treatment of him while he did live scarcely suggested such esteem, and later he apparently did not lift a hand to help Smith publish Cotes's mathematical papers.[171]

During the whole period of the second edition, Newton's position at the Mint, by which, for better or for worse, he now primarily defined his life, steadily grew more difficult. On 1 May 1708, Craven Peyton succeeded Sir John Stanley as warden of the Mint. Peyton was the son and heir of Sir Robert Peyton, an erratic and violent figure of Restoration politics who earned the enmity of both sides by supporting the exclusion of James Stuart from the throne while he tried secretly to negotiate an accommodation with him.

---

[166] *Ibid.*, p. 249. Cotes to Newton, 22 Dec. 1713; *Corres 6*, 48–9.
[167] Cotes to Newton, 29 April and 13 May 1715; *Corres 6*, 218–21, 223.
[168] Newton to Cotes, 8 Nov. 1715; *Corres 6*, 248.
[169] Cotes to Jones, 5 May 1716; *Corres 6*, 335–6.   [170] *Edleston*, p. lxxvii.
[171] Smith to Newton, 23 Dec. 1718; *Corres 7*, 28–9. There is no evidence of a reply from Newton. Smith's letter to Newton in August 1720, which told how the edition was progressing, implied that he had received no help from Newton during the intervening two years (*Corres 7*, 98).

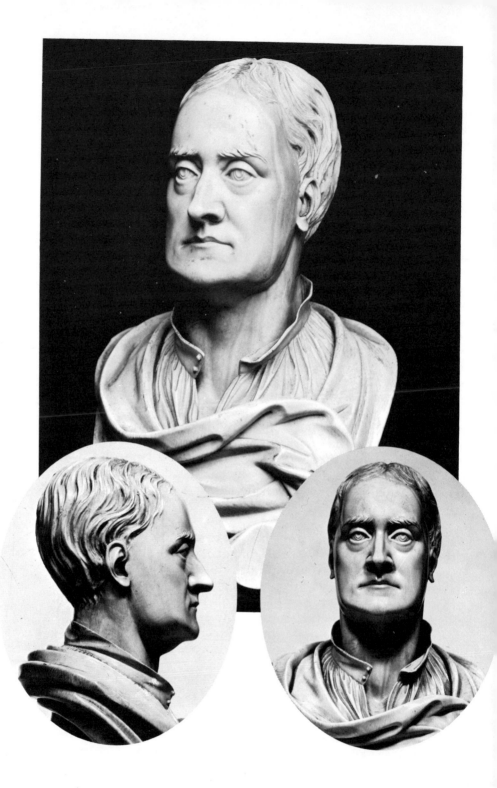

Sir Robert also wasted the family fortune; his son recouped what he could by marrying Catherine Granville, daughter of the Earl of Bath. He entered Parliament in 1705 as the nominee of the Duke of Newcastle at Boroughbridge, Yorkshire. Strong willed like his father and well connected, Peyton discovered as quickly as an earlier warden that he was no more master of the Mint in fact than he was in name. Apparently he resented his subordination to an erstwhile Cambridge don. The earliest evidence of friction is found in a letter of roughly March 1709 from Newton to Allardes, master of the Edinburgh Mint, in response to some questions about procedures. Newton apologized to Allardes for the tardiness of his answer. He had feared that some of the questions might be referred to other officers of the Mint, "with whom I find it sometimes difficult to agree . . .," and he stressed that he was finally replying, not publicly as an officer, but privately as a friend.[172] With each year, Newton found it more difficult to agree. Hostility was overt by 1711, adding a further load to his burden of troubles at this time. Among the more poignant documents that Newton left are some at the Mint on which his multiple worries struggled with each other for their share of the page. On the letter from John Chamberlayne about the election of the Royal Society in 1713, when according to Flamsteed the society nearly turned Newton out, he drafted a letter on Mint business, a letter to Flamsteed written in his capacity as Visitor to the Observatory, and paragraphs about Leibniz. Another sheet with Mint business contains as well a draft of the final paragraph of the General Scholium and two furious passages directed against Leibniz.[173] The man who said – or may have said, and could have said – that he discovered the law of universal gravitation by thinking on it continually found himself so distracted that he could hardly focus his mind long enough to compose a full page.

One of Peyton's first acts was to use his position in Parliament to arrange a modification of the Coinage Act, upon its renewal in 1708, to give him an account of £400 per annum, independent of the master, for his function as watchdog over coiners. At least, Newton attributed this alteration of the law directly to Peyton's agency.[174] In his analysis of the Mint, when he was in Peyton's

---

[172] *Corres 4*, 534–5.   [173] *Mint* 19.2, ff. 334–5, 88.
[174] In two drafts of a letter to the lords commissioners in early 1716; *Mint* 19.1, ff. 440, 448.

Figure 14.6. Newton at seventy-five. An ivory bust sculpted by David Le Marchand, 1718. George Vertue mentioned a second bust by Le Marchand in 1718, this one with a wig. (By permission of the Trustees of the British Museum.)

shoes, Newton had seen financial control as the key to the master's power over the warden. He still saw it that way, and he resented Peyton's partial escape from his domination. In 1714, when this issue was a factor in bringing Newton's relations with Peyton to a head, he drafted a memorandum stating his opinion that, in the pursuit of counterfeiters, "it would have been better that all the Coynage Duty including the 400^{li} had been imprested to ye M^r for the time being, so that he might advance to ye Warden any reasonable summs of ready money for that service upon receipts for the same & be discharged in his yearly Acc^{ts} by those Receipts."[175] Beginning with the accounts for 1708 meanwhile, the annual entries of money passing through his hands to the warden for that service ceased to appear.

Peyton may have precipitated another contest of will, over the control of melting. Among Newton's papers are thirty-four folios of Elizabethan documents about an attempt of a warden to assert his authority over melting and the exertions of the master to repel him. Nothing in the papers indicates their relevance to Newton, but usually he had a point in mind when he collected historical precedents.[176]

There is no doubt about an angry confrontation in 1711. Concerned with the unremitting shortage of silver coin, Parliament legislated a bounty for plate brought into the Mint to be coined. In its address to her majesty, the House of Commons asked that "the Officers of the Mint" be directed to receive plate and give receipts. It was Newton's opinion that this was the function, not of the officers, but of the master, and he told the others that he assumed Commons to mean merely that "the proper Officer or Officers" be so directed. One draft of a memorandum on the issue was explicit about his motive; letting the other officers give receipts would tend to take the control of the Mint's finances out of his hands.[177] In his customary manner, he even prepared a draft of the warrant for the Treasury to issue authorizing him as master to receive plate, and he left blanks in which the lords commissioners could insert the names of the other officers if they wished. "But the Warden of the Mint fell into a passion at the blanks & said he would not go into the Lords unless the blanks were first filled up, & at his desire they were filled up. Then the Master prepared a distinct Warrant for himself as Master to coin the Plate, but the Warden opposed it."[178] Peyton prevailed on both accounts. The warrant directed the officers to receive the plate, and a separate warrant to coin it was not

[175] *Mint* 19.1, f. 484. Cf. two warrants for payments to Weddell on bills certified by the warden alone (*CTB 24 (2)*, 193, 401).

[176] *Mint* 19.1, ff. 28–61. Cf. John Craig, *Newton at the Mint*, (Cambridge, 1946), p. 35.

[177] *Mint* 19.2, f. 513.   [178] *Corres 5*, 180–1.

issued to the master.[179] The scene occurred between 5 May, the date of the address by the House of Commons, and 10 May, the date of the warrant, at exactly the time when the reply to Leibniz's first letter was in preparation.

Soon a further complication arose. By the terms of the address of Commons, those who brought plate into the Mint received receipts that were acceptable as payments into any loan that Parliament might later pass that session. When the Two Million Act was passed in June, however, it limited payments to receipts issued before 15 May. The Act made no provision for receipts issued on or after 15 May, and the Mint, which by law received the plate at a higher price than its value as coin, faced a loss in redeeming those receipts. Needless to say, it was Newton who spotted the defect in the legislation, to Peyton's apparent annoyance.

> When the two Million Act was published [he explained in a memorandum to the Earl of Oxford, now Lord Treasurer] & the Master alone (after a stay of some days for the concurrence of his fellow Officers) acquainted the Lord H. Treasurer wth the defect of the Act & in a second memorial, laid the state of the plate before his Lord[p], & in order to a third memorial was informing himself whether 5s per ounce would content the Importers [of plate into the Mint] till the Parliament met, & told the Warden that he found that it would: the Warden declared against it . . .[180]

If Newton can be believed, Peyton opposed every resolution of the dilemma he suggested. Finally, late in July, the Lord Treasurer summoned the officers of the Mint to the Treasury. Newton devoted considerable effort to his memorandum on the problem, composing at least four drafts of it. He prevailed; Oxford accepted his proposed solution.[181]

Early in January 1713, Daniel Brattel, the assaymaster at the Mint, died. Newton petitioned Oxford for the appointment of his brother Charles Brattel to succeed him.[182] Signed by Newton alone, the petition is peculiar, for the assaymaster stood on the warden's account. In the events which followed, Peyton's name did not appear as an antagonist. Nevertheless, it appears that he was involved. One Catesby Oadham contested the position and claimed to be the better assayer. On order from the Treasury, the officers of the Mint arranged a trial between them in April. Newton was on sure ground here. If Peyton did stand behind Oadham, he could not begin to match Newton's technical knowledge, and all three officers signed the report endorsing Brattel.[183] The matter did not

---

[179] *Corres 5*, 126. Newton did draft a separate warrant for the master (*Mint* 19.2, f. 529[v]).
[180] *Corres 5*, 180–1.
[181] *CTP 1708–14*, p. 292. *CTB 25 (2)*, 85, 87. Newton's drafts of the memorandum he presented are in *Mint* 19.2, ff. 513, 532, 534, 537. The final warrant issued on 30 July is printed in *Corres 5*, 182–3.   [182] *Corres 5*, 374.   [183] *Corres 5*, 403–4.

end there, however. Oadham cried foul, and it was not at the other officers but solely at Newton that he directed his complaint. The trial had been unfair; Newton had already decided in Brattel's favor.[184] Newton's reply, the multiple drafts of which again suggest some emotional intensity, loans credence to the suggestion that Peyton stood behind Oadham, for Newton claimed that Oadham was attempting to have him expelled from his post.[185] One version of his memorandum accompanied a draft of the *Principia*'s proposition on the approximation of area for Kepler's second law. Another lapsed from his defense against Oadham into his defense against Leibniz.[186]

In 1714, the mounting tension in the Mint reached the breaking point. Two issues became the focus of conflict: the warden's independent account for the prosecution of coiners and the annual accounts for 1712 and 1713, but they appear to have been pretexts for the deeper question of ultimate authority. Two items from 1713 bore on the warden's independent account. In the years immediately after its establishment, charges on it passed on the signature of Peyton alone. Early in 1713, however, a memorial by Peyton for the payment of his assistant, Robert Weddell, was referred to Newton and Edward Phellips, the comptroller, for their opinion on its propriety. Moreover, a further question hangs over the matter, since Peyton had submitted his memorial nearly a full year earlier. Newton and Phellips recommended payment in full, but if Peyton's account had to pass Newton's scrutiny, it was no longer fully independent.[187] The year 1713 also witnessed the termination of the warden's special clerk for the business of coiners, an appointment which went back to Newton's days as warden. The clerk had continued to appear as an annual charge on the master's account after the new arrangement of 1708, but a replacement was not appointed when the incumbent died in August 1713. Then in 1714, Peyton suddenly found his entire conduct of the responsibility to suppress coiners challenged. One Henry Smithson, an old coiner himself who had learned to win a safer living by pursuing coiners, composed several memorials to the Lord Treasurer charging that Peyton's laxity had permitted a number of counterfeiters to go free. Smithson's memorials involved Newton. Newton had encouraged him and said he was ready to impress money to the warden to carry on the prosecution of coiners Smithson had apprehended, but the warden had refused to furnish the necessary expenses. The second memorial, 16 April 1714, indicated that the Lord Treasurer could

---

184 *Mint* 19.1, f. 94.
185 *Mint* 19.1, ff. 91, 95, 98–9, 100–1. F. 90 is an amanuensis's copy. Cf. the short memorandum ultimately sent on 26 August (*Corres 6,* 25).
186 *Mint* 19.1, ff. 98–9, 91ᵛ.   187 *Corres 5,* 338, 401–2.

leave an answer for him with Newton.[188] The Lord Treasurer did, on the same day, and on 26 April, at Newton's house, Smithson confronted Peyton. Phellips, the comptroller, was also present. He emerged as Peyton's ally. Smithson left the only account of the meeting. According to him, when he tried to present his case, Peyton and Phellips brought up other frivolous matters, so that he was not able to proceed effectively. His specific goal was the payment of a bill for £96 expenses. Peyton and Phellips won; Newton had to agree that Smithson had not proved Peyton to have authorized the expenses.[189]

A political tone crept into the issue, for Smithson later alleged that Peyton was a special favorite of Oxford. This appears to have been true. Though returned to Parliament three times in a pocket borough of the Duke of Newcastle, Peyton had followed his father's example, and influenced perhaps by the extent of the Tory victory in 1710, perhaps by the threat of losing his position at the Mint, he had gone over to Oxford. Phellips, who had been appointed to the Mint in 1711 by the Tory administration, undoubtedly shared Peyton's new loyalties. Political motives probably lay behind the next move in the game. In May, Philip Bertie supported Smithson's claims in a letter to the Lord Treasurer charging that Smithson had not received a fair hearing and that the warden was neglecting his duty.[190] The Berties were a powerful political family, although Philip Bertie was not their most prominent member. Nothing appears to have come of his letter, but the letter itself indicates that more merely than the apprehension of coiners was at stake.

Also in 1714, Newton's accounts for 1713 were challenged. Finances in the Mint were always a difficult problem. The Coinage Act allowed £3,000 per annum for salaries and upkeep. Special warrants had added £500 in temporary annual expenditures. The expenses of coining were separate and varied of course from year to year. Since salaries consumed most of the £3,500, only about £405 was left each year for upkeep. These expenses fell to the warden's account and responsibility. The sum available was insufficient, however, and Newton, who felt himself to be in command, tended to manage both accounts to squeeze the maximum out of them by juggling sums between upkeep and coining expenses. In the 1712

[188] *Mint* 1.7, pp. 64–6 (these are not part of Newton's Mint papers). Cf. depositions by Smithson and his petition of 1717 (Public Record Office, T1/175, no. 28, ff. 85–7; T1/208, no. 19, ff. 86–7).

[189] T1/175, no. 28, f. 89. Smithson remained in touch with Newton. On 26 May of what appears to be a later year he wrote to Newton about coiners (*Corres 6*, 389).

[190] Public Record Office, T1/175, no. 48, ff. 155–6. Cf. P.B. to Newton, 8 May 1714; *Corres 6*, 125–6.

accounts, a bill of £130 for repairs for which the warden was responsible had to be deferred a year since the total would otherwise have exceeded the limit; this maneuver appears to have nettled Peyton, who resented Newton's iron control in any case. In 1714, several items on Newton's account connected with assaying furnaces were deemed to be repairs rather than coining expenses and disallowed since they pushed the total above £3,500.[191] We do not know the exact timing, but the accounts finally cleared on 6 August, one week after the political climax that witnessed Oxford's dismissal by the dying queen to insure the Hanoverian succession. It may be relevant that Auditor Harley, who challenged the account, was Oxford's brother.

It is hard to imagine that the incident was unconnected either with the political situation or with Newton's conflict with Peyton. The Tory regime had been notorious for its purge of office holders to insure that its own followers might feed on the incomes. Peyton's shift of loyalties had probably been influenced by a threatened removal. Newton's response to Oadham's charges in 1713 included the intriguing assertion that Oadham was trying to have him expelled from the Mint. The story that Conduitt heard from his wife fits into the context. Bolingbroke, the chief agent of the Tory purge, sent Dr. Swift to Catherine Barton to let Newton know he held it a sin for his thoughts to be distracted by the Mint. The queen, Bolingbroke suggested, would happily settle a pension of £2,000 on him, which was, he added, nearly double the value of the position. Before the upsurge in coining after the war, this assessment was correct. Perhaps it indicates what one is inclined to believe in any case, that the approach was made before 1714. Newton curtly refused, saying his place was at their service but he would have no pension.[192] The challenge to his accounts suggests that the Tories were becoming more aggressive in their determination to seize the position.

As for Peyton, the renegade Whig, Newton went out of his way in his defense to implicate him and to vent his spleen. All of his own expenses were fair and reasonable. The major item on his account was £2,004 9s. impressed to the warden. "And the Warden is to discharge himself of what has been imprest to him, & in his next Accompt to charge himself with the surplus, if any there be."[193] The specific nature of the challenge, of course, involved their unending conflict over authority, for if repairs to assaying furnaces were part of normal upkeep, they fell within the warden's competence. In the same account of 1713, Peyton received the pay-

[191] *Mint* 19.1, ff. 332–3. *Mint* 19.4, f. 35.
[192] *Keynes MS* 130.7, Sheet 1.   [193] *Mint* 19.1, f. 335.

Figure 14.7. Newton at seventy-five. An ivory plaque in high relief
carved by David Le Marchand. (By kind permission of
Sotheby Parke Bernet & Co.)

ment of £130 deferred from 1712, while Newton's charges were
disallowed; in recompense Peyton found his special clerk terminated. There is every indication that their struggle for control
reached its climax with the political upheaval that surrounded
Anne's death.

The succession of the Hanoverian dynasty and the return of the Whigs assured Newton's victory and sealed Peyton's fate. Indeed Newton found himself the unwilling host of the Earl of Oxford, who was imprisoned in the Tower and lodged in a house belonging to the Mint. Newton duly complained of this further encroachment of the Tower on his prerogatives.[194] Both Peyton and Phellips departed from the Mint at the end of the year to be replaced by loyal Whigs. Nevertheless, Newton could not permit such a challenge to his honor to go unanswered. He devoted immense energy to a reply, of which at least seven drafts and two copies by amanuenses exist, fully exonerating himself of course.[195] With his command of relevant fact, he justified all the items on his account by abundant reference to accepted practice, what he called "the custome or course of the Mint," all stated with all the passion of which he was capable. The burden of his argument was the necessary distinction between fixed charges to keep a Mint in being, and charges resulting from coinage which must vary with the level of coinage.

> The gold furnaces were necessary to be repaired for carrying on the coynage & the charges thereof & those of Assaying & reducing the bullion to standard were free from extravagance & just & unavoidable & the fees of the Exchequer & Treasury & other Offices were customary & necessary to be paid, & all these expences are placed in my accompt according to the course of the Mint & the Vouchers are good. And therefore all these charges are I think to be allowed at present by the article of the Indenture above recited.[196]

It appears that he had finally to cover the challenged expenditures of 1713 himself—only £27, a miniscule sum in comparison with the passion it generated—but in the future similar charges regularly passed on his accounts. Stung by the hint of extravagance, he imposed a puritanical regime on the Mint, forcing the other officers to consume a spartan diet, if the expenditures are a gauge, and denying future juries at the trials of the pyx the splendors to which they had formerly accustomed themselves. As for the warden, the independent account for prosecuting coiners was immediately suppressed, and Mint finances returned once more fully to Newton's authority.

If the crisis in the Mint resolved itself, the dispute with Leibniz did not. On 29 July 1713, a sheet without the name of its author, its printer, or even the city in which it was printed, appeared and spread quickly through interested circles on the Continent. Known

---

[194] Newton to the lords commissioners, 20 July 1715; *Corres 6*, 235.
[195] *Mint* 19.1, ff. 5, 87, 334–40, 342, 352, 354. Copies by amanuensis are on ff. 87, 334–5.
[196] *Mint* 19.1, f. 335.

as the Charta volans or (as Newton anglicized it) flying sheet, it was Leibniz's reply to *Commercium epistolicum*. Leibniz's continuing use of anonymity can no more be praised than Newton's practice of hiding behind John Keill or a committee of the Royal Society. Equally, it can no more be blamed. The Charta volans said little in detail about the correspondence of the 1670s. Rather it pointed out that Newton had published nothing on the calculus before Leibniz did and went on to assert that when the English began to ascribe everything to Newton, Leibniz, who had hitherto been inclined to believe Newton's claim of independent discovery, reexamined the matter more carefully and became convinced that Newton had developed the fluxional method in imitation of his calculus. In support of the last, the Charta quoted a letter of 7 June 1713 from a "leading mathematician" who expressed the opinion that in the 1670s Newton had invented only his method of infinite series. The "leading mathematician" also quoted a "certain eminent mathematician" on Newton's error in regard to second-order differentials in the *Principia*.

> From these words it will be gathered [the Charta continued] that when Newton took to himself the honour due to another of the analytical discovery or differential calculus first discovered by Leibniz . . . he was too much influenced by flatterers ignorant of the earlier course of events and by a desire for renown; having undeservedly obtained a partial share in this, through the kindness of a stranger, he longed to have deserved the whole – a sign of a mind neither fair nor honest. Of this Hooke too has complained, in relation to the hypothesis of the planets, and Flamsteed because of the use of his observations.[197]

Apparently no one was ever deceived as to the author of the Charta volans. Bernoulli told Leibniz that it was openly attributed to him, and Newton gave two good reasons why Leibniz was the author to John Arnold, who relayed them to Leibniz.[198]

The "leading mathematician" quoted by the Charta volans was Johann Bernoulli, who with characteristic modesty quoted himself as an eminent mathematician. Though suspected, Bernoulli's authorship of the passage proved to be a better-kept secret. Written as soon as Bernoulli had seen a copy of *Commercium epistolicum,* the letter probably gave Leibniz his first account of the volume. It was a scathing account, since Bernoulli was outraged by the injustice of the whole proceeding. Nevertheless, he was not anxious to make his outrage public. "I do indeed beg you [he concluded the letter] to use what I now write properly and not to involve me with Newton

[197] *Corres 6,* 18–19; original Latin, p. 15–17.
[198] Bernoulli to Leibniz, 23 May 1714; *Mathematische Schriften, 3/2,* 931. Arnold to Leibniz, 5 Feb. 1716; *Corres 6,* 274.

and his people, for I am reluctant to be involved in these disputes or to appear ungrateful to Newton who has heaped many testimonies of his goodwill upon me."[199] In the continuing correspondence, Bernoulli never failed to repeat his request for secrecy. One can understand his desire to avoid a bruising struggle which was not his, but he might have done better not to supply Leibniz with ammunition and not to fan his resentment in the same letters. Bernoulli did not cover himself with glory, or even credit, in the calculus dispute. We regret the dishonesty and the subterfuge of Newton and Leibniz, but we also understand how two proud men goaded by perceived injustices could resort to such. Bernoulli arouses only disgust. While he damned Newton to Leibniz (and complained of not receiving a presentation copy of the *Principia!*), he assured Newton's friends of his neutrality and of his regard for Newton. He feared "that it may be imagined among you that by the remarks on second differences, I wished to furnish arms to the enemies of Mr. Newton to use against him; imagination ridiculous and amusing!"[200] Leibniz indicated to Bernoulli that though he did not wish to embroil him in quarrels, "I expect from your honesty and sense of justice that you will as soon as possible make it evident to our friends that in your opinion Newton's calculus was posterior to ours, and say this publicly when opportunity serves . . ."[201] By immediately printing his letter in the Charta volans, albeit anonymously, Leibniz gave himself the means of making Bernoulli's support public should Bernoulli hold back. Apparently he knew him well.

If Newton was furious, Leibniz was no less so as he read Bernoulli's description of *Commercium epistolicum:*

> those idiotic arguments which (as I gather from your letter) they have brought forward deserve to be lashed by a satirical wit. They would maintain Newton in the possession of his own invented calculus and yet it appears that he no more knew our calculus than Apollonius knew the algebraic calculus of Viète and Descartes. He knew fluxions, but not the calculus of fluxions which (as you rightly judge) he put together at a later stage after our own was already published. Thus I have myself done him more than justice, and this is the price I pay for my kindness.[202]

Such Leibniz undoubtedly believed: that Newton's method in the 1670s, whatever it was, was not yet the calculus. Again in August

---

[199] Bernoulli to Leibniz, 7 June 1713; *Mathematische Schriften, 3/2,* 912; translation in *Corres 6,* 5.

[200] Bernoulli to DeMoivre, 4 Aug. 1714; Wollenschläger, "Briefwechsel," p. 296.

[201] Leibniz to Bernoulli, 28 June 1713; *Mathematische Schriften, 3/2,* 914; translation in *Corres 6,* 9.

[202] *Mathematische Schriften, 3/2,* 913; translation in *Corres 6,* 8.

he repeated substantially the same thing. There was "unmistakeable evidence" that Newton did not understand the calculus in the 1670s. He was therefore compelled to blame Newton for insincerity since, knowing Leibniz could not have taken the calculus from him, he still encouraged his followers to accuse Leibniz. Had Newton not protested so loudly, he would have believed his claim that he had himself stumbled upon "our" calculus, which, when the matter was more diligently sifted out, appeared clearly to be untrue.[203] Nevertheless, haunted by the skeleton in his closet, he chose not to mention "De analysi" to Bernoulli, and to Christian Wolf he wrote with careful precision that "I have indeed observed no shadow of an argument whence it appears that notice of an infinitesimal calculus invented by Newton came to me."[204] By a similar careful choice of words, he also denied to Bernoulli that he wrote the offensive review of "De quadratura."

While Newton perceived Leibniz's authorship of the Charta volans, Leibniz received intelligence of his maneuvers via Newton's enemies in England. Dr. Woodward sent word "that everything done against you here has proceeded solely from Mr. Newton and thus he hopes that you will not attribute it to the Royal Society." Woodward also promised to try to get a copy of *Commercium epistolicum* for Leibniz. Flamsteed also sent his compliments – together with a list of errors in Newton's lunar theory.[205]

With two men angry beyond willingness to hear reason, the dispute could only grow. The inaugural issue, for May and June 1713, of the *Journal literaire,* a new journal launched by a group of Dutch savants, carried an anonymous letter from London, written by Keill, which presented the Newtonian version of the priority question together with French translations of the report of the committee of the Royal Society and of Newton's tangent letter of December 1672. Leibniz felt he must answer. Hence the issue for November and December of the same journal printed a French translation of the Charta together with anonymous "Remarks" on the difference between Newton and Leibniz.[206] In the "Remarks" Leibniz, who composed them, repeated at greater length the case he had made in his letters to Bernoulli, laying emphasis on the absence of the calculus from the *Principia* where it was needed and the lack of understanding shown when the work did try to use it. Newton's error on Proposition X and the proof it seemed to offer that he did

[203] Leibniz to Bernoulli, 19 Aug. 1713; *Mathematische Schriften, 3/2,* 919.
[204] Leibniz to Wolf, 2 April 1715; Gerhardt, ed. *Leibniz-Wolf Briefwechsel,* p. 162.
[205] Arnold to Leibniz, 22 Dec. 1715; Hasperg to Leibniz, 16 April 1714; *Corres 6,* 260, 102.
[206] "Extrait d'une lettre de Londres," *Journal literaire,* 2nd ed., *1* (May and June 1713), 206–14. "Remarques sur le different entre M. de Leibnitz, & M. Newton," *Journal literaire,* 2nd ed., *2* (Nov. and Dec. 1713), 445–53.

not understand second-order differentials in 1687 loomed ever larger in an argument seriously short of empirical content.

By now, Newton was also wholly swept up in the quarrel and its growing ramifications. At some time, probably in 1713 or 1714, he saw the anonymous review of "De analysi" in the *Acta*. Undoubtedly he recognized it as Leibniz's. He started to draw up a letter to the editor of the *Acta* in reply, and pursued it through at least two drafts before he decided to drop it.[207] When he finally saw the Charta volans, late in 1713 or early in 1714, it exercised him greatly. Although he had a printed copy of it, he copied it out in his own hand at least twice and drew up detailed comments on it. Later he did the same for Leibniz's "Remarks."[208] The first intimation he had of the "Remarks" was a letter from Keill in February 1714 enclosing one from a Scottish bookseller in the Hague, T. Johnson, the publisher of the *Journal literaire*. Like journalists of all ages, the editors of the *Journal* did not want a sure source of racy copy to dry up too quickly. Through Johnson they informed Keill and Newton of their good will, of Leibniz's recent retort, and of their readiness to print a reply. The issue itself with the "Remarks" he did not send, however. Keill asked Newton to forward a copy to him with his opinion of what should be done, "and I will observe your orders as far as I can. . . I am of opinion that Mr. Leibnits should be used a litle smartly and all his Plagiary and Blunders showed at large. However S^r I expect your directions . . ."[209] Bernoulli referred to Keill as Newton's ape. Had he needed confirmation, he would have found Keill's correspondence a rich source to mine. It was April before Newton finally saw the "Remarks" and rushed a copy to Keill with the opinion that they required an answer. In tones more of command than petition, he suggested that Keill send his thoughts on a reply; he would send his own to Keill "that you may compare them wth your own sentiments & then draw up such an Answer as you think proper. You need not set your name to it."[210] By the middle of April, he had received further the *Acta* of February-March 1713 and learned, as he must have guessed from the events of 1712, that Bernoulli was the "eminent mathematician" who had pointed out the error in Proposition X.[211]

As for his own thoughts on a reply to Leibniz's "Remarks," Newton wrote seven drafts of a letter to the bookseller Johnson and had DeMoivre translate it into French, though he probably never

[207] *Corres 5*, 383.

[208] *Add MSS* 3968.34, ff. 478–9, 480–5; 3968.35, ff. 487–504; 3968.41, ff. 60–7.

[209] Johnson to Keill, 9 Feb. 1714 (N.S.); Keill to Newton, 8 Feb. 1714 (O.S.); *Corres 6*, 63, 62.   [210] *Corres 6*, 79–80.

[211] Newton to Keill, 20 April 1714; *Corres 6*, 108.

sent it.[212] Meanwhile, throughout May, he corresponded intensely with Keill about an answer. Clearly, the argument on Proposition X pricked an exposed nerve. Newton's drafted reply devoted considerable attention to a rebuttal, and Keill's answer did also. Keill even rose from his usual tone of brutal assault to unaccustomed irony when he concluded that Bernoulli ought to acknowledge publicly that he was mistaken, "wch he is more particularly obliged to doe because our tuo Authors [Leibniz and the 'leading mathematician' quoted in the Charta] (who doe not seem to understand much of that matter themselves) have made use of his great name to defame Mr Newton on that score."[213] But Leibniz was vulnerable on the same count. Newton had once told Keill that in his *Tentamen* Leibniz confused second fluxions. Would he please send particulars? Indeed Newton would. His detailed notes on the *Tentamen,* showing the error on second differences, concluded that Leibniz could not have found the principal results by his calculus since his demonstration was fallacious. Hence this too he stole from Newton.[214]

> Honored S^r [Keill wrote on 17 May]
> The papers I have sent I intirely submitt to you, and shall be glad you'l take any pains about them to change add or leave out anything . . . [And again on 25 May:] I leave my whole paper to You and Dr Halley to change or take away what your please . . .[215]

Newton's faithful ape did insist on putting his name to the "Answer," which appeared in the *Journal literaire* for July and August.[216] It was no matter. His style was beyond concealment. Other aspects of the "Answer" did not suggest Keill; in France and no doubt elsewhere the learned thought it too powerful to have been Keill's own. Newton must have provided the arguments.[217] To silence the suggestion that *Commercium epistolicum* had distorted the letters it cited, or worse, the "Answer" concluded with a notice that the originals were available for inspection at the Royal Society. Newton and Keill also issued the "Answer" as a separate pamphlet.

Although both Wolf and Bernoulli strongly urged him to reply to Keill's latest attack lest it appear that he could not, Leibniz sensed that he was being forced into battle on the least advantageous ground; made to discuss the correspondence of the early 1670s,

[212] *Add MSS* 3968.35, ff. 494–504; 3985.20, ff. 1–5.

[213] Keill to Newton, 17 May 1714; *Corres 6,* 138–9.

[214] Keill to Newton, 2 May 1714; *Corres 6,* 113–14. Newton's notes on the *Tentamen; Corres 6,* 116–17.   [215] *Corres 6,* 138–9, 142.

[216] "Réponse de M. Keill, M.D. Professeur d'Astronomie Savilien, aux auteurs des Remarques sur le différent entre M. de Leibnitz & M. Newton," *Journal literaire,* 4 (July and August 1714), 319–58.

[217] Wolf to Leibniz, 4 May 1716; Gerhardt, ed., *Leibniz-Wolf Briefwechsel,* p. 164.

made to demean himself by debating an underling. "I cannot bring myself to make a reply to that crude man Keill," he explained to Wolf.[218] In the end, however, he did insert a letter addressed to René-Joseph Tournemine in the *Mémoires de Trévoux* which confined itself to insisting that in the 1670s Newton merely had a version of the infinitesimal method of Fermat and others, not the "new calculus, which he does not appear to have understood well."[219]

As the dispute deepened, it also broadened to include more men. On the Continent, Bernoulli and Wolf labored on Leibniz's behalf, one more covertly than the other. In England, Newton coopted not only Keill but also the Huguenot refugee DeMoivre, who became the translator of whatever went to the *Journal literaire* for publication. Bernoulli and DeMoivre struggled to maintain their personal correspondence, writing deliberately misleading letters which attempted to conceal their participation.[220] Almost no British mathematician or natural philosopher remained unmoved by the cause. It even worked to heal some of the wounds opened in the battle within the Royal Society in 1710. Soon after *Commercium epistolicum* appeared, William Derham informed Newton that he had some letters from Collins to Towneley which confirmed the account of Newton's mathematical development before 1669.[221] After an absence of three years, Derham reappeared on the council of the Royal Society in the election of 1713.

By now, well-meaning men had begun to worry about the impact of such a battle to the death on the world of learning. Among them was John Chamberlayne, a member of the Royal Society as well as a minor political figure connected with the court and well aware that Leibniz was a prominent adviser to the Hanoverian dynasty which would soon be installed in England. As early as 1710 he had entered into correspondence with Leibniz. He had made his support of Newton in the Royal Society evident. In February 1714, he wrote to Leibniz to express his dismay over "the Differences Fatal to Learning between two of the greatest Philosophers & Mathematicians of Europe . . ." Perhaps he should know better than to involve himself, "yet as it would be very Glorious to me, as well as Advantageous to the common Wealth of Learning, if I could bring such an Affair to a happy end, I humbly offer my Poor Meditation . . ."[222] Leibniz's reply did not outdo itself in generosity or candor. He delivered a brief history of the dispute, which in his version began with Keill's unprovoked attack, and complained both

[218] Leibniz to Wolf, 2 April 1715; *ibid.*, p. 162; translation in *Corres 6*, 211.
[219] Leibniz to Tournemine, 28 Oct. 1714; *Mémoires de Trévoux*, January 1715, pp. 154–6.
[220] Bernoulli to DeMoivre, 20 March and 4 Aug. 1714; DeMoivre to Bernoulli, 28 June 1714; Wollenschläger, "Briefwechsel," pp. 286–9, 289–93, 294–9.
[221] Derham to Newton, 20 Feb. 1713; *Corres 5*, 379–80.   [222] *Corres 6*, 71.

of the "chicanery & foul play" by which someone led the Royal Society to think he submitted himself to its judgment, and of the injustice of its proceeding.

Also I do not at all beleive that the judgment wch is given can be taken for a final judgment of the Society. Yet Mr Newton has caused it to be published to the world by a book printed expresly for discrediting me, & sent it into Germany, into France, & into Italy in the name of the Society. . . .

As for me I have always carried my self with the greatest respect that could be towards Mr Newton. And tho it appears now that there is great room to doubt whether he knew my invention before he had it from me; yet I have spoken as if he had of himself found something like my method; but being abused by some flatterers ill advised, he has taken the liberty to attaque me in a manner very sensible. Judge now S^r, from what side that should principally come wch is requisite to terminate this controversy.[223]

As he read Leibniz's reply, Chamberlayne no doubt wished he had followed his better sense and forgotten the glory of settling the dispute. There was no backing out now, however, so he took the letter to Newton. Recall that Newton received it early in May 1714, when tension with Peyton at the Mint was reaching its height and when he had engaged himself with Keill in formulating an answer to Leibniz's "Remarks." Newton's reaction did not exceed Leibniz's in generosity or candor. For nine years now Leibniz had been attacking his reputation, he said, an arithmetic that set the blame from his point of view on the review in the *Acta*. Keill defended him before Newton even knew of the attack. Leibniz's response to Keill's defense virtually demanded that he, Newton, retract what he had published in the *Principia*'s scholium on fluxional method and in the introduction to "De quadratura": "If you can show me something in which I have injured him, I shall endeavor to give him satisfaction; but I do not wish to retract what I know to be true, and I believe that the Committee of the Royal Society did not do him any wrong."[224] Nor did Newton stop with this reply. Greatly agitated by Leibniz's letter, he translated it into English himself with the purpose of laying it before the Royal Society as an attack on their honor. Chamberlayne was beside himself. Thoughts of glory long forgotten, he could think now only of Queen Anne's imminent death and the prudence of using some moderation "with

---

[223] Leibniz to Chamberlayne, 28 April 1714; Pierre Des Maizeaux, *Recueil de diverses pièces, sur la philosophie, la religion naturelle, l'histoire, les mathematiques, &c. par Mrs. Leibniz, Clarke, Newton & autres auteurs célèbres*, 2nd ed., 2 vols. (Amsterdam, 1740), *2*, 120–5. The English is Newton's own translation (*Corres 6*, 105–6).

[224] Newton to Chamberlayne, 11 May 1714; *Corres 6*, 126–7.

a Gentleman that is in the Highest Esteem at the Court of Hanover . . ." He pleaded with Newton not to treat a private letter as a public document.[225] Newton had long since passed the boundary of prudence. Despite Chamberlayne's plea, he presented the letter at the meeting on 20 May. Several years later, Newton described part of the proceedings.

> Upon the reading of this Letter to the R. Society Mr Newton represented that he was so far from printing the Commercium Epistolicum himself that he did not so much as deliver to the Committee of the Society the Letters wch he had in his own custody. & to prove this, produced a letter of Mr Leibnitz dated   & another of Dr Wallis dated   both wch after the hands had been examined were deposited in the Archives of the Society.[226]

Thus, apparently, he used the occasion to put on record Leibniz's letter of 17 March 1693 and Wallis's letter to him of 10 April 1695, which he could henceforth use, as part of the public record, in anonymous attacks. If he did bring up those letters at the meeting, as he asserted above, it was a full year later, after two sessions devoted to public collations of the letters on deposit in the society with the extracts printed in *Commercium epistolicum,* when he brought them in for deposit.[227] In fact, the Royal Society disappointed Newton by failing to rise to Leibniz's challenge. Since the letter was not addressed to them, they refused to concern themselves with it. Nor were they desired to do so, the minutes add, though Newton presumably had something in mind when he brushed Chamberlayne's qualms aside to read the letter to them.[228]

Chamberlayne found an excuse not to attend the meeting on 20 May. It was 30 June before he screwed up his resolution far enough to tell Leibniz about it. He explained weakly that "I did presume your consent to show it to all whom it concern'd . . ." and reported his failure both with Newton and (as he saw it) with the Royal Society. Later he sent a transcript of the minutes.[229] The report did not distress Leibniz as Chamberlayne had feared, for it bolstered his contention that the Royal Society itself had not passed judgment upon him. Nevertheless, he did not let Chamberlayne escape so easily. The Royal Society had letters concerning him; he wanted copies for a *Commercium epistolicum* of his own which he was composing.[230] Poor Chamberlayne had to lay this request before the society in November after the annual break, and this time Newton

---

[225] Chamberlayne to Newton, 20 May 1714; *Corres 6,* 140.

[226] *Add MS* 3968.41, a scrap after f. 152. Cf. Newton to ?DeMoivre, ca. early 1720; *Corres 7,* 83. The blanks are in the original.

[227] 5 May 1715; *JBC 11,* 63.   [228] *JBC 10,* 571.

[229] Chamberlayne to Leibniz, 30 June and 27 July 1714; *Corres 6,* 152–3, 158–9.

[230] Leibniz to Chamberlayne, 25 Aug. 1714; Des Maizeaux, *Recueil, 2,* 128–9.

received more satisfactory support. The Society thought rather "that M^r. Leibnitz ought either to make good his Charge against D^r. Keill or to ask pardon of the Society for Suspecting their Judgm^t & Integrity in the *Commercium epistolicum* already published by their Order & Approbation."[231] Blessed are the peacemakers, for they shall inherit the enmity of both sides. As far as extant correspondence indicates, Chamberlayne did not communicate with Newton again.

By the end of 1714 the controversy with Leibniz had obsessed Newton for nearly four years. During that time, he had pored over the manuscript record of the early years and the published record beginning with 1684. As he assisted Keill in composing his various answers and himself put together *Commercium epistolicum,* he gradually formulated his case against Leibniz, worked out its exposition, and polished its paragraphs endlessly so that they would perfectly express his rage. Vicarious expression through Keill or a committee could not finally suffice to convey his indictment. He worried as well that the message had not reached its intended audience, for *Commercium epistolicum* had been a restricted publication. Toward the end of 1714, he began to compose his own essay on the dispute, "An Account of the Book entituled *Commercium Epistolicum.*" He could not, of course, put his name to it and appear openly in the lists. He could publish it anonymously in the *Philosophical Transactions* for January–February 1715, however, filling all but three pages of the whole issue.[232] He could have DeMoivre translate it into French for publication in the *Journal literaire.* He could have a review of it sent to the *Nouvelles littéraires.* He could have the French version of the "Account" printed as a separate pamphlet. He could have copies of the pamphlet spread about the Continent by whatever means presented themselves.[233] And if he could, he did. Later he also published it in Latin.

Newton left an immense manuscript record of the "Account," testimony to the extent and the intensity of the effort that he put into it.[234] Its content repeated at greater length what he had been

[231] 11 Nov. 1714; *JBC 11,* 25.

[232] *Philosophical Transactions, 29* (1714–16), 173–224.

[233] "Extrait du livre intitulé *Commercium epistolicum* . . ." *Journal literaire,* 7 (1715), 114–58, 344–65. Cf. Halley to Keill, 3 Oct. 1715; *Corres 6,* 242. "An Account of the Book &c.," *Nouvelles littéraires, 2* (21 Sept. 1715), 184–5. Cf. Henri du Sauzet to Leibniz, 9 Nov. 1715; *Corres 6,* 247. Arnold to Leibniz, 5 Feb. 1716 (*Corres 6,* 275), and Conti to Leibniz, March 1716 (Gerhardt, ed., *Briefwechsel,* p. 269), both mentioned a pamphlet form of the French version. The English envoy to Venice, Alexander Cunningham, had copies of the pamphlet to pass out where he could (Cunningham to Newton, 21 Feb. 1716; *Corres 6,* 278–9).

[234] *Add MSS* 3968.8, ff. 67–96; 3968.41, ff. 34–7, 77–82, 94–127, 153–8. There are many other folios in *Add MS* 3968 devoted to things very similar to the "Account," which it is difficult readily to distinguish from it. *Add MS* 3968.20, ff. 296–311, is an amanuensis's copy of the Latin translation, called the *Recensio.*

saying through Keill for some time. Building on the materials in *Commercium epistolicum,* he presented a history of his own mathematical development by the year 1676, the year of the two *Epistolae.* "And by these improvements," he concluded, "Mr. Newton had in those Days made his Method of Fluxions much more universal than the Differential Method of Mr. Leibnitz is at present."[235] Leibniz made the absence of fluxions from the *Principia* the cornerstone of his case. Newton threw truth to the winds to insist on its presence.

> By the help of the new *Analysis* Mr. Newton found out most of the Propositions in his *Principia Philosophiae:* but because the Ancients for making things certain admitted nothing into Geometry before it was demonstrated synthetically, he demonstrated the Propositions synthetically, that the Systeme of the Heavens might be founded upon good Geometry. And this makes it now difficult for unskilful Men to see the Analysis by which those Propositions were found out.[236]

Wholly partisan, wholly uncharitable, wholly unfair, the "Account" delivered a searing indictment of Leibniz, first for plagiary, second for inciting conflict. For all its vicious hypocrisy, it was a powerful argument in a genre of which Newton was master. For years he had trained himself in empirical historical research connected with the prophecies. In the "Account," he turned the skills he had developed to his own advantage. Leibniz sensed that he could not compete effectively on these grounds and largely avoided the issues Newton posed in order to pitch his case in more philosophical terms. As history the "Account" does not pass inspection, whatever the skill of its presentation. The passion with which he delivered it, however, leading him to sacrifice factual truth for the higher truth he never doubted, tells much about Newton.

At about the same time when he composed the "Account," Newton also worked on at least two other expositions of his case. When Lemma II, near the beginning of Book II, was printed for the second edition, Leibniz's initial letter had not yet arrived, and Newton allowed the scholium about the correspondence of 1676 to appear in the edition unchanged. Now he wrote several versions of an alternate scholium that would insert his story of the controversy into the *Principia.*[237] In the third edition a further version of the scholium (which did not mention Leibniz) cited the tangent letter of 1672 and insisted on the treatise of 1671 as the foundation of the lemma. At one point, he thought of much more. He also planned to issue a reprint of the second edition with a full version of the 1691 "De quadratura" and a reply to Bernoulli's criticisms of his expertise appended.[238] Nothing came of either plan.

---

[235] "Account," p. 194.   [236] *Ibid.,* p. 206.   [237] *Var Prin 2,* 794–9.
[238] Cohen, *Introduction,* p. 346.

Not long after the publication of the "Account" a French delegation, led by Pierre Rémond de Monmort, arrived in England to observe a solar eclipse. With Monmort was a Venetian cleric of noble descent, the Abbé Antonio Schinella Conti, who stayed on for a number of years. Abbé Conti had lived in Paris for two years while he gathered material for a history of philosophy. He had made the acquaintance of all the leaders of French science, and he had entered into correspondence with Leibniz on the eve of his departure for England. Manifestly he was both an impressive and a charming man. In London he repeated his French conquests anew. Leibniz welcomed his correspondence, took care to secure his introduction to Princess Caroline, and regarded him, at first, as a valuable philosophic ally. Newton was taken with him at once, called on him soon after he arrived, and invited him to dinner.[239] Very quickly they became frequent companions, and it was not long before Conti was dreaming of the glory that Chamberlayne had pursued in vain. Some years later, he complained people had convinced Newton that he had involved himself in the quarrel "to give me a name."[240] There is no serious reason to doubt it, though undoubtedly the name that he, like Chamberlayne before him, sought was a worthy one. He pursued it more tenaciously and thereby won for himself the same reward in greater proportion.

On 6 December 1715 Leibniz sent a letter to Conti with a long postscript about Newton which was probably intended for his eyes. Leibniz had no intention of engaging in a contest with a nonentity like Keill. Though he had not been excessively open himself, his letter sought to draw Newton onto the field in person. He had also decided that Bernoulli would have to surrender the cloak of anonymity and appear publicly on his behalf. Leibniz opened the postscript with a brief résumé of the priority dispute from his point of view. "It does not appear to me," he stated, "that Mr. Newton had the Essence and the Algorithm of the infinitesimal method before me, as Mr. Bernoulli has rightly judged . . ."[241] This was to repeat the heart of the opinion of the "leading mathematician" quoted in the *Charta volans*, using his key word "algorithm," and to attribute the opinion to Bernoulli. Newton may have been an old man, but he was not too old to catch that hint. Nor had Leibniz intended him to miss it. At almost the same time, in the 28 December 1715 issue of the *Nouvelles littéraires*, Leibniz's anonymous reply to the "Ac-

---

[239] Antonio Schinella Conti, *Prose e poesi*, 2 vols. (Venice, 1739–56), 2, 24.

[240] Conti to Des Maizeaux, 1 Sept. 1721; *Corres 7*, 149–50. Evidently this charge struck a vulnerable spot. Conti complained of it in almost identical words in a letter to Brook Taylor, 22 May 1721 (Brook Taylor, *Contemplatio philosophica* [London, 1793], p. 125).

[241] Gerhardt, ed., *Briefwechsel*, pp. 263–4.

count" also identified Bernoulli and translated the letter in the Charta into French.[242]

In the letter to Conti, Leibniz risked a sally onto Newton's chosen ground, the events of the early 1670s, to sustain his contention that *Commercium epistolicum* presented only selected and partial excerpts from the correspondence. On his second visit to London, he mentioned, Collins allowed him to see part of his correspondence, "and in it I noted that Mr. Newton also confessed his ignorance about several things, and among others said that he had not discovered anything on the measurement of the well known Curves except for the Cissoid. But all of that has been suppressed."[243] He could not have ventured a more disastrous gambit. Newton's reply pointed out the passage in question in *Commercium epistolicum*. Embarrassed, Leibniz clutched at another example in his second letter only to have it destroyed in the same way. He could not have demonstrated more effectively the folly of trying to attack Newton through the empirical record of the early events. He also revealed more about the visit in London than he had ever said before. Newton grasped the implication, and as he pondered Leibniz's two examples, he convinced himself that Leibniz must also have read "De analysi" in 1676. By the time he reached this conclusion, Leibniz was dead, but in his later expositions of the early events he began to assert as much. Keill made the same charge in 1716.[244] Page 32 in volume 2 of Newton's copy of Des Maizeaux's *Recueil* is folded down in the manner Newton used to remind himself of important passages. The corner of the page points directly at Leibniz's admission, in his letter to the Baroness von Kilmansegge, that Collins showed him papers by Newton when he visited London in 1676.[245]

Leibniz always sought to push the discussion onto broader philosophical grounds. A month before his letter to Conti, a set of criticisms of Newtonian philosophy in a letter to Princess Caroline had inaugurated a celebrated and extended correspondence with Samuel Clarke on points of difference in their natural philosophies. The letter to Conti quickly moved beyond the priority dispute to the same issues. "His Philosophy seems rather strange to me," he began, "and I do not think it can be established." The concept of attraction, either an occult quality or a perpetual miracle, furnished the centerpiece of his critique. He objected to the void and wondered at Newton's belief that he could prove empirically that a

---

[242] A piece said to come from Amsterdam (*Nouvelles littéraires, 2*, 413–15).

[243] Gerhardt, ed., *Briefwechsel*, p. 264.

[244] The draft of a letter to Varignon, late 1718; *Corres 7*, 18. *Add MSS* 3968.26, f. 372^v; 3968.41, f. 55^v, "Lettre de Monsieur Jean Keill . . . à Monsieur Jean Bernoulli," *Journal literaire, 10* (1719), 286. The introduction to the letter stated that Keill wrote it in 1716.

[245] John Harrison, *The Library of Isaac Newton* (Cambridge, 1978), p. 27.

void exists. He wondered more at Newton's attribution of parts, including a sensorium, to God, and his questioning of God's intelligence by suggesting the creation was so imperfect as to need continual repairs. "Thus I find the Metaphysic of these Gentlemen *a narrow one* . . ."[246] For Newton, the raising of philosophic questions became a fresh grievance. Leibniz was trying to avoid the matter of fact "by running the dispute into a squabble about a Vacuum, & Atoms, & universal gravity, & occult qualities, & Miracles, & the Sensorium of God, & the perfection of the world, & the nature of time & space, & the solving of Problemes, & the Question whether he did not find the Differential Method *proprio marte:* all which are nothing to the purpose."[247] He had given his answers to the philosophic questions elsewhere. By and large he refused here to be diverted by them.

Leibniz's letter to Conti contained one other matter. For some time, Bernoulli had been urging on him the desirability of posing a new challenge problem that would reveal the superiority of the differential calculus. Indeed, at Leibniz's request, Bernoulli furnished the problem used. Thus the letter to Conti ended with a paragraph containing a challenge: "To test the pulse of our English Analysts . . . ," as he put it. The problem was to find a line *BCD* that cuts at right angles all of a determined family of curves, for example all the hyperbolas *AB, AC, AD* with the same apex and the same center.[248] Alas for Leibniz, it was another disaster; English mathematicians read the problem, not unnaturally, as one posed specifically about hyperbolas, and answers quickly poured in, from Halley, from Keill, from Pemberton, from Taylor, even one from a student at Oxford, James Stirling. Newton also tried his hand at it, and since he was Newton, the *Philosophical Transactions* published it, though anonymously. It was just as well. Approaching seventy-five now, he failed even to get a handle on it, and his solution wandered from the point.[249] Leibniz had meant to pose a general problem calling for a general method of solution, not a specific one about hyperbolas, which he used only as an example. He tried to amend his statement of the problem to make it clear that a general solution was demanded.[250] The English mathematicians had already submit-

---

[246] Gerhardt, ed., *Briefwechsel*, pp. 264–5. He emphasized "a narrow one" by putting the phrase in English.

[247] The draft of a letter to Des Maizeaux; *Add MS* 3968.27, f. 390.

[248] Gerhardt, ed., *Briefwechsel*, p. 267.

[249] Several drafts of his attempt to solve the problem are printed in *Corres 6*, 290–2. One of them was published anonymously in the *Philosophical Transactions, 29* (1716), 399–400.

[250] He sent the amended statement of the problem no less than four times to be sure it arrived. Leibniz to Rémond, 27 Jan. and 27 March 1716; G. W. Leibniz, *Die Philosophischen Schriften*, ed. C. I. Gerhardt, 7 vols. (Berlin, 1875–90), *3*, 669–70, 674. Leibniz to Arnold, 7 Feb. and 17 March 1716; *Corres 6*, 270, 299.

ted their solutions before the amplification arrived. In consternation he wrote to Bernoulli for assistance and received the formulation of a general problem, of which the former was a specific case, which he forwarded in turn to Conti. He also identified Bernoulli as the author of the problem.[251] The English retorted in disdain that Leibniz had effectively set a new problem after they had foiled his first attempt. They also noted a further instance of Bernoulli's active engagement against Newton.

Leibniz had an English friend and supporter in London, John Arnold, and a letter from him to Leibniz early in February 1716 offers a precious glimpse behind the scene provided by the formal correspondence. Arnold had gone to visit Conti and found Newton there. The conversation turned to the calculus dispute. When Newton asked, Arnold admitted that he had a copy of the *Charta volans*, which he agreed to show to Newton. Hence he and Conti had gone to Newton's house together. Newton placed the blame entirely on Leibniz and Keill. He had refused to let Keill write until Leibniz accused him of plagiary – an ambiguous statement without dates, probably a reference to Keill's "Answer" to Leibniz's "Remarks," though it could refer to earlier events, which it would then put in a wholly new light. Newton agreed that Keill had been too brusque. He resented the challenge problem sent in defiance of a whole nation. Newton went out of his way to woo Arnold with many civilities. He had the two men stay to dinner. He gave Arnold the pamphlet with the French version of the "Account" for his edification. Arnold remarked that similar civilities had succeeded in winning Conti over to Newton's side, and he indicated that Newton was beginning to turn his attention to the Hanoverian court. He had arranged to demonstrate his optical experiments to the Baronness von Kilmansegge, the mistress of King George.[252]

Arnold also informed Leibniz that Newton was preparing to enlist even the diplomatic corps in the struggle. What he referred to was a reception that Newton arranged at the Royal Society about 20 February 1716. We know about it primarily from a letter Conti wrote to Taylor five years later. Leibniz's charge that the selection of letters in *Commercium epistolicum* was slanted had stung. A year earlier Newton had devoted two sessions of the Royal Society to a public scrutiny of the letters. A month later, when the French delegation led by Monmort arrived to view the eclipse, Newton immediately entertained them with an exhibition of the letters and manuscripts while Halley in the background softly commented on the age of the paper and the faded ink, which proved they were not

[251] Bernoulli to Leibniz, 11 March 1716; *Die Mathematische Schriften, 3/2*, 957–8. Leibniz to Conti, 14 April 1716; Gerhardt, ed., *Briefwechsel*, pp. 294–5.
[252] Arnold to Leibniz, 5 Feb. 1716; *Corres 6*, 274–5.

recent fabrications. And later in 1716 he revived the committee to examine Leibniz's charge that "the committee" unfairly omitted from the letters passages that were prejudicial to Newton.[253] He wanted testimony beyond the Royal Society, however, and he managed, according to Conti, to assemble a number of ambassadors at Crane Court to collate the originals with *Commercium epistolicum*. The plan went awry. The Baron von Kilmansegge, the Hanoverian minister and husband of George's mistress, loudly asserted that this procedure would not suffice. The only way to end the quarrel was a direct correspondence between Newton and Leibniz. Everyone present agreed, and the king later indicated his agreement as well.[254] Until this moment, it was by no means clear that Newton intended to reply to Leibniz's letter. Now he had no choice. Later, at least twice, Newton claimed that Conti had pressed him to reply so that the two letters, Leibniz's and his, might be shown to the king.[255] Conti had made himself a popular figure at the court; popular figures must work continually at making a "name." There is good reason to suspect that Conti set Kilmansegge up for his decisive intervention.

In 1720, Newton asserted that when George I came to England, Leibniz's friends tried to reconcile them in order that he too might come over, "but they could not get me to yeild."[256] Certainly there was little conciliation in the letter he now sent to Leibniz. His immense capacity for feeling grievance amply displayed itself again. Without even blinking at his own prevarication as he referred to the compilation of *Commercium epistolicum* by an international committee, and without any sense of the injuries he had inflicted on Leibniz, he complained loudly of Leibniz's injury to him in not answering the published report but referring it to an eminent mathematician.

> And the Answer of the Mathematician (or pretended Mathematician) dated 7 June 1713, was inserted into a defamatory Letter dated 29 July following, & published in Germany without the name of the Author or Printer or City where it was printed. And the whole has since been translated into French & inserted into another abusive Letter (of the same Author, as I suspect) . . .[257]

Since Leibniz had identified Bernoulli, Newton undoubtedly inserted the insulting reference to a "pretended Mathematician" as a pointed retort to one whom Newton regarded, after his offer of

---

[253] 3 and 10 March 1715; 12 July 1716; *JBC 11*, 53, 54, 135. Conti, *Prose e poesi, 2*, 23.

[254] Conti recounted the events in a letter to Taylor on 22 May 1721 (Taylor, *Contemplatio philosophica*, pp. 121–2).

[255] Newton to Des Maizeaux, ca. Aug. 1718; *Corres 6*, 455. Newton to Varignon, 29 Aug. 1718; *Corres 7*, 3.

[256] The draft of a letter from about early 1720; *Corres 7*, 83.

[257] Newton to Conti (for Leibniz), 26 Feb. 1716; *Corres 6*, 285.

friendship in 1712, as a traitor. Leibniz immediately forwarded the letter to Bernoulli to provoke him further. Thus introduced, Newton's letter proceeded on to a bitter and powerful statement of Newton's case. He mentioned the recent ambassadorial endorsement of the letters' accuracy. To stimulate Leibniz's memory, he cited what Leibniz had said about Newton's mathematics in 1676, 1677, 1684, 1693, 1696, and 1700.

> But as he has lately attaqued me with an accusation wch amounts to plagiary: if he goes on to accuse me, it lies upon him by the laws of all nations to prove his accusation on pain of being accounted guilty of calumny. He hath hitherto written Letters to his correspondents full of affirmations complaints & reflexions without proving any thing. But he is the agressor & it lies upon him to prove his charge.[258]

Conti treated the letter as public property. It passed around the court for a month before it was sent. Pierre Coste, another French refugee, who had translated Locke into French and would soon translate Newton's *Opticks,* put it into that language for the king and for transmission to Leibniz.

Like all such exchanges, this one fed on itself and grew. Newton's reply was longer than Leibniz's letter. Leibniz wrote a new one twice as long. Newton responded with a set of savagely furious "Observations" just as long, and they could have continued indefinitely had Leibniz not died.[259] Leibniz expressed satisfaction to be dealing with Newton directly, and he sent copies of the correspondence to Paris to be shown about. Clearly he thought his letters placed him in a good light. And yet his second letter, his response to Newton's implacable rage, is defensive and apologetic, puny in comparison to Newton's power. Modern scholarship has vindicated Leibniz. His own attempt to give an account of his independent development of the calculus, expounded without corroborating evidence, appears weak beside Newton's wealth of citations from Collins's correspondence. Nothing could vindicate Leibniz from the charge on which so much of the dispute hinged, that he failed to mention the correspondence of 1676 when he first published, and his efforts to elude the issue were especially unconvincing. To modern eyes, his crime was no worse than Newton's in suppressing his method altogether for thirty years. Each committed a sin of omission. Newton's injured the whole world, which had no voice to speak its grievance. Leibniz's omission injured Newton, who certainly did.

Shortly after his second letter to Newton via Conti, Leibniz also wrote two letters to Germans prominent in the Hanoverian court,

---

[258] *Corres 6,* 285–8.
[259] Leibniz to Conti, 9 April 1716; Gerhardt, ed., *Briefwechsel,* pp. 274–84. Newton's "Observations," *Corres 6,* 341–9.

the Baroness von Kilmansegge and Count Bothmar.[260] Newton had begun to pay serious attention to wooing the Hanoverians. As Arnold mentioned, he arranged a demonstration of his optical experiments for the baroness; in May he had himself presented at court.[261] Leibniz wanted to be sure that the king and his retinue did not forget their faithful servant in Hanover. The letters gave accounts of the dispute, partisan accounts of course in the best tradition of the controversy.

After Leibniz's death, Newton had his correspondence with Leibniz, plus five items from the 1690s that he was accustomed to cite on his own behalf, published as an appendix to a reissue of Raphson's *Historia fluxionum.* The publication has been roundly and almost universally denounced. Newton did many reprehensible things during the conflict, but I see no reason to list this act among them. He justified himself by claiming he wanted to forestall an imperfect – that is, partial – publication abroad. He had good reason to fear such. From the beginning, when he passed it around the court, Conti had treated the correspondence as public, and Leibniz had sent copies of it all to Paris for dissemination. There is some cause to think that the announcement and preparation of Pierre Des Maizeaux's *Recueil,* to which Conti contributed the correspondence, stimulated Newton's publication. Newton considered Des Maizeaux to be a Leibnizian partisan, objected to the manner in which he ordered the correspondence, and later held up the publication.[262] He may have rushed his own version into print in 1718 to beat the *Recueil,* which did not finally appear for another three years. Newton has enough crimes to answer for in the dispute. There is no need to blame him for a justified act of self-defense.

While Newton's own correspondence with Leibniz, which he kept focused on the priority controversy, continued, another correspondence pursued the philosophic issues, which Newton refused to discuss. In November 1715, replying to a question about the theology of Samuel Clarke put to him by the Princess of Wales, Leibniz sent a famous challenge, which could not be ignored.

Natural religion itself, seems to decay (in England) very much. Many will have human souls to be material: others make God himself a corporeal being. . . . Sir Isaac Newton says, that space is an organ, which God makes use of to perceive things by. . . . Sir Isaac Newton, and his followers, have also a very odd opinion concerning the work of God. According to their doctrine, God Almighty wants

260 Pierre Des Maizeaux, *Recueil, 2,* 33–52.

261 Cf. Robert Balle to Newton, 28 May 1716; *Corres 6,* 357.

262 Newton was correct in this. Cf. Leibniz to Princess Caroline, 31 July 1716, and Leibniz to Des Maizeaux, 21 Aug. 1716 (*Die Werke von Leibniz,* ed. Onno Klopp, 11 vols. [Hannover, 1864–88], *11,* 129–30, 178–80) for the origins of the *Recueil.*

to wind up his watch from time to time: otherwise it would cease to move. He had not, it seems, sufficient foresight to make it a perpetual motion.[263]

Leibniz addressed the challenge more to Clarke, who at that time was attending upon the princess assiduously, than to Newton, and Clarke (who was a dedicated Newtonian in any case) undertook to reply. The exchange lasted through five rounds, ten letters in all, each one longer than the last as every point made required more words for its refutation, until Leibniz's death finally terminated the correspondence. Leibniz's original letter defined perhaps the central issue, the divine governance of the universe. Inevitably the discussion also spread out into natural philosophy and embraced questions such as attractions and voids. Although it did not figure as a central issue, Leibniz could not bring himself to ignore Newton's conception of space as the sensorium of God. Once he saw it in Query 28 of the *Opticks,* it appeared to him an absurdity which cast doubt on Newton's competence in philosophy as a whole. "And so this man has little success with Metaphysics," he told Bernoulli in commenting on the passage.[264] It remains an open question how extensively Newton participated in the composition of Clarke's side of the correspondence. Unlike the case of Keill's letters, extensive manuscript evidence of his active role does not exist. On the other hand, Clarke was one of his close adherents. He was the rector of the chapel in Golden Square of which Newton was a trustee, two Arians masquerading as orthodox. Among Newton's papers is a copy in his own hand of the insulting passage on atoms and voids printed as a postscript to Leibniz's fourth letter. At the bottom Newton wrote, "Received of ye Princess May 7th 1716, & copied May 8."[265] Though attached to Leibniz's letter of 2 June in the published correspondence, this passage was a postscript to Leibniz's letter of 12 May (that is, 1 May in the Julian calendar); the princess must have dispatched it as soon as it arrived.[266] At the least, New-

---

[263] H. G. Alexander, ed., *The Leibniz-Clarke Correspondence* (New York, 1956), p. 11. See Princess Caroline to Leibniz, 15 Nov. 1715 (Klopp, ed., *Werke von Leibniz, 11,* 53) for the query that provoked Leibniz's initial challenge. On the Clarke-Leibniz correspondence and the general issue of the philosophical differences between Newton and Leibniz, see Ernst Cassirer, "Newton and Leibniz," *Philosophical Review, 52* (1943), 366–91; F. E. L. Priestley, "The Clarke-Leibniz Controversy," in Butts and Davis, eds., *Methodological Heritage,* pp. 34–56; Margula R. Perl, "Physics and Metaphysics in Newton, Leibniz, and Clarke," *Journal of the History of Ideas, 30* (1969), 507–26; and I. Bernard Cohen, "Newton and Keplerian Inertia: An Echo of Newton's Controversy with Leibniz," in Allen Debus, ed., *Science, Medicine and Society in the Renaissance,* 2 vols. (New York, 1972), *2,* 199–211.

[264] Leibniz to Bernoulli, 9 April 1715; *Mathematische Schriften, 3/2,* 939; translation in *Corres 6,* 213.   [265] Add MS 3968.36, f. 517.

[266] Klopp, ed., *Werke von Leibniz, 11,* 102–3.

ton's copy of it indicates his active interest in the correspondence, which touched on most of the issues that had been central to his own lifelong enquiry into the nature of things. Indeed, Princess Caroline assured Leibniz that Clarke's letters were not written "without the advice of the Chevalier Newton."[267] Clarke published the correspondence in 1717. For some reason, no one has found this a reprehensible act on his part.

Though he refused to be diverted by philosophical issues in his correspondence with Leibniz, Newton did take up the differences between them. The General Scholium to the *Principia* had already done so, and the "Account of *Commercium Epistolicum*" also closed with three pages devoted to a forceful exposition of the differences between their philosophies. Newton's philosophy was experimental, Leibniz's hypothetical. He quoted anew the various reservations he had inserted in the *Principia* and the *Opticks* about forces and their possible causes. "And after all this, one would wonder that Mr. Newton should be reflected upon for not explaining the Causes of Gravity and other Attractions by Hypotheses; as if it were a Crime to content himself with Certainties and let Uncertainties alone."[268] Can the constant and universal laws of nature, whether derived from the power of God or from a cause yet unknown, be called miracles and occult qualities, that is wonders and absurdities?

> Must all the Arguments for a God taken from the Phaenomena of Nature be exploded by *new hard Names?* And must Experimental Philosophy be exploded as *miraculous* and *absurd,* because it asserts nothing more than can be proved by Experiments, and we cannot yet prove by Experiments that all the Phaenomena in Nature can be solved by meer Mechanical Causes? Certainly these things deserve to be better considered.[269]

The manifesto, written in the twilight of Newton's scientific career, voiced his continued support of the principles that had guided him to his revolution in natural philosophy.

"Mr. Leibniz is dead; and the dispute is finished."[270] Newton received this word in December 1716 from Conti, who wrote from Hanover, where he had gone in the hope of meeting Leibniz. He had died on 4 November, well before Conti arrived. As to the second clause in his announcement, Conti could not have been more mistaken. The passions generated had reached a pitch that required another six years for their dissipation. Nevertheless, Leib-

---

[267] Princess Caroline to Leibniz, 30 Dec. 1715; *ibid.,* *11,* 71. For arguments that Newton did participate actively in the correspondence see Alexandre Koyré and I. Bernard Cohen, "Newton and the Leibniz-Clarke Correspondence," *Archives internationale d'histoire des sciences,* 15 (1962), 63–126; and A. R. and Marie Boas Hall, "Clarke and Newton," *Isis,* 52 (1961), 583–5.   [268] "Account," p. 223.

[269] *Ibid.,* p. 224.   [270] *Corres 6,* 376–7.

niz's death removed the object of Newton's wrath, and with time even he tired of the repetition of stale taunts. If the dispute was not finished, at least its conclusion had been announced. Newton had probably celebrated his seventy-fourth birthday before Conti's message arrived. With it ended the last passionate episode of his life. Though he had more than ten years to live, they were, inevitably for a man of his age, years of decline.

15

Years of decline

THE priority dispute dragged on with diminished intensity for another six years and during that time continued to occupy a major part of Newton's consciousness. He had never been able to lay a project down easily. Wound up as tightly as he was now, and with his honor at stake, he could not put the dispute aside simply because his antagonist had died. When Fontenelle's *Éloge* of Leibniz appeared, he was upset by the author's unwillingness to believe that Leibniz was a plagiarist, and he made extensive notes on the *Éloge* by way of correction, though he never put them to use.[1] Although Newton had invested an immense effort in the "Account of *Commercium epistolicum*," he was not satisfied with its exposition of his case – as indeed he was never finally satisfied with any work. Sometime after Leibniz's death, he composed a *Historia methodi fluxionum* (*The History of the Method of Fluxions*).[2] Mercifully, he eventually forebore to impose this further recital of a barren chronology upon his loyal public. In 1717, when Desaguliers dug out of an old *History of the Academy* a discussion in which Leibniz attempted to show that the fall of the barometer in rainy weather derives from the condensation of vapor into drops which do not gravitate while they fall, Newton quickly seized it and regaled the next meeting of the Royal Society with a demonstration that Leibniz's explanation could account at most for a small fraction of observed barometric variations.[3]

His own personal involvement was no longer the most powerful factor in prolonging the struggle, however. Both Newton and Leibniz had drawn others into the fray as they had attempted to conceal their own participation. John Keill had happily assumed the mantle of Newton's defender. Johann Bernoulli had done the same for Leibniz, though more reluctantly and with every effort to keep it secret. Quasi-anonymity made him a tempting target, for one could say things about an unnamed "leading mathematician" that one would not say about Johann Bernoulli, not only a leading mathematician but the leading one in Europe at that time. Though he had only

---

[1] Newton saw the *Éloge* only in 1718. His notes on it are in *Add MS* 3968.26, ff. 372–80. Probably near the end of 1718, Newton drafted a letter about it to Varignon, which he apparently did not finally send (*Corres 7*, 17–19).

[2] He used a number of different but similar titles on various drafts. *Add MSS* 3968.12, ff. 146–72; 3968.13, ff. 173–218, 223–35.

[3] 28 Feb. and 7 March 1717; *JBC 11*, 162–4.

himself to thank for his inglorious posture, he took offense at every barb. John Keill, whose only tactic was to flail about wildly in all directions, readily gave more offense than Bernoulli could stomach. Leibniz might be dead; Keill and Bernoulli were not. Their quarrel, which took on an independent life of its own, did more than anything else to prolong the calculus controversy.

The proximate cause of their quarrel was an "Epistola pro eminente mathematico, Dn. Johanne Bernoullio, contra quendam ex Anglia antagonistam scripta" (Letter Written on Behalf of the Eminent Mathematician, Mr. Johann Bernoulli, against a Certain Adversary from England) which Bernoulli composed a few months before Leibniz's death. A reply to Keill's rejoinders to Bernoulli's published criticisms of deficiencies in the *Principia*, the letter rehearsed anew the argument that Newton had not known the calculus when he wrote his work. The second clause of the first sentence advanced the unequivocal assertion that Leibniz had invented the differential calculus. The first clause preceded that assertion with the equally unequivocal one that Johann Bernoulli had invented the integral calculus, a claim in which Bernoulli thus implicitly forced Leibniz's acquiescence as the price of his support. Bernoulli dispatched the "Epistola" to Christian Wolf on 8 April 1716 with a covering letter which permitted and indeed begged its publication in full. Wolf should submit it first to Leibniz for approval:

> however I hope it will turn out that he will in particular not reject the very powerful arguments which I have drawn together from which, whatever Keill or his supporters whine, it is firmly established that Newton, at the time when he wrote his *Principia Philos. Mathematica*, still had not understood the method of differentiating differentials. As regards the actual form in which I would choose that the contents of this letter appear, they can keep the form of a letter, but, please, so altered as though it had been written by an anonymous writer, or indeed by someone of another either real or fictitious name; in a word, you should arrange the whole thing with as much discretion as you can, lest Keill suspect this letter to have been written by me. It would be exceedingly unpleasant for me to have Keill vent his spleen upon me, and to be rudely exposed to ridicule as his opponents usually are, after he has hitherto treated me quite politely.[4]

Thus the *Acta* for July 1716 carried the anonymous "Epistola" in which Johann Bernoulli covertly defended and praised the eminent mathematician Johann Bernoulli. Wolf and Leibniz, however, had in mind for Bernoulli a stance more heroic than the one that he

---

[4] Bernoulli to Wolf, 8 April 1716; Johann (III) Bernoulli, "Anecdotes pour servir à l'histoire des mathématiques," *Histoire de l'Académie royale des sciences et belle-lettres, 1799–1800* (Berlin), p. 44. Translation in *Corres 6*, 303.

himself projected. In transforming the letter from the first person to the third, they contrived to overlook one phrase in which Bernoulli referred to "my formula" for the inverse problem of central forces.[5] It was picked up at once and Bernoulli's authorship proclaimed, though his hand was transparently obvious in the shape of the argument in any case.

By the time the July issue of the *Acta* reached England in the spring of 1717, Leibniz was dead. Keill was incensed by the "Epistola." "I believe there was never such apeice for falshood malice envie and ill nature published by a Mathematician before . . ." He also pointed out the phrase that showed Bernoulli had written it.[6] Recalling how Bernoulli had spurned his advances of 1712, Newton was no less incensed than Keill. He went over Keill's draft of a reply amending it extensively and venting his own spleen, which was no less venomous than Keill's, on Bernoulli.

> But since you make a practice of writing controversial abusive papers without setting your name to them & of applauding your self in them [he made Keill taunt Bernoulli]: whenever I meet with such anonymous papers wherein you are applauded or cited as a witness or your enemies abused: I shall for the future look upon them as written by your self or at least by your procurement, unless the contrary appears to me.[7]

Newton decided to contain his anger, however. He was nearly seventy-five years old. He was beginning to tire of the battle and probably reflected that the reply to Bernoulli's "Epistola" could only provoke another retort, which would demand still another reply from Keill, and so the battle would continue without end. Newton held onto Keill's "Letter to Bernoulli" and did not release it until Bernoulli offered further provocation. It finally appeared, with Newton's alterations, in 1720.[8]

Perhaps Newton paused in 1717 because of a letter shown to him by Brook Taylor about the same time the *Acta* with Bernoulli's "Epistola" arrived. Early in April, Bernoulli wrote a letter to Rémond de Monmort in Paris, a letter intended for transmission to Brook Taylor who was to show it to Newton, as in fact he did. With Leibniz dead, Bernoulli was taking new counsel about his own best interests, and peace with Newton appeared desirable. At

[5] *Acta eruditorum*, July 1716, pp. 296–315. The telltale phrase is on p. 314.

[6] Keill to Newton, 17 May 1717; *Corres 6*, 385–6.

[7] *Add MSS.* 3968.23, ff. 363–7; 3968.34, ff. 474–7. The phrase quoted is published in *Corres 6*, 387.

[8] Cf. Inglis to Keill, 19 Dec. 1717 and 14 Jan. 1718; *Corres 6*, 425, 429. "Lettre de Monsieur Jean Keyll, . . . à Monsieur Jean Bernoulli . . . ," *Journal literaire, 10*, Part II, (1719) (The Hague, 1720), pp. 261–87. Keill also published it in Latin as a separate pamphlet, *Epistola ad Bernoulli* (London, 1720).

that time, Bernoulli had not yet seen his published "Epistola" in the *Acta* and did not know about the incriminating phrase. As far as he was aware, the only mark against him in Newton's book was the challenge problem of 1715, and he explained at length how Leibniz had pried it from him and advertised his connection with it against his will.

> You will then have the goodness [he wrote to Monmort] to disabuse Mr. Newton of the opinion he holds in this regard, and to assure him on my part that I never intended to test the English by challenges of this sort, and that I desire nothing so much as to live in good amity with him, and [you will have the goodness] to find the occasion to make him see how much I value his rare merit; in fact I never speak of him without much praise.[9]

One might search some time to find a more repellent missive. Since Newton was convinced that Bernoulli had composed the letter in the Charta volans, he also, according to his own account, received the message with skepticism and made no response. Nevertheless, as I have indicated, he did hold up Keill's reply to Bernoulli. A year later, he received a second overture from Bernoulli via Monmort. Again Bernoulli laid the entire blame for the challenge problem on Leibniz and expressed his great respect for the discoveries of Newton, "whose esteem and friendship are very precious to me." By 1718, the "Letter on Behalf of Bernoulli" could not be ignored, and Monmort quoted to Newton Bernoulli's categorical denial that he had participated in the composition of a piece the tone of which he found objectionable.[10] Newton sent an account of the letter to Keill; again he did not respond to Bernoulli, however.

Late in 1718, two letters from France intervened. Rémond de Monmort wrote one of them in answer to a letter from Brook Taylor. With the ardor characteristic of Newton's supporters, Taylor had asserted that no one but Newton had contributed to the invention and improvement of the calculus. Monmort, who was not a slavish follower of Leibniz, let Taylor know why he would have none of such exaggerated claims. On the invention of the calculus he would not comment, he said, but Leibniz and the Bernoullis had been its true and almost sole promoters.

> It is they and they alone who taught us the rules of differentiation and integration, the way to use the calculus to find tangents to curves, their points of inflection and reversal, their extrema, evolutes, caustiques by reflection and refraction, the quadratures of curves, centers of gravity, of oscillation, and of percussion, problems of the inverse method of tangents such as that, for example, which

[9] Bernoulli to Monmort, 8 April 1717; *Corres 6*, 383. Cf. Newton to ?DeMoivre, ca. early 1720; *Corres 7*, 82.
[10] Monmort to Newton, 27 March 1718; *Corres 6*, 435–8.

excited so much admiration by Huygens in 1693, *to find the curve the tangent of which has a given ratio to the subtangent.* It is they who first expressed mechanical curves by equations, who taught us to separate the unknowns in differential equations, to decrease their power, and to construe them by logarithms or by the rectification of curves when that is possible, & who finally, by many and beautiful applications of the calculus to the most difficult problems of mechanics, such as the catenary, the sail, the spring, the quickest descent, and the paracentric, have set us and our descendants on the path of the most profound discoveries.[11]

A word of candor spoken forthrightly, Monmort's letter was a breath of fresh air in a miasma of deceit and hypocrisy. If Taylor showed Newton the letter, as he had shown him the earlier one, Newton never formulated a reply to it. With his conviction that second inventors have no rights, he had long since placed himself beyond the call of reason and equity. Even Taylor, who had no personal commitment to defend, failed to heed Monmort's message enough to refrain from undertaking his own imbroglio with Bernoulli. Monmort's letter did not alter the controversy in the least, but it does serve us as a timely reminder of historical reality, so easily forgotten in the flood of polemics.

The second letter from France, which did alter the course of events, came from Pierre Varignon, a member of the Académie and a respected student of mathematics and mechanics. For over ten years, until the priority dispute began in earnest, membership in the Académie had left no visible impression on Newton's life. Suddenly, the good opinion of academicians in Paris acquired importance in his eyes, and he left no stone unturned in his efforts to woo them. He was careful in distributing copies of the *Principia*'s second edition to the Académie. He outdid himself in entertaining French visitors. Deslande dined with him in 1713, and later a French mathematician, Nicole, returned to Paris from a visit in London with the report that Newton had entertained him at dinner, where he drank to the whole Académie and to Fontenelle and Varignon in particular.[12] In 1714, after Varignon had indicated his desire to be a fellow of the Royal Society, Newton himself proposed him at the special meeting called to elect Prince Menshikov.[13] Upon receiving his copy of the *Mémoires* of the Académie that contained Fontenelle's *Éloge* of Leibniz, Newton chose to thank Varignon for sending the *Mémoires* annually, and he enclosed three copies of the second English edition of the *Opticks*. Two months later, he sent Varignon

---

[11] Monmort to Taylor, 18 Dec. 1718; *Corres 7*, 21–2.
[12] A.-F. Deslande, *Histoire critique de la philosophie*, 3 vols. (Amsterdam, 1737), *2*, 264–5. Varignon to Newton, 28 Nov. 1720; *Corres 7*, 105.
[13] *CM 2*, 284, *JBC 11*, 17. Cf. Varignon to Newton, 18 Nov. 1714; *Corres 6*, 187.

Figure 15.1. Newton at seventy-seven. Portrait by Sir Godfrey
Kneller, painted for Abbé Varignon, 1720. (Courtesy of
Lord Egremont and the Petworth Estate.)

five copies of the second Latin edition of the *Opticks*. Thus began an
extended exchange which continued with mounting esteem on both
sides until Varignon's death late in 1722, a correspondence which
provided one of the happier interludes in Newton's old age.

From the beginning, the priority controversy figured in it. In his
first letter, Newton thanked Varignon for not acting against him in

the dispute. Along with the information that Varignon had sent a copy of the *Opticks* to Bernoulli as a gift from Newton, his reply confirmed that he had done nothing to injure Newton. He had always maintained silence about the dispute in his letters to Leibniz and Bernoulli, he said, "only bewailing, to myself and privately, that such great men are harassed by it, whom, if I had any influence, I would have brought back to their former cordiality; and this was the sole object I had in mind when I sent to Johann Bernoulli from yourself the new edition of your book on colours."[14] When the copies of the new Latin edition finally arrived in Paris in 1719, he also presented one of them to Bernoulli as though from Newton.

As Varignon had hoped, the second gift to Bernoulli finally provoked him to write a letter to Newton himself, sent via Varignon, in July 1719. However thankless, peacemaking is a sainted activity; it was not Varignon's fault that Bernoulli's letter induces nausea. After the opening flourishes, in which he called Newton "a man of divine genius of whom our age has no equal," Bernoulli went on to lament the fact that he had fallen from Newton's favor. Such a misfortune could only have been the work of toadies about Newton who misrepresented others.

> Therefore I do not doubt, great Sir, that you have also been told many lies and inventions about me . . . Beyond all doubt they are mistaken who have reported me to you as the author of certain of those flying sheets in which perhaps you were not treated with sufficient respect; but I earnestly entreat you, famous Sir, and implore you in the name of everything sacred to humanity, that you convince yourself thoroughly that whatever has been published anonymously in this way is falsely imputed to me. For it has not been my custom to issue anonymously what I neither wish nor dare to acknowledge as my own.[15]

Newton did not love Bernoulli. Only a few months before he received the letter, William Jones had shown him Brook Taylor's "Apologia contra Bernoullium," which Newton had ordered to be printed in the *Philosophical Transactions*. "I am very sensible that Sir Isaac Newton is very glad that any thing should be publisht which affects Bernoulli . . . ," Taylor had confided to Jones.[16] Not only did he not love Bernoulli, Newton also knew that Bernoulli's denial was an outright lie, as his earlier denial of the "Epistola pro Bernoullio" had been. Nevertheless, Newton accepted it. He even worked over the first draft of his reply to Varignon to eliminate passages that might give offense if sent on to Basel.[17] "Now that I am old," he

---

[14] Varignon to Newton, 17 Nov. 1718; *Corres 7*, 16; original Latin, p. 15.
[15] *Corres 7*, 44–6; original Latin, pp. 42–4.
[16] Taylor to Jones, 5 May 1719; *Corres 7*, 39–40.
[17] Newton to Varignon, 29 Sept. 1719; *Corres 7*, 62–4. Prior drafts, pp. 66–7.

wrote directly to Bernoulli, "I take very little pleasure in mathematical studies . . ."[18] Manifestly, he took even less pleasure in the continuing dispute, of which he had grown weary. He would rather accept a patent lie than prolong the controversy further.

As it turned out, he was not able simply to snuff the dispute out. Taylor was now engaged in his own quarrel with Bernoulli. Keill was eager to enter the fray once more to take his revenge for the insults Bernoulli continued to offer him. Bernoulli would not believe that Newton was unable to control their pens. Another publication in 1720, which also emanated from London, Pierre Des Maizeaux's *Recueil de diverses pièces,* aroused his anger even more, for Des Maizeaux published Newton's correspondence with Leibniz via Conti together with other letters of Leibniz that named Bernoulli as the author of the anonymous letter in the Charta volans. Des Maizeaux's preface also drew attention to Bernoulli's participation. A Huguenot refugee resident in England, Des Maizeaux had begun to assemble his *Recueil* in 1716, not as Newtonian propaganda, which it never was, but as a collection of Leibniz's papers and letters. Leibniz himself supplied him with the Clarke correspondence, and encouraged him, and Conti furnished him with the Newton-Leibniz correspondence as well as some other items. Only in 1718 did Newton learn of the project, when Des Maizeaux submitted proofs for his inspection. He was not entirely pleased. The order in which Des Maizeaux printed the letters did not present the message that Newton wished to draw from them.[19] He was unhappy enough that he contemplated his own publication of the letters, which included several that he had not printed earlier that year in Raphson's *History of Fluxions.*[20] When Bernoulli's denial arrived in 1719, Newton realized that the *Recueil,* now finally ready to appear, might shatter the fragile armistice, and he attempted to halt the publication. He paid the Dutch bookseller who was printing it twelve guineas to delay it until the spring.[21] According to DeMoivre, who must have been quoting Newton, he attempted to suppress the work entirely by buying it out, offering even to pay Des Maizeaux his expected profit.[22] We can measure in these efforts the extent of his weariness with the dispute.

His fears about Bernoulli's reaction to the *Recueil* were well founded. Newton had mentioned its imminent publication in his reply to Bernoulli, adding that pieces in it claimed Bernoulli had

---

[18] Newton to Bernoulli, 29 Sept. 1719; *Corres 7,* 70; original Latin, p. 69.

[19] Newton to Des Maizeaux, ca. August 1718; *Corres 6,* 462. He did acquiesce in the publication, however (Newton to Des Maizeaux, late 1718; *Corres 6,* 463).

[20] Cf. a draft of the preface to his intended publication (*Corres 6,* 463).

[21] Newton to Des Maizeaux, Nov. 1719; *Corres 7,* 73–4.

[22] DeMoivre to Varignon as quoted in Varignon to Bernoulli, 16 Feb. 1721; *Corres 7,* 130.

written the anonymous letter in the Charta volans. Early in 1720, he received the response to this information. Bernoulli began by hedging his denial with ambiguity.

I am not certain of what kind that letter addressed to Mr. Leibniz is of which you speak, which is dated 7 June 1713. I do not remember having written to him myself that day, yet I would not deny it altogether, since I have not kept copies of all the letters I have written. But if perhaps, among the innumerable letters which I had written to him, one were found which bore the very day and year mentioned, I would have dared to assert with complete confidence that nothing was contained in it which would in any way weaken your reputation for honesty, nor did I ever give him leave to make public certain of my letters, particularly such as would not please you, albeit against my hope and wish.

With the subject introduced, he proceeded to tell Newton that the publication in question would not contribute to peace. He implied that Newton could halt it if he wished and that he could prevent Taylor and Keill from attacking him. None too subtly, he added that he had some letters in his possession that would tell a different story. And if all that was not enough, he closed by repeating two pieces of information he had recently received that seemed to indicate that he had been expelled from the Royal Society.[23]

DeMoivre delivered the letter, which Newton read at once. At first, he saw the repeated disavowal of the Charta volans, which for all its lame evasions seemed still to desire peace. Only a second reading impressed the letter's true meaning upon him.[24] He may have been too old to find pleasure in continuing disputes. He was not so far gone that he would stomach crude threats. A letter about this time to an unnamed correspondent, that he had restrained Keill for two years but could not forever hinder him from defending himself, suggests that he now finally gave Keill permission to publish his reply to Bernoulli's "Epistola pro eminente mathematico."[25] Keill even included in it a malicious reference to Bernoulli's denial of the Charta volans – to Newton's dismay, real or pretended. To Bernoulli, Newton drafted a cold reply which he probably never sent.[26] Bernoulli continued to believe that Newton could staunch the flow of abuse from London if he would. Varignon's truce had survived only a few months.

Abbé Varignon was a man dedicated to peace, however, and he refused to give up. He devoted the next two years, until he died at the end of 1722, to attempting to revive the collapsed reconciliation. A rejected Conti, by then resident once more in Paris and

23 Bernoulli to Newton, 21 Dec. 1719; *Corres 7*, 77–9; original Latin, pp. 75–7.
24 DeMoivre to Varignon as quoted in Varignon to Bernoulli, 8 March 1720; *Corres 7*, 80–1.
25 The draft from ca. early 1720; *Corres 7*, 83.   26 *Corres 7*, 80.

given to increasingly cynical comments as he watched events in which he no longer participated, remarked that the reconciliation made him laugh since it was a forced one on both sides.[27] Newton and Bernoulli had stopped communicating directly. Varignon resumed his role as mediator. With supreme gall, Bernoulli seized on the *Recueil* to pretend that he was the injured party. Nothing speaks more eloquently of Newton's wish to be done with the controversy than his willingness to explain, via Varignon, how little he had had to do with the *Recueil*.[28] Bernoulli wanted more. Letters of Newton in the *Recueil* referred unflatteringly to the anonymous mathematician quoted in the *Charta volans* as a pretended mathematician, a novice, and a knight errant. Even while denying that he was the anonymous mathematician, Bernoulli chose to take offense at the epithets. Newton submitted to this as well and patiently explained what he had meant by the phrases and why they were not insults aimed at Bernoulli.[29] While Varignon was carrying on the mediation, Newton received two letters from James Wilson about early mathematical manuscripts of Newton which Wilson wanted to publish to establish his priority. Newton refused to cooperate because their publication might "occasion Disputes concerning their Antiquity."[30] When Wilson asked him to check his copies of the papers against the originals, he chose instead not to return the copies.

As always, however, his patience had limits. He expressed them to Varignon in the same letter that explained the three epithets.

Mr. DeMoivre has told me that Mr. Bernoulli desires a picture of me: but he still has not publicly acknowledged that I had the method of fluxions and moments in the year 1672, as was admitted in the 'Éloge' of Mr. Leibniz published in the *Histoire* of your Academy. He still has not acknowledged that I gave a correct rule for differentiating differentials in the first proposition of the book on quadratures, published by Wallis in 1693, and demonstrated synthetically in 1686 in Lemma 2, Book 2 of the *Principia,* and that I had that rule in 1672, by means of which I did, beyond question, at that time determine the curvatures of curves. He has still not acknowledged that in 1669, when I wrote my *Analysis per quantitatum series,* I had a method for accurately squaring curved lines, if it can be done, just as I explained it in my Letter to Oldenburg of 24 October 1676 and in the fifth proposition of the book on quadratures; and that the Tables of curved lines that can be compared with conic sections were composed by me at that time. If he will

---

[27] Conti to Des Maizeaux, 1 Sept. 1721; *Corres* 7, 149.
[28] Newton to Varignon, 19 Jan. and early Aug. 1721; *Corres* 7, 119–20, 141.
[29] Newton to Varignon, early Aug. and 26 Sept. 1721; *Corres* 7, 141, 160–2.
[30] Wilson's summary of his objection in Wilson to Newton, 15 Dec. 1720 and 21 Jan. 1721; *Corres* 7, 107–10, 125–7.

submit on those points so as entirely to remove the cause of disputes, I will not readily refuse him my picture.[31]

To ask Bernoulli to submit to the third point was to ask him to surrender his pretension that he had discovered the integral calculus.

Varignon relayed Newton's explanation of the three epithets to Bernoulli in the fall of 1721. By now, even he was beginning to tire of his mission, and he decided to omit Newton's demands to Bernoulli. Varignon no longer dared to submit Bernoulli's complaints directly to Newton but sent them instead to DeMoivre, who placed them before Newton and transmitted Newton's replies. DeMoivre also was sick of the burden, as he indicated in a passage which Varignon repeated to Bernoulli.

If you are not content with this satisfaction from Mr. Newton [Varignon told him], I shall be very annoyed, since Mr. Newton does not appear inclined to proceed any further, nor (as you have just seen) is Mr. DeMoivre inclined to ask more of him, nor I either, who have only Mr. DeMoivre to carry your complaints since I do not dare to do it myself at this distance.[32]

Bernoulli did want more, however. He insisted that Newton's explanation be published. Varignon gave in and forwarded the new demand; DeMoivre gave in and read it to Newton. "Since my youth I have hated disputes," Newton replied, and DeMoivre repeated; "now that I am eighty years old I detest them. I do not know how Mr. Bernoulli would use that letter, or an extract of it, in print; hence I cannot consent to its printing." Newton said this with enough feeling, DeMoivre reported, that he was afraid to argue with him. He did dare to suggest that at least Newton could write that much to Varignon. He would not. He made DeMoivre send the message for him.[33] Varignon made a last effort. Would Newton allow him to summarize the explanation in a letter to Bernoulli which he could print? When Newton did not reply, he even composed the letter and submitted it for Newton's approval. He wrote again when he still did not hear.[34] Eventually, through DeMoivre, Newton sent his agreement, and in June 1722 Varignon

---

[31] *Corres* 7, 165; original Latin, p. 162. I have altered the translation.

[32] Varignon to Bernoulli, 21 Oct. 1721; *Corres* 7, 167–9.

[33] DeMoivre to Varignon, Jan. 1722, as quoted in Varignon to Bernoulli, 4 Feb. 1722; *Corres* 7, 185. There may have been some deliberate deception by Newton and DeMoivre in fostering a pretense that Newton only responded to DeMoivre's mediations. Among Newton's papers is a scrap with two drafts in his hand of a reply by DeMoivre apparently to Varignon. It seems to concern Bernoulli's letter of July 1719; perhaps he meant it for Taylor's reply to Monmort in 1717. Thus it may not suggest anything about the later correspondence (*Add MS* 3965.13, f. 432). Who knows what deceptions Bernoulli and Varignon practiced in their turn?

[34] Varignon to Newton, early February, 4 April, 28 April 1722; *Corres* 7, 183, 193–4, 199–200. A draft of Varignon to Bernoulli, ca. June 1722; *Corres* 7, 196.

sent the text to Bernoulli, who was, predictably, still unsatisfied. In December, death finally released the gentle Varignon from his thankless task. DeMoivre refused to have more to do with it.

Newton was not quite done, however. Faced with Bernoulli's endless complaints and demands, he republished the *Commercium epistolicum* with the anonymous letter from the Charta volans and a detailed rebuttal to it appended.[35] Whereas the original publication had been distributed by the Royal Society, he placed the new one on sale to insure its circulation. It did not contain Bernoulli's name, but there could not have been many interested readers who did not know that he had composed the letter in the Charta.

Bernoulli, who was a stranger to both shame and propriety, was not done either. In February 1723, he wrote to Newton to report his indignation at a recent book by Nicholas Hartsoeker which accused Newton of plagiary in the *Opticks.* As he proceeded, Bernoulli's indignation increased, for Hartsoeker had also attacked him. Indeed, Hartsoeker had used Newton's words about the Charta volans, as found in Des Maizeaux's *Recueil,* to assault Bernoulli, twisting the words of course, as Bernoulli explained at length, to serve his purpose.

> Meanwhile, whatever the case, Hartsoeker's accusation redounds more on you than on me, for he tries to produce it from your words, wickedly distorted. If you should consider me worthy of a reply, I would very gladly learn from your own pen what you decide to do, in order to defend your innocence . . .[36]

Newton allowed his innocence to fend for itself and treated Bernoulli's letter with the contempt it deserved. And with his silence, the priority dispute finally came to rest.

Among its other effects, the controversy served to remind Newton that he needed to give attention to his intellectual legacy. As a result, he devoted considerable attention during his old age to new editions of his works. Newton was well aware of the connection with the quarrel. A letter that accompanied a gift of the *Opticks* to Fontenelle rang changes on a theme he had explored in distinguishing his philosophy from Leibniz's. "Here I cultivate the experimental philosophy," he informed Fontenelle, "as that which is worthy to be called philosophy, and I treat hypothetical philosophy, not as knowledge, but by means of queries."[37]

In 1717, Newton turned his attention to a new edition of the *Opticks.* He did not touch the body of the treatise, which continued to set forth conclusions as he had established them forty-five years earlier. The Queries, where he dealt with hypothetical philosophy

[35] *Comm epist* (Biot-Lefort ed.), pp. 185–9.
[36] *Corres 7,* 221; original Latin, pp. 219–20.
[37] Newton to Fontenelle, autumn 1719; *Corres 7,* 72.

with a greater air of assurance than he seems to have realized, offered room for change.

As we have seen, Newton had concluded the second edition of the *Principia* with a paragraph which referred to a spirit found in the pores of bodies. This paragraph, and his more extended unpublished speculations on the electric spirit, reflected at once the impact of Hauksbee's electrical experiments on Newton and perhaps also the effect of unrelenting criticism by mechanical philosophers on an aging man no longer able to sustain alone a revolutionary position in the face of general opposition. With the electrical spirit, which he allowed to account for the various short-range attractions and repulsions of particles, Newton retreated toward a mode of explanation which at least appeared acceptable to conventional mechanical philosophers, though in its rarity and elasticity his spirit embodied features alien to their gross fluids.

As he began to work on the new *Opticks,* Newton expanded the paragraph in the General Scholium into two new Queries, 24 and 25, on the "very subtle active potent elastic spirit by wch light is emitted refracted & reflected, electric attractions & fugations are performed, & the small particles of bodies cohaere when contiguous, agitate one another at small distances & regulate almost all their motions amongst them selves. For electric bodies could not act at a distance without a spirit reaching to that distance."[38] Associated papers "De vita & morte vegetabili (On Vegetable Life and Death) and "De Motu & sensatione Animalium" (On the Motion and Sensation of Animals), carried the speculations into further areas, some of which recalled themes of alchemy.[39] At some point, he felt confident enough about the electrical spirit that instead of relegating it to the Queries he thought of including it in the body of the treatise, where it would be presented, in his terminology, as philosophy rather than speculation. He sketched in a Part II, Book III, of "Observations concerning the Medium through which light passes, & the Agent which emits it."[40]

---

[38] *Add MS* 3970.3, f. 235. Cf. ff. 235^v, 240, 241^v, 293^v. See Joan L. Hawes, "Newton and the 'Electrical Attraction Unexcited'," *Annals of Science, 24* (1968), 121–30.

[39] *Add MS* 3970.3, ff. 236–7. Cf. ff. 235, 241.

[40] *Add MS* 3970.9, ff. 623–9. Cf. *Add MS* 3970.3, ff. 427–8. See Henry Guerlac, "Newton's Optical Aether," *Notes and Records of the Royal Society, 22* (1967), 45–57; Joan L. Hawes, "Newton's Revival of the Aether Hypothesis and the Explanation of Gravitational Attraction," *Notes and Records of the Royal Society, 23* (1968), 200–12; R. W. Home, "Newton on Electricity and the Aether," forthcoming, and " 'Newtonianism' and the Theory of the Magnet," *History of Science, 15* (1977), 252–66; Léon Rosenfeld, "Newton's Views on Aether and Gravitation," *Archive for History of Exact Sciences, 6* (1969), 29–37; P. M. Heimann, " 'Nature is a Perpetual Worker': Newton's Aether and Eighteenth-Century Natural Philosophy," *Ambix, 20* (1973), 1–25; and N. R. Hanson, "Waves, Particles, and Newton's 'Fits'," *Journal of the History of Ideas, 21* (1960), 370–91.

In the end, Newton suppressed the explicit discussion of the electric spirit, although he did refer to it briefly in two passages added to the new edition.[41] Instead, he composed a set of eight new Queries, which proposed a similar but distinct cosmic aethereal fluid. He inserted them as Queries 17–24 between the original set of sixteen Queries in the first edition and the set of seven added to the Latin edition. From the beginning of his speculations on the electric spirit, Newton had insisted on the difference between the interparticulate forces, which it explained, and gravity, which was always proportional to the quantity of matter and obeyed the inverse-square law, and was by implication nonmechanical. Swept on by the retreat from the radical stance of earlier years, he now postulated a cosmic aether to explain gravity itself. To be sure, the aether had so little in common with conventional mechanical fluids that the retreat was more apparent than real, and Newton may have intended it more as a sop than a concession. The aether, he said, was "exceedingly more rare and subtile than the Air, and exceedingly more elastick and active." He concluded indeed that the ratio of its elastic force to its density must be more than 490 billion times that of air. Was such a medium possible? It was, Newton argued, if one supposed that the aether, like the air, is composed of "Particles which endeavour to recede from one another . . ."[42] That is, Newton's new aether, like the electrical spirit, embodied the very problem it seemed to explain, action at a distance in the form of a mutual repulsion between aethereal particles. Standing rarer in the pores of bodies than in free space, the aether caused the phenomena of gravity by its pressure.

Newton also used it to explain the periodic phenomena in thin films which he had investigated nearly fifty years before. When he rejected the aether in 1679, Newton had found himself bereft of a convincing explanation of these phenomena. One of the advantages of a subtle elastic medium was its ability to vibrate periodically. He had insisted on this with the electric medium and had discussed its role in optical phenomena. Now he transferred optical phenomena to the aether and devoted more space to them in the new Queries than to gravity.

The *Opticks* was in English, of course. Only a Latin edition could effectively reach the Continental audience. In 1719, Newton published a second Latin edition that included the new Queries. Two years later, in 1721, a third English edition, which did not differ significantly from the second, appeared.

In fact, Newton's optics had so far had little impact on the Conti-

---

[41] An addition to Query 8 and part of Query 22; *Opticks*, pp. 340–1, 352.
[42] *Opticks*, pp. 349, 352.

nent, where efforts to repeat his experiments had uniformly failed. Years before, Mariotte had denied in print that the experiments revealed what Newton claimed. In 1715, when a French delegation led by Monmort came to England to view an eclipse, Newton decided to seize the occasion. The very day after they met Newton, he arranged for Desaguliers to demonstrate the experiments for their benefit in a manner that pointedly reminded the French of Mariotte's assertions against him.[43] Only with that demonstration did his theory of colors begin to spread significantly outside Great Britain. The second Latin edition helped its progress. It prompted an honorary member of the Académie, Sebastien Truchet, to perform them in Paris, this time with success.[44] (Dortous de Mairan had performed them successfully a couple of years earlier but in a place far removed from Paris, the center of French scientific life.) By now interest ran high. Brother Sebastien's demonstration drew a large audience of influential people, including Varignon. So did a later demonstration by Gauger. When Pierre Coste's French translation of the *Opticks* was published in Amsterdam in 1720, the Académie chose Varignon to review it. By 1721, Varignon found himself in charge of a second French edition in Paris. The Chancellor of France, Henri François Daguesseau, who had witnessed Gauger's demonstration of the experiments, took a personal interest in the edition. Daguesseau examined the first sheet, rejected it, ordered the use of a new font of type, and rejected that font as well because it was not sufficiently elegant, before he would allow the edition to proceed.[45] He even wrote to Newton to explain his concern for the edition.

> But, by your leave, most famous Sir, I should have said that I saw fit to act in this particular way not for your sake but my own, or rather to speak more truly not for my own sake but for that of all France, who freely and with applause offers her tongue and her speech to you; nor for the sake of France alone, but on behalf of all philosophers and mathematicians of every nation everywhere, who already look up to you and venerate you as obvious master of them all.[46]

Varignon asked Newton to furnish a sketch for an engraving to set at the head of each book. To represent his work in optics, Newton chose the *experimentum crucis,* and on his drawing he summarized

---

43 Antonio Schinella Conti, *Prose e poesi,* 2 vols. (Venice, 1739–56), *2,* 23–4. The news of the demonstration spread quickly; *cf.* "Notanda circa theoriam colorum Newtonianam," *Acta eruditorum,* May 1717, pp. 232–4.

44 Cf. his account of the experiments and his failure to understand the theory (Truchet to Newton, ca. 1721; *Corres 7,* 111–13).

45 Pierre Coste to Newton, 16 Aug. 1721; Varignon to Newton, 18 Sept. 1721; *Corres 7,* 147–8, 152–4.

46 Daguesseau to Newton, 28 Sept. 1721; *Corres 7,* 157–8; original Latin, p. 157.

his theory in a motto: *Nec variat lux fracta colorem,* Light does not change color when it is refracted (Figure 15.2). The edition required a year to complete. On 4 August 1722, Varignon sent a copy to Newton with the hope that it would please him.[47] It was the last letter Newton received from Varignon. He did have time to send his thanks before Varignon died in December.

While the French edition was in preparation, Newton received a challenge to his optics. Despite the successful demonstrations in France, one Giovanni Rizzetti denied some of his experiments, his interpretation of them, and his theory of colors.[48] Newton, who had long since lost his ardor for Conti, convinced himself that Rizzetti was Conti's friend and that Conti had instigated the challenge.[49] His papers reveal that he took Rizzetti seriously enough to compose three drafts of a reply. Desaguliers performed the critical experiment one more time before the Royal Society to assure them of its validity.[50] Rizzetti did not make any impression on scientific opinion, and Newton's optics remained in secure supremacy both in England and elsewhere. Not long before Newton's death, an undergraduate in Trinity addressed Latin verses to him on the topic "color est connata lucis proprietas" (color is a connate property of light) and received a sumptuously bound copy of the *Principia* for his pains.[51]

Newton also gave one last effort to the *Principia.* It is clear that he regarded the *Principia* rather than the *Opticks* as his masterwork. He had thrown the *Opticks* together largely from previously written papers in the 1690s and had scarcely touched the text, as opposed to the Queries, through three English and two Latin editions. In contrast, he worked over the *Principia* without end to hone its language to a perfect expression of his ideas. As he had done with the first edition, he had a special copy of the second bound for him with interleaved sheets on which he could enter corrections. He also entered such into the margins of another copy. Into these two he copied the list of corrigenda and addenda that he sent to Cotes at the end of 1713, and apparently he continued to add to them.[52] In

[47] *Corres 7,* 206–7.
[48] Rizzetti to Martinello (copied in Newton's hand) and Rizzetti to the Royal Society; *Keynes MS* 103. Rizzetti later published two books directed against Newton: *De luminis affectionibus specimen physico mathematicum* (Treviso, 1727) and *Saggio dell'antinevvtonianismo sopra le leggi del moto e dei colori* (Venice, 1741).
[49] Newton's "Remarks upon the Observations made upon a Chronological Index of Sir Isaac Newton . . . publish'd at Paris," *Philosophical Transactions, 33* (1724–5), 315–21. I cite from p. 321.
[50] Manuscripts concerned with his reply are in *Add MSS* 3970.3, ff. 481–2; 3970.5, ff. 585–8; 3970.8, ff. 608–9. 6 Dec. 1722; *JBC 12,* 290–2.
[51] John Nichols, *Literary Anecdotes of the Eighteenth Century,* 9 vols. (London, 1812–15), *5,* 280.
[52] I. Bernard Cohen, *Introduction to Newton's 'Principia'* (Cambridge, 1971), pp. 258–64.

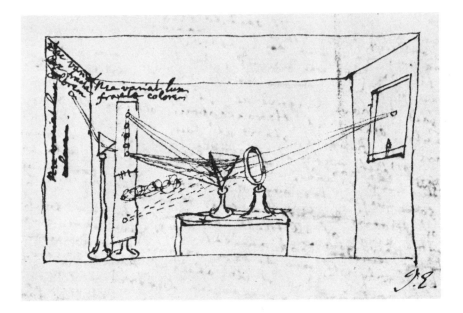

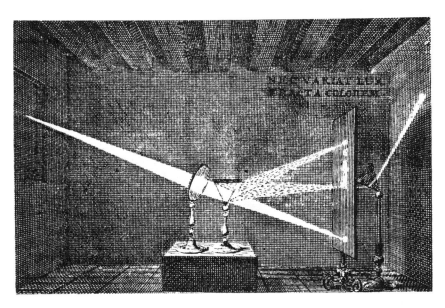

Figure 15.2. The *experimentum crucis* according to Newton's drawing for the second French edition of the *Optique*. Below, the vignette as it was used in the edition. (Courtesy of the Warden and Fellows of New College, Oxford.)

1719, to Varignon, he mentioned a copy with corrections, which looked forward to another edition.[53] His manuscripts contain a considerable number of papers with intended corrections, which in the end he omitted. Probably he stood behind the new experiments on the resistance of the air, performed in St. Paul's in 1719 by Desaguliers, which compared the fall of lead balls with light hollow spheres five inches in diameter molded from hogs' bladders.[54] When Conduitt conversed with Newton on 7 March 1725, a copy of the *Principia,* which Newton consulted during the talk, lay on the table.[55] To be sure, the third edition was then in progress. Nevertheless, the *Principia,* the embodiment of his legacy to philosophy, was constantly in his mind in a way the *Opticks* never was.

Perhaps the appearance of a reprint of the second edition in Amsterdam in 1723 stimulated Newton to put his plan for a new edition into action. Perhaps a serious illness in 1722 reminded him that he could not delay forever. We only know that printing of an edition more sumptuous than either of the others began in the fall of 1723. As editor, Newton had the services of a young member of the Royal Society, Henry Pemberton, who had recently returned to England after completing a medical degree in Leyden in 1719. Pemberton had met the *Principia* during the course of his medical studies. Although his thesis, on the mechanism by which the eye accommodates itself in viewing objects at various distances, was a work of enduring if minor importance, the impact of the *Principia* on Pemberton, like its impact on a generation of young English scientists, was such as to shape his future career. He had returned to England determined to enroll himself among Newton's epigoni. An attempt to approach the great man through Keill had failed, but he persevered with an essay that criticized Leibniz's measure of force. It was an obsequious essay punctuated with references to "the great Sir Isaac Newton." Dr. Richard Mead performed the favor of showing it to the great Sir Isaac Newton, who approved and had it printed in the *Philosophical Transactions.*[56] Pemberton followed it in 1723 with an "Introduction, Concerning the Muscles and their Actions" to Mead's edition of Cowper's *Myotomia reformata* (published in 1724), which Mead probably also showed to Newton. Muscular motion may seem to have little enough to do with Newton, but

---

[53] Newton to Varignon, 29 Sept. 1719; *Corres 7,* 63.

[54] Reported to the Royal Society on 22 Oct. 1719; *JBC 11,* 381–2. Earlier in 1719, Desaguliers had performed experiments with lead, glass, and pasteboard balls (*JBC 11,* 339–40). On 6 April 1721, Desaguliers described experiments with the gold bob of a pendulum oscillating in water and mercury (*JBC 12,* 104–6).   [55] *Keynes MS* 130.11.

[56] "A Letter to Dr. Mead . . . Concerning an Experiment, Whereby It Has Been Attempted to Shew the Falsity of the Common Opinion, in Relation to the Force of Bodies in Motion," *Philosophical Transactions, 32* (1722–3), 57.

Pemberton converted the essay into a panegyric on Newtonian method as obsequious in its references as the piece on force. It is necessary to add that Pemberton was a competent mathematician who had also published a pamphlet on mathematics in the form of a letter to his friend James Wilson, and that he was already engaged in writing a popular exposition of Newtonian philosophy. For whatever reason, Newton entrusted primary responsibility for the edition to Pemberton, who, in the fall of 1723, addressed to him the first of thirty-one communications which stretched over the following two and a half years while the edition passed through the press.

By universal agreement, Pemberton performed his function with less understanding and skill than Cotes had shown with the second edition, and he left little of himself impressed upon the work.[57] His letters tended to be short and uninteresting, raising only small points of style. To be fair to Pemberton, it is necessary to recall that Newton was now more than eighty years old and no longer capable of the serious exchange that occurred with Cotes. On occasion Pemberton did try to press beyond details, and he brought up some basic issues that had been among the more hotly debated topics in the second edition: the flow of water from a tank, and aspects of the treatment of tides that bore on the ratio of the sun's force to the moon's and thus on the precession of the equinoxes.[58] Newton refused to become engaged and simply swept the arguments aside.

Other evidence confirms that Newton could no longer sustain a penetrating reexamination of his work. In 1719, James Stirling sent him objections raised by Nikolaus Bernoulli (whom Stirling took to be the mouthpiece for Johann Bernoulli) to the treatment of pendulums in resisting media.[59] As far as we know, Newton never answered. Certainly he did not take advantage of the criticism to correct the passage. While the edition was in progress, Brook Taylor informed Pemberton that the derivation of precession was incorrect.[60] We do not know if Pemberton passed this on to Newton, but the treatment of precession did not change. Molyneux told Conduitt that he, together with Graham and Bradley, had accidentally discovered a nutation of the earth's axis while they were attempting to observe stellar parallax. Unable to account for the nutation, they were convinced that it undermined the whole Newtonian philosophy. Molyneux did not know how to break such

---

[57] Cohen, *Introduction*, pp. 270–80. *Var Prin 2*, 827–47.

[58] Pemberton to Newton, ca. Sept. 1724, and a paper of queries on parts of Book III, ca. April 1725; *Corres 7*, 289–91, 313–14. Cf. another sheet of queries on Book III, ca. May or June 1725, which brought up a delicate problem about the tails of comets (*Corres 7*, 324).

[59] Stirling to Newton, 6 Aug. 1719; *Corres 7*, 53–4; Cf. *Math 6*, 441–2.

[60] Taylor to Pemberton, 27 May 1725; *Corres 7*, 321.

news to Newton and did so very gradually and gently, only to meet the same lack of interest which Stirling had found; "all Sʳ I. said in answer, was, it may be so, there is no arguing against facts & experiments."[61] It is true that Newton's treatment of precession implied nutations whatever Molyneux thought, but the story expresses indifference rather than quiet confidence.

Nevertheless, Newton's relations with Pemberton cannot be summarized wholly in terms of his age. Pemberton wrote to Newton. Newton did not write to Pemberton. To be sure, Pemberton's notes accompanied proofs so that Newton's replies took the form of corrections on the printed pages. Since they both lived in London, Pemberton could drop by for discussion when the occasion demanded, and his letters mentioned such. For all of that, one catches a note approaching contempt in the way that Newton frequently brushed aside reasonable suggestions.[62] After Newton's death, when Pemberton published his *View of Sir Isaac Newton's Philosophy* with the claim that Newton had read and approved it, Conduitt recalled that Newton said he had only glanced through it hastily.[63]

In view of his age, it is not surprising that the third edition did not embody major changes. Newton inserted Desaguliers's results in the scholium that treated the experimental investigation of resistance. He added new observations of the shape of Jupiter and the elongation of its satellites made by James Pound with a telescope that Newton had helped him erect. Following Halley's computations, he treated the orbit of the comet of 1680–1 as an ellipse (with a period of 575 years) instead of a parabola. Here and there, miscellaneous paragraphs attempted to elaborate on his concepts in a minor way. In 1724, when the French astronomer Delisle visited him, Newton filled his ear with complaints that Halley had not paid sufficient attention to Newton's various corrections for the moon when he constructed his lunar tables, which could otherwise have removed all sensible discrepancy between theory and observation.[64] Nevertheless, the amendments of lunar theory in the third edition were minimal.

The most important addition appeared near the beginning of Book III, a new Rule IV for reasoning in philosophy which carried on the argument with Leibniz. Among Newton's papers are a number of drafts which testify to plans which were at times much more extensive. He drafted definitions of "phenomena," "body," "vac-

---

[61] *Keynes MS* 130.5, Sheet 3.

[62] For example, about April 1725 Pemberton made a couple of good points about minor turns of phrase in Book III that could be improved; Newton paid no attention (*Corres 7*, 312).

[63] *Keynes MSS* 130.6, Book 2; 130.5, Sheet 1.   [64] *Edleston*, p. lxxviii.

uum," and "hypothesis." He composed, not one new rule, but two. In all of the papers one theme predominates, the same theme that he had pursued in the final pages of the "Account of *Commercium epistolicum*." As he put it to Fontenelle, only experimental philosophy deserves the name philosophy. The new definitions and rules rejected hypotheses from the precincts of philosophy, especially hypotheses of aetherial fluids that carry the planets in vortices.[65] Eventually he settled for a single Rule IV, a restrained statement from which he succeeded in eliminating the passion that had mounted anew within the octogenarian as he thought once more about Leibniz.

> Rule IV. In experimental philosophy we are to look upon propositions inferred by general induction from phenomena as accurately or very nearly true, notwithstanding any contrary hypotheses that may be imagined, till such time as other phenomena occur, by which they may either be made more accurate, or liable to exceptions.
>
> This rule we must follow, that the argument of induction may not be evaded by hypotheses.[66]

Through 1724 and 1725 the edition made its slow but steady progress toward completion with none of the delays that stopped the press during the second edition. Pemberton, who was staking his future on it, was anxious to make himself conspicuous in the edition if he could. He suffered a cruel disappointment in Book III. Newton had received alternative treatments of the motion of the lunar nodes from both Pemberton and John Machin. For reasons which remain unknown, he decided to insert one of them in Book III after his own derivation, which concluded with Proposition XXXIII. He chose Machin's instead of Pemberton's! Ever eager, Pemberton put in a reference to an integral he had published which was useful in the solution of the problem.[67] Newton struck it out without a word of explanation though he did allow a mention of Pemberton's derivation to appear in the introduction to Machin's.

A revealing incident occurred in February 1725. Early in December, Newton had asked Halley to compute the place of the comet of 1680–1 on a given day in the proximate parabolic orbit. On 16 February, two and a half months later, Halley wrote in what can only be called alarm.

> Honour^d S^r
>
> A mistake I committed in considering the scheme of your Comets Orb, which was no less than my taking the Suns motion the contrary way, made me conclude that no other than an Elliptick Orb could suffice to represent the first observations therof with the de-

---

[65] *Add MS* 3965.13, ff. 419–20, 422–2^v.   [66] *Prin*, p. 400.

[67] Pemberton to Newton, 17 May 1725; *Corres 7*, 321.

sired exactness, and you being indisposed out of town, I waited for
your return to consult you. Being yesterday at London I guessed by
some Symptoms that you take it ill that I have not dispatcht the
Calculus I undertook for you, but the afores^d mistake made me
despair of pleasing you in it. Being got home last night I was aston-
isht to find my self capable of such an intollerable blunder, for which
I hope it will be easier for you to pardon me, than for me to pardon
my self, who hereby run the risk of disobliging the person in the
Universe I most esteem. I entreat therefore that you would not think
of any other hand for this computus, and that you please to allow me
the rest of this week to do it in, being desirous to approve my self in
all things

> Honr^d S^r.
> Your most faithfull serv^t
> Edm: Halley.[68]

It was the Astronomer Royal, a man sixty-eight years old and
possessed of a considerable reputation of his own, who cowered
thus before the aging despot. Nor did his plea suffice. The table in
question appeared in the third edition without that particular calculated place.

Pemberton finally fared better than Cotes had done. Newton
forgot him as he first drafted his preface to the new edition, but he
recollected in time to insert his regards for "Henry Pemberton,
M.D., a man of the greatest skill in these matters . . ."[69] Although
he also rewarded him handsomely with two hundred guineas, Pemberton told his friend James Wilson that he valued the recognition
more.[70] Whatever he said to Conduitt about Pemberton's *View of
Newton's Philosophy,* he subscribed for no less than twelve copies at
one pound each.[71]

An engraving of the well-known Vanderbank portrait, done in
1725, graced the new edition. Newton dated the preface 12 January
1726. The edition contained a royal privilege to the publishers,
William and John Innys, in effect a copyright for fourteen years,
that bore the date 25 March.[72] Six days later, Martin Folkes presented a copy "richly Bound in morocco Leather" to the Royal
Society in Newton's name.[73] In all, 1,250 copies were printed, fifty
of them on superfine paper. Newton intended at least some of these

---

[68] *Corres 7*, 302.   [69] *Prin*, p. xxxv. Cf. *Var Prin 2*, 848–50.
[70] James Wilson's biographical sketch prefaced to Henry Pemberton, *A Course of Chemistry*
(London, 1771).
[71] The receipt appeared in the inventory of his estate (*Villamil*, p. 56). See I. Bernard Cohen,
"Pemberton's Translation of Newton's *Principia*, with Notes on Motte's Translation,"
Isis, 54 (1963), 319–51.   [72] *Var Prin 1*, 3, 36.   [73] *JBC* 12, 643.

Figure 15.3 Newton at eighty-two. Portrait by John Vanderbank, 1725. (Courtesy of the Royal Society.)

as presentation volumes. He gave one of them to his friend and associate of thirty years at the Mint, John Francis Fauquier. He was lavish in his concern for the Paris Académie; they received no fewer than six copies.[74]

---

[74] Newton to Fontenelle, ca. June 1726; *Corres* 7, 349. *Edleston*, p. lxxix. Cohen, *Introduction*, p. 284.

With the calculus controversy still fresh in his mind, Newton also devoted some attention to the publication of his mathematical work. In 1720, Raphson's English translation of the *Universal Arithmetick* finally appeared, and two years later a second edition, somewhat revised, of the original Latin. Conduitt recorded an anecdote about the latter publication. "Machin overlooked the press for w^{ch} [Sr I.] intended to have given him 100 Guineas but he made him wait 3 years for a preface & then did not write one . . ."[75] Machin's unsatisfactory performance with the algebra may explain why Newton chose Pemberton to edit the *Principia* despite his belief that Machin understood the work better.[76] In 1723, the year after the second edition of *Commercium epistolicum*, Jones's volume with *De analysi* was reprinted in the Netherlands.

The mathematical and scientific works did not, however, exhaust the Newtonian legacy, and in his old age Newton turned again to the theological interests that had burned with consuming intensity during the years of his early manhood. Although he had never completely abandoned theological studies, a hiatus of about twenty years beginning with the publication of the *Principia* had interrupted their vigorous pursuit. Only a small number of theological papers can be placed with assurance in this period.[77] In the years 1705–10, he returned to theology with renewed vigor, and for all his editions of the *Principia* and *Opticks,* theology was the primary occupation of his old age.[78] Some of the papers, in which information on early heresies appears on the same page with material on heathen religions or the nature of Christ or the chronology of the prophecies, suggest that during his final years his mind was frequently a chaos of undifferentiated religious preoccupation.[79] Newton's conclusions in theology had been as radical as his conclusions in natural philosophy. So far they had seen the light of

---

[75] *Keynes MSS* 130.5, Sheet 1; 130.6, Book 1.

[76] *Keynes MSS* 130.6, Book 2; 130.5, Sheet 2.

[77] Two chapters related to the "Origines": "The original of Monarchies" (*Keynes MS* 146) and "The original of religions" (*Yahuda MS* 41). There are a number of sheets in New College MSS 361.2 and 361.3 (Bodleian Library) from this period that are also related to the "Origines." Cf. Gregory's reply to Newton's request for information about manuscripts of Papias in Oxford (Gregory to Newton, 16 May 1703; *Corres 7,* 431).

[78] As far as I can tell from the manuscripts, he began to write again on Revelation in the period 1705–10, and for the first time he began to devote serious attention to Daniel (*Yahuda MSS* 6, 7.1, 7.2, 7.3 and some sheets in 15.7. *Keynes MS* 5, ff. 7–134. *Keynes MS* 5 is not one continuous treatise. The first twelve folios [ff. I–VI, 1–6], printed in *Theological Manuscripts,* ed. H. McLachlan (Liverpool, 1950) pp. 119–26, belonged to an earlier period). Cf. ———— to William Lloyd, 7 Nov. 1713, and Newton to Pierre Allix, ca. 1713; *Corres 6,* 34–5, and *Corres 7,* 357.

[79] *Yahuda MS* 15.7, ff. 108^v, 109, 185^v.

day only within a limited circle of trusted confidants. As the years mounted and he became conscious that the inevitable end approached, Newton gave thought to his responsibilities. He said when he died "he should have the comfort of leaving Philosophy less mischievous" than he found it, Conduitt reported; "he might say the same of the revealed religion then mention his Irenicum his creed."[80] What Conduitt did not mention were the factors that made it even more difficult for an old man fond of his position and respect to publish heretical views than it had been for a rebellious young don fifty years earlier.

Newton had been at work revising his interpretation of the prophecies and other theological works for about ten years when the arrival of Abbé Conti set in motion a train of events which forced him into partial publication. Already in the 1690s he had begun to recast the "Origines" into a different form which, after further evolution, ultimately became his *Chronology of Ancient Kingdoms Amended*. In order to square the chronology of ancient kingdoms with his account of the development and diffusion of arts and of idolatry among the descendants of Noah, he had found it necessary to lop off the antiquity to which most ancient societies had pretended, and among other things he had cut about four hundred years out of the accepted chronology of Greek history. When Conti arrived and won Newton's friendship, chronology was one of their topics of conversation.[81] Conti, who had cultivated the art of cultivating influential people, had also made himself welcome at court. There, in 1716, he mentioned Newton's new principles of chronology to Caroline, Princess of Wales. Interested, she summoned Newton and asked for a copy of what he had written.

Newton never lightly surrendered one of his compositions. He had even less desire to hand over to the Princess of Wales a treatise which might still have contained assertions heretical enough to secure his instant dismissal from the Mint. Well schooled as he was in the art of delay, he pleaded that the work was "imperfect and confus'd," but he knew very well that one did not dally with a royal command. In haste he drew up an "Abstract" of his chronology, what was later called the "Short Chronology," which put the work into "that shape the properest for her Perusal . . . ," and delivered it to the princess in a few days.[82] Conti wanted a copy also. Apparently Newton refused, but Conti arranged for the prin-

---

[80] *Keynes MS* 130.6, Book 3. When Conduitt copied this anecdote into 130.7 (Sheet 1), he decided to omit the second half of it. [81] Conti, *Prose e poesi, 2,* 27.

[82] So Newton described it to Zachary Pearce. Pearce to Hunt, 10 Aug. 1754; *The Lives of Dr. Edward Pocock . . . of Dr. Zachary Pearce . . . of Dr. Thomas Newton . . . and of the Rev. Philip Skelton,* 2 vols. (London, 1816), *1,* 430.

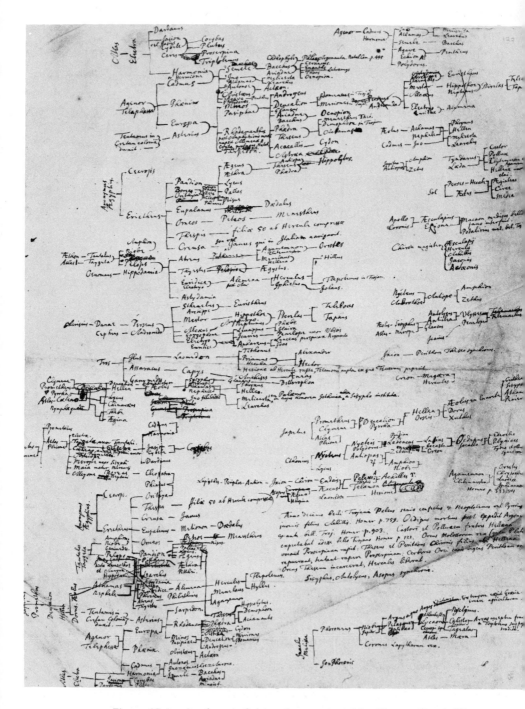

Figure 15.4. A scheme of chronology prepared by Newton, probably between 1710 and 1715. (Courtesy of the Warden and Fellows of New College, Oxford.)

cess to request one for him. It was another request Newton could not refuse, though he did extract a pledge that Conti would not communicate it.[83] Such at least was Newton's story nine years later. In fact, a number of copies of the "Abstract" were made. In 1725, Zachary Pearce, the Rector of St. Martins-in-the-Fields, had one and knew of many others who did.[84] Newton had also talked about his ideas on chronology with others, at least one of whom wrote to him that he had spread the word still further, and that all hoped he would publish his scheme.[85]

Despite Newton's desire to keep the "Abstract" secret, there was nothing in it to excite theological objection. A list of dates beginning with the flight of the Canaanites from Joshua, which he placed shortly before 1125 B.C., and extending to the death of Darius Codomannus in 331 B.C., the "Abstract" might, and eventually did, arouse the objections of professional classicists to its arbitrary truncation of ancient chronology. As far as theologians were concerned, Newton had sanitized it so thoroughly that no heterodox odor likely to tease an ecclesiastical nostril remained from its earlier model, the "Origines." Some hints did, and to one familiar with the "Origines" they offer unmistakable clues to the source of Newton's chronology. In 963, the "Abstract" asserted, Amphictyon brought the twelve gods of Egypt, "the *Dii magni majorum gentium,* to whom the Earth and Planets and Elements are dedicated," into Greece. In 956, the Egyptian king Sesac was slain and after his death deified, one of a number of deifications which Newton included in the chronological table.

Sesac from his making the river Nile useful, by cutting channels from it to all the cities of Egypt, was called by its names, Sihor or Siris, Nilus and Egypt. The Greeks, hearing the Egyptians lament, O Siris and Bou Siris, called him Osiris and Busiris. The Arabians from his great acts called him Bacchus, that is, the Great. The Phrygians called him Ma-sors or Mavors, the valiant, and by contraction Mars. Because he set up pillars in all his conquests, and his army in his father's Reign fought against the Africans with clubs, he is painted with pillars and a club: and this is that Hercules who, according to Cicero, was born upon the Nile; and according to Eudoxus, was slain by Typhon; and according to Diodorus, was an Egyptian, and went over a great part of the world, and set up the pillars in Afric. He seems to be also the Belus who, according to Diodorus, led a colony of Egyptians to Babylon, and there instituted Priests

83 Newton, "Remarks," p. 317. Newton to Cavelier, 27 May 1725; *Corres 7,* 322. Cf. Conduitt's dedication of the *Chronology of Ancient Kingdoms Amended* (London, 1728), p. v.

84 Pearce to Hunt, 10 Aug. 1754; *Lives, 1,* 430.

85 Alexander Cunningham to Newton, 1 May 1716; *Corres 6,* 331. Frank E. Manuel, *Isaac Newton, Historian* (Cambridge, Mass., 1963), p. 22.

called Chaldeans, who were free from taxes, and observed the stars, as in Egypt.[86]

By themselves, cut off from the "Origines," there was nothing very novel in these ideas and nothing to excite odium. By disguising radical theology as chronology, Newton had made it safe enough even for royal consumption.

Perhaps the need to produce the "Abstract" stimulated Newton's interest anew. Since many of the manuscripts on chronology appear to come from the years shortly before the "Abstract," it is more likely that Newton had already taken the topic up again, as the conversation with Conti, the letter of 1713 to Bishop Lloyd, and the letter of about the same time to Pierre Allix suggest. Whenever he began, the manuscript remains testify to extensive work on a longer treatise on chronology which occupied much of Newton's time and attention during the last ten or fifteen years of his life.[87] Conduitt indicated that he was still at work revising it during his final days.[88]

Newton had not heard the last of the "Abstract," however. When Conti left England late in 1716, he carried his copy with him and showed it around in the learned circles of Paris where he settled down. A work on ancient chronology by the great Sir Isaac Newton was an item sure to generate interest, but Newton's name was not enough to silence objections when it transpired that he had seriously foreshortened the accepted span of history. One expert on the subject who saw it, Father Etienne Souciet, wanted to learn the principles on which Newton based his argument, for the "Abstract" presented little more than a table of dates. Through Conti he attempted in 1720 to inquire into the system. His queries reached John Keill who received from Newton the oral reply that the manuscript Conti had was the abstract of a longer work, and that he had relied on astronomical calculations, which he described briefly, in his revision of Greek chronology. Keill transmitted the answer to Brook Taylor, then in Paris, who relayed it on to Souciet.[89] Souciet chose not to cross swords with an unpublished paper and held his peace. By 1724, Conti had loaned his manuscript to M. de Pouilly of the Académie des inscriptions et belles-lettres, who allowed Nicolas Fréret to take a copy of it. Like Souciet, Fréret was a scholar of antiquities. Unlike Souciet, he did not propose to let the unpublished state of the manuscript silence his objections. Fréret trans-

---

[86] *Chronology of Ancient Kingdoms*, pp. 23–4. I quote from the "Abstract," which was also published in the *Chronology*.

[87] *Yahuda MSS* 25 and 26. New College MSS 361.1–3.

[88] New College MS 361.1, ff. 88–90.

[89] Etienne Souciet, *Recueil de dissertations critiques sur des endroits difficiles de l'Écriture sainte*, 3 vols. (Paris, 1725–36), *2*, 56.

lated the "Abstract" into French as the *Abrégé de la chronologie,* composed a refutation of it, and arranged for a joint publication of the two.[90]

In the spring of 1724, Newton received a letter from Guillaume Cavelier, a Parisian bookseller, who informed him that a small manuscript said to be Newton's work had come into his hands and he wanted to publish it. He was told, he continued, that it had faults which would pain Newton should they appear under his name. Hence he asked Newton to supply him with a correct copy of his chronology.[91] Total silence greeted this shameless piece of duplicity, which sought to trick Newton into owning the opinions that Fréret had already prepared to pounce on. After waiting several months, Cavelier apparently wrote a second time, although the letter has not survived. Silence followed once more. Finally, on 20 March 1725 (N.S.), he wrote one last time. Would Newton tell him if there was anything he wanted to change; "if I do not receive any News from you I will take your silence for your Consent that it appear as it is and I will give it to the public with some Remarks."[92] Even the threat did not sting Newton to sudden action, but eventually, on 27 May (O.S.), he did reply to Cavelier's third letter.

> I remember that I wrote a Chronological Index for a particular friend, on condition that it should not at be communicated. As I have not seen the manuscript which you have under my name, I know not whether it be the same. That which I wrote was not at all done with design to publish it. I intend not to meddle with that which hath been given you under my name, nor to give any consent to the publishing of it.[93]

Cavelier claimed that the letter did not arrive until after the *Abrégé* had appeared. The dates hardly matter. It is impossible to imagine that he would have allowed the letter to deter him at that point.

Newton received a copy of the *Abrégé* on 11 November 1725. If Cavelier's letter had not aroused him, the publication itself did. Outraged, he seized his pen and composed a short essay, "Remarks upon the Observations made upon a Chronological Index of Sir Isaac Newton . . . publish'd at Paris." He was sufficiently agitated to write out seven full drafts before he was satisfied.[94] The issue of the *Philosophical Transactions* for July and August was just then going to press. He thrust the "Remarks" in, placing them, not at the end where they might be missed, but at the beginning. Newton may have been nearly eighty-three years old, but he had lost none

[90] Chap. 1, "Academic Piracy," in Manuel, *Newton, Historian,* pp. 21–36. Monmort heard of the intended publication in January 1724 (Monmort to Taylor, 15 Jan. 1724; Brook Taylor, *Contemplatio philosophica,* [London, 1793], p. 140).
[91] *Corres 7,* 279.   [92] *Corres 7,* 311.   [93] *Corres 7,* 322.   [94] *Yahuda MS 27.*

Figure 15.5.   Newton at eighty-two. Portrait by John Vanderbank, 1725. (Courtesy of the Master and Fellows of Trinity College, Cambridge.)

of his capacity for indignation or for expressing it in pungent prose. The "Remarks" recounted the history of the edition from his point of view. Cavelier had never shown him the translation, named the translator, or told him that the same man had written a confutation of the "Abstract" to be published with it—"as if," he remarked, "any Man could be so foolish as to consent to the publishing of an unseen Translation of his Papers, made by an unknown Person, with a Confutation annexed, and unanswered at their first Appearance in Publick." He took up Fréret's criticisms one by one and was able to show that Fréret had misconstrued his arguments.

> So then the Observator [that is, the author of the "Observations" on his "Abstract"] hath mistaken my Meaning, in the two main Arguments on which the Whole is founded, and hath undertaken to translate and to confute a Paper which he did not understand, and been zealous to print it without my Consent; tho' he thought it good for nothing, but to get himself a little Credit, by translating it to be confuted, and confuting his own Translation.

But Newton stored up his major reservoir of bitterness to pour out on Conti, his quondam friend and present Judas, whom he took to be the moving force behind the whole proceeding. Once aroused against Conti, Newton made him the scapegoat for every incident that had recently troubled his calm, for the exchange with Leibniz, for Rizzetti's challenge to his optical experiments, and for the tirades of a German, Johann von Hatzfeld, whose letters against him on perpetual motion had recently been read to the Royal Society.[95]

> Abbé Conti came into England in Spring 1715, and, while he staid in England, he pretended to be my Friend, but assisted Mr. Leibnitz in engaging me in new Disputes, and hath since acted in the same Manner in France. . . . And what he hath been doing in Italy, may be understood by the Disputes raised there by one of his Friends, who denies many of my Optical Experiments, though they have been all tried in France with Success. But I hope that these Things, and the perpetual Motion, will be the last Efforts of this Kind.[96]

Nor were the *Abrégé* and the "Remarks" the end of the business. Once they appeared, Father Souciet no longer felt constrained, and in 1726 he published five dissertations against Newton's system of

---

95 In 1724 Hatzfeld published *The Case of the Learned Represented According to the Merit of the Ill Progress hitherto made in Arts and Sciences shewing I. The Cause of Gravity and Attraction. II. What Nature is and the Effects it is Capable of, etc. . . . Contained in Two Letters to the Royal Society.* On 27 May and 21 Oct. 1725, on both of which occasions Newton presided, the Royal Society considered letters from von Hatzfeld related to his book, in which he took a very high tone about their neglect of what he said about perpetual motion (*JBC 12*, 568–9, 587). Cf. von Hatzfeld to Newton, probably 1725; *Corres 7*, 253–4.

96 Newton, "Remarks," pp. 316–21.

dates.[97] Remembering the calculus dispute, Conduitt was so worried about the manner in which Souciet attacked the "Abstract" that he arranged for a friend to prepare an extract of the objections "stript of the extraordinary ornaments wth which they are cloathed . . ." Their only effect, Conduitt added, was to convince Newton of Souciet's ignorance.[98] The "Remarks" indicate a considerably less impassive Newton than Conduitt tried to present, however. Zachary Pearce, also a student of antiquity, whose acquaintance Newton therefore began to cultivate about this time, urged him to publish the full work. Newton replied "that at his time of life it was too late to enter into a controversy, which might perhaps arise from his publishing his thoughts on Ancient Chronology, as they differed so much from the common opinion . . ."[99] Eventually he changed his mind and decided that he must defend himself by bringing the treatise out. He was at work revising it when he died. We should not from that conclude that the work was unfinished. Endless revision, however minute the alterations, was Newton's fate. Whiston asserted that when Newton composed the chronology, he "wrote out 18 Copies of its first and principal Chapter with his own Hand, but little different one from another . . . ," and Pearce mentioned sixteen copies of the whole work.[100] Conduitt published the full volume, *The Chronology of Ancient Kingdoms Amended,* in 1728, the year after Newton's death. The book stimulated the debate anew, and within a confined circle it continued for another fifty years.[101]

Newton opened the *Chronology* with the assertion of one of its basic themes: "All Nations, before they began to keep exact accounts of Time, have been prone to raise their Antiquities; and this humour has been promoted, by the Contentions between Nations about their Originals."[102] Hence with every ancient people, the Greeks, the Egyptians, the Assyrians, the Babylonians, the Medes, and the Persians, it was necessary to eliminate the extra years, stretching in some cases to hundreds of thousands of years, which in their vanity they had feigned to enhance their antiquity. Only one people escaped Newton's razor: the Israelites, whose written record, the oldest such extant in Newton's belief, gave their history a solidity by which the others' could be amended.

In the "Abstract," Newton had said he intended "to make Chronology suit with the Course of Nature, with Astronomy, with

[97] In vol. 2 of Souciet, *Recueil de dissertations critiques.*   [98] *Keynes MS* 129A, pp. 16–17.

[99] Pearce to Hunt, 10 Aug. 1754; *Lives, 1,* 433.

[100] William Whiston, *Memoirs of the Life and Writings of Mr. William Whiston* (London, 1749), p. 39. Pearce to Hunt, 10 Aug. 1754; *Lives, 1,* 430.

[101] Chap. 10, "The Battle of the Dates," in Manuel, *Newton, Historian,* pp. 166–93.

[102] *Chronology of Ancient Kingdoms,* p. 43.

Sacred History, with Herodotus the Father of History, and with it self . . ."[103] The last three criteria, the historical books of the Old Testament, Herodotus' *History,* and the principle of consistency, were straightforward enough. The novelty of the work lay in the first two. Astronomy referred to the use of the precession of the equinoxes to place the Argonauts' expedition forty-three or forty-four years after the death of Solomon, or about 936 B.C. Newton convinced himself that Chiron had been an astronomer and that on the first celestial sphere, made for the Argonauts, he had placed the four cardinal points of the ecliptic in the middle of the constellations Aries, Cancer, Chelae [Libra], and Capricorn. With the equinoxes moving at a rate of fifty seconds a year or one degree in seventy-two years, it was then a simple calculation to date the expedition.[104] The course of nature referred to his second principle of dating, the average length of the reigns of kings. Ancient chronologers figured them at the same rate with generations, about three to a century. From the records of Israel, Persia, the Hellenistic monarchs, England, and France, Newton calculated that the average length of a reign is rather eighteen to twenty years.[105] Hence chronologies needed to be shortened to about four-sevenths their asserted length when phantom monarchs had not stretched them yet farther. The reigns he counted were the reigns of figures in ancient mythology, such as Chiron, who were taken to have been real men. Euhemerism was not novel with Newton; neither was it confined to him in his age. Be that as it may, by counting years, Newton arrived at the same date for the Argonauts' expedition that he had established by astronomy.[106] Using the expedition as a benchmark, he fixed other dates mostly by the course of nature, that is, by the length of reigns. He could not place any event outside of Greek history by the equinoxes, but the principle of the average length of reigns, and the synchronism and interconnection of all ancient history, which he accepted as obvious, helped him to date events in the other kingdoms. Using these devices, he drastically shortened the accepted chronologies of the ancient world as we have seen, in order (though he did not state his purpose) to fit them to the pattern of the multiplying offspring of Noah, which he had sought to establish in the "Origines."

Here and there, his real interest briefly asserted its presence. He had much to say about the deification of kings and heroes and the spread of that custom from Egypt and Chaldea. At the end of the first chapter, on the Greeks (which constituted nearly half of the whole book), Newton permitted himself a few words about the descendants of Noah and how all mankind lived together in Chal-

---

[103] *Ibid.,* p. 8.   [104] *Ibid.,* pp. 82–95.   [105] *Ibid.,* pp. 52–3.   [106] *Ibid.,* p. 64.

dea under his sons until five generations before Abraham, when they divided the earth and began to spread out, taking their laws, customs, and religion with them. After the days of Melchizedek, the nations revolted to the worship of false gods; even the Israelites were prone to revolt. Some of the laws of the original religion survived, however; they were mentioned in Job. It was the religion of Moses and the prophets comprehended in the two commandments to love God and to love your neighbor, "and this is the primitive religion of both Jews and Christians, and ought to be the standing religion of all nations, it being for the honour of God, and good of mankind." Older than Moses, it was the religion of Noah to whom it was given.[107] Such was the most venturesome passage in a most careful work.

At one point, however, Newton had envisaged a bolder project. Among his papers, apparently in the hand of the period 1710–15, are two drafts of "The Introduction. Of the times before the Assyrian Empire."

> Idolatry had its rise [Newton began] from worshipping ye founders of Cities kingdoms & Empires, & began in Chaldea a little before the days of Abraham, most probably by ye worship of Nimrod the founder of several great cities. Till Abrahams days the worship of the true God propagated down from Noah to his posterity continued in Canaan as is manifest by the instance of Melchizedeck but in a little time the Canaanites began to imitate the Chaldees in worshipping the founders of their dominions, calling them Baalim & Melchom & Asteroth Lords & Kings & Queens, & sacrificing to them upon their gravestones & in their sepulchres & directing their worship to their statutes as their representatives, & instituting colleges of Priests with sacred rites to perpetuate their worship.[108]

At the time when Newton wrote this passage he was master of the Mint, with an annual income of about £2,000 per year, a trusted adviser to the government, and the honored president of the Royal Society who was privileged to refer in print to the Princess of Wales as "a particular friend." He has not told us why he suppressed the projected introduction and disguised the central theme of his chronological interests, but it is not farfetched to speculate that he feared to endanger his position by revealing too much. A dramatic slip of the pen about this time portrays his situation. Intending to write "St John," he put down instead "Sr John."[109] In truth, the Sir Johns of Augustan England bulked larger in his life by 1715 than the St. Johns of the primitive church. The man who had once prepared to surrender his fellowship so as not to accept the mark of the beast now cultivated the odor of orthodox sanctity by serving

---

[107] *Ibid.*, p. 188–9.   [108] New College MS 361.3, f. 242.   [109] *Keynes MS* 3, p. 39.

as a trustee of the Tabernacle on Golden Square and as a member of the Committee to Build Fifty New Churches in London. He was not likely to publish a tract that would place him beyond the pale.

Isaac Newton historian was Isaac Newton heretic engaged in one of his characteristic lifelong activities, the concealment of his heterodox views. In this he was eminently successful. In the continuing comments on the *Chronology* in the eighteenth century, only Arthur Young appears to have perceived Newton's drift. "I am sorry," he remarked, "to see Principles so favouring the Schemes of the Deists, with so great a Name affix'd to them."[110] The most recent student of his *Chronology*, writing before the manuscripts of the "Origines" were known or accessible, interpreted the work as a standard exercise in Christian piety.[111] There was indeed only one slip in the whole performance. By concealing his true purpose so effectively, Newton produced a book with no evident point and no evident form.[112] A work of colossal tedium, it excited for a brief time the interest and opposition of the handful able to get excited over the date of the Argonauts before it sank into oblivion. It is read today only by the tiniest remnant who for their sins must pass through its purgatory.

After Newton's death, William Whiston asserted that Newton and Samuel Clarke had given up the good fight for the restoration of primitive Christianity because Newton's interpretation of the prophecies led them to expect a long age of corruption before it would take place. Whiston quoted Dr. Sykes on Clarke's "Expression of his Fears, that the Face of Protestantism would once more be covered by as foul a Corruption as ever was that of Popery, before the happy Liberty and Light of the Gospel should take place." Whereas Sykes ascribed that view to Clarke, Whiston insisted that the idea came from Newton.[113] Whiston's assertion pointed correctly to Newton's resumption of his study of the prophecies, which together with his allied interest in ancient chronology became his principal study. Whiston also suggested a connection between Newton's suppression of his radical theological views, his interpretation of the prophecies,

[110] Arthur Young, *Historical Dissertation on Idolatrous Corruptions in Religion*, 2 vols. (London, 1734), 2, 269.

[111] See especially chap. 2, "The Learning of a Christian Chronologist," and chap. 6, "Israel Vindicated", in Manuel, *Newton, Historian*, pp. 37–49, 89–102.

[112] Cf. *ibid.*, p. 48. I believe Manuel is simply mistaken, however, in his assertion that only the insistence of Whiston, who wanted to refute it, secured the publication of the *Chronology of Ancient Kingdoms* (p. 36). The records of the estate indicate that the heirs were determined to realize on every merchantable item, and they sold the manuscript of the *Chronology* to a bookseller at once (Keynes MS 127A).

[113] William Whiston, *Historical Memoirs of the Life of Dr. Samuel Clarke* (London, 1730), p. 156.

and his growing conviction that not years but centuries had yet to pass before the second coming.

The papers on the prophecies contain a number of calculations on the Day of Judgment. One of them, apparently from the 1720s, concluded that it could not come before 2060 and that it might be later yet.

> I mention this period [he continued] not to assert it, but only to shew that there is little reason to expect it earlier, & thereby to put a stop to the rash conjectures of Interpreters who are frequently assigning the time of the end, & thereby bringing the sacred Prophesies into discredit as often as their conjectures do not come to pass. It is not for us to know the times & seasons which God has put in his own breast.[114]

The published *Observations upon the Prophecies* contained a passage similar to the final sentence above, perhaps the most quoted lines in the work. "The folly of Interpreters has been, to foretel times and things by this Prophecy, as if God designed to make them Prophets." God's design was much different, however. He delivered the prophecies, not to gratify men's curiosity about the future, "but that after they were fulfilled they might be interpreted by the event, and his own Providence, not the Interpreters, be then manifested thereby to the world."[115] Nevertheless, the internal chronology of the prophecies remained. If the second coming lay in the future, the beginnings of various delimited periods lay in the past, and whatever the folly of other interpreters, Newton could not suppress his own curiosity. The same published *Observations* implicitly indulged in the folly of interpreters at least twice, once placing the beginning of the 1,260 years about A.D. 800, another time placing the beginning of a period of 2,300 years, found in Daniel, in the first or second century A.D.[116] A scrap which seems to date from his final years reduced the matter in typical fashion to a set of propositions.

> Prop. 1. The 2300 prophetick days did not commence before the rise of the little horn of the He-Goat.
>
> 2 Those days did not commence after the destruction of Jerusalem & ye Temple by the Romans A.C. 70.
>
> 3 The time times & half a time did not commence before the year 800 in wch the Popes supremacy commenced
>
> 4 They did not commence after the reigne of Gregory the 7th. 1084.
>
> 5 The 1290 days did not commence before the year 842.

---

[114] *Yahuda MS* 7.3g, f. 13. Cf. f. 13ᵛ. There are at least two passages substantially the same which came from the period 1700–10 (*Yahuda MSS.* 7.3i, f. 54; 7.3l, f. 5).

[115] Newton, *Observations upon the Prophecies*, p. 251.   [116] *Ibid.*, pp. 113–14, 125–6.

6 They did not commence after the reign of Pope Greg. 7th.
1084
7 The diffence between the 1290 & 1335 days are a part of the
seven weeks.
Therefore the 2300 years do not end before the year 2132 nor after
2370.
The time times & half time do not end before 2060 nor after [2344]
The 1290 days do not begin [*sic;* end] before 2090 [*sic;* 2132] nor
after 1374 [*sic;* 2374][117]

The net result of the calculations in his old age was to move the
second coming back to the end of the twenty-first century at the
earliest, and probably later.

Other changes also appeared in Newton's study of the prophecies
during his old age. Whereas earlier he had concentrated almost exclusively on Revelation, now Daniel became the main object of his
interest and dominated the manuscript that was published eventually.[118] The shift of attention was not unrelated to the alterations by
which the "Origines" became the *Chronology*. In Newton's interpretation, Revelation focused attention on the early years of Christianity
and the rise of trinitarianism; although Daniel did not eliminate that
theme, it diffused attention by bringing in the pre-Christian era. The
early years of Christianity continued to loom large, but Newton
carried further the changes he had introduced years before and
couched his exposition of the great apostasy in terms Protestants
would have understood as a condemnation of Roman Catholicism.

For if placing supernatural vertue in words & figures & ceremonies &
things consecrated & reliques & images & invocations of dead men,
be abominable actions, actions of the same kind with the heathen
charms & magic & sorcery; if the invocation of dead men be idolatry, & if Christian sorcerers & Christian idolaters be the worst of
hereticks, then the Roman Empire before ye end of the fourth century became very heretical . . .[119]

The "homousians," who had dominated the original exposition,
had vanished. Newton had cleansed his *Observations upon the Prophecies* as effectively as he had the *Chronology*. Conduitt called it
"Revelation & Prophecies without *Enthusiasm or superstition* . . ."[120]
Unfortunately, we must add, as we did with the *Chronology*, prophecies without focus or point, but his heirs could publish the manuscript without concern.

---

[117] *Yahuda MS* 7.3o, f. 8. Cf. a manuscript, which can be dated with assurance to 1708,
which seems to place the Day of Judgment in 2132 (*Yahuda MS* 7.3n, f. 1ᵛ).
[118] See especially *Yahuda MS* 7.1.
[119] *Yahuda MS* 7.3i, f. 46ᵛ. I could greatly expand his treatment of analogous themes;
passages are legion throughout the manuscript. [120] *Keynes MS* 130.7, Sheet 4.

Figure 15.6.   Newton at eighty-three. Portrait by John Vanderbank, 1726. (Courtesy of the Royal Society.)

After Newton's death, Conduitt sought advice about the work. The replies he received indicate that Newton was still at work on it when he died. They referred to two manuscripts, an old one completed and a new one left incomplete by his death.[121] From the

[121] *Yahuda MSS* 7.2b, ff. 1–9; 7.2j, ff. 95–6.

memoranda of advice that Conduitt received, it is evident that the work eventually published in 1733 was an artificial product composed by the advisers by melding the two manuscripts together so as to use the old, finished manuscript to round out and complete the unfinished new one. They also inserted three chapters which did not appear in either of Newton's versions, chapters 1 of both Parts I and II (the chapters on the composition of the books in the Old Testament and of Revelation) and Chapter XI, Part I, "Of the Times of the Birth and Passion of Christ."[122] The two chapters 1 do not seriously interrupt what minimal flow the work possesses. The chapter on the life of Christ, which is utterly unrepresentative of Newton's theological interests and finds almost no repetition in his very extensive papers, sits in the belly of the work like an undigested lump of wax. Nevertheless, the major failures of continuity and coherence in the book cannot be attributed to the editors. Newton himself must bear the blame. Neither the old nor the new manuscript, which one can effectively reconstruct from the chapter titles in the memoranda and extant versions of such chapters, had an argument or an organization. They were the meandering products of an old man. As I have suggested, the suppression of their real argument completed the decomposition of whatever senescence had left intact.

Both the *Chronology* and the *Observations upon the Prophecies* appeared in print after Newton's death. Other theological papers, which were also products of his old age, did not. They provide a perspective by which we can judge the extent to which Newton edited those which he prepared for publication. Among them are many sheets that took up again the history of the early church.[123] A codicil to Catherine Conduitt's will, in which she expressed her strong desire that Newton's theological papers be published, mentioned "the history of the Creed or criticism of it, and a Church History compleat . . ."[124] Although the papers mentioned above contain fragments that appear to belong to such works, the works themselves do not appear among the papers now accessible.[125] At least ten different chapter headings of a history of the church do survive among them, one of which is numbered Chapter XVI.[126]

---

[122] Cf. *Yahuda MS* 7.2f, f. 11.

[123] *Yahuda MSS* 7.3c, ff. 28–30; 7.3d, ff. 1–2; 7.3g, ff. 1–2; 7.3i, ff. 31–2, 42–3, 52; 7.3m, ff. 1–5, 7, 10, 12; 15.1, ff. 1–22, 25–44; 15.2, ff. 23–4; 15.3, ff. 55–9; 15.4, ff. 71–6; 15.5, ff. 92–5, 15.7, f. 124. I do not care to assert that this list exhausts those on the history of the church. It does indicate those I identified as such.

[124] New College MS 361.4, f. 139.

[125] There is one extensive batch of theological papers, in the Bodmer Library, Geneva, Switzerland, which is not accessible.

[126] *Yahuda MS* 15.7, f. 124. Other chapter titles in *Yahuda MSS* 7.3m, f. 5ᵛ; 15.2, f. 23; 15.3, ff. 55, 57, 59; 15.5, ff. 91–2.

In so far as I can judge from the fragments, the history of the early church underwent a change analogous to those in the *Chronology* and the *Observations* but less complete. Whereas Newton had once centered his attention on the fourth century and the triumph of trinitarianism, he now concentrated more on the earlier period of primitive Christianity and attempted (as Catherine Conduitt's codicil suggests) to reconstruct the creed of the early church. Perhaps we can see in the papers the reason why her expressed desire to publish them was not carried out. The distance between Newton's religion and the established faith of the Church of England emerges in these and allied papers far more clearly than it did in the published works.

Among the allied papers, one of the most significant, which exists in multiple drafts like everything to which Newton attached importance, bore the title "Irenicum." It was the paper Conduitt called "his Irenicum his creed," which would have made revealed religion less mischievous – that is, it would have, had Newton worked up his courage to publish it.[127] "Irenicum" returned to the theme of the "Origines": "All nations were originally of one religion & this religion consisted in the Precepts of the sons of Noah . . ." The principal heads of the primitive religion were love of God and love of neighbor. This religion descended to Abraham, Isaac, and Jacob. Moses delivered it to Israel. Pythagoras learned it in his travels and taught it to his disciples. "This religion," Newton concluded, "may be therefore called the Moral Law of all nations."[128] Chapter I, "Of the Church of God," in his history of the early church began with the same theme.

> The true religion was propagated by Noah to his posterity, & when they revolted to yᵉ worship of their dead Kings & Heros & thereby denied their God & ceased to be his people, it continued in Abraham & his posterity who revolted not. And when they began to worship the Gods of Egypt & Syria, Moses & the Prophets reclaimed them from time to time till they rejected the Messiah from being their Lord, & he rejected them from being his people & called the Gentiles, & thenceforward the beleivers both Jews & Gentiles became his people.[129]

To the two great commandments of the primitive religion, to love God and to love one's neighbor, the Gospels added the further doctrine that Jesus was the Christ foretold in prophecy. The risen Christ

---

[127] No item among Newton's theological manuscripts suffered more at McLachlan's hands than "Irenicum." What he printed from *Keynes MS 3*, where several drafts are collected, contains in my opinion the least typical parts. The paper that comes closest to capturing the sense of "Irenicum" (pp. 31–5 from *Keynes MS 3*, pp. 9–14) has been rearranged and distorted by the insertion of passages from other sheets of *Keynes MS 3*.

[128] *Keynes MS 3*, p. 27.   [129] *Yahuda MS 15.3*, f. 57.

appeared to his disciples and commanded them to preach repentence and remission of sins in his name. Repentence and remission of sins, "Irenicum" insisted, refer to transgressions of the two basic laws, sins such as idolatry, a failure in the love of God, and covetousness, a failure in the love of neighbor.[130] When Jesus was asked what was the great commandment of the law, he answered that it was to love God, and he added that the second commandment was to love your neighbor. "This was the religion of the sons of Noah established by Moses & Christ & still in force."[131]

Thus, Newton argued, we are to believe in one God and in one Lord, Jesus Christ, who is next to him in power and glory. All this was taught from the beginning in the primitive church. It is what Paul called milk for babes. Those things taught after baptism were meat for men of full age. After baptism, men were urged to study the Scriptures, "& especially the Prophesies," to learn as much as they could, and they were to teach others what they learned. Beyond the minimal requirements for baptism, however, they were, in the primitive church, to proceed in the spirit of charity. They were not to break communion, to excommunicate, or to persecute over matters beyond the requirements for baptism. To impose now any article of communion that was not such from the beginning was to preach another gospel. To persecute Christians for not receiving that Gospel was to make war on Christ.[132] The two great commandments, he insisted over and over, "always have & always will be the duty of all nations & The coming of Jesus Christ has made no alteration in them." As often as mankind has turned from them, God has made a reformation – through Noah, Abraham, Moses, the Jewish prophets, and Jesus. Now that the gentiles had corrupted themselves, men must expect a new reformation.

And in all the reformations of religion hitherto made the religion in respect of God & our neighbor is one & the same religion (barring ceremonies & forms of government wch are of a changeable nature) so that this is the oldest religion in the world . . .[133]

"These are the laws of nature," he added in another paper, "the essential part of religion wch ever was & ever will be binding to all nations, being of an immutable eternal nature because grounded upon immutable reason."[134]

The law was ancienter then the days of Moses [he wrote on the back of a bill dated May 1719] being given to Noah & all his posterity, & therefore w^{n} the Apostles & Elders in the Council at Jerusalem declared that the Gentiles were not obliged to be circumcised & observe

130 *Keynes MS* 3, p. 1.   131 *Keynes MS* 3, pp. 5–7.
132 *Keynes MS* 3, pp. 3, 14. The second of these passages is printed (incorrectly) in McLachlan, *Theological Manuscripts*, p. 35. Cf. similar passages in *Yahuda MSS* 7.1m, f. 2; 7.3b, f. 1.   133 *Keynes MS* 3, p. 35.   134 *Yahuda MS* 15.5, f. 91.

the law of Moses, they excepted this law as being imposed on all nations not as the sons of Araham [*sic*] but as the sons of Noah not by circumcision but by an earlier law of God not by conversion to the Christian religion but even before they were Christians. And of the same kind is the law of absteining from meats offered to Idols. & from fornication.

    not as Christians but as Gentiles

    – as being imposed on all nations not by the law of Moses but by an earlier law of God, not as sons of Abraham but as sons of Noah, not as Christians but even as Gentiles. And of the same kind is the law of absteining from meats offered to Idols & from fornication.[135]

Newton returned to the same theme in "A short schem of the true Religion," in which he argued that the precepts that fill out the command to love your neighbor were taught to the heathens by Socrates, Cicero, Confucius, and other philosophers, to the Israelites by Moses, and to the Christians more fully by Christ. "Thus you see there is but one law for all nations the law of righteousness & charity dictated to the Christians by Christ to the Jews by Moses & to all mankind by the light of reason & by this law all men are to be judged at the last day."[136]

One cannot miss the filiation of such concepts, stemming from the "Origines," with Newton's Arianism. In other papers he continued to explore the nature of God and Christ and to reaffirm in his old age the Arian stance he had taken originally in his young manhood. Some of the fragments on the early creed contain the most important statements of his theological position that he ever set on paper. His research on the early church led him to a statement of the true creed.

    I believe in one God, the Father Almighty, maker of heaven & earth & of all things visible & invisible, & in one Lord Jesus Christ, the Son of God, who was born of the Virgin Mary, suffered under Pontius Pilate, was buried, the third day rose again from the dead. He ascended into heaven & from thence shall come to judge the quick & the dead whose kingdom shall have no end. And I believe in the holy Ghost who spoke by the Prophets.

Such a creed, easy to understand even to those of limited capacity, was fit to be proposed as a statement of the first principles of religion. It contained no theories but only practical truths on which the whole practice of religion depended. Passages such as this, from the time of the calculus controversy, reflected aspects of the philosophic stance that Newton assumed vis-à-vis Leibniz. The primitive religion, easily understood by the meanest of people, was handed

[135] *Yahuda MS* 7.4, n. f.
[136] *Keynes MS* 7, pp. 2–3; printed in McLachlan, *Theological Manuscripts*, p. 52. Cf. a sheet with nine "Positions": *Keynes MS* 3, pp. 17–18; *Theological Manuscripts*, pp. 28–31.

down in simplicity "untill men skilled in the learning of heathens Cabbalists & Schoolmen corrupted it with metaphysicks, straining the scriptures from a moral to a metaphysical sence & thereby making it unintelligible."[137] To cap his other crimes, Leibniz repeated the intellectual stance of the archfiend Athanasius.

Newton proceeded to explicate the meaning of his creed. We must beleive that he is the father Almighty, or first author of all things by the almighty power of his will, that we may thank & worship him & him alone for our being and for all the blessings of this life. . . . the only invisible God & the only God whom we are to worship & therefore we are not to worship any visible image picture likeness or form. We are not forbidden to give the name of Gods to Angels & Kings but we are forbidden to worship them as Gods. 'For tho there be that are called Gods whether in heaven or in earth (as there are Gods many & Lords many) yet to us there is but one God the Father of whom are all things & we in him & one Lord Jesus Christ by whom are all things & we by him,' that is, but one God & one Lord in our worship: 'One God & one mediator between God & man the man Christ Jesus.' We are forbidden to worship two Gods but we are not forbidden to worship one God & one Lord: One God for creating all things & one Lord for redeeming us with his blood. We must not pray to two Gods but we may pray to one God in the name of one Lord. . . . We must beleive that he was crucified being slain at the Passover as a propitiary sacrifice for us, that in gratitude We may give him honour & glory & blessing as the Lamb of God wch was slain & hath redeemed us & washed us from our sins in his own blood & made us Kings & Priests unto God his Father. We must believe that he rose again from the dead that we may expect the like resurrection & that he ascended into heaven to prepare a place or mansion for the blessed that by the expectation of such a glorious & incorruptible inheritance we may endeavour to deserve it. We must beleive that he is exalted to the right hand of God (Acts 2   ) or is next in dignity to God the Father Almighty, the first begotten the heir of all things & Lord over all the creation next under God, & we must give him suitable worship. . . . The worship wch we are directed in scripture to give to Jesus Christ respects his death & exaltation to the right hand of God & is given to him as our Lord & King & tends to the glory of God the Father. Should we give the father that worship wch is due to the Son we should be Patripassians, & should we give the Son all that worship which is due to the father we should make two creators & be guilty of polytheism & in both cases we should practically deny the father & the Son. . . . We must also beleive that Jesus Christ shall come to judge

---

[137] *Yahuda MS* 15.5, f. 97^v. Similar passages were very common among the theological papers that dated from the years after 1715.

the quick & the dead, that is to reign over them with justice & judgment untill he shal subdue all rule & all authority & power, & all
enemies be put under his feet the last of wch is death, & by consequence untill all the dead be raised & judged. For he sits at the right
hand of God not only in this world but also in that which is to come
untill all enemies be put under his feet. And this his coming to judgm^t
we must beleive that we may with understanding pray for the coming
of his kingdom & fit our selves to stand before him in that day, & to
deserve an early resurrection. . . .[138]

The Arian features of Newton's christology continued to be evident. Although we are to worship Christ as Lord, "yet we are to do
it without breaking the first commandment." The true manhood of
Christ was important to Newton, who believed that trinitarianism
effectively denied his manhood and with it the reality of his suffering
on the cross. However, "he was not an ordinary man but incarnate
by the almighty power of God & born of a Virgin without any other
father then God himself."[139] That is, Newton had reached back to
the primitive church to resurrect a concept of Christ as a human
body animated by a divine or semidivine spirit. He rejected any
notion of a unity of substance between God the Father and Christ the
Son, and asserted instead what he called a monarchical unity –

> an unity of Dominion, the Son receiving all things from the father,
> being subject to him, executing his will, sitting in his throne &
> calling him his God, & so is but one God wth the Father as a king &
> his viceroy are but one king. For the word God relates not to the
> metaphysical nature of God but to his dominion.[140]

Though created by God in time, Christ existed before the world
began. As the spirit of prophecy, he was the angel of God who
appeared to Abraham, Jacob, and Moses and governed Israel in
the days of the judges. After Israel rejected him and desired a
king, the angel appeared no more but rather sent his messenger to
the prophets.[141]

For Newton, to whom prophecy was the very essence of revelation, the most essential thing about Christ was not his special relation
to God but his special relation to prophecy. To the original true worship of the sons of Noah, he asserted, "nothing more has been added . . . with relation to Jesus Christ then to belive in the predictions

---

[138] *Yahuda MS* 15.3, ff. 45–6. Cf. a short restatement of this position on f. 46^v. Also *Yahuda MS* 15.5, f. 95^v, and *Keynes MS* 3, pp. 43, 45. The space is in the original.

[139] *Keynes MS* 3, p. 45. *Yahuda MS* 15.3, f. 46.

[140] *Yahuda MS* 15.7, f. 154. Cf. a number of other manuscripts on the nature of Christ; all of them post-1715; *Yahuda MSS* 15.3, ff. 58, 66^v; 15.4, ff. 67–8; 15.5, ff. 87, 90^v, 96–8; 15.7, f. 108; *Keynes MS* 11, printed in McLachlan, *Theological Manuscripts*, pp. 44–7.

[141] *Yahuda MS* 15.5, f. 96^v.

of the holy Ghost by the Prophets concerning him, vizt that he is the Messiah & the son of man predicted by Daniel & ye son of God predicted by David in the 2^d Psalm, & the Lamb of God predicted in ye Pascal Lamb by Moses, &c."[142] Whenever Newton attempted to summarize the true religion in a series of articles, a further aspect of the special relation of Jesus to prophecy always appeared.

> That all foreknowledge of things is originally in the breast of him that sitteth upon the throne, & that the Lamb received this prophesy from him & was the only being in heaven earth or under the earth who was worthy to receive it from him, & by his death obteined this worthiness . . .[143]

As Newton also insisted, giving ear to the prophets is the mark of the true church.

When Voltaire came to England in the 1720s, he interested himself in Newton's religion as well as his philosophy. The latter he could find in the published works. He learned what he could about the former from Newton's friends, such as Samuel Clarke. "Newton was firmly persuaded of the Existence of a God, and by that word he understood not only a Being infinite, omnipotent, and eternal, who is the creator, but a master who has made a relation between himself and his creatures . . ."[144] Clarke instructed Voltaire well. Among Newton's theological papers, many passages emphasized the very point.

> We are therefore to acknowledge one God infinite eternal omnipresent, omniscient omnipotent, the creator of all things most wise, most just, most good most holy: & to have no other Gods but him. We must love him feare him honour him trust in him pray to him give him thanks praise him hollow his name obey his commandments & set times apart for his service as we are directed in the third & fourth commandments. For this is the love of God that we keep his commandments & his commandments are not grievous 1 John 5.3. These things we must do not to any mediators between him & us but to him alone, that he may give his Angels charge over us who being our fellow servants are pleased with the worship wch we give to their God. And this is the first & principal part of religion. This always was & always will be the religion of all Gods people, from the beginning to the end of the world.[145]

---

[142] *Yahuda MS* 7.3k, n. f. before f. 1.

[143] *Yahuda MS* 7.2e, f. 4^v. Cf. another list on the same sheet and a sheet with twelve articles, *Keynes MS* 8; printed in McLachlan, *Theological Manuscripts*, pp. 56–7.

[144] Voltaire, *Oeuvres complètes de Voltaire* (Beuchot), 72 vols. (Paris, 1834–40) *38*, 11.

[145] *Keynes MS* 7, p. 2; printed in McLachlan, *Theological Manuscripts*, p. 51. Cf. *Keynes MS* 3, p. 35, and *Yahuda MSS* 15.3, f. 59; 15.5, f. 98. The last manuscript, a theological paper, is interesting in the virtual identity of part of it to the General Scholium to the *Principia*.

It is impossible to mistake the intense affective quality of such passages, which one can only describe as worshipful, and we should be rash indeed to challenge their sincerity. The piety they express can be found more readily in the General Scholium to the *Principia* and the Queries attached to the *Opticks*. Shortly after Newton's death, John Craig, in summarizing his achievement for Conduitt, asserted that "the reason of his showing the errors of Cartes's philosophy, was because he thought it was made on purpose to be the foundation of infidelity."[146] Newton's papers from over the years support Craig's statement. Such had been the heritage he received before his long pilgrimage through Christianity. It has supported the picture frequently presented of a religiously traditional Newton largely wedded to the forms of established Protestantism.[147]

Newton was a far more complex man, however, than such a picture allows. Pious he undoubtedly was, but his piety had been stained indelibly by the touch of cold philosophy. It is impossible to wash the Arianism out of his religious views. Newton set out at an early age to purge Christianity of irrationality, mystery, and superstition, and he never turned from that path. His study of the prophecies, the work most frequently cited in support of a contrary interpretation of his religion, was in fact one of the cornerstones of his program. True, he undertook to purge Christianity in the name of Gospel purity, but in the light of the role that Arianism played in the early church and the role that it and its offspring played in the eighteenth century, one cannot view Newton's Arianism in isolation from the intellectual currents of his day. Rather we do him more justice and acknowledge anew his manifest genius by allowing that here too he stood in the van, although the very reform of Christianity he sought to foster was already, in his old age, surging far beyond the limits he had envisaged. He justified himself in terms of the Bible, but the Bible as he understood it was far removed from the Bible of traditional belief. Where that Bible contained truths beyond reason, Newton summed up true religion in terms that effectively dispensed with all of revelation beyond the prophecies. Christians for centuries had understood divine revelation in terms of a new dispensation foretold in the Old Testament and fulfilled in the New; divine revelation as Newton understood it centered on two books, Daniel and Revelation, which revealed the almighty dominion of God over history as natural philosophy re-

---

[146] *Keynes MS* 132.

[147] Richard S. Brooks, "The Relationships between Natural Philosophy, Natural Theology and Revealed Religion in the Thought of Newton and Their Historiographic Relevance" (dissertation, Northwestern University, 1976). William H. Austin, "Isaac Newton on Science and Religion," *Journal of the History of Ideas, 31* (1970), 521–40. Leonard Trengrove, "Newton's Theological Views," *Annals of Science, 22* (1966), 277–94.

vealed His dominion over nature. Newton questioned the plenary inspiration of the received canon of books and regarded the historical books of the Old Testament as the compilations of men.[148]

Though he wrote at some length about Christ, his interest largely exhausted itself in proving that Christ was not God. His soteriology, the focus of traditional Christian concern, was uninspired and jejune, substituting a mere legal pact for the reconciliation of fallen man to the majesty of God which generations of theologians had explored. The two fundamental duties of true religion, to love God and to love one's neighbor, seem to present the opportunity for spiritual insight. Alas, when Newton sought to give them content, he could do so only in negative terms.

> We are to forsake the Devil, that is, all fals God & all manner of idolatry, this being a breach of the first & great commandment. And we are to forsake the flesh & the world, or as the Apostle John expresseth it, the lust of the flesh, the lust of the eye, & the pride of life, that is, unchastity, covetousness pride & ambition; these things being a breach of the second of the two great commandments.[149]

The new covenant in Christ's blood promising life renewed had become in his hands a list of peremptory thou-shalt-not's.

Somehow Newton blended this desiccated vision of Christianity with a living faith in the almighty God which suffused his life.

> We must beleive that there is one God or supreme Monarch that we may fear & obey him & keep his laws & give him honour & glory. We must beleive that he is the father of whom are all things, & that he loves his people as his children that they may mutually love him & obey him as their father. We must beleive that he is the παντο-κρατωρ Lord of all things with an irresistible & boundles power & dominion that we may not hope to escape if we rebell & set up other Gods or transgress the laws of his monarchy, & that we may expect great rewards if we do his will. We must beleive that he is the God of the Jews who created the heaven & earth all things therein as is exprest in the ten commandments that we may thank him for our being, & for all the blessings of this life, & forbear to take his name in vain or worship images or other Gods. We are not forbidden to give the name of Gods to Angels & Kings, but we are forbidden to have them as Gods in our worship. For tho there be that are called God whether in heaven or in earth (as there are Gods many & Lords many, yet to us there is but one God the father of whom are all things & we in him & one Lord Jesus Christ by whom are all things & we by him: that is, but one God & one Lord in o^r worship.[150]

[148] New College MS 361.2, ff. 132–3. Newton, *Observations upon the Prophecies*, pp. 4–13. Newton, *Chronology*, pp. 357–8. Manuel, *Newton, Historian*, pp. 59–60.

[149] *Keynes MS* 3, p. 1. Cf. Manuel, *Newton, Historian*, pp. 137–8, on Newton's analogous dehumanized treatment of history.   [150] *Yahuda MS* 15.3, f. 46^v.

The concept of pantocrator caught Newton's imagination and held it. The word appeared repeatedly throughout the theological papers from his final years. Autocrat over all that is, He dictated the form of the natural world and the course of human history. Newton did not meet him in the intimacies of watchful providence, a point related to his Arianism. Rather he found Him in the awful majesty of universal immutable laws–an austere God, one perhaps whom only a philosopher could worship.

Very few items indicate a more personal side to Newton's religion. One that does, a letter from Joseph Morland, a member of the Royal Society, probably written in 1716, the year Morland died, should not be omitted.

> Sʳ
>
> I have done and will do my best while I live to follow your advice to repent and believe I pray often as I am able that god would make me sincere & change my heart. Pray write me your opinion whether upon the whole I may dye with comfort. This can do you no harm written without your name. God knows I am very low & uneasie & have but little strength
>
> > Yours most humble servᵗ
> > Jos. Morland
>
> Pray favour me with one line because when I parted I had not your last words to me you being in hast[151]

Even in a rare venture in the cure of souls, it appears, Newton could not lay aside his fear of disclosing too much.

Whereas Newton published statements of his belief in God, he not only kept the unorthodox aspects of his religion to himself, but he exercised some care in London to mask his heterodoxy behind a facade of public conformity. He continued to act as a trustee of Archbishop Tenison's chapel on Golden Square until 1722.[152] When Parliament passed an act in 1711 to finance the construction of fifty new churches in the expanding suburbs of London, Newton became one of the commissioners appointed to implement Parliament's will, and he sat on the commission until at least 1720.[153] Likewise he accepted membership on the new commission to supervise the completion of St. Paul's cathedral, and attended meetings of it in the period 1715–21.[154] In view of such assiduous atten-

---

[151] *Corres 7*, 382.   [152] Cf. Warren to Newton, 19 Dec. 1721; *Corres 7*, 182.

[153] Summons to meetings of the commission in 1717 and 1720; *Corres 6*, 406–7; *Corres 7*, 483–4.

[154] A record shows his attendance at a total of twelve meetings in the period 1715–21 ("Minute Book. H. M. Commission for Rebuilding St. Paul's Cathedral," *The Wren Society, 16* [1939], 33–137). There is a notice of a meeting in New College MS 361.2, f. 77ᵛ.

tion to the proprieties, it is not surprising that William Stukeley, noting his care to attend Sunday services and his well-thumbed Bible, called him "an intire Christian" and opined that his steady support of the Church of England was the product of true philosophy. Stukeley also mentioned that Arians tried to claim Newton.[155] He was probably sincere in his vigorous denial of their claim, for Newton did not lightly lay his soul bare.

A few did know better. In his letter to Conduitt after Newton's death, John Craig mentioned Newton's extensive religious study and said that he had not published his theological writings "because they show'd that his thoughts were some times different from those which are commonly receiv'd, which would ingage him in disputes, & this was a thing which he avoided as much as possible."[156] Conduitt, who scarcely needed Craig's instruction in the matter, made his reference, in the sketch of Newton's life that he sent to Fontenelle, even more elliptical. Newton believed firmly in revealed religion, he told Fontenelle, "but his notion of the Xtian religion was not founded on a narrow bottom, nor his charity & morality so scanty as to shew a coldness to those who thought otherwise than he did in matters indifferent . . ."[157] A few rumors did circulate. Thomas Hearne picked one up in 1732.

> Sir Isaac Newton, tho' a great Mathematician, was a man of very little Religion, in so much that he is ranked with the Heterodox men of the age. Nay they stick not to make him, with respect to belief, of no better principles than M[r]. Woolaston [corrected later to Wolston], who hath written so many vile books and made so much noise.[158]

Andrew Michael Ramsay, who knew friends of Newton such as Fatio and Clarke, asserted in a letter that Newton had wanted to revive Arianism by means of his disciple Clarke, though Clarke also confessed shortly before his death how much he regretted the publication of his arianizing work.[159]

Newton himself did not make Clarke's mistake. Well concealed behind circumlocutions such as Craig's and Conduitt's, his heterodoxy slid into virtual oblivion, not to be uncovered until the twentieth century or to be fully revealed until the Yahuda papers became available quite recently.

His conclusions functioned only vicariously in the religious ferment of the eighteenth century. When Joseph Hallet, alarmed by the spread of Arianism, published in 1735 *An Address to Conform-*

---

[155] *Stukeley*, pp. 69–71.   [156] *Keynes MS* 132.   [157] *Keynes MS* 129A, p. 22.
[158] Thomas Hearne, *Remarks and Collections*, ed. C. E. Doble *et al.*, 11 vols. (Oxford, 1885–1921), *11*, 100–1. Thomas Woolston (Hearne's Wolston) was a freethinker whose works attracted some notoriety in the 1720s.
[159] Joseph Spence, *Anecdotes, Observations, and Characters, of Books and Men*, ed. Samuel Singer (London, 1820), p. 379.

ing Arians to convince them of their hypocrisy and to lead them
to repent, he named two men as the source of the infection,
William Whiston and Samuel Clarke.[160] Both were Newton's
disciples and known as such. Later another disciple, Hopton
Haynes, would publish unitarian tracts, and a more aggressive
unitarian, Richard Baron, would lament that Samuel Clarke, who
had performed good work in purging Christianity of much absurdity and rubbish, had stopped short in Arianism when a fully
rational Christianity lay only another step beyond.[161] But Newton's extended quest, barely hinted at in his published works,
·had to enter the stream of religious controversy through disciples
more daring than he. He carefully laundered what he himself
prepared for publication. The rest he locked away. It is wholly
unlikely that his views, formulated a generation before similar
ones became widespread, had a significant causal role in the religious history of the Enlightenment.

Though study formed the main substance of Newton's life in his
declining years, as it had always done, he continued until the end to
exercise the public functions he had assumed. He presided over the
Royal Society, attending its meetings with great regularity. After
1713, with both secretaryships filled by his followers, the society
was fully under his control, and he assumed an almost regal bearing
in its affairs. When it lost £600 in the South Sea Bubble, Newton
offered to make the sum up from his own pocket.[162] In 1722, he
presented to the society a manuscript by Tycho Brahe with unpublished observations of four comets and, as president, ordered that
they be printed.[163] Even more than in the past, he assumed the role
of patron to young scientists. In 1717, he subscribed for two copies
of James Stirling's *Liniae tertii ordinis newtonianae,* a work which
added four species of cubics which Newton had missed in his own
pioneering *Enumeration,* though to be sure in the context of a Newtonian manifesto vis-à-vis Leibniz.[164] A few years later, according
to Conduitt, Newton gave Stirling money to return home from
Venice so that he might recommend him for a professorship.[165] In
1725, he thought of offering £20 per annum toward the salary of a
young Scottish mathematician, Colin Maclaurin, if he were elected
in Edinburgh to a joint professorship with James Gregory, who had

---

[160] Reprinted in Thomas Gordon, *A Cordial for Low Spirits,* 2 vols. (London, 1751), 2,
321–49.

[161] In his preface to *ibid.,* pp. xv–xvi.   [162] *Stukeley,* p. 13.

[163] *JBC 12,* 271. They did not, however, appear in the *Philosophical Transactions,* or elsewhere as far as I can find.

[164] Stirling, *Liniae tertii ordinis newtonianae* (Oxford, 1717). Cf. *Math 7,* 574.

[165] *Keynes MS* 130.7, Sheet 1. Cf. Stirling to Newton, 17 Aug. 1719; *Corres 7,* 53–4.

Figure 15.7. Newton at eighty-three. He holds a copy of the third edition. This portrait may be a copy of the Vanderbank of 1725. Artist unknown. George Vertue mentioned a portrait of Newton "in his hair" by Michael Dahl which hung at Wimpole in the eighteenth century. This could be the Dahl portrait. Trinity College has a nearly identical portrait attributed to Vanderbank. The attribution appears to rest solely on tradition. (Courtesy of the Trustees of the National Portrait Gallery.)

become too ill to fill the position he held. It does not appear that he ever made the offer formally, however.[166] He did contribute more than £100 to the support of the astronomer James Pound. He also purchased Huygens's great lens, which had a focal length of 123 feet, and had it set up at Wanstead for Pound's use. He even purchased the maypole that had been erected in the Strand and had it transported to Wanstead to support the glass. The lens had a special Newtonian story attached to it. When Newton imported it into England, the customs officer asked him its value. The officer would have accepted its value as glass; Newton supplied instead its value

---

[166] Newton to Maclaurin, 21 August, and Newton to John Campbell, ca. November 1725; Maclaurin to Newton, 25 Oct. 1725; *Corres 7*, 329, 336, 338.

as a lens and paid as a consequence a duty of £20.[167] Pound repaid
him with observations of Jupiter and Saturn and their satellites and
of fixed stars near the path of the comet of 1680–1, observations
added to the third edition of the *Principia*. To promising young
scientists such as the men above Newton was accessible. He held
his distance from others. An irate German noble (real or pretended), Johann von Hatzfeld, a perpetual-motion fanatic, wrote to
Newton in high dudgeon about 1725 that when he tried to call to
show him an experiment demonstrating that a perpetual motion is
possible, Newton's servant refused to admit him without a "Recommendation."[168] An Englishman whose circumstances did not
permit him to assume such airs indicated the same trouble when he
came for the humble purpose of begging.[169]

Not only president of the Royal Society but Europe's most re-
nowned scientist, whose fame multiplied daily, Newton found
himself more than ever before almost a tourist attraction to traveling literati, who attempted to meet him if they could. Conduitt
heard that Monseigneur Branchini, the pope's chamberlain, came
all the way to London for the express purpose of seeing Newton,
that Count Marsigli did the same, and that the Marquis de l'Hôpital
(who was a mathematician of note) wanted to do so but died before
he could.[170] Conti indicated that the French party with Monmort
came to London in 1715 as much to meet Newton as to see the
eclipse.[171] From the time of the priority dispute, when he began
seriously to woo the support of the Académie, Newton seems to
have opened his door hospitably to French visitors. He received
Monmort's party the day after they arrived. Joseph-Nicolas Delisle,
an astronomer and member of the Académie who briefly replaced
Varignon as Newton's contact with that body, wrote enthusiastically about his reception by Newton in 1724. Newton even gave

---

[167] Pound's "Account Book," quoted in James Bradley, *Miscellaneous Works and Correspondence,* ed. S. P. Rigaud, reprint ed. (New York, 1972), p. iii. Stukeley heard about the
customs duty at the breakfast with Newton and Halley on 22 Feb. 1721 (Stukeley to
Conduitt, 15 July 1727; *Keynes MS* 136, p. 10). Newton may also have had a hand in the
first significant Newtonian telescope. In October 1720, the Reverend Dr. Hill sent to the
Royal Society the design of a reflecting telescope "vastly Different" from Newton's.
Although the society found it much inferior to Newton's, it was less than three months
later when John Hadley showed up at a meeting on 12 Jan. 1721 with a reflector six feet
long which magnified 200 times and was made according to Newton's plan. He presented it to the society, which placed it under Halley's care (*JBC 12,* 48, 67–8. Cf. pp.
108–10). See Hadley, "An account of a Catadioptrick Telescope," *Philosophical Transactions,* 32 (1722–3), 303–12; and "A Letter from the Rev. Mr. James Pound . . . Concerning Observations made with Mr. Hadley's Reflecting Telescope," *ibid.,* 32 (1722–3),
382–4.   [168] *Corres 7,* 253.   [169] *Corres 7,* 383.
[170] *Keynes MSS* 130.6, Book 3; 130.5, Sheet 2.   [171] Conti, *Prose e poesi, 2,* 23.

Delisle a portrait of himself as he had done for Varignon before.[172] Some of the visits produced continuing consequences. Delisle reminded Newton of Varignon, and he proceeded to have an engraving made of Varignon's portrait for inclusion in the publication of his mechanics. To Delisle he insisted that the engraving was his gift to Varignon's friends.[173] The consequences of Marsigli's visit were less pleasant. Marsigli presumed that a friendship had been established, and he kept trying to force his correspondence on Newton. In 1724, he mentioned a general work he was writing on the organic structure of the earth which he intended to dedicate to the Royal Society, "provided that you will so amend it in your kindly and diligent way, that it may seem not altogether unworthy of so great an honour."[174] Be it said that Newton refused to bow to suasion of this sort. Marsigli reaped silence as his reward; he never published the intended work.

The Abbé Alari, another traveler from France, left a full account of his visit to Newton on 1 July 1725. He called at nine in the morning. Newton began by informing Alari that he was eighty-three years old (awarding himself the benefit of six unearned months). They talked about ancient history. When the abbé gave a satisfactory account of his own learning, Newton invited him to dine – the miserable meal with cheap wine, as the abbé perceived it, of which we have heard already. In the afternoon, Newton took him to the Royal Society, where he seated him on his right hand at the table. The meeting began. Newton promptly dozed off. As the Journal Book summarizes the meeting – a letter on drugs from the Netherlands, an account of some French bottles that spoiled wine stored in them, a letter with a record of the weather in Zurich during 1724 – one only wonders how Alari stayed awake. The meeting over, Newton took Alari back to his home and kept him until nine in the evening – rather a longer visit, one gathers from the account, than Alari had bargained for.[175]

Members of the scientific community also had other ways to demonstrate their esteem. Thus in 1723, Pieter van Musschenbroek presented Newton with a copy of his inaugural lecture in Utrecht suitably inscribed.

[172] Delisle to Newton, 10 Dec. 1724; *Corres 7*, 296–7.

[173] Newton to Delisle, ca. January 1725; *Corres 7*, 300–1. Cf. Jombert to Newton, 12 Sept. 1725; *Corres 7*, 332.

[174] Marsigli to Newton, ca. May 1724; *Corres 7*, 283–4. Cf. Marsigli to Newton, 29 Feb. 1724; *Corres 7*, 264–5. See also the letter of Philippe Naudé, a member of the Berlin Academy, to Newton, 6 Feb. 1723; *Corres 7*, 223–31.

[175] Général Grimoard, "Essai historique sur Bolingbroke," *Lettres historiques, politiques, philosophiques et particulières de Henri Saint-John, Lord Vicomte Bolingbroke, depuis 1710 jusqu'en 1736*, 3 vols. (Paris, 1808), *1*, 155–6. *JBC 12*, 583–6.

The Author sends [this book] to the Very Great and Noble Man Isaac Newton, Most Eminent of Geometers and Restorer of Solid Philosophy.[176]

No other scientist in Europe would have received – or deserved – such a dedication.

The presidency of the Royal Society combined with Newton's reputation to involve him in time-consuming advice to the government on a technical matter during the final thirteen years of his life. With British commerce as well as the British empire growing apace, the problem of determining a ship's longitude at sea was becoming acute. The wreck of Sir Cloudesley Shovell's fleet in the Scilly Islands in 1707, with the loss of many lives and much treasure, served to underline the need. Aspiring navigators and inventors, aware of the reward likely to follow the discovery of a workable method, were turning their thoughts to it. In 1711, Denis Papin tried to get the Royal Society to endorse a clock he claimed to have developed for that purpose.[177] The following year, the Earl of Oxford sought the advice of Newton and Halley on a proposal by a Mr. Cawood, not to find longitude, but to aid navigation by a compass that he claimed would point to the true geographic north pole.[178]

Such was the background to a petition of 25 May 1714 to the House of Commons by a group of captains and London merchants. The problem of determining longitude at sea had become so necessary to England, they argued, that Parliament should establish a reward to encourage the discovery of a method.[179] The House of Commons appointed a committee which established a precedent for later generations by summoning technical experts to instruct them. William Whiston and Humfrey Ditton appeared before the committee as the joint authors of a proposal that among other things would help ships near shore determine their location by means of shots which could be heard at sea. Newton, Halley, Cotes, and Samuel Clarke all commented on the proposal sometime between 25 May and 11 June when the committee gave in its report. Whereas Halley, Cotes, and Clarke all spoke briefly, Newton arrived with his testimony written out in full, and the committee duly incorporated his opinion into its report. There were four possible methods of determining longitude, he said: a watch that kept time perfectly, eclipses of the satellites of Jupiter, the positions of the moon, and the Whiston-Ditton project. The first was rendered difficult by the motion of the ship and variations of temperature and weather. The second required a long telescope, which was hard to use on a toss-

---

[176] John Harrison, *The Library of Isaac Newton* (Cambridge, 1978), p. 198.
[177] 10 Oct. 1711; *JBC 10*, 321–2.   [178] *Corres 5*, 332.
[179] *Journal of the House of Commons*, 17, 641.

ing ship. The theory of the moon, which was accurate enough only to fix longitude within two or three degrees, stood in the way of the third. The fourth method, the Whiston-Ditton proposal, was limited to certain situations.[180]

Years later Whiston described the hearing. After he read his paper, Newton sat down. The committee had not understood him, and a silence ensued. The chairman of the committee, who opposed the concept of a prize, said that unless Newton declared his opinion that the proposal of Whiston and Ditton was apt to be useful he thought there would not be a prize. Newton said nothing, and no one else on the committee spoke. The whole design of a prize stood in jeopardy. At the last moment Whiston spoke in desperation. The problem was just Newton's caution, he urged; Newton knew that their method was useful near the shore, the place of greatest danger. With this, Newton rose and said that he thought the bill should pass because the method was useful near the shore, and the committee then reported unanimously in favor of the prize.[181]

J. B. Biot, who was convinced of Newton's mental incompetence after the breakdown of 1693, seized on this incident (or rather on Whiston's description of it) as evidence of conduct that was "almost puerile."[182] Such a judgment ignores known facts about Newton. He never trusted himself to express anything extemporaneously. The repeated drafts on even minor letters bear witness to this trait. Whereas the other three spoke, Newton read a prepared statement, tried to avoid speaking again, and allowed Whiston to put words in his mouth. That is, he did so if Whiston's account was accurate. There is good reason to doubt its accuracy, however. Newton's testimony makes it clear that he did not regard the Whiston-Ditton proposal highly. It was a method "rather for keeping an Account of the Longitude at Sea, than for finding it, if at any time it should be lost, as it may easily be in cloudy Weather."[183] The committee understood him well enough to propose a prize for a workable method without any reference to that of Whiston and Ditton. Far from finding Newton almost puerile, they incorporated his suggestion that the prize offered vary in size with the degree of accuracy that the method gave: £10,000 for a method of determining longitude to an accuracy of one degree, £15,000 for an accuracy of forty minutes, and £20,000 for an accuracy of half a degree.[184] In

---

[180] *Ibid.*, 17, 677–8. Newton's own manuscript of his testimony is in *Add MS* 3972, f. 32.

[181] Whiston, Preface, *Longitude Discovered by the Eclipses, Occultations and Conjunctions of Jupiter's Planets* (London, 1738), pp. ii–vii. The historical preface, which appears only in some copies, is dated 1742.

[182] "Newton (Isaac)," *Biographie universelle, ancienne et moderne*, 52 vols. (Paris, 1811–28), *31*, 127–95. The phrase quoted is on p. 193.

[183] *Journal of the House of Commons*, 17, 678.   [184] *Ibid.*, 17, 678, 715–16.

a separate communication, he urged that Parliament strike from the bill a clause that recommended a watch made by Mr. Hutchinson since he doubted that any watch could ever reach such a degree of perfection.[185]

The eventual act of Parliament not only offered the prize but established a Board of Longitude to judge proposals submitted for it. Among others, the board included the Savilian professors from Oxford, the Lucasian and Plumian professors from Cambridge, the Astronomer Royal, and the president of the Royal Society, who thus found himself in the company of John Flamsteed once more. The *Journal littéraire* carried a notice of the reward even before Parliament enacted it, and before July passed Newton had received at least three proposals from France.[186] They were the first installment of a large number to follow, a burden Newton had to carry until the end of his life.[187] In November 1714, Sir Christopher Wren, then eighty-two years old, but interested in the renown if not in the money, sent Newton, in a cipher, the descriptions of three instruments suitable to determine longitude.[188] In June 1720, Whiston, who had participated in launching the board, presented before the Royal Society a method dependent on magnetic dip, as measured by needles six to eight feet long, coordinated with maps that charted lines of dip.[189] Even before a year had passed, Newton had grown impatient with some of the fantastic schemes put before him.

Sr

I have received your Letter (dated yesterday) by the hands of Mr John Vat & can acquaint you that his Project for the Longitude is as impracticable as to make a perpetual motion like that of the heart but much more uniform or to observe the Sun's meridional altitude to a second or to deduce the Longitude from the complement of the Latitude, or to find that complement by burning brandy. . . . And I have told you oftner then once that it is not to be found by Clockwork alone. Clock work may be subservient to Astronomy but without Astronomy the longitude is not to be found. Exact instruments for keeping of time can be usefull only for keeping the Longitude while you have it. If it be once lost it cannot be found again by such Instruments. Nothing but Astronomy is sufficient for this purpose. But if you are unwilling to meddle with Astronomy (the only right method & the method pointed at by the Act of Parliament) I

[185] *Yahuda MS 7.3g*, f. 21.
[186] "Extrait d'une lettre de Londres," *Journal littéraire*, 4 (1714), 235–6. Jordan to ?Oxford, 22 July 1714; LeMuet to the Longitude Commission, 10 July 1714; d'Alesme to ?Newton, ca. late July 1714; *Corres 7*, 163–5.   [187] *Corres 6, 7, passim*.
[188] Wren to Newton, 30 Nov. 1714; *Corres 6*, 193.
[189] *JBC 12*, 32–5. Cf. 22 March and 5 April 1722; *JBC 12*, 219, 224.

am unwilling to meddle with any other methods then the right one.[190]

Later, he refused to assemble the board to consider a clock until it had been made and tested by acknowledged experts.[191]

In letters to Josiah Burchett, the secretary of the Admiralty who forwarded proposals, Newton frequently repeated his negative opinion of clocks for the determination of longitude.

A good Watch may serve to keep a recconing at Sea for some days & to know the time of a celestial Observation: & for this end a good Jewel watch may suffice till a better sort of Watch can be found out. But when the Longitude at sea is once lost, it cannot be found again by any watch.[192]

His deprecation of clocks may have helped later to delay the acceptance of Harrison's chronometers, which did in fact offer a practical determination of longitude at sea.[193]

Newton also held onto his position at the Mint during these years, and the position at the Mint in turn continued to impose its burden upon him. Year in and year out, the overvaluation of gold in relation to silver had furnished his constant refrain. Because England undervalued silver, merchants could turn a profit by melting down silver coin and exporting the bullion to the Continent, where it fetched a higher price. In a paper of April 1714, "Observations upon the valuation of Gold and Silver in proportion to one another," he returned to the theme once more. The current value of the guinea was 21s 6d, a rate which made a pound of gold worth slightly more than fifteen and a half pounds of silver. On the Continent, Spain and Portugal excepted, the rate was fifteen to one. Britain ought therefore, he argued, to devalue the guinea to 20s 7d or at most 20s 8d.[194] In 1714, no one heeded his song. By 1717, however, as silver coin continued to melt away, an audience had appeared. On 10 August, the lords commissioners of the Treasury directed Newton to report on the state of the gold and silver coin and to propose what he thought their relative values should be in order to discourage the melting down of silver.[195] He was ready enough to do that. He had only to dust off arguments now well

---

[190] Newton to —— [French], 22 March 1715; *Corres 6*, 211–12. For the reference to burning brandy, see French to the Royal Society, 9 Dec. 1714; *Corres 6*, 194.

[191] *Corres 7*, 173.

[192] Newton to Burchett, ca. October 1721; *Corres 7*, 172–3. Cf. Newton to the Admiralty, 26 Aug. 1725; *Corres 7*, 330–2.

[193] So the editors of the Correspondence suggest (*Corres 6*, 212).

[194] *Mint* 19.1, f. 6. See C. R. Fay, "Newton and the Gold Standard," *Cambridge Historical Journal, 5* (1935–7), 109–17; and William Stanley Jevons, "Sir Isaac Newton and Bimetallism," in *Investigations in Currency and Finance*, ed. H. S. Foxwell (London, 1884), pp. 330–60. [195] *CTB 31 (2)*, 44.

polished, though in his usual fashion he devoted immense effort to his memorandum and went through his standard quota of drafts before he sent in his finished version of 21 September.[196]

Commerce among nations established the value of gold in relation to silver, he argued; nothing England might do could alter it. "And therefore the lowering of Gold in proportion to our silver money will also lower forreingn [*sic*] silver in proportion to the same silver money. Or to speak more truly it will raise the value of our silver coin to that of foreign silver."[197] Hence he urged that the government reduce the legal value of the guinea by as much as 10d or 12d; even a reduction of 6d would help.[198] The final suggestion was a mistake. A government eager to do something but equally anxious not to do too much seized on it as an appropriate compromise and on 22 December issued a proclamation setting the value of the guinea at 21s. The reduction was not enough to save the silver coinage. It was enough to injure the interests that had been profiting from the old arrangement, and they reacted angrily.

Interest in the coinage and in the related problem of foreign exchange ran high. Newton's memorandum was printed in the *Daily Courant* on 30 December and in the monthly résumé of events, *The Political State of Great Britain*. On 21 January 1718, the House of Lords summoned him to report to them on the quantities of gold and silver coined annually since the beginning of the recoinage.[199] When the devaluation of the guinea led to no apparent improvement in the supply of silver, the Treasury began plans to coin quarter-guineas to facilitate daily commerce. The Mint did indeed produce these tiny gold wafers for a brief period, until their impracticality became obvious.[200] The furor did not die away immediately. In the autumn of 1718, the Treasury ordered Newton to lay before them whatever thoughts he had on the state of the coinage since the proclamation of December, and he responded with a set of eight "Observations upon the State of the Coins of Gold and Silver" which defended the effect of the action.[201] Nevertheless, it was politically impossible to lower the value of the guinea further to its true exchange rate. It remained fixed permanently at 21s, which became the effective definition of the word "guinea" after the gold coin disappeared from circulation much later. The realities of British trade in the Far East, where silver fetched a much higher price than it did in Europe, make it unlikely that any reasonable action could have saved the silver coinage.

Meanwhile, new problems internal to the Mint had arisen. With

[196] *Mint* 19.2, ff. 38, 65, 67, 69, 96, 98, 100, 111–115, 236.
[197] *Mint* 19.2, f. 98. This passage was not in the final memorandum.
[198] *Corres 6*, 415–18.   [199] *Journal of the House of Lords, 20*, 579.
[200] *CTB 32 (2)*, 7, 37, 107, 120.   [201] *Corres 7*, 5, 8–10.

the end of 1714, Newton's nemesis, Craven Peyton, had disappeared from the scene forever. By the end of 1715, however, a completely different issue, the finances of the Mint, had become a matter of concern. With the end of the War of the Spanish Succession, the rate of coinage had surged forward, and the expense of coinage, always a fixed percentage of the amount coined, increased in proportion. The receipts of the duty on imported liquors, which financed the Mint, did not keep pace. Newton calculated that over the three previous years the coinage duty had produced an average income of £9,600 per year, while the Mint had expended an average of £14,380. Moreover, the annual expenditure appeared to be growing. Although the Mint had held a large balance at the end of the war, Newton was worrying by the end of 1715 that it would soon be exhausted. With the Coinage Act, which came up for renewal every seven years, due to expire in March 1716, he began to lobby the Treasury for a change in its provisions. His first proposal, in February 1716, called for an increase in the coinage duty of 50 percent.[202] He must have discovered quickly that such a proposal had no prospects whatever. It disappeared from his papers in favor of an alternative by which the general fund would supply supplementary appropriations to the Mint, to cover a maximum annual expenditure of £15,000, whenever the income from the coinage duty failed to meet expenses.[203]

Newton's papers make it apparent that another question bothered him at least as much as the Mint's income. By the Act of Union, a separate mint was maintained in Edinburgh, independent of the London Mint but drawing on the same coinage duty for its support, which was set at £1,200 per year. From the beginning, Newton had seen the Scottish Mint as a threat to his bureaucratic domain. Now the Scots began to propose an additional £500 per year for the expenses of coining. Once the Scottish recoinage had been completed, the Edinburgh Mint had ceased to operate. Newton would have preferred to abolish it. All of its positions were now sinecures. Furthermore, it had an extra official, the general of the Mint, who had no function even when the Mint was active but nevertheless drew £300 per annum. Before union, the Scottish Mint had operated and coined on £1,200 per year. Now the £1,200 went entirely into salaries and maintenance, and they wanted £500 more. "And thus the coinage moneys are taken away from the Mint in England."[204] If Newton could not abolish the Scottish Mint, at least he could keep it inactive, and he sought to use the new Coinage Act to control its threat once and for all. His first proposal was

---

[202] *Corres 6*, 276–7.
[203] At least seven different drafts of this provision exist among Newton's papers; *Mint* 19.1, ff. 369–78; 19.3, f. 206.   [204] *Mint* 19.3, f. 215.

a complete separation of the two mints, each to operate on the proceeds from the coinage duty within the respective borders of England and Scotland. It was Newton's impression that the coinage duty would not produce as much as £1,200 per year in Scotland. One draft of his proposal suggested that the Scottish Mint also be allowed 6 percent of the English income; another draft magnanimously proposed that appropriations from the general fund, within the total limit of £15,000 for the two mints, might raise its income to the established £1,200.[205]

In the upshot, Newton got an amendment, though not exactly the one he sought. It did authorize additional appropriations from the general fund when needed, to cover a maximum annual expenditure of £15,000.[206] In fact, the Mint never drew a penny from the general fund during his lifetime. The rate of coinage fell off for a couple of years from the peak of the postwar period. When it rose again in 1719 and 1720, Newton's balance at the Mint shrank to £10 18s 6½d at the end of 1720. However, coinage declined once more, while the income from the duty suddenly shot up about 50 percent. The balance recovered quickly and stood at a healthy £22,278 at the end of 1726, a few months before Newton's death.[207] His aspirations in regard to the Scottish Mint, however, suffered shipwreck on the rocks of political reality. No word of the amendments he proposed to limit its income crept into the final act. He did succeed in the years ahead in keeping the Scottish Mint inactive and even in further restricting its income. The Scottish Mint never coined again, either in Newton's life or after. In 1718, when the general of the Mint tried to collect the £2,400 due for the last two years, Newton advised the Treasury that £1,000 per year was enough to cover their salaries and maintenance, and the Treasury apparently accepted his advice.[208]

A minor incident in 1716, which was related to the level of silver coinage, required extended attention by Newton. About March or April, one James Hamilton showed up in London with stories of a silver mine on the property of Sir John Erskine near Stirling, Scotland. Erskine had discovered the mine in 1714 and employed Hamilton to smelt the ore. In 1715, however, Erskine had joined the Jacobite rebellion, and Hamilton, scenting the possibility of gain,

---

[205] *Mint* 19.1, ff. 370, 376. *Mint* 19.3, f. 206–9. Cf. Newton to the Treasury and Newton to ———— [Lauderdale?], both late 1715; *Corres 6*, 263–4, 264–5.

[206] *Acts 1 George*, pp. 623–4.

[207] *Mint* 19.4. Craig asserts that the fiscal crisis of the Mint continued for another four years and that payment of salaries had sometimes to be postponed (John Craig, *The Mint: A History of the London Mint from A.D. 287 to 1948* [Cambridge, 1953], p. 211). I saw no evidence of this in the accounts. As I say, the Mint never drew on the extra allowance, as it could have done, to pay salaries.

[208] *Corres 7*, 6–7. Cf. Newton and Bladen to the lords commissioners, 5 Sept. 1717; *Corres 6*, 410–11. *CTP, 1720–28*, p. 86.

had come to London with information about the mine.[209] Since Erskine's property had been forfeited by rebellion, the mine held out the promise of a steady flow of bullion into the Mint. A government cognizant of the technical expertise at its command turned immediately to Newton. The Mint assayed the sample of ore that Hamilton had brought with him and found it very promising.[210] The Treasury now proposed to send Newton north to survey the mine. Seventy-three years old and hardly eager for such an excursion, Newton pled a lack of technical knowledge, but he furnished instructions to the German expert for George's mines in Hanover, Dr. Justus Brandeshagen, whom the Crown did employ. The complete public servant, ever conscious of the possibilities of fraud, he outlined careful procedures, with the constant use of witnesses, to insure that no one imposed on the Crown.[211] It had good cause to be concerned. Nothing came of the mine, but Brandeshagen and Hamilton deliberately loitered in Scotland to inflate their expenses as much as possible. In 1717, Newton spent considerable time advising the Treasury about their accounts, and six and seven years later he had to deal with a pair of law suits about expenses denied.[212]

In 1717, Newton acquired a major new responsibility that had long been under discussion at the Mint, the coinage of copper half-pennies and farthings. A similar coinage carried out by separate patentees rather than the Mint had been underway when Newton came to London in 1696. Between 1694 and 1700, they had coined 700 tons of coins, leaving the nation oversupplied, in Newton's opinion. New proposals to resume small coinage again under similar arrangements had begun in 1703 and continued thereafter in regular succession. Newton had found a reason to oppose every project. He did not believe there was a shortage of copper coins, and generally he did not favor the existence of coinage operations outside the Mint, although he wavered on this principle at times.[213] Under the regime of the Earl of Oxford, the pattern of consultation and delay suddenly changed. On 2 May 1713, Newton received a summons to a meeting about copper coinage from a Lord Treasurer

---

[209] Newton's history of the Erskine mine; *Mint* 19.3, f. 246.

[210] Newton to Townshend, April 1716; *Corres 6*, 316.

[211] Newton to Treasury, 25 Aug. and 14 Nov. 1716; *Corres 6*, 367–9, 373–4. Cf. Newton to Brandeshagen, autumn 1716; *Corres 6*, 378–9.

[212] Newton to Treasury, 27 June 1717; *Corres 6*, 395–6. Notes on the accounts, drafts, and copies are found in *Mint* 19.3, ff. 235, 237–8, 246, 251, 261, 263. In 1723 James Hamilton brought a suit against Newton over the payment (*Mint* 19.3, ff. 240–1). The following year John Walker, a creditor of Brandeshagen, who was dead by then, petitioned the Treasury for money supposedly owed to Brandeshagen (Scrope to Newton 11 May; Newton to the Treasury, 18 May 1724; *Corres 7*, 280–2).

[213] Newton to Godolphin 13 July 1703; *Corres 4*, 408–9. Newton to the Treasury, 13 Dec. 1710; *Corres 5*, 81–2. Newton to Oxford, Dec. 1712; *Corres 5*, 357–60.

who professed to be "much Importuned about the Business of Coyning Farthings . . ."[214] Various copper merchants and undertakers, as they were called, eager to profit from such coinage, were present with the officers of the Mint at the Treasury on 8 May. Oxford left no doubt that he intended to press forward, and the merchants present were delighted with the prospects of prosperity that he offered to the copper business.

For his part, Newton insisted that the coinage be carried out in the Mint "because of the danger of Trusting proper Tools in any other hands . . . ," and he urged as well that the coins issued bear the value of the metal plus the expense of coinage to prevent counterfeiting. The question of assaying copper came up. Newton indicated that the Mint's test of copper good enough to coin was its capacity to be hammered red hot into a thin sheet without cracking.[215] Newton would later claim credit for developing this method of assaying where before there had been no accepted standards for the purity of copper.[216] The Mint continued to use it until bronze replaced copper in the middle of the nineteenth century.

The meeting at the Treasury touched off a flurry of activity. Newton rapidly made himself the master of the various technical details involved in order to be able to speak from his customary position of authority. Some trial coins were struck early in 1714. A flood of proposals by various people eager to turn a dishonest pound poured into the Treasury and found their way to the Mint for commentary.[217] Newton composed at least six technical memoranda to the Treasury about the various problems that had to be solved before the coinage could be put in motion.[218] Then Anne died, Oxford fell, and after a short hesitation the project collapsed.

The need for small coins was such that the issue refused to die, however, and in 1717 the same forces that had impelled Oxford to action breathed life into it once more. On 29 April, the Treasury decided to proceed and placed an advertisement in the *Gazette* for bids to supply copper to the Mint. There was no question any longer of contracting with patentees to coin outside the Mint. Twelve proposals to supply copper came in by July and were passed on to the Mint, which sifted them and reported early in August. On 12 August a warrant was drawn for £500 to start the coinage and on 14 August another warrant for Newton to proceed;

[214] *Corres 5*, 404–5.
[215] *Corres 5*, 405–6. Cf. Newton to Oxford, Feb. 1713; *Corres 7*, 407; and "Considerations about the Coynage of Copper Moneys"; *Corres 5*, 415–16.
[216] Newton to the Treasury, 12 Aug. 1719; *Corres 7*, 58.
[217] *Corres 6*, 23, 58–9, 74–5, 95–6, 97, 99, 99–100, 111. *CTB 29 (2)*, 228, 230.
[218] *Corres 6*, 51–2, 52–3, 55–7, 75–6, 98, 190–2.

the final signing of the last warrant waited a month while Newton reached agreement with the moneyers.[219]

Although he was nearly seventy-five years old when the copper coinage finally began, the negotiations in the summer of 1717 to arrange its details showed Newton the civil servant at his best. On the one hand, the technical command he insisted on acquiring allowed him to assess the various ways in which copper might be supplied and to recommend the one most likely to provide copper suitable for coining. On the other hand, twenty years at the Mint had taught him to specify careful procedures to insure the quality of the coinage and to protect the Crown from fraud, even when he was the one most likely to profit from laxity.[220] He also calculated the elements of the total cost with care. The copper bars would cost 18d or 19d per pound weight avoirdupois; in the end he recommended that the Treasury accept an offer by Messrs. Hines and Appleby to supply them for 18d. As master, he himself received 3½d per pound weight, out of which he had to defray all the expenses of coining. That set the total cost at 21½d per pound weight. Each pound was coined into 23d worth of coins, 1½d of which the Crown received. Thus in its finance the copper coinage returned to the old practice of seigneurage after the Crown had coined precious metals free as a public service for half a century. Most of the 3½d that Newton received went to the moneyers. In 1711, they had asked for 3¾d per pound weight to coin the copper. He bargained them down to 2d, though eventually they got an extra ¾d from the copper supplier. The smith and the graver received ¼d each. Charges for clerks and delivery consumed about ⅜d more, leaving Newton about ⅝d per pound weight for supervising the whole process. Over a period of seven years, until the Treasury judged the demand fulfilled, the Mint coined a total of 140 tons of copper coins, an average of 20 tons a year. At this rate, Newton pocketed about £117 per year in return for a great deal of trouble.[221] One of the earlier proposals he had analyzed for the Treasury had offered to inundate the country with 1,500 tons of copper money while the undertakers reaped a pleasant profit of about £85,000.[222]

Troubles did not end with the official warrant to coin in Septem-

---

[219] *CTB 31 (2)*, 13, 30, 40, 43, 45, 69, 450, 509. *CTB 31 (3)*, 571, 574, 575–7. Newton to the Treasury, 3 Aug., and Newton to Stanhope, 17 Aug. 1717; *Corres 6*, 404–5, 408. The warrant to coin; *Corres 6*, 412–14.

[220] *Corres 6*, 390–1. Cf. drafts of this document in *Mint* 19.2, ff. 346–7, 356, 361, 444.

[221] *Mint* 19.2, f. 426. *Corres 6*, 404–5. A note by Conduitt summarized the various payments (*Mint* 19.2, f. 428). The final accounts of the copper coinage are in the Public Record Office, A.O.1/1635, Roll 281.

[222] *Corres 6*, 99–100. Elsewhere he estimated the profit at £97,000 (*Mint* 19.2, ff. 322.3).

ber. As Newton informed the Treasury in November, coining copper was different from smelting copper. It was an art not hitherto
practiced in England, which had formerly imported bars prepared
abroad for the coining operations, and they had to grope forward
by trial and error.[223] Refining the copper presented difficulties.
Fumes from coal rendered it impossible to coin. Cracks appeared in
the coins, and they had to alter the method of rolling the bars and
annealing.[224] Complaints about the delay made themselves heard.

> To be concerned in this sort of coinage I never desired but it falls to
> my lot [Newton replied with some warmth]. And things are upon
> such a foot that I can get nothing but discredit by coyning the money
> ill. . . . I am very willing to lend the Copper rooms in the Mint to
> any body who may be authorized to take care of this coinage &
> content my self wth the coinage of ye gold & silver; but if it be your
> Lord^{ps} pleasure that I go on with it, I will take the best care I can to
> have it well performed.[225]

It was their lordships' pleasure, and as always Newton persevered
to a successful conclusion. In January 1718 the Mint finally began to
coin. By April it had produced six tons of halfpennies.[226]

More trouble was brewing. When the government advertised in
1717 for offers to supply copper, twelve bids came in. Only one
could be accepted, of course, and the disappointed projectors, their
dreams of profit dissolving, tried other means to seize the prize.
One of them, Richard Jones, published a pamphlet, "A Letter
from a Livery-Man of the City of London to a Member of the
Honble House of Commons relating to ye Coinage of Copper
farthings & Half pence," which made various allegations about the
quality of the copper. According to Newton, coarse copper was
spirited into the Mint, stamped without his knowledge, and
shown about as a sample of the coinage. Others did the same with
very fine copper and exhibited these coins as examplars of what
they could do.[227] By February the agitation by disappointed bid-
ders was enough to make the Treasury summon Newton to a
meeting, where he was instructed to accept five tons of copper
from each of two disappointed projectors.[228] The man who had
battered Leibniz did not propose to surrender to a parcel of copper
merchants, however. The Treasury had specified British copper
and British copper alone, and he insisted on proof that the copper
did not come from Barbary. This drove off some of the agitators.
His insistence on his own assays drove off more. Some argued

[223] Cf. Newton to the lords commissioners, 23 Nov. 1717; *Corres 6*, 422.
[224] Newton to the Treasury, 12 Aug. 1719; *Corres 7*, 56–8.
[225] *Mint* 19.2, ff. 441^v–2.   [226] *CTB 32 (2)*, 37.
[227] Newton to the lords commissioners, ca. July 1718; *Corres 6*, 451–4. *CTP*, 1714–19, p. 360.   [228] *CTB 32 (2)*, 17. *Corres 6*, 434.

that part of the work should be done outside the Mint. Newton resisted with all his might, demanding that the job be performed in the Mint, where his mandate carried weight. "I cannot undertake absolutely that in the copper imported there shall be no faulty barrs wch may escape the assays, but I am safest in people that are afraid of me."[229]

There were problems also of another sort, for the copper merchants sometimes represented more powerful patrons. Newton complained to William Derham of the interruptions caused by petitions, usually from people of quality.

> Amongst others he told me y^t an Agent of one had made him an offer of above 6000^{li}. W^{ch} S^r Is: refusing on account of its being a bribe; the Agent said, He saw no dishonesty in the acceptance of the Offer, & y^t S^r Is: understood not his own Interest. To wch S^r Is: replied y^t he knew well enough w^t was his Duty, & y^t no Bribes should corrupt him. The Agent then told him y^t he came from a great Dutchesse, & pleaded her Quality & Interest. To wch S^r Is: roughly answered, I desire you to tell the Lady, y^t if she was here her self, & had made me the Ofter [*sic*], I would have desired her to go out of my house, & so I desire you, or you shall be turned out.[230]

Later, Derham added, Newton found out who the duchess was, but he did not pass that information on to Derham, who thus did not repeat it for us.

The sniping continued without respite through 1718. There were enough problems still in process of being solved that telling criticism could be made. Early in 1719, when the method of rolling copper bars still left them rough and the coins therefore imperfect, the critics succeeded in winning an order from the Treasury to halt coinage for about four months.[231] Newton had still to reply to criticisms of disappointed copper merchants as late as November.[232] For all their efforts, when Newton submitted his final accounts for the copper coinage, he had purchased copper only from Hines and Appleby, the partners whose bid he originally selected in August 1717.[233] It is worth recording as well that at a trial of the pyx in 1722 six halfpennies made by his critics were tried along with the Mint's. Their coins all shattered when hammered red hot. All of Newton's stood the test.[234]

[229] *Mint* 19.2, f. 442.   [230] *Keynes MS* 133, pp. 12–13.
[231] A memorial on 13 April 1719 said coinage had been stopped for three months (*CTP, 1714–19*, p. 446). Lowndes to Newton, 15 April, and Newton to the Treasury, 21 April 1719; *Corres 7*, 32, 35–6. Cf. *Corres 6*, 424, 445–6, 446, 449.
[232] He was called into the Treasury with Briggs and Nicholson on 20 Nov. 1719 (Public Record Office, T29/24, Pt. I, f. 230). Cf. an earlier meeting with the same two in June (f. 183) and Newton's reply to their complaints on 12 August (*Corres 7*, 56–8).
[233] Public Record Office, A.O.1/1635, Roll 281.   [234] *Mint* 19.2, f. 317.

In 1717, while Newton wrestled with the copper coinage, wrote memoranda on the value of gold, and brought out a new edition of his *Opticks,* and while the priority dispute temporarily slumbered following Leibniz's death, a young man who would figure prominently in his declining years entered his life. On 26 August 1717, John Conduitt married Newton's niece, Catherine Barton. Born in London in 1688 and educated at Westminster School, Conduitt had entered Trinity College in 1705 as one of the Westminster scholars.[235] After leaving Cambridge without a degree, he next appeared in 1711 as judge advocate and provost marshal of the British forces in Portugal and secretary to the Earl of Portmore, the general of that army, during the War of the Spanish Succession. He was always "Mr. Conduitt" in the documents that mentioned him, not Capt. or Maj. Conduitt, though it is said that he later served with a regiment of dragoons.[236] On 6 April 1713, Conduitt received appointment as commissary to the British forces on the new base at Gibraltar and held that post at least until July 1715, more probably until early 1717.[237] When he was the husband of Catherine Barton, Conduitt was a wealthy man. Clearly he had been born to prosperity, and he may have inherited his wealth. Positions such as commissary had traditionally yielded immense profits, however, and it is not implausible to speculate that the years at Gibraltar enhanced Conduitt's means at the least. He did more than accumulate money while he was there. He also identified the site of the Roman city, Carteia. On 13 December 1716 and on 14 March 1717, the Royal Society heard communications from Conduitt, still apparently at Gibraltar, about Carteia. On 6 June, a meeting which Newton missed, Conduitt received permission to attend the society, and two weeks later, this time with Newton in the chair, he read his paper on Carteia.[238] As it turned out, Carteia was the only string on Conduitt's learned bow. In 1719, he sent the Royal Society yet another discourse on the location of Carteia, a discourse now closely enough related to the president of the society to find its way into the *Philosophical*

---

[235] His memorial plaque on the wall of Westminster Abbey states that he was forty-four when he died in 1737. This would place his birthday in 1692 or 1693. Unless there was a second John Conduitt, whose records of baptism and admission to Westminster and Trinity have been confused with Catherine Barton's husband, such a date seems very implausible. He was admitted to Trinity in 1705. He became judge advocate of the British forces in Portugal in 1711 – at eighteen or nineteen if he born in 1692 or 1693.

[236] *CTB 25 (2),* 115, 595; *CTB 26 (2),* 51, 371.

[237] *CTB 32 (2),* 231–4. Cf. *CTB 28 (2)* 31; *31 (2),* 365, 387–9; *CTP,* 1708–14, pp. 568, 575; *CTP,* 1714–19, pp. 44, 45, 165, 256. Conduitt submitted his accounts on 18 Nov. 1717 (*CTB 31 (3),* 672). Petitions about his accounts continued through 1718 (*CTP,* 1714–19, p. 425).

[238] *JBC 11,* 147, 166–7, 187, 190.

Figure 15.8. Newton in his old age. The portrait has been attributed (incorrectly) to William Hogarth. Another opinion holds that Vanderbank executed it, but it differs markedly in aspect from his others. It may be a copy of the Seeman portrait (Figure 15.11) or of a Vanderbank of 1725 or of Figure 15.7, all of which it resembles in some respects. It could also be the Dahl portrait mentioned above. (Courtesy of W. Heffer and Sons, Ltd., Cambridge, England.)

Transactions. He even had a priority dispute with a Mr. Breval over the identification of Carteia.[239] It so happened that Newton was also interested in Carteia, a city built by the Tyrians, as he believed, during mankind's expansion through the Mediterranean basin in the first millennium B.C.[240] We do not know the course of events following Conduitt's appearance at the Royal Society. Newton was working on his *Chronology* in that period, however, and he might well have spoken to the author of a paper that fit in with his current studies. Three months later Conduitt did for Catherine Barton what Halifax had not; he gave her a new name and married respectability. Conduitt was then twenty-nine years old, and Mrs. Catherine Barton was thirty-eight. The evidence indicates that she was still beautiful and charming. Her uncle may nevertheless have possessed at least as much attraction as the bride, for Conduitt worshiped him unabashedly as a hero. Though Conduitt's own capabilities may have been limited, he recognized that he stood in the presence of one of the geniuses of all ages, and he vowed to respond adequately to the opportunity. He wrote down accounts of their conversations. He collected anecdotes about Newton. When he died twenty years after his wedding, Conduitt arranged to have his memorial plaque begin with a statement, not about himself, not about his wife, not about his parents, but about Isaac Newton, to whom he was related and near whose remains he had contrived to have his own placed.

The Conduitts stayed on with Newton when they were in London. That was only part of the time, since they were frequently at Conduitt's country seat, Cranbury Park near Winchester. In 1719, Catherine wrote to Newton from Cranbury, thanking him for a letter and explaining that she had instructed her servant to enquire about Newton's health and send her an account.[241] There is a fragment from 1724 addressed to her in Cranbury, and a letter from John Conduitt there on 21 May 1724.[242] Even when they were in London, Newton probably did not see much of them. His niece was an energetic woman, and beginning in 1721 her husband was a member of Parliament. In 1719, the couple presented Newton with a new grandniece (far from his first), whom they named Catherine. There is no evidence that the child entered deeply into the life of the old philosopher. In a letter to Conduitt in 1726, after he himself had removed to Kensington, Newton did at least acknowledge her exis-

---

[239] 26 Feb., 5 and 19 March 1719; *JBC 11*, 301, 302, 310. "A Discourse tending to shew the Situation of the ancient Carteia, and some other Roman Towns near it," *Philosophical Transactions, 30* (1717–19), 903–22.

[240] Newton, *Chronology*, pp. 108, 111–12.

[241] Catherine Conduitt to Newton, 16 Nov. 1719; *Corres 7*, 74.

[242] *Mint* 19.3, f. 471. *JBC 12*, 477.

tence. "I hope your Lady and Kitty are well. My service to them."[243] Kitty was well; she went on, after her parents' deaths, to marry into the nobility.

About the time of Catherine Barton's wedding, according to one story, Newton brought her cousin, Benjamin Smith, a racy profligate young man about eighteen years old at that time, to live with him.[244] If the story is true, Ben Smith left singularly little trace of his presence, though he does not appear to have been self-effacing.

Early in 1718, Newton made the acquaintance of William Stukeley, who was practicing medicine in London and joined the Royal Society at that time. Stukeley was regular in his attendance, and he derived from Lincolnshire as well. Newton became friendly with him. When he later moved to Grantham, Stukeley, like Conduitt, made it his business to collect information about Newton. From the two of them comes much of our knowledge of Newton's characteristics in his final years. Newton's life, Conduitt wrote, was "one continued series of labour, patience, humility, temperance, meekness, humanity, beneficence & piety without any tincture of vice . . ."[245] Such are the fruits of hero worship. Fortunately, in addition to beatifying Newton as a plaster saint, he also recorded a few details. Newton was of middle stature and, in his later years, plump. He had "a very lively piercing eye" and a gracious aspect. His head of hair, as white as snow, was full with no baldness. Even as an old man he retained the bloom and color of youth and all his teeth except one.[246] In contrast to Conduitt, Bishop Atterbury, who could not have known him well, denied that Newton had a piercing eye, at least during his last twenty years when he was acquainted with him. "Indeed, in the whole air of his face and make, there was nothing of that penetrating sagacity which appears in his composures. He had something rather languid in his look and manner, which did not raise any great expectation in those who did not know him."[247] Thomas Hearne thought much the same. Newton was "of no promising Aspect. He was a short, well set man. He was full of thought, and spoke very little in company, so that his conversation was not agreeable. When he rode in his Coach, one arm would be out of the Coach on one side, and the other on the other."[248] In reporting Humphrey Newton's story that he saw Newton laugh only once, Stukeley commented that his own experience was otherwise, though as he filled in the details it appeared less different than he claimed.

---

[243] *Corres 7*, 349.

[244] John Nichols, *Illustrations of the Literary History of the Eighteenth Century*, 8 vols. (London, 1817–58), *4*, 32.   [245] *Keynes, MS* 130.3, p. 2.   [246] *Keynes MS* 129A, pp. 23–4.

[247] Francis Atterbury, *Epistolary Correspondence*, 5 vols. (London, 1783–90), *1*, 180.

[248] Hearne, *Remarks and Collections, 9*, 294.

According to my own observation, tho' Sir Isaac was of a very serious and compos'd frame of mind, yet I have often seen him laugh, and that upon moderate occasions. . . . He usd a good many sayings, bordering on joke and wit. In company he behavd very agreably; courteous, affable, he was easily made to smile, if not to laugh. . . . He could be very agreable in company, and even sometime talkative.[249]

That is, Stukeley agreed with Hearne and Atterbury but spoke as a friend. Percival, the tenant at Woolsthorpe, told Spence that Newton was a man of very few words; "that he would sometimes be silent and thoughtful for above a quarter of an hour together, and look all the while almost as if he was saying his prayers: but that when he did speak, it was always very much to the purpose."[250] All four remind us more of the Newton of Cambridge days than Conduitt's panegyric does.

Perhaps the most prominent feature in Conduitt's recollection of Newton was studiousness. Newton's age of creativity had ended twenty years before Conduitt first met him. After the move to London, he did nothing but reshuffle ideas and themes from his years in Cambridge. Nevertheless, the pattern of a lifetime remained intact. If he could only reshuffle old ideas, at least he could do that, and a life which found adventure in exploring the seas of thought held true to itself until the end. He gave all of his time not devoted to business and what Conduitt called the civilities of life to study, "& he was hardly ever alone without a pen in his hand & a book before him . . ."[251] Conduitt noticed that his eyes never grew tired from reading. He added that Newton was shortsighted in his age but had not been so as a boy.[252] Stukeley testified to the exact opposite – that he was shortsighted in his youth but not in his age. In November 1726, before the annual meeting of the society, he saw Newton cast up the accounts without spectacles and without pen and ink, a testimony both to his sight and to his power of concentration.[253]

Stukeley also recalled Newton's moderation in his style of living. His breakfast consisted only of bread and butter and a tea made by boiling a bit of orange peel in water which he sweetened with sugar. He partook freely of wine only with dinner, and for the most part drank only water.[254] Whiston spread it about that Newton refused to eat rabbits because they were strangled and black puddings because they were made of blood. Catherine Conduitt told her husband that this was a matter of ethics rather than taste;

[249] *Stukeley,* pp. 57, 68.   [250] Spence, *Anecdotes,* p. 362.
[251] *Keynes MS* 129A, pp. 15–16. Cf. p. 28.   [252] *Keynes MS* 130.10, f. 4ᵛ.
[253] Stukeley to Conduitt, 15 July 1727; *Keynes MS* 136, p. 11.
[254] *Keynes MS* 136, pp. 10–11.

Newton thought that strangling was a cruel way to kill an animal and that eating blood excited men to brutality.[255] Apparently he did regard abstention from cruelty to animals as a moral command almost on a par with love of neighbor.[256]

Stukeley had the good fortune, at the request of both men involved, to attend all the sittings when Sir Godfrey Kneller painted Newton for Varignon in 1720.

> Tho' it was Sir Isaac's temper to say little, yet it was one of Sir Godfrys arts to keep up a perpetual discourse, to preserve the lines and spirit of a face. I was delighted to observe Sir Godfry, who was not famous for sentiments of religion, sifting Sir Isaac to find out his notions on that head, who answered him with his usual modesty and caution.[257]

Stukeley asked Newton to let Kneller do a profile of him; "what says S^r: Isaac, would you make a medal of me? & refus'd it, tho' I was then in highest favor with him."[258] Nevertheless, at some time or other, Stukeley, who was something of an amateur artist, sketched a profile of Newton for himself. For that matter, Conduitt had a medal struck.

Coming from Lincolnshire himself, Stukeley noticed that Newton remained fond of his native county and liked to join in its annual feast in London. On 20 February 1721, Stukeley attended such a feast at the Ship Tavern in Temple Bar. He went to the dining room upstairs, "where the better sort of company was," and found them talking about an old gentleman belowstairs who they thought was Sir Isaac Newton. Stukeley went down at once, found that it was Newton, and took a seat with him. Those upstairs sent word that they should come up to the chief room. The chief room, Stukeley informed them, was where Sir Isaac Newton was, whereupon the upper room was left to the ordinary company and the better sort came down.[259]

As he remained fond of Lincolnshire, so Newton remained concerned with his native village and his estate in it. Less than two years before his death, he wrote to his tenant about a plan to stint the common, so that each property would have a definite number of sheep that were allowed to graze. Nearly sixty-five years after his departure from Woolsthorpe, he remained aware of the details of his estate and determined to protect its rights.[260] His concern for

---

[255] *Keynes MS* 130.7, Sheet 4.   [256] *Chronology*, p. 190.

[257] *Stukeley*, p. 12. Cf. Varignon's account of receiving the portrait. He added that Brook Taylor, who visited him soon after it arrived, said that it was a good likeness (Varignon to Newton, 28 Nov. 1720; *Corres* 7, 104).

[258] Stukeley to Conduitt, 22 July 1727; *Keynes MS* 136.   [259] *Stukeley*, pp. 13–14.

[260] Newton to his tenant Percival, 12 May 1725; *Corres* 7, 317. Cf. an earlier letter to Henry Ingle, 13 Oct. 1712, on the same topic (*Corres* 5, 346–7), and one to John Newton with no date (*Corres* 7, 364).

Figure 15.9. Newton at about seventy-seven. Sketch by William
Stukeley. (Courtesy of the Royal Society.)

Colsterworth extended beyond his estate, however. He contributed
£12 to erect a gallery in the church there, and later £3 to repair the
floor. He even talked about founding a school in Woolsthorpe,
though he never took any definite steps in that direction.[261]

---

[261] Newton to Mason, ca. February 1725, 12 May 1725, 10 May 1726, 4 Feb. 1727; Mason
to Conduitt, 23 March 1727; *Corres 7*, 303, 318, 347, 355, 355–6.

Woolsthorpe was not the only old connection he maintained. Intermittent chance references suggest that Bentley made it a point to see him whenever he came to London, and Newton valued the friendship enough that he commissioned a jeweled watch for Bentley.[262] As agent for the commission, Newton turned to Fatio, whose continued advocacy of jeweled watches appears to have been his one line of communication with Newton. In 1724, in the second of the two letters he is known to have written to Newton after 1693 (both about watches), Fatio sought Newton's permission to use his name in an advertisement for timepieces.[263] There is no indication that Newton replied; if the advertisement appeared, it also disappeared, and quickly. No mention of it exists. Friendships established in the early days at the Mint also continued. John Francis Fauquier served as Newton's deputy and as his private financial agent until his own death not long before Newton's, and Newton signified the value he placed on him by presenting him with a copy of the third edition of the *Principia*.[264] When Thomas Hall, the effective master of the Mint during the recoinage, died in 1718, his will named Newton (together with Hopton Haynes) an executor of his estate.[265]

He had other friends as well. Conduitt informed Fontenelle that George II and his wife (the former Princess Caroline) showed Newton favor and often admitted him to their presence for hours together. The queen liked to hear arguments on questions of philosophy and divinity and sought Newton's company for that purpose. She even claimed to consider his "Abstract" of chronology, which he had written out for her in his own hand, one of her choicest treasures.[266] Such attentions ranked among the hazards that fame entailed. A letter from Pierre Coste, written two days after he had seen the Princess Caroline, does tend to confirm Conduitt's account.[267]

By inference we know something else about Newton during his years of decline, something that harmonizes well with his evident pleasure in being a familiar acquaintance of royalty. He was greatly concerned to leave his image behind him. It is sometimes asserted that Newton did not like to sit for portraits. If that is true, he was a masochist as well, since few if any men of his age were more frequently painted. Not just in his old age, but during the whole of

---

[262] Robert Smith to Newton, 12 Aug. 1720; *Corres 7*, 98. Fatio to Newton, 15 June 1717; *Corres 6*, 391–2.   [263] *Corres 7*, 270.

[264] *A Descriptive Catalogue of the Grace K. Babson Collection of the Works of Sir Isaac Newton* (New York, 1950), p. 12.

[265] *Keynes MS* 148 contains the records of the executors. Cf. Sir Charles Godolphin to Newton, 17 Dec. 1718, and Francis Hall to Newton, 2 Sept. 1719; *Corres 7*, 24–7, 59–60.

[266] *Keynes MS* 129A, pp. 20–1.   [267] Coste to Newton, 16 Aug. 1721; *Corres 7*, 148.

his residence in London, he sat constantly to portraits such that following the Kneller of 1702 (already Kneller's second), no more than four years passed without a new one. During his final decade, portraiture appears to have become almost a mania. After Kneller painted him in 1702, Jervas did so in 1703, Gandy in 1706, and Thornhill twice in 1709–10. In 1714 he sat for a miniature by Richter, and that same year Le Marchand sculpted a bust in ivory. Four years later, in 1718, Le Marchand did a second bust plus a number of reliefs and Murray a portrait. Kneller executed his third portrait (for Varignon) in 1720, and in the three years before his death in 1723 two more (of which only one apparently survives) for Conduitt (Figure 15.10). Vanderbank did two portraits in 1725 and a third in 1726, and Seeman another in 1726. There is testimony to a portrait by Dahl, probably from Newton's final years. Two other portraits of him in old age by unknown artists exist, one in the National Portrait Gallery and one in the possession of W. Heffer and Sons. One or both may be copies; one of them may be the Dahl. Many or perhaps most of these were commissioned by other men, but they could only have been carried out with Newton's cooperation. By any reckoning it is a considerable record; obsession does not seem too strong a word.

Another characteristic does not reduce readily to his concern with his image. Charity supplied a constant background to Newton's final years. He dispensed much of it to various branches of his family, for he was by far the most prosperous man in the clan, the other members of which looked to him. The tendency of descendents of Barnabas Smith and at least one Ayscough to name children Isaac or Newton – Isaac Warner, Newton Smith, Newton Barton, Newton Chapman – speaks eloquently of his position in the family. In the early eighteenth century their troubles outnumbered their joys, and they brought the troubles to rich Sir Isaac. When Thomas Pilkington, the husband of his half-sister, died and left Mary Smith Pilkington a widow like her younger sister Hannah, Newton came to her support, and later he was making regular quarterly payments of £9 to sustain her daughter Mary.[268] He stood surety for a loan of £20 to his sister's son, Thomas Pilkington.[269] According to Conduitt's testimony at the time of Newton's death, he had also loaned money, which was not yet repaid at that time, to John Warner, the husband of Catherine Conduitt's sister. Conduitt

[268] Mary Pilkington to Newton, 26 April 1709, 22 March 1712; and 30 Oct. 1712; *Corres 7,* 473; *Corres 5,* 251–2, 353. Cf. *Stukeley,* pp. 68–9.
[269] Augustine Tampyan to Newton, 30 April 1723; *Corres 7,* 242.

Figure 15.10. Newton at about eighty. Portrait by Kneller. Stukeley stated that Kneller did three portraits of Newton in his old age, one for Varignon in 1720 and two for Conduitt, one in his own hair and one in a wig. The last of these is not known to have survived. This portrait appears to belong nearer to Kneller's death in October 1723 than to the beginning of Newton's acquaintance with Conduitt. (Courtesy of Lord Portsmouth and the Trustees of the Portsmouth Estates.)

did not, however, know the size of the debt involved.[270] There were moments of joy as well, and according to Stukeley Newton savored them.

[270] *Villamil*, p. 61.

He was generally present at the marriages of his relations, when conveniently he could be. He would on those occasions lay aside gravity, be free, pleasant, and unbended. He generally made a present of £100 to the females, and set up the men to trade and business.[271]

A surviving list of expenditures seems to indicate £25 instead of £100 on the marriage of the daughter of his half-brother Benjamin to Carrier Thompson.[272] A letter from Margaret Barton Warner, Catherine's sister, thanking him for a present of wine, suggests a graceful continuing relation with that branch of the family.[273]

When Colonel Robert Barton, the brother of Catherine and Margaret, died in 1711, in the shipwreck that brought Hill's expedition against Quebec to a disastrous end, Newton stepped in. As far as Colonel Barton was concerned, his death does not appear to have grieved anyone. Swift reported that Catherine mourned him only as a matter of form.[274] He left a widow and three children whom Newton could not ignore, however. He drafted the widow's application for a pension, whereby she secured a bounty of £30 in 1712 and two and a half years later a pension of £80 in the military establishment of Ireland.[275] Eventually she married a Colonel Robert Gardner, and a letter from Newton to Gardner indicated that he did not lose sight of her.[276]

The onslaughts of fortune brought a constant stream of destitute relations to beg at his door. Richard Pindar, the grandson of Newton's uncle Richard Newton, needed to borrow fifteen or twenty pounds.[277] For the most part, however, the Newtons seem to have stayed on the land near Colsterworth, often renting fields from Sir Isaac and farming them successfully. It was the offspring of his stepfather who needed assistance, and also the Ayscoughs, who were numerous and penurious. In 1714, Katherine Rastall, the daughter of his uncle William Ayscough, wrote in desperation from Basingthorpe after her landlord had seized her husband's goods. "S^r I humbley disire you that you will be pleased to give the bearer [of the letter] sumthing for me . . . Sr Humbley beging the faver that you will be pleased to Answer this I remain S^r your humble sarvant." Clearly he did respond to her need, for seven years later she sent him "a thousand thanks for ye many civilities y^t I received from You I have been ill & can never get well since, I had ye happiness to bounteously hear from you & how ye Lord designs to

---

[271] *Stukeley*, pp. 68–9.   [272] Printed in *Corres 7*, 368.   [273] *Corres 7*, 383.

[274] Swift, *The Works of Jonathan Swift*, 19 vols. (Edinburgh, 1814), *2*, 375.

[275] Newton to George Greenwood, 9 Oct. 1711, and Newton to Oxford, ca. October 1712; *Corres 5*, 199, 345. *CTB 26 (2)*, 546. *CTB 29 (2)*, 594. *CTB 31 (2)*, 536.

[276] *Corres 7*, 359.   [277] Pindar to Newton, 9 June 1725; *Corres 7*, 325.

deall with me, He alone knows."[278] Trailing her anguish behind her, Katherine Rastall disappeared, but Hannah Tonstall, a granddaughter of William Ayscough, equally in debt and more illiterate, wrote to beg assistance. Whatever else he did, Newton solicited the help of Colonel Gardner in an unsuccessful effort to secure her husband a place among the Poor Knights of Windsor.[279] In 1716, he gave £500 to Ralph Ayscough, and in 1720 loaned £100 to Thomas Ayscough, a debt he later remitted.[280]

There was also another Ayscough, Newton Chapman, the grandson of Sir Isaac's aunt, who solicited his assistance for young Chapman while Newton still lived on Jermyn Street. By 1719, Newton Chapman had grown into a servile young man who fawned over Newton as "ye only patron and benefactor I have in the world to whom I am oblig'd (under God) for ye preservation of my life and my future fortune . . ." The source of Chapman's concern was his failure to keep an appointment with Newton. His excuse showed promise: "I was at Church offering my thanks and praise, for my recovery from my late dangerous indisposition of body." Apparently Newton had to help him over some rocky stretches of the road, but his final appearance in 1725, when he came to London to give an account of himself, sounded hopeful.

> My affairs (I thanke god and your Honours Bounty) are not so perplext and Cumbersome as of late they have been. I have made the Burden lighter and supported my family (my wife and Infant son) comfortably and Reputably by an Indefatigable Incessant Industry not presuming to Attend your Honour, till I had Reduced things to an Easie narrow Compass.[281]

Benjamin Smith, the son of Newton's half-brother and hence Catherine Barton's cousin, presented a different problem. A hellion who brooked no restrictions, Benjamin later earned a modest notoriety as the most profligate clergyman of a profligate age. Indeed it was none other than William Stukeley, Newton's friend who had turned to the ministry, who ordained him, an act Bishop Warburton called a "furious scandal." Stukeley, who had his wits about him, told the bishop he had refused to give Smith a testimonial of character but had granted him only a title of office, which had reference to his livelihood and not to his morals, an explanation

---

[278] *Corres 6*, 183. *Corres 7*, 166–7.   [279] *Corres 7*, 359, 383.
[280] *Sotheby Catalogue*, p. 50. *Babson MS* 426. According to the recollections of a granddaughter of William Ayscough, perhaps the same Hannah Tonstall, Newton was generous to the Ayscoughs, giving £800 to one, £200 to another, £100 to a third, and many other sums (I.H. [James Hutton], "New Anecdotes of Sir Isaac Newton," *Annual Register*, 19 (1776), p. 25).   [281] *Corres 7*, 33–4, 335.

which sufficed to gain Stukeley congratulations from the bishop for quick thinking. But this happened after Newton's death. When Benjamin was still a young man, his conduct was bad enough to arouse the ire of his aged uncle. There is a story, seemingly well based, that Newton wrote him angry letters denouncing his morals in very direct terms. A clergyman who came upon the letters later in the century destroyed them lest their forthright language reflect, not upon Smith, but upon Newton.[282] Hooke and Flamsteed might have recognized their author.

As far as charity was concerned, Newton did not confine his alms to his family. Among his papers are a large number of letters begging assistance. Their very number implies that he was known in circles that cared as a charitable man. The letters contain internal evidence that he answered many appeals. They present heart-rending pictures of destitution and suffering. No doubt they exaggerated their authors' plight, but they serve nevertheless to remind us that charity was the only resource of the unfortunate in an age which restricted public welfare to the rich. The letters could not have been wholly mendacious; Newton usually gave something. Everyone named Newton considered him as a natural resource. James Newton, lately recovered from illness, sought further assistance from one who, he said, had earlier saved his life. George Needham solicited his help for a boy named Robert Newton, whose aunts were cheating him of an inheritance. William Newton, who claimed that his father was named Isaac, thanked him profusely for help in 1716, then wrote eight months later from the Marshalsea, where he was imprisoned for debt, begging for another pound beyond the £3 4s 6d Newton had already given him. "I could not live on the air," he explained in regard to his expenditures, "& my bedroom tho God knows very bad paymt for 6 weeks twenty shillings."

> God knows it is a very dismal thing to perish for hunger. Dear S^r pray pardon my Importunities, my life being in danger such a weakness I have upon me, that my health is much impair'd. If I dye must end my life miserably here God knows, Dear S^r pray let not y^r good intention be now diverted, my liberty may be now had . . .

Newton seems to have met his need; two years later he heard from him again in Whitby, where he had hoped to obtain a post with the Customs, and where he still needed money.[283]

The name Newton was not a necessary condition for an appeal to his charity. John Corker laid a long tale of woe before him in 1717.

[282] John Nichols, *Illustrations of the Literary History of the Eighteenth Century,* 8 vols, (London, 1817–58), *4,* 34.
[283] *Corres 5,* 410. *Corres 6,* 364, 381–2. *Corres 7,* 30–1, 318–19.

S^r

My misfortune is so great which makes me trouble you at this time is that I have been out of Bisness so long and all my mony spent by Resonn that my famaly fell ill when they Came to Town and then my Wife Dying, my Daughter falling ill of the Small Pox and not fitt for servis yeat wherfore I humbly Crave your pardon in Takeing this freedom with you S^r as Letting my Case be known to you and Dow humbly Crave your Asisdance . . . [284]

Four years later, Elizabeth Johnson appealed for help, and not for the first time.

Hono[ure]d Sir

With humble submission i beg leave to trouble You once more being in ye greatest trouble Imagenable. my Son, who has layn Sick these two Years, have Reduced me to ye lowest Extremity Yesterday he Departed, haveing tasted of Y^r Honours Charity [I] beg once more to Consider my Deploarable Condition. [285]

In 1723, continuing a connection that must have stretched back thirty years, Mary or Ann Davies wrote to Newton from Cambridge.

Honoured Sir

I have made bould to troubel your Honour with these few lines to return your Honour thankes for the too ginnes that your Honour whas pleas'd to send ous by the gentelman that whated upon your Honour with the letter Honoured Sir we hope your Honour will pardon our rudeness in not riting before but my Mother and I have bin very bad and that was the caus of our not riting to return your Honour thankes before now[286]

Along with the others above, John Arnold wrote to him at least five times, acknowledging help as he appealed for more; Peter Gardner twice, Isaac Banastre, Daniel Harrison, Edmund Longbridge, Joseph Trevor, and someone named Hickstan at least once. Robert Corbey asked for a chance to thank him personally for favors received.[287] William Howard had a more novel approach to soliciting alms. "The Author of the Poem on the Resurrection to You humbly Dedicated now humbly waits in hopes of Your favourable acceptance of it."[288] These letters survive primarily because Newton used them as scratch paper to draft passages he was working on. Who knows how many others he received–and answered?

Without extensive evidence about the charitable habits of others at the time, it is impossible to assess the meaning of Newton's giving

[284] *Corres* 6, 394. [285] *Corres* 7, 152. [286] *Corres* 7, 243.
[287] *Corres* 7, 373–83. [288] *Corres* 7, 380.

Figure 15.11. Newton at eighty-three. Portrait by Enoch Seeman. The originality of the portrait, which closely resembles Figure 15.7, has been questioned. (Courtesy of the Master and Fellows of Trinity College, Cambridge.)

with assurance. Nevertheless, there is relevant evidence. Conduitt asserted that his charity had no limits; "he used to say they who gave away nothing till they died never gave . . ." He added his opinion that no one in Newton's circumstances ever gave away so much in alms, in the encouragement of learning, and in support of relatives.[289] Of course, Conduitt tended to glorify his every move. As we have seen, however, Stukeley also praised his generosity. There is reason to think that it was well above average. Together with the abiding friendships of his London years, his charity works to soften the image left by the quarrels with Flamsteed and Leibniz. The quarrels were real. So was the charity to unfortunates, as though he hoped to compensate for his own shortcomings.

The extensive charity should not be taken to mean that Newton was unconcerned with his own material condition. When he died, he left a considerable estate, and we must assume that he watched over its accumulation carefully during all the years in London. Among other things, his period in London embraced the time when the speculative fever known as the South Sea Bubble led many to dream of wealth and to taste of disaster. In the eighteenth century the story was told on the stated authority of Catherine Conduitt that Newton also caught fever and lost £20,000 as a result.[290] Villamil has offered an extended rebuttal of the story based on two things: a memorandum addressed to Fauquier on 27 July 1720, when the speculation was at its peak, to subscribe annuities of the value of £650 per annum in South Sea stock in Newton's name, and second, the amount of South Sea stock in Newton's estate at his death.[291] His argument rests on the assumptions that we know enough about Newton's finances to compute his average annual accumulation of wealth from the time he came to London, and that the one memorandum to Fauquier was Newton's only transaction in South Sea stock. The first assumption is highly dubious. The investigation of Newton's papers since Villamil wrote has shown the second one to be false. As early as 1713, well before the Bubble, Newton owned £2,500 of South Sea stock. On 19 April 1720, when the Bubble was swelling, Newton signed a power of attorney to Fauquier to sell £3,000 of his South Sea stock, indicating in the document that it was only part of the stock he held in the company. He also subscribed £1,000 to the company early in June 1720, and in a letter of August 1722 stated that he then owned South Sea stock worth £21,696 7s 4d.[292] It is impossible to know whether or not Newton lost £20,000 when the Bubble burst, as the story has it. Clearly he invested heavily in the company, however, and a loss is

---

[289] *Keynes MS* 179A, p. 19.
[290] William Seward, *Anecdotes of Distinguished Men*, 5th ed., 4 vols. (London, 1804), 2, 295.
[291] *Villamil*, pp. 19–35. [292] *Corres 6*, 27. *Corres 7*, 96, 210, 358.

not impossible. Spence later quoted Lord Radnor's recollection of Newton's comment on the soaring value of South Sea stock when the speculative fever was running high – "that he could not calculate the madness of the people."[293] He may have been reflecting ruefully on himself. According to the story about his loss, he did not like to hear it mentioned.

Whether he participated in the South Sea mania or not, it is clear that Newton, who was never poor, became a wealthy man during his years in London. In 1724, there were forty-eight proprietors of the East India Company who held £10,000 or more of stock. Newton, with £11,000, was one of them.[294] When he died three years later, this stock was no longer part of his estate. The documented changes in his holdings tend to support the story that he was among those who tasted of the Bubble's madness.

During his final years, Newton liked to reminisce over the various topics that had formed the substance of his life. As I remarked earlier, the stories about the apple and the law of gravity all came from his old age. Stukeley engaged him at times on chronology and the prophecies, although Newton never let him glance into the depths of his theological reflections. Conduitt occasionally heard some of his wider-ranging speculations. On 7 March 1725 they had a long conversation, which Conduitt recorded in a memorandum, about circulations in the cosmos. Newton told him his belief that there was a sort of revolution of the heavenly bodies. Light and vapors from the sun gather together to make secondary bodies like the moon, which continue to grow as they gather more matter and become primary planets and ultimately comets, which in turn fall into the sun to replenish its matter. He thought that the great comet of 1680, after five or six more orbits, would fall into the sun, increasing its heat so much that life on earth would cease. Mankind was of recent date, he continued, and there were marks of ruin on the earth which suggested earlier cataclysms like the one he predicted. Conduitt asked how the earth could have been repeopled if life had been destroyed. It required a creator, Newton answered. Why did he not publish his conjectures as Kepler had done? "I do not deal in conjectures," he replied. He picked up the *Principia* and showed Conduitt hints of his belief which he had put in the discussion of comets. Why did he not own it outright? He laughed and said that he had published enough for people to know his meaning.[295]

Not long before his death, Newton looked back over his life and summarized it for some unnamed companion, a magnificient reflec-

[293] Spence, *Anecdotes*, p. 368.
[294] P. G. M. Dickson, *The Financial Revolution in England* (New York, 1967), p. 279.
[295] Keynes MS 130.11.

tion which catches the essence of a life devoted to the pursuit of Truth.

> I don't know what I may seem to the world, but, as to myself, I seem to have been only like a boy playing on the sea shore, and diverting myself in now and then finding a smoother pebble or a prettier shell than ordinary, whilst the great ocean of truth lay all undiscovered before me.[296]

Signs of senescence though never senility began to appear toward the end. Conduitt, who was usually careful not to reveal any hint of decay in Newton, mentioned in his memorandum of 7 March 1725 that Newton's head was clearer and his memory stronger that day than it had been for some time.[297] Pemberton also noted that Newton's memory was much decayed, and Newton himself complained of it to Pearce not long before his death when he was unable to recall the name of an ancient king, though he knew the exact date of an event that happened during his reign.[298] Already the Royal Society had begun to feel the effects of his infirmities. During the last years of Newton's presidency, the meetings ran down seriously. One can sense the absence of the guiding hand which had earlier rescued the meetings from formless vapidity; and if they did not decline again to the level of the 1690s, they did move in that direction. Desaguliers stopped presenting weekly demonstrations; the council cut his remuneration back to £30 per annum but did not succeed in stirring him to renewed action. The meetings tended to center on whatever letters arrived. Newton himself largely ceased to be a positive presence in the sessions and became instead another source of undirected garrulity, which always threatened the intellectual coherence of the society. When Stukeley read an account of a woman sixty-six years old who grew a set of new teeth, Newton recalled a similar case near Cambridge. He regaled the society one day with an account of his bitch, which had lately grown blind with cataracts, and another day he suggested that freshly raised well water, which was considered poor for watering vegetables, might lack the aerial particles present in rivers and ponds exposed to the air. An experiment, which also reminded him of John Mayow's theory that certain particles in the air sustain combustion and life, led Newton to recall a "Direfull Accident" in Grantham thirty years before, when the town clerk and his whole family went to bed in a freshly plastered room in which a pan of charcoal and a lighted candle were both present. The family was found dead the

---

[296] Spence attributed this anecdote to Andrew Michael Ramsay (*Anecdotes*, p. 54).
[297] *Keynes MS* 130.11.
[298] Henry Pemberton, *A View of Sir Isaac Newton's Philosophy* (London, 1728), preface. Pearce to Hunt, 10 Aug. 1754; *Lives, 1,* 434–5.

following morning. Newton surmised that a damp arose from the sulfur of the charcoal, mixed with the lime in the plaster, and consumed the air, and he went on to speculate that some particles in the air were a necessary cause of the heart's motion. He cited experiments with dogs in vivisection to support his speculation and rambled on to tell about "a very remarkable Experiment he made formerly in Trinity Colledge Kitchin at Cambridge upon the heart of an Eel which he Cutt into three peices and observed every One of them Beat at the same Instant & Interval putting Spittle upon any of the Sections had no Effect but a Drop of Viniger utterly Extinguished its Motion."[299] In fact, he had already recited the account of the eel's heart to the society once before. On the whole, Newton entered less and less into the discussions at the meetings. When he did, his comments, like those above, tended to reminisce over the past – interestingly, nearly always a recollection of Cambridge, not of Woolsthorpe.

The council also ran down during his final years. It met infrequently and did little of substance when it did meet. As though the society recognized his decline and need of assistance, they began to elect Conduitt to the council, though he was inactive in the society and attended only one council meeting. It was not to Conduitt and not to Sloane that Newton turned for help but to Martin Folkes. After a serious illness, which would keep him away from the society much of the time for two and a half months, struck early in 1723, he appointed Folkes as his "Deputy or Vice President."[300] Things did not pick up following the appointment, however; and when Newton's death finally cleared the way for a new administration, Sir Hans Sloane, elected to succeed him, imposed a rigorous schedule of meetings on the council as he attempted to revive what he clearly thought was a moribund society. Sloane himself later made a point of retiring before he declined to a similar level of decrepitude.

Affairs at the Mint went much like affairs at the Royal Society, except that the Mint, with an articulated bureaucratic structure and largely routine functions, could better survive a debilitated leadership. Newton never entirely lost his grip on affairs. When the Ordnance tried to seize more of the Mint's domain in 1720, he reacted with the vigor of old.[301] The records available do not indicate the outcome, but one hopes, at least, that the Ordnance got the space, unused by the Mint, that they needed. The following year, when a John Rotherham informed the government that he had a method of

[299] 10 April and 19 June 1718, 24 Dec. 1719, 31 March 1720; *JBC 11,* 234, 252, 421, 473–5.
[300] 17 Jan. 1723; *CM 2,* 337.
[301] The Ordnance to the Mint, 23 Feb. 1720; *Corres 7,* 86–7. Newton to the lords justices, ca. July 1720; *Corres 7,* 94–5. There are at least four drafts of this reply from Newton (*Mint* 19.3, ff. 425–8).

coining which would prevent counterfeiting, Newton was still able to summon up the tartness with which he had learned to greet such proposals. "I take him to be a trifler," Newton replied, "more fit to embroyle the coinage then to mend it."[302] With the likes of Rotherham he could prevail, and no more was heard of his proposal. Similarly, Newton did not modify his views on handling coiners.

I know nothing of Edmund Metcalf convicted at Derby assizes of counterfeiting the coyne [he advised Lord Townshend in 1724]; but since he is very evidently convicted, I am humbly of opinion that its better to let him suffer, then to venture his going on to counterfeit the coin & teach others to do so untill he can be convicted again, For these people very seldom leave off. And its difficult to detect them.[303]

Newton allowed himself to be named as comptroller both of William Wood's mint in Bristol, which coined copper money for Ireland, and of Wood's copper coinage for the West Indies.[304] He was permitted to act by deputy in both cases and appears to have employed one Matthew Barton (of no known relation to Catherine Barton) for both.

Nevertheless, although he did not become wholly inactive, the quantity of Mint papers from the period after he instituted the copper coinage is very small. The Treasury Papers for these years contain very little about the Mint, suggesting that it was functioning smoothly and that officials in the Treasury realized the master had passed the age of useful advice on broader issues. He tended now to collect what orders of reference he did receive and to reply to several on the same day, a practice which implied less frequent attendance at the Mint.[305] When the copper coinage terminated in 1725, Fauquier prepared Newton's accounts, something that would never have happened at an earlier time.[306] Conduitt stated that after 1725 Newton hardly ever went to the Mint and that he himself took over the duties. At least two documents, which appear to show Conduitt as Newton's legal deputy after Fauquier's death in 1726, support his account.[307]

In the last five years, Newton's health began visibly to deteriorate. He was ill during the spring of 1722; DeMoivre mentioned the difficulty he had in seeing him at this time because of his sick-

302 Newton to Delafaye, Oct. 1720; *Corres* 7, 102. Cf. various relevant papers in *Mint* 19.1, ff. 488–94.
303 Newton to Townshend, 25 Aug. 1724; *Corres* 7, 289.   304 *CTP* 1720–28, p. 190.
305 Cf. three miscellaneous items, all on 26 June 1722 (*Corres* 7, 202–3) and replies on 13 Sept. 1723 to two petitions referred on 3 August and 12 September (*Corres* 7, 245–6).
306 Public Record Office, A.O.1/1636, Roll 281.
307 *Keynes MS* 129A, p. 26. *Mint* 3.168, p. 2. Newton to Conduitt, 25 June 1726; *Corres* 7, 349. At the end of 1726, Newton paid Conduitt £200 for "his trouble & Charges" (Huntington Library MS).

ness.[308] Nevertheless, he missed only four meetings of the Royal Society during six months of 1722 before the summer break. In 1723, however, he was seriously ill, and the Royal Society saw him only twice between the New Year and 21 March. He placed himself in the care of Richard Mead and William Cheselden, two of the most prominent physicians in London. At first they thought his problem was the stone, and in 1724 he did in fact void a stone, without serious pain according to Conduitt. His basic problem, perhaps a result of the sickness, turned out to be a weakness of the sphincter. From this time on, Newton suffered from incontinence of urine. Since motion excited the affliction, he gave up his carriage and took to a chair. He quit dining abroad and seldom entertained at home. He stopped eating flesh in any quantity and lived chiefly on broth, vegetables, and soup.[309]

One of Conduitt's anecdotes concerned a visit by Cheselden. As Cheselden prepared to leave, Newton took a handful of guineas from his pocket to give him for a fee. The doctor refused; a guinea or two was the utmost he ought to receive. Newton laughed and said, "why suppose I do give you more than your fee."[310] Apparently he finally accepted the coins. There was no break as there had been when Cheyne refused money twenty years earlier, and Cheselden continued as his physician to the end.

Another illness, though not his own, very nearly complicated Newton's case in 1724. All we know about it is the outcome. On 22 December, Thomas Hill congratulated John Conduitt and his wife on their recovery from alarming sickness.[311]

The following month, January 1725, Newton suffered a violent cough and inflammation of the lungs. An attack of gout further compounded his problems. After 7 January, he did not occupy the chair of the Royal Society again until 22 April, and from that time until his death he missed more meetings than he attended. After much ado, the Conduitts persuaded him to take a house in Kensington in Orbell's Buildings, which stood to the west of Church Lane (now Kensington Church Street), and a little to the north of the present Duke's Lane. Halley wrote to him there in the middle of February.[312] The air proved to be good for him. Conduitt noted

---

[308] DeMoivre to Varignon, ca. May 1722; *Corres 7*, 198. Cf. Newton to Varignon, ca. July 1722; *Corres 7*, 204.

[309] *Keynes MS 129A*, p. 24. However, the estate settled bills of £10 16s 4d with a butcher, £1 14s 0d with a poulterer, and £0 14s 9d with a fishmonger, though only £1 1s 0d with a grocer and £3 1s 9½d with a baker. With the brewer, it is true, there was a bill of £7 10s 0d for fifteen barrels of beer (Huntington Library MS).

[310] *Keynes MSS 130.6*, Book 2; 130.7, Sheet 1.

[311] *Eighth Report of the Historical Manuscripts Commission*, Appendix, p. 61b.

[312] *Keynes MS 129A*, pp. 24–5. *Corres 7*, 302. Constructed at the end of the seventeenth century, Orbell's Buildings were later known as Pitt's Buildings; the two surviving ones were known as Bullingham House and Newton House when they were demolished in

that he was "visibly better" than he had been for some years, an implicit admission of greater decay than Conduitt ever stated openly. James Stirling visited him in Kensington and confirmed his decline. "S Isaac Newton lives a little way of in the country. I go frequently to see him, and find him extremely kind and serviceable in every thing I desire but he is much failed and not able to do as he has done."[313] His spirit never deserted him. Conduitt tried to get him not to walk to church, to which he replied, "use legs & have legs."[314] Conduitt also remarked that he continued to study and write until the very end.

Eighty-two years old, Newton was beyond reasonable expectation of restoration to full health. The visits of the doctors must have been constant. Mead informed Conduitt about one examination in August 1726. "Mr. Cheselden has waited on Sr Isaac and examined into the complaint which made him suspect a Fistula; he finds nothing but a little Relaxation of the inward coat of the Gut, of no consequence; and he has therefore made our old Friend very easy in the matter."[315] Like all doctors, Mead rather exaggerated the confidence the medical profession inspires.

Realizing that the end must be near, Newton began to dispose of his estate. He had always been generous to his relatives. Now he began to make specific gifts, virtually endowments, for many of them who would not share in the inheritance. To his godson, Isaac Warner, the son of Margaret Barton Warner, he gave £100 in the form of rents worth £25 per annum for four years.[316] He purchased a farm for his second cousin Robert Newton and gave land worth £30 per annum to John Newton, the grandson of Sir Isaac's uncle Robert.[317] When Newton died intestate, the son of this man, another John Newton, who was Sir Isaac's heir at law, inherited the estate in Colsterworth. The gift of land to John Newton in 1723 appears to indicate an intention at that time of a will to dispose of the remaining land otherwise, perhaps to support the school he

---

the 1890s to make way for Bullingham Mansions, which presently occupy the site (I owe this information to the great kindness of Miss R. J. Ensing of the Central Library of the Royal Borough of Kensington and Chelsea). In his memorandum of a conversation on 7 March 1725, Conduitt also mentioned that Newton had gout in his eighty-third year (i.e., during 1725) and that he had it that day (*Keynes MS* 130.11). He described the illness and more in his draft of a biography (*Keynes MS* 129A, pp. 24–5). Cf. Newton to Mason, ca. February 1725, and Newton to Colonel Armstrong, n.d.; *Corres 7*, 303, 309. He paid rent of £70 per annum for the house in Kensington but also retained occupancy of the one on St. Martin's Street, in which he installed a maid to look after it. (Huntington Library MS).

[313] Charles Tweedie, *James Stirling* (Oxford, 1922), p. 13.
[314] *Keynes MS* 130.6, Book 2.   [315] *Sotheby Catalogue*, p. 45.
[316] Huntington Library MS. Cf. British Library, *Add MS* 5017*, f. 73 (cited in *Corres 7*, xliv).
[317] Stukeley to Conduitt, 15 July 1727; *Keynes MS* 136, p. 11. Cf. *Corres 7*, 487.

talked about. He gave £100 to an unidentified Ayscough and no less than £500 to wild Ben Smith (who was in addition one of the heirs).[318] For Kitty Conduitt, the daughter of Catherine, he purchased an estate in Kensington worth £4,000, and he paid the same price for another one at Boyden in Berkshire for the three children of Robert Barton and their half sister, Joannah Gardner. On the last purchase, made only ten days before his death, he was cheated. "When he had been imposed upon in buying an estate at Boyden" Conduitt recalled, "& given double the value, & might have vacated the bargain in equity, he said he would not for the sake of £2000 go to West^r Hall to prove he had been made a fool of."[319]

He did one other thing as he prepared for death. He burned a number of papers; at least Conduitt testified to such.[320] When one considers the papers he left behind, multiple drafts and scraps on every topic known to have interested him, even sheets covered with nothing but raw calculations, it is difficult to imagine what he destroyed, and why. Professor Manuel thinks it was correspondence, some of it with his mother. In view of his mother's virtual illiteracy, this is hard to accept, and there is no evidence of a lost body of correspondence. Conduitt's assertion was straightforward, however, though made in passing in a way that did not imply that significant papers were involved. Elsewhere Conduitt mentioned the burning of depositions of coiners; they could have contributed the sum of what Newton destroyed.[321]

After the summer break in 1726, Newton attended only four meetings of the Royal Society and only one meeting of the council. He presided for the last time on 2 March 1727. Appropriately, a letter addressed to Newton and the fellows of the society from the new Academy of Science in St. Petersburg had arrived in time for the meeting. The letter gave a brief account of the founding of the academy and went on to hope that regular communication between the two bodies would be acceptable to the Royal Society, "which they say they are the more Inclined to Make their Address to, and desire most to have the Approbation of, as being the first of the kind and that which gave Rise to all the Rest."[322]

A month earlier, Newton had written his last surviving letter – that too an appropriate one since he addressed it to Thomas Mason,

---

[318] Stukeley to Conduitt, 15 July 1727; *Keynes MS* 136, p. 11. Nichols, *Illustrations of Literary History, 4,* 33. In fact the Reverend William Sheepshanks, who supplied the information in Nichols, said that Newton left Benjamin Smith estates worth £500 per annum. I find this unbelievable.

[319] *Sotheby Catalogue,* p. 50. *Keynes MSS* 129A, p. 29; 130.7, Sheet 1. He also gave £100 to Kitty Conduitt (Huntington Library MS).

[320] *Keynes MS* 129B, pp. 11–12; quoted in *Math 5,* xii–xiii.

[321] *Keynes MS* 130.7, Sheet 2.   [322] *JBC 13,* 53.

rector of the church in his native village of Colsterworth. He reported the negative result of an assay of ore sent him from Woolsthorpe.[323] Although he could not grant his village wealth on his deathbed, he could still serve it.

A few days before his death, Zachary Pearce, rector of Newton's home parish, St. Martins-in-the-Fields, visited him.

> I found him writing over his *Chronology of Ancient Kingdoms,* without the help of spectacles, at the greatest distance of the room from the windows, and with a parcel of books on the table, casting a shade upon the paper. Seeing this, on my entering the room, I said to him, "Sir, you seem to be writing in a place where you cannot so well see." His answer was, "A little light serves me." He then told me that he was preparing his Chronology for the press, and that he had written the greatest part of it over again for that purpose. He read to me two or three sheets of what he had written, (about the middle, I think, of the work) on occasion of some points in Chronology, which had been mentioned in our conversation. I believe, that he continued reading to me, and talking about what he had read, for near an hour, before the dinner was brought up.[324]

The meeting of the Royal Society on 2 March exhilarated Newton. He stayed on in London that night, and the next day Conduitt thought he had not seen him better in many years. But the strain both of the meeting and of the visits he received on the morrow brought his violent cough back, and he returned to Kensington on 4 March. Conduitt sent for Mead and Cheselden, who diagnosed the distemper as a stone in the bladder and offered no hope of recovery. Newton was in great pain. In his anguish sweat ran from his face. The spectacle of Christian dying was more than Stukeley's poetic gift could resist.

> It [the pain] rose to such a height that the bed under him, and the very room, shook with his agonys, to the wonder of those that were present. Such a struggle had his great soul to quit its earthly tabernacle! All this he bore with a most exemplary and remarkable patience, truly philosophical, truly Christian . . .[325]

Conduitt felt obliged to note, not without embarrassment, what was surely Newton's most significant act as he lay dying, though Stukeley would have found it difficult to reconcile with his account. Newton refused to receive the sacrament of the church.[326] He must have been planning the gesture for some time, the personal declaration of belief he had not dared to utter in public for more than fifty years. Even in death it was compromised. He had, after all, spent

[323] *Corres 7,* 355. [324] Pearce to Hunt, 10 Aug. 1754; *Lives 1,* 434–5.
[325] *Stukeley,* pp. 82–3. [326] *Keynes MSS* 130.6, Book 1; 130.7, Sheet 1.

his final fifteen years purging objectionable opinions from the theological works he left behind for publication. Similarly, he made his gesture to the limited audience of Catherine and John Conduitt, who did not care to jeopardize his memory by making it known. As far as one can tell, Stukeley never heard a whisper about it.

On 15 March Newton was somewhat better, and the watchers began to hope. He declined again immediately, however, was insensible on Sunday, 19 March, and died the following morning about one o'clock without further pain.[327]

He may not have given the Royal Society much leadership during his final years, but they were aware of what they had lost.

March 23d. 1726.

The Chair being Vacant by the Death of Sir Isaac Newton there was no Meeting this Day.[328]

Death has its unavoidable practical dimension. The estate was considerable, and the various heirs intended to enjoy it. As soon as the news reached Colsterworth, the next surviving Newton, John, who became the owner of the patrimonial estate, set out for London bearing a letter from the Reverend Thomas Mason to Conduitt. He did not know presumably that Mason described him bluntly as "God knows a poor Representative of so great a man."[329] Stukeley, who claimed to know this "idle fellow" after he moved to Grantham, said that he quickly wasted the whole inheritance, "by cocking, horse racing, drinking and folly."[330] In less than six years he was forced to sell the manor to one Thomas Alcock, having previously mortgaged it to James Cholmeley. Edmund Turnor purchased it from Alcock in 1733 for £1,600.[331] According to the story, John Newton killed himself accidently when he fell down drunk with a pipe in his mouth.

The liquid estate, made up primarily of stock and annuities in the Bank of England and the South Sea Company, amounted to £32,000, not a princely sum but a very substantial one. Eight nieces and nephews of half blood – three Pilkingtons, three Smiths, and two Bartons (including Catherine Barton Conduitt, of course) – shared it. Among them, only Catherine Conduitt, seconded by her husband, appreciated Newton's achievement. To the others he was only rich Uncle Isaac, and they determined to realize as much as they could from their good fortune. Most of them came to town for the funeral and immediately fell to bickering. Conduitt was the only man of the world in the group, and he tried initially simply to take over the administration of the estate. The others became suspi-

---

[327] *Keynes MS* 130.15.   [328] *JBC 13*, 62.   [329] *Corres 7*, 355–6.
[330] *Stukeley*, p. 36.   [331] Turnor Papers, Lincolnshire Archives.

cious. On 1 April, four of them sent him a letter proposing an immediate inventory of Newton's effects, for which he would have to give them access to the house. They asked if he would join the rest in granting the administration to three of them "or whether you resolve to contest it . . ." They suggested further that the manuscripts be sealed up until all the rest was settled.[332]

The manuscripts became the focus of controversy. In an account of the matter, Conduitt mentioned "several contests & disputes in the Prerogative court about the Administration & the disposal of the Manuscripts . . ." As far as the stocks, annuities, and cash (which ultimately constituted about 95 percent of the estate) were concerned, there was no problem, beyond the desire of the other nephews and nieces for an immediate distribution. The library was an item of possible value. The inventory eventually placed a value of £270 on its 1,896 books plus a hundredweight of pamphlets and waste books. They found a buyer for them in July: John Huggins, warden of Fleet Prison, who was willing to put down £300 to purchase the library for his son, Charles, the rector of Chinnor near Oxford.[333] The manuscripts were the unknown quantity. The heirs were aware that Newton's name would sell almost anything, and they were anxious to capitalize on whatever he had left. One gains the impression that the others, except the Conduitts, greatly overestimated their potential value. The manuscripts were also the focus of Conduitt's concern, not as items of possible profit but as records of the great man whose memory he intended to preserve. As Catherine Conduitt would later say, with reference to the theological manuscripts, he wanted to insure that "the labour and sincere search of so good a Xtian and so great a Genius, may not be lost to the world . . ."[334]

Conduitt had certain advantages. Before any proceeds of the estate could be distributed, Newton's final account would have to be passed and his final coinage seen through a trial of the pyx. As master, Newton drew the entire income from the coinage duty each year. Since 1720, the income had exceeded expenses substantially; according to Conduitt, the balance Newton held stood at £34,330, a debt owed by the estate to the Crown. The eagerness of the heirs strongly implies that this sum was separate from the £32,000 at

---

[332] Most of the documents relating to the estate are in *Keynes MS* 127A. The accounts of the executors are in the Huntington Library MS. A draft of the agreement among the heirs is printed in the *Eighth Report of the Historical Manuscripts Commission*, Appendix, p. 63b. The inventory of the estate is printed in *Villamil*, pp. 49–61; Cf. Whiteside's account of the settlement in *Math 1*, xvii–xxi, and see also *Villamil*, p. 3, 54.

[333] *Villamil*, pp. 3, 54. On 6 May the estate had paid £21 for an appraisal of the library plus £6 6s for a catalogue and two additional copies (Huntington Library MS).

[334] Bodleian Library, New College MS 361.4, f. 139.

which the inventory valued his estate, though there is no evidence of other assets. Stukeley's account implied the same. In any event, before the courts would allow the estate to be distributed, someone would have to assume the obligation of the debt, assuring the Crown against loss. Among the heirs, only Conduitt was in a position to shoulder such a burden. Hence a bargain was struck, embodied in a court order on 3 April and confirmed by a further agreement on 27 April when seven of the eight were present. Conduitt backed off from his effort to control the administration of the estate, and a set of three administrators was appointed: Thomas Pilkington, Benjamin Smith, and Catherine Conduitt. All parties would peruse the papers, and Dr. Thomas Pellet, who was a member of the Royal Society, would also examine them. All of the manuscripts deemed worthy of publication would be printed and sold "to the best advantage." In the meantime, Catherine Conduitt would keep them. Conduitt agreed to pass the accounts or to get a release within a month of the time the administrators were appointed and to oversee the trial of the pyx. In return, he would have possession of all the papers not found worthy of publication, on condition that he post a bond (eventually set at £2,000) that if he ever should "publish any or make any advantage thereof" he would be accountable to the heirs. They agreed also, no doubt at Conduitt's prodding, to set aside £500 from the estate for a monument. They spent £87 on one hundred and one funeral rings, sent £20 to the poor in Colsterworth parish, and gave every servant a year's wages.[335]

On 18 April, the Prerogative Court of Canterbury ordered an inventory of the estate, and on 5 May, when it was submitted, appointed the three administrators. By 17 May, Conduitt had obtained a warrant from the lords commissioners of the Treasury directing the auditor to give an allowance to the administrators of the estate for Newton's balance as soon as Conduitt gave in a receipt for it. He did so at once, freeing the estate from any claim by the Crown. All the heirs dined together at Doctors Commons on 18 May, presumably to celebrate.[336] It took Conduitt nearly two more years fully to clear the accounts through the six offices that had to sign them. Immediately after he gave in the receipt, however, the administrators distributed the liquid assets and called in Dr. Pellet, who spent three days (20, 21, and 26 May) judging manuscripts to which an army of scholars from a less heroic age have devoted decades. His bold "Not fit to be printed" must have been a blow to the hopes of some of the heirs. Indeed, he found only one manuscript unquestionably suitable for publication, the

[335] Huntington Library, MS.   [336] *Ibid.*

Chronology, and two booksellers snapped it up at once for £350. He did think four others worthy of reconsideration. Two of these, Newton's original draft of the final book of the *Principia* and his *Observations upon the Prophecies,* did in fact see print, the first under the title *De mundi systemate* (with a return to the estate of £31 10s) in 1728, the second (with no recorded price) in 1733. Conduitt posted the bond as soon as Pellet was done, thus obtaining effective ownership of the papers.

Conduitt was adroit enough to secure much the richest inheritance, the mastership of the Mint, for himself. In 1730, Whiston asserted that the Crown offered the post to Samuel Clarke, who consulted with friends and finally rejected it. He also mentioned rumors that Conduitt had bought Clarke out, though he did not believe them.[337] His story has generally been accepted, and it may be true. It is worth noting, however, that Georgian governments, which were not excessively sentimental, generally reserved such plums for supporters in Parliament, a qualification Clarke did not possess but Conduitt did. Moreover, Newton had been actively easing him into the post for at least two years. It may also be relevant that Conduitt's warrant for the mastership was dated 31 March.[338] If the Crown offered the position first to Clarke, the entire transaction took a remarkably short period. Conduitt could be confident in the obligations he undertook for the heirs early in April, since he was directly involved with the accounts in any case.

Stukeley asserted that each of the eight nieces and nephews inherited about £3,500 (which understated it by nearly £500), "but all soon found a period."[339] Little is known about seven of them, but Stukeley's summary judgment does not suffice for Catherine Conduitt. She died in 1739, two years after her husband John. Their daughter Catherine married the Hon. John Wallop, Viscount Lymington, in 1740, and their son became the second Earl of Portsmouth. Through Catherine Conduitt, the daughter, Newton's papers, preserved by her father's farsightedness, passed into the possession of the Portsmouth family and, most of them, eventually into the Cambridge University Library.

Newton's death scarcely passed unnoticed. The various gazettes carried announcements of it. *The Political State of Great Britain* for March devoted three pages to an encomium which adequately summarized the position Newton held in English learning by proclaiming him "the greatest of Philosophers, and the Glory of the British Nation."[340] James Thomson's "Poem Sacred to the Memory of Sir Isaac Newton" reached a fifth edition before the year was out. The

[337] Whiston, *Clarke,* pp. 135–6.
[338] Public Record Office, T52/27, ff. 409–10.   [339] *Stukeley,* p. 84.
[340] *The Political State of Great Britain,* March 1727, p. 328.

nation that had honored him with knighthood honored him more lavishly in death. On 28 March he lay in state in the Jerusalem Chamber in Westminster Abbey, whence he was interred in a prominent position in the nave. Conduitt informed Fontenelle that the dean and chapter had often refused that place to the greatest noblemen.[341] According to the proprieties of the day, a Knight of the Bath named Sir Michael Newton performed as chief mourner followed by some other relations, as they were called, and eminent persons acquainted with Newton. The Lord Chancellor, the dukes of Montrose and Roxburgh, and the earls of Pembroke, Sussex, and Macclesfield, all of them members of the Royal Society, supported the pall; the Bishop of Rochester attended by the prebends and choir performed the office. The monument specified by the heirs was finally erected in 1731, a baroque monstrosity with cherubs holding emblems of Newton's discoveries, Newton himself in a reclining posture, and a female figure representing Astronomy, the Queen of the Sciences, sitting and weeping on a globe that surmounts the whole. Twentieth-century taste runs along simpler lines, and the monument is now roped off in the Abbey so that one can scarcely even see it. A similar tone marked the inscription, which concluded with the exhortation, "Let Mortals rejoice That there has existed such and so great an Ornament to the Human Race." In this case, baroque extravagance struck the proper note. Faults Newton had in abundance. Nevertheless, only hyperbole can hope to express the reality of the man who returned to dust in the early spring of 1727.

[341] *Keynes MS* 129A, p. 29.

Bibliographical essay

SINCE Peter and Ruth Wallis have published an exhaustive and thoroughly professional bibliography, *Newton and Newtoniana, 1672–1975* (Folkestone, 1977), less than two years ago, there is no point in my repeating, in a less exhaustive form, what they have done. I intend rather in this essay to offer a brief guide to the location of Newton manuscripts and, less briefly, an indication of what appears to me to be the best literature about Newton.

I shall begin by calling attention to a number of articles on what were the most important recent studies of Newton at the time the respective articles were written at various points during the past two decades. I. Bernard Cohen, whose name will appear frequently below, published the first of these: "Newton and Recent Scholarship," *Isis, 51* (1960), 489–514. The same year Clelia Pighetti published "Cinquant'anni di studi newtoniani (1908–1959)," *Rivista critica di storia della filosofia, 15* (1960), 181–203, 295–318, a chronological list of works with brief descriptions of many of them. D.T. Whiteside (another name which will appear frequently) followed them with "The Expanding World of Newtonian Research," *History of Science, 1* (1962), 16–29. More recently I added a further article in this genre to the group: "The Changing World of the Newtonian Industry," *Journal of the History of Ideas, 37* (1976), 175–84, and earlier a more restricted one, "Newton's Optics: The Present State of Research," *Isis, 57* (1966), 102–7.

Although the Wallis volume includes, as its name promises, a bibliography of Newton's own publications, they can also be found in *A Descriptive Catalogue of the Grace K. Babson Collection of the Works of Sir Isaac Newton* (New York, 1950; supplement, Babson Park, Mass., 1955).

Although Newton's papers are now widely scattered throughout the world, five major collections contain the overwhelming bulk of them. Much the most important of these, not least because it contains the great majority of the papers that Newton devoted to mathematics, mechanics, and optics, are the Portsmouth Papers in the Cambridge University Library. There is a published *Catalogue of the Portsmouth Collection* (Cambridge, 1888), which does not in fact give a very good indication of their richness. The Portsmouth family gave the papers to the university library in the late nineteenth century. The manuscripts they retained at that time were scattered widely after they were sold at auction in 1936; three of the major collections resulted from that sale. The Keynes MSS, King's College, Cambridge, contain a considerable variety, many letters, John Conduitt's collection of materials about Newton, a few theological manuscripts, Newton's papers relevant to the crisis in the university in 1687, and the largest single group of alchemical manuscripts now to be found in one place. The greatest number of Newton's theological papers reside now in Yahuda MS Var. 1, The Jewish National and University Library, Jerusalem. Nearly all of Newton's own papers from his years at the Mint returned to the Mint and went from there to the Public Record Office (Mint Papers, 19.1–5). The fifth important collection was already separated from the rest of Newton's papers in the eighteenth century. Concerned mostly with Newton's chronological and associated theological work, it reposes in the Bodleian Library, Oxford (New College MSS 361. 1–4).

Four lesser collections follow the five great ones. The Royal Society has a number of papers (never associated with the Portsmouth family), mostly concerned with the *Principia,* plus quite a few letters. They also hold, of course, their own journals and records from the more than fifty years of Newton's membership, including over twenty years as president. The Trinity College Library has a number of miscellaneous papers and letters, including an important notebook with accounts from Newton's student years, in addition to vast institutional records from the forty years when he was a member of the college. In the United States, Babson College, Massachusetts, includes within its Grace K. Babson Collection a number of papers of several sorts, all stemming from the auction, as does the Dibner Collection in the Smithsonian Institution Libraries.

The remaining papers are dispersed among a very large number of locations. In Great Britain, the most important are the papers in the possession of the Earl of Macclesfield (separated from the main collection by William Jones before Newton's death and not now available to the public) and a notebook from Newton's undergraduate days in the Fitzwilliam Museum, Cambridge. Lesser papers, usually single items from the auction, are in the possession of the Museum Collection of the Bank of London; the British Library; the Clifton College Library, Bristol; the Iron and Steel Institute Library, London; G.L. Keynes; the University of Keele; Lord Lymington; Magdalen College, Oxford; and St. Andrews University.

Elsewhere in Europe, M. Emmanuel Fabius of Paris purchased thirteen lots at the auction; they are not available to the public. The Martin Bodmer Library, Geneva, possesses a long theological manuscript that in the auction was part of a lot the rest of which, now in the Yahuda MSS, consists of miscellaneous scraps. It is possible, however, that the Bodmer manuscript is a connected history of the church, which would be important. Unfortunately, the Bodmer Library, in contrast to every other library with Newton material, chooses to withhold its possession from scholarly use. The Soviet Academy of Science also has a paper from the auction, which is more accessible than the Bodmer manuscript.

In the United States there are several secondary collections of manuscripts. Eleven manuscripts of various sorts are in the Joseph Halle Schaffner Collection in the Library of the University of Chicago. The Pierpont Morgan Library, New York, has an early notebook. Stanford University Library has several items as does the Humanities Research Center of the University of Texas. The Yale University Library and the Yale Medical Library have several alchemical manuscripts plus some letters. In addition there are manuscripts in the American Philosophical Society, Philadelphia; the Clark Memorial Library, Los Angeles; the Columbia

University Library; the Countway Medical Library, Boston; the Historical Society of Pennsylvania, Philadelphia; the Huntington Library, San Marino, California; the Lehigh University Library, Bethlehem, Pennsylvania; the library of the Massachusetts Institute of Technology; the library of the Mount Wilson Observatory, Pasadena, California; the New York Public Library; and the Princeton University Library.

The *Catalogue of the Newton Papers Sold by Order of the Viscount Lymington* (London, 1936), published by Sotheby and Co., gives a good indication of the papers sold and dispersed. Mr. Peter Spargo has been collecting information on the present location of the papers, which he intends soon to publish with a reissue of the catalogue.

With the activity in Newtonian publication of late, a considerable number of his manuscripts are now available in print. References to these volumes are spread through the pages of this book, and full bibliographical citations appear in the list of abbreviations at the beginning.

Newton has been the subject of many biographers. The reigning biography is still David Brewster, *Memoirs of the Life, Writings, and Discoveries of Sir Isaac Newton*, 2 vols. (Edinburgh, 1855). Augustus De Morgan published a number of pieces criticizing Brewster's excessive hero worship. Three of them were collected together and edited by Philip E. B. Jourdain in *Essays on the Life and Work of Newton* (Chicago, 1914). De Morgan also published *Newton: His Friend: and His Niece* (London, 1885), an examination of the relation of Halifax and Catherine Barton and its bearing on Newton, in which he demonstrated that he was not as free from hero worship as he liked to think. Louis Trenchard More, *Isaac Newton: A Biography* (New York, 1934), probably the biography most readily available at this time, failed to add anything new and also failed in its attempt to supplant Brewster as the leading biography. J. W. N. Sullivan, *Isaac Newton, 1642–1727* (London, 1938), a shorter work, contains general interpretive ideas that continue to merit attention. Among the large number of popular biographies, unquestionably the best is that by E. N. da C. Andrade which appeared in two slightly different versions: *Isaac Newton* (London, 1950) and *Sir Isaac Newton* (London, 1954). And as introduction to a reprint of Brewster's *Memoirs* (New York, 1965), I composed a detailed historical essay on the biographies, beginning with Fontenelle's *Éloge*.

No survey of biographies of Newton can omit Frank E. Manuel's *Portrait of Isaac Newton* (Cambridge, Mass., 1968). Since it does not attempt to deal with Newton's scientific career, it is not a biography in the full sense of the word. On every other aspect of Newton's life it offers insights no student of Newton can afford to ignore. It also offers a Freudian analysis of the roots of Newton's character which may or may not be true but can be separated from the portrait of Newton, on which it is empirically based.

On Newton's family and boyhood, the biographies generally offer the most information. In addition to them, consult the authoritative genealogical study, C. W. Foster, "Sir Isaac Newton's Family," *Reports and Papers of the Architectural Societies of the County of Lincoln, County of York, Archdeaconries of Northampton and Oakham, and County of Leicester, 39*, Part I

(1928), 1–62. On Newton's student days, see in addition to the biographies D.T. Whiteside, "Newton's Marvellous Year: 1666 and All That," *Notes and Records of the Royal Society, 21,* (1966), 32–41.

There are many studies of Newton's work and thought in general. I chose to begin with Ferdinand Rosenberger, *Isaac Newton und seine physikalischen Principien* (Leipzig, 1895), a book now approaching the age of one hundred but in my opinion not yet surpassed. Not long after Rosenberger, Léon Bloch published *La philosophie de Newton* (Paris, 1908), another work with many insights but marred by the author's determination to force Newton into his own positivistic mold. E. A. Burtt, *The Metaphysical Foundations of Modern Physical Science* (New York, 1925), concludes with a long discussion of Newton which, along with the book as a whole, deservedly continues to exercise influence. I. Bernard Cohen, the dean of practicing Newtonian scholars, effectively began the Newtonian aspect of his career with his important *Franklin and Newton, an Inquiry into Speculative Newtonian Experimental Science and Franklin's Work in Electricity as an Example Thereof* (Philadelphia, 1956). As the title indicates, the primary focus of Cohen's work falls beyond Newton; nevertheless, it begins with a long discussion of him as the basis of the rest. I cannot, in a bibliographical essay of this sort, begin to list all of Cohen's contributions to the understanding of Newton. Many of them will appear in later parts of the essay, especially that devoted to mechanics and the *Principia.* Here let me add *Isaac Newton. The Creative Scientific Mind at Work,* the Wiles Lectures for 1966, and, on a popular plane, *The Birth of a New Physics* (Garden City, N. Y., 1960). Alexandre Koyré, who did so much to shape the modern discipline of the history of science, turned his attention to Newton in his late years. See especially his *Newtonian Studies* (Cambridge, Mass., 1965). One of the *Newtonian Studies* deserves separate mention: "The Significance of the Newtonian Synthesis," originally published in *Archives internationales d'histoire des sciences, 11* (1950), 291–311, and since then in other places as well. See also the parts devoted to Newton in Koyré's *From the Closed World to the Infinite Universe* (Baltimore, 1965), and an article published with Cohen, "The Case of the Missing *Tanquam,*" *Isis, 52* (1961), 555–66. Henry Guerlac has also contributed to our understanding of Newton during the later part of his career. Under this category see his *Newton et Epicure,* Conférence donnée au Palais de la Découverte le 2 Mars 1963 (Paris, 1963), and "Newton's Optical Aether," *Notes and Records of the Royal Society, 22* (1967), 45–57. In addition to A. R. and Marie Boas Hall's "Newton's Theory of Matter," *Isis, 51* (1960), 131–44, see their introductions in *Unpublished Scientific Papers of Isaac Newton* (Cambridge, 1962), and also A. R. Hall's introductions (with Laura Tilling) to the final three volumes of the *Correspondence* and the chapters on Newton in his *Scientific Revolution* (London, 1954) and *From Galileo to Newton* (London, 1963).

To the extent that they relied on the published works of Newton, all of the studies done before 1945 have been somewhat superseded by more recent ones (such as most of those in the paragraph above), which are based primarily on his manuscripts. One of the earliest of these, and one that broke decisively with the established pattern of apotheosizing Newton, was Lord Keynes's essay drawn from his own collection of papers:

"Newton, the Man," in the Royal Society's volume, *Newton Tercentenary Celebrations* (Cambridge, 1947), pp. 27–34. During the last few years, J. E. McGuire has utilized the manuscript sources extensively in a number of influential articles on Newton which have emphasized the influence of the Cambridge Platonists on him and have also examined deeply various philosophical issues involved in his thought: "Atoms and the 'Analogy of Nature': Newton's Rules of Philosophizing," *Studies in History and Philosophy of Science, 1* (1970), 3–58; "Body and Void in Newton's De Mundi Systemate: Some New Sources," *Archive for History of Exact Sciences, 3* (1966), 206–48; "Existence, Actuality and Necessity: Newton on Space and Time," *Annals of Science, 35* (1978), 463–508; "Force, Active Principles, and Newton's Invisible Realm," *Ambix, 15* (1968), 154–208; "Neoplatonism and Active Principles: Newton and the *Corpus Hermeticum*," in Robert S. Westman and J. E. McGuire, *Hermeticism and the Scientific Revolution* (Los Angeles, 1977); "Newton's 'Principles of Philosophy': An Intended Preface for the 1704 *Opticks* and a Related Draft Fragment," *British Journal for the History of Science, 5* (1970), 178–86; and (with P. M. Rattansi) "Newton and the 'Pipes of Pan'," *Notes and Records of the Royal Society, 21* (1966), 108–43. Analogous to McGuire in its use of Newton's manuscripts as a means of examining philosophical questions in his science is Ernan McMullin, *Matter and Activity in Newton* (Notre Dame, Ind., 1977).

Augusto Guzzo has produced an extended study of the major aspects of Newton's scientific thought in a series of articles: "Newton, Leibniz e l'analysi infinitesimale," "Meccanica e cosmologia newtoniane," "Ottica e atomistica newtoniane," *Filosofia, 5* (1954), 13–36, 229–66, 383–419. Another influential article is David Kubrin, "Newton and the Cyclical Cosmos: Providence and the Mechanical Philosophy," *Journal of the History of Ideas, 28* (1967), 325–46.

Among the older studies from the period before the intensive exploitation of Newton's manuscripts are at least four others that require mention: C. D. Broad, *Sir Isaac Newton* (London, 1927); A.J. Snow, *Matter & Gravity in Newton's Physical Philosophy* (London, 1926); E. W. Strong, "Newton's 'Mathematical Way'," *Journal of the History of Ideas, 12* (1951), 90–110; and Ernst Cassirer, "Newton and Leibniz," *Philosophical Review, 52* (1943), 366–91.

At the present rate of research on Newton, developments are rapid. The most recent, though already aging, summary is the collection of essays covering every aspect of his career delivered at a conference at the University of Texas to commemorate Newton's early years of discovery and collected by Robert Palter in *The 'Annus Mirabilis' of Sir Isaac Newton* (Cambridge, Mass., 1970). I do not list any of the articles in this volume separately.

One other book that defies categorization but cannot be omitted fits best under the heading of general works: John Harrison, *The Library of Isaac Newton* (Cambridge, 1978). A catalogue of books known to have been in Newton's library, with full bibliographical details plus an indication of notes that Newton entered and much other useful information, Harrison's volume is invaluable.

D. T. Whiteside has established himself as the unquestioned authority on Newton's mathematics. He should be consulted first of all in the introductory essays and editorial apparatus of his great edition of the *Mathematical Papers,* and in the introduction to a reprint edition of the six mathematical papers published during Newton's life or shortly thereafter: *The Mathematical Works of Isaac Newton,* 2 vols. (New York, 1964). Beyond these, he has published a number of important works and articles: "Patterns of Mathematical Thought in the Later Seventeenth Century," *Archive for History of Exact Sciences, 1* (1961), 179–388; "Newton's Mathematical Method," *Bulletin of the Institute of Mathematics and its Applications, 8* (1972), 173–8; "Isaac Newton: Birth of a Mathematician," *Notes and Records of the Royal Society, 19* (1964), 53–62; "Newton's Discovery of the General Binomial Theorem," *Mathematical Gazette, 45* (1961), 175–80. An older work, which is also more accessible to the reader who is not expert in mathematics, is H. W. Turnbull, *The Mathematical Discoveries of Newton* (London, 1945). One of the masters of the history of mathematics, Joseph E. Hofmann, examined Newton's mathematics up to 1676: "Studien zur Vorgeschichte des Prioritätstreites zwischen Leibniz und Newton um die Entdeckung der höheren Analysis," *Abhandlungen der Preussischen Akademie der Wissenschaften,* Mathematisch-naturwissenschaftliche Klasse, 1943. Carl B. Boyer, *The Concepts of the Calculus: A Critical and Historical Discussion of the Derivative and the Integral* (New York, 1939), also contains an excellent discussion of Newton.

The best narratives of the priority dispute with Leibniz are in the biographies. J. O. Fleckenstein, *Der Prioritätstreit zwischen Leibniz und Newton* (Basel and Stuttgart, 1956) contains a very brief account. The *Correspondence,* especially volume 6, contains very full information on the controversy, to the extent that it constitutes the best treatment of it. A. R. Hall, who oversaw the compilation and editing of volume 6, has published (too recently for me to have used) a monograph, *Philosophers at War: The Quarrel between Newton and Leibniz* (Cambridge, 1980), which complements the *Correspondence* and will surely become the authoritative treatment of the episode. Volume 8 of the *Mathematical Papers* will also devote itself primarily to the dispute.

Among early studies of Newton's optics, see Léon Rosenfeld, "La Théorie des couleurs de Newton et ses adversaires," *Isis, 9* (1927), 44–65, and Michael Roberts and E. R. Thomas, *Newton and the Origin of Colours: A Study of One of the Earliest Examples of Scientific Method* (London, 1934). A. R. Hall effectively inaugurated the recent heavy exploitation of the Newtonian manuscripts with his article, "Sir Isaac Newton's Note-book, 1661–65," *Cambridge Historical Journal, 9* (1948), 239–50, which he followed with "Further Optical Experiments of Isaac Newton," *Annals of Science, 11* (1955), 27–43. A. I. Sabra, the authority on optics in the seventeenth century, is the author of two articles: "Explanations of Optical Reflection and Refraction: Ibn-al-Haytham, Descartes, Newton," *Proceedings of the Tenth International Congress of the History of Science* (Paris, 1964), *1,* 551–4, and "Newton and the 'Bigness' of Vibrations," *Isis, 54* (1963), 267–8. There is also an extensive passage on Newton in his *Theories of Light from Descartes to Newton* (London, 1967). Thomas S. Kuhn's introductory essay to "Newton's Optical Papers," in I. B. Cohen's volume of *Papers & Letters* is excellent. Cohen, "Versions of Isaac Newton's First Published Paper," *Archives internationales d'histoire des sciences, 11* (1958), 357–75, contains important material. N. R. Hanson, "Waves, Particles, and Newton's 'Fits'," *Journal of the History of Ideas, 21* (1960), 370–91, discusses Newton's theory in relation to the existing state of knowledge.

Three scholars still active have produced a number of recent significant articles on Newton's optics: J. A. Lohne, "Isaac Newton: The Rise of a Scientist, 1661–1671," *Notes and Records of the Royal Society, 20* (1965), 125–39; "Experimentum crucis," *ibid., 23* (1968), 169–99; "Newton's 'Proof' of the Sine Law and his Mathematical Principles of Colors," *Archive for History of Exact Sciences, 1* (1961), 389–405; and "Newton's Table of Refractive Powers: Origins, Accuracy, and Influence," *Sudhoffs Archiv, 61* (1977), 229–47. Alan E. Shapiro, "Newton's Definition of a Light Ray and the Diffusion Theories of Chromatic Dispersion," *Isis, 66* (1975), 194–210; "Light, Pressure, and Rectilinear Propagation: Descartes' Celestial Optics and Newton's Hydrostatics," *Studies in History and Philosophy of Science, 5* (1974), 239–96; and "The Evolving Structure of Newton's Theory of White Light and Color: 1670–1704," forthcoming in *Isis*. Shapiro has embarked on an edition of Newton's optical manuscripts. Zev Bechler, "Newton's Search for a Mechanistic Model of Colour Dispersion: A Suggested Interpretation," *Archive for History of Exact Sciences, 11* (1973), 1–37; " 'A less agreeable matter': The Disagreeable Case of Newton and Achromatic Refraction," *British Journal for the History of Science, 8* (1975), 101–26; and "Newton's 1672 Optical Controversies: a Study in the Grammar of Scientific Dissent," in Y. Elkana, ed., *The Interaction between Science and Philosophy* (Atlantic Highlands, N.J., 1974), pp. 115–42. Maurizio Mamiani has recently published a book on Newton's early work, *Isaac Newton, filosofo della natura. Le Lezioni giovanili di ottica e la genesi del metodo newtoniano* (Florence, 1976), which I have seen praised highly but have not been able to put my hands on.

The *Principia* and related questions in mechanics have been the subject of historical study for longer than the optics. In the nineteenth century, two important studies published basic documents plus interpretative essays: S. P. Rigaud, *Historical Essay on the First Publication of Sir Isaac Newton's Principia* (Oxford, 1838), and W. W. Rouse Ball, *An Essay on Newton's 'Principia'* (London, 1893). I. Bernard Cohen has made the *Principia* his special province in works too extensive to be cited here in entirety. His *Introduction to Newton's 'Principia'* (Cambridge, 1971), is an invaluable history of the book itself. Among his many articles, see especially "Newton's Theory versus Kepler's Theory and Galileo's Theory," in Elkana, ed., *Interaction*, pp. 299–338; " 'Quantum in se est': Newton's Concept of Inertia in Relation to Descartes and Lucretius," *Notes and Records of the Royal Society, 19* (1964), 131–55; "The Concept and Definition of Mass and Inertia as a Key to the Science of Motion: Galileo-Newton-Einstein," *Colloquium: Principle Stages in the Evolution of Classical Dynamics (17th–20th Centuries)*, XIII International Congress of the History of Science; and "Galileo and Newton," in *Saggi su Galileo Galilei* (Florence, 1972). In *Isaac*

Newton's 'Theory of the Moon's Motion' (Folkestone, 1975), he has published a number of texts, which stem ultimately from the 1690s, relevant to that subject.

John Herivel is the leading student of the early development of Newton's mechanics. His articles on this subject are effectively embodied in the various essays included in *The Background to Newton's Principia* (Oxford, 1965), which also publishes all of the documents on Newton's mechanics before the *Principia*. With his command of Newton's mathematics, D. T. Whiteside has also had important things to say about the *Principia* (see "The Mathematical Principles Underlying Newton's *Principia Mathematica*," *Journal for the History of Astronomy, 1* [1970], 116–38, which was also published separately under that title [Glasgow, 1970]: "Newton's Early Thoughts on Planetary Motion: a Fresh Look," *British Journal for the History of Science, 2* [1964], 117–37: "Before the *Principia*: The Maturing of Newton's Thoughts on Dynamical Astronomy, 1664–84," *Journal for the History of Astronomy, 1* [1970], 5–19; and "Newton's Lunar Theory: From High Hope to Disenchantment," *Vistas in Astronomy, 19* [1975–6], 317–28).

In addition to his general work on Newton, Alexandre Koyré also had things to say that were relevant directly to the *Principia*: "A Documentary History of the Problem of Fall from Kepler to Newton," *Transactions of the American Philosophical Society*, n. s., *45* (1955), 329–95; "La gravitation universelle de Kepler á Newton," *Archives internationales d'histoire des sciences, 4* (1951), 638–53; and "An Unpublished Letter of Robert Hooke to Isaac Newton," *Isis, 43* (1952), 312–37.

Among the many other items under this heading, the following have seemed the most important to me: W. H. Macauley, "Newton's Theory of Kinetics," *Bulletin of the American Mathematical Society, 3* (1896–7), 363–71; Brian D. Ellis, "The Origin and Nature of Newton's Laws of Motion," in Robert G. Colodny, ed., *Beyond the Edge of Certainty* (Englewood Cliffs, N.J., 1965), pp. 29–68; Alan Gabbey, "Force and Inertia in Seventeenth-Century Dynamics," *Studies in History and Philosophy of Science, 2* (1971), 1–67; Curtis A. Wilson, "From Kepler's Laws, So-called, to Universal Gravitation: Empirical Factors," *Archive for History of Exact Sciences, 6* (1970), 89–170; and A. N. Kriloff, "On Sir Isaac Newton's Method of Determining the Parabolic Orbit of a Comet," *Monthly Notices of the Royal Astronomical Society, 85* (1925), 640–56. Among the things written on Hooke's charge of plagiary, see J. A. Lohne, "Hooke *versus* Newton," *Centaurus, 7* (1960), 5–52, for a judicious assessment of the question.

Already in the eighteenth century, aids to the study of the *Principia* and its difficult mathematics began to appear. The so-called Jesuit edition of the *Principia* by Thomas Le Seur and Francis Jacquier (Geneva, 1739–42), remains one of the best. In the nineteenth century, Henry Brougham and E. J. Routh published *Analytical View of Sir Isaac Newton's Principia* (London, 1855), which set the demonstrations over from Newton's geometric form into analytic form, though one must add that their analytic notation differed from that now generally employed, making the book less useful to us. In volume 6 of his edition of the *Mathematical Papers*, D. T. Whiteside makes a distinguished contribution to this genre.

Newton's chemistry (and alchemy) have been much less studied than other aspects of his science. B. J. T. Dobbs, *The Foundations of Newton's Alchemy: The Hunting of the Greene Lyon* (Cambridge, 1975), supplants all earlier work and, for the first time, offers real guidance into the comprehension of the extensive alchemical papers he left behind. Karin Figala has written a dissertation on Newton's alchemy that unfortunately remains unpublished. She did review Dobbs's book in an extensive article, "Newton as Alchemist," *History of Science*, 15 (1977), 102–37. P. M. Rattansi is the author of two general articles which insist on the importance of the alchemical papers: "Newton's Alchemical Studies," in Allen G. Debus, ed., *Science, Medicine and Society in the Renaissance* (New York, 1972), pp. 167–82, and "Some Evaluations of Reason in Sixteenth- and Seventeenth-Century Natural Philosophy," in Mikulas Teich and Robert Young, ed., *Changing Perspectives in the History of Science* (London, 1973), pp. 148–66. Among works that are not concerned primarily with alchemy, see Hélène Metzger, *Newton, Stahl, Boerhaave et la doctrine chimique* (Paris, 1930) (the second volume, as it were, of her pioneering work in the history of chemistry), A. R. and Marie Boas Hall, "Newton's Chemical Experiments," *Archives internationales d'histoire des sciences*, 11 (1958), 113–52, and the relevant early parts of Arnold Thackray, *Atoms and Powers: An Essay on Newtonian Matter-Theory and the Development of Chemistry* (Cambridge, Mass., 1970).

With one exception, all existing studies of Newton's religious thought are based on his published works and on the very small number of manuscripts that were available at an earlier time. The very recent acquisition of the Yahuda MSS by the Jewish National and University Library has made them all partially obsolete. Herbert McLachlan edited a volume of *Theological Manuscripts* (Liverpool, 1950), from the Keynes MSS. His publication took great liberty with the originals so that it cannot be relied upon. Moreover, the theological manuscripts in the Keynes collection are not typical of the whole.

Among older works, Hélène Metzger, *Attraction universelle et religion naturelle chez quelques commentateurs anglais de Newton* (Paris, 1938), argues the close connection between the concept of universal gravitation and religious considerations. Frank E. Manuel, *The Religion of Isaac Newton* (Oxford, 1974), is the only work of which I am aware that draws upon the Yahuda MSS. Manuel concerns himself entirely, however, in extending the Freudian themes of his *Portrait*, so that the book does not give any insight into the theological content of the papers. Manuel's *Isaac Newton, Historian* (Cambridge, Mass., 1963), is the only significant study of Newton's chronology, a topic intimately allied with his theology. Among other works, see also Perry Miller, "Bentley and Newton," the introduction to the theologically related pieces in Cohen's *Papers & Letters;* William H. Austin, "Isaac Newton on Science and Religion," *Journal of the History of Ideas*, 31 (1970), 521–42; Leonard Trengove, "Newton's Theological Views," *Annals of Science*, 22 (1966), 277–94; and Richard S. Brooks, "The Relationships between Natural Philosophy, Natural Theology and Revealed Religion in the Thought of Newton and their Historiographic Relevance," dissertation, Northwestern University, 1976. Margaret Jacob, *The*

Newtonians and the English Revolution, 1689–1720 (Ithaca, N.Y., 1976), which sums up a number of earlier articles, examines the interrelation of Newtonian natural philosophy, the practical theology of the Latitudinarians, and the political situation in England at the time of the Glorious Revolution.

For Newton's activities at the Mint, the unquestioned authority is John Craig, *Newton at the Mint* (Cambridge, 1946), a work based directly on Newton's own papers. In addition, Craig is the author of *The Mint: A History of the London Mint from A.D. 287 to 1948* (Cambridge, 1953), which is also useful for understanding Newton's tenure there. See also Craig's "Issac Newton and the Counterfeiters," *Notes and Records of the Royal Society*, 18 (1963), 136–45.

Newton's presidency of the Royal Society has not been much studied except in histories of the society and in Manuel's *Portrait*. The best account of his administrative effort in my opinion is that in Henry Lyons, *The Royal Society 1660–1940: a History of Its Administration Under Its Charters* (Cambridge, 1944).

List of illustrations

Index of Newton's life and works

For all other subjects, see the General Index.

General index

For all subjects relating directly to Newton, see Index of Newton's Life and Works.